# Advances in Phage Display

## A LABORATORY MANUAL

## ALSO FROM COLD SPRING HARBOR LABORATORY PRESS

**RELATED TITLES**

*Machine Learning for Protein Science and Engineering*

*T-Cell Memory*

*Immune Memory and Vaccines: Great Debates*

*A Cure Within: Scientists Unleashing the Immune System to Kill Cancer*

**OTHER LABORATORY MANUALS**

*Antibodies: A Laboratory Manual,* Section Edition

*Budding Yeast: A Laboratory Manual*

*CRISPR—Cas: A Laboratory Manual*

Drosophila *Neurobiology: A Laboratory Manual,* Section Edition

*Experiments in Bacterial Genetics: A Laboratory Manual*

*Molecular Cloning: A Laboratory Manual,* Fourth Edition

*Mouse Phenotypes: Generation and Analysis of Mutants: A Laboratory Manual,* Section Edition

Xenopus: *A Laboratory Manual*

**WEBSITE**

www.cshprotocols.org

# Advances in Phage Display

A LABORATORY MANUAL

EDITED BY

**Gregg J. Silverman**
*New York University School of Medicine*

**Christoph Rader**
*The Scripps Research Institute*

**Sachdev S. Sidhu**
*University of Toronto*

COLD SPRING HARBOR LABORATORY PRESS
Cold Spring Harbor, New York • www.cshlpress.org

## Advances in Phage Display
A LABORATORY MANUAL

All rights reserved
© 2026 by Cold Spring Harbor Laboratory Press, Cold Spring Harbor, New York

| | |
|---|---|
| Publisher | John Inglis |
| Executive Editor | Alejandro Montenegro-Montero |
| Acquisition Editor | Maria Smit |
| Assistant Editor | Christin Munkittrick |
| Project Managers | Inez Sialiano and Barbara Acosta |
| Editorial Assistant | Danett Gil |
| Permissions Coordinator | Carol Brown |
| Production Editors | Kathleen Bubbeo and Cynthia Blaut |
| Production Manager and Cover Designer | Denise Weiss |
| Director of Product Development & Marketing | Wayne Manos |

*Cover art:* Modes of phage display. The structure of Ff phage (fd, f1, or M13), depicted in monochromatic or rainbow colors, was determined by Mathew McLaren, Rebecca Conners, and Vicki Gold. The Ff-derived nanorods used to determine the structure were produced by Rayén León-Quezada. Proteins, displayed as fusions to full-length or truncated pIII or to pVIII, include a larger single-chain antibody and a smaller affibody. Red coloring along a selection of the monochromatic phage structures shown represents peptides displayed on the major coat protein, pVIII. The cover image was designed by Jasna Rakonjac and Vicki Gold, with input from Gregg Silverman.

*Library of Congress Cataloging-in-Publication Data*

Names: Silverman, Gregg J., 1955– editor | Rader, Christoph, 1965– editor | Sidhu, Sachdev S., 1968– editor
Title: Advances in phage display: a laboratory manual / edited by Gregg J. Silverman, New York University School of Medicine, Christoph Rader, The Scripps Research Institute and Sachdev S. Sidhu, University of Toronto.
Description: Cold Spring Harbor, New York : Cold Spring Harbor Laboratory Press, [2025] | Includes bibliographical references and index. | Summary: "Phage display is a powerful technique for mining peptide and protein libraries that has greatly accelerated drug discovery, structural biology, biomarker development, antigen identification, and protein-domain engineering. Designed for researchers at all levels, this Laboratory Manual provides expert reviews on principles and applications of phage display, along with detailed laboratory protocols"-- Provided by publisher.
Identifiers: LCCN 2025007161 (print) | LCCN 2025007162 (ebook) | ISBN 9781621824251 paperback | ISBN 9781621824268 epub
Subjects: LCSH: Bacteriophages--Research--Laboratory manuals | Pharmaceutical biotechnology--Laboratory manuals | LCGFT: Handbooks and manuals
Classification: LCC QR342 .A38 2025 (print) | LCC QR342 (ebook) | DDC 579.2/6--dc23/eng/20250609
LC record available at https://lccn.loc.gov/2025007161
LC ebook record available at https://lccn.loc.gov/2025007162

Users following the procedures in this manual do so at their own risk. Cold Spring Harbor Laboratory makes no representations or warranties with respect to the material set forth in this manual and has no liability in connection with the use of these materials. All registered trademarks, trade names, and brand names mentioned in this book are the property of the respective owners. Readers should please consult individual manufacturers and other resources for current and specific product information.

All World Wide Web addresses are accurate to the best of our knowledge at the time of printing.

Certain experimental procedures in this manual may be the subject of national or local legislation or agency restrictions. Users of this manual are responsible for obtaining the relevant permissions, certificates, or licenses in these cases. For experiments involving animals or human subjects, users must follow the ethical standards of the relevant institutional and national committees for such matters and obtain all necessary approvals and permissions. Neither the authors of this manual nor Cold Spring Harbor Laboratory assume any responsibility for failure of a user to do so.

The materials and methods in this manual may infringe the patent and proprietary rights of other individuals, companies, or organizations. Users of this manual are responsible for obtaining any licenses necessary to use such materials and to practice such methods. COLD SPRING HARBOR LABORATORY MAKES NO WARRANTY OR REPRESENTATION THAT USE OF THE INFORMATION IN THIS MANUAL WILL NOT INFRINGE ANY PATENT OR OTHER PROPRIETARY RIGHT.

Authorization to photocopy items for internal or personal use, or the internal or personal use of specific clients, is granted by Cold Spring Harbor Laboratory Press, provided that the appropriate fee is paid directly to the Copyright Clearance Center (CCC). Write or call CCC at 222 Rosewood Drive, Danvers, MA 01923 (978-750-8400) for information about fees and regulations. Prior to photo-copying items for educational classroom use, contact CCC at the above address. Additional information on CCC can be obtained at CCC Online at www.copyright.com. For a complete catalog of all Cold Spring Harbor Laboratory Press publications, please visit our website at www.cshlpress.org.

# Contents

Preface ix

## SECTION 1. FOUNDATIONS OF PHAGE DISPLAY

**CHAPTER 1**

OVERVIEW

Advances in Phage Display—A Perspective 1
*George P. Smith*

**CHAPTER 2**

OVERVIEW

Structure, Biology, and Applications of Filamentous Bacteriophages 27
*Jasna Rakonjac, Vicki A.M. Gold, Rayén I. León-Quezada, and Catherine H. Davenport*

**CHAPTER 3**

OVERVIEW

Principles of Affinity Selection 49
*George P. Smith*

**CHAPTER 4**

OVERVIEW

The pComb3 Phagemid Family of Phage Display Vectors 63
*Christoph Rader*

## SECTION 2. POSTIMMUNE ANTIBODY PHAGE DISPLAY LIBRARIES

**CHAPTER 5**

INTRODUCTION

Generation and Selection of Phage Display Antibody Libraries in Fab Format 77
*Christoph Rader*

PROTOCOLS

1 Generation of Antibody Libraries for Phage Display: Human Fab Format 86
*Haiyong Peng and Christoph Rader*

2 Generation of Antibody Libraries for Phage Display: Chimeric Rabbit/Human Fab Format 102
*Haiyong Peng and Christoph Rader*

3 Generation of Antibody Libraries for Phage Display: Preparation of Electrocompetent E. coli 116
*Haiyong Peng and Christoph Rader*

4 Generation of Antibody Libraries for Phage Display: Preparation of Helper Phage 122
*Haiyong Peng and Christoph Rader*

5 Generation of Antibody Libraries for Phage Display: Library Reamplification 125
*Haiyong Peng and Christoph Rader*

| | 6 Phage Display Selection of Antibody Libraries: Panning Procedures | 129 |
| | *Haiyong Peng and Christoph Rader* | |
| | 7 Phage Display Selection of Antibody Libraries: Screening of Selected Binders | 139 |
| | *Haiyong Peng and Christoph Rader* | |
| | 8 Cloning, Expression, and Purification of Phage Display-Selected Fab for Biophysical and Biological Studies | 149 |
| | *Matthew G. Cyr, Haiyong Peng, and Christoph Rader* | |

## SECTION 3. CHICKEN ANTIBODY PHAGE DISPLAY LIBRARIES

### CHAPTER 6

**INTRODUCTION**

Generation of Chicken Antibody Libraries and Selection of Antigen Binders — 159
*Hyunji Yang, Jisu Chae, Hyori Kim, Jinsung Noh, and Junho Chung*

**PROTOCOLS**

1 Chicken Immunization Followed by RNA Extraction and cDNA Synthesis for Antibody Library Preparation — 166
*Hyunji Yang, Jisu Chae, Hyori Kim, Jinsung Noh, and Junho Chung*

2 Generation of a Phage Display Chicken Single-Chain Variable Fragment Library — 172
*Hyunji Yang, Jisu Chae, Hyori Kim, Jinsung Noh, and Junho Chung*

3 Preparation of VCSM13 Helper Phage for Display Library Reamplification and Bio-Panning — 185
*Hyunji Yang, Jisu Chae, Hyori Kim, Jinsung Noh, and Junho Chung*

4 Selection of Antigen Binders from a Chicken Single-Chain Variable Fragment Library — 188
*Hyunji Yang, Jisu Chae, Hyori Kim, Jinsung Noh, and Junho Chung*

## SECTION 4. SYNTHETIC HUMAN ANTIBODY PHAGE DISPLAY LIBRARIES

### CHAPTER 7

**OVERVIEW**

Structural Survey of Antigen Recognition by Synthetic Human Antibodies — 201
*Maryna Gorelik, Shane Miersch, and Sachdev S. Sidhu*

### CHAPTER 8

**INTRODUCTION**

Beyond Natural Immune Repertoires: Synthetic Antibodies — 223
*Gianluca Veggiani and Sachdev S. Sidhu*

**PROTOCOL**

1 Generation and Selection of Synthetic Human Antibody Libraries via Phage Display — 226
*Gianluca Veggiani and Sachdev S. Sidhu*

## SECTION 5. SEMISYNTHETIC scFv PHAGE DISPLAY LIBRARIES

### CHAPTER 9

**INTRODUCTION**

Considerations for Using Phage Display Technology in Therapeutic Antibody Drug Discovery — 245
*Mary Ann Pohl and Juan C. Almagro*

## PROTOCOLS

1 Semisynthetic Phage Display Library Construction: Design and Synthesis of Diversified Single-Chain Variable Fragments and Generation of Primary Libraries  260
*Juan C. Almagro and Mary Ann Pohl*

2 Semisynthetic Phage Display Library Construction: Generation of Filtered Libraries  271
*Juan C. Almagro and Mary Ann Pohl*

3 Semisynthetic Phage Display Library Construction: Generation of Single-Chain Variable Fragment Secondary Libraries  281
*Juan C. Almagro and Mary Ann Pohl*

## SECTION 6. HIGH-THROUGHPUT ANTIBODY REPERTOIRE LIBRARIES

### CHAPTER 10

#### INTRODUCTION

Beyond Single Clones: High-Throughput Sequencing in Antibody Discovery  293
*Ahmed S. Fahad, Matías F. Gutiérrez-Gonzalez, Bharat Madan, and Brandon J. DeKosky*

#### PROTOCOLS

1 Clonal Variant Analysis of Antibody Engineering Libraries  298
*Ahmed S. Fahad, Matías F. Gutiérrez-Gonzalez, Bharat Madan, and Brandon J. DeKosky*

2 Antibody Data Analysis from Diverse Immune Libraries  305
*Ahmed S. Fahad, Matías F. Gutiérrez-Gonzalez, Bharat Madan, and Brandon J. DeKosky*

3 Clonal Lineage and Gene Diversity Analysis of Paired Antibody Heavy and Light Chains  311
*Ahmed S. Fahad, Matías F. Gutiérrez-Gonzalez, Bharat Madan, and Brandon J. DeKosky*

## SECTION 7. B-CELL EPITOPE PHAGE DISPLAY LIBRARIES

### CHAPTER 11

#### INTRODUCTION

Insights from the Study of B-Cell Epitopes of a Microbial Pathogen by Phage Display  317
*Gregg J. Silverman*

#### PROTOCOL

1 Cloning and Selection from Antigen Fragment Libraries for Epitope Identification  334
*Gregg J. Silverman*

## SECTION 8. DOMAIN ENGINEERING IN PHAGE DISPLAY LIBRARIES

### CHAPTER 12

#### OVERVIEW

Use of Phage Display and Other Molecular Display Methods for the Development of Monobodies  357
*Akiko Koide and Shohei Koide*

| CHAPTER 13 | OVERVIEW | |
|---|---|---|
| | Phage-Displayed SH2 Domain Libraries: From Ultrasensitive Tyrosine Phosphoproteome Probes to Translational Research | 367 |
| | *Gregory D. Martyn and Gianluca Veggiani* | |

| CHAPTER 14 | OVERVIEW | |
|---|---|---|
| | Generation and Characterization of Engineered Ubiquitin Variants to Modulate the Ubiquitin Signaling Cascade | 381 |
| | *Chen T. Liang, Olivia Roscow, and Wei Zhang* | |

| CHAPTER 15 | INTRODUCTION | |
|---|---|---|
| | Engineering of Affibody Molecules | 403 |
| | *Stefan Ståhl, Hanna Lindberg, Linnea Charlotta Hjelm, John Löfblom, and Charles Dahlsson Leitao* | |
| | **PROTOCOLS** | |
| | 1 Cloning of Affibody Libraries for Display Methods | 409 |
| | *Stefan Ståhl, Linnea Charlotta Hjelm, Charles Dahlsson Leitao, John Löfblom, and Hanna Lindberg* | |
| | 2 Selection of Affibody Molecules Using Phage Display | 422 |
| | *Linnea Charlotta Hjelm, Charles Dahlsson Leitao, Stefan Ståhl, John Löfblom, and Hanna Lindberg* | |
| | 3 Selection of Affibody Molecules Using *Escherichia coli* Display | 436 |
| | *Charles Dahlsson Leitao, Linnea Charlotta Hjelm, Stefan Ståhl, John Löfblom, and Hanna Lindberg* | |
| | 4 Selection of Affibody Molecules Using Staphylococcal Display | 443 |
| | *John Löfblom, Linnea Charlotta Hjelm, Charles Dahlsson Leitao, Stefan Ståhl, and Hanna Lindberg* | |

| Appendix | General Safety and Hazardous Material Information | 451 |
|---|---|---|

Index 457

# Preface

Phage display technology, a groundbreaking method for studying protein interactions, was first introduced by George Smith in his seminal 1985 paper (Smith 1985), following his sabbatical in the laboratory of Robert Webster at Duke University. The technology harnesses the elegant molecular properties of a type of bacteriophage—viruses that infect bacteria and contain only a handful of genes—as a powerful platform for experimental investigation. This technology has since evolved and expanded, shaped by countless contributions from researchers in basic and applied sciences across a variety of disciplines.

This collection of review articles and protocols features a cross-section of the many research applications that have blossomed from the foundational work in phage display. A unifying thread among these contributions is that they are all derived from instructors, students, and collaborators of our Cold Spring Harbor Laboratory (CSHL) course, originally launched in 1992 under the title "Monoclonal Antibodies from Combinatorial Libraries," with minor variations in naming over the years. In 1993, Gregg Silverman, formerly at University of California San Diego and now at New York University, joined the course leadership and has remained a central figure, providing continuity and guiding its evolution.

Cold Spring Harbor Laboratory, a 100+-year-old research base that sprawls along the shores of an idyllic harbor off the Long Island Sound, has long served as a unique center for learning and scientific discourse. Originally a 19th-century whaling station and later a marine biology outpost, CSHL has hosted, since the 1930s, the renowned *Symposium on Quantitative Biology*, which has annually brought together leading investigators on a variety of different fields, including those working to unravel the foundations of molecular biology, structural biology, and principles of diversification of immune repertoires. In parallel, for 25 years and until 1970, CSHL also offered an annual course on the emerging field of phage biology (Susman 1995).

Our current phage display course was born in the early 1990s from a conversation between Richard Lerner, then Director of what would later become The Scripps Research Institute, and James Watson, then Director of CSHL. At the time, Scripps had become a vibrant center for early antibody engineering studies, which quickly morphed to adopt vector systems to display peptides and proteins on the surface of the minuscule particle of an engineered phage. Innovations like the phagemid vectors introduced in 1991 by Carlos Barbas et al. (1991) and Angray Kang et al. (1991) made the system more flexible and easier to use, helping the field take off. As interest in the approach was growing fast, Lerner found himself fielding a growing number of requests from candidate visiting scientists who were eager to learn the method firsthand. To both meet the increasing demand and share the technology more broadly, Lerner proposed to Watson to continue the legacy of the early CSHL phage course by starting an annual course, to spread the new gospel of phage display technology. There could be no more appropriate home base than CSHL, with purpose-built facilities for our course nestled among the many historic "Houses of Science" that have been the homes for renowned research groups and symposia for years (Watson 1991).

Scientific advances have always been accelerated by collaborations—and competition. In addition to the research done at Scripps, work in phage display of antibodies was also pioneered at Cambridge, UK, by brilliant protégées under the leadership of Sir Greg Winter (Winter et al. 1994; Winter 2019), who, in 2018, shared the Nobel Prize in Chemistry with George Smith (Smith 2019) for these contributions, along with Francis Arnold, for her work on evolutionary protein design

(Arnold 2019). The competing teams at Scripps and Cambridge (which morphed into Cambridge Antibody Technology [CAT]) reported several remarkably parallel ideas in their publications. Notably, CAT developed the first commercial therapeutic antibody derived from phage display, adalimumab, which was the first fully human anti-TNFα cytokine antibody (den Broeder et al. 2002) and is now used in the treatment of a range of inflammatory diseases. In parallel, James Wells and collaborators at Genentech were combining phage display with other protein engineering technologies (Clackson and Wells 1995). Since then, the products of this approach have become major antibody-based therapeutics for cancer and autoimmune diseases, with many other applications subsequently developed in academic centers and innovative biotechnology companies around the world that now each year bring in tens of billions of dollars.

As such, and based on the exciting advances in the field in laboratories around the world and the increased interest from the community, Dennis Burton and Carlos Barbas launched the CSHL phage display course in 1992, with key aspects of the original curriculum rumored to have been sketched out during their flight to the Laboratory. Since then, we have refined and updated the course curriculum to a practical but rigorous 14-day schedule that combines lectures and hands-on laboratory work. Participants become sequestered, and intensely focus on their assigned tasks, fully immersed in the somewhat monastic atmosphere at CSHL. Aside from only two instances, once in 2011 due to a financial hiccup, and more recently in 2020 because of the global COVID pandemic, our CSHL course has been held annually ever since. A product of this work and collaboration—and after quite a bit of cobbling and recobbling—our earlier compendium book, "*Phage Display: A Laboratory Manual*," was published by CSHL Press in 2001 (Barbas et al. 2001). This manual compiled protocols and concepts from the early years of the course.

In each edition of the course, and for over three decades, our lecture series has been kicked off by a presentation on the biology of the filamentous phage that infects *Escherichia coli*, as elucidated by Marjorie Russel and Peter Model (Russel and Model 1982) at Rockefeller University. Their contributions are now presented at our course by one of their leading disciples, Jasna Rakonjac (Rakonjac et al. 2017; see Chapter 2, Overview: Structure, Biology, and Applications of Filamentous Bacteriophages [Rakonjac et al. 2024]). Each year, the course hosts 16 participants—graduate students, postdocs, junior faculty, and some young industry scientists (with a few exceptions)—who work in pairs in our purpose-built laboratory. Applicants are selected based on their educational background and work experience, and on an essay describing how they plan to apply the technology in their own work. We encourage participants to not only immerse themselves in the course, to ask hard questions, and to later use their network of new colleagues—and our class alumni and lecturers—to help them pressure test new ideas and applications, but also, and most notably, to then share their know-how in their networks and with their own students and colleagues.

This collection represents the fruit of the contributions of many individuals. We thank our contributors and coauthors. We also thank our technical assistants, and we hope their efforts, practical assistance, and support, which have been key for the success of our demonstration studies in library construction and selection, have also helped their own professional development. Among our hosts, we especially thank David Stewart, CSHL Director of Courses and Meetings, who has always been our stalwart advocate, and Barbara Zane, who ensures we have everything we need in the laboratory to be successful.

We also especially thank Don Siegel at the University of Pennsylvania, who was enlisted as a course Instructor in 1995 and later served as Co-Director, who has contributed to the course as much as anyone. We also thank our more recently recruited faculty, Johannes Yeh, who leads a phage display research group at CSHL, and our current course Co-Director, Gianluca Veggiani, formerly at the University of Toronto, now at Louisiana State University. We also thank all of our colleagues who have served as instructors for the course, and who continue to carry the phage display flame.

Finally, we loudly express our appreciation to all of our students, who have always been a source of great inspiration.

We dedicate this volume to our close friend, Carlos Barbas, a world-renowned pioneering antibody engineer and inspired tireless innovator, who suffered a much too premature ending. Carlos was appropriately eulogized by invoking the spirit of the fictional Roy Batty "The candle that burns twice as bright, burns half as long" (Fancher 1981) (which had been borrowed from I Chin, credited to Lao Tzu).

Gregg Silverman
(Course Instructor: 1993–present; Co-Director: 1995–present)

Christoph Rader
(Course Instructor: 2000–present; Co-Director: 2019–2022)

Sachdev Sidhu
(Course Instructor: 2001–present)

## REFERENCES

Arnold FH. 2019. Innovation by evolution: bringing new chemistry to life (Nobel lecture). *Angew Chem Int Ed Engl* **58:** 14420–14426. doi:10.1002/anie.201907729

Barbas CF III, Kang AS, Lerner RA, Benkovic SJ. 1991. Assembly of combinatorial antibody libraries on phage surfaces: the gene III site. *Proc Natl Acad Sci* **88:** 7978–7982.

Barbas CF III, Burton DR, Scott JK, Silverman GJ. 2001. *Phage display: a laboratory manual.* Cold Spring Harbor Laboratory Press, Cold Spring Harbor, NY.

Clackson T, Wells JA. 1995. A hot spot of binding energy in a hormone–receptor interface. *Science* **267:** 383–386.

den Broeder A, van de Putte L, Rau R, Schattenkirchner M, Van Riel P, Sander O, Binder C, Fenner H, Bankmann Y, Velagapudi R, et al. 2002. A single dose, placebo controlled study of the fully human anti-tumor necrosis factor-alpha antibody adalimumab (D2E7) in patients with rheumatoid arthritis. *J Rheumatol* **29:** 2288–2298.

Fancher H (writer), Peoples DW (writer), Dick PK (writer). 1981. *Blade runner* [film]. The Ladd Company; Shaw Brothers; Blade Runner Partnership; Warner Bros.

Kang AS, Barbas CF, Janda KD, Benkovic SJ, Lerner RA. 1991. Linkage of recognition and replication functions by assembling combinatorial antibody Fab libraries along phage surfaces. *Proc Natl Acad Sci* **88:** 4363–4366.

Rakonjac J, Russel M, Khanum S, Brooke SJ, Rajic M. 2017. Filamentous phage: Structure and biology. *Adv Exp Med Biol* **1053:** 1–20.

Rakonjac J, Gold VAM, Leon-Quezada RI, Davenport CH. 2024. Structure, biology, and applications of filamentous bacteriophages. *Cold Spring Harb Protoc* **2024:** pdb.over107754.

Russel M, Model P. 1982. Filamentous phage pre-coat is an integral membrane protein: Analysis by a new method of membrane preparation. *Cell* **28:** 177–184.

Smith GP. 1985. Filamentous fusion phage: Novel expression vectors that display cloned antigens on the virion surface. *Science* **228:** 1315–1317.

Smith GP. 2019. Phage display: Simple evolution in a petri dish (Nobel Lecture). *Angew Chem Int Ed Engl* **58:** 14428–14437.

Susman M. 1995. The Cold Spring Harbor Phage Course (1945–1970): A 50th anniversary remembrance. *Genetics* **139:** 1101–1106.

Watson EL. 1991. *Houses of science: a pictorial history of Cold Spring Harbor Laboratory.* CSHL Press, Cold Spring Harbor, NY.

Winter G. 2019. Harnessing evolution to make medicines (Nobel Lecture). *Angew Chem Int Ed Engl* **58:** 14438–14445.

Winter G, Griffiths AD, Hawkins RE, Hoogenboom HR. 1994. Making antibodies by phage display technology. *Annu Rev Immunol* **12:** 433–455.

## GENERAL SAFETY AND HAZARDOUS MATERIAL INFORMATION

Cold Spring Harbor Laboratory Manuals should be used by laboratory personnel with experience in laboratory and chemical safety or students under the supervision of such trained personnel. The procedures, chemicals, and equipment referenced in these manuals are hazardous and can cause serious injury unless performed, handled, and used with care and in a manner consistent with safe laboratory practices. Students and researchers using the procedures in this manual do so at their own risk. It is essential for your safety that you consult the appropriate Material Safety Data Sheets, the manufacturers' manuals provided with the relevant equipment, and your institution's Environmental Health and Safety Office, as well as the General Safety and Disposal Cautions in the Appendix for proper handling of hazardous materials. Cold Spring Harbor Laboratory makes no representations or warranties with respect to the material set forth in its manuals and has no liability in connection with the use of these materials.

All registered trademarks, trade names, and brand names mentioned in this book are the property of the respective owners. Readers should please consult individual manufacturers and other resources for current and specific product information.

Appropriate sources for obtaining safety information and general guidelines for laboratory safety are provided in the General Safety and Hazardous Material Information Appendix.

CHAPTER 1

# Advances in Phage Display—A Perspective

George P. Smith[1]

*Division of Biological Sciences, University of Missouri, Columbia, Missouri 65211, USA*

Phage display technology is enabled by genetic fusion of a foreign protein domain to a phage coat protein, without interfering with the phage's ability to replicate by infecting bacterial host cells. The displayed domain is exposed on the phage particle (virion) surface, where it can interact with molecules or other substances in the surrounding medium; in this regard, it acts like a normal protein. However, it possesses a superpower that is unavailable to ordinary proteins: It is easily replicated in great abundance because it is attached to a replicating virion whose genome includes its coding sequence. The main way this technology is exploited is construction of huge phage display "libraries," comprising billions of phage clones, each displaying a different protein domain, and each represented by thousands, millions, or billions of genetically identical virions—all mixed together in a single vessel. Surface display allows exceedingly rare virions whose displayed protein domains happen to bind a user-defined molecule or other substance—generically called the "selector"—to be isolated from such libraries by an affinity selection process. The yield of selector-binding virions is much too low to be of practical use, but their number is readily increased by many orders of magnitude by propagating the virions in host bacteria in culture. This overview is a critical review of recent developments of this technology. It does not review the entire arena of contemporary phage display; there is special emphasis on phage display's most prominent application, phage antibodies, in which the displayed domain is an antibody domain, and the selector is an antigen of interest.

## INTRODUCTION

The centerpiece of phage display is a type of engineered phage whose genome includes the coding sequence for a foreign protein domain genetically fused to one of the phage's coat protein genes. As a consequence of coat protein fusion, the foreign domain is exposed on the surface of the phage particle, called the virion, where it is accessible to molecules and other substances in the surrounding medium. The foreign domain is said to be "displayed" on the virion surface—hence the term "phage display." In this sense, the displayed domain acts like an ordinary protein. However, at the same time, it can be replicated by many orders of magnitude, simply by infecting a culture of host bacteria with the virion that displays the domain and carries the domain's coding sequence in its genome.

Phage display's core processes are library construction and affinity selection. A typical phage display library contains billions of phage clones, displaying billions of different foreign domains on the virion surface. Affinity selection (explained briefly in the next section and more fully in a later section) isolates rare library virions displaying domains that happen to bind a "selector"—a molecule, collection of molecules, or other substance chosen by the investigator. The ability of virions displaying a protein domain to replicate is essential to affinity selection, as explained in the next section.

---

[1]Correspondence: smithgp194@gmail.com

© 2026 Cold Spring Harbor Laboratory Press
Cite this overview as *Cold Spring Harb Protoc*; doi:10.1101/pdb.over107753

# Chapter 1

The articles discussed in this review emphasize phage display's most conspicuous enterprise: phage antibody technology. Here, the displayed entity is an antibody domain such as a single-chain variable fragment (scFv) or an antigen-binding fragment (Fab), and the selector is a target antigen or mixture of antigens. Many of the articles featured in this review are from an accompanying collection, called *Advances in Phage Display*, edited by Gregg Silverman of New York University School of Medicine, Christoph Rader of the University of Florida, and Sachdev Sidhu of The Anvil Institute in Ontario. The collection is a successor to *Phage Display: A Laboratory Manual*, edited by Carlos Barbas and Dennis Burton of the Scripps Research Institute, Jamie Scott of Simon Fraser University, and Silverman, and published by the Cold Spring Harbor Laboratory Press in 2001 (Barbas et al. 2001). The entire *Advances in Phage Display* collection is available online at *Cold Spring Harbor Protocols* and can be accessed at https://cshprotocols.cshlp.org/.

This overview is divided into seven sections:

- **What Is a Phage Antibody?**—a brief preview of phage antibodies
- **Filamentous Phage Structure and Function**—essential information about the predominant phage species serving as phage display vectors
- **Phagemid Vectors for Phage Antibody Display**—discussion of the predominant type of filamentous phage vector for display of antibody domains
- **Phage Antibody Libraries**—review of design and construction of libraries with billions of phage clones, each displaying a different antibody domain with different binding specificity, and each represented by thousands, millions, or billions of genetically identical phages
- **Affinity Selection**—explanation of exposing a solid surface coated with an antigen to a phage antibody library, in order to isolate very rare phages whose displayed antibody domains happen to bind the antigen with high affinity
- **Display of Peptides and Protein Domains Other than Antibodies**—discussion of libraries displaying antigenic peptides, antibody-like binding domains, and other binding domains
- **Postscript: The C.A.R.L.O.S. Project**—account of an emergency project using phage antibody display to develop chimeric antigen receptor T-cell (CAR T) therapy against a cancer afflicting Carlos Barbas, a key contributor to phage antibody technology from its earliest days

The articles in the collection serve not only individually as contributions to phage display methodology, but also collectively as a laboratory manual for students of an advanced phage display course that has been taught at the Cold Spring Harbor Laboratory since 1992, when phage display in its contemporary form, including phage antibody technology, was just emerging in a growing community of researchers. The course was first offered under the title "Phage Display of Combinatorial Antibody Libraries," with Barbas and Burton as the founding instructors. It is now called "Antibody Engineering and Display Technologies," with Silverman and Gianluca Veggiani of Louisiana State University as the current instructors. I refer to it here simply as the "Phage Display Course."

In large measure, the Phage Display Course has been a joint project of the Cold Spring Harbor Laboratory's Courses Program and the Scripps Research Institute, which has been a center of phage antibody development from its earliest days and the source of key resources used in the course's laboratory investigations. Many of the instructors and lecturers, including authors represented in this collection, have worked at Scripps during their careers. These include in particular Barbas, Burton, Silverman (the longest serving instructor), and Rader (who, along with Barbas, assumed a major role in developing the pComb family of phagemids, leading vectors for phage antibody technology). I have tried not only to engage with the articles' experimental data and logic, but also to touch on the broader intellectual culture that they are embedded in.

I assume that the reader is already familiar with the basic structure of antibodies and their interaction with antigens. Given only that background, I review the logic of the articles included in the collection, in the broader context of progress in phage antibody technology over the past 25 years.

## WHAT IS A PHAGE ANTIBODY?

A phage antibody (Fig. 1) is an engineered phage in which an antibody domain—most often an scFv or Fab—is genetically fused to a phage protein that is exposed on the surface of the virion. The virion-borne antibody domain is thus accessible to antigens or other potential ligands in the surrounding medium. At the same time, the antibody domain is physically attached to a virion whose chromosome includes its coding sequence. The antibody domain is said to be "displayed" on the surface of the virion. Peptides and protein domains other than antibodies can be displayed on virions in the same way.

Physical linkage of the antibody domain to its coding sequence in the virion means that the domain can be propagated to prodigious numbers, simply by infecting host bacteria with virions displaying the domain. In effect, the antibody domain can replicate. That is what makes it feasible to isolate exceedingly rare phage clones whose displayed antibody domains happen to bind a chosen antigen from enormous phage antibody libraries comprising many billions of phage clones, displaying billions of different antibody domains. All the virions in the library are jumbled together in a single vessel. Rare antigen-binding clones are obtained, not via clone-by-clone screening, but rather via an affinity selection process, which is outlined in Figure 2. As explained in the legend, virions that have been affinity-selected must be propagated ("amplified") to be useful.

## FILAMENTOUS PHAGE STRUCTURE AND FUNCTION

The phage display concept has been realized with a number of phage platforms other than Ff filamentous phage, including λ, T4, T7, and Qβ (Istomina et al. 2024); each has key advantages. Nevertheless, the great majority of phage display applications still use the Ff family of filamentous phages (the very closely related natural strains M13, f1, and fd), and those phages have also been the platform for the Phage Display Course ever since its inception in 1992. This overview accordingly focuses on Ff filamentous phages.

The first chapter of the 2001 *Phage Display: A Laboratory Manual* was an authoritative review of filamentous phage structure and function by Robert Webster of Duke University (Webster 2001). Much of what we know about these phages was already clear then. Virions can easily be prepared at high purity. Electron microscopy (Fig. 3) reveals filamentous particles 6 nm wide and 900 nm long in the case of wild-type virions (the length of the virion can vary widely depending on the length of the DNA chromosome it contains). There are 11 phage proteins: five structural proteins that comprise the virion's cylindrical outer coat, and six proteins required for replication of the phage DNA and assembly of the virion. All but the tips of the cylindrical outer coat are formed by a fivefold helical array of thousands of copies of the 50-amino-acid major coat protein pVIII. The two tips of the virion are physically different: one pointy and one rounded, as depicted in high-resolution electron micrographs (Fig. 3). By 2001, it had been established that there are three to five subunits each of minor coat proteins pVII and pIX at the rounded tip, and three to five subunits each of minor coat proteins pIII

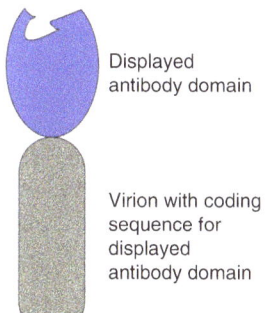

FIGURE 1. Schematic diagram of a phage antibody (not to scale). An antibody domain—most often a single-chain variable fragment (scFv) or an antigen-binding fragment (Fab)—is displayed on the surface of a virion (phage particle). The displayed antibody domain is exposed in the surrounding medium, where it can bind to an antigen. The virion's chromosome contains the coding sequence for the antibody domain it displays.

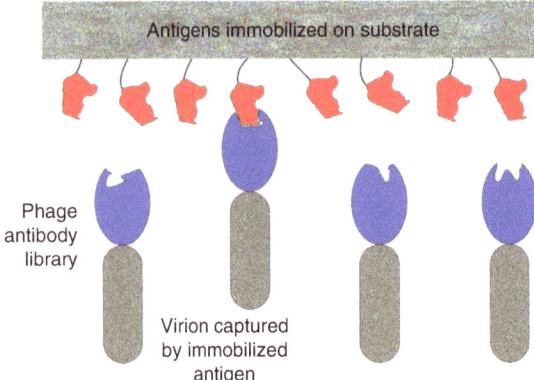

FIGURE 2. Isolation of rare antigen-binding antibody domains by affinity selection. An antigen of interest (red in the figure) is immobilized on a solid substrate of some kind. The antigen-coated substrate is exposed to the library—the "input" to the affinity selection process. Rare virions whose displayed antibody domains happen to bind the immobilized antigen are captured on the substrate surface, while all other virions—the overwhelming majority—remain free in solution. Virions that have not been captured are thoroughly washed away, leaving only the captured virions on the surface. The captured virions are released from the substrate in some manner, resulting in an "output" virion population that is greatly enriched for virions displaying antigen-binding antibody domains. No clone in the output population is present in sufficient numbers to permit meaningful characterization, or to serve as input to another round of capture and release. Instead, taking advantage of the physical link between the displayed antibody domain and its coding sequence in the phage genome, the output virions are greatly "amplified" (propagated) by infecting fresh bacterial host cells with output virions, and culturing the infected cells in growth medium. The amplified output population is then analyzed to identify clones of interest (those displaying antibody domains of interest), or used as input to another round of capture, release, and amplification.

and pVI at the pointy tip. All four minor coat proteins are now known to be present in exactly five copies each.

Table 1 summarizes the domains of minor coat protein pIII, one of the minor coat proteins at the pointy tip, and a key component of phage display technology. An N-terminal domain of each pIII subunit is flexibly connected to the main body of the virion. A few of these domains can sometimes be discerned in electron micrographs, as exemplified by the four small, irregular bodies visible at the tip of the arrow pointing to the pointy tip in Figure 3. An X-ray crystallographic structure of the N-terminal domain has been determined (Lubkowski et al. 1999), revealing two subdomains, now called N1 and N2, attached to each other and to the C-terminal domain through flexible linkers. Domain N2 binds to the tip of a bacterium's F pilus to initiate infection; domain N1 subsequently interacts with the secondary receptor TolQRA in the bacterium's inner membrane, mediating entry of the phage DNA into the cytoplasm.

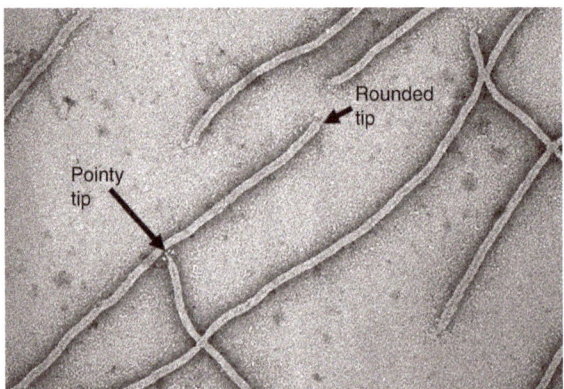

FIGURE 3. High-resolution negative-stained electron micrograph of filamentous phage virions. Putative pointy and rounded tips are indicated. Courtesy of Lee Makowski, Northeastern University, with permission.

TABLE 1. Domains of filamentous phage Ff minor coat protein pIII

| Domain | Amino acids | Function |
| --- | --- | --- |
| SS | 18 | Signal peptide, targeting the polypeptide to the Sec secretion system; cleaved off during secretion |
| N1 | 67 | Interacts with the inner membrane protein TolQRA during infection |
| L1 | 19 | Flexible linker |
| N2 | 131 | Binds F pilus to initiate infection |
| L2 | 39 | Long flexible linker connecting the N-terminal domains to the C-terminal domain C embedded in the pointy tip |
| C (including TM) | 150 | C-terminal domain: five copies intertwined with five copies of pVI and the C-terminal ends of five copies of pVIII to form the pointy tip. The final 27 residues of C comprise the transmembrane domain TM; prior to incorporation into the virion, the first 21 residues of TM act as a transmembrane helix anchoring pIII in the inner membrane of the host bacterium, with the final six residues of TM exposed in the cytoplasm. |

N1, L1, and N2 together comprise the Exposed N-terminal domains.

For further details, see Pellegri et al. (2023), and see also Chapter 2, Overview: Structure, Biology, and Applications of Filamentous Bacteriophages (Rakonjac et al. 2024).

Several features of minor coat protein pIII make it the most commonly used viral host protein for phage display, especially of antibody domains. Its flexible attachment to the main body of the virion ensures that peptides or proteins displayed on pIII are exposed in the surrounding medium and are thus able to react with antibodies, antigens, or other substances there. Like the other phage coat proteins, pIII is an inner membrane protein prior to being incorporated into the nascent virion. Its signal peptide, SS, targets it to the Sec secretion system (filamentous phages are secreted through the bacterial envelope without lysing the cell, as described below). The unfolded polypeptide is transported through the inner membrane until the first 21 residues of its 27-amino-acid transmembrane domain, TM, anchor it in the membrane, leaving only the last six amino acids exposed in the cytoplasm. Meanwhile, the 18-amino-acid signal peptide SS is cleaved from the polypeptide during secretion. The protein therefore folds into its functional conformation in the oxidizing environment of the periplasm between the inner and outer membranes. The periplasm is a putatively propitious environment for antibody domains displayed on pIII to likewise fold into their functional disulfide-stabilized conformations.

Jasna Rakonjac of Massey University and her colleagues have written a successor to Webster's review, reporting some stunning achievements since 2001 (see Chapter 2, Overview: Structure, Biology, and Applications of Filamentous Bacteriophages [Rakonjac et al. 2024]). In particular, two advances have allowed the structures at the tips to be defined in detail (Conners et al. 2023). First, short, rigid "nanorods" have been constructed that have the same tip structures as wild-type virions. The viral DNA in these virions is much shorter than in wild-type virions, and the length of the particle is reduced in proportion. Instead of constituting ∼2.6% of a flexible wild-type virion, the tips constitute 28% of a rigid nanorod, making imaging much easier. Second, spectacular improvements in cryo-electron microscopy (cryoEM) allow detailed structures to be extracted computationally from great numbers of individually blurry images of nanorods in random orientations. These advances have allowed a detailed composite structure of the filamentous phage virion to be assembled (Fig. 4).

Short peptides are typically displayed pentavalently (on all five pIII subunits), fused to the N terminus at the beginning of domain N1. As explained in the next section, in most phage antibodies, the antibody domain is displayed monovalently (on only one of the five pIII subunits); it can be fused either to the N terminus of N1 or to domain C through all or part of flexible linker L2, replacing domains N1, L1, and N2.

Progeny filamentous virions are not released into the medium by lysing the host cell, as in the majority of well-studied phage species. Instead, the circular single-stranded viral DNA, twisted into a long, thin helix, is secreted through the envelope, losing its intracellular coat of pV subunits and acquiring its extracellular coat of virion structural proteins from the inner membrane as it passes. Host bacteria survive and continue to divide, albeit at a reduced rate. The rounded tip of the virion, assembled at the beginning of the twisted DNA, emerges first through the envelope; the pointy tip, assembled at the end of the twisted DNA, emerges last. The length of the virion is determined by the length of the viral DNA, rather than by constraints imposed by the geometry of the coat. Domains N1,

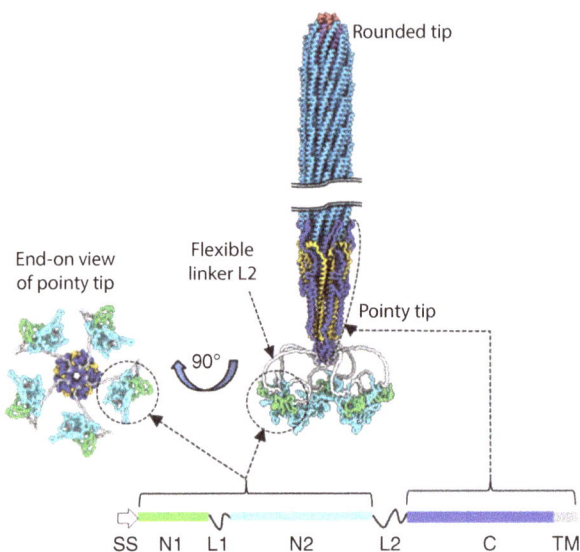

FIGURE 4. Composite structure of the Ff filamentous phage virion. Short segments of the virion near the rounded and pointy tips are depicted at the *right* (the wild-type virion is 150 times longer than its diameter). The circular single-stranded viral DNA, twisted into a long, thin helix, lies inside the protein coat; the length of the virion is variable, depending on the length of the DNA. The major coat protein pVIII subunits, thousands forming the bulk of the cylindrical outer coat, are bright blue; the minor coat protein pVII and pIX subunits (five each at the rounded tip) are salmon and purple, respectively; and minor coat protein VI subunits (five at the pointy tip) are gold. The domains of the minor coat protein pIII subunits (five at the pointy tip), which are mapped at the *lower right* and summarized in Table 1, are colored as follows: SS (absent from the virion) is white, N1 is green, flexible linkers L1 and L2 are gray, N2 is cyan, C is dark blue, and TM is gray. TM becomes part of domain C when incorporated into the virion, where it is colored dark blue like the rest of the domain. An end-on view of the pointy tip is shown at the *left*, with the five pIII subunits artificially arranged symmetrically around the central axis; in reality, these domains have highly variable orientations because of the flexibility of linker L2. Adapted from Chapter 2, Overview: Structure, Biology, and Applications of Filamentous Bacteriophages [Rakonjac et al. 2024]) with permission from Cold Spring Harbor Laboratory Press.

L1, and N2 of the five pIII subunits are connected to the pointy tip through their flexible L2 linkers, as depicted in Figure 4.

The channel through which the nascent virion is secreted is composed of phage proteins pIV, pI, and pXI. CryoEM has been deployed to determine the detailed structure of the outer membrane component of the channel, formed by a 15-member ring of pIV subunits (Conners et al. 2021; see also Chapter 2, Overview: Structure, Biology, and Applications of Filamentous Bacteriophages [Rakonjac et al. 2024]). Figure 5 depicts a nascent progeny virion emerging, rounded tip first, through the channel. The near half of the ring of pIV subunits (colored green in Fig. 5) has been cut away, revealing the tight fit between the channel and the emerging virion. It is remarkable that the N-terminal domains of the pIII subunits at the pointy tip of the nascent virion are able to pass through this channel, especially when they display extra antibody domains. Despite this remarkable research, it is not yet possible to formulate clear rules governing limitations on foreign peptides or protein domains that can be successfully displayed on the virion surface.

Foundational investigation of Ff phage replication by Kensuke Horiuchi and Norton Zinder of Rockefeller University (Zinder and Horiuchi 1985) defined a 508-nt segment of the genome, called the intergenic region (IR), that orchestrates phage replication and assembly. It contains no coding sequence, but three essential *cis*-acting functions: a packaging signal required for efficient virion secretion, the minus-strand origin of replication, and the plus-strand origin of replication (Fig. 6). As explained in the next section, most phage antibody vectors are plasmids containing the IR (often referred to as the "origin") and a recombinant pIII gene that includes the coding sequence for the antibody domain, but no other phage sequences.

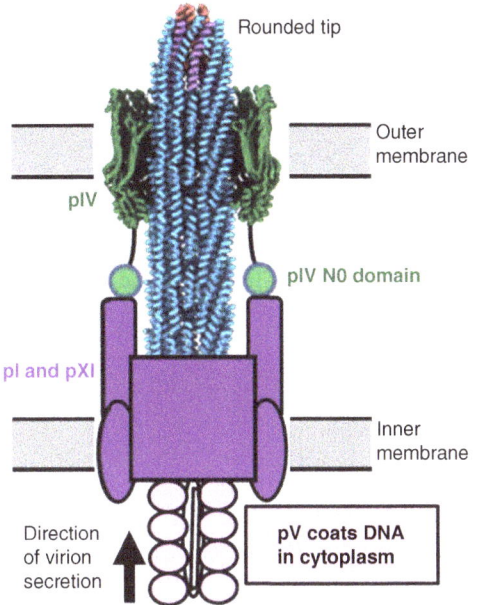

FIGURE 5. Schematic diagram of a nascent virion in the process of being secreted through the cell envelope. The outer membrane component of the channel, composed of a ring of 15 pIV subunits, is colored dark green. The front half of the outer membrane channel has been cut away in the image, in order to reveal the tight fit between the emerging virion and the interior of the channel. The inner membrane component of the channel, colored dark purple, is composed of pI and pXI subunits. A small N-terminal domain, N0, of each pIV subunit interacts with the inner-membrane component through a flexible linker (only two of the 15 N0 domains are depicted). As the nascent virion passes through the inner membrane, its DNA chromosome loses its cytoplasmic coat of phage pV subunits (light purple in the image) and acquires its sheath of coat proteins from the inner membrane. The chromosome is oriented with its packaging signal (hairpin [A] in Fig. 6) at the rounded tip. When the end of the chromosome is reached, five subunits each of minor coat proteins pIII and pVI are added to form the pointy tip of the virion. Adapted from Chapter 2, Overview: Structure, Biology, and Applications of Filamentous Bacteriophages (Rakonjac et al. 2024) with permission from Cold Spring Harbor Laboratory Press.

## PHAGEMID VECTORS FOR PHAGE ANTIBODY DISPLAY

Phagemid vectors such as pBluescript (Alting-Mees and Short 1989) are hybrids between ordinary recombinant DNA plasmids and filamentous phages. Like standard recombinant DNA plasmids, pBluescript has a plasmid origin of replication, a gene conferring resistance to an antibiotic (ampicillin), and a short multiple cloning site (MCS) with unique cleavage sites for several restriction enzymes. The MCS facilitates inserting a foreign DNA payload into the plasmid. The feature that defines pBluescript as a phagemid is inclusion of the filamentous phage IR (Fig. 6). The IR is ordinarily quiescent, but if the phagemid-bearing cell is infected by a fully functional filamentous phage (called the helper phage—helper for short), phage proteins act not only on the helper's own IR, but on the phagemid IR as well. Helpers usually include a gene conferring resistance to kanamycin; culturing the helper-infected cells in medium containing both ampicillin and kanamycin ensures that only cells bearing both genomes replicate. Two different single-stranded DNAs are produced and packaged into virions by the doubly resistant cells: the helper chromosome and the phagemid chromosome.

Phagemids provide an attractive platform for phage display (Qi et al. 2012), as exemplified by the pComb3 family of phagemids, introduced by Barbas and his colleagues (Barbas et al. 1991). They installed an expression cassette driven by the *Escherichia coli lac* promoter operator in a pBluescript-based backbone. The expressed coding sequence has cloning sites that allow a foreign protein or peptide, particularly an antibody domain, to be fused to pIII domain C, through all or part of pIII's flexible linker L2 (the domains of pIII are explained in Table 1 and Fig. 4).

There is an important advantage to fusing the antibody domain to pIII domain C, bypassing pIII's N-terminal domains N1 and N2. Expression of N2 suppresses synthesis of the F pilus, the primary receptor for filamentous phage infection; N1 suppresses synthesis of the secondary receptor TolQRA. This means that if the antibody–pIII fusion protein includes N1 and N2, its expression must be repressed before the host bacteria can be infected with helper phages. Adequate repression is difficult to achieve in the case of the *lac* promoter, especially at the optimal growth temperature of 37°C. Eliminating N1 and N2, as in the pComb3 family of vectors, avoids this problem. Nevertheless, expression of the fusion protein from the *lac* promoter is seldom derepressed by adding the inducer isopropyl β-D-1-thiogalactopyranoside (IPTG) to the culture medium. The usual reason given for thus relying on leaky expression is to reduce expression of the antibody domain so that it is displayed predominantly monovalently (on only one of the five pIII subunits). However, a more important reason is probably that even modest overexpression of the protein is highly toxic to the cell.

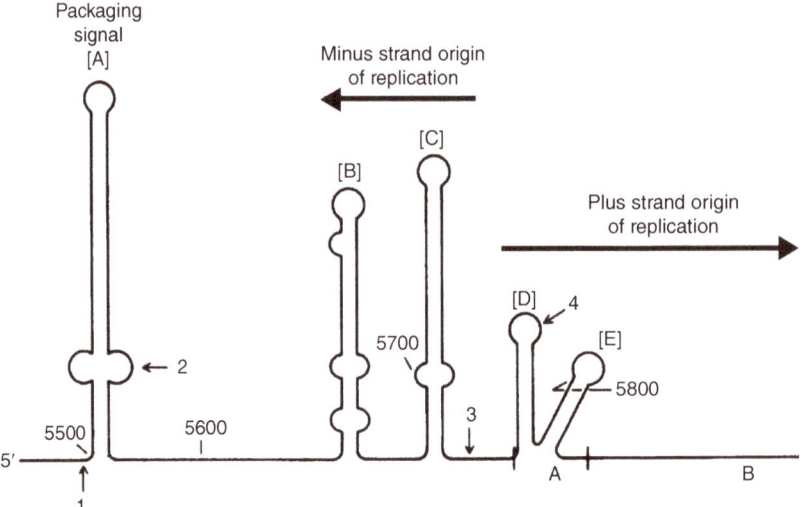

FIGURE 6. Intergenic region (IR) of phage f1 single-stranded viral DNA (the plus strand), comprising three cis-acting elements: packaging signal, minus-strand origin, and plus-strand origin. The plus-strand origin is divided into functional domains A and B (only part of domain B is included). [A]–[E] indicate the secondary structure hairpins in the plus strand. Four-digit numbers indicate nucleotide positions in the f1 genome. Arrows point to the end of the pIV coding sequence (1), the transcription terminator (2), the initiation site of the minus-strand RNA primer (3), and the origin of plus-strand synthesis (4). Modified from Zinder and Horiuchi (1985) with permission from the American Society for Microbiology.

Most phage antibody libraries today, including those created in the Phage Display Course, are constructed in pComb3 family vectors, in conjunction with improved helper phages such as VCSM13 (Russel et al. 1986). The pComb3 family is reviewed by Christoph Rader, a leader in the development of phage antibody libraries, starting when he was a colleague of Barbas at Scripps (see Chapter 4, Overview: The pComb3 Phagemid Family of Phage Display Vectors [Rader 2024b]).

When a pComb3-bearing cell is infected with a helper phage, both the wild-type pIII encoded by the helper and the recombinant pIII encoded by the phagemid are produced and incorporated into both helper and phagemid virions. In the recombinant protein, the cloned foreign protein or peptide replaces the normal N-terminal domains N1, L1, and N2. Even though the recombinant pIII is missing the domains that mediate the infection process, the virions are still infective because some of their pIII subunits are the wild-type protein encoded by the helper.

Antibody domains (either Fabs or scFvs) are almost always displayed monovalently (on only one pIII subunit) or not at all, for two reasons. First, wild-type pIII subunits encoded by the helper phage greatly outcompete recombinant pIII subunits encoded by the phagemid for incorporation into nascent virions. Second, only a minority of the phagemid-encoded pIII subunits that are incorporated into completed virions have intact antibody domains, because of the susceptibility of foreign proteins to degradation in the periplasm or cytosol.

Figure 7 diagrams an example of a coding sequence whose expression is driven by the *lac* promoter in a pComb3 family phagemid. In this construct, it is a chicken scFv that is fused to domain C of pIII. This format of the scFv–pIII genes in the phage antibody library is used by students in the Phage Display Course today.

## PHAGE ANTIBODY LIBRARIES

A phage antibody library is a mixture of millions or billions of phage clones, each clone displaying a different antibody domain, with different antigen-binding specificity. Each clone is represented by

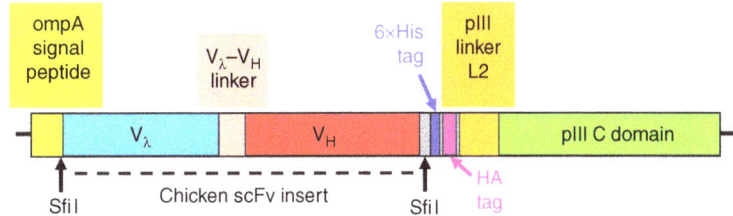

FIGURE 7. Schematic diagram of the coding sequence in a phage antibody construct in a pComb3 family phagemid vector (Andris-Widhopf et al. 2000). The coding sequence for a chicken single-chain variable fragment (scFv) is inserted between the SfiI cloning sites. The colored blocks together constitute the coding sequence for a recombinant protein in which a chicken scFv is fused to the C domain of pIII (including TM) through the last 27 amino acids of pIII's flexible linker, L2. The N-terminal domains N1, L1, and N2 are absent. The gray blocks are spacers with no specified function in the protein. The entire coding sequence spans 463 amino acids if $V_\lambda$ and $V_H$ are the germline genes (the number can be slightly different if $V_\lambda$ or $V_H$ have been modified by somatic gene conversion, as explained in the next section). The 21-amino-acid ompA signal peptide (yellow), like pIII's natural signal peptide, targets the protein to the Sec secretion system, and is cleaved off as the recombinant polypeptide passes through the inner membrane into the periplasmic space. The 6×His and HA tags allow virions to be specifically purified by immobilized metal affinity chromatography (IMAC) or affinity chromatography with an antibody against influenza hemagglutinin protein, respectively.

hundreds, thousands, or millions of genetically identical virions. The whole population is mixed together in a single vessel.

Antibody domains can be displayed on phages in different formats (Istomina et al. 2024):

- Fab fragments are natural compact domains of antibodies, each including a light chain and the first two segments ($V_H$ and $C_H1$) of a heavy chain (see Chapter 5, Protocol 8: Cloning, Expression, and Purification of Phage Display-Selected Fab for Biophysical and Biological Studies [Cyr et al. 2024], Chapter 5, Protocol 2: Generation of Antibody Libraries for Phage Display: Chimeric Rabbit/Human Fab Format [Peng and Rader 2024a], Chapter 5, Protocol 1: Generation of Antibody Libraries for Phage Display: Human Fab Format [Peng and Rader 2024b], and Chapter 5, Introduction: Generation and Selection of Phage Display Antibody Libraries in Fab Format [Rader 2024a]).

- scFv constructs consist of only the $V_H$ and $V_L$ segments of an Fab, connected in either order by a flexible linker (they are illustrated in Fig. 7 and discussed in the first subsection below).

- Diabodies consist of two scFv domains (identical or different) connected by a short linker.

- Single-domain antibodies (sdAbs) consist of the heavy-chain variable region $V_H$ from a class of camelid antibodies that contain no light chain and no heavy-chain $C_H1$ domain.

In almost all cases but sdAbs, heavy- and light-chain V region repertoires are created separately and joined in random combinations to make the final constructs, containing both heavy- and light-chain V regions. The number of possible combinations of heavy- and light-chain V regions is orders of magnitude greater than in the component V region repertoires.

The larger and more diverse the library, the more likely it is that a few of its displayed antibody domains will have high affinity for a target antigen of interest (called the selector). Such clones are isolated from the library by affinity selection (see Fig. 2), which is detailed in the next section. The three subsections of the current section review a few examples of phage antibody libraries, displaying antibody domains with various formats.

## Phage Antibody Libraries from Immunized Humans, Chickens, and Other Animals

In many phage antibody projects, antibody domains are obtained from animals that are immune to one or more antigens of interest, either because they have been artificially immunized with a vaccine or other immunogen or because they have been naturally infected by a pathogen. For example, Jing Yi Lai and Theam Soon Lim of Universiti Sains Malaysia have reviewed projects in which the donors are humans who have been infected by viruses, bacteria, or parasites (Lai and Lim 2020).

Junho Chung of Seoul National University and his colleagues introduce the chicken immune system (see Chapter 6, Introduction: Generation of Chicken Antibody Libraries and Selection of Antigen Binders [Yang et al. 2025c]) and detail the construction and use of a chicken phage antibody library displaying scFv domains against a chosen antigen (see Chapter 6, Protocol 1: Chicken Immunization Followed by RNA Extraction and cDNA Synthesis for Antibody Library Preparation [Yang et al. 2025a], Chapter 6, Protocol 2: Generation of a Phage Display Chicken Single-Chain Variable Fragment Library [Yang et al. 2025b], Chapter 6, Protocol 3: Preparation of VCSM13 Helper Phage for Display Library Reamplification and Bio-Panning [Yang et al. 2025d], and Chapter 6, Protocol 4: Selection of Antigen Binders from a Chicken Single-Chain Variable Fragment Library [Yang et al. 2025e]). The project that the students carry out in the Phage Display Course largely follows these articles, using a pComb3 family phagemid vector developed at Scripps (Andris-Widhopf et al. 2000; Cary et al. 2000).

The chicken antibody system differs from the human and mouse systems in several respects that make library generation simpler, and therefore particularly appropriate for the Phage Display Course (see Chapter 6, Introduction: Generation of Chicken Antibody Libraries and Selection of Antigen Binders [Yang et al. 2025c]). There are only two immunoglobulin loci in chickens, one for the heavy (H) chain and one for the λ light chain; there is no κ light chain. There is only a single functional germline $V_H$ gene at the H locus, followed by four diversity (D) segments and a single joining (J) segment. Diversity is generated by VDJ joining and multiple somatic gene conversion events with more than 80 upstream $V_H$ pseudogenes. There is also only a single functional $V_\lambda$ gene at the λ locus, followed by a single J segment. Diversity is generated by VJ joining and multiple somatic gene conversion events with 25 upstream $V_\lambda$ pseudogenes. At both loci, there is no sequence variation near the end of assembled V genes in antibody-producing B cells, and very limited diversity near the beginning. The entire repertoire of $V_H$ genes can be amplified with only three forward primers combined with a single reverse primer, and the same is true for the $V_\lambda$ gene repertoire (see Chapter 6, Protocol 2: Generation of a Phage Display Chicken Single-Chain Variable Fragment Library [Yang et al. 2025b]). Most of the repertoire can be captured with a single forward primer for $V_\lambda$ genes and a single forward primer for $V_H$ genes, as is done in the Phage Display Course. This simplicity is in sharp contrast to the profusion of primers required to capture the repertoire of $V_H$, $V_\lambda$, and $V_\kappa$ genes in human B cells (see Chapter 5, Protocol 1: Generation of Antibody Libraries for Phage Display: Human Fab Format [Peng and Rader 2024b]). Constructing an scFv library from chickens is therefore relatively straightforward, making it a practical exercise in the Phage Display Course.

Chicken scFv libraries are constructed from laying hens that have been hyperimmunized with the antigen of interest (Andris-Widhopf et al. 2000; Cary et al. 2000; see also Chapter 6, Protocol 2: Generation of a Phage Display Chicken Single-Chain Variable Fragment Library [Yang et al. 2025b]). The progress of the hens' immune response to the antigen is assayed by titering immunoglobulin in blood or in egg yolks, each of which contains up to 100 mg of IgY (the chicken homolog of mammalian IgG). When the titer is adequate, the hens are euthanized, and RNA is extracted from lymphoid organs, such as spleen and bone marrow. The RNA is reverse-transcribed to create cDNA molecules that include $V_H$ and $V_\lambda$ coding sequences. The cDNAs in turn serve as templates for PCR amplification of the hens' entire $V_H$ and $V_\lambda$ repertoires, using the simple primer sets described in the previous paragraph. The primers are designed to create a flexible linker between the $V_H$ and $V_\lambda$ amino acid sequences so that they fold into a functional scFv domain, and to install suitable flanking restriction sites so that the scFv construct can be inserted into the phagemid vector in the correct reading frame, as illustrated for SfiI cloning sites in Figure 7. Vector DNAs bearing the scFv inserts are transfected into *E. coli* host cells; when the transfected cells are infected with helper phages, they release virions bearing scFv domains on their recombinant pIII subunits. These virions are the input to the first round of affinity selection.

Chicken phage antibody libraries can be smaller than the all-purpose libraries discussed in the next subsection. All-purpose libraries, either synthetic or representing large numbers of individual donors, are intended to include antibody domains that bind almost any chosen antigen of interest. Hundreds of millions to hundreds of billions of independent phage clones are required to meet this goal. A large portion of the library must be used to ensure that individual clones are represented by a sufficient number of virions. A chicken phage antibody library, in contrast, represents the antibody repertoire of chickens that have been hyperimmunized with the antigen of interest, so that a substantial number of their circulating antibodies are specific for the immunogen. A few million independent clones usually suffice to include high-affinity antibodies specific for the antigen.

More than 20 years ago, a few hens were hyperimmunized with the dye fluorescein conjugated to bovine serum albumin (FL-BSA), as described previously (Andris-Widhopf et al. 2000). The hens made not only antibodies that reacted with BSA, but also antibodies that reacted with the fluorescein dye. Small molecules like fluorescein that provoke an antibody response when conjugated to a protein antigen are called haptens; the proteins they are conjugated to are called carriers. RNA was extracted from the hens' spleens and bone marrows and frozen in aliquots for long-term storage. Aliquots have served as the starting point for the Phage Display Course's student projects ever since. The availability of that RNA stock, as well as the small size of the library required to represent the repertoire of anti-fluorescein scFv domains, have made it practical for the students to complete their project in the 2 weeks available.

More generally, hyperimmunizing a few hens with a chosen antigen of interest, euthanizing them and harvesting their spleens and bone marrow, and constructing a small phage antibody library and affinity-selecting antibodies against the antigen or hapten using easily obtained supplies and standard protocols—that can be an attractive alternative to constructing a large all-purpose library (next subsection), if such a resource is unavailable or too costly to buy or construct.

## All-Purpose Phage Antibody Libraries: Naive and Synthetic

All-purpose libraries are designed to represent such a diversity of independent clones that they are likely to include high-affinity antibody domains specific for almost any antigen of interest. There are two major types (Almagro et al. 2019): natural naive libraries, meant to express the entire natural antibody repertoire of many (sometimes as many as 200) donors (human or nonhuman) who have not been purposely immunized with an antigen of interest, and synthetic libraries, in which randomized synthetic complementarity-determining regions (CDRs) are installed in the backbone of an animal's (usually a human's) natural antibody variable regions.

A 2017 review summarizes anti-pathogen antibodies that have been affinity-selected from dozens of natural naive libraries assembled from RNA pooled from many healthy human donors (Chan et al. 2017). A series of articles by Haiyong Peng (University of Florida) and Rader details construction of naive human libraries in Fab format (see Chapter 5, Protocol 2: Generation of Antibody Libraries for Phage Display: Chimeric Rabbit/Human Fab Format [Peng and Rader 2024a], Chapter 5, Protocol 1: Generation of Antibody Libraries for Phage Display: Human Fab Format [Peng and Rader 2024b], Chapter 5, Protocol 5: Generation of Antibody Libraries for Phage Display: Library Reamplification [Peng and Rader 2024c], Chapter 5, Protocol 3: Generation of Antibody Libraries for Phage Display: Preparation of Electrocompetent *E. coli* [Peng and Rader 2024d], Chapter 5, Protocol 4: Generation of Antibody Libraries for Phage Display: Preparation of Helper Phage [Peng and Rader 2024e], Chapter 5, Protocol 6: Phage Display Selection of Antibody Libraries: Panning Procedures [Peng and Rader 2024f], and Chapter 5, Protocol 7: Phage Display Selection of Antibody Libraries: Screening of Selected Binders [Peng and Rader 2024g]).

Like natural naive libraries, synthetic libraries do not entail immunizing live animals or humans. However, they also have a critical advantage: Antibodies against self-antigens are not censored by immunological tolerance. Tolerance might block discovery of high-affinity human antibody domains against human antigens, for instance. Human anti-human antibodies are an increasingly important class of pharmaceuticals.

Three articles by Sidhu, Veggiani, Maryna Gorelik (University of Waterloo), and Shane Miersch (University of Waterloo) review the "minimalist" approach to synthetic human libraries that the Sidhu laboratory has been pursuing for more than two decades (see Chapter 7, Overview: Structural Survey of Antigen Recognition by Synthetic Human Antibodies [Gorelik et al. 2025], Chapter 8, Introduction: Beyond Natural Immune Repertoires: Synthetic Antibodies [Veggiani and Sidhu 2024a], and Chapter 8, Protocol 1: Generation and Selection of Synthetic Human Antibody Libraries via Phage Display [Veggiani and Sidhu 2024b]). The overall goal is to develop a practical way to construct libraries from which promising therapeutic antibodies for any chosen antigen can be affinity-selected. They focus on two desirable physical properties of such antibodies: high affinity for the antigen and high "developability," meaning possession of properties such as solubility and stability that facilitate manufacture and favorable pharmacokinetic behavior. The defining feature of the minimalist approach is using an already existing therapeutic antibody with favorable developability as the sole scaffold for mutagenesis, thus greatly simplifying library construction and hopefully increasing the chance that antibody domains selected from the library will share the favorable developability of the chosen scaffold antibody. The scaffold for many minimalist projects has been the anti-HER2 antibody trastuzumab. Minimalism's key assumption is that high affinity can be achieved without exploiting the entire natural diversity of scaffolds in natural antibodies. This assumption has been largely vindicated in the sequel.

Minimalist libraries are constructed by introducing random mutations in strategically chosen codons in the scaffold antibody's six CDRs. One of the three articles mentioned above provides a general protocol for constructing such a library (see Chapter 8, Protocol 1: Generation and Selection of Synthetic Human Antibody Libraries via Phage Display [Veggiani and Sidhu 2024b]):

- A "stop template" is cloned in a phagemid vector. The stop template encodes the scaffold sequence (e.g., trastuzumab), but with a stop codon at each position chosen for mutagenesis.

- The resulting vector is transfected into a mutant *E. coli* host that incorporates deoxy-U (dU) in place of many T nucleotides. The transfected cells are infected with helper phages and incubated in medium that selects for both phagemid and helper. Virions are prepared from the culture supernatant (there are no phagemid-encoded pIII molecules on these virions).

- Single-stranded plus-strand DNA is prepared from the virions; that DNA has numerous dU nucleotides in place of T nucleotides.

- Synthetic minus-strand mutagenic oligonucleotides, one for each of the six CDRs, are hybridized to the plus-strand virion DNA. These oligonucleotides have randomized sense codons opposite each of the stop codons in the plus strand. The minus strand is then closed by incubation with DNA polymerase and ligase. This reaction creates double-stranded heteroduplex DNA whose plus strand has numerous dU nucleotides and stop codons at the sites chosen for mutagenesis, and whose minus strand has no dU nucleotides and has randomized sense codons at the sites chosen for mutagenesis. Many (but not all) of the heteroduplexes will be covalently closed circles. This DNA is stored frozen in aliquots for use in preparing fresh libraries, as described in the next bullet.

- To prepare a new library, an aliquot of the heteroduplex DNA is electroporated into electrocompetent *E. coli* prepared from helper-infected cells. The electroporated cells are cultured in medium that selects for both helper and phagemid genomes. The cells do not incorporate dU nucleotides into DNA, and have an intact dU repair system that cuts the plus strand into small segments. The great majority of replicating DNAs, and of the DNAs in the secreted virions, are therefore derived from covalently closed circular minus strands, in which the stop codons have been replaced with randomized sense codons. Phages displaying antibody domains that bind a chosen antigen can then be affinity-selected from the resulting library.

By now, many human antibodies that have been affinity-selected from minimalist libraries are available for comparison to natural antibodies (see Chaper 7, Overview: Structural Survey of Antigen Recognition by Synthetic Human Antibodies [Gorelik et al. 2025]). These authors compared 50 antibodies affinity-selected from minimalist libraries with 50 natural antibodies from the same $V_H$ and $V_\kappa$ subgroups. High-resolution structures of the antibody–antigen interaction were available for all 100 antibodies. By several broad criteria, the structures of the CDRs in the two types of antibody and the interaction of the CDRs with the antigen were strikingly similar. Figure 8 shows an example of the authors' analysis. All CDR amino acids defined as "core" in the figure are represented in at least 60% of the structures analyzed. To facilitate graphical comparison, those core amino acids are mapped onto

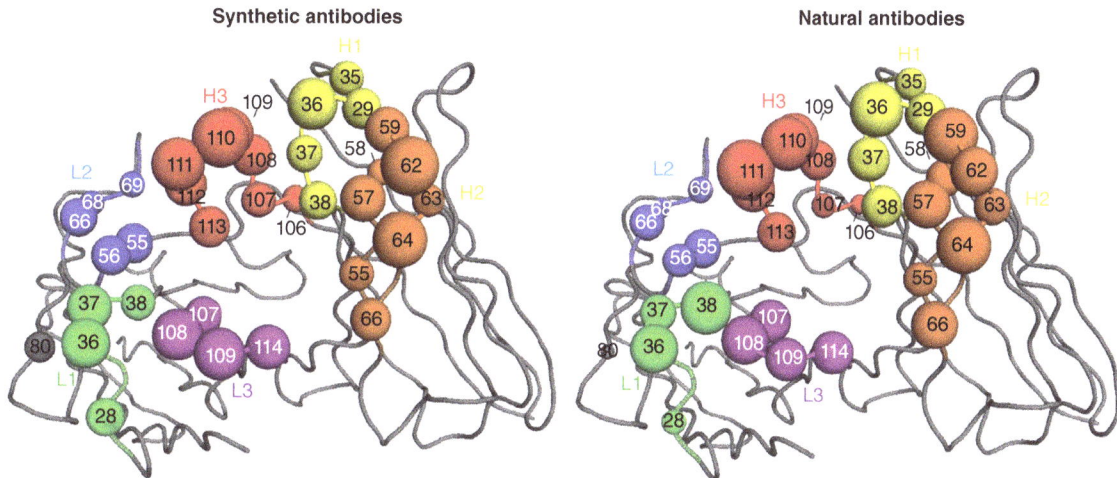

FIGURE 8. Graphical representation of the interaction between "core" complementarity-determining region (CDR) amino acids and antigen in synthetic and natural antibodies. The definition of core amino acids and their representation in the figure are explained in the text. Figure reprinted from Chapter 7, Overview: Structural Survey of Antigen Recognition by Synthetic Human Antibodies (Gorelik et al. 2025) with permission from Cold Spring Harbor Laboratory Press.

the backbone of the trastuzumab Fab (even though the backbone conformation differs significantly from one antibody to another). The size of each colored circle in the figure is proportional to the percent of structures in which that core amino acid participates significantly in the CDR–antigen interface; amino acids for which this percent was <20 are omitted. The striking similarity of the graphs for synthetic and natural antibodies implies that synthetic and natural CDRs interact with antigen in very similar ways. The authors argue that their results suggest considerable commonality of CDR structure, regardless of the backbone in which they reside, and that high-affinity antibodies can be selected from libraries in which only one of the natural antibodies' myriad scaffolds is represented.

A series of four articles by Mary Ann Pohl of Ailux Biologics and Juan Almagro of GlobalBio, Inc., describe an elaboration of the minimalist concept that they call semisynthetic libraries (see Chapter 9, Protocol 1: Semisynthetic Phage Display Library Construction: Design and Synthesis of Diversified Single-Chain Variable Fragments and Generation of Primary Libraries [Almagro and Pohl 2024a], Chapter 9, Protocol 2: Semisynthetic Phage Display Library Construction: Generation of Filtered Libraries [Almagro and Pohl 2024b], Chapter 9, Protocol 3: Semisynthetic Phage Display Library Construction: Generation of Single-Chain Variable Fragment Secondary Libraries [Almagro and Pohl 2024c], and Chapter 9, Introduction: Considerations for Using Phage Display Technology in Therapeutic Antibody Drug Discovery [Pohl and Almagro 2024]).

Such libraries are created in three stages:

- Construction of four primary libraries: Each of the four primary scFv libraries is constructed on one of four human $V_\kappa$ scaffolds in the N-terminal position, connected by a flexible linker to a common human $V_H$ scaffold in the C-terminal position. The $V_H$ scaffold's V gene segment is followed by a mixture of 90 combinations of human diversity (D) and joining (J) gene segments. The sequence from the beginning of CDR3 to the end of FR4 (corresponding to the last codon of the V gene segment through the D and J gene segments) is a temporary placeholder that will be replaced in construction of the secondary libraries (last bullet). Hopefully, the very limited variation in the placeholder's sequence and length is sufficient for the purposes of "filtration" (next bullet). Meanwhile, random variation is introduced at strategically chosen codons in all three CDRs of the $V_\kappa$ scaffolds and CDR1 and CDR2 of the $V_H$ scaffold. The four primary libraries are assembled by chemical DNA synthesis and stored as frozen aliquots.

- Construction of four filtered libraries: The "filtered" libraries consist of members of the primary libraries whose scFv domains are resistant to heat shock and are therefore presumably well folded into a stable native three-dimensional structure. Construction begins by PCR-amplifying an aliquot of each of the four primary libraries, splicing the amplified DNA into the SfiI cloning sites of a pComb3-like phagemid vector, and propagating the library by electroporating the spliced DNA into an *E. coli* host and infecting with a helper phage. Partially purified phagemid virions are prepared from the four culture supernatants. The virions are then heat-shocked and cooled to allow renaturation of the displayed scFvs (the virions themselves are impervious to the heat shock). Superparamagnetic beads coated with protein L are used to affinity-purify virions from each of the four heat-shocked virion preparations, discarding virions that are not captured on the beads. Protein L binds FR1 of all four $V_\kappa$ scaffolds in well-folded scFvs. The affinity-purified virions are therefore enriched for virions displaying particularly stable scFvs. Virions from each of the four libraries are released from the beads and propagated separately by infecting *E. coli* host cells. Aliquots of phagemid-bearing cells are frozen to serve as input to construction of the secondary libraries (next bullet).

- Construction of four secondary libraries: An aliquot of frozen phagemid-bearing cells from each of the four filtered libraries is thawed and cultured in medium that selects for the phagemid (the cells are not infected with helper phages, and the medium does not select for helpers). Phagemid DNA is extracted from the four cultures and partially purified. A secondary library is prepared from each of the four phagemid DNAs by replacing the 90 placeholder CDR3-FR4 sequences at the end of the $V_H$ regions with a mixture of hundreds of millions of natural CDR3-FR4 sequences prepared from 200 adult human donors. Replacement is mediated by a highly conserved nucleotide sequence at the end of human $V_H$ FR3. The conserved sequence serves as a PCR priming site for fusing the natural and filtered library sequences. Each of the four fused DNAs is spliced into the SfiI cloning sites of the pComb3-like phagemid vector, which is electroporated into *E. coli* host cells. The phagemid-bearing cells are infected with helper phages and cultured. Phagemid virions prepared from the culture supernatants comprise the four secondary libraries.

The resulting ensemble of four semisynthetic libraries is called "ALTHEA Gold Plus Library." It is a successor to the "ALTHEA Gold Library," an ensemble of two semisynthetic libraries, representing two, rather than four, human $V_\kappa$ scaffolds (Valadon et al. 2019); for the ALTHEA Gold Library, filtration was accomplished with protein A rather than protein L.

High-affinity, conformationally stable scFvs have been affinity-selected from both the ALTHEA Gold Library (Valadon et al. 2019; Pedraza-Escalona et al. 2021; Dao et al. 2022a,b) and the ALTHEA Gold Plus Library (Guzmán-Bringas et al. 2023; Mata-Cruz et al. 2025). Mata-Cruz and her colleagues, for example, sought human scFvs that bind human CD36, a multifunctional membrane protein that is implicated in numerous diseases and is the target of drug development projects (Guerrero-Rodríguez et al. 2022). After three rounds of affinity selection using the extracellular domain of human CD36 as selector, they analyzed phagemids from 90 randomly selected clones. Fifty-seven of the phagemids, representing four distinct scFvs, bound strongly and specifically to the CD36 selector. All but 17 of the 90 phagemids, including all 57 that bound strongly to the selector, bound strongly to protein L, indicating that their displayed scFvs were predominantly in a natural immunoglobulin conformation. One of the four distinct scFvs, called D11, was shown to mimic the anti-metastatic effects of a mouse monoclonal antibody that binds both mouse and human CD36 (Pascual et al. 2017). These results commend D11 as a possible pharmaceutical lead. Whether or not that promise is realized, the results demonstrate the effectiveness of the Althea Gold Plus Library as a source of well-folded, high-affinity scFvs for a chosen selector.

## Affinity Maturation

In the course of a natural antibody response, the affinity of the antibodies for the antigen increases progressively as a result of somatic mutation of the V regions in the responding B cells, along with clonal competition and increasingly stringent selection for high affinity by the antigen. An artificial analog of natural affinity maturation can be imposed on phage antibody clones that emerge from affinity selection with the antigen of interest (next section). DNA sequences from the selected clones are mutagenized at strategic positions and used to create a library of mutagenized phages. The mutagenized library is subjected to increasingly stringent affinity selection, in the hope of creating phage antibody clones with even higher affinity (Tiller et al. 2017).

## AFFINITY SELECTION

In the phage antibody context, as diagrammed in Figure 2, affinity selection is accomplished by immobilizing the antigen of interest on a substrate of some kind, and then reacting the antigen-coated substrate with a library of phage antibodies (see Chapter 3, Overview: Principles of Affinity Selection [Smith 2024]). Virions displaying antibody domains that happen to bind the immobilized antigen are captured on the substrate surface, while other virions—the overwhelming majority—remain free in solution. Virions that have not been captured are washed away, and the captured virions are released from the surface; the released virions are the unamplified output of affinity selection. The output virions are then used to infect fresh bacterial cells, which are cultured in growth medium; the resulting progeny virions constitute the amplified output. Each virion in the unamplified output is represented by millions or billions of genetically identical progeny virions in the amplified output. The amplified output serves as the input virions for another round of affinity selection or is analyzed to identify promising clones for further investigation.

A generic vocabulary is introduced here in order to summarize the principles of affinity selection in general, not just in the particular case of affinity selection from phage antibody libraries. The immobilized species (an antigen in the case of phage antibody projects) is referred to generically as the "selector." The species that is displayed on the surface of a virion, and potentially binds the immobilized selector (an antibody domain in the case of phage antibody projects), is referred to generically as a "peptide" regardless of size; whether the peptide is a protein, a protein domain, or a small peptide is to be understood in context. Using this vocabulary, a few general principles of affinity selection were explained (see Chapter 3, Overview: Principles of Affinity Selection [Smith 2024]) and are discussed here.

### Yield Versus Stringency

Theoretically, the yield of a phage clone from a round of affinity selection is the ratio of the number of virions of that clone in the input to that round to the number of virions of that clone in the unamplified output of that round. It is not generally possible to measure yield directly, but it is possible to estimate the relative yields of different phage clones bioinformatically, as explained in the final subsection of this section.

There are two types of yield:

- Specific yield is dependent on the identity of the displayed peptide. The obvious reason for high specific yield is that the clone in question displays a peptide with high affinity for the selector—exactly the desired property in an affinity selection project. Such specific clones are selector-specific, as are their yields. However, some clones owe their relatively high specific yields to something other than high affinity of their displayed peptides for the selector. One example is a clone whose displayed peptide binds the substrate, rather than the selector. Such clones are examples of selector-unrelated phages, or SUPs, which are the subject of the next subsection.

- Nonspecific background yield is independent of the identity of the displayed peptide. Some virions, whatever peptide they display, end up in the output just by chance, because no physical separation process is perfect. Nonspecific background yield is very low in affinity selection, but it is not zero.

Stringency is the degree to which a round of affinity selection favors phage clones whose displayed peptides bind the immobilized selector with high affinity over clones whose displayed peptides bind the immobilized selector with low affinity.

# Chapter 1

High yield and high stringency are both desirable characteristics of a round of affinity selection, but they are mutually antagonistic: High stringency almost always entails low yield and vice versa. High yield is mainly promoted by a high surface density of selectors on the substrate, while high stringency is promoted by a low surface density of selectors on the substrate. In the Phage Display Course, for example, stringency of selection for scFvs that bind the fluorescein (FL) hapten selector could be adjusted by immobilizing mixtures of FL-conjugated ovalbumin (FL-Ova) and unconjugated Ova in different proportions on the substrate surface. The smaller the proportion of FL-Ova, the higher the stringency.

High yield is imperative in the first round of affinity selection, when the input is a large library with billions of clones, even though that entails low stringency. Each individual clone in such a library is represented by a limited number of genetically identical virions—10 or 100, say. That includes the most desired clones: those displaying peptides that bind the selector with high affinity. If stringency is so high that not even a single virion from such a clone is represented in the unamplified output of the first round of affinity selection, that clone cannot magically reappear in subsequent rounds.

Stringency can be greatly increased in the second and subsequent rounds. That is because even when yield in the first round is as high as practical (i.e., stringency is as low as practical), the overall yield is still quite low. The number of clones represented in the output is almost never more than 0.01% of the number of clones represented in the starting library, and usually is much lower. That means that each clone can be represented by at least 10,000 times more identical virions in the input to the second round than in the input to the first round, assuming that the total number of virions in the inputs to the first and second rounds are the same.

## Selector-Unrelated Phages, or SUPs

SUPs are unwanted phage clones that increase in prevalence with successive rounds of affinity selection for reasons other than affinity of their displayed peptides for the selector (Thomas et al. 2010). They can be classified into two types.

- Some SUPs reflect high specific yield during the capture or release stages of affinity selection (the capture, release, and amplification stages of affinity selection are described in the first paragraph of this section). For instance, the displayed peptide might bind some component of the affinity selection apparatus other than the immobilized selector—the substrate itself, for example.

- Other SUPs reflect overreplication during the amplification stage in each round of affinity selection. This unwanted overreplication is specific for the displayed peptide, but it is seldom caused by the displayed peptide. Much more often, it is the result of a property that is determined by genetic alterations that are linked to the coding sequence for the displayed peptide in the same phagemid chromosome.

I argue that with a few exceptions, measures to eliminate SUPs from a library have limited success (see Chapter 3, Overview: Principles of Affinity Selection [Smith 2024]). A more promising general approach today is to identify SUPs bioinformatically, with the aid of high-throughput sequencing (HTS) of the clones that emerge from an affinity selection project. HTS is already increasingly favored as the "primary readout" in affinity selection projects; that is, the first stage in analysis of the selected clones. The role of HTS in identifying phage clones of interest, while eliminating SUPs from consideration, is the subject of the next subsection. That subsection discusses HTS in the context of affinity selection from phage antibody libraries, but its application to other types of libraries and selectors should be apparent.

## High-Throughput Sequencing (HTS) as Primary Readout in Affinity Selection Projects

A series of articles by Brandon DeKosky and his colleagues at the Massachusetts Institute of Technology introduces HTS (see Chapter 10, Introduction: Beyond Single Clones: High-Throughput Sequencing in Antibody Discovery [Fahad et al. 2025b]) and details its use in antibody engineering (see Chapter 10, Protocol 2: Antibody Data Analysis from Diverse Immune Libraries [Fahad et al.

2025a], Chapter 10, Protocol 3: Clonal Lineage and Gene Diversity Analysis of Paired Antibody Heavy and Light Chains [Fahad et al. 2025c], and Chapter 10, Protocol 1: Clonal Variant Analysis of Antibody Engineering Libraries [Fahad et al. 2025d]).

HTS can deliver partial sequences of the antibody domains displayed by millions of virions in the output of a round of affinity selection. That is usually enough for a thorough census of the selected antibody domains in the outputs of the first and second rounds of affinity selection, giving the number of times each clone occurs; that number is referred to as the clone's reads. A clone's reads divided by the sum of the reads for all clones in the output is a measure of the clone's prevalence. The amplified output of the first round of affinity selection is the input to the second round. Therefore, the ratio of a clone's prevalence in the amplified output of the second round of affinity selection to its prevalence in the amplified output of the first round is a measure of the clone's yield in the second round of affinity selection relative to the yields of the other clones in that round. If, following the advice in the first subsection above, stringency is greatly increased in the second round, clones whose displayed antibody domains bind the selector with high affinity are strongly favored over clones displaying low-affinity antibody domains. This profile of the clones' relative yields in the second round may be enough to choose the most promising clones for further investigation. Promising clones can be recovered from the second-round output by PCR with clone-specific primers.

Not all clones with high second-round yields necessarily display antibody domains with high affinity for the selector. Some of them might be SUPs, as explained in the previous subsection. Unwanted SUPs can be distinguished from desirable high-affinity clones if affinity selections with multiple selectors are carried out in parallel from a common library. The parallel selections should be as identical as possible, apart from the identity of the selectors. SUPs reveal themselves because their yield is high in all (or almost all) the parallel selections, regardless of the selector.

## DISPLAY OF PEPTIDES AND PROTEIN DOMAINS OTHER THAN ANTIBODIES

Although phage antibodies are the most conspicuous arena of phage display technology and the focus of the Phage Display Course's laboratory experiments, many other types of peptides and proteins have been displayed on filamentous phages, a few of which are described below.

### Epitope Discovery

"Epitope discovery"—identification of peptides or protein domains that are recognized by antibodies—has been an important application of phage display technology from its earliest days (Cortese et al. 1994). Antibodies—either monoclonal or polyclonal—are used to affinity-select antibody-binding peptides from two types of phage display library: random peptide libraries (RPLs) and natural peptide libraries (NPLs). RPLs display synthetic peptides with randomized sequences; NPLs display the entirety or fragments of natural proteins.

Two articles by Gregg Silverman (see Chapter 11, Introduction: Insights from the Study of B-Cell Epitopes of a Microbial Pathogen by Phage Display [Silverman 2025a] and Chapter 11, Protocol 1: Cloning and Selection from Antigen Fragment Libraries for Epitope Identification [Silverman 2025b]), summarizing recent work by his group at New York University (Hernandez et al. 2020a, b), exemplify epitope discovery from NPLs. His laboratory has developed a suite of techniques for rapid identification of short peptide epitopes, for potential incorporation into modular vaccines against infectious diseases. To this end, genes for target pathogen proteins are randomly fragmented into small DNA pieces, which are cloned into a pComb-derived vector called pComb-Opti8. This vector displays the cloned peptides on subunits of the major coat protein pVIII through a flexible linker, rather than on subunits of minor coat protein pIII (as illustrated in Fig. 4). The resulting NPLs include DNA inserts in all possible reading frames in both orientations, but there is sufficient redundancy for many thousands of in-frame inserts, spanning the entire protein-coding sequence, to be represented. Monoclonal or polyclonal antibodies of interest are then used to affinity-select phage clones displaying protein fragments that bind those antibodies. Individual affinity-selected clones can be sequenced to identify the selected epitope-containing peptides. Alternatively, and increasingly, outputs of affinity selection are analyzed en masse by HTS, as outlined in the final subsection of the previous section.

The investigators' "proof-of-principle" project targets the leucocidins, which are major virulence factors in invasive infection by *Staphylococcus aureus*, including by methicillin-resistant *S. aureus* (MRSA). The leucocidins, reviewed by Spaan et al. (2017), are eight-member β-barrel rings of alternating type S and type F subunits. Such rings kill cells in infected hosts by creating pores in the cell membrane, thus breaching the barrier between inside and outside. There are multiple leucocidins, with different type S and type F pairs. In most cases, the two types of subunits are secreted separately. The type S subunit binds a receptor on a target cell; then, additional type S subunits and type F subunits assemble into the β-barrel pore at the receptor site. Different leucocidins target different cell types, depending on the receptors they bind. Receptors have been identified for the major leucocidins, and X-ray crystallographic structures are available for many leucocidins and individual type S and type F subunits.

Two illustrative articles from Silverman's group report construction of gene fragment libraries from two type S subunits in the pComb-Opti8 vector, and affinity selection of phage-borne peptides from each library, using as selectors a mouse monoclonal antibody elicited by immunization with the respective unfragmented parent type S subunit (Hernandez et al. 2020a,b). Clone-by-clone sequencing (Hernandez et al. 2020b) or HTS analysis (Hernandez et al. 2020a) of the affinity-selected clones from each library identified a minimal epitope recognized by the selecting monoclonal antibody. Both epitopes were also recognized by antibodies from patients recovering from invasive *S. aureus* infection. The availability of the X-ray crystallographic structure of the type S subunits allows the minimal epitopes to be located in the native folded proteins. One epitope corresponds to a solvent-exposed β-turn at the tip of a β-hairpin (Hernandez et al. 2020a); the other corresponds to an entire solvent-exposed β-hairpin (Hernandez et al. 2020b).

Remarkably, when mice were immunized with these small peptides conjugated to keyhole limpet hemocyanin, they produced antibodies that bind not only the immunizing peptides, but also the intact parent proteins from which the peptides derive. This ability to elicit antibodies that react with the target pathogen protein in its native state has been called immunogenic fitness (Matthews et al. 2002); it is presumably a necessary condition for almost any effective vaccine component. It is not clear how general strong immunogenic fitness will be for short peptide epitopes affinity-selected from fragment libraries. For some parent proteins, many or most epitopes may be discontinuous, composed of amino acids from distant parts of the linear amino acid sequence. In any case, the logic of these investigators' strategy does not depend on the size distribution of the protein fragments. A gene could be randomly fragmented into much larger pieces, some of which happen to encode peptides that fold into natural subdomains that are present on the intact protein. Discontinuous epitopes on such subdomains are likely to be shared by the pathogen or pathogenic protein itself, and therefore have good prospects of strong immunogenic fitness.

## Display of Nonantibody Antigen-Binding Domains

Antibody V regions are folded β-sheets supporting three loops with diverse amino acid sequences at one end. The β-sheet is conformationally stable, serving as a secure "scaffold" that keeps the loops in place despite their diversity. It is the diversity of the loops that accounts for the antibodies' ability to bind specifically to an enormous diversity of antigens. The loops that mediate binding by an antibody are called complementarity-determining regions (CDRs), because it is their sequence diversity that confers an almost unlimited diversity of complementarity (i.e., antigen-binding specificity) on antibodies.

Several stable nonantibody scaffolds have been explored for construction of "phage nonantibody libraries" (as I call them here) representing a great diversity of binding sites. From such libraries, specific, high-affinity ligands for a great diversity of selectors can be affinity-selected. I call the selectors "antigens" here to highlight the parallelism between nonantibody binding domains and canonical antibodies. Such nonantibody antigen-binding domains, like antibody V regions, are composed of a structurally stable scaffold holding in place CDRs with highly diverse sequences and antigen-binding specificities.

Akiko Koide and Shohei Koide of New York University Grossman School of Medicine summarize the construction and use of monobodies, whose scaffolds are fibronectin type III (FN3) domains (see Chapter 12, Overview: Use of Phage Display and Other Molecular Display Methods for the Development of Monobodies [Koide and Koide 2024]). The FN3 motif is widespread in nature: Humans produce 4104 of them in 673 proteins, a few of which have served as scaffolds for monobodies. Like antibody V regions, FN3 domains are folded β-sheets, but with seven rather than nine β-strands. As in the case of antibody V regions, there are three loops at one end of the folded sheet, connecting pairs of anti-parallel β-strands. The amino acid sequences in these loops can vary greatly without weakening the folded β-sheet structure; that sequence diversity gives monobodies the ability to bind specifically to a great diversity of antigens. Unlike an antibody V region, a monobody is not paired with another monobody in a stable superstructure comparable to a Fab. They are expressed as stand-alone antigen-binding domains, like the VHH domains in some camelid antibodies (Muyldermans 2013).

The investigators set out to display monobodies fused to the C-terminal domain of filamentous phage pIII protein, using a phagemid vector in the pComb3 family (see "Phagemid Vectors for Phage Antibody Display"), and have explored other display formats as well. Efficient display required changing pComb3's ompA signal peptide to a signal peptide dependent on the signal recognition particle (SRP), so that the protein was secreted through the inner membrane cotranslationally, rather than post-translationally. They argue that this change was necessitated by one of the FN3 domain's chief attractions: its extraordinary noncovalent stability, which is not reinforced by an intradomain disulfide bond. Rapid formation of the folded structure in the cytoplasm would arguably interfere with post-translational secretion.

Stefan Ståhl and his colleagues at KTH Royal Institute of Technology in Stockholm have authored articles introducing affibodies, another type of nonantibody antigen-binding domain (see Chapter 15, Protocol 3: Selection of Affibody Molecules Using *Escherichia coli* Display [Dahlsson Leitao et al. 2024], Chapter 15, Protocol 2: Selection of Affibody Molecules Using Phage Display [Hjelm et al. 2024], Chapter 15, Protocol 4: Selection of Affibody Molecules Using Staphylococcal Display [Löfblom et al. 2024], Chapter 15, Protocol 1: Cloning of Affibody Libraries for Display Methods [Ståhl et al. 024a], and Chapter 15, Introduction: Engineering of Affibody Molecules [Ståhl et al. 2024b]). Here, the scaffold is a three-helix bundle, and the CDRs are not loops protruding from one end of the bundle, but rather 13 amino acids in helices 1 and 2 that protrude from one side of the bundle and are not required for its structural integrity. The investigators explain how to display affibodies on a pComb3 family phagemid vector, on the surface of *E. coli* bacteria, and on the surface of *Staphylococcus carnosus* bacteria. With only 58 amino acids, affibodies are smaller than any other antigen-binding domain, yet they can have subnanomolar and even subpicomolar affinities for their target antigens. These features commend them for biomedical applications requiring rapid and long-lasting binding to antigens in human subjects, along with rapid clearance of unbound molecules. For example, affibody ABY-25 has 76 pM affinity for the breast cancer marker HER2 (Feldwisch et al. 2010). When conjugated to the chelator DOTA and labeled with the positron-emitting radionuclide $^{68}$Ga, whose half-life is only 68 min, it allows metastatic lesions to be imaged at high resolution in HER2-positive breast cancer patients by positron emission tomography (Velikyan et al. 2019; Alhuseinalkhudhur et al. 2020).

## Display of Specialized Protein Domains

Gregory Martyn of the University of Toronto and Veggiani describe libraries of modified Src homology 2 (SH2) domains, which bind phosphorylated tyrosine (pTyr) residues on specific receptor tyrosine kinases or other proteins (see Chapter 13, Overview: Phage-Displayed SH2 Domain Libraries: From Ultrasensitive Tyrosine Phosphoproteome Probes to Translational Research [Martyn and Veggiani 2024]). The human genome includes 122 of these ~100-amino-acid domains, present on intracellular signal transduction proteins. When an SH2 domain binds pTyr, the protein that it is part of is triggered to set in train a signal transduction cascade. Being able to exploit this specificity to detect or modify the behavior of specific regulatory pTyr residues could have many potential biomedical applications. However, although SH2 domains generally bind specifically to their cognate pTyr residues, their affinity is generally too weak for most applications. The article describes procedures for randomizing strategic codons in an SH2-coding sequence, using phage display to affinity-select variants with higher affinity or altered specificity.

An article by Chen T. Liang, Olivia Roscow, and Wei Zhang summarizes affinity selection of high-affinity ligands for chosen selectors from libraries of ubiquitin (Ub) variants (see Chapter 14, Overview: Generation and Characterization of Engineered Ubiquitin Variants to Modulate the Ubiquitin Signaling Cascade [Liang et al. 2024]). Ub is a highly conserved 76-amino-acid protein present in virtually all eukaryotic cells. The article emphasizes how detailed knowledge of the structure and properties of Ub can guide affinity selection projects. Ub might well have been included in the previous subsection, as yet another attractive scaffold for "phage nonantibody libraries," as I call them there. Following the pioneering work of the Sidhu laboratory (Ernst et al. 2013), however, almost all proteins chosen as selectors for Ub variants already interact specifically with Ub (Tang et al. 2023).

Ub is central to a sprawling network of cellular functions.

- Ubiquitination—enzymatic covalent conjugation of Ub to target proteins, which are called "substrates"—is at the heart of the Ub system. It is carried out by an enzyme cascade composed of three types of enzymes acting in succession: E1 activating enzymes, followed by E2 conjugating enzymes, followed by E3 ligating enzymes. The human proteome includes two E1s (Barghout and Schimmer 2021), about 40 E2s (Stewart et al. 2016), and well over 600 E3s (Yang et al. 2021). It is the E3 ligating enzymes that are responsible for conjugating Ub to specific substrate proteins, creating conjugates with single Ubs, or branched or unbranched chains of Ubs. The combinations of substrates that can be ubiquitinated by the E3 ligating enzymes have only begun to be mapped (Suiter et al. 2025).

- Countering ubiquitination are about 100 deubiquitinating enzymes with different specificities (Snyder and Silva 2021).

- Ubiquitinated proteins are recognized by a plethora of Ub-binding effector proteins, triggering myriad physiological responses. Prominent among the effector proteins is the proteasome, which degrades the ubiquitinated proteins it recognizes (Ciechanover and Stanhill 2014; Grice and Nathan 2016); that indeed was the context in which Ub was first discovered. However, there are an unknown number of other effector proteins, mediating a largely unknown array of responses. Recognition of ubiquitinated proteins by effector proteins is primarily mediated by numerous small Ub-binding domains (UBDs), falling into 10 or 11 classes (Hurley et al. 2006). Effector proteins typically have multiple UBDs. Using Ub variants to probe specific effector functions is a major goal of this arena of investigation.

Affinity selection of Ub variants that bind specific elements of the Ub network is a promising new way to interrogate a great diversity of cellular functions. Binding of Ub-binding proteins to wild-type Ub is specific but weak; it seldom interferes with affinity selection of high-affinity ligands for specific Ub-binding proteins from libraries of Ub variants.

A project by Veggiani and his colleagues (Veggiani et al. 2022) exemplifies application of the principles outlined (see Chapter 14, Overview: Generation and Characterization of Engineered Ubiquitin Variants to Modulate the Ubiquitin Signaling Cascade [Liang et al. 2024]). They prepared a panel of 60 different selectors, representing the Ub-interacting motif (UIM) class of UBDs (last bullet in the previous paragraph) from human proteins. Such UIMs bind weakly, but specifically, to wild-type Ub, with dissociation equilibrium constants in the range $K_D = 100–1000$ μM. The selectors were used to affinity-select ligands from a library of Ub variants displayed on phages. The library was designed in light of the contact amino acids in the known structure of a Ub variant in complex with a UIM from yeast. Altogether 23 Ub amino acids were chosen for "soft" randomization, such that at each randomized position, ~50% of the codons would encode the wild-type amino acids, while the other 50%

would encode other amino acids. The resulting library contained 17 billion different Ub variants. A total of 138 distinct Ub variants, affinity-selected by 42 of the 60 UIM selectors, bound their selectors well above background levels. Further analysis of the most selective variant for each of the 42 selectors revealed high affinity and specificity for their cognate selectors:

- Phage ELISA showed that 26% of the 42 variants recognized only their cognate selectors, and 52% recognized three or fewer selectors.

- Competitive phage ELISA showed that 57% of the variants gave $IC_{50}$ values <100 nM with their cognate selectors, while wild-type Ub gave $IC_{50}$ values between 100 and 500 µM.

These results indicate that many UIMs, involved in a wide variety of physiological functions, can be specifically addressed with affinity-selected Ub variants. It seems likely that this approach will apply to UBDs other than UIMs as well. Design of a Ub variant library for each class of UBD can be guided by available structures of Ub in complex with members of that UBD class. A vast array of cellular effector functions would thus be opened up for detailed investigation.

The UIM project seems tailor-made for using HTS, rather than clone-by-clone analysis, as the first step in analyzing the output of affinity selections, as recommended above in "High-Throughput Sequencing (HTS) as Primary Readout in Affinity Selection Projects." Thus, two rounds of affinity selection with all 60 UIM selectors would be carried out in parallel, stringency being greatly increased in the second round. HTS of the 120 first- and second-round outputs would allow clones of interest, with high specific affinity for their cognate UIM selectors, to be distinguished from the expected selector-unrelated phages (SUPs): phages that bind the carrier or other elements of the selection apparatus, phages that overreplicate during amplification, and phages that bind off-target UIMs. This line of experimentation would arguably be a streamlined way to choose a small number of clones for detailed individual investigation.

## POSTSCRIPT: THE C.A.R.L.O.S. PROJECT

Carlos Barbas was one of the founders of the Phage Display Course, as explained above, and was a prime contributor to the phage antibody technology that is practiced worldwide in research laboratories and industry, and taught in the Phage Display Course to the present day.

In October 2013, Carlos sent an apology to the other instructors of the Phage Display Course. He would not be able to participate that year because he had just been diagnosed with metastatic medullary thyroid cancer (MTC). The terrible news galvanized Don Siegel at the University of Pennsylvania (Penn), codirector of the Phage Display Course from 1997 to 2019, to launch an emergency campaign to save Carlos's life. He recruited Christoph Rader, then still at Scripps, and Siegel's Penn colleagues Michael Milone and Vijay Bhoj to the cause. Here, I call them the MTC team and their campaign the MTC project.

Penn is a major center for development of a promising approach to cancers like Carlos's, called chimeric antigen receptors (CARs), in which some of a patient's own T cells are artificially hyperweaponized to eliminate the cancer. Milone, Carl June, and their coworkers had recently described a CAR (Milone et al. 2009) that would eventually be the first to be approved by the U.S. Food and Drug Administration (FDA) in 2017. It is now called tisagenlecleucel. Here, it is referred to as the anti-CD19 CAR because it targets a B-cell cell surface marker called CD19 that is also present on leukemia and lymphoma cells. The MTC plan was to develop a CAR that targets a cell surface protein marker (yet to be identified) on Carlos's cancer cells. They referred to the project informally as C.A.R.L.O.S.: Chimeric Antigen Receptor Lymphocytes with Oncolytic Specificity.

A CAR is an artificial T-cell receptor that mimics a native T-cell receptor (TCR); T cells expressing a CAR are called CAR T cells. Like a native TCR, a CAR is a membrane protein whose extracellular domain recognizes a specific antigen (protein marker) on the surface of target cells. A native TCR's extracellular domain recognizes specific peptide epitopes presented by MHC class I or II proteins on a target cell's surface. TCRs that recognize peptides from self-proteins are eliminated by immunological tolerance, severely limiting the repertoire of antigens that can be targeted. In contrast, the extracellular domain of a CAR is an scFv that binds directly to a specific surface antigen on the target cell. The repertoire of target antigens is not limited by tolerance. The scFv in the anti-CD19 CAR was constructed from a conventional monoclonal antibody. The antibody domain for the MTC project was to be a rabbit scFv obtained via phage antibody technology, once a target antigen had been identified.

In both TCRs and CARs, engagement of the extracellular domain with its specific peptide epitope or antigen on a target cell triggers a signal transduction cascade that activates the T cell, enabling several effector functions that result in death of the target cell. In the case of a native TCR, activation is triggered by accessory membrane proteins, including CD3 ζ, that are associated with the TCR noncovalently in a TCR complex. In the case of a CAR, the cascade is triggered by the intracellular domain of CD3 ζ, which is part of the CAR's intracellular segment. The intracellular segment also includes the intracellular domain of a costimulatory protein called 4-1BB, which triggers functions that enhance CAR T-cell persistence and performance (Philipson et al. 2020). A schematic diagram of the planned CAR's coding sequence is shown in Figure 9.

Producing the scFv component of a CAR for Carlos's MTC required identification of a suitable target surface antigen specific for MTC cells. Carlos had already arranged for a gene expression profile on a sample of his tumor. That profile was compared to a publicly available normal gene expression profile, pinpointing a few proteins that were overexpressed in the tumor. One of the overexpressed proteins was GFRα4, a receptor on the surface of MTC cells and their noncancerous precursors, but absent from almost all other cells in the body. The GFRα4 protein was used to affinity-select anti-GFRα4 Fabs from a 10-billion-clone naive rabbit Fab library supplied by Rader (Peng et al. 2017). The scFv form of one of those Fabs serves as the extracellular domain of the MTC project's CAR, which was constructed as described in Box 1.

The anti-GFRα4 CAR was tested in a mouse model of MTC (Bhoj et al. 2021). The "patients" were immunodeficient mice that would not reject engrafted human tumor cells. In the experiment described below, the mice were engrafted with cells of a human leukemia cell line that naturally expresses the CD19 surface protein, and that had been genetically engineered to express the GFRα4 surface protein and a luciferase enzyme in the cytoplasm. The purpose of the luciferase was to cause tumor cells to luminesce (emit bright light) when the luciferase substrate was administered to the mice, so the tumor could be imaged optically. The engrafted cells formed tumors within days. Five days after engraftment, the mice were infused with human CAR T cells expressing either the FDA-approved anti-CD19 CAR or the MTC project's anti-GFRα4 CAR, prepared as described in Box 1. As a negative control, mice were also infused with human T cells expressing no CAR at all. Two days before T-cell infusion (day 2) and on several days after T-cell infusion, groups of five mice were injected with the luciferase substrate, and 6 min later, the luminescence from their tumors was imaged. Figure 10 shows false-color images of tumor luminescent intensity, with blue for the least intense and red for the most intense.

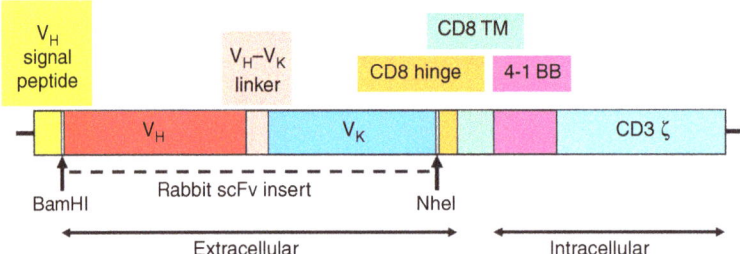

FIGURE 9. Schematic diagram of the planned chimeric antigen receptor (CAR)-coding sequence. This sequence encodes a rabbit scFv that specifically binds an antigen on the surface of Carlos's tumor cells. It is connected through a CD8 hinge and transmembrane (TM) to the intracellular domains of coreceptor protein CD3 ζ and costimulatory protein 4-1BB. The functions of the intracellular domains are explained in the text. CAR T cells (T cells expressing the CAR gene) deploy the CAR on the cell surface, with the single-chain variable fragment (scFv) exposed extracellularly, and the 4-1BB and CD3 ζ domains exposed intracellularly. Data are from Bhoj et al. (2021).

## BOX 1. CONSTRUCTION OF A CAR AND PREPARATION OF PATIENT CAR T CELLS

A chimeric antigen receptor (CAR) is constructed by inserting the coding sequence for the single-chain variable fragment (scFv) into the cloning sites (BamHI and NheI in the MTC project) of a lentiviral transfer plasmid, as diagrammed in Figure 9. The plasmid supplies the promoter and all elements of the CAR-coding sequence, except the scFv. Human producer cells in culture are transiently transfected with the transfer plasmid, along with a packaging plasmid encoding lentivirus proteins GagPol and Rev, and an envelope plasmid encoding the vesicular stomatitis virus (VSV) G surface glycoprotein. When the triply transfected producer cells are cultured, they release artificial enveloped virions with the structure of a lentivirus inside the envelope membrane, but with the VSV G protein deployed on the outside. The viral RNA molecules contain the CAR gene, but no viral genes. Lentiviral proteins GagPol (source of reverse transcriptase, integrase, and other proteins) and Rev are present as in a natural lentiviral virion, though without their genes. The artificial virions are harvested from the culture medium and frozen in aliquots for future treatment of patients.

To create patient CAR T cells, a preparation of the patient's T cells is infected with the artificial virions (the VSV G glycoprotein mediates infection of almost any cell type, including T cells). Inside an infected cell, the artificial CAR-bearing viral RNA is reverse-transcribed to DNA by the retroviral reverse transcriptase, with the help of the retroviral Rev protein. The resulting DNA is trafficked into the nucleus, and with the help of retroviral integrase is integrated into a random position in one of the T cell's chromosomes. The CAR gene in the integrated DNA is transcribed into CAR mRNAs, which in turn are translated to CAR proteins. The CAR proteins are trafficked to the cell membrane, with the scFv domain exposed extracellularly and the 4-1BB and CD3 ζ domains exposed intracellularly, as depicted in Figure 9. The virus cannot be propagated because no viral genes are present, and because the artificial CAR-bearing viral RNA is engineered to prevent the production of progeny artificial viral RNA by transcription of the integrated viral DNA.

The results in Figure 10 are striking: The tumors grew aggressively in the mice infused with the control T cells that did not express a CAR, but were rapidly eliminated in mice infused with human CAR T cells expressing either the FDA-approved anti-CD19 CAR or the anti-GFRα4 CAR developed in the MTC project. Based on these and other results reported in the article (Bhoj et al. 2021), the FDA

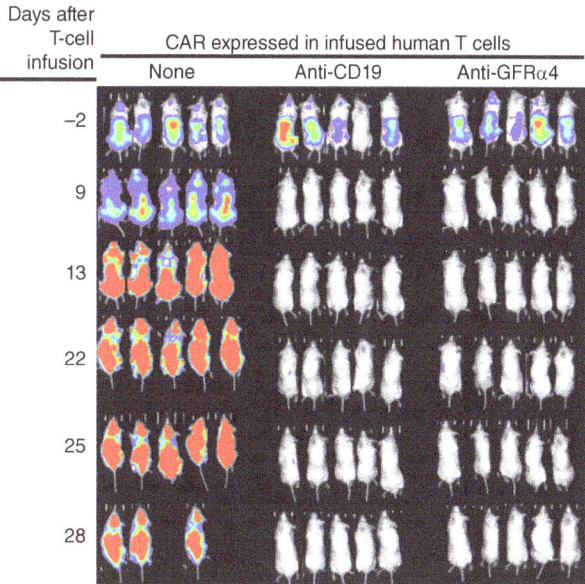

FIGURE 10. False-color images of luminescence from mice engrafted with a human tumor cell line expressing luciferase, and infused with human T cells expressing an anti-CD19 chimeric antigen receptor (CAR), an anti-GFRα4 CAR, or no CAR at all. False-color images of tumor luminescent intensity are shown, with blue for the least intense and red for the most intense. Adapted from Bhoj et al. (2021), under a Creative Commons License Attribution (CC BY-NC-ND); https://creativecommons.org/licenses/by-nc-nd/4.0/

has approved a Phase 1 trial of the anti-GFRα4 CAR in about 18 patients with metastatic or recurrent MTC (NCT no. 04877613).

Carlos will not participate in the trial. He died at age 49 on June 24, 2014, just as the mouse experiments were getting under way. It was only 8 months after his diagnosis. However, his legacy will live on—in the expanding realm of phage antibody technology to which he contributed, in the Phage Display Course he cofounded at the Cold Spring Harbor Laboratory, and perhaps in the anti-GFRα4 CAR if the FDA approves it for treatment of the cancer that cut short his life. Christoph Rader and two other colleagues of Carlos wrote a fitting appreciation of his many contributions to science (Rader et al. 2014).

## COMPETING INTEREST STATEMENT

The author declares no competing interests.

## ACKNOWLEDGMENTS

I am very grateful to Gregg Silverman, Don Siegel, Gianluca Veggiani, Brandon DeKosky, Mary Ann Pohl, and Juan Carlos Almagro for their generous help in writing this perspective.

## REFERENCES

Alhuseinalkhudhur A, Lubberink M, Lindman H, Tolmachev V, Frejd FY, Feldwisch J, Velikyan I, Sörensen J. 2020. Kinetic analysis of HER2-binding ABY-025 affibody molecule using dynamic PET in patients with metastatic breast cancer. *EJNMMI Res* **10:** 21. doi:10.1186/s13550-020-0603-9

Almagro JC, Pohl MA. 2024a. Semisynthetic phage display library construction: design and synthesis of diversified single-chain variable fragments and generation of primary libraries. *Cold Spring Harb Protoc* doi:10.1101/pdb.prot108614

Almagro JC, Pohl MA. 2024b. Semisynthetic phage display library construction: generation of filtered libraries. *Cold Spring Harb Protoc* doi:10.1101/pdb.prot108615

Almagro JC, Pohl MA. 2024c. Semisynthetic phage display library construction: generation of single-chain variable fragment secondary libraries. *Cold Spring Harb Protoc* doi:10.1101/pdb.prot108616

Almagro JC, Pedraza-Escalona M, Arrieta HI, Pérez-Tapia SM. 2019. Phage display libraries for antibody therapeutic discovery and development. *Antibodies (Basel)* **8:** 44. doi:10.3390/antib8030044

Alting-Mees MA, Short JM. 1989. pBluescript II: gene mapping vectors. *Nucleic Acids Res* **17:** 9494. doi:10.1093/nar/17.22.9494

Andris-Widhopf J, Rader C, Steinberger P, Fuller R, Barbas CF III. 2000. Methods for the generation of chicken monoclonal antibody fragments by phage display. *J Immunol Methods* **242:** 159–181. doi:10.1016/S0022-1759(00)00221-0

Barbas CF III, Kang AS, Lerner RA, Benkovic SJ. 1991. Assembly of combinatorial antibody libraries on phage surfaces: the gene III site. *Proc Natl Acad Sci* **88:** 7978–7982. doi:10.1073/pnas.88.18.7978

Barbas CF III, Burton DR, Scott JK, Silverman GJ, eds. 2001. *Phage display: a laboratory manual*. Cold Spring Harbor Laboratory Press, Cold Spring Harbor, NY.

Barghout SH, Schimmer AD. 2021. E1 enzymes as therapeutic targets in cancer. *Pharmacol Rev* **73:** 1–58. doi:10.1124/pharmrev.120.000053

Bhoj VG, Li L, Parvathaneni K, Zhang Z, Kacir S, Arhontoulis D, Zhou K, McGettigan-Croce B, Nunez-Cruz S, Gulendran G, et al. 2021. Adoptive T cell immunotherapy for medullary thyroid carcinoma targeting GDNF family receptor α 4. *Mol Ther Oncolytics* **20:** 387–398. doi:10.1016/j.omto.2021.01.012

Cary SP, Lee J, Wagenknecht R, Silverman GJ. 2000. Characterization of superantigen-induced clonal deletion with a novel clan III-restricted avian monoclonal antibody: exploiting evolutionary distance to create antibodies specific for a conserved VH region surface. *J Immunol* **164:** 4730–4741. doi:10.4049/jimmunol.164.9.4730

Chan SK, Rahumatullah A, Lai JY, Lim TS. 2017. Naïve human antibody libraries for infectious diseases. *Adv Exp Med Biol* **1053:** 35–59. doi:10.1007/978-3-319-72077-7_3

Ciechanover A, Stanhill A. 2014. The complexity of recognition of ubiquitinated substrates by the 26S proteasome. *Biochim Biophys Acta* **1843:** 86–96. doi:10.1016/j.bbamcr.2013.07.007

Conners R, McLaren M, Łapińska U, Sanders K, Stone MRL, Blaskovich MAT, Pagliara S, Daum B, Rakonjac J, Gold VAM. 2021. CryoEM structure of the outer membrane secretin channel pIV from the f1 filamentous bacteriophage. *Nat Commun* **12:** 6316. doi:10.1038/s41467-021-26610-3

Conners R, León-Quezada RI, McLaren M, Bennett NJ, Daum B, Rakonjac J, Gold VAM. 2023. Cryo-electron microscopy of the f1 filamentous phage reveals insights into viral infection and assembly. *Nat Commun* **14:** 2724. doi:10.1038/s41467-023-37915-w

Cortese R, Felici F, Galfre G, Luzzago A, Monaci P, Nicosia A. 1994. Epitope discovery using peptide libraries displayed on phage. *Trends Biotechnol* **12:** 262–267. doi:10.1016/0167-7799(94)90137-6

Cyr MG, Peng H, Rader C. 2024. Cloning, expression, and purification of phage display-selected Fab for biophysical and biological studies. *Cold Spring Harb Protoc* doi:10.1101/pdb.prot108604

Dahlsson Leitao C, Hjelm LC, Ståhl S, Löfblom J, Lindberg H. 2024. Selection of affibody molecules using *Escherichia coli* display. *Cold Spring Harb Protoc* **2024:** pdb.prot108400. doi:10.1101/pdb.prot108400

Dao T, Mun S, Korontsvit T, Khan AG, Pohl MA, White T, Klatt MG, Andrew D, Lorenz IC, Scheinberg DA. 2022a. A TCR mimic monoclonal antibody for the HPV-16 E7-epitope p11-19/HLA-A*02:01 complex. *PLoS One* **17:** e0265534. doi:10.1371/journal.pone.0265534

Dao T, Mun SS, Molvi Z, Korontsvit T, Klatt MG, Khan AG, Nyakatura EK, Pohl MA, White TE, Balderes PJ, et al. 2022b. A TCR mimic monoclonal antibody reactive with the "public" phospho-neoantigen pIRS2/HLA-A*02:01 complex. *JCI Insight* **7:** e151624. doi:10.1172/jci.insight.151624

Ernst A, Avvakumov G, Tong J, Fan Y, Zhao Y, Alberts P, Persaud A, Walker JR, Neculai AM, Neculai D, et al. 2013. A strategy for modulation of enzymes in the ubiquitin system. *Science* **339:** 590–595. doi:10.1126/science.1230161

Fahad AS, Gutiérrez-Gonzalez MF, Madan B, DeKosky BJ. 2025a. Antibody data analysis from diverse immune libraries. *Cold Spring Harb Protoc* **2025**: pdb.prot108627. doi:10.1101/pdb.prot108627

Fahad AS, Gutiérrez-Gonzalez MF, Madan B, DeKosky BJ. 2025b. Beyond single clones: high-throughput sequencing in antibody discovery. *Cold Spring Harb Protoc* **2025**: pdb.top107772. doi:10.1101/pdb.top107772

Fahad AS, Gutiérrez-Gonzalez MF, Madan B, DeKosky BJ. 2025c. Clonal lineage and gene diversity analysis of paired antibody heavy and light chains. *Cold Spring Harb Protoc* **2025**: pdb.prot108628. doi: 10.1101/pdb.prot108628

Fahad AS, Gutiérrez-Gonzalez MF, Madan B, DeKosky BJ. 2025d. Clonal variant analysis of antibody engineering libraries. *Cold Spring Harb Protoc* **2025**: pdb.prot108626. doi:10.1101/pdb.prot108626

Feldwisch J, Tolmachev V, Lendel C, Herne N, Sjöberg A, Larsson B, Rosik D, Lindqvist E, Fant G, Höidén-Guthenberg I, et al. 2010. Design of an optimized scaffold for affibody molecules. *J Mol Biol* **398**: 232–247. doi:10.1016/j.jmb.2010.03.002

Gorelik M, Miersch S, Sidhu SS. 2025. Structural survey of antigen recognition by synthetic human antibodies. *Cold Spring Harb Protoc* **2025**: pdb.over107759. doi:10.1101/pdb.over107759

Grice GL, Nathan JA. 2016. The recognition of ubiquitinated proteins by the proteasome. *Cell Mol Life Sci* **73**: 3497–3506. doi:10.1007/s00018-016-2255-5

Guerrero-Rodríguez SL, Mata-Cruz C, Pérez-Tapia SM, Velasco-Velázquez MA. 2022. Role of CD36 in cancer progression, stemness, and targeting. *Front Cell Dev Biol* **10**: 1079076. doi:10.3389/fcell.2022.1079076

Guzmán-Bringas OU, Gómez-Castellano KM, González-González E, Salinas-Trujano J, Vázquez-Leyva S, Vallejo-Castillo L, Pérez-Tapia SM, Almagro JC. 2023. Discovery and optimization of neutralizing SARS-CoV-2 antibodies using ALTHEA Gold Plus Libraries. *Int J Mol Sci* **24**: 4609. doi:10.3390/ijms24054609

Hernandez DN, Tam K, Shopsin B, Radke EE, Kolahi P, Copin R, Stubbe FX, Cardozo T, Torres VJ, Silverman GJ. 2020a. Unbiased identification of immunogenic *Staphylococcus aureus* leukotoxin B-cell epitopes. *Infect Immun* **88**: e00785-19. doi:10.1128/IAI.00785-19

Hernandez DN, Tam K, Shopsin B, Radke EE, Law K, Cardozo T, Torres VJ, Silverman GJ. 2020b. Convergent evolution of neutralizing antibodies to *Staphylococcus aureus* γ-hemolysin C that recognize an immunodominant primary sequence-dependent B-cell epitope. *mBio* **11**: e00460-20. doi:10.1128/mBio.00460-20

Hjelm LC, Dahlsson Leitao C, Ståhl S, Löfblom J, Lindberg H. 2024. Selection of affibody molecules using phage display. *Cold Spring Harb Protoc* **2024**: pdb.prot108399. doi:10.1101/pdb.prot108399

Hurley JH, Lee S, Prag G. 2006. Ubiquitin-binding domains. *Biochem J* **399**: 361–372. doi:10.1042/BJ20061138

Istomina PV, Gorchakov AA, Paoin C, Yamabhai M. 2024. Phage display for discovery of anticancer antibodies. *N Biotechnol* **83**: 205–218. doi:10.1016/j.nbt.2024.08.506

Koide A, Koide S. 2024. Use of phage display and other molecular display methods for the development of monobodies. *Cold Spring Harb Protoc* **2024**: 107982. doi: 10.1101/pdb.over107982

Lai JY, Lim TS. 2020. Infectious disease antibodies for biomedical applications: a mini review of immune antibody phage library repertoire. *Int J Biol Macromol* **163**: 640–648. doi:10.1016/j.ijbiomac.2020.06.268

Liang CT, Roscow O, Zhang W. 2024. Generation and characterization of engineered ubiquitin variants to modulate the ubiquitin signaling cascade. *Cold Spring Harb Protoc* **2024**: 107784. doi: 10.1101/pdb.over107784

Löfblom J, Hjelm LC, Dahlsson Leitao C, Ståhl S, Lindberg H. 2024. Selection of affibody molecules using staphylococcal display. *Cold Spring Harb Protoc* **2024**: pdb.prot108401. doi:10.1101/pdb.prot108401

Lubkowski J, Hennecke F, Plückthun A, Wlodawer A. 1999. Filamentous phage infection: crystal structure of g3p in complex with its coreceptor, the C-terminal domain of TolA. *Structure* **7**: 711–722. doi:10.1016/S0969-2126(99)80092-6

Martyn GD, Veggiani G. 2024. Phage-displayed SH2 domain libraries: from ultrasensitive tyrosine phosphoproteome probes to translational research. *Cold Spring Harb Protoc* **2024**: 107981. doi:10.1101/pdb.over107981

Mata-Cruz C, Guerrero-Rodríguez SL, Gómez-Castellano K, Carballo-Uicab G, Almagro JC, Pérez-Tapia SM, Velasco-Velázquez MA. 2025. Discovery and in vitro characterization of a human anti-CD36 scFv. *Front Immunol* **16**: 1531171. doi:10.3389/fimmu.2025.1531171

Matthews LJ, Davis R, Smith GP. 2002. Immunogenically fit subunit vaccine components via epitope discovery from natural peptide libraries. *J Immunol* **169**: 837–846. doi:10.4049/jimmunol.169.2.837

Milone MC, Fish JD, Carpenito C, Carroll RG, Binder GK, Teachey D, Samanta M, Lakhal M, Gloss B, Danet-Desnoyers G, et al. 2009. Chimeric receptors containing CD137 signal transduction domains mediate enhanced survival of T cells and increased antileukemic efficacy in vivo. *Mol Ther* **17**: 1453–1464. doi:10.1038/mt.2009.83

Muyldermans S. 2013. Nanobodies: natural single-domain antibodies. *Annu Rev Biochem* **82**: 775–797. doi:10.1146/annurev-biochem-063011-092449

Pascual G, Avgustinova A, Mejetta S, Martín M, Castellanos A, Attolini CS, Berenguer A, Prats N, Toll A, Hueto JA, et al. 2017. Targeting metastasis-initiating cells through the fatty acid receptor CD36. *Nature* **541**: 41–45. doi:10.1038/nature20791

Pedraza-Escalona M, Guzmán-Bringas O, Arrieta-Oliva HI, Gómez-Castellano K, Salinas-Trujano J, Torres-Flores J, Muñoz-Herrera JC, Camacho-Sandoval R, Contreras-Pineda P, Chacón-Salinas R, et al. 2021. Isolation and characterization of high affinity and highly stable anti-chikungunya virus antibodies using ALTHEA Gold Libraries. *BMC Infect Dis* **21**: 1121. doi:10.1186/s12879-021-06717-0

Pellegri C, Moreau A, Duché D, Houot L. 2023. Direct interaction between fd phage pilot protein pIII and the TolQ–TolR proton-dependent motor provides new insights into the import of filamentous phages. *J Biol Chem* **299**: 105048. doi:10.1016/j.jbc.2023.105048

Peng H, Rader C. 2024a. Generation of antibody libraries for phage display: chimeric rabbit/human Fab format. *Cold Spring Harb Protoc* doi:10.1101/pdb.prot108598

Peng H, Rader C. 2024b. Generation of antibody libraries for phage display: human Fab format. *Cold Spring Harb Protoc* doi:10.1101/pdb.prot108597

Peng H, Rader C. 2024c. Generation of antibody libraries for phage display: library reamplification. *Cold Spring Harb Protoc* doi:10.1101/pdb.prot108601

Peng H, Rader C. 2024d. Generation of antibody libraries for phage display: preparation of electrocompetent *E. coli*. *Cold Spring Harb Protoc* doi:10.1101/pdb.prot108599

Peng H, Rader C. 2024e. Generation of antibody libraries for phage display: preparation of helper phage. *Cold Spring Harb Protoc* doi:10.1101/pdb.prot108600

Peng H, Rader C. 2024f. Phage display selection of antibody libraries: panning procedures. *Cold Spring Harb Protoc* doi:10.1101/pdb.prot108602

Peng H, Rader C. 2024g. Phage display selection of antibody libraries: screening of selected binders. *Cold Spring Harb Protoc* doi:10.1101/pdb.prot108603

Peng H, Nerreter T, Chang J, Qi J, Li X, Karunadharma P, Martinez GJ, Fallahi M, Soden J, Freeth J, et al. 2017. Mining naïve rabbit antibody repertoires by phage display for monoclonal antibodies of therapeutic utility. *J Mol Biol* **429**: 2954–2973. doi:10.1016/j.jmb.2017.08.003

Philipson BI, O'Connor RS, May MJ, June CH, Albelda SM, Milone MC. 2020. 4-1BB costimulation promotes CAR T cell survival through non-canonical NF-κB signaling. *Sci Signal* **13**: eaay8248. doi:10.1126/scisignal.aay8248

Pohl MA, Almagro JC. 2024. Considerations for using phage display technology in therapeutic antibody drug discovery. *Cold Spring Harb Protoc* doi:10.1101/pdb.top107757

Qi H, Lu H, Qiu HJ, Petrenko V, Liu A. 2012. Phagemid vectors for phage display: properties, characteristics and construction. *J Mol Biol* **417**: 129–143. doi:10.1016/j.jmb.2012.01.038

Rader C. 2024a. Generation and selection of phage display antibody libraries in Fab format. *Cold Spring Harb Protoc* doi:10.1101/pdb.top107764

Rader C. 2024b. The pComb3 phagemid family of phage display vectors. *Cold Spring Harb Protoc* **2024**: pdb.over107756. doi:10.1101/pdb.over107756

Rader C, Segal DJ, Shabat D. 2014. Carlos F. Barbas III (1964–2014): visionary at the interface of chemistry and biology. *ACS Chem Biol* **9**: 1645–1646.

Rakonjac J, Gold VAM, León-Quezada RI, Davenport CH. 2024. Structure, biology, and applications of filamentous bacteriophages. *Cold Spring Harb Protoc* **2024**: pdb.over107754. doi:10.1101/pdb.over107754

Russel M, Kidd S, Kelley MR. 1986. An improved filamentous helper phage for generating single-stranded plasmid DNA. *Gene* **45**: 333–338. doi:10.1016/0378-1119(86)90032-6

Silverman GJ. 2025a. Insights from the study of B-cell epitopes of a microbial pathogen by phage display. *Cold Spring Harb Protoc* doi:10.1101/pdb.top107777

Silverman GJ. 2025b. Cloning and selection from antigen fragment libraries for epitope identification. *Cold Spring Harb Protoc* doi:10.1101/pdb.prot108660

Smith GP. 2024. Principles of affinity selection. *Cold Spring Harb Protoc* **2024**: pdb.over107894. doi:10.1101/pdb.over107894

Snyder NA, Silva GM. 2021. Deubiquitinating enzymes (DUBs): regulation, homeostasis, and oxidative stress response. *J Biol Chem* **297**: 101077. doi:10.1016/j.jbc.2021.101077

Spaan AN, van Strijp JAG, Torres VJ. 2017. Leukocidins: staphylococcal bi-component pore-forming toxins find their receptors. *Nat Rev Microbiol* **15**: 435–447. doi:10.1038/nrmicro.2017.27

Ståhl S, Hjelm LC, Dahlsson Leitao C, Löfblom J, Lindberg H. 2024a. Cloning of affibody libraries for display methods. *Cold Spring Harb Protoc* **2024**: pdb.prot108398. doi:10.1101/pdb.prot108398

Ståhl S, Lindberg H, Hjelm LC, Löfblom J, Dahlsson Leitao C. 2024b. Engineering of affibody molecules. *Cold Spring Harb Protoc* **2024**: pdb.top107760. doi:10.1101/pdb.top107760

Stewart MD, Ritterhoff T, Klevit RE, Brzovic PS. 2016. E2 enzymes: more than just middle men. *Cell Res* **26**: 423–440. doi:10.1038/cr.2016.35

Suiter CC, Calderon D, Lee DS, Chiu M, Jain S, Chardon FM, Lee C, Daza RM, Trapnell C, Zheng N, et al. 2025. Combinatorial mapping of E3 ubiquitin ligases to their target substrates. *Mol Cell* **85**: 829–842.e826. doi:10.1016/j.molcel.2025.01.016

Tang JQ, Marchand MM, Veggiani G. 2023. Ubiquitin engineering for interrogating the ubiquitin-proteasome system and novel therapeutic strategies. *Cells* **12**: 2117. doi:10.3390/cells12162117

Thomas WD, Golomb M, Smith GP. 2010. Corruption of phage display libraries by target-unrelated clones: diagnosis and countermeasures. *Anal Biochem* **407**: 237–240. doi:10.1016/j.ab.2010.07.037

Tiller KE, Chowdhury R, Li T, Ludwig SD, Sen S, Maranas CD, Tessier PM. 2017. Facile affinity maturation of antibody variable domains using natural diversity mutagenesis. *Front Immunol* **8**: 986. doi:10.3389/fimmu.2017.00986

Valadon P, Pérez-Tapia SM, Nelson RS, Guzmán-Bringas OU, Arrieta-Oliva HI, Gómez-Castellano KM, Pohl MA, Almagro JC. 2019. ALTHEA Gold Libraries: antibody libraries for therapeutic antibody discovery. *MAbs* **11**: 516–531. doi:10.1080/19420862.2019.1571879

Veggiani G, Sidhu SS. 2024a. Beyond natural immune repertoires: synthetic antibodies. *Cold Spring Harb Protoc* **2024**: 107768. doi:10.1101/pdb.top107768

Veggiani G, Sidhu SS. 2024b. Generation and selection of synthetic human antibody libraries via phage display. *Cold Spring Harb Protoc* **2024**: 108347. doi:10.1101/pdb.prot108347

Veggiani G, Yates BP, Martyn GD, Manczyk N, Singer AU, Kurinov I, Sicheri F, Sidhu SS. 2022. Panel of engineered ubiquitin variants targeting the family of human ubiquitin interacting motifs. *ACS Chem Biol* **17**: 941–956. doi:10.1021/acschembio.2c00089

Velikyan I, Schweighöfer P, Feldwisch J, Seemann J, Frejd FY, Lindman H, Sörensen J. 2019. Diagnostic HER2-binding radiopharmaceutical, [(68)Ga]Ga-ABY-025, for routine clinical use in breast cancer patients. *Am J Nucl Med Mol Imaging* **9**: 12–23.

Webster R. 2001. Filamentous phage biology. In *Phage display: a laboratory manual* (ed. Barbas CF III, et al.), pp. 1.1–1.37. Cold Spring Harbor Laboratory Press, Cold Spring Harbor, NY.

Yang Q, Zhao J, Chen D, Wang Y. 2021. E3 ubiquitin ligases: styles, structures and functions. *Mol Biomed* **2**: 23. doi:10.1186/s43556-021-00043-2

Yang H, Chae J, Kim H, Noh J, Chung J. 2025a. Chicken immunization followed by RNA extraction and cDNA synthesis for antibody library preparation. *Cold Spring Harb Protoc* **2025**: pdb.prot108568. doi:10.1101/pdb.prot108568

Yang H, Chae J, Kim H, Noh J, Chung J. 2025b. Generation of a phage display chicken single-chain variable fragment library. *Cold Spring Harb Protoc* **2025**: pdb.prot108213. doi:10.1101/pdb.prot108213

Yang H, Chae J, Kim H, Noh J, Chung J. 2025c. Generation of chicken antibody libraries and selection of antigen binders. *Cold Spring Harb Protoc* **2025**: pdb.top108210. doi:10.1101/pdb.top108210

Yang H, Chae J, Kim H, Noh J, Chung J. 2025d. Preparation of VCSM13 helper phage for display library reamplification and biopanning. *Cold Spring Harb Protoc* **2025**: pdb.prot108569. doi:10.1101/pdb.prot108569

Yang H, Chae J, Kim H, Noh J, Chung J. 2025e. Selection of antigen binders from a chicken single-chain variable fragment library. *Cold Spring Harb Protoc* **2025**: pdb.prot108211. doi:10.1101/pdb.prot108211

Zinder ND, Horiuchi K. 1985. Multiregulatory element of filamentous bacteriophages. *Microbiol Rev* **49**: 101–106. doi:10.1128/mr.49.2.101-106.1985

CHAPTER 2

# Structure, Biology, and Applications of Filamentous Bacteriophages

Jasna Rakonjac,[1,2,5] Vicki A.M. Gold,[3,4] Rayén I. León-Quezada,[1,2] and Catherine H. Davenport[1,2]

[1]School of Natural Sciences, Massey University, Auckland 0632, New Zealand; [2]Nanophage Technologies Ltd., Palmerston North, Manawatu 4474, New Zealand; [3]Living Systems Institute; [4]Faculty of Health and Life Sciences, University of Exeter, Exeter, EX4 4QD, United Kingdom

The closely related *Escherichia coli* Ff filamentous phages (f1, fd, and M13) have taken a fantastic journey over the past 60 years, from the urban sewerage from which they were first isolated, to their use in high-end technologies in multiple fields. Their relatively small genome size, high titers, and the virions that tolerate fusion proteins make the Ffs an ideal system for phage display. Folding of the fusions in the oxidizing environment of the *E. coli* periplasm makes the Ff phages a platform that allows display of eukaryotic surface and secreted proteins, including antibodies. Resistance of the Ffs to a broad range of pH and detergents facilitates affinity screening in phage display, whereas the stability of the virions at ambient temperature makes them suitable for applications in material science and nanotechnology. Among filamentous phages, only the Ffs have been used in phage display technology, because of the most advanced state of knowledge about their biology and the various tools developed for *E. coli* as a cloning host for them. Filamentous phages have been thought to be a rather small group, infecting mostly Gram-negative bacteria. A recent discovery of more than 10 thousand diverse filamentous phages in bacteria and archaea, however, opens a fascinating prospect for novel applications. The main aim of this review is to give detailed biological and structural information to researchers embarking on phage display projects. The secondary aim is to discuss the yet-unresolved puzzles, as well as recent developments in filamentous phage biology, from a viewpoint of their impact on current and future applications.

## INTRODUCTION

Ff (f1, fd, and M13) phages of *Escherichia coli* replicate extremely efficiently, reaching, under optimal laboratory conditions, titers of $10^{12}$–$10^{13}$ per mL of culture. These phages have been used for over the past 60 years in molecular biology, protein evolution, and for the generation of monoclonal antibodies, and have even been used in nanotechnology and material science research (Smith 1985; McCafferty et al. 1990; Scott and Smith 1990; Mao et al. 2004). This article is part of a collection focused on phage display—a powerful combinatorial technology that allows affinity selection of rare protein variants using a selector molecule of interest. The entire collection is available online at https://cshprotocols.cshlp.org/. The power of phage display technology is based on a physical connection between polypeptides displayed on the surface of the virion and their coding sequences packaged within the virion. For a discussion, see Chapter 3, Overview: Principles of Affinity Selection (Smith 2023).

For our discussion on Ff phages, we have organized the content in various sections—namely, structure, genome, genes and transcription, stages of the life cycle (including the mechanism of

[5]Correspondence: j.rakonjac@massey.ac.nz

© 2026 Cold Spring Harbor Laboratory Press
Cite this overview as *Cold Spring Harb Protoc*; doi:10.1101/pdb.over107754

infection, replication, and assembly), and effect on *E. coli* physiology—and we end with a discussion of the limitations imposed by the host on phage display. Details of the biology of the Ff phage (i.e., its life cycle and assembly) give a basis on which the protocols for phage display library construction, amplification, and screening have been developed. Knowledge of the virion structure and correct subcellular targeting, folding, and assembly of the virion proteins into an infectious Ff virion is critical for designing fusions that will be folded correctly and displayed on the virion surface. Finally, understanding replication will contribute to successful use of phage display setups, including more complex ones that do not solely rely on the phage vectors but are composed of a phagemid vector and a helper phage. The effect of Ff on the physiology of its host *E. coli* provides a context to phage production; a better understanding of the balance between the host and Ff may result in innovations that will increase virion production.

## FILAMENTOUS PHAGE STRUCTURE

Filament-like virions of the inoviruses are composed of five proteins: one major coat protein (pVIII or g8p or gp8) that forms the shaft and is present in thousands of copies, and two pairs of minor coat proteins that cap the filaments at two asymmetrical ends (pVII and pIX at the blunt end, and pIII and pVI that form the pointy end), each present in five copies (Fig. 1; Conners et al. 2023). The circular single-stranded DNA (ssDNA) genome forms the filament backbone as a double-stranded helix. The diameter of filamentous phage is 6–7 nm (depending on the species), and the length is determined by the size of the genome, 860 nm in the case of wild-type Ff phages (f1, fd, or M13). Virion length can be further extended if the initiation or termination of the assembly–secretion process is impaired. The

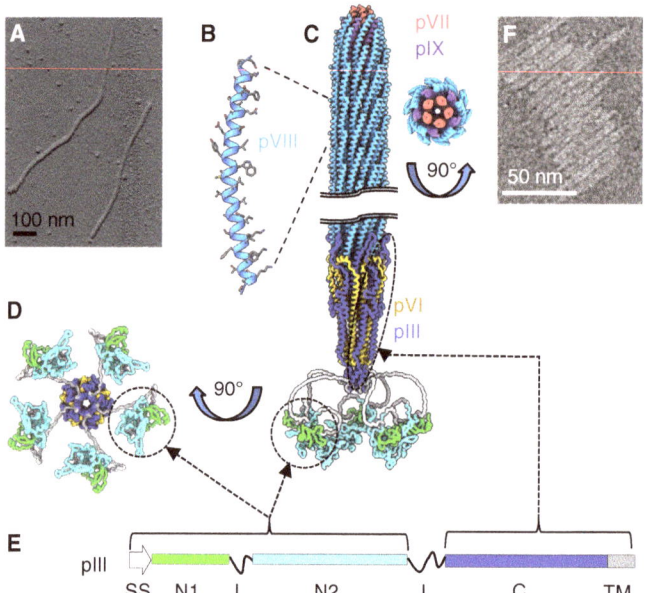

FIGURE 1. The Ff virion structure. (*A*) Helper phage R408 and a phagemid imaged by atomic force microscopy (J Rakonjac, M Russel, and P Model, unpubl.). (*B*) Major coat protein pVIII subunit structure (minus the four residues at the amino terminus) in the virion extracted from the cryo-electron microscopy (cryoEM) structure (PDB 8B3Q; Conners et al. 2023). (*C*) Composite structure of an f1 phage virion obtained from the cryoEM structures PDB 8B3O, 8B3P, and 8B3Q, and a prediction of the N1–N2 domains and linkers, based on the data in Conners et al. (2023), pVII is shown in salmon, pIX in purple, pVIII in bright blue, pVI in gold, the C domain of pIII in dark blue, the pIII linkers in gray, the pIII N1 domain in green, and the pIII N2 domain in cyan. (*D*) The structure in C, viewed from the pIII–pVI end. (*E*) A schematic of the nascent pIII protein. (SS) Signal sequence, (N1, N2, C) domains of pIII; (L) glycine-rich linkers; (TM) transmembrane helix. (*F*) f1-derived nanorods (50 nm in length) (Adapted from Sattar et al. 2015, under a Creative Commons License Attribution (CC BY); http://creativecommons.org/licenses/by/4.0/.)

longest Ff filaments observed, up to 20 μm, have been those released after infection with a phage missing genes encoding one of the minor virion proteins (Lopez and Webster 1983; Rakonjac and Model 1998). The rise per nucleotide ($h$) is another determinant of filamentous phage virion length. The $h$, in turn, depends on the mode of interaction between the DNA and the major coat protein (pVIII), in which the positively charged residues at the pVIII carboxyl terminus interact with the negatively charged phosphates of DNA. For the same phage, $h$ can be modified by changing the number of positive charges at the pVIII carboxyl terminus which changes the ratio of pVIII subunits to nucleotides (Marvin et al. 2014). The length of the virion can be calculated as a product of $h$ and the genome length in nucleotides divided by 2. The $h$ value varies between different phages. For instance, in Ff, it is 0.28 nm, whereas it is as much as 0.61 nm in the *Pseudomonas* phage Pf1. The structure of the DNA helix in the Pf1 phage is much more extended (P-DNA-like form) in comparison to that of the f1 phage, in which the B form of DNA is the best fit (Conners et al. 2023).

Only the packaging signal forms a proper helix because of the self-complementarity of its inverted repeat. Over the remaining length the two strands are not inverted repeats, hence by chance 25% of nucleotides are expected to form complementary pairs. The atomic-level structure of DNA within the virion has not been resolved to date. In contrast to Ff, DNA encapsulated in the Pf1 phage of *Pseudomonas aeruginosa* forms a helix in which phosphates are in the center, with the bases pointing outward (Day 2011).

## Major Coat Protein Structure

The Ff major coat protein pVIII is synthesized with a signal sequence, which is cleaved as pVIII is targeted to the inner membrane of *E. coli* by the translocon YidC (Samuelson et al. 2000). Mature pVIII is 50 residues in length and forms a slightly curved α-helix (Fig. 1B), apart from the amino-terminal five residues, which are flexible (Fig. 1B,C; Marvin et al. 2006, 2014; Goldbourt et al. 2007; Conners et al. 2023). The α-helix is amphipathic up to the 20th residue and then hydrophobic to residue 39, ending with a 10-residue positively charged helix that interacts with encapsulated DNA (Marvin et al. 2006). Thousands of pVIII helices are packed together in a helical array to coat the length of the viral DNA genome (Fig. 1C). The negatively charged amino termini of pVIII subunits are exposed on the surface, whereas their positively charged carboxyl termini face the negatively charged DNA.

Filamentous phages are separated into two classes according to their symmetry (Day et al. 1988). The filaments of Class I phages have pentameric symmetry and include Ff, Ike, and If1. Class II phage filaments have C1 symmetry and include Pf 1, Pf4, and Xf (Marvin et al. 1974a,b; Tarafder et al. 2020). Although the atomic-level structures of pVIII and the shaft of the Ff, Ike, Pf1, Pf4, and several other *E. coli* and *Pseudomonas* phages have been determined over the past 40 years (Goldbourt et al. 2010; Morag et al. 2011; Sergeyev et al. 2011; Xu et al. 2019), the structure of the virion caps has only been determined very recently (Conners et al. 2023).

## Minor Virion Proteins Forming the End Caps of the Filament

The high-resolution structure of a short version of the Ff virion, including the end caps, demonstrates unequivocally that each of the minor proteins is present in five copies (Fig. 1C; Conners et al. 2023). Furthermore, the "plug" of the flat cap is formed of a five-helix bundle of pVII, surrounded by a ring of pIX subunits. The pointy end of the virion is a cone formed of five intertwined copies each of pIII and pVI (Fig. 1C).

pVII and pIX are small hydrophobic proteins, 32 and 33 residues in length, respectively. They each form a single α-helix in the assembled virion (Fig. 1C). After translation and before assembly into the virion, they are inserted in the inner membrane. Neither has a signal sequence, and it has been reported that pIX insertion into the membrane is mediated by the YidC translocon, whereas pVII appears to insert spontaneously (Ploss and Kuhn 2011). The assembly of filamentous phages is initiated by pVII and pIX (Grant et al. 1980; Endemann and Model 1995), in which the two proteins seem to bind DNA (the packaging signal) via their carboxyl termini (Russel and Model 1989).

The second pair of minor virion proteins, pIII and pVI, form the conical or "pointy" cap of the filament (Fig. 1C). These two proteins are incorporated last into the virion particle on egress from the infected cell. With their incorporation into the virion, assembly is finished, and the filament is released from the cell (Rakonjac and Model 1998; Rakonjac et al. 1999). pIII is composed of three domains, N1, N2, and C (Fig. 1C–E). The amino-terminal domains (N1 and N2) are required for receptor binding at the start of infection but not assembly, whereas the C domain is required for both assembly and the entry stage of infection. Like other filamentous phage proteins, pIII and pVI are targeted to the inner membrane after translation. From there, they are translocated into the virion at the end of assembly (Boeke and Model 1982; Endemann and Model 1995).

pVI is a 112-residue protein that, in the virion, forms one long α-helix and two shorter α-helices that interact extensively with pIII (Fig. 1C). In the membrane-inserted state, two out of the three transmembrane α-helices predicted by TMHMM2.0 (Krogh et al. 2001) correspond to the folded-up long α-helix (Conners et al. 2023). The amino termini and carboxyl termini are predicted to face the periplasm and the cytoplasm, respectively (Endemann and Model 1995).

Nascent pIII is 424 residues long, whereas the mature protein, after signal sequence cleavage, is 406 residues in length. The amino-terminal signal sequence (signal peptide or leader peptide) mediates translocation of pIII across the inner membrane in a SecYEG–SecA-dependent manner, and a carboxy-terminal hydrophobic helix anchors the protein to the membrane. The carboxy-terminal five residues are exposed to the cytoplasm (Boeke and Model 1982). Most of pIII (apart from the membrane anchor at the carboxy-terminal end) folds in the periplasm in the host cell (Davis and Model 1985). Amino-terminal fusions to pIII, which are the most common mode of phage display, therefore fold in the oxidizing environment of the periplasm.

Structures of the receptor-binding pIII N1 and N2 domains, expressed as fragments in the absence of the carboxy-terminal domain, have been determined (Fig. 1D; Lubkowski et al. 1998; Holliger et al. 1999). An AlphaFold prediction based on these structures was used to generate a model of the N1 and N2 receptor binding domains linked to the carboxyl terminus of pIII (Fig. 1C,D; Conners et al. 2023).

## Physical Properties of the Ff Filamentous Phages

Ff is very stable, resisting exposure to pH extremes, detergents, and heat (Branston et al. 2011, 2013). Given that phage display library screening by biopanning typically requires detergents and low pH at the wash and elution steps, respectively, these properties are precious, and they place the Ff as preferred particles for phage display, ahead of tailed phages like λ, which are sensitive to low pH (see Chapter 3, Overview: Principles of Affinity Selection [Smith 2023]). Not only polar but also ionic detergents below the critical micellar concentration are harmless to Ff phages (Stopar et al. 2002). Ff filaments are, however, disassembled by bleach and chloroform. Using a laboratory evolution approach, a variant of pVIII was identified that prevents Ff disassembly by organic solvents (Petrenko et al. 1996).

Another of filamentous phages' properties of interest, specifically for applications as nanoparticles, is in that they form liquid crystals at high concentration or in solutions of high osmolarity (Dogic 2016). The Ff liquid crystalline arrays can form shapes such as disc, sphere, ribbon, stacks, or star-like disc-ribbon. These properties make them a suitable model for nanoscale rod-shaped objects in soft matter physics (Sharma et al. 2014; Dogic 2016). The various assemblies of filamentous phages, when scaled up and combined with inorganic or organic molecules, have been shown to form various types of nanomaterials, such as films, fibers, or hydrogels (Lee et al. 2002; Lee and Belcher 2004; Souza et al. 2010). The liquid crystalline nature of the Ff phages is also the basis of "phage litmus," in which spectral changes are used to monitor an analyte (Oh et al. 2014). Changes in the orientation of the filamentous phage liquid crystals can also be measured by changes in linear dichroism (LD) values and are used in immunodetection of analytes (e.g., pathogenic bacteria; Pacheco-Gomez et al. 2012). The length of the filament is important for increased LD signal; hence, there is interest to produce filamentous phage particles that are much longer than the "regular" Ff phages. The physical properties of filamentous phages can be modified not only by recombinant display of peptides or antibodies, but

## GENOME, GENES, AND TRANSCRIPTS

The Ff (f1, fd, M13) phage genome is 6407 nt in length (Fig. 2A). DNA sequence identity between f1, fd, and M13 is 98% and, in phage display technology, they are used interchangeably. The genome contains nine open reading frames (ORFs) encoding 11 proteins (Table 1). The Ff genes are all transcribed in the same direction and are organized into two operons (Fig. 2A; Model and Russel 1988). Genes encoding replication proteins (gII, gX, and gV) form an operon together with genes gVII, gIX, and gVIII, which encode virion proteins. The second operon is composed of genes encoding the remaining two virion proteins (gIII, gVI) and those encoding the secretion/assembly machinery (gI/gXI and gIV). Both operons harbor strong transcriptional terminators. Two genes, gX and gXI, encode truncated—but functional—translational products of gII and gI, respectively, because of the translation start within the latter genes (Fulford and Model 1984). These shorter proteins are required for phage replication and assembly, respectively (Model and Russel 1988; Haigh and Webster 1999). A ~500-nt intergenic sequence (Fig. 2A,B) located downstream from gIV and upstream of gII contains the origins for positive (+) and negative (−) strand replication, as well as a packaging signal (Fig. 2B; Zinder and Horiuchi 1985).

A large body of published work has given a good qualitative and quantitative picture of Ff phage transcription and translation (for review, see Webster 2001). Despite the simple two-operon organization discussed above, there are vast differences in the number of each of the proteins encoded within the same operon per cell. This is achieved by the presence of multiple nested promoters, resulting in a number of overlapping transcripts (Edens et al. 1978; Moses et al. 1980). These multiple transcripts

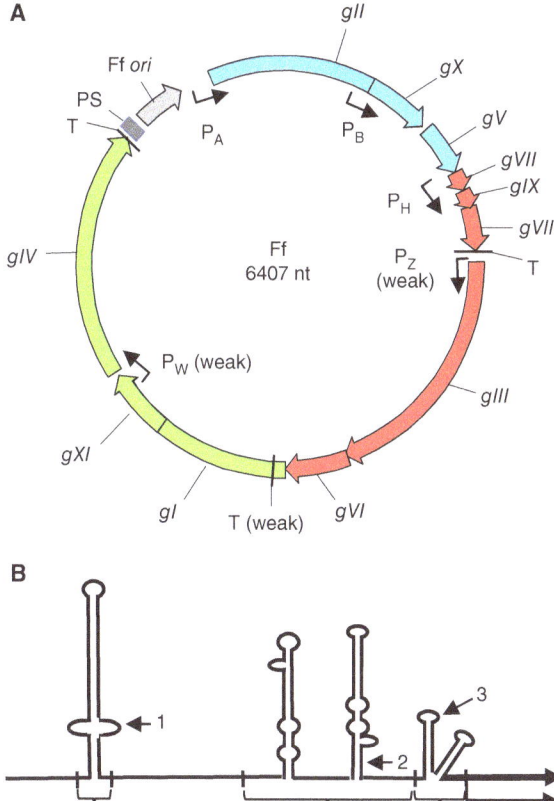

FIGURE 2. The Ff genome and the intergenic sequence. (A) The Ff genome. Open reading frames are indicated by block arrows that are color-coded based on the functions of the encoded proteins: blue, gII/gIX-gV, DNA replication; red, gVII, gIX, gVIII, gIII, and gVI, virion; green, gI/gXI, gIV, assembly. (PS) Packaging signal, (Ff ori) origin of replication, ($P_A$, $P_B$, $P_H$, $P_W$, $P_Z$) promoters, (T) terminators. (B) Predicted secondary structure of the single-stranded DNA (ssDNA) at the Ff origin of replication. Components of the intergenic sequence are (PS) packaging signal; followed by origins of replication: (−) ori, (+) ori region I (A); (+) ori region II, negative and positive origins of replication; (1) gIV stop codon; (2) gIV mRNA terminator; (3) initiation site of (−) strand primer synthesis; (4) initiation site of (+) strand synthesis.

## Chapter 2

TABLE 1. Genes and proteins of Ff phage

| Gene/Protein | Function |
|---|---|
| $gI \rightarrow$ pI (g1p) | Morphogenesis (phage assembly-export)<br>Inner membrane component of the *trans*-envelope assembly/secretion system. Interacts with pIV. |
| $gII \rightarrow$ pII (g2p) | Replication<br>Replication protein (rolling circle replication).<br>Cleaves the (+) strand of the dsDNA replicative form (RF).<br>Joins the ends of the displaced (+) strand to generate a circular single-stranded DNA. |
| $gIII \rightarrow$ pIII (g3p) | Structural<br>Forms the pIII/pVI virion cap.<br>Interacts with pVI, pVIII, the F pilus, and domain III of TolA protein.<br>Essential role in all the stages of infection and in termination of Ff assembly, and release of new virions from the host membranes. |
| $gIV \rightarrow$ pIV (g4p) | Morphogenesis (phage assembly-export)<br>Forms a gated channel across the host outer membrane through which virions are exported. |
| $gV \rightarrow$ pV (g5p) | Replication<br>ssDNA binding protein.<br>Coats (+) strand ssDNA to form an assembly substrate.<br>Negatively regulates the translation of *gII* mRNA |
| $gVI \rightarrow$ pVI (g6p) | Structural<br>Forms the pIII/pVI virion cap.<br>Essential role in termination of assembly and release of new virions from the host membranes. Role in the infection and entry has not been examined. |
| $gVII \rightarrow$ pVII (g7p) | Structural<br>Forms the pVII/pIX virion cap.<br>Initiates the virion assembly-export process, along with pIX, by interacting with the packaging signal of the viral genome and the pI–pXI complex. |
| $gVIII \rightarrow$ pVIII (g8p) | Structural<br>Major coat protein.<br>Assembles to form a helical filament-like capsid along the DNA backbone. |
| $gIX \rightarrow$ pIX (g9p) | Structural<br>Forms the pVII/pIX virion cap.<br>Initiates the virion assembly-export process, along with pVII, by interacting with the packaging signal of the viral genome and the pI–pXI complex. |
| $gX \rightarrow$ pX (g10p) | Replication<br>Promotes the RF synthesis and depresses the accumulation of ssDNA.<br>Inhibits the activity of pII. |
| $gXI \rightarrow$ pXI (g11p) | Morphogenesis (phage assembly-export)<br>Part of a transmembrane complex with pI and pIV to protect pI from cleavage by endogenous proteases. |

undergo differential turnover; furthermore, translation of individual ORFs occurs at different rates because of the differences in the ribosome strength of the ribosome binding site (RBS) and translation repression by the phage-encoded ssDNA-binding protein pV (La Farina and Model 1983; Model and Russel 1988; Michel and Zinder 1989; Goodrich and Steege 1999). No dedicated transcriptional regulators of Ff gene expression have been identified; nevertheless, these existing modes of regulation suffice to ensure very high levels of phage production, including the synthesis of $\sim 10^6$ copies of pVIII per cell per generation (Lerner and Model 1981; Smeal et al. 2017a).

Published quantitative and qualitative data on the transcripts and proteins of Ff phages during infection have been used recently for developing a mathematical model of the Ff life cycle, followed by simulations of single-generation and multigeneration infection cycles (Smeal et al. 2017a,b). Simulations correctly predicted a drastic decrease in Ff production after seven generations of *E. coli* postinfection. Host DNA polymerase III and the Ff-encoded replication protein pII were found to be limiting factors. Persistent Ff replication and assembly over 20 *E. coli* generations was predicted to be possible without these limitations. Such sustained production is modeled to increase the Ff number of assembled virions by about threefold over 20 generations, from $\sim$7500 to up to $\sim$23,000 phages per cell (Smeal et al. 2017a,b).

## FILAMENTOUS PHAGE INFECTION

The Ff filamentous phages were originally isolated from plaques on the $F^+$ (but not $F^-$) *E. coli* strains (Loeb 1960; Hofschneider 1963; Marvin and Hoffmann-Berling 1963). The F pilus is the primary receptor, the tip of which interacts with the N2 domain of pIII (Fig. 3A; Riechmann and Holliger 1997; Deng and Perham 2002). The secondary receptor is the periplasmic protein TolA, of which the C (or III) domain interacts with the N1 domain of pIII (Fig. 3B; Riechmann and Holliger 1997; Deng and Perham 2002). A number of structural and biochemical analyses have resulted in a model whereby binding to the F pilus induces *cis–trans* isomerization of $Pro_{213}$ within the N2 domain of pIII, resulting in detachment of the N1 domain and exposure of the TolA binding site (Riechmann and Holliger 1997; Lubkowski et al. 1999; Eckert et al. 2007). The F pilus, however, is not absolutely required for infection. It has been shown that Ff can infect the F-negative host in the presence of $Ca^{2+}$ ions at a frequency of $10^{-6}$ relative to the $F^+$ hosts (Russel et al. 1988).

Filamentous phage must cross the outer membrane to gain access to its secondary receptor, TolA. It is unclear, however, how this is achieved mechanistically (Fig. 3B,C). In contrast to the F pilus, the secondary receptor TolA and the cognate N1 domain of pIII are absolutely required for infection (Russel et al. 1988; Click and Webster 1997, 1998). TolA belongs to the TolQRA inner membrane complex. The TolQR component, a proton motive force (pmf)-powered motor, was also thought to be essential for filamentous phage infection. However, it was recently shown that, provided TolA is present, TolQR can be replaced by overexpression of a homologous ExbBC motor (Samire et al. 2020). In this setup, ExbBD must form a complex with TolA to allow phage entry. The ExbBD complex usually works with TonB, a protein involved in energizing the uptake of large nutrient molecules or powering protein export by TonB-dependent outer membrane gated channels (Gómez-Santos et al. 2019).

Once the N1 domain of pIII binds to TolA, it has been proposed that the TolQR motor pulls the TolA-bound phage toward the inner membrane (Cascales et al. 2007; Gerding et al. 2007). This notion, however, has been challenged by the recent finding that strains harboring TolQR mutations that prevent the complex's pmf-dependent conformational changes but allow formation of the TolQRA complex are still susceptible to Ff infection (Samire et al. 2020).

The entry phase of filamentous phage infection also requires the pIII C-domain to be linked covalently to N1 and N2. Long linkers or noncovalent interaction between N1/N2 and C domains decreases infectivity of the Ff-derived virions by several orders of magnitude (Krebber et al. 1997). Entry into the cell results in phage ssDNA being translocated into the cytoplasm and pVIII inserting into the inner membrane (Fig. 3B–D; Trenkner et al. 1967; Smilowitz 1974; Click and Webster 1997;

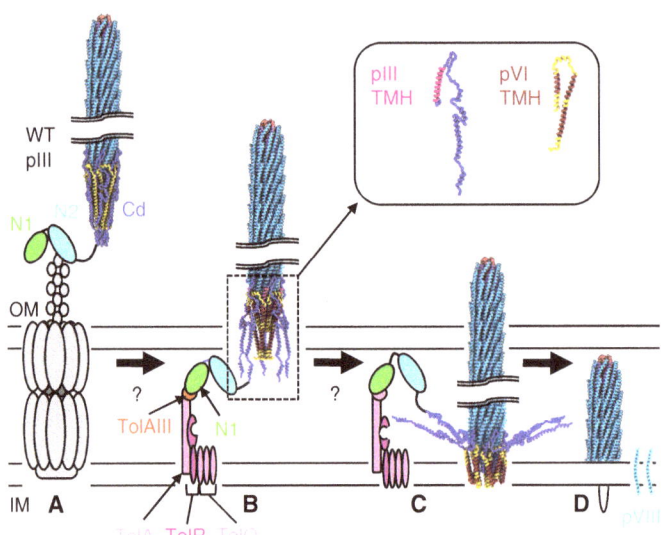

FIGURE 3. Model of Ff phage infection. (*A*) The pIII N2 domain (cyan oval) binds to the F-pilus tip (gray circles), and the pilus retracts. The question mark below the arrow pointing to *B* indicates an unknown mechanism by which the phage penetrates the outer membrane. (*B*) The pIII N1 domain (green oval) binds the TolAIII domain (orange circle). The pIII C domain (Cd, dark blue) goes through conformational changes that begin to expose the pVI and pIII transmembrane helices (see box; a single subunit of pIII is shown in dark blue with its predicted transmembrane helix (THM) pink, and pVI in gold with its predicted transmembrane helices cherry red). The question mark pointing to *C* indicates an unknown mechanism by which the interaction with TolA results in "opening" of the virion to allow phage DNA entry into the host cell. (*C*) Hydrophobic helices of pIII and pVI insert into the inner membrane and/or interact with the transmembrane helices of TolQRA (pink ovals). (*D*) The single-stranded DNA (ssDNA) crosses the inner membrane as pVIII (bright blue) is peeled off and inserts into the membrane. (OM) Outer membrane, (IM) inner membrane. The phage structure is obtained from the cryoEM structures PDB 8B3O and 8B3P; pVII, salmon; pIX, purple; F pilus assembly/retraction system, gray ovals. For simplicity only one N1 and N2 domain of pIIII is shown. The structural transitions shown on phage insertion into the membrane are based on images and the model in Conners et al. (2023).

Bennett and Rakonjac 2006). The high-resolution structure of the pIII–pVI virion cap gives some hints as to how the pointy cap could disassemble at this stage. The C domain of pIII contains a hydrophobic membrane anchor, joined to the remainder of the protein by a β-hairpin stabilized by a disulfide bridge (Conners et al. 2023). Not only N1 and N2 domains of pIII are required for phage entry, but also at least 121 residues at the carboxyl terminus, including the membrane anchor, the β-hairpin, and the upstream sequence (Bennett et al. 2011; Conners et al. 2023). The pIII membrane anchor is buried in the assembled virion (box in Fig. 3B), thus infection is likely dependent on major conformational changes in pIII and pVI, exposing the hydrophobic sequences of both proteins (Conners et al. 2023).

The pIII N2 and N1 domains, when expressed on their own or in infected cells, cause resistance to superinfection by incapacitating the primary and secondary receptors, respectively. The pIII N1 domain binds and blocks TolA, whereas N2 domain prevents assembly of the F pilus (Boeke et al. 1982).

## FILAMENTOUS PHAGE REPLICATION

Ff filamentous phage DNA is replicated as an episome, by a rolling circle mechanism, one strand at a time. Other nonintegrative filamentous phages also use the same replication mechanism, as do "induced" lysogenic filamentous phages. Lysogenization upon ssDNA entry does not apply to the Ff phage; for a discussion on filamentous phages in general, the reader is referred to relevant reviews (Das 2014; Mai-Prochnow et al. 2015).

The most detailed knowledge of the filamentous phage rolling-circle replication mechanism comes from Ff (for reviews, see Model and Russel 1988; Webster 2001). Both the positive (+) and negative (−) strand origins are located within the intergenic sequence (Fig. 2B). The Ff genome that enters the host cytoplasm corresponds to the (+) strand. The first step after entry is synthesis of the (−) strand, which results in a dsDNA genome (Fig. 4). Replication of the (−) strand is primed by a short RNA primer which, in turn, is synthesized by *E. coli* RNA polymerase. Transcription at the (−) origin

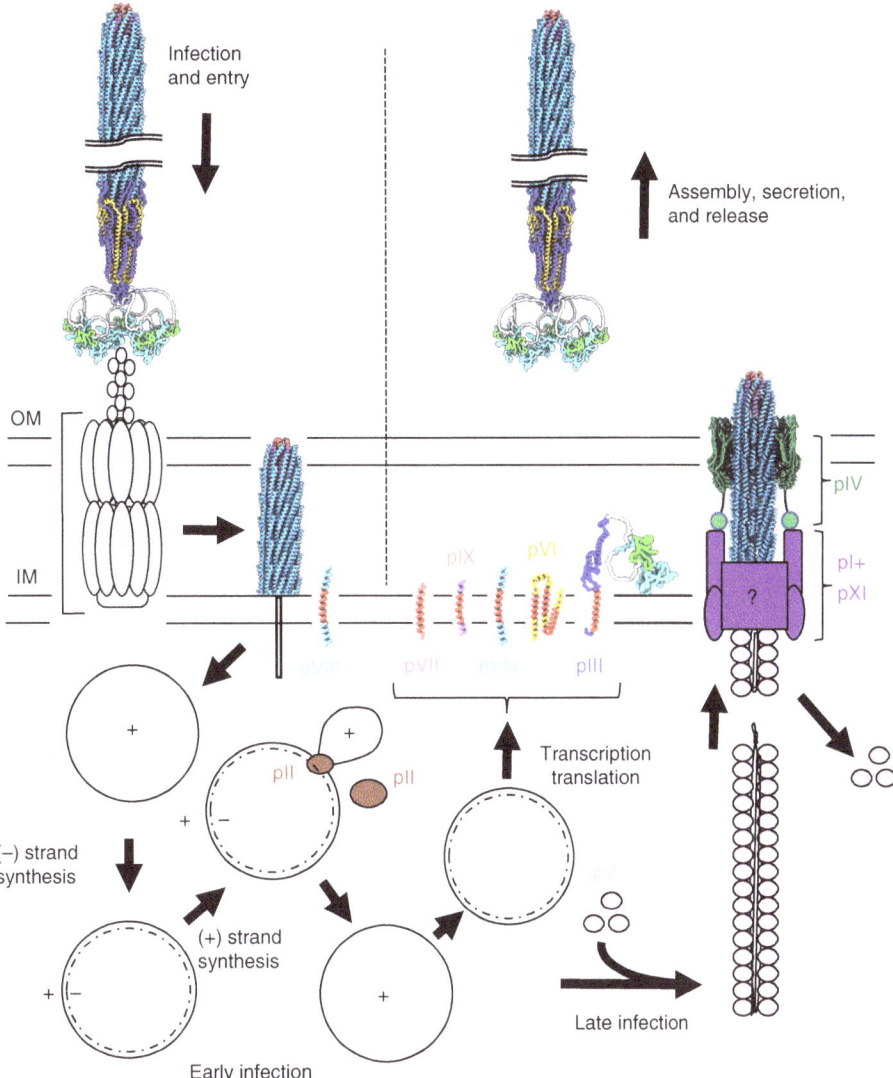

FIGURE 4. The Ff phage life cycle. After infection, (+) strand ssDNA enters the cytoplasm, and pVIII, the major coat protein, integrates into the inner membrane. The (−) strand replication is initiated by an RNA primer synthesized by RNA polymerase (Zenkin et al. 2006). Replication of the (−) strand by DNA polymerase III proceeds until the (−) strand is completely synthesized, resulting in the complete double-stranded RF (replicative form). Positive strand synthesis is initiated in supercoiled RF by pII (brown oval), which creates a nick in the (+) strand of the dsDNA at the initiator site within the (+) origin of replication. The RF is a template for further (+) strand replication and production of phage proteins. Proteins pII, pX (not shown), and pV (pink ovals) are involved in the replication and preparation of the (+) strand for assembly. Proteins pI and pXI (purple) form a ring-shaped multimer in the inner membrane, interacting with the 15-mer pIV channel (green) in the outer membrane. The open state of pIV was modeled as described in Conners et al. (2021). Both the inner and outer membrane complexes have domains protruding into the periplasm. Based on genetic evidence, the N0 domains of pIV (green circles) interact with the periplasmic domains of the pI–pXI complex, where they interact to form the *trans*-envelope Ff assembly complex. Virion proteins pIX, pVII, pVIII, pVI, and pIII are fed from the inner membrane into the growing phage filament. The composite structure of an f1 phage virion was generated as described in Figure 1. The structures of individual subunits are extracted from the cryoEM maps (PDB 8B3O, 8B3P, and 8B3Q; Conners et al. 2023) as per Figure 1 and are colored in the same way, with additional coloring of the predicted transmembrane helices in red. The ssDNA binding protein pV dissociates from the DNA and returns to cytoplasm. (OM) Outer membrane; (IM) inner membrane.

of replication is made possible by formation of a secondary structure, the stem of which mimics a typical −35 and −10 promoter sequence (Higashitani et al. 1996). The synthesized primer is extended by DNA polymerase III, to complete the (−) strand synthesis. The resulting double-stranded circular DNA (replicative form or RF) becomes supercoiled by the activity of the host enzyme DNA gyrase.

The (+) strand replicates using the supercoiled RF as a template. The phage-encoded replication protein pII creates a primer by nicking the (+) strand origin of replication at a specific sequence, GTTCTTT↓AATA. Replication terminates when pII encounters the same sequence, after the complete (+) strand is replicated. A pII-catalyzed strand-transferase reaction joins the 5′ and 3′ ends of the newly synthesized (+) strand to regenerate the intact double-stranded template (or the RF), while the "old" (+) strand is released (Fig. 4; Model and Russel 1988).

The (+) origin of replication is 147-nt-long, but the core (A or I segment) that is absolutely essential for replication is much shorter, only 65 nt in length. The B or II segment is not essential, but its absence or mutation results in a 100-fold decrease in replication efficiency. This decrease can be compensated for by mutations that increase production of pII (see below).

The Ff phage (+) origin acts as an initiator as well as a terminator for the (+) strand replication. If the Ff (+) origins are duplicated, short ssDNA circles are produced. They are packaged into short "defective interfering particles" (Enea et al. 1977; Ravetch et al. 1979). The short replicons result in a decrease in the RF copy number of their large counterparts through interference with pII-mediated replication. Low copy number of the phage RF in turn decreases the number of templates available for transcription and translation of phage proteins. This scarcity of the virion building blocks limits the number of full-length phage and defective interfering particles that can be produced. Propagation of these phages was found to lead to the appearance of interference-resistant mutants that contained mutations in the 5′ untranslated portion of *gII* mRNA and in the *gII* coding sequence, which were named interference-resistance mutations (Enea and Zinder 1982). Similar mutations are engineered in the helper phages used in phage display, to boost the production of particles containing phagemid vectors (Russel et al. 1986; Vieira and Messing 1987). The phagemid vectors containing an expression cassette for construction of the virion protein fusions were key for broadening the size and display modes used in phage display libraries in comparison to phage vectors (Barbas et al. 1991, 2001; Marks et al. 1991; Sblattero and Bradbury 2000; see also Chapter 4, Overview: The pComb3 Phagemid Family of Phage Display Vectors [Rader 2023]).

Plasmids containing two Ff (+) origins of replication have been shown to result in the generation of circular ssDNA between the two origins (Dotto et al. 1982). Constructs resulting in 221-nt ssDNA assembling nanorods of 50 nm in length have been produced (Fig. 1E; Specthrie et al. 1992). Very short Ff-derived particles have been functionalized and used in lateral flow assays, as vaccine carriers (Sattar 2013; Sattar et al. 2015), or to form aerogel nanofoam applied to Ni-MnOx cathode design (Cha et al. 2021).

Depending on the time point after infection, (+) strands can either serve as templates for additional (−) strand synthesis (in early infection) or form a complex with the ssDNA binding protein pV (20 min or longer after entry). The former fate of ssDNA results in the increase of the double-stranded RF to ∼70 per cell, whereas the latter results in formation of the pV–ssDNA complex, which is assembled into the new virions (Fig. 4). The virion production essentially ceases seven cell divisions after infection (Lerner and Model 1981).

The pV–ssDNA complex (packaging substrate) is formed by pV binding to the (+) strand DNA, commencing at the base of the packaging signal hairpin loop. pV binds to DNA as a dimer, with each subunit binding one of the two antiparallel strands. In this way, they zip up the single-stranded circular DNA into a rod-like pV–ssDNA filament (Guan et al. 1995).

Besides its role in forming the packaging substrate, pV also serves as a regulatory protein that inhibits translation of the mRNA encoding replication proteins pII and pX (Michel and Zinder 1989), and (−) strand synthesis (Fulford and Model 1988). Replication defects due to the interruption of the (+) strand origin B (II) segment or domain in a number of Ff phage vectors (e.g., R218, fdTet, M13KE, and M13mp1; Yanisch-Perron et al. 1985; Michel and Zinder 1989; Scott and Smith 1990; Marks et al. 1991; Nguyen et al. 2014) is compensated for by *trans* mutations in pV or *cis* mutations in the *gII* 5′

UTR that prevent the translational attenuation. pV attenuates translation of several other Ff genes, including its own. Recent analyses have shown that in mutants that decreased pV translation, Ff phage production is increased by up to 5.7-fold (Lee et al. 2021). Modeling and simulation studies suggested that lowering the amount of pV would increase not only the production of pII due to decreased translational suppression, but also conserve the cells' resources for production of the major coat protein pVIII (Smeal et al. 2017a,b). Taken together, it appears as if there is a balance between the formation of a ssDNA–pV packaging substrate and repression of the Ff mRNA's translation by pV that is tipped toward lowering overall phage production in the wild-type Ff. This effect can be overcome by increasing pII production and lowering the overall amount of pV in cells, leading to an increased production of phage (Lee et al. 2021).

## FILAMENTOUS PHAGE ASSEMBLY

Newly synthesized Ff phage virion proteins are targeted to the inner membrane (Fig. 4). From there, they are fed to an assembly complex that catalyzes their translocation from the inner membrane and onto the DNA that is being pushed out of the cell through the assembly channel. The major coat protein pVIII is one of the most highly expressed proteins in infected cells. This is consistent with the hourly production of approximately 1000 phage per cell, each of which contains ~3000 copies of pVIII.

Filamentous phage assembly at the bacterial envelope is a secretion-like process. It is related to bacterial pilus assembly or toxin secretion through dedicated *trans*-envelope protein "secretion systems." Assembly is initiated with minor virion proteins pVII and pIX, interacting with the pI–pXI complex, which forms the inner membrane component of the assembly complex (Fig. 4; Russel and Model 1989). During the elongation phase of assembly, pV is progressively released from ssDNA on the cytoplasmic side of the inner membrane, whereas pVIII associates with the DNA in an energy-dependent process that occurs at the inner membrane and continues until the entire ssDNA is covered (Feng et al. 1997).

Despite being an integral membrane protein before assembly, the subunits of pVIII interact directly with each other, and no lipids are observed in between them (Day 2011). Protein–phospholipid interactions, therefore, must be replaced with protein–protein interactions at the point of integration into the growing virion (Park et al. 2010). Although little is known about the mechanistic and structural details of how the pI–pXI complex feeds the major coat protein pVIII onto the DNA, this must be a highly coordinated process, in which the translocation of DNA across the inner membrane is coupled to association of the major coat protein subunits. During this process, DNA serves as an axis around which the helical array of the coat protein is assembled (Russel et al. 1997).

Ff assembly is finished once the DNA is completely covered with pVIII. The two minor phage proteins, pIII and pVI, are required to assemble into the virion in order to form the terminating cap, at the same time releasing the phage from the cell (Rakonjac et al. 1999). In the absence of pIII or pVI in infected cells, hundreds of cell-trapped phage particles can be detected. Depending on the time point after infection, the cell-associated Ff filaments may contain one or multiple sequentially packaged Ff genomes (Rakonjac and Model 1998).

Gene I encodes for pI, the inner membrane component of the filamentous phage assembly machinery, which contains an essential ATP-binding Walker motif. This gene is the most characteristic signature of a filamentous phage genome. The requirement of ATP and a pmf for Ff assembly was shown in a semipermeable assembly system (Feng et al. 1997). Very recently, the interactions before initiation of assembly have been shown in vitro, by combining pI–pXI proteoliposomes, Ff ssDNA, and pV. These experiments showed that the ssDNA–pV complex, but not pV alone, associates with pI–pXI-containing proteoliposomes (Haase et al. 2022). The size of the purified pI–pXI complex was estimated to be 320 kDa, corresponding approximately to a 12-mer containing six pI and six pXI subunits. Low-resolution electron microscopy imaging of the purified pI–pXI complex suggested a ring-like multimer structure with a sixfold rotational symmetry, similar to the inner membrane–associated ATPases in other secretion systems (Haase et al. 2022).

The outer membrane component of the Ff assembly machinery is a large gated outer membrane channel protein called pIV, which interacts with the pI–pXI complex (Fig. 4). Incidentally, the outer diameters of the pI/pXI and pIV multimers are similar, ~12 nm. The pI–pXI–pIV *trans*-envelope complex is thought to be necessary and sufficient for assembly (Russel 1993). pIV belongs to a family of outer membrane proteins called secretins. Such secretin channels share a high level of identity in their channel-forming domain, whereas domains that interact with the periplasmic and inner membrane components of the secretion machineries are less conserved at the protein level (Majewski et al. 2018).

Near-atomic resolution structures of several secretins revealed that they are composed of multiple copies of radially arranged identical subunits forming a β-barrel, with large internal diameters (6.4–8 nm). A second inner β-barrel contains two loops that project into the channel lumen at the level of the periplasm (Worrall et al. 2016; Yan et al. 2017; Conners et al. 2021). These two loops correspond to the "Gate 1" and "Gate 2" regions previously identified in a mutagenesis study, positively selecting for leaky mutants in pIV that could take up maltopentaose in the absence of the maltoporin LamB (Spagnuolo et al. 2010). Most leaky mutations map to the gates, with the severe "leaky" phenotype exhibited by strains harboring mutations of key inter- and intrasubunit interactions (Conners et al. 2021). Modeling pIV into an open conformation (based on the homologous type III secretion system secretin InvG; Hu et al. 2019) showed that the gates undergo substantial rearrangements to allow channel opening and the passage of the assembling phage filament (Fig. 5A; Conners et al. 2021).

When the filamentous phage assembly complex is compared to that of its closest relatives (i.e., the bacterial type II secretion system and type IV pilus biogenesis system), its simplicity is contrasted by their complexity. The inner membrane components of the bacterial secretion systems are particularly complex, with at least five different proteins taking part in feeding prepilin subunits into an assembling pseudopilus (type II secretion system) or pilus (type IV pilus) (Chang et al. 2016; Thomassin et al. 2017; Luna Rico et al. 2019; Naskar et al. 2021).

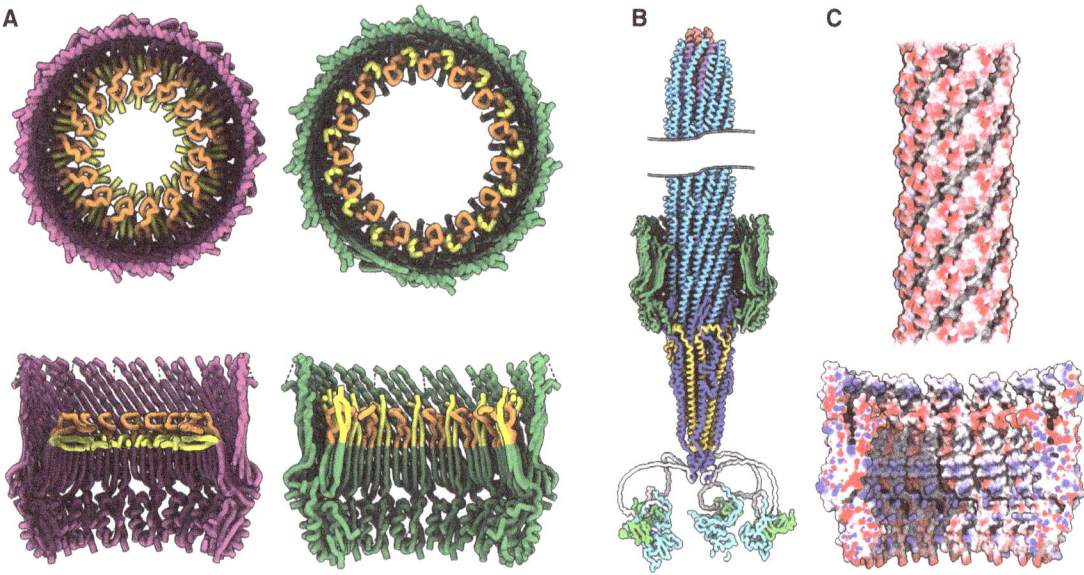

FIGURE 5. The pIV channel gate opening and interaction with the phage. (*A*) Top and side views of the closed (purple) and open (green) states of the pIV channel. The closed state is obtained from the cryoEM structure (PDB 7OFH), and a model of the open state was generated as described in Conners et al. (2021). Gate 1 is shown in yellow and Gate 2 in orange. (*B*) The composite structure of an f1 phage virion (generated as described in Fig. 1) docked into the open state of pIV. (*C*) Electrostatic surface potential of the fd virion (PDB 2C0W; Marvin et al. 2006) and the open state of pIV, colored red for negative and blue for positive. The figure is based on data from Conners et al. (2021).

Thioredoxin, a bacterial protein that normally reduces disulphide bonds in proteins, is required for assembly of the Ff phage f1, but not fd or M13 (Russel and Model 1986). The reduced rather than oxidized conformation of thioredoxin is required for f1 (but not fd or M13) assembly, albeit the reducing activity of thioredoxin is dispensable. Thioredoxin's role in the f1 phage assembly may be to improve processivity of pV dissociation from the ssDNA at the pI–pXI assembly site. A mutant of f1 phage that propagates normally in the absence of thioredoxin has a single mutation, in pI—Asn 142 to Tyr. In M13 and fd, residue 142 is a histidine (Russel and Model 1983, 1986). The requirement for thioredoxin is, therefore, determined by the properties of the inner membrane assembly protein pI.

## EFFECT OF Ff ON *E. COLI* PHYSIOLOGY

The high production of Ff phages is beneficial for their application. Even though the production of the wild-type Ff phage is high, resulting in titers of up to $10^{13}$ per mL, phage display recombinant phages and phagemids can be produced at 100-fold lower titers. Knowledge of limitations in Ff replication or assembly imposed by the host cell physiology will allow higher production of phage-derived particles for various applications.

In wild-type Ff infection, high productivity is a result of extremely efficient use of host resources, as recently modeled (Smeal et al. 2017a,b). The first seven *E. coli* generations after infection (3–5 h) is the window during which most progeny phages are produced. In subsequent generations, production drops due to a low level of Ff DNA replication in the infected cells (Lerner and Model 1981). Mathematical modeling has predicted that, among the host resources, the bottleneck for phage replication over an extended number of generations is the amount of DNA polymerase III allocated to Ff replication, but not the transcription (RNA polymerase) or the protein translation machinery (Smeal et al. 2017a,b). If this is the case, increasing the total amount of DNA polymerase III in the cells could conceivably increase replication and production of particles over 20 generations while minimizing the effect on host cell replication.

During the peak of Ff production, as many as four million copies of pVIII per infected cell are targeted to and anchored in the inner membrane (Lerner and Model 1981) before assembly into the virion. There is an "eclipse" period of 10–12 min postinfection before commencement of the Ff assembly, during which pVIII accumulates in the cells. The composition of the inner membrane changes drastically during that period, by increasing cardiolipin and decreasing phosphoethanolamine in comparison to uninfected cells (Woolford et al. 1974). When cells are infected with Ff mutants that lack pI and pIV and, hence, cannot assemble the virion but produce pVIII, cardiolipin increases to 30% of the total membrane lipids, in contrast to 5% in uninfected cells. The cells undergo major changes in morphology, with cristae-like inner membrane structures increasing with accumulation of pVIII (Schwartz and Zinder 1968; Onishi 1971). Membrane accumulation is pVIII-dependent (it does not occur in infection with a *gVIII* mutant). The increase in internal membranes could be a result of a decrease in the cardiolipin turnover, possibly due to being sequestered by millions of pVIII copies that accumulate in the cells, combined with increased synthesis (Chamberlain and Webster 1976, 1978).

Electron microscopy of cells infected with the Ff *gIII* deletion mutant (where phage stay attached to the cell surface) allowed the number of the assembly complexes to be estimated at 400 per cell (Rakonjac and Model 1998). Such a high number of *trans*-envelope phage assembly complexes is bound to cause stress in the host (Weiner and Model 1994). An inner membrane stress-response pathway, the phage shock protein (Psp) response, is triggered by phage infection (Model et al. 1997). Expression of type II and type III secretion systems also induces the Psp response, as reported for *Salmonella enterica* Sv. Typhimurim and *Yersinia enterocolitica* (Darwin 2005; Karlinsey et al. 2010). In Ff infection, it appears that the Psp response is mainly induced by mistargeting of pIV (the outer membrane channel) to the inner membrane (Daefler et al. 1997), rather than by the massive production of the major coat protein.

Another cellular response to Ff infection is increased phosphorylation of several proteins, including chaperone DnaK (Rieul et al. 1987). This observation is in agreement with the highly productive replication of phage DNA (1000-nt × 6047-nt ssDNA per cell per generation).

Wild-type Ff infection does not kill *E. coli*; however, as mentioned above, infection in which pVIII is produced but phage assembly is prevented is lethal (Pratt et al. 1966). In contrast, mutations affecting phage release due to the lack of terminating proteins pIII and pVI that still allow filament extrusion do not kill the cells (Rakonjac et al. 1999). Similarly, mutations in *gII* that do not support phage replication are not lethal (Pratt et al. 1966).

Lethality of pVIII in the absence of filament assembly, however, is not observed in cells containing "helper plasmids." These plasmids contain a theta-replicating origin of replication and all Ff genes, but lack the Ff origin of replication and the packaging signal (Chasteen et al. 2006). Given the constitutive expression of pVIII, which is toxic to *E. coli* (Pratt et al. 1966), survival of cells containing these plasmids is surprising and unexpected, but it could be a result of translational repression by pV due to the absence of ssDNA.

Even though phage assembly ceases seven generations after infection, the loss of the Ff replicon from the cells is minimal, only in one cell per 10 generations (Lerner and Model 1981). These cells are immediately reinfected by the phage from the medium; hence, the 100% infection of the cells in culture is maintained.

pIII also has a major effect on cell physiology through retracting or blocking assembly of the F pilus. This effect is beneficial to the host, because it prevents superinfection by lytic phages such as f2, MS2, R17, or Qβ that bind along the side of the F pilus (Loeb 1960; Olsthoorn and van Duin 2011). The second effect of pIII expression is through blocking the TolQRA secondary receptor, causing tolerance to colicins and resistance to infection by other filamentous phages. Another consequence of pIII expression is a leaky outer membrane. In cells that are transformed with ampicillin resistance plasmids that express pIII, it is observed that the release of periplasmic β-lactamase increases, resulting in protection of the sensitive cells in the culture (Boeke et al. 1982).

## LIMITATIONS TO PHAGE DISPLAY IMPOSED BY Ff ASSEMBLY AND BACTERIAL PHYSIOLOGY

This final section describes modifications of Ff phages that enable phage display applications and how these modifications affect the phage life cycle. It also describes phage display improvements based on the knowledge of Ff replication, assembly, and interactions with the host.

Polypeptides can be displayed on all Ff virion proteins (Fig. 6), although most common fusions are to pIII (Fig. 6B–E), pVII/pIX (Fig. 6F), and pVIII (Fig. 6G,H). For pIII and pVIII fusions to be inserted into a virion, they have to first be targeted to the membrane by a signal sequence. The coding sequence of the polypeptides to be displayed is, therefore, inserted in the same reading frame, and between the sequences coding the signal sequence and the mature portion of these proteins. A detailed review of membrane targeting and folding of the protein fusions in the periplasm can be found in Webster (2001).

Minor virion proteins pVII and pIX (at the round cap) are used in phage display *via* fusions with one or both proteins. Double fusions to both pVII and pIX allow display of proteins that are normally heterodimers (Gao et al. 1999). There is some controversy, however, about the use of these two minor virion proteins as display platforms. Earlier publications postulated a requirement for inclusion of a SecA-dependent signal sequence upstream of the inserted protein-coding sequence in order for the scFv fusion to the amino termini of pVII and pIX to be displayed efficiently (Gao et al. 1997, 2002; Huang et al. 2005). In contrast, more recent reports show that display on pVII and pIX is more effective without a signal sequence (Loset et al. 2011a,b). pVI, the virion protein that is least frequently used in phage display, only displays proteins that are fused to its carboxyl terminus. Display at the amino terminus of pVI is likely not successful because of potentially critical interactions with pIII, observed in the high-resolution structure of the pIII–pVI virion cap (Conners et al. 2023). Furthermore, wild-type

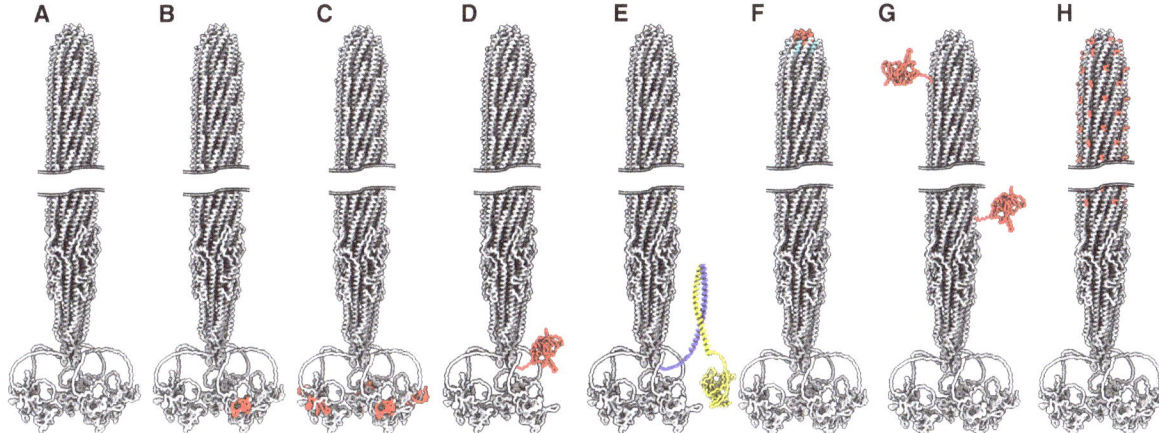

FIGURE 6. Positions of displayed proteins on the Ff virion. The wild-type Ff is shown in *A* and various sites and modes of display are shown in *B–H*. (*B*) Monovalent and (*C*) polyvalent display at the amino terminus of full-length pIII; (*D*) monovalent display on the C domain of pIII; (*E*) indirect phage display where the leucine zipper dimerization domain of Fos (*blue*) is fused to the pIII C domain, and the proteins to be displayed (*yellow*) are fused to the carboxyl terminus of its partner Jun (Jun and Fos Leu zipper coordinates are obtained from 5VPE [Yin et al. 2017]); (*F*) pVII (*red*) and /or pIX display (*cyan*); (*G*) mosaic pVIII display, where the fusions are assembled into the virion in combination with wild-type pVIII; (*H*) uniform pVIII display of short peptides on each pVIII subunit. The composite structure of an f1 phage virion was generated as described in Figure 1.

copies of pVI are also required to allow the virion to assemble; as a result, the frequency of the fusions is very low, about one per 10 phage particles (Jespers et al. 1996). The reason for inefficient display could be that the carboxyl terminus of pVI is not exposed on the surface of the virion but rather "encapsulated" within a dome formed by the C domain of pIII (Conners et al. 2023). The proteins or peptides fused to pVI are, therefore, expected to interfere with assembly of the pIII–pVI "pointy" end of the virion, resulting in overall poor incorporation into the phage.

Among the Ff proteins, the major coat protein pVIII is used for multicopy display. High avidity due to large copy numbers allows selection of peptides that have low affinity for the "selector" or target or bait molecule. Like the minor virion proteins pVII and pIX, pVIII was reported to be used for display of heterodimeric proteins. In this display mode, two different pVIII fusions are designed, each with a cysteine-containing linker. Dimerization of two fusions occurs by formation of an S–S bridge in the linker (Zwick et al. 2000).

The recent near-atomic structure of the secretin channel pIV and modeling of the egressing Ff within it (Fig. 5B) suggest that the virion fits snugly within the inner diameter of the open pIV channel (Conners et al. 2021). The tight fit of the virion inside the open pIV channel is bound to result in interactions of the virion with the open channel during extrusion, whereby the overall negative charge of the virion may be important (Fig. 5C; Conners et al. 2021). A study examining the correlation between diversity of peptides displayed on pVIII and their length showed that six residues can be displayed on each pVIII without a sequence bias. Longer peptide fusions to pVIII are more restricted, with 40%, 80%, and 99% of all possible sequences absent from the libraries of, respectively, 8-, 10-, or 16-residue-pVIII fusions (Iannolo et al. 1995). Display bias against specific sequences of these longer peptides may be caused by interference with packing of the pVIII copies into the virion, or by increased diameter combined with unfavorable interactions with the inner wall of the pIV channel. Display of even longer peptides is possible by combining the pVIII-peptide fusions with wild-type copies of pVIII. Large proteins, such as alkaline phosphatase (a homodimer, ∼47 KDa per subunit), are reported to be displayed in such a mosaic manner. The assembled virions, however, were shown to display no more than one copy of the fusion out of approximately 3000 pVIII subunits (Weichel et al. 2008). Curiously, mutations in *gVIII* were found to increase the number of human growth hormone (20-kDa monomer) and streptavidin (a homotetramer, ∼15 kDa per subunit) fusions per virion

(Sidhu et al. 2000). Modeling of the pVIII fusions displaying these larger proteins within the pIV channel would be of interest, to explain how they are able to be extruded from the cells.

Display on pIII has a number of modalities. Proteins can be fused to the amino terminus of either full-length pIII or to the amino terminus of the C domain, which is sufficient for assembly of the virion (Fig. 6D). Irrespective of the length of pIII fragment used to construct the fusions, an amino-terminal signal sequence is required (Barbas et al. 1991; Griffiths et al. 1993). If a truncated pIII is used for display, a wild-type (full-length) pIII must also be included to provide the N1 and N2 domains of pIII, to allow phage infection and amplification of the phage.

Display of peptides and proteins in five copies on pIII can be achieved by exclusively expressing the fusions (Griffiths et al. 1993). The copy number, however, is often lower than 5 because of proteolytic degradation of the fusion counterpart. Fusions to the amino terminus of pIII somewhat attenuate infection of the host, so those fusions that are susceptible to proteolysis have an amplification advantage during library screening. This means that cleaved fusions can be enriched during biopanning even if they do not bind to the immobilized selector molecule (Matochko and Derda 2013). This issue can be alleviated if the insert is placed downstream from the coding sequence for N1 or N2 domains of pIII. In this way, peptide degradation destroys the phage's ability to infect cells, and the recombinant clone encoding this variant is eliminated from the library at the stage of amplification (Tjhung et al. 2015).

pIII fusions can be expressed from the phage or phagemid vectors. Phagemid vectors contain two origins of replication: one is a plasmid *ori* that replicates in the absence of the helper phage, and one is an Ff origin of replication. The vector also contains the Ff packaging signal, an antibiotic resistance gene, and a multiple cloning site between sequences encoding the signal sequence and mature pIII (Barbas et al. 2001). Phagemid-transformed cells infected with a helper phage switch to rolling circle replication from the Ff origin and are packaged into phage-like phagemid particles that are termed PPs or transducing particles (TDPs).

A truncated pIII, containing only the carboxy-terminal domain, is sufficient for assembly into the virion, as long as a signal sequence is present. Phage display phagemid vectors encoding the carboxy-terminal domain of pIII are used as a platform for display of antibody variable domains (see Chapter 4, Overview: The pComb3 Phagemid Family of Phage Display Vectors [Rader 2023]). The sequence encoding the fusion counterpart is inserted between a signal sequence and the pIII C domain (Barbas et al. 2001). Given that these fusions lack the receptor-binding domains of pIII, the full-length *gIII* is provided in *trans* from a helper phage or a helper plasmid. Fusions to truncated pIII expressed from phagemid vectors are displayed only in one or two copies per virion, resulting in low avidity of the phage or phagemid particles. This mode of display is used to screen phage display libraries for high-affinity ligands, such as antibodies, from the libraries derived from an immunized source (Barbas et al. 1991).

Phagemid vectors offer flexibility in that the expression of pIII fusion for display can be tuned by placing it under the control of an inducible promoter. This helps overcome potential toxicity of pIII fusions to the host and thereby avoid counterselection against the recombinant clones expressing fusions that impair host growth. The small size of the phagemid is also advantageous over the phage vectors because of an increased efficiency of transformation; *E. coli* transformation limits the diversity (or primary size) of phage display libraries.

The number of displayed pIII fusions per phagemid particle can be tuned by changing the amount of pIII expressed from a helper phage. If a wild-type helper phage is used, the copy number of the fusions per virion is low because the wild-type pIII encoded by the helper phage is preferentially assembled into the phagemid particles. There are a few ways the number of fusions per particle can be increased. One option is to use a helper containing a *gIII* with an amber mutation (*gIII$^{am}$*). If the particles are assembled in a host that has a suppressor mutation, pIII encoded by the helper is produced at a lower amount in comparison to the expression from the wild-type *gIII*, allowing an increased proportion of the fusion (Oh et al. 2007). Another option to allow multivalent display is to use a *gIII$^{am}$* helper phage in a nonsuppressor host, or a helper that has a deletion of *gIII* (Griffiths et al. 1993; Rakonjac et al. 1997; de Wildt et al. 2002). If so, all five copies of pIII will be fusions, allowing for selection of low-affinity binders due to the increased avidity of the particles.

Another variant of the phagemid system uses a helper plasmid instead of a helper phage (Chasteen et al. 2006). This setup eliminates a need for superinfection with a helper phage, and was introduced to simplify the procedures of biopanning used in phage display library screening. However, transduction of the helper-plasmid-containing cells with the phagemid particles was achieved at a rather low efficiency, 1%–18% relative to transduction of the host strain without a helper plasmid (Chasteen et al. 2006). This low efficiency is likely due to the constitutive expression of pIII from the helper plasmid, which is known to incapacitate the primary and secondary Ff receptors, the F pilus, and TolQRA (Boeke et al. 1982). Helper plasmids that contain mutations in *gIII* are susceptible to phagemid transduction at an efficiency that is equal to that of a host that is devoid of helper plasmid (Chasteen et al. 2006).

A functional signal sequence is required not only for membrane insertion of pIII, but also, consequently, for assembly and release of the virion. This property has been used as a trap for selective display of bacterial surface and secreted proteins at the genomic and metagenomic scale. A vector that encodes the pIII C domain without a signal sequence was used to select for virions that formed the intact pointy caps based on resistance to the ionic detergent sarkosyl (Jankovic et al. 2007). Interestingly, when applied on the metagenomic scale, Gram-negative signal sequences were favored over the Gram-positive ones for targeting pIII to the inner membrane of *E. coli* and assembly into the virion. However, the bar for membrane targeting and display was shown to be reasonably low, with transmembrane helices and archaeal signal sequences being able to target the pIII fusions to the SecA–SecYEG translocon and enable display (Ciric et al. 2014; Gagic et al. 2016; Ng et al. 2016).

In general, the periplasmic folding of fusions to Ff virion proteins is not suitable for display of cytosolic proteins. The reason for this is in that the periplasm, where the amino-terminally displayed fusions to pIII and pVIII fold, is an oxidizing environment, where disulphide bonds are formed between the cysteine residues. In contrast, in the reducing environment of the cytoplasm, disulphide bonds do not form, and reduced cysteines are often essential for correct folding and function of cytoplasmic proteins (Bardwell et al. 1991).

Options for display of cytoplasmic proteins have been explored, including use of mutants that decrease formation of disulphide bonds or anaerobic growth (Rebar and Pabo 1994). Another approach to display cytoplasmic proteins is construction of fusions to the termini of phage proteins that locate to the cytoplasm during assembly. The issue with this strategy is that the termini of the Ff proteins that are exposed to the cytoplasm also serve as contacts with ssDNA during assembly, or are buried within the virion caps; hence, incorporation of such fusions into the virions may be poor. It was reported that it was possible to display proteins at the carboxyl terminus of pIII; however, the fusion had to be combined with wild-type pIII copies (Fuh and Sidhu 2000). An alternative strategy applied to allow folding of cytosolic proteins in the cytoplasm was to construct fusions containing a Tat signal sequence. The Tat secretion pathway allows export of folded protein domains through the inner membrane (Velappan et al. 2010).

Phage display libraries of antibody or peptide repertoires contain inserts of fixed length, and thus construction of fusions in-frame with the upstream signal sequence and downstream mature portion of the protein is not an issue. However, when cDNA libraries are constructed, where there are no fixed ends to the inserts, the probability of correct frame and orientation of the insert is 1 in 18 ($18 = 3 \times 3 \times 2$). The functional size of the library is, therefore, nearly 20-fold lower than its primary size. Furthermore, many fusions in the library display incorrect reading frames, with the resulting peptides being degraded, giving an advantage to "noise" clones during amplification steps because of higher infectivity in comparison to "legitimate" fusions.

One method to increase the odds of the in-frame clones in a cDNA library is "indirect" display, in which the library is constructed as a carboxy-terminal fusion to a protein that interacts with a cognate partner displayed on pIII. This strategy means that the inserts must be in-frame only with the upstream sequence, increasing the effective size of the library from 1/18 to 1/6. The interacting partners tested using this were leucine zipper domains of the transcription factors Jun (fused to the library) and Fos (fused to pIII; Fig. 6E). Jun and Fos pair within the *E. coli* periplasm, and the released phagemid particles display the cDNA fusion proteins (Crameri et al. 1994).

Ff production in the context of phage display implies expression of additional proteins in comparison to the wild-type Ff infection. These fusions present an additional burden to the host cell, because of interference with phage assembly or overexpression from inducible promoters in the phagemid vectors. This can be particularly problematic with fusions to pIII, whose membrane targeting is SecA–SecYEG-dependent, and which may form channels in the inner membrane when overexpressed, as was shown for wild-type pIII (Glaser-Wuttke et al. 1989). Because of the toxicity of fusions (or diminished infectivity), there are often "censored" variants that are lost in the course of amplification (Derda et al. 2011).

## CONCLUSIONS

This overview provides background information on Ff filamentous phage biology and structure. It also discusses the effect of Ff infection and production on *E. coli* and how this impacts the production of phage during the time course of culture growth. At a protein level, this overview discusses how Ff protein processing and assembly into the virion determines the folding, size, and copy number of displayed polypeptides. Overall, this overview gives a "beyond the protocols" look at the main character of phage display, the filamentous phage Ff, with brief reference to its applications as nanorods or nanofibers.

It is also worth mentioning that recent bioinformatic analyses of prokaryotic genomes and shotgun metagenomes have revealed 10,295 novel filamentous phage-derived prophages in nearly all bacterial phyla and in some archaea (Roux et al. 2019). From this new vantage point, it becomes clear that thousands of filamentous phage species growing on a range of diverse hosts are a treasure trove of novel properties that will undoubtedly help overcome some limitations of Ff-based phage display and other applications in the future.

## ACKNOWLEDGMENTS

We are indebted to Marjorie Russel and to the editors and reviewers for comments on the manuscript. Funding to the V.A.M.G. laboratory from the Wellcome Trust (Seed Award in Science) (210363/Z/18/Z) and to the J.R. laboratory from the Callaghan Innovation Grants MAUX1915 and BNANO2101, Marsden Fund (the Royal Society of New Zealand), Bridgewest Ventures New Zealand, Palmerston North Medical Foundation, Massey University Research Fund, School of Natural Sciences, and donations by Anne and Bryce Carmine and Anonymous Donor are acknowledged.

## REFERENCES

Barbas CF III, Kang AS, Lerner RA, Benkovic SJ. 1991. Assembly of combinatorial antibody libraries on phage surfaces: the gene III site. *Proc Natl Acad Sci* **88:** 7978–7982. doi:10.1073/pnas.88.18.7978

Barbas CF III, Burton DR, Scott JK, Silverman GJ. 2001. *Phage display: a laboratory manual*. Cold Spring Harbor Laboratory Press, Cold Spring Harbor, NY.

Bardwell JCA, McGovern K, Beckwith J. 1991. Identification of a protein required for disulfide bond formation *in vivo*. *Cell* **67:** 581–589. doi:10.1016/0092-8674(91)90532-4

Bennett NJ, Rakonjac J. 2006. Unlocking of the filamentous bacteriophage virion during infection is mediated by the C domain of pIII. *J Mol Biol* **356:** 266–273. doi:10.1016/j.jmb.2005.11.069

Bennett NJ, Gagic D, Sutherland-Smith AJ, Rakonjac J. 2011. Characterization of a dual-function domain that mediates membrane insertion and excision of Ff filamentous bacteriophage. *J Mol Biol* **411:** 972–985. doi:10.1016/j.jmb.2011.07.002

Bernard JM, Francis MB. 2014. Chemical strategies for the covalent modification of filamentous phage. *Front Microbiol* **5:** 734. doi:10.3389/fmicb.2014.00734

Boeke JD, Model P. 1982. A prokaryotic membrane anchor sequence: carboxyl terminus of bacteriophage f1 gene III protein retains it in the membrane. *Proc Natl Acad Sci* **79:** 5200–5204. doi:10.1073/pnas.79.17.5200

Boeke JD, Model P, Zinder ND. 1982. Effects of bacteriophage f1 gene III protein on the host cell membrane. *Mol Gen Genet* **186:** 185–192. doi:10.1007/BF00331849

Branston S, Stanley E, Ward J, Keshavarz-Moore E. 2011. Study of robustness of filamentous bacteriophages for industrial applications. *Biotechnol Bioeng* **108:** 1468–1472. doi:10.1002/bit.23066

Branston SD, Stanley EC, Ward JM, Keshavarz-Moore E. 2013. Determination of the survival of bacteriophage M13 from chemical and physical challenges to assist in its sustainable bioprocessing. *Biotechnol Bioprocess Eng* **18:** 560–566. doi:10.1007/s12257-012-0776-9

Cascales E, Buchanan SK, Duche D, Kleanthous C, Lloubes R, Postle K, Riley M, Slatin S, Cavard D. 2007. Colicin biology. *Microbiol Mol Biol Rev* **71:** 158–229. doi:10.1128/MMBR.00036-06

Cha TG, Tsedev U, Ransil A, Embree A, Gordon DB, Belcher AM, Voigt CA. 2021. Genetic control of aerogel and nanofoam properties, applied to Ni-MnO$_x$ cathode design. *Adv Funct Mater* **31:** 2010867. doi:10.1002/adfm.202010867

Chamberlain BK, Webster RE. 1976. Lipid-protein interactions in *Escherichia coli*. Membrane-associated f1 bacteriophage coat protein and phospholipid metabolism. *J Biol Chem* **251:** 7739–7745. doi:10.1016/S0021-9258(19)56996-4

Chamberlain BK, Webster RE. 1978. Effect of membrane-associated f1 bacteriophage coat protein upon the activity of *Escherichia coli* phosphatidylserine synthetase. *J Bacteriol* **135:** 883–887. doi:10.1128/jb.135.3.883-887.1978

Chang YW, Rettberg LA, Treuner-Lange A, Iwasa J, Søgaard-Andersen L, Jensen GJ. 2016. Architecture of the type IVa pilus machine. *Science* **351:** aad2001. doi:10.1126/science.aad2001

Chasteen L, Ayriss J, Pavlik P, Bradbury AR. 2006. Eliminating helper phage from phage display. *Nucl Acids Res* **34:** e145. doi:10.1093/nar/gkl772

Chung WJ, Lee DY, Yoo SY. 2014. Chemical modulation of M13 bacteriophage and its functional opportunities for nanomedicine. *Int J Nanomed* **9:** 5825–5836.

Ciric M, Moon CD, Leahy SC, Creevey CJ, Altermann E, Attwood GT, Rakonjac J, Gagic D. 2014. Metasecretome-selective phage display approach for mining the functional potential of a rumen microbial community. *BMC Genomics* **15:** 356. doi:10.1186/1471-2164-15-356

Click EM, Webster RE. 1997. Filamentous phage infection: required interactions with the TolA protein. *J Bacteriol* **179:** 6464–6471. doi:10.1128/jb.179.20.6464-6471.1997

Click EM, Webster RE. 1998. The TolQRA proteins are required for membrane insertion of the major capsid protein of the filamentous phage f1 during infection. *J Bacteriol* **180:** 1723–1728. doi:10.1128/JB.180.7.1723-1728.1998

Conners R, McLaren M, Lapinska U, Sanders K, Stone MRL, Blaskovich MAT, Pagliara S, Daum B, Rakonjac J, Gold VAM. 2021. CryoEM structure of the outer membrane secretin channel pIV from the f1 filamentous bacteriophage. *Nat Commun* **12:** 6316. doi:10.1038/s41467-021-26610-3

Conners R, León-Quezada RI, McLaren M, Bennett NJ, Daum B, Rakonjac J, Gold VAM. 2023. Cryo-electron microscopy of the f1 filamentous phage reveals insights into viral infection and assembly. *Nat Commun* **14:** 2724. doi:10.1038/s41467-023-37915-w

Crameri R, Jaussi R, Menz G, Blaser K. 1994. Display of expression products of cDNA libraries on phage surfaces. A versatile screening system for selective isolation of genes by specific gene-product/ligand interaction. *Eur J Biochem* **226:** 53–58. doi:10.1111/j.1432-1033.1994.tb20025.x

Daefler S, Guilvout I, Hardie KR, Pugsley AP, Russel M. 1997. The C-terminal domain of the secretin PulD contains the binding site for its cognate chaperone, PulS, and confers PulS dependence on plV(f1) function. *Mol Microbiol* **24:** 465–475. doi:10.1046/j.1365-2958.1997.3531727.x

Darwin AJ. 2005. Genome-wide screens to identify genes of human pathogenic Yersinia species that are expressed during host infection. *Curr Issues Mol Biol* **7:** 135–149.

Das B. 2014. Mechanistic insights into filamentous phage integration in Vibrio cholerae. *Front Microbiol* **5:** 650. doi:10.3389/fmicb.2014.00650

Davis NG, Model P. 1985. An artificial anchor domain: hydrophobicity suffices to stop transfer. *Cell* **41:** 607–614. doi:10.1016/S0092-8674(85)80033-7

Day LA. 2011. Family inoviridae. In *Virus taxonomy: classification and nomenclature of viruses: Ninth report of the International Committee on Taxonomy of Viruses* (ed. King AMQ, Adams MJ, Carstens EB, Lefkowitz EJ), pp. 375–384. Elsevier Academic, San Diego.

Day LA, Marzec CJ, Reisberg SA, Casadevall A. 1988. DNA packing in filamentous bacteriophages. *Annu Rev Biophys Biophys Chem* **17:** 509–539. doi:10.1146/annurev.bb.17.060188.002453

Deng LW, Perham RN. 2002. Delineating the site of interaction on the pIII protein of filamentous bacteriophage fd with the F-pilus of *Escherichia coli*. *J Mol Biol* **319:** 603–614. doi:10.1016/S0022-2836(02)00260-7

Derda R, Tang SK, Li SC, Ng S, Matochko W, Jafari MR. 2011. Diversity of phage-displayed libraries of peptides during panning and amplification. *Molecules* **16:** 1776–1803. doi:10.3390/molecules16021776

de Wildt RM, Tomlinson IM, Ong JL, Holliger P. 2002. Isolation of receptor-ligand pairs by capture of long-lived multivalent interaction complexes. *Proc Natl Acad Sci* **99:** 8530–8535. doi:10.1073/pnas.132008499

Dogic Z. 2016. Filamentous phages as a model system in soft matter physics. *Front Microbiol* **7:** 1013. doi:10.3389/fmicb.2016.01013

Dotto GP, Horiuchi K, Zinder ND. 1982. Initiation and termination of phage f1 plus-strand synthesis. *Proc Natl Acad Sci* **79:** 7122–7126. doi:10.1073/pnas.79.23.7122

Eckert B, Martin A, Balbach J, Schmid FX. 2007. Prolyl isomerization as a molecular timer in phage infection. *Nat Struct Mol Biol* **12:** 619–623. doi:10.1038/nsmb946

Edens L, Konings RN, Schoenmakers JG. 1978. A cascade mechanism of transcription in bacteriophage M13 DNA. *Virology* **86:** 354–367. doi:10.1016/0042-6822(78)90076-4

Endemann H, Model P. 1995. Location of filamentous phage minor coat proteins in phage and in infected cells. *J Mol Biol* **250:** 496–506. doi:10.1006/jmbi.1995.0393

Enea V, Zinder ND. 1982. Interference resistant mutants of phage f1. *Virology* **122:** 222–226. doi:10.1016/0042-6822(82)90395-6

Enea V, Horiuchi K, Turgeon BG, Zinder ND. 1977. Physical map of defective interfering particles of bacteriophage f1. *J Mol Biol* **111:** 395–414. doi:10.1016/S0022-2836(77)80061-2

Feng JN, Russel M, Model P. 1997. A permeabilized cell system that assembles filamentous bacteriophage. *Proc Natl Acad Sci* **94:** 4068–4073. doi:10.1073/pnas.94.8.4068

Fuh G, Sidhu SS. 2000. Efficient phage display of polypeptides fused to the carboxy-terminus of the M13 gene-3 minor coat protein. *FEBS Lett* **480:** 231–234. doi:10.1016/S0014-5793(00)01946-3

Fulford W, Model P. 1984. Gene X of bacteriophage f1 is required for phage DNA synthesis. Mutagenesis of in-frame overlapping genes. *J Mol Biol* **178:** 137–153. doi:10.1016/0022-2836(84)90136-0

Fulford W, Model P. 1988. Bacteriophage f1 DNA replication genes. II. The roles of gene V protein and gene II protein in complementary strand synthesis. *J Mol Biol* **203:** 39–48. doi:10.1016/0022-2836(88)90089-7

Gagic D, Ciric M, Wen WX, Ng F, Rakonjac J. 2016. Exploring the secretomes of microbes and microbial communities using filamentous phage display. *Front Microbiol* **7:** 429. doi:10.3389/fmicb.2016.00429

Gao C, Lin CH, Lo CH, Mao S, Wirsching P, Lerner RA, Janda KD. 1997. Making chemistry selectable by linking it to infectivity. *Proc Natl Acad Sci* **94:** 11777–11782. doi:10.1073/pnas.94.22.11777

Gao C, Mao S, Lo CH, Wirsching P, Lerner RA, Janda KD. 1999. Making artificial antibodies: a format for phage display of combinatorial heterodimeric arrays. *Proc Natl Acad Sci* **96:** 6025–6030. doi:10.1073/pnas.96.11.6025

Gao C, Mao S, Kaufmann G, Wirsching P, Lerner RA, Janda KD. 2002. A method for the generation of combinatorial antibody libraries using pIX phage display. *Proc Natl Acad Sci* **99:** 12612–12616. doi:10.1073/pnas.192467999

Gerding MA, Ogata Y, Pecora ND, Niki H, de Boer PA. 2007. The trans-envelope Tol-Pal complex is part of the cell division machinery and required for proper outer-membrane invagination during cell constriction in E. coli. *Mol Microbiol* **63:** 1008–1025. doi:10.1111/j.1365-2958.2006.05571.x

Glaser-Wuttke G, Keppner J, Rasched I. 1989. Pore-forming properties of the adsorption protein of filamentous phage fd. *Biochim Biophys Acta* **985:** 239–247. doi:10.1016/0005-2736(89)90408-2

Goldbourt A, Gross BJ, Day LA, McDermott AE. 2007. Filamentous phage studied by magic-angle spinning NMR: resonance assignment and secondary structure of the coat protein in Pf1. *J Am Chem Soc* **129:** 2338–2344. doi:10.1021/ja066928u

Goldbourt A, Day LA, McDermott AE. 2010. Intersubunit hydrophobic interactions in Pf1 filamentous phage. *J Biol Chem* **285:** 37051–37059. doi:10.1074/jbc.M110.119339

Gømez-Santos N, Glatter T, Koebnik R, Świątek-Połatyńska MA, Søgaard-Andersen L. 2019. A TonB-dependent transporter is required for secretion of protease PopC across the bacterial outer membrane. *Nat Commun* **10:** 1360. doi:10.1038/s41467-019-09366-9

Goodrich AF, Steege DA. 1999. Roles of polyadenylation and nucleolytic cleavage in the filamentous phage mRNA processing and decay pathways in *Escherichia coli*. *RNA* **5:** 972–985. doi:10.1017/S1355838299990398

Grant R, Lin T, Webster R, Konigsberg W. 1980. Structure of filamentous bacteriophage: isolation, characterization, and localization of the minor coat proteins and orientation of the DNA. In *Bacteriophage assembly* (ed. DuBow M), pp. 413–428. Alan. R. Liss, New York.

Griffiths AD, Malmqvist M, Marks JD, Bye JM, Embleton MJ, McCafferty J, Baier M, Holliger KP, Gorick BD, Hughes-Jones NC, et al. 1993. Human anti-self antibodies with high specificity from phage display libraries. *EMBO J* 12: 725–734. doi:10.1002/j.1460-2075.1993.tb05706.x

Guan Y, Zhang H, Wang AH. 1995. Electrostatic potential distribution of the gene V protein from Ff phage facilitates cooperative DNA binding: a model of the GVP-ssDNA complex. *Protein Sci* 4: 187–197. doi:10.1002/pro.5560040206

Haase M, Tessmer L, Köhnlechner L, Kuhn A. 2022. The M13 phage assembly machine has a membrane-spanning oligomeric ring structure. *Viruses* 14: 1163. doi:10.3390/v14061163

Haigh NG, Webster RE. 1999. The pI and pXI assembly proteins serve separate and essential roles in filamentous phage assembly. *J Mol Biol* 293: 1017–1027. doi:10.1006/jmbi.1999.3227

Henry KA, Arbabi-Ghahroudi M, Scott JK. 2015. Beyond phage display: non-traditional applications of the filamentous bacteriophage as a vaccine carrier, therapeutic biologic, and bioconjugation scaffold. *Front Microbiol* 6: 755. doi:10.3389/fmicb.2015.00755

Higashitani N, Higashitani A, Guan ZW, Horiuchi K. 1996. Recognition mechanisms of the minus-strand origin of phage f1 by *Escherichia coli* RNA polymerase. *Genes Cells* 1: 829–841. doi:10.1046/j.1365-2443.1996.d01-279.x

Hofschneider PH. 1963. Untersuchungen uber kleine *E. coli* K 12 bakteriophagen 1 und 2 mitteilung. *Z Naturforsch Pt B* 18: 203–210. doi:10.1515/znb-1963-0306

Holliger P, Riechmann L, Williams RL. 1999. Crystal structure of the two N-terminal domains of g3p from filamentous phage fd at 1.9 A: evidence for conformational lability. *J Mol Biol* 288: 649–657. doi:10.1006/jmbi.1999.2720

Hu J, Worrall LJ, Vuckovic M, Hong C, Deng W, Atkinson CE, Brett Finlay B, Yu Z, Strynadka NCJ. 2019. T3S injectisome needle complex structures in four distinct states reveal the basis of membrane coupling and assembly. *Nat Microbiol* 4: 2010–2019. doi:10.1038/s41564-019-0545-z

Huang Y, Chiang CY, Lee SK, Gao Y, Hu EL, De Yoreo J, Belcher AM. 2005. Programmable assembly of nanoarchitectures using genetically engineered viruses. *Nano Lett* 5: 1429–1434. doi:10.1021/nl050795d

Iannolo G, Minenkova O, Petruzzelli R, Cesareni G. 1995. Modifying filamentous phage capsid: limits in the size of the major capsid protein. *J Mol Biol* 248: 835–844. doi:10.1006/jmbi.1995.0264

Jankovic D, Collett MA, Lubbers MW, Rakonjac J. 2007. Direct selection and phage display of a Gram-positive secretome. *Genome Biol* 8: R266. doi:10.1186/gb-2007-8-12-r266

Jespers LS, De Keyser A, Stanssens PE. 1996. λZLG6: a phage lambda vector for high-efficiency cloning and surface expression of cDNA libraries on filamentous phage. *Gene* 173: 179–181. doi:10.1016/0378-1119(96)00217-X

Karlinsey JE, Maguire ME, Becker LA, Crouch MLV, Fang FC. 2010. The phage shock protein PspA facilitates divalent metal transport and is required for virulence of *Salmonella enterica* sv. Typhimurium. *Mol Microbiol* 78: 669–685. doi:10.1111/j.1365-2958.2010.07357.x

Krebber C, Spada S, Desplancq D, Krebber A, Ge L, Pluckthun A. 1997. Selectively-infective phage (SIP): a mechanistic dissection of a novel in vivo selection for protein-ligand interactions. *J Mol Biol* 268: 607–618. doi:10.1006/jmbi.1997.0981

Krogh A, Larsson B, von Heijne G, Sonnhammer EL. 2001. Predicting transmembrane protein topology with a hidden Markov model: application to complete genomes. *J Mol Biol* 305: 567–580. doi:10.1006/jmbi.2000.4315

La Farina M, Model P. 1983. Transcription in bacteriophage f1-infected *Escherichia coli*. Messenger populations in the infected cell. *J Mol Biol* 164: 377–393. doi:10.1016/0022-2836(83)90057-8

Lee S, Belcher A. 2004. Virus-based fabrication of micro- and nanofibers using electrospinning. *Nano Lett* 4: 387–390. doi:10.1021/nl034911t

Lee S, Mao C, Flynn C, Belcher A. 2002. Ordering of quantum dots using genetically engineered viruses. *Science* 296: 892–895. doi:10.1126/science.1068054

Lee BY, Lee J, Ahn DJ, Lee S, Oh MK. 2021. Optimizing protein V untranslated region sequence in M13 phage for increased production of single-stranded DNA for origami. *Nucl Acids Res* 49: 6596–6603. doi:10.1093/nar/gkab455

Lerner TJ, Model P. 1981. The "steady state" of coliphage f1: DNA synthesis late in infection. *Virology* 115: 282–294. doi:10.1016/0042-6822(81)90111-2

Loeb T. 1960. Isolation of a bacteriophage specific for the F$^+$ and Hfr mating types of *Escherichia coli* K-12. *Science* 131: 932–933. doi:10.1126/science.131.3404.932

Lopez J, Webster RE. 1983. Morphogenesis of filamentous bacteriophage f1: orientation of extrusion and production of polyphage. *Virology* 127: 177–193. doi:10.1016/0042-6822(83)90382-3

Loset GA, Bogen B, Sandlie I. 2011a. Expanding the versatility of phage display I: efficient display of peptide-tags on protein VII of the filamentous phage. *PLoS ONE* 6: e14702. doi:10.1371/journal.pone.0014702

Loset GA, Roos N, Bogen B, Sandlie I. 2011b. Expanding the versatility of phage display II: improved affinity selection of folded domains on protein VII and IX of the filamentous phage. *PLoS ONE* 6: e17433. doi:10.1371/journal.pone.0017433

Lubkowski J, Hennecke F, Pluckthun A, Wlodawer A. 1998. The structural basis of phage display elucidated by the crystal structure of the N-terminal domains of g3p. *Nat Struct Biol* 5: 140–147. doi:10.1038/nsb0298-140

Lubkowski J, Hennecke F, Pluckthun A, Wlodawer A. 1999. Filamentous phage infection: crystal structure of g3p in complex with its coreceptor, the C-terminal domain of TolA. *Structure* 7: 711–722. doi:10.1016/S0969-2126(99)80092-6

Luna Rico A, Zheng W, Petiot N, Egelman EH, Francetic O. 2019. Functional reconstitution of the type IVa pilus assembly system from enterohaemorrhagic *Escherichia coli*. *Mol Microbiol* 111: 732–749. doi:10.1111/mmi.14188

Mai-Prochnow A, Hui JG, Kjelleberg S, Rakonjac J, McDougald D, Rice SA. 2015. Big things in small packages: the genetics of filamentous phage and effects on fitness of their host. *FEMS Microbiol Rev* 39: 465–487. doi:10.1093/femsre/fuu007

Majewski DD, Worrall LJ, Strynadka NC. 2018. Secretins revealed: structural insights into the giant gated outer membrane portals of bacteria. *Curr Opin Struct Biol* 51: 61–72. doi:10.1016/j.sbi.2018.02.008

Mao C, Solis D, Reiss B, Kottmann S, Sweeney R, Hayhurst A, Georgiou G, Iverson B, Belcher A. 2004. Virus-based toolkit for the directed synthesis of magnetic and semiconducting nanowires. *Science* 303: 213–217. doi:10.1126/science.1092740

Marks JD, Hoogenboom HR, Bonnert TP, McCafferty J, Griffiths AD, Winter G. 1991. By-passing immunization. Human antibodies from V-gene libraries displayed on phage. *J Mol Biol* 222: 581–597. doi:10.1016/0022-2836(91)90498-U

Marvin DA, Hoffmann-Berling H. 1963. A Fibrous DNA phage (Fd) and a spherical Rna phage (Fr) specific for male strains of *E. coli* Ii. physical characteristics. *Z Naturforsch B* 18: 884–893. doi:10.1515/znb-1963-1106

Marvin DA, Pigram WJ, Wiseman RL, Wachtel EJ, Marvin FJ. 1974a. Filamentous bacterial viruses. XIL. Molecular architecture of the class I (fd, If1, IKe) virion. *J Mol Biol* 88: 581–598. doi:10.1016/0022-2836(74)90409-4

Marvin DA, Wiseman RL, Wachtel EJ. 1974b. Filamentous bacterial viruses. XI. Molecular architecture of the class II (Pf1, Xf) virion. *J Mol Biol* 82: 121–138. doi:10.1016/0022-2836(74)90336-2

Marvin DA, Welsh LC, Symmons MF, Scott WR, Straus SK. 2006. Molecular structure of fd (f1, M13) filamentous bacteriophage refined with respect to X-ray fibre diffraction and solid-state NMR data supports specific models of phage assembly at the bacterial membrane. *J Mol Biol* 355: 294–309. doi:10.1016/j.jmb.2005.10.048

Marvin DA, Symmons MF, Straus SK. 2014. Structure and assembly of filamentous bacteriophages. *Prog Biophys Mol Biol* 114: 80–122. doi:10.1016/j.pbiomolbio.2014.02.003

Matochko WL, Derda R. 2013. Error analysis of deep sequencing of phage libraries: peptides censored in sequencing. *Comput Math Methods Med* 2013: 491612. doi:10.1155/2013/491612

McCafferty J, Griffiths AD, Winter G, Chiswell DJ. 1990. Phage antibodies: filamentous phage displaying antibody variable domains. *Nature* 348: 552–554. doi:10.1038/348552a0

Michel B, Zinder ND. 1989. Translational repression in bacteriophage f1: characterization of the gene V protein target on the gene II mRNA. *Proc Natl Acad Sci* **86:** 4002–4006. doi:10.1073/pnas.86.11.4002

Model P, Russel M. 1988. Filamentous bacteriophage. In *The bacteriophages* (ed. Calendar R), pp. 375–456. Plenum, New York.

Model P, Jovanovic G, Dworkin J. 1997. The *Escherichia coli* phage shock protein operon. *Mol Microbiol* **24:** 255–261. doi:10.1046/j.1365-2958.1997.3481712.x

Morag O, Abramov G, Goldbourt A. 2011. Similarities and differences within members of the Ff family of filamentous bacteriophage viruses. *J Phys Chem B* **115:** 15370–15379. doi:10.1021/jp2079742

Moses PB, Boeke JD, Horiuchi K, Zinder ND. 1980. Restructuring the bacteriophage f1 genome: expression of gene VIII in the intergenic space. *Virology* **104:** 267–278. doi:10.1016/0042-6822(80)90332-3

Naskar S, Hohl M, Tassinari M, Low HH. 2021. The structure and mechanism of the bacterial type II secretion system. *Mol Microbiol* **115:** 412–424. doi:10.1111/mmi.14664

Ng F, Kittelmann S, Patchett ML, Attwood GT, Janssen PH, Rakonjac J, Gagic D. 2016. An adhesin from hydrogen-utilizing rumen methanogen *Methanobrevibacter ruminantium* M1 binds a broad range of hydrogen-producing microorganisms. *Environ Microbiol* **18:** 3010–3021. doi:10.1111/1462-2920.13155

Nguyen KT, Adamkiewicz MA, Hebert LE, Zygiel EM, Boyle HR, Martone CM, Melendez-Rios CB, Noren KA, Noren CJ, Hall MF. 2014. Identification and characterization of mutant clones with enhanced propagation rates from phage-displayed peptide libraries. *Anal Biochem* **462:** 35–43. doi:10.1016/j.ab.2014.06.007

Oh MY, Joo HY, Hur BU, Jeong YH, Cha SH. 2007. Enhancing phage display of antibody fragments using *gIII*-amber suppression. *Gene* **386:** 81–89. doi:10.1016/j.gene.2006.08.009

Oh JW, Chung WJ, Heo K, Jin HE, Lee BY, Wang E, Zueger C, Wong W, Meyer J, Kim C, et al. 2014. Biomimetic virus-based colourimetric sensors. *Nat Commun* **5:** 3043. doi:10.1038/ncomms4043

Olsthoorn R, van Duin J. 2011. Bacteriophages with ssRNA. In *eLS*, pp. 778. John Wiley & Sons, New York.

Onishi Y. 1971. Phospholipids of virus-induced membranes in cytoplasm of *Escherichia coli*. *J Bacteriol* **107:** 918–925. doi:10.1128/jb.107.3.918-925.1971

Pacheco-Gomez R, Kraemer J, Stokoe S, England HJ, Penn CW, Stanley E, Rodger A, Ward J, Hicks MR, Dafforn TR. 2012. Detection of pathogenic bacteria using a homogeneous immunoassay based on shear alignment of virus particles and linear dichroism. *Anal Chem* **84:** 91–97. doi:10.1021/ac201544h

Park SH, Marassi FM, Black D, Opella SJ. 2010. Structure and dynamics of the membrane-bound form of Pf1 coat protein: implications of structural rearrangement for virus assembly. *Biophys J* **99:** 1465–1474. doi:10.1016/j.bpj.2010.06.009

Petrenko VA, Smith GP, Gong X, Quinn T. 1996. A library of organic landscapes on filamentous phage. *Protein Eng* **9:** 797–801. doi:10.1093/protein/9.9.797

Ploss M, Kuhn A. 2011. Membrane insertion and assembly of epitope-tagged gp9 at the tip of the M13 phage. *BMC Microbiol* **11:** 211. doi:10.1186/1471-2180-11-211

Pratt D, Tzagoloff H, Erdahl WS. 1966. Conditional lethal mutants of the small filamentous coliphage M13. I. Isolation, complementation, cell killing, time of cistron action. *Virology* **30:** 397–410. doi:10.1016/0042-6822(66)90118-8

Rader C. 2023. The pComb3 phagemid family of phage display vectors. *Cold Spring Harb Protoc* doi:10.1101/pdb.over107756

Rakonjac J, Model P. 1998. Roles of pIII in filamentous phage assembly. *J Mol Biol* **282:** 25–41. doi:10.1006/jmbi.1998.2006

Rakonjac J, Jovanovic G, Model P. 1997. Filamentous phage infection-mediated gene expression: construction and propagation of the *gIII* deletion mutant helper phage R408d3. *Gene* **198:** 99–103. doi:10.1016/S0378-1119(97)00298-9

Rakonjac J, Feng J, Model P. 1999. Filamentous phage are released from the bacterial membrane by a two-step mechanism involving a short C-terminal fragment of pIII. *J Mol Biol* **289:** 1253–1265. doi:10.1006/jmbi.1999.2851

Ravetch JV, Horiuchi K, Zinder ND. 1979. DNA sequence analysis of the defective interfering particles of bacteriophage f1. *J Mol Biol* **128:** 305–318. doi:10.1016/0022-2836(79)90090-1

Rebar EJ, Pabo CO. 1994. Zinc finger phage: affinity selection of fingers with new DNA-binding specificities. *Science* **263:** 671–673. doi:10.1126/science.8303274

Riechmann L, Holliger P. 1997. The C-terminal domain of TolA is the coreceptor for filamentous phage infection of *E. coli*. *Cell* **90:** 351–360. doi:10.1016/S0092-8674(00)80342-6

Rieul C, Cortay JC, Bleicher F, Cozzone AJ. 1987. Effect of bacteriophage M13 infection on phosphorylation of DnaK protein and other *Escherichia coli* proteins. *Eur J Biochem* **168:** 621–627. doi:10.1111/j.1432-1033.1987.tb13461.x

Roux S, Krupovic M, Daly RA, Borges AL, Nayfach S, Schulz F, Sharrar A, Matheus Carnevali PB, Cheng JF, Ivanova NN, et al. 2019. Cryptic inoviruses revealed as pervasive in bacteria and archaea across Earth's biomes. *Nat Microbiol* **4:** 1895–1906. doi:10.1038/s41564-019-0510-x

Russel M. 1993. Protein-protein interactions during filamentous phage assembly. *J Mol Biol* **231:** 689–697. doi:10.1006/jmbi.1993.1320

Russel M, Model P. 1983. A bacterial gene, *fip*, required for filamentous bacteriophage fl assembly. *J Bacteriol* **154:** 1064–1076. doi:10.1128/jb.154.3.1064-1076.1983

Russel M, Model P. 1986. The role of thioredoxin in filamentous phage assembly—construction, isolation, and characterisation of mutant thioredoxins. *J Biol Chem* **261:** 4997–5005. doi:10.1016/S0021-9258(18)66819-X

Russel M, Model P. 1989. Genetic analysis of the filamentous bacteriophage packaging signal and of the proteins that interact with it. *J Virol* **63:** 3284–3295. doi:10.1128/jvi.63.8.3284-3295.1989

Russel M, Kidd S, Kelley MR. 1986. An improved filamentous helper phage for generating single-stranded plasmid DNA. *Gene* **45:** 333–338. doi:10.1016/0378-1119(86)90032-6

Russel M, Whirlow H, Sun TP, Webster RE. 1988. Low-frequency infection of F⁻ bacteria by transducing particles of filamentous bacteriophages. *J Bacteriol* **170:** 5312–5316. doi:10.1128/jb.170.11.5312-5316.1988

Russel M, Linderoth NA, Sali A. 1997. Filamentous phage assembly: variation on a protein export theme. *Gene* **192:** 23–32. doi:10.1016/S0378-1119(96)00801-3

Samire P, Serrano B, Duche D, Lemarie E, Lloubes R, Houot L. 2020. Decoupling filamentous phage uptake and energy of the TolQRA motor in *Escherichia coli*. *J Bacteriol* **202:** e00428-19. doi:10.1128/JB.00428-19

Samuelson JC, Chen M, Jiang F, Moller I, Wiedmann M, Kuhn A, Phillips GJ, Dalbey RE. 2000. YidC mediates membrane protein insertion in bacteria. *Nature* **406:** 637–641. doi:10.1038/35020586

Sattar S. 2013. "Filamentous phage-derived nano-rods for applications in diagnostics and vaccines." PhD thesis, Massey University, Palmerston North, New Zealand.

Sattar S, Bennett NJ, Wen WX, Guthrie JM, Blackwell LF, Conway JF, Rakonjac J. 2015. Ff-nano, short functionalized nanorods derived from Ff (f1, fd, or M13) filamentous bacteriophage. *Front Microbiol* **6:** 316. doi:10.3389/fmicb.2015.00316

Sblattero D, Bradbury A. 2000. Exploiting recombination in single bacteria to make large phage antibody libraries. *Nat Biotechnol* **18:** 75–80. doi:10.1038/71958

Schwartz FM, Zinder N. 1968. Morphological changes in *Escherichia coli* infected with the DNA bacteriophage f1. *Virology* **34:** 352–355. doi:10.1016/0042-6822(68)90246-8

Scott JK, Smith GP. 1990. Searching for peptide ligands with an epitope library. *Science* **249:** 386–390. doi:10.1126/science.1696028

Sergeyev IV, Day LA, Goldbourt A, McDermott AE. 2011. Chemical shifts for the unusual DNA structure in Pf1 bacteriophage from dynamic-nuclear-polarization-enhanced solid-state NMR spectroscopy. *J Am Chem Soc* **133:** 20208–20217. doi:10.1021/ja2043062

Sharma P, Ward A, Gibaud T, Hagan MF, Dogic Z. 2014. Hierarchical organization of chiral rafts in colloidal membranes. *Nature* **513:** 77–80. doi:10.1038/nature13694

Sidhu SS, Weiss GA, Wells JA. 2000. High copy display of large proteins on phage for functional selections. *J Mol Biol* **296:** 487–495. doi:10.1006/jmbi.1999.3465

Smeal SW, Schmitt MA, Pereira RR, Prasad A, Fisk JD. 2017a. Simulation of the M13 life cycle I: assembly of a genetically-structured deterministic chemical kinetic simulation. *Virology* **500:** 259–274. doi:10.1016/j.virol.2016.08.017

Smeal SW, Schmitt MA, Pereira RR, Prasad A, Fisk JD. 2017b. Simulation of the M13 life cycle II: investigation of the control mechanisms of M13 infection and establishment of the carrier state. *Virology* **500**: 275–284. doi:10.1016/j.virol.2016.08.015

Smilowitz H. 1974. Bacteriophage f1 infection: fate of the parental major coat protein. *J Virol* **13**: 94–99. doi:10.1128/jvi.13.1.94-99.1974

Smith GP. 1985. Filamentous fusion phage: novel expression vectors that display cloned antigens on the virion surface. *Science* **228**: 1315–1317. doi:10.1126/science.4001944

Smith GP. 2023. Principles of affinity selection. *Cold Spring Harb Protoc* doi:10.1101/pdb.over107894

Souza GR, Molina JR, Raphael RM, Ozawa MG, Stark DJ, Levin CS, Bronk LF, Ananta JS, Mandelin J, Georgescu MM, et al. 2010. Three-dimensional tissue culture based on magnetic cell levitation. *Nat Nanotechnol* **5**: 291–296. doi:10.1038/nnano.2010.23

Spagnuolo J, Opalka N, Wen WX, Gagic D, Chabaud E, Bellini P, Bennett MD, Norris GE, Darst SA, Russel M, et al. 2010. Identification of the gate regions in the primary structure of the secretin pIV. *Mol Microbiol* **76**: 133–150. doi:10.1111/j.1365-2958.2010.07085.x

Specthrie L, Bullitt E, Horiuchi K, Model P, Russel M, Makowski L. 1992. Construction of a microphage variant of filamentous bacteriophage. *J Mol Biol* **228**: 720–724. doi:10.1016/0022-2836(92)90858-H

Stopar D, Spruijt RB, Wolfs CJ, Hemminga MA. 2002. Structural characterization of bacteriophage M13 solubilization by amphiphiles. *Biochim Biophys Acta* **1594**: 54–63. doi:10.1016/S0167-4838(01)00281-3

Tarafder AK, von Kugelgen A, Mellul AJ, Schulze U, Aarts D, Bharat TAM. 2020. Phage liquid crystalline droplets form occlusive sheaths that encapsulate and protect infectious rod-shaped bacteria. *Proc Natl Acad Sci* **117**: 4724–4731. doi:10.1073/pnas.1917726117

Thomassin JL, Santos Moreno J, Guilvout I, Tran Van Nhieu G, Francetic O. 2017. The *trans*-envelope architecture and function of the type 2 secretion system: new insights raising new questions. *Mol Microbiol* **105**: 211–226. doi:10.1111/mmi.13704

Tjhung KF, Deiss F, Tran J, Chou Y, Derda R. 2015. Intra-domain phage display (ID-PhD) of peptides and protein mini-domains censored from canonical pIII phage display. *Front Microbiol* **6**: 340. doi:10.3389/fmicb.2015.00340

Trenkner E, Bonhoeffer F, Gierer A. 1967. The fate of the protein component of bacteriophage fd during infection. *Biochem Biophys Res Commun* **28**: 932–939. doi:10.1016/0006-291X(67)90069-1

Velappan N, Fisher HE, Pesavento E, Chasteen L, D'Angelo S, Kiss C, Longmire M, Pavlik P, Bradbury AR. 2010. A comprehensive analysis of filamentous phage display vectors for cytoplasmic proteins: an analysis with different fluorescent proteins. *Nucl Acids Res* **38**: e22. doi:10.1093/nar/gkp809

Vieira J, Messing J. 1987. Production of single-stranded plasmid DNA. *Methods Enzymol* **153**: 3–11. doi:10.1016/0076-6879(87)53044-0

Webster RE. 2001. Filamentous phage biology. In *Phage display: a laboratory manual* (ed. Barbas CF III, Burton DR, Scott JK, Silverman GJ), pp. 1–37. Cold Spring Harbor Laboratory Press, Cold Spring Harbor, NY.

Weichel M, Jaussi R, Rhyner C, Crameri R. 2008. Display of *E. coli* alkaline phosphatase pIII or pVIII fusions on phagemid surfaces reveals monovalent decoration with active molecules. *Open Biochem J* **2**: 38–43. doi:10.2174/1874091X00802010038

Weiner L, Model P. 1994. Role of an *Escherichia coli* stress-response operon in stationary-phase survival. *Proc Natl Acad Sci* **91**: 2191–2195. doi:10.1073/pnas.91.6.2191

Woolford JL Jr, Cashman JS, Webster RE. 1974. F1 coat protein synthesis and altered phospholipid metabolism in f1 infected *Escherichia coli*. *Virology* **58**: 544–560. doi:10.1016/0042-6822(74)90088-9

Worrall LJ, Hong C, Vuckovic M, Deng W, Bergeron JRC, Majewski DD, Huang RK, Spreter T, Finlay BB, Yu Z, et al. 2016. Near-atomic-resolution cryo-EM analysis of the Salmonella T3S injectisome basal body. *Nature* **540**: 597. doi:10.1038/nature20576

Xu J, Dayan N, Goldbourt A, Xiang Y. 2019. Cryo-electron microscopy structure of the filamentous bacteriophage IKe. *Proc Natl Acad Sci* **116**: 5493–5498. doi:10.1073/pnas.1811929116

Yan Z, Yin M, Xu D, Zhu Y, Li X. 2017. Structural insights into the secretin translocation channel in the type II secretion system. *Nat Struct Mol Biol* **24**: 177–183. doi:10.1038/nsmb.3350

Yanisch-Perron C, Vieira J, Messing J. 1985. Improved M13 phage cloning vectors and host strains: nucleotide sequences of the M13mp18 and pUC19 vectors. *Gene* **33**: 103–119. doi:10.1016/0378-1119(85)90120-9

Yin Z, Machius M, Nestler EJ, Rudenko G. 2017. Activator Protein-1: redox switch controlling structure and DNA-binding. *Nucl Acids Res* **45**: 11425–11436. doi:10.1093/nar/gkx795

Zenkin N, Naryshkina T, Kuznedelov K, Severinov K. 2006. The mechanism of DNA replication primer synthesis by RNA polymerase. *Nature* **439**: 617–620. doi:10.1038/nature04337

Zinder ND, Horiuchi K. 1985. Multiregulatory element of filamentous bacteriophages. *Microbiol Rev* **49**: 101–106. doi:10.1128/mr.49.2.101-106.1985

Zwick MB, Shen J, Scott JK. 2000. Homodimeric peptides displayed by the major coat protein of filamentous phage. *J Mol Biol* **300**: 307–320. doi:10.1006/jmbi.2000.3850

CHAPTER 3

# Principles of Affinity Selection

George P. Smith[1]

*Division of Biological Sciences, University of Missouri, Columbia, Missouri 65211, USA*

The most common application of phage-display technology is the discovery of peptides or proteins that specifically bind some molecule or other substance of interest—for example, antibodies that specifically bind an antigen. The discovery process starts with a library encompassing a very large array of proteins or peptides with a great diversity of binding specificities—for example, single-chain antibodies with a great diversity of antigen-binding sites. Each member of the array is displayed on the surface of hundreds to billions of identical virus particles (virions) belonging to a single-phage clone; the library as a whole comprises millions to billions of such clones, all mixed together in a single vessel. Affinity selection is the process by which a molecule or substance of interest—generically called the selector—is used to select very rare clones in the library displaying proteins or peptides that happen to bind the selector with high affinity and selectivity. Here, I explain general principles guiding a successful affinity-selection project—principles grounded in phage biology, kinetics of reversible binding, technological advances, and the practical experience of thousands of investigators around the globe.

## BACKGROUND

Here I summarize fundamental principles to consider in designing an affinity-selection project: that is, a project with the goal of discovering peptides or proteins that bind specifically to some chosen molecule or other substance, which is called the selector. Affinity selection is often called "panning" in laboratory jargon, reflecting the phage display's origin in immunology (Wysocki and Sato 1978), but that is much too narrow a term to cover the common realizations of the selection process. The principles explained here are grounded in basic reversible binding kinetics, phage biology, technological advances, and practical experience.

There are numerous realizations of the general phage-display concept, involving different phages or phage proteins; most of the principles apply to all of them. But here the principles are shown specifically for systems in which the peptides or proteins are displayed on coat protein III (pIII) of the Ff family of filamentous phages: derivatives of wild-type strains fd, f1, and M13 (Rakonjac et al. 2017). Even within that category, innumerable innovative variations have been investigated by the global phage-display community. Here, however, only the most common, "mainstream" procedures are covered explicitly.

## THE AFFINITY-SELECTION PROCESS

The upper part of Figure 1 shows a side-view cartoon of the long, thin phage particle—called the virion—showing the basic structural features involved in the affinity-selection process

---

[1]Correspondence: smithgp@missouri.edu

© 2026 Cold Spring Harbor Laboratory Press
Cite this overview as *Cold Spring Harb Protoc*; doi:10.1101/pdb.over107894

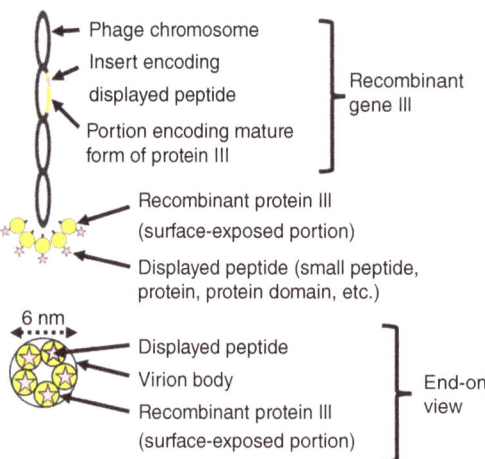

FIGURE 1. Schematic cartoons of virion structure, with a side view on *top* and an end-on view on the *bottom*. See text for further explanation.

(Rakonjac et al. 2017). There is a coiled, single-stranded, circular chromosome inside the particle, represented by a black chain with colored segments. The colored segments correspond to one of the chromosome's genes: a recombinant form of phage gene III, encoding pIII. The parts of the recombinant gene that derive from the wild-type gene III are represented by yellow segments, whereas the foreign (i.e., nonphage) insert that encodes the displayed peptide is represented by a pink segment. The shorter yellow segment encodes the amino-terminal signal peptide of pIII, which is removed in the process of virion assembly. The pink segment plus the longer yellow segment encode the mature form of the recombinant pIII in the completed virion, with the displayed foreign peptide at the amino terminus.

I use the term "displayed peptide" (or simply "peptide") generically for the amino acids encoded by the insert, regardless of the number of amino acids. So, for example, single-chain antibodies (∼240 amino acids; see Box 1) and short random peptides (typically 6–15 amino acids) are both called peptides, despite their disparity in size.

### BOX 1. DISPLAY SYSTEMS

An important detail is suppressed in this and all upcoming cartoons for the sake of simplicity. In most affinity-selection projects, the recombinant gene III resides in a complete phage genome that bears no other gene III: a so-called type 3 display system (Smith 1993; Smith and Petrenko 1997). This system is well-adapted for short displayed peptides, which are likely to be displayed on all five recombinant pIII molecules at the tip of the virion. This is what is shown in Figure 1, and what is assumed here unless otherwise indicated. In many important projects, however, the displayed peptide is a ∼240-amino-acid single-chain antibody (Kang and Seong 2020) or other antibody domain; phages displaying such constructs are referred to generically as phage–antibodies. Almost all phage–antibody projects, along with other projects involving relatively large displayed peptides, use a type 3 + 3 display system (Smith 1993; Smith and Petrenko 1997) in which there are two gene III sequences in two different genomes: a special type of plasmid called a phagemid that bears the recombinant gene III encoding the displayed peptide and a helper phage that bears a wild-type or engineered gene III that does not encode a displayed peptide. A cell containing both genomes releases two types of virions: virions containing the phagemid chromosome and virions containing the helper genome. Both types of pIII molecules can be incorporated into both types of virions. When the displayed peptide is a relatively large domain such as a single-chain antibody, it is only occasionally expressed intact on either type of virion, for two reasons. First, the pIII encoded by the helper phage almost always outcompetes the recombinant pIII encoded by the phagemid for incorporation during virion assembly. Second, only a few large displayed foreign peptide domains escape degradation in the cytosol or periplasm during virion assembly, leaving only the rest of the protein intact in the completed virion. Consequently, only a minority of completed virions display an intact, functional foreign peptide domain; virtually no virions display two or more intact foreign peptide domains; and many virions display no intact foreign peptide domains at all. On the few occasions when antibody-bearing phagemid virions are mentioned in the following text, I assume (contrary to the cartoons) that an intact, functional antibody domain is displayed monovalently or not at all.

pIII molecules form a five-member ring that caps off the end of the long filamentous virion as it emerges from the cell (Rakonjac et al. 2017). This five-member ring is represented schematically in the end-view cartoon in the lower part of Figure 1; in the upper, side-view cartoon, it is represented less realistically as a fan. The carboxy-terminal half of the recombinant pIII is not depicted; it is buried within the tight architecture of the virion body and is not exposed to the surrounding medium. The amino-terminal half, with the displayed peptide at the very amino terminus, is flexibly attached to the virion, where it is exposed to the medium. The part of the exposed amino-terminal half that corresponds to wild-type pIII is represented by the yellow balls in both the side-view and end-view cartoons. The displayed peptide, represented here by the pink star, is flexibly attached to the phage-derived domain (the yellow ball), and is fully exposed to the medium. Virions belonging to other phage clones bear different gene-III inserts (they are colored differently in the following cartoons), and thus display different peptides (they have different shapes and the same color as the insert in the following cartoons).

There are two starting resources in an affinity-selection project:

- A library with millions or billions of different phage clones, each displaying a different peptide (represented by different shapes with different colors in Fig. 2), and each represented by a relatively small number of identical virions. A large-scale affinity-selection project might start with 1 mL of buffer containing $10^{10}$ clones displaying $10^{10}$ different peptides, each represented by 1000 identical virions on average.

- A selector, such as an antibody, antigen, or receptor molecule. In the case of phage–antibody libraries (see Box 1), for instance, the selector is a single antigen of interest. (In some projects, the selector is not a molecule. In other projects, the selector is a collection of thousands of different molecules represented in varying proportions—for example, all the molecules on the surface of some cell type. I will not consider such cases explicitly here.)

The goal of affinity selection is to isolate rare library clones with virions displaying selector-binding peptides—for example, rare phage–antibody clones with virions displaying antibodies that bind the antigen acting as the selector.

The cartoons in Figure 2 show the steps in a single round of affinity selection; almost all affinity-selection projects include two or more successive rounds. The steps below are numbered to correspond to the step numbers in the figure. There are countless innovative variations on the basic procedure, some of which bear little resemblance to the cartoons. But the standard process covers the majority of affinity-selection experiments, and most of the principles here apply even to such procedural variants.

- **Step 1.** The selector is immobilized on a solid surface that is called the substrate. Typical examples are the surface of a polystyrene Petri dish, the surface of the well in a multiwell polystyrene dish, and the impermeable hydrophilic surfaces of 1-µm paramagnetic beads (e.g., SpeedBeads from Cytiva Life Sciences).

    Most protein selectors can be immobilized by direct absorption onto the polystyrene surface of a Petri or multiwell dish; this is thought to entail local denaturation of a small patch of protein surface, exposing hydrophobic groups that bond with the hydrophobic polystyrene. Alternatively, the selector can be covalently coupled to a hydrophilic surface derivatized with a reactive group such as amine-reactive N-hydroxysuccinimide or sulfhydryl-reactive N-alkyl maleimide; this method does not ordinarily entail denaturation of the selector.

    A third, very popular immobilization mode is to react a biotinylated selector with a hydrophilic substrate coated with streptavidin or neutravidin (an engineered derivative of hen's egg avidin), each molecule of which can bind four biotin molecules; the selector is not ordinarily denatured. Small (1-µm in diameter) paramagnetic beads densely coated ($\sim$0.2 molecules per $nm^2$) with streptavidin or neutravidin are commercially available (SpeedBeads from Cytiva Life Sciences), as are 96-well dishes coated with streptavidin at $\sim$0.2 molecules per $nm^2$ or with neutravidin at $\sim$0.024 molecules per $nm^2$ (Thermo Fisher Scientific). Binding of biotin to streptavidin or neutravidin is extraordinarily fast, and dissociation is so slow that the bond is effectively

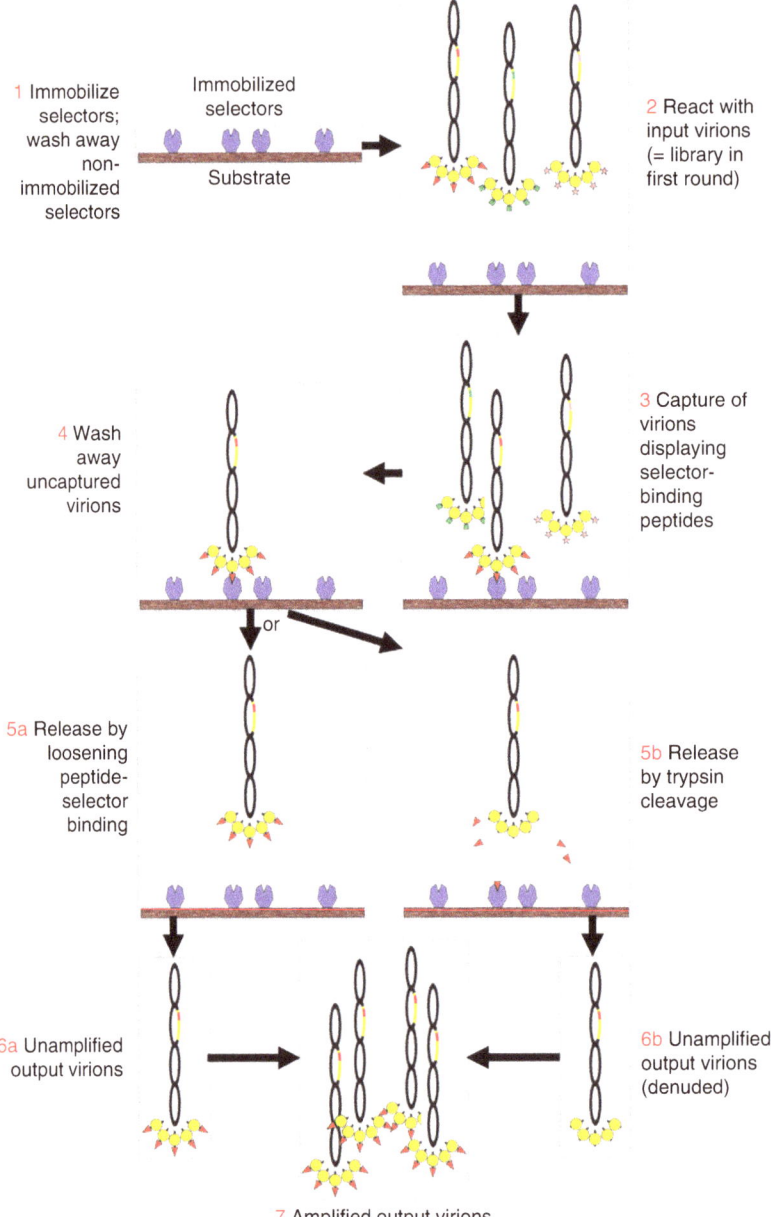

FIGURE 2. Schematic representation of the steps in a single round of affinity selection. The numbered steps are explained in the text.

covalent. The dissociation equilibrium constant, which is the ratio of the dissociation rate constant to the association rate constant, is roughly 1–10 fM. Essentially all selector molecules that are exposed to the surface are immobilized until the binding capacity of streptavidin or neutravidin is saturated. Streptavidin and neutravidin can themselves act as selectors, necessitating measures to avoid selecting unwanted virions that display streptavidin- or neutravidin-binding peptides. One such measure is to alternate streptavidin and neutravidin in successive rounds of affinity selection, because many peptides that bind one do not bind the other.

Whichever immobilization mode is chosen, the substrate must be thoroughly washed to remove all nonimmobilized selectors before proceeding.

- **Steps 2 and 3.** The selector-coated substrate is reacted with the input virions. In the first round, the input virions are the library; in subsequent rounds, the input virions are the amplified output virions from the previous round (Step 7). During the reaction, virions displaying selector-binding peptides are captured on the substrate surface by immobilized selectors, whereas other virions (the overwhelming majority) remain free in solution. Often unwanted virions are also captured for reasons other than binding of their displayed peptides to the selector—for example, virions displaying peptides that bind streptavidin or neutravidin. These unwanted virions are examples of selector-unrelated phages (SUPs), which are the subject of a later section.

- **Step 4.** The substrate surface is then thoroughly washed, with the aim of removing all virions that have not been captured—again, the overwhelming majority. The captured virions that remain after washing are ready for the next step: release from the substrate surface. I will discuss two alternatives for release.

- **Steps 5a and 6a.** One common way of releasing captured virions is to elute them with an eluent that loosens peptide–selector binding without permanently damaging the virions. Examples of such eluents are acidic (e.g., 1 M glycine, pH adjusted to 2.2 with HCl) and basic (e.g., 100 mM triethylamine, pH 12) solutions. Once the eluate is transferred from the substrate to another vessel (Step 6a), the solution is readjusted to physiological conditions (e.g., neutral rather than acidic or basic pH) to allow the virions to recover infectivity and the displayed peptides to refold into their native conformation (if any). These are the unamplified output virions.

    A potential disadvantage of release by loosening selector–peptide binding is that it is possible that virions displaying peptides with particularly high affinity for the selector might be particularly resistant to release. This is obviously very undesirable if the goal is to discover peptides with particularly high affinity for the selector. It should be noted, however, that there is little danger of this undesirable bias when acid or alkali is used to loosen the binding between a virion-borne, single-chain antibody and an immobilized antigen selector, or between a virion-borne peptide and an immobilized antibody selector. This is because acid or alkali loosens antigen–antibody binding not only directly, by interfering with the molecular interactions between the antibody's binding site and the antigen's epitope (the part of the antigen that makes direct contact with the antibody's binding site), but also indirectly, by loosening the three-dimensional folded framework that holds the segments of the antibody's antigen-binding site together.

- **Steps 5b and 6b.** A second common way of releasing captured virions is cleaving the peptide from the virion with the intestinal protease trypsin (Thomas and Smith 2010). This requires that one or more trypsin-sensitive bonds (lysine or arginine followed by any amino acid other than proline) be engineered or exist naturally between the peptide and pIII. Most single-chain antibody domains (see Box 1) have a trypsin-sensitive bond at the carboxy-terminal end. Filamentous virions, the natural habitat of which is the mammalian intestine, are themselves highly resistant to trypsin and other intestinal proteases (Salivar et al. 1964; Sieber et al. 1998). (The cartoon for Step 5b shows the detached peptides intact in solution or still bound to the selector, but either peptide or selector or both may also be cleaved by trypsin.) Once the released virions are transferred to another vessel (Step 6b), trypsin can be neutralized with a commercially available trypsin inhibitor (e.g., lima bean trypsin inhibitor from Worthington Biochemical). These are the unamplified output virions; unlike the unamplified output virions in Step 6a, they are "denuded" (stripped of their displayed peptides).

    A key advantage of trypsin release is that it does not depend at all on the nature and strength of peptide–selector interaction. Therefore, there is no possibility of bias that depends on the strength or nature of that interaction—a distinct possibility when release is accomplished by loosening peptide–selector interaction, as explained in the text for Steps 5a and 6a. Another advantage is that denuded virions have uniform, relatively high infectivity (ability to infect susceptible cells); this means that they are approximately uniformly represented in the first cohort of infected cells when the unamplified output virions are amplified (Step 7). In contrast, large displayed peptides can have large and highly variable inhibitory effects on infectivity (Løset et al. 2008).

- **Step 7.** The unamplified output virions (Step 6a or 6b) are used to infect fresh bacterial host cells, which are then cultured, producing enormous numbers of copies of each of the infecting virions. This process is called amplification, and the resulting population constitutes the amplified output virions. Amplification is absolutely necessary to prepare output virions in sufficient numbers for another round of affinity selection or initial analysis of the output clones—what I will call readout. For the purposes of readout, amplification is often performed with individual output clones rather than en masse. In a later section ("NGS as Primary Readout"), however, I will advocate for en masse readout.

## YIELD VERSUS STRINGENCY

Yield is the probability that a virion from a given clone in the input will be represented in the unamplified output. Technically, it is an abstract property intrinsic to the clone under specified conditions of affinity selection; the actual ratio of unamplified output virions to input virions is randomly distributed about the intrinsic yield.

It is important to distinguish between specific yield, which is highly dependent on the identity of the displayed peptide, and nonspecific background yield, which is largely independent of the identity of the displayed peptide. The obvious reason for a relatively high specific yield is that the clone in question displays a peptide with relatively high affinity for the selector. Such specific clones are called selector-specific, as are their yields. But some specific clones owe their relatively high specific yields to something other than the affinity of their displayed peptides for the selector. Such clones are examples of SUPs, which are the subject of the next section.

Nonspecific background yield is never zero (no selection process is perfect), but is generally very low. It can depend on the physical setup (e.g., the geometry and composition of the substrate) and the details of the selection procedure. Unless selector-specific yield is considerably higher than nonspecific background yield, there is no way to distinguish the desired selector-specific clones, which are very rare in the starting library, from the background clones, which are overwhelmingly abundant in the starting library, and which emerge randomly from imperfection of the selection process.

Stringency is the degree to which affinity selection favors virions with high selector-specific yield over those with low selector-specific yield (I will not attempt an algebraic definition). High stringency too is desirable.

Increasing stringency generally reduces selector-specific yield relative to nonspecific background yield. Conversely, increasing selector-specific yield generally reduces stringency and may also increase nonspecific background yield. Choosing which desirable characteristic to emphasize is a critical strategic consideration in the design of the affinity-selection projects.

- In the first round of affinity selection, high selector-specific yield is of the highest priority, even though that means sacrificing stringency. The reason is that the input library contains a very large number of clones, each represented by relatively few virions. Consider one of the desired clones: a clone with a displayed peptide that binds the selector with high affinity. If we fail to capture at least one of that clone's virions in the first round, the clone obviously cannot reappear magically in subsequent rounds.

- In the second and subsequent rounds of affinity selection, selector-specific yield can be sacrificed in the interest of high stringency. This is because there are orders of magnitude fewer input clones in these rounds, and each clone is represented by orders of magnitude more virions. If selection is too stringent, however, clones that appear in the output because of their peptides' high affinity for the selector cannot be distinguished from nonspecific background clones that appear randomly in the output because the selection process is not perfectly discriminatory.

The obvious way to maximize selector-specific yield in the first round is to maximize the surface density of the selector on the substrate. This promotes high yield in two ways. First, the higher the surface density of selector, the faster the rate at which virions displaying selector-binding peptide will be captured on the surface, and the more likely that virions that disengage from the surface will reengage.

But there is a second important way that high selector surface density increases selector-specific yield: the avidity effect. Recall that five copies of the displayed peptide form a ring at one tip of the virion (cartoon in the lower part of Fig. 1). If the selector is densely arrayed on the substrate surface, it is possible for two peptides on a single virion to bind simultaneously to two neighboring immobilized selectors: bivalent binding. Bivalent binding greatly increases the strength with which the virion is captured on the substrate surface. This is because, if one peptide temporarily disengages from its selector, the virion is not free to diffuse away because it is held in place by the other peptide–selector interaction; the temporarily disengaged peptide is thus in position to rapidly reengage its selector. The result is that the virion's effective dissociation rate from the substrate is orders of magnitude slower than the underlying monovalent dissociation rate of an individual peptide from an individual selector molecule.

The avidity effect is largely irrelevant in the case of single-chain antibodies (and most other protein-sized peptides) displayed on phagemid virions. This is because display of these large "peptides" is almost always monovalent (Box 1).

How can stringency be increased in the second and subsequent rounds? Three approaches are considered.

- The obvious approach is to reduce the surface density of the selector. This will favor displayed peptides with fast association rates and slow dissociation rates. It will also tend to decrease the avidity effect. The avidity effect is a disadvantage when stringency is at a premium, because even virions displaying peptides with weak monovalent affinity for the selector can be strongly captured on the substrate when two of their peptides are bound to selectors. Avoiding the avidity effect thus favors displayed peptides with slow monovalent dissociation rates. But there is a limit to how far stringency can be increased before selector-specific yield descends to the level of nonspecific background yield. For some libraries and selectors, it may not be possible to reduce surface density sufficiently to avoid the avidity effect without erasing the difference between selector-specific and nonspecific background yields. This is an issue that is addressed in the section on "How to Regulate Stringency."

- A second approach is to use stringent washing conditions (Step 4 in Fig. 2) that partially loosen the binding between displayed peptides and the selector. The hope is that virions displaying weakly binding peptides will be preferentially washed away, whereas virions displaying strongly binding peptides will remain captured on the substrate surface. Perhaps this can be accomplished simply by longer and more vigorous washing. Alternatively, the substrate can be washed with nonphysiological solutions that partly loosen peptide–selector binding. It should be borne in mind, though, that the peptides with binding favored under such nonphysiological conditions may not be the same as the peptides with the strongest affinity under physiological conditions.

- A third approach is competition with a known selector ligand. This option is available only in very special circumstances and will not be discussed further.

## THE SCOURGE OF SUPs

A selector-unrelated phage (SUP) is a phage clone that increases in prevalence over the course of affinity selection for some reason other than binding of its displayed peptide to the selector (Menendez and Scott 2005; Brammer et al. 2008; Thomas et al. 2010). Many affinity-selection projects are plagued by such SUPs.

SUPs have previously been called TUPs, which originally stood for target-unrelated peptides. There are two problems with this nomenclature. First, "target" is an ambiguous term for the selector, because it might equally be taken to refer to the peptide. Second, the problem may not have anything to do with a clone's displayed peptide, as will be explained below.

# Chapter 3

There are two categories of SUPs (some SUPs may fall into both categories):

- Capture- or release-related SUPs. Their increased prevalence reflects increased yield in the unamplified output virion population (Step 6a or 6b in Fig. 2) relative to the input population (Step 2). Such increased prevalence presumably involves some characteristic of the displayed peptide, as this is the only difference between clones at this stage; it is thus an example of high specific yield, but it does not depend on the identity of the selector. Perhaps the peptide binds the substrate or some other element of the selection apparatus. A common example of capture-related SUPs are clones displaying peptides that bind streptavidin or neutravidin when these biotin-binding proteins are used to mediate immobilization of biotinylated selectors (see the explanation for Step 1 in Fig. 2).

- Propagation-related SUPs. Their increased prevalence reflects overrepresentation in the amplified output virion population (Step 7 in Fig. 2) relative to the unamplified output virion population (Step 6a or 6b). Overrepresentation could reflect increased infectivity, increased replication inside the infected cell, increased growth of infected cells, or increased production of completed virions in the medium. Some propagation-related SUPs can be related somehow to the displayed peptide. As a particularly perverse example, in type 3 + 3 systems (see Box 1), deletion or lack of expression of the recombinant gene III (which encodes the peptide) can give the infected cell a growth advantage, if that particular recombinant protein is somewhat toxic. Other propagation-related SUPs are unrelated to the displayed peptide. For example, the clone may harbor a mutation elsewhere in the genome that increases replication of phage DNA in the infected cell (Brammer et al. 2008; Thomas et al. 2010).

There are several countermeasures that can be taken against SUPs:

- Unnecessary propagation steps should be avoided to minimize propagation-related SUPs (Thomas et al. 2010). Of course, some propagation steps are unavoidable, including amplification of unamplified output virions. But others are avoidable, including, for example, multiple serial reamplifications of the library. Obviously, avoiding unnecessary propagation can have no effect on capture- or release-related SUPs.

- Counterselection with a mock selector (or no selector at all) may remove unwanted virions that bind to some element of the selection apparatus other than the selector (for example, the substrate). The virions that escape capture in the counterselection (Step 3 of Fig. 2) are retained; there is no need for the subsequent steps depicted in Figure 2. Meanwhile, the counterselected virions serve as input to the subsequent round of ordinary (positive) affinity selection (Steps 2 and 3). Obviously, counterselection can only affect capture- or release-related SUPs; it can have no effect on propagation-related SUPs.

    A common example is counterselection with streptavidin and/or neutravidin in the absence of biotinylated selector when those biotin-binding proteins are used to mediate immobilization of biotinylated selectors for the subsequent round of positive selection (see the explanation for Step 1 in Fig. 2). This counterselection is a useful complement to alternating between streptavidin and neutravidin in successive rounds of positive selection.

    Counterselection is almost never 100% effective; two or more consecutive rounds of counterselection are typically required to sufficiently reduce the concentration of unwanted SUPs. It should also be borne in mind that, in many cases, especially with large displayed peptides, counterselection cannot permanently remove undesired clones from the phage population. This is because some virions representing such an undesired clone may not display the unwanted peptide in intact form, and thus will not be subject to counterselection. In such cases, amplification subsequent to counterselection will partially restore representation of the unwanted clone.

- SUPs may be identified bioinformatically without being removed. Thus, a series of control affinity selections with a control selector (or no selector at all) is performed in parallel with the real series of positive affinity selections, using exactly the same procedures in both cases. The frequencies of the clones that emerge from each selection are compared. Clones that are abundant in the output of both selection series are SUPs; both capture- or release-related and propagation-related SUPs can be detected in this way. Clones that are abundant in the output of the positive selection series but rare in the output of the control selection series may well display a selector-binding peptide—exactly what is sought in an affinity-selection project. Clones that are rare in the output of the positive selection series but abundant in the output of the control selection series may display a peptide that binds the control selector (if any). This approach requires analyzing large numbers of output clones, which is one of the arguments for next-generation sequencing (NGS) as the primary readout in affinity-selection projects, the subject of the next section.

## NGS AS PRIMARY READOUT

Next-generation sequencing (NGS) has become an essential tool in life sciences today. In the context of affinity-selection projects, it can provide a thorough census of the sequences of the displayed peptides, including single-chain antibodies, in the output virion population of an affinity-selection project (Brinton et al. 2016; Yang et al. 2017; Rouet et al. 2018; Braun et al. 2020; Zambrano et al. 2022). Tens or even hundreds of millions of output virion sequences from positive and control affinity selections become available for bioinformatic analysis. These results allow enumeration of the number of times that the sequence from any given clone appears among those tens or hundreds of million sequences; this number is called the clone's reads. The number of a clone's reads divided by the total number of reads is a measure of the clone's prevalence in the population. Such population prevalence data can pinpoint clones that are particularly likely to display peptides with high affinity for the selector, as explained in this section. Moreover, the displayed peptide sequences can be subjected in turn to further bioinformatic analysis—for example, to identify common sequence motifs (O'Shea et al. 2013). For these reasons, NGS is increasingly favored as the primary readout in affinity-selection projects: the first level of analysis of output clones.

NGS delivers sequence information, not physical clones. But modern molecular biology provides abundant means for using sequence information to recover any particular clone of interest from the output phage population for further analysis and development.

NGS cannot deliver a census of a large starting library with billions of clones, each displaying a different peptide. But affinity selection reduces the number of clones so drastically that 10 million NGS reads will ordinarily provide a nearly complete census of the first-round amplified output, and thus a measure of each clone's prevalence in that output population.

First-round prevalence data by themselves do not suffice to identify promising clones for two reasons. First, only for a tiny minority of the clones in the first-round amplified output can their prevalence in that output be compared to their prevalence in the first-round input (the large starting library); thus, the yield during the first round of affinity selection (essential for assessing the promise of a clone's displayed peptide as a high-affinity ligand for the selector) cannot be calculated except in extremely rare cases. Second, as explained in the section on "Yield Versus Stringency" above, the first round of affinity selection must prioritize yield, severely compromising its ability to discriminate among clones on the basis of the affinity of their displayed peptides for the selector.

The full value of a thorough first-round census becomes apparent only when a thorough second-round census is also available. This allows each clone's yield in the second round of affinity selection to be estimated by comparing its prevalence in the second-round output population to its prevalence in the second-round input population (the same as the first-round amplified output population). If, again following the principles outlined in the section on "Yield Versus Stringency" above, the stringency of the second round has been adjusted such that the yields of selector-specific clones reflect the affinity of their displayed peptides for the selector, this information suffices to identify clones that are promising candidates for further investigation.

It may be necessary in a few cases to analyze the third or even fourth round of affinity selection. Eventually, however, information will decline in succeeding rounds, as clones with the highest yield crowd out clones with lesser yields. This is a severe disadvantage even if the goal of the project is to identify the clone with the highest affinity for the selector. This is because the specific clones with the very highest yields may be undesired SUPs rather than the desired clones displaying peptides with high affinity for the selector. It is vital to collect data for all clones with specific yields substantially higher than the nonspecific background yield.

Figure 3 is a schematic diagram showing NGS analysis in the context of a two-round affinity-selection project from a type 3 + 3 phage–antibody library (see Box 1): for instance, a library displaying single-chain antibodies (Andris-Widhopf et al. 2000) derived from chicken immunoglobulin Y (Carlander et al. 2000). The recombinant pIII displaying the single-chain antibody domain is encoded on a phagemid chromosome, which also bears a phage origin of replication and a selectable marker (typically ampicillin resistance), but no additional phage elements. Only when phagemid-bearing cells are superinfected with helper phage virions, which encode an alternative pIII (either wild-type or engineered) as well as all the other phage proteins necessary for virion assembly, do the cells produce virions: both phagemid and helper virions. In phage–antibody projects, it is the antigen of interest that serves as the selector. The steps highlighted in red in Figure 3 are discussed below.

In each round, the input (the initial library in round 1 and the amplified first-round output in round 2) is a mixture of phagemid and helper phage virions (the former predominating because of the design of the helper). Some of each type of virion display a single functional single-chain antibody; almost no virions display more than one. Capture and release (Steps 1–6a or Steps 1–6b in Fig. 2) proceed as usual, resulting in the unamplified output (Step 6a or 6b).

Amplification in type 3 + 3 systems is broken down into distinct steps. First, the output is used to infect cells. Second, the infected cells are cultured under conditions that select for the phagemid's selectable marker (again, usually ampicillin resistance). When the culture has grown sufficiently to ensure that all infecting phagemids are abundantly represented in the population, part of the culture is split off and allowed to continue growth; this culture is the source of the phagemid DNA that is subjected to NGS, as symbolized in the figure by the leftward-branching arrows. Meanwhile (third),

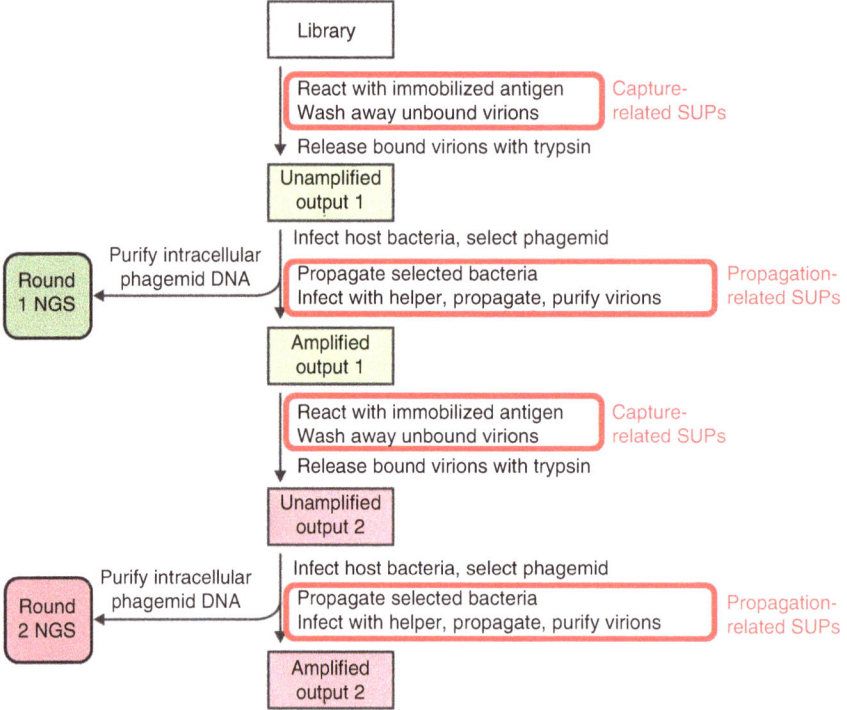

FIGURE 3. Flow diagram for two rounds of affinity selection, starting with a library of single-chain antibodies displayed from a phagemid vector. The output of each round is analyzed by next-generation sequencing (NGS). Further explanation can be found in the text. (SUPs) Selector-unrelated phages.

the remainder of the phagemid culture is superinfected with a large excess of fresh helper virions, and the doubly infected cells are cultured to yield the amplified output virions: both phagemid and helper phage, with the former predominating. As usual, the amplified output virions from the first round are the input virions for the second round.

The red rectangles in Figure 3 highlight the steps in each round at which bias in favor of SUPs can occur:

- The capture stage of affinity selection can favor capture-related SUPs—for example, virions with a displayed peptide that binds some component of the affinity-selection apparatus other than the selector.
- If release is accomplished by trypsin cleavage, there is little room for bias of any kind, including bias in favor of SUPs during the release stage.
- As trypsin release results in denuded virions (including phagemid virions) in the unamplified output, there is also little room for bias during infection of bacterial cells with phagemid virions.
- However, there is opportunity for bias in favor of propagation-related SUPs during propagation of phagemid-infected cells and during virion assembly after superinfection with helper virions.

Because bias in favor of SUPs is independent of the selector (the antigen of interest in the case of phage–antibody libraries), the very same SUPs should be favored in parallel affinity selections from the same initial library with different antigens as selectors. Comparisons of the clones emerging from the parallel selections should allow likely SUPs of all kinds to be identified, as explained in the section on "The Scourge of SUPs" above. In light of this consideration, it makes sense to carry out affinity selections in batches, with each affinity selection in a batch using a different selector. Apart from the choice of selector, the different affinity selections should be as identical as possible. That way, the different affinity selections serve as controls for one another.

## HOW TO REGULATE STRINGENCY

The section on "Yield Versus Stringency" above explains that stringency can be increased in the second and subsequent rounds of affinity selection by decreasing the surface density of the selector. This favors virions that are captured at a fast rate, and reduces the avidity effect, which would allow unwanted virions displaying peptides with weak monovalent affinity for the selector to contribute to the output. However, there is a limit to this approach. It may not be possible to reduce selector surface density sufficiently to significantly discourage the avidity effect without simultaneously severely reducing the ratio of specific yield to nonspecific background yield. If the ratio is reduced too much, the desired clones (those displaying selector-binding peptides) are barely distinguishable in frequency from random background clones that appear only because the selection process is not perfectly discriminatory.

This is not a problem for phage–antibody libraries (Box 1), in which display is monovalent and avidity is therefore irrelevant. This section applies instead to projects in which five peptides are displayed on each virion, as is likely to be the case for type 3 libraries (Box 1) displaying short peptides.

This section outlines a possible alternative selection process in which the avidity effect might actually be exploited to enhance selection in favor of clones with displayed peptides that bind the selector with particularly strong monovalent affinity. It should be kept in mind that the proposal is speculative at this time.

The proposed alternative affinity-selection process differs from the usual one in the capture stage (Steps 1–3 in Fig. 2). It begins by reacting the library with selector molecules free in solution, not immobilized on a substrate. During this liquid-phase reaction, selectors bind reversibly to displayed peptides; the system is allowed to come to equilibrium.

The selector molecules are assumed to have been biotinylated. As described in the explanation for Step 1 in Figure 2, biotinylated molecules bind streptavidin or neutravidin with an extraordinarily fast association rate and an extraordinarily slow dissociation rate—exactly what is required for the next stage of the proposed procedure.

The next stage is to remove unbound selectors (biotinylated selector molecules that are not bound to displayed peptides) and immediately deliver the virions, along with their still-bound selectors, directly onto a substrate surface that is densely coated with streptavidin or neutravidin, where they are rapidly captured on the substrate surface. This might be accomplished in a matter of minutes (before selectors that are strongly bound to displayed peptides have a chance to dissociate) by gel filtration through a spin column from which the effluent flows directly into a suspension of streptavidin- or neutravidin-coated 1-μm paramagnetic beads (e.g., SpeedBeads from Cytiva Life Sciences; 0.2 streptavidin or neutravidin molecules per $nm^2$).

Spin columns (for example, Zeba desalting columns from Thermo Fisher Scientific) are widely used in molecular biology to remove small solutes from macromolecules such as proteins and nucleic acids. In that case, the gel filtration medium is beads with small pores that only small solutes can penetrate, allowing macromolecules to emerge from the column free of these solutes. Because the solution is driven through the column by centrifugal force, filtration is finished in a few minutes. In the process proposed here, the gel filtration medium would be beads such as Sephacryl S-500 HR (MilliporeSigma) with pores that allow macromolecules such as selectors to penetrate, whereas particles like virions are excluded (Zakharova et al. 2005). It remains to be determined whether such a column would suffice to free virions from almost all selectors that are not bound to displayed peptides.

From this point on, the proposed alternative affinity-selection procedure is the same as the usual procedure (Steps 4–7 in Fig. 2). In particular, the substrate is thoroughly washed to remove virions that are not captured on the surface. The substrate is then ready for the usual release stage, either by loosening peptide–selector binding (Steps 5a and 6a in Fig. 2) or by trypsin cleavage (Steps 5b and 6b), as described in the section "The Affinity-Selection Process."

The attraction of the proposed alternative capture procedure lies in the kinetics of reversible solution-phase selector–peptide binding. It is to those kinetics that I now turn, making the following simplifying assumptions:

- Selectors bind monovalently to displayed peptides. Either one or two selectors can be bound to a single virion; in the latter case, the two selectors are bound to nonvicinal peptides (see bottom cartoon in Fig. 1).
- All selector–peptide interactions for a given clone occur independently with the same reversible monovalent kinetics.
- All reversible-binding interactions for a given clone are governed by the same monovalent kinetic parameters:
  - an association rate constant (dimensional units $time^{-1}$ $concentration^{-1}$),
  - a dissociation rate constant (dimensional unit $time^{-1}$), and
  - the dissociation equilibrium constant $K_D$, which is the dissociation rate constant divided by the association rate constant. $K_D$ has the dimension unit of concentration.

At selector concentrations at which the selector can plausibly bind to a displayed peptide at significant levels, the selector will be in vast molar excess over all selector-binding peptides. Effectively, therefore, the free selector concentration can be assumed to be the same as the total selector concentration, which is abbreviated as "$S$."

Each virion is assumed to display five peptides in a ring at one tip of the particle (lower part of Fig. 1). There are therefore five ways that a virion can associate with a selector, and only one way for a virion with one bound selector to dissociate. The overall dissociation equilibrium constant for virions with no versus one bound selector will thus be the monovalent dissociation rate constant divided by five times the monovalent association rate constant, or $K_D/5$. Similarly, there are two ways that a virion with one bound selector can associate with another selector to create a virion with two bound selectors, and there are two ways that a virion with two bound selectors can dissociate to create a virion with one bound selector. The overall dissociation equilibrium constant for virions with one versus two bound selectors will thus be two times the monovalent dissociation rate constant divided by two times the monovalent association rate constant, or $K_D$. The two types of reversible binding reactions are diagrammed in the upper part of Figure 4.

At equilibrium, the equilibrium equations for both reversible reactions (between virions with no bound selector and virions with one bound selector [governed by the dissociation equilibrium constant $K_D/5$] and between virions with one bound selector and virions with two bound selectors [governed by the dissociation equilibrium constant $K_D$]) must be simultaneously satisfied for every phage clone displaying a selector-binding peptide.

Let us consider a single phage clone with a displayed peptide that binds the selector with a monovalent dissociation equilibrium constant of $K_D$ (the lower the value of $K_D$, the stronger the binding affinity), and define the ratio $R \equiv K_D/S$, the peptide's dissociation equilibrium constant relative to the prevailing selector concentration. In this case, following the simplifying kinetic assumptions and their consequences in the preceding paragraphs, we can calculate the fraction of that clone's virions that have a selector molecule bound to one of its displayed peptides at equilibrium as $5R/(5 + 5R + R^2)$. Similarly, the fraction of the clone's virions with selector molecules bound to two of its displayed peptides at equilibrium is $5/(5 + 5R + R^2)$. The graph in the lower part of Figure 4 plots the percent of a clone's virions with one or two bound selectors as a function of $R$, equal to that clone's peptide dissociation equilibrium constant $K_D$ divided by the selector concentration $S$.

Once equilibrium has been established, the solution is processed using the alternative capture phase presented earlier in this section. During this process, almost all free biotinylated selector molecules will (hopefully) be removed by the spin column before the solution arrives at the streptavidin- or neutravidin-coated substrate. Virions with peptides that have high dissociation rate constants will lose one or both of their bound selectors to the gel filtration medium in the spin column. But the most desirable virions—those with one or especially two selectors still bound to their peptides—will be captured monovalently or bivalently on the substrate surface. If that substrate surface is then washed vigorously, it may be possible to eliminate most virions that are not captured

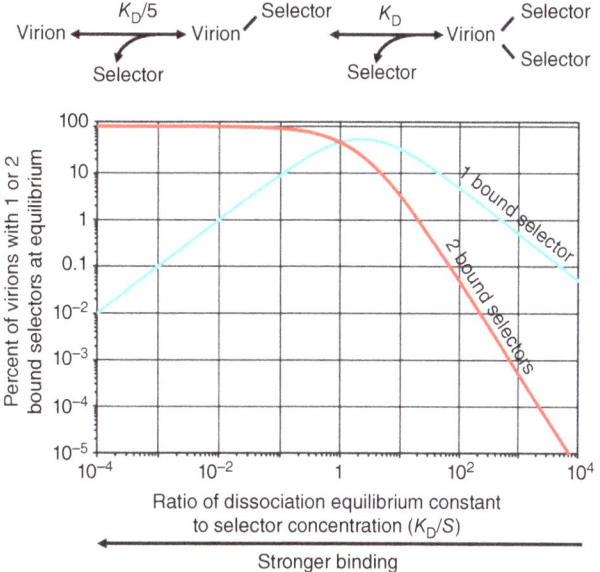

FIGURE 4. (*Top*) Schematic representation of the equilibria governing the reaction of the selector with a single-phage clone with a displayed peptide that binds the selector with a monovalent dissociation equilibrium constant of $K_D$. The simplifying kinetic assumptions are explained in the text. (*Bottom*) Graph showing the fraction of a clone's virions with one or two selector molecules bound to the virion's displayed peptides at equilibrium, as a function of the ratio of the dissociation equilibrium constant for these peptides to the concentration of the selector. The simplifying kinetic assumptions underlying the calculation are explained in the text.

bivalently through two bound selectors. As the red curve in the graph indicates, this will result in sharply progressive discrimination against virions with displayed peptides that have monovalent dissociation equilibrium constants $K_D$ progressively greater than the prevailing selector concentration $S$. In this context, the avidity effect from bivalent binding works for, rather than against, the goal of discovering peptides with strong monovalent binding to the selector.

The effectiveness of this approach depends critically on nearly complete removal of free selector molecules by gel filtration through spin columns. Whether this is possible with macromolecular solutes, with their small diffusion constants, has yet to be addressed to my knowledge.

## REFERENCES

Andris-Widhopf J, Rader C, Steinberger P, Fuller R, Barbas CF III. 2000. Methods for the generation of chicken monoclonal antibody fragments by phage display. *J Immunol Methods* 242: 159–181. doi:10.1016/S0022-1759(00)00221-0

Brammer LA, Bolduc B, Kass JL, Felice KM, Noren CJ, Hall MF. 2008. A target-unrelated peptide in an M13 phage display library traced to an advantageous mutation in the *gene II* ribosome-binding site. *Anal Biochem* 373: 88–98. doi:10.1016/j.ab.2007.10.015

Braun R, Schönberger N, Vinke S, Lederer F, Kalinowski J, Pollmann K. 2020. Application of next generation sequencing (NGS) in phage displayed peptide selection to support the identification of arsenic-binding motifs. *Viruses* 12: 1360. doi:10.3390/v12121360

Brinton LT, Bauknight DK, Dasa SS, Kelly KA. 2016. PHASTpep: analysis software for discovery of cell-selective peptides via phage display and next-generation sequencing. *PLoS ONE* 11: e0155244. doi:10.1371/journal.pone.0155244

Carlander D, Kollberg H, Wejåker PE, Larsson A. 2000. Peroral immunotherapy with yolk antibodies for the prevention and treatment of enteric infections. *Immunol Res* 21: 1–6. doi:10.1385/IR:21:1:1

Kang TH, Seong BL. 2020. Solubility, stability, and avidity of recombinant antibody fragments expressed in microorganisms. *Front Microbiol* 11: 1927. doi:10.3389/fmicb.2020.01927

Løset GA, Kristinsson SG, Sandlie I. 2008. Reliable titration of filamentous bacteriophages independent of pIII fusion moiety and genome size by using trypsin to restore wild-type pIII phenotype. *Biotechniques* 44: 551–552. doi:10.2144/000112724

Menendez A, Scott JK. 2005. The nature of target-unrelated peptides recovered in the screening of phage-displayed random peptide libraries with antibodies. *Anal Biochem* 336: 145–157. doi:10.1016/j.ab.2004.09.048

O'Shea JP, Chou MF, Quader SA, Ryan JK, Church GM, Schwartz D. 2013. pLogo: a probabilistic approach to visualizing sequence motifs. *Nat Methods* 10: 1211–1212. doi:10.1038/nmeth.2646

Rakonjac J, Russel M, Khanum S, Brooke SJ, Rajič M. 2017. Filamentous phage: structure and biology. *Adv Exp Med Biol* 1053: 1–20. doi:10.1007/978-3-319-72077-7_1

Rouet R, Jackson KJL, Langley DB, Christ D. 2018. Next-generation sequencing of antibody display repertoires. *Front Immunol* 9: 118. doi:10.3389/fimmu.2018.00118

Salivar WO, Tzagoloff H, Pratt D. 1964. Some physical-chemical and biological properties of the rod-shaped coliphage M13. *Virology* 24: 359–371. doi:10.1016/0042-6822(64)90173-4

Sieber V, Pluckthun A, Schmid FX. 1998. Selecting proteins with improved stability by a phage-based method. *Nat Biotechnol* 16: 955–960. doi:10.1038/nbt1098-955

Smith GP. 1993. Preface. Surface display and peptide libraries. *Gene* 128: 1–2. doi:10.1016/0378-1119(93)90145-S

Smith GP, Petrenko VA. 1997. Phage display. *Chem Rev* 97: 391–410. doi:10.1021/cr960065d

Thomas W, Smith G. 2010. The case for trypsin release of affinity-selected phages. *Biotechniques* 49: 651–654. doi:10.2144/000113489

Thomas WD, Golomb M, Smith GP. 2010. Corruption of phage display libraries by target-unrelated clones: diagnosis and countermeasures. *Anal Biochem* 407: 237–240. doi:10.1016/j.ab.2010.07.037

Wysocki LJ, Sato VL. 1978. "Panning" for lymphocytes: a method for cell selection. *Proc Natl Acad Sci* 75: 2844–2848. doi:10.1073/pnas.75.6.2844

Yang W, Yoon A, Lee S, Kim S, Han J, Chung J. 2017. Next-generation sequencing enables the discovery of more diverse positive clones from a phage-displayed antibody library. *Exp Mol Med* 49: e308. doi:10.1038/emm.2017.22

Zakharova MY, Kozyr AV, Ignatova A, Vinnikov IA, Shemyakin IG, Kolesnikov AV. 2005. Purification of filamentous bacteriophage for phage display using size-exclusion chromatography. *Biotechniques* 38: 194, 196, 198.

Zambrano N, Froechlich G, Lazarevic D, Passariello M, Nicosia A, De Lorenzo C, Morelli MJ, Sasso E. 2022. High-throughput monoclonal antibody discovery from phage libraries: challenging the current preclinical pipeline to keep the pace with the increasing mAb demand. *Cancers (Basel)* 14: 1325. doi:10.3390/cancers14051325

CHAPTER 4

# The pComb3 Phagemid Family of Phage Display Vectors

Christoph Rader[1]

*The Herbert Wertheim UF Scripps Institute for Biomedical Innovation & Technology, University of Florida, Jupiter, Florida 33458, USA*

A phagemid is a plasmid that contains the origin of replication and packaging signal of a filamentous phage. Following bacterial transformation, a phagemid can be replicated and amplified as a plasmid, using a double-stranded DNA origin of replication, or it can be replicated as single-stranded DNA for packaging into filamentous phage particles. The use of phagemids enables phage display of large proteins, such as antibody fragments. Phagemid pComb3 was among the first phage display vectors used for the generation and selection of antibody libraries in the 50-kDa Fab format, a monovalent proxy of natural antibodies. Affording a robust and versatile tool for more than three decades, phage display vectors of the pComb3 phagemid family have been widely used for the discovery, affinity maturation, and humanization of antibodies in Fab, scFv, and single-domain formats from naive, immune, and synthetic antibody repertoires. In addition, they have been used for broadening phage display to the mining of nonimmunoglobulin repertoires. This review examines conceptual, functional, and molecular features of the first-generation phage display vector pComb3 and its successors, pComb3H, pComb3X, and pC3C.

## INTRODUCTION

The generation and selection of antibody libraries by phage display was introduced in the early 1990s and has remained a key tool for the de novo discovery and in vitro evolution of monoclonal antibodies for research and medicine. Its success is grounded in the robustness and versatility of phage display in general and the use of phagemids as phage display vectors in particular. With a focus on the pComb3 phagemid family, this review examines features of phagemids at a conceptual, functional, and molecular level and provides a thorough technical understanding of their utility for the mining of antibody and other protein repertoires by phage display. Properties of filamentous phage that are relevant to phage display will be discussed first, followed by classifying phage display vectors, examining phagemids, and dissecting the three components of phagemid-based phage display (i.e., phagemid library, host bacterial cells, and helper phage). This is followed by a closer look at the recombinant fusion of the protein of interest to a coat protein of filamentous phage, affording the linkage of phenotype and genotype that defines phage display. Finally, with this background in place, the review turns to phagemid pComb3 and its descendants.

## FILAMENTOUS PHAGE IN THE CONTEXT OF PHAGE DISPLAY

Filamentous phage that infect gram-negative bacteria including *Escherichia coli* (*E. coli*) belong to the genus *Inovirus*. Fertility factor (F)-specific filamentous phage, also known as Ff phage, infect F+ *E. coli*

---

[1]Corresponding author: rader33458@gmail.com

© 2026 Cold Spring Harbor Laboratory Press
Cite this overview as *Cold Spring Harb Protoc*; doi:10.1101/pdb.over107756

and stress—but do not kill—the host bacterial cell. Unlike plaques formed on bacterial lawns by lytic phage, the plaque-like structures on bacterial lawns infected with Ff phage are a result of slower growth. Ff phage comprise the highly homologous strains f1, fd, and M13. On their nearly identical genome (6.4 kb), which is a circular single-stranded DNA (ssDNA) packaged as a supercoiled double helix with ∼25% complementary base pairs, the Ff phage strains share interchangeable coding and noncoding DNA elements essential for phage display (Rakonjac et al. 2011). These elements include 11 encoded proteins that comprise the replication proteins pII, pV, and pX; the major coat protein pVIII; the minor coat proteins pIII, pVI, pVII, and pIX; and the export proteins pI, pIV, and pXI. They also include origins of replication for +strand and −strand, to enable replication in the host bacterial cell, as well as a phage packaging signal.

The structure and function of all components of Ff phage and their utility and limitations for phage display have been described in detail (Rakonjac et al. 2011). Briefly, Ff phage are elongated protein tubes, with ∼2700 copies of major coat protein pVIII encapsulating the ssDNA and five copies of two pairs of minor coat proteins pIII/pVI and pVII/pIX forming a spiky and a flat cap, respectively, on opposite ends. They do not contain any lipids. The ssDNA serves as a helical array for deposition of inner membrane-accumulated pVIII during its push from the cytoplasm through the inner membrane channel (formed by pI and pXI) to the periplasm. As such, the length of the protein tube is defined by the length of the ssDNA, and it can accommodate recombinant genomes smaller or larger than the natural 6.4 kb. Once the ssDNA is completely coated with pVIII, the pIII/pVI pair forms a terminating cap and releases the phage particle through the outer membrane channel (formed by pIV). Filamentous phage particles morphologically resemble F pili of *E. coli*, which are encoded on a conjugative plasmid known as the fertility factor (F) and, intriguingly, also serve as the primary receptor for Ff phage infection. This interaction is mediated by the N2 domain of pIII and triggers a subsequent interaction of the N1 domain of pIII with the periplasmic domain of inner membrane protein TolA, the secondary receptor for Ff phage infection. The C domain of pIII is involved in forming a pore in the inner membrane, through which the ssDNA enters the host bacterial cell. Thus, all three domains of pIII—N1, N2, and C—are required for Ff phage infection. Notably, their expression by the host bacterial cell incapacitates primary and secondary receptors and causes resistance to superinfection by Ff phage.

A key feature of filamentous phage is that their assembly takes place on the periplasmic side of the inner membrane of *E. coli*, where the major and minor coat proteins fold in an oxidizing environment after export. Thus, recombinant fusion of the protein of interest to a major or minor coat protein permits its folding under conditions similar to those of the endoplasmic reticulum of mammalian cells, where immunoglobulin domains fold into disulfide bridge (cystine)-stabilized β-sandwiches. Conversely, cytoplasmic proteins of interest that require unpaired cysteines for folding or function and are easily oxidized, necessitate modifications to conventional phage display systems. Another key feature of Ff phage is their high resistance to harsh environments with respect to solvents, solutes, pH, and temperature, affording a broad range of binding and elution conditions as long as these do not affect the protein of interest. The robustness of phage also facilitates the chemical or enzymatic modification of coat proteins and displayed peptides or proteins.

## CLASSIFICATION OF PHAGE DISPLAY VECTORS

Phage display vectors have been classified (Smith and Petrenko 1997) based on (i) the filamentous phage coat protein used for display of the peptide or protein by recombinant fusion (i.e., predominantly minor coat protein pIII or major coat protein pVIII), (ii) whether the recombinant fusion involves all or only a fraction of pIII or pVIII copies, and (iii) whether the recombinant fusion is encoded on a filamentous phage genome or on a plasmid that contains the origin of replication and packaging signal of a filamentous phage. Such plasmids are known as phagemids. For example, a phage display vector encoding a recombinant fusion to all copies of pIII on a phage genome is referred to as "type 3" phage display vector and was used in the original report of phage display (Smith 1985). M13KE, a widely used "type 3" phage display vector, forms the basis of the commercially available "Ph.D.-7," "Ph.D.-12," and "Ph.D.-C7C" peptide libraries from New England Biolabs (Noren and Noren 2001). Unlike phage display of peptide libraries, which requires only short DNA insertions to encode the recombinant fusion, phage display of protein libraries necessitates long DNA insertions and more complex systems, in which only a fraction of the expressed pIII or pVIII copies involve the recombinant fusion. As a larger fusion protein can interfere with phage assembly, display, and infectivity, these undesirable effects need to be mitigated by expression of an excess of wild-type pIII or pVIII in each phage particle.

The focus of this article, the pComb3 phagemid family of phage display vectors (Barbas et al. 1991), are "type 3 + 3" phage display vectors that encode the recombinant fusion of the protein of interest to a carboxy-terminal fragment of pIII that becomes mixed in the assembled phage particle with several copies of wild-type pIII from a phage genome. Thus, only a fraction of pIII copies on the phage particle that is produced displays the protein of interest.

The length and composition of peptides that can be fused to major coat protein pVIII (∼2700 copies per phage particle; 5.2 kDa) without compromising phage assembly is much more limited compared to minor coat protein pIII (five copies per phage particle; 3.6 kDa). Thus, the more common mode of peptide and protein display by recombinant fusion to pVIII is "type 8 + 8" compared to "type 8," with a small fraction (<10%) of phagemid-encoded pVIII copies displaying the peptide or protein of interest and a large fraction (>90%) of phage genome-encoded wild-type pVIII copies. Phagemid pComb8 (Addgene plasmid #63889) is a "type 8 + 8" phage display vector that is identical to pComb3 (Addgene plasmid #63888) except for its pVIII expression cassette, which permits multivalent pVIII rather than monovalent pIII display.

In the following sections, key concepts and features of phagemid-based phage display that were used to inform the architecture of the pComb3 phagemid family are discussed.

## PHAGEMIDS AS PHAGE DISPLAY VECTORS

A phagemid is a hybrid of a phage and a plasmid. It denotes a plasmid that contains an Ff intergenic region comprising origin of replication and packaging signal. Following transformation of bacteria, a phagemid can be replicated as double-stranded DNA (dsDNA), like a plasmid, or replicated as ssDNA, for packaging into phage particles. The latter requires infection with a suitable filamentous phage that encodes three proteins required to change the mode of replication—namely, pII, pV, and pX. Phagemids are attractive tools for DNA cloning and sequencing, as they combine the durability and convenience of plasmids with the ability to switch from dsDNA replication to ssDNA replication (Sambrook et al. 1989). Although DNA cloning and sequencing became less reliant on ssDNA with the advance of molecular biology techniques, phagemids experienced renewed interest and broad use as vectors for phage display of proteins, such as the 25-kDa scFv and 50-kDa Fab antibody fragments. As such, they are built to encode the physical linkage of phenotype (protein of interest displayed on the phage particle) and genotype (protein-encoding ssDNA encapsulated in the phage particle) by fusing the protein of interest with a coat protein of filamentous phage, such as major coat protein pVIII and minor coat protein pIII. The ease of plasmid cloning paired with the physical, chemical, and biological robustness of phage facilitated the generation of large protein libraries that can be stringently selected for de novo clones or in vitro evolved clones that recognize large or small molecules of interest with high affinity and specificity. For example, phage display vectors belonging to the pComb3 phagemid family have been widely used for the discovery, affinity maturation, and humanization of antibodies in various formats (Table 1). In addition, they have been used for selecting nonimmunoglobulin repertoires, such as synthetic zinc finger libraries (Segal et al. 1999; Dreier et al. 2001, 2005). As protein library diversity is dependent on the transformation efficiency of host bacterial cells which, in turn, inversely correlates with plasmid size, phagemids are generally kept small. For example, pComb3H

TABLE 1. Representative examples of the utilization of the pComb3 phagemid family for mining antibody repertoires

| | pComb3 (see Fig. 3A) | pComb3H (see Fig. 3B) | pComb3X (see Fig. 3C) | pC3C (see Fig. 3D) |
|---|---|---|---|---|
| Repertoire | Immune human (Barbas et al. 1991; Burton et al. 1991; Duchosal et al. 1992; McIntosh et al. 1996); Synthetic human (Barbas et al. 1992, 1993, 1994); Immune macaque (Samuelsson et al. 1995); Immune mouse (Williamson et al. 1996); Synthetic mouse (Gram et al. 1992) | Naive human (Rader et al. 1998); Immune human (Roben et al. 1996; Steinberger et al. 1996; Siegel et al. 1997; Roark et al. 2002; Zhang et al. 2003); Synthetic human (Yang et al. 1995; Rader et al. 2002); Immune chimpanzee (Schofield et al. 2000); Immune macaque (Glamann et al. 1998); Naive bovine (O'Brien et al. 1999); Immune rabbit (Lang et al. 1996; Rader et al. 2000); Immune mouse (Nathan et al. 2005); Immune chicken (Cary et al. 2000) | Naive human (Steinberger et al. 2000; Tanaka et al. 2004); Immune human (Payne et al. 2005); Synthetic human (Chung et al. 2004; Chen et al. 2008; Cao et al. 2020; Li et al. 2020); Immune macaque (Kuwata et al. 2011); Naive canine (Mason et al. 2021); Naive camel (Hong et al. 2022); Immune alpaca (Kim et al. 2012); Immune rabbit (Steinberger et al. 2000; Popkov et al. 2003); Immune rat (Norbury et al. 2019); Immune mouse (Moreland et al. 2012); Immune chicken (Andris-Widhopf et al. 2000; Yoo et al. 2020); Naive shark (Feng et al. 2019) | Naive human (Kwong et al. 2008); Immune human (Baskar et al. 2009); Synthetic human (Kwong et al. 2008); Immune rabbit (Hofer et al. 2007; Yang et al. 2011); Naive rabbit (Peng et al. 2017; Wilson et al. 2018); Synthetic rabbit (Goydel et al. 2020) |
| Format | Fab (Barbas et al. 1991, 1992, 1993, 1994; Burton et al. 1991; Duchosal et al. 1992; Gram et al. 1992; Samuelsson et al. 1995; McIntosh et al. 1996; Williamson et al. 1996) | Fab (Yang et al. 1995; Lang et al. 1996; Roben et al. 1996; Steinberger et al. 1996; Siegel et al. 1997; Glamann et al. 1998; Rader et al. 1998, 2000, 2002; O'Brien et al. 1999; Schofield et al. 2000; Roark et al. 2002; Zhang et al. 2003; Nathan et al. 2005); scFv (Cary et al. 2000) | Fab (Andris-Widhopf et al. 2000; Steinberger et al. 2000; Popkov et al. 2003; Chung et al. 2004; Tanaka et al. 2004; Kuwata et al. 2011; Moreland et al. 2012); scFv (Andris-Widhopf et al. 2000; Steinberger et al. 2000; Payne et al. 2005; Norbury et al. 2019; Yoo et al. 2020; Mason et al. 2021); Single domain (Chen et al. 2008; Kim et al. 2012; Feng et al. 2019; Cao et al. 2020; Li et al. 2020; Hong et al. 2022) | Fab (Hofer et al. 2007; Kwong et al. 2008; Baskar et al. 2009; Yang et al. 2011; Peng et al. 2017; Wilson et al. 2018; Goydel et al. 2020) |
| Application | Discovery (Barbas et al. 1991, 1992, 1993; Burton et al. 1991; Duchosal et al. 1992; Samuelsson et al. 1995; McIntosh et al. 1996; Williamson et al. 1996); Affinity maturation (Gram et al. 1992; Barbas et al. 1994) | Discovery (Lang et al. 1996; Roben et al. 1996; Steinberger et al. 1996; Siegel et al. 1997; Glamann et al. 1998; O'Brien et al. 1999; Cary et al. 2000; Rader et al. 2000; Schofield et al. 2000; Roark et al. 2002; Zhang et al. 2003; Nathan et al. 2005); Affinity maturation (Yang et al. 1995; Rader et al. 2002); Humanization (Rader et al. 1998, 2000) | Discovery (Andris-Widhopf et al. 2000; Popkov et al. 2003; Chung et al. 2004; Tanaka et al. 2004; Payne et al. 2005; Chen et al. 2008; Kuwata et al. 2011; Feng et al. 2012; Moreland et al. 2012; Norbury et al. 2019; Cao et al. 2020; Li et al. 2020; Yoo et al. 2020; Mason et al. 2021; Hong et al. 2022); Affinity maturation (Chung et al. 2004); Humanization (Steinberger et al. 2000) | Discovery (Hofer et al. 2007; Kwong et al. 2008; Baskar et al. 2009; Yang et al. 2011; Peng et al. 2017; Wilson et al. 2018; Goydel et al. 2020); Affinity maturation (Kwong et al. 2008; Goydel et al. 2020) |

and pComb3X, the two most common members of phagemid family pComb3, have a size of 4.7 kb when harboring a Fab-encoding cassette.

## PHAGEMID-BASED PHAGE DISPLAY

Phagemid systems comprise three components that are essential for phage display: (i) phagemid library, (ii) host bacterial cells, and (iii) helper phage. For phagemid-based phage display, these components have to be carefully tailored and matched to act in concert (Fig. 1). The three components are discussed below.

### Phagemid Library

In addition to its defining origin of replication and packaging signal from a filamentous phage, the phagemid bears a plasmid origin of replication (e.g., ColE1), an antibiotic resistance gene (e.g., β-lactamase for ampicillin or carbenicillin resistance), and an expression cassette that encodes the protein library of interest fused to a coat protein or coat protein fragment of filamentous phage.

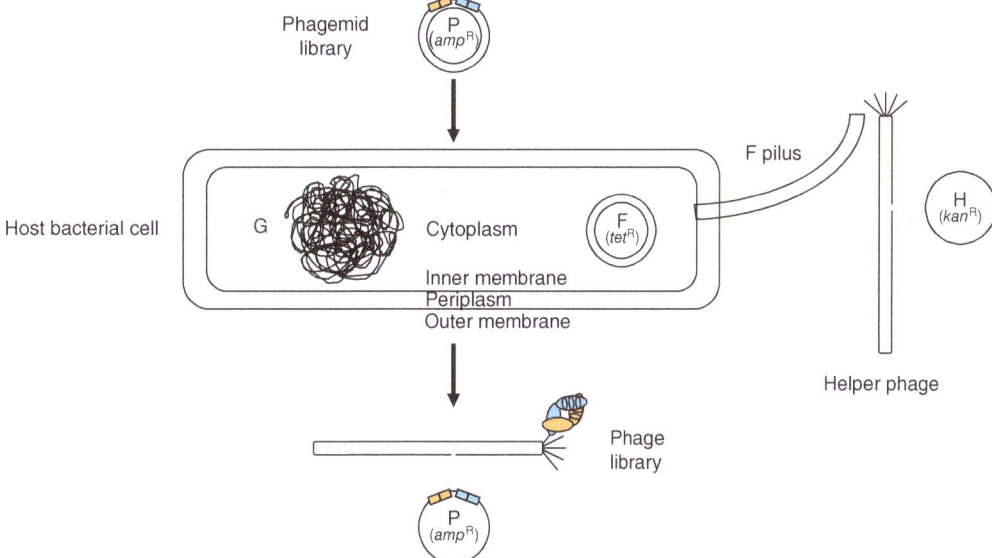

FIGURE 1. Phagemid-based phage display. In phagemid systems, the three components essential for phage display comprise the phagemid library, the host bacterial cell, and the helper phage. The cloned phagemid library is a circular double-stranded (ds) DNA ("P"; ~5 kb) that encodes the recombinant fusion of a protein library of interest (here shown as Fab, orange and blue) to a carboxy-terminal segment of pIII (see also Fig. 2 for details) and an antibiotic resistance gene (e.g., $amp^R$), and additionally harbors both an origin of replication and packaging signal from an Ff phage and a plasmid (bacterial) origin of replication. *Escherichia coli* serves as the host bacterial cell. Upon *E. coli* transformation with the phagemid library (typically by electroporation for higher transformation efficiency), phagemid-harboring *E. coli* clones are selected with ampicillin, or the more stable carbenicillin. In addition to their genome ("G"; ~5000 kb), the host bacterial cell harbors an episomal circular dsDNA ("F"; ~100 kb) that encodes information for the conjugative F pilus and an orthogonal antibiotic resistance gene (e.g., $tet^R$). As such, selection with tetracycline renders the host bacterial cell (F+ or male) susceptible to infection with helper phage. Like all Ff filamentous phage, the helper phage harbors a circular single-stranded (ss) DNA ("H"; ~6 kb) that forms a supercoiled double helix. "H" encodes pI, pII, pIII, pIV, pV, pVI, pVII, pVIII, pIX, pX, and pXI along with a different orthogonal resistance gene (e.g., $kan^R$). Following infection of the host bacterial cell by the helper phage, "H" follows the life cycle of Ff phage through DNA replication, mRNA transcription, protein translation, and assembly of phage particles containing ssDNA surrounded by major and minor coat proteins. In the presence of "P," phagemid ssDNA is preferentially packaged and the phagemid-encoded protein of interest fused to pIII is displayed at low frequency. The resulting physical linkage of phenotype and genotype of the protein of interest, here shown as a Fab, and the monoclonality of the transformed host bacterial cell are the essential elements of phage display that enable the selection of large libraries of naturally diverse or synthetically diversified proteins of interest.

# Chapter 4

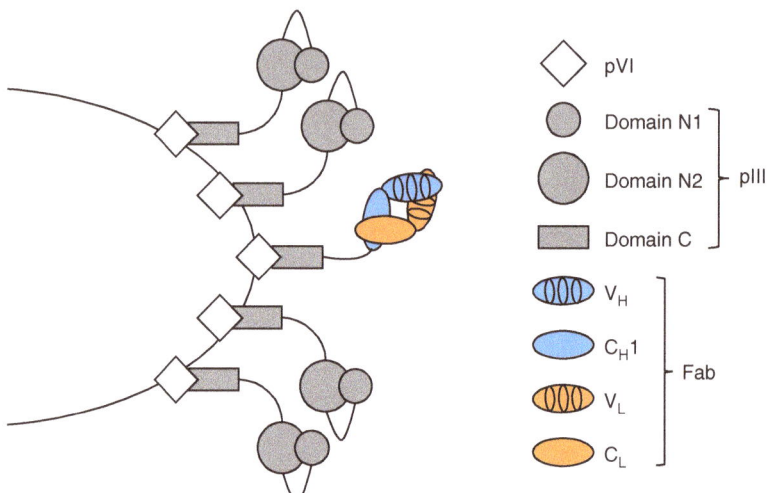

FIGURE 2. Monovalent Fab-phage display. The spiky cap of filamentous phage particles comprises five pIII/pVI copies. Wild-type pIII consists of three domains (N1, N2, and C) separated by glycine-rich linkers. All three domains are required for host bacterial cell infection. Phagemids of the pComb3 family harbor a fusion of the protein of interest to domain C of pIII. A typical protein of interest is a Fab (blue and orange), shown here in its heterodimeric assembly of Fd fragment ($V_H$–$C_H1$) and light chain (LC) ($V_L$–$C_L$), which are further stabilized by a carboxy-terminal disulfide bridge that forms in the periplasm of the host bacterial cell. Low expression of the Fab-ΔpIII fusion protein in the host bacterial cell leads to its incorporation in infectious phagemid-harboring phage particles that display one copy of Fab-ΔpIII and four copies of wild-type pIII. Monovalent Fab-phage display is preferred for affinity-driven selection.

The phagemid library is assembled by polymerase chain reaction (PCR) amplification of an expression cassette with diverse or diversified sequences, followed by restriction digestion of the amplified DNA, ligation into the correspondingly cut phagemid backbone, transformation of the ensuing phagemid library into host bacterial cells, and selection with the appropriate antibiotic. Using optimized reagents and protocols, this process typically yields $10^7$–$10^{10}$ independent transformants and can be scaled up to >$10^{11}$.

## Host Bacterial Cells

An *E. coli* strain that can be superinfected with filamentous phage typically serves as host for the phagemid library. In the absence of filamentous phage, the phagemid library is propagated as a plasmid library, taking advantage of the antibiotic resistance gene incorporated into the vector to enable selection and replication. Superinfection with filamentous wild-type or helper phage initiates replication of the (+) strand of the phagemid and its packaging into phage particles. Infection by filamentous phage requires the expression of F pili by the host bacterial cells encoded by the F plasmid. F plasmid-harboring bacteria are referred to as F+ or male. The F plasmid typically harbors a different antibiotic (e.g., tetracycline) resistance gene. Under antibiotic pressure, all host bacterial cells are F+ and, as such, are susceptible to infection by filamentous phage. Importantly, infection by a phage particle renders the host bacterial cell resistant to superinfection by another phage particle, ensuring monoclonality. As a result, host bacterial cells are not infected by more than one phage particle within the library and are, therefore, monoclonal with respect to the protein of interest. This ensures a match of genotype and phenotype, which is a prerequisite for selecting from protein libraries by phage display.

## Helper Phage

Helper phage are derived from wild-type Ff phage. The helper phage strains M13K07 (Vieira and Messing 1987), and its descendant VCSM13, are derivatives of wild-type Ff phage M13, whereas helper phage strain R408 is a derivative of wild-type Ff phage f1 (Russel et al. 1986). The helper phage delivers all 11 Ff phage genome-encoded functional and structural proteins (pI–pXI) required for ssDNA replication and packaging into infectious phage particles. The functional proteins of the helper phage act in *trans* on the origins of replication harbored by both the phagemid and helper phage. Helper phage can be engineered to harbor an antibiotic resistance gene (e.g., kanamycin for M13K07 and VCSM13) that is different from the antibiotic resistance genes of the phagemid and F plasmid. Thus, in pComb3 phagemids, three different antibiotics (e.g., carbenicillin, tetracycline, and kanamycin) are required for the selection of host bacterial cells that contain the phagemid and helper phage ssDNA. The presence of two Ff origins of replication and Ff packaging signals results in the host bacterial cell producing infectious phage particles that encapsulate either the phagemid or helper phage ssDNA. Both types of phage particles display the protein of interest, but only the one harboring the phagemid ssDNA provides the physical linkage of phenotype and genotype required for protein library selection. The helper phage genome contains a debilitated Ff packaging signal, which allows the preferential packaging of phagemid (>90%) over helper phage ssDNA when both "payloads" are present in the host bacterial cell. Several helper phage variants have been developed that further improve the efficiency of protein display. For example, impairing the ability of the helper phage genome to express pIII in the host bacterial cell facilitates higher efficiency of incorporation of the phagemid-encoded recombinant fusions of pIII with the protein of interest (Rondot et al. 2001; Baek et al. 2002). This higher efficiency is of particular interest for selection strategies that may benefit from multivalent over monovalent display on each phage particle, such as avidity-driven selection of low (micromolar) affinity binders. On the other hand, monovalent display is preferred for affinity-driven selection of high (nanomolar) affinity binders.

## PHYSICAL LINKAGE OF PHENOTYPE AND GENOTYPE IN PHAGEMID-BASED PHAGE DISPLAY

To understand "type 3 + 3" phage display vectors, one needs to consider the function of minor coat protein pIII and the consequences of mixing, in a host bacterial cell, copies of wild-type pIII with copies of the recombinant fusion of the protein of interest to a carboxy-terminal fragment of pIII (Rakonjac and Model 1998). As discussed above and depicted in Figure 2, pIII, which comprises 406 amino acids (aa) after signal peptide cleavage, can be functionally and structurally dissected into three domains, termed N1 (aa 1–68), N2 (aa 87–217), and C (aa 257–406), with two intersecting glycine-rich linkers (aa 69–86 and aa 218–256). Domain N2 mediates the attachment of filamentous phage to the tip of the F pilus of male *E. coli*, which emanates from the inner membrane and protrudes through the outer membrane. Following attachment, the F pilus retracts into the periplasm, triggering the interaction of domain N1 with the inner membrane protein TolA. This interaction promotes the domain C-assisted penetration of the inner membrane of the host bacterial cell by the filamentous phage. Domains N1 and N2, whether delivered by a phage or expressed by a phage-infected or phagemid-transformed host bacterial cell, render *E. coli* immune to superinfection. Domain C is also used as an anchor to display the protein of interest on the phage particle. In "type 3 + 3" phage display vectors, the phagemid can harbor a fusion of the protein of interest to pIII with or without domains N1 and N2.

Phagemids of the pComb3 family exclude the N1 and N2 domains and fuse the protein of interest directly to the carboxy-terminal segment of pIII. With such phage display vectors, the host bacterial cell remains nonimmune to superinfection by helper phage. Other "type 3 + 3" phage display vectors use phagemids that fuse the protein of interest to the amino terminus of full-length pIII (Hoogenboom et al. 1991). The inclusion of N1 and N2 domains in these systems requires the use of a recombinant fusion-driving promoter that can be switched off to allow superinfection by the helper phage and then switched on to allow phage display. In contrast, in pComb3 phagemids, the recombinant fusion is under control of the *lac* promoter and the overlapping *lac* operator (*lacO*). Host bacterial cells that express the *lac* repressor (encoded by *lacI*) suppress expression of the recombinant fusion. Binding of the lactose metabolite allolactose, or of its synthetic mimic isopropyl-β-D-1-thiogalactopyranoside (IPTG), to the *lac* repressor inhibits this suppression. In addition to high lactose levels, the *lac* promoter is triggered by low glucose levels. Decreased glucose leads to increased cAMP concentration which, in turn, allosterically activates the catabolic activator protein (CAP) to bind to the *lac* promoter and trigger transcription by RNA polymerase. Thus, high glucose concentrations in the medium (e.g., 2%, w/v, or 20 mg/mL) suppress expression and display of the protein of interest. At normal glucose concentrations in the medium (e.g., 0.1%, w/v, or 1 mg/mL), the *lac* promoter is leaky, even in the absence of allolactose or IPTG. Although assembling phage particles at this basic activity of the *lac* promoter favors monovalent over multivalent phage display and can mitigate potential toxicity of the recombinant fusion toward the host bacterial cell, it comes at the expense of the majority of phage particles being "bald" (i.e., not displaying the protein of interest).

## "TYPE 3 + 3" PHAGE DISPLAY VECTOR pComb3

Phagemid pComb3 (Fig. 3A) was first described in 1991 by Carlos F. Barbas III and colleagues (Barbas et al. 1991). This phage display vector was generated by modification of the phagemid pBluescript SK (-) (Short et al. 1988), which includes an f1 intergenic region (comprising f1 origin of replication and f1 packaging signal), a ColE1 origin of replication, a β-lactamase gene, and a multiple cloning site (MCS). Phagemid pComb3 was designed for phage display of antibodies in Fab format. Fab molecules pair the $V_H$–$C_H1$ portion of the heavy chain (HC), known as Fd fragment, with the $V_L$–$C_L$ LC. To assemble Fab and Fab library-encoding phagemids, Fd and LC cDNAs are cloned sequentially using XhoI/SpeI and SacI/XbaI restriction sites, respectively. Whereas the "Comb" of pComb3 signifies the random recombination of Fd and LC resulting in combinatorial Fab libraries, the "3" denotes the fusion of Fd to a carboxy-terminal segment of the minor coat protein pIII encompassing aa 198–406 —that is, a carboxy-terminal fragment (aa 198–217) of domain N2 (aa 87–217), the second glycine-rich linker (aa 218–256), and domain C (aa 257–406). At their amino termini, Fd and LC are each fused to a pelB signal peptide (each comprising 22 aa; MKYLLPTAAAGLLLLAAQPAMA), which directs the export of the recombinant proteins across the cytoplasmic membrane to the periplasm through the general secretory (SEC) pathway of *E. coli*. Note that other phagemid systems use the cotranslational signal recognition particle (SRP) pathway instead of the post-translational SEC pathway to export and display fast-folding recombinant proteins without disulfide bridges, such as DARPins, more efficiently (Steiner et al. 2006).

The expression of pelB-Fd-ΔpIII and pelB-LC is driven by the promoter/operator element of the *lacZ* gene of *E. coli*. The constitutively expressed *lacI* gene in the *E. coli* genome generates the lac repressor that controls the promoter through binding to the operon. As discussed above, leaky expression is sufficient for monovalent phage display, but can be further constrained in the presence of high concentrations of glucose or can be fully released in the presence of IPTG. Unlike later generations of the pComb3 phagemid family, pComb3 features two promoter/operator elements for the two expression cassettes. However, by inclusion of the two pelB encoding sequences, this design incorporates repetitive elements prone to deletion by homologous recombination and, therefore, requires the use of *E. coli* strains deficient in RecA as host bacterial cells (e.g., XL1-Blue). Another liability of pComb3 is the requirement of sequential XhoI/SpeI cloning of Fd and of SacI/XbaI cloning of LC. As a consequence, in addition to the introduction of cloning biases due to the requirement of an intermittent transformation and amplification step, the $V_L$- and $V_H$-encoding cDNAs that contain at least one of these restriction sites are excluded from the final Fab library. Nonetheless, pComb3 enabled the successful generation and selection of numerous Fab libraries from a variety of different antibody repertoires (Table 1).

Another feature of pComb3 is the ability to remove ΔpIII, which is adjacent to the cloning site, by cutting its flanking restriction sites SpeI and NheI and self-ligating the phagemid through their compatible cohesive ends. This effectively generates an expression cassette for soluble Fab expression following Fab library selection (Barbas et al. 1991).

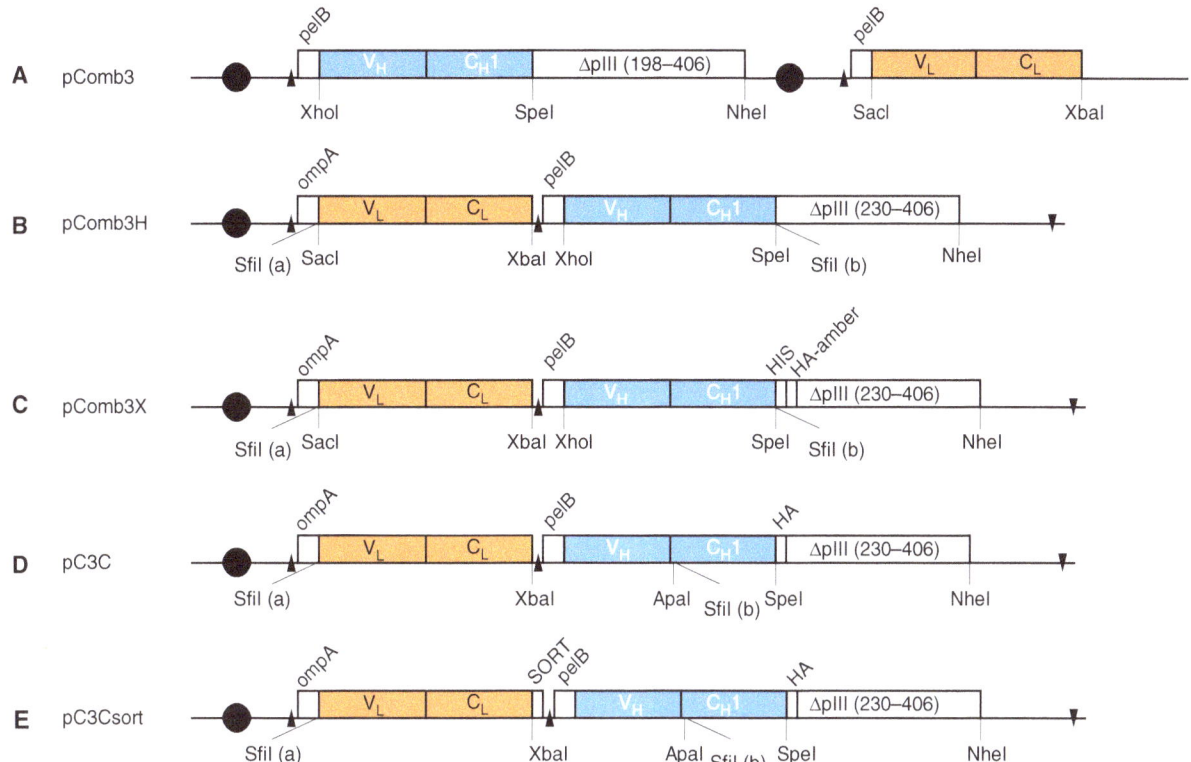

FIGURE 3. Expression cassettes of the pComb3 phagemid family. First-generation phagemid pComb3 (A) features separate lac promoter (black circle)-driven pelB–$V_H$–$C_H$1–ΔpIII and pelB–$V_L$–$C_L$ expression cassettes. Shine–Dalgarno sequences are indicated as black triangles. Fd fragment and LC libraries are generated by sequential XhoI/SpeI and SacI/XbaI cloning, respectively. SpeI/NheI self-ligation eliminates ΔpIII and converts pComb3 to a plasmid for soluble Fab expression. Second-generation phagemids pComb3H (B) and pComb3X (C) feature a single lac promoter that drives bicistronic ompA–$V_L$–$C_L$ and pelB–$V_H$–$C_H$1–ΔpIII expression cassettes. The reverse black triangle indicates a transcriptional terminator sequence. Although sequential SacI/XbaI and XhoI/SpeI cloning of LC and Fd fragment libraries is still possible, asymmetric SfiI cloning via SfiI (a) and SfiI (b) facilitates library generation. The SpeI/NheI self-ligation feature of pComb3 is conserved in pComb3H and pComb3X. Compared to pComb3H, pComb3X adds a HIS tag and an HA tag terminating in an amber stop codon between $C_H$1 and ΔpIII. Phagemid pC3C (D) moves the downstream SfiI (b) site upstream between $V_H$ and $C_H$1, eliminating one PCR amplification step in library generation. Phagemid pC3Csort (E) adds a SORT tag to the carboxyl terminus of the LC to enable Fab biotinylation and capture selections.

Phagemid pComb8, a "type 8 + 8" phage display vector that has also been used for Fab libraries (Kang et al. 1991; Dinh et al. 1996), is nearly identical to pComb3 but fuses Fd to major coat protein pVIII, resulting in multivalent Fab display along the sides of the phage particle. Maps and the complete nucleic acid sequence of the pComb3 phagemid are available online (https://www.addgene.org/63888).

## pComb3 DESCENDANTS pComb3H AND pComb3X

The second-generation of pComb3 phagemids addressed the noted shortcomings of the first-generation, including the use of redundant sequences and the requirement for separate Fd and LC cloning steps. Shown in Figure 3B,C are phagemids pComb3H and pComb3X, respectively. These phage display vectors do not contain a lacZ promoter/operon and pelB repeat, but use a single lacZ promoter/operon and a combination of signal peptides ompA and pelB for LC and Fd-ΔpIII, respectively. The bicistronic mRNA contains ribosome binding sites (Shine–Dalgarno sequences) upstream of the ompA and pelB start codons. A transcriptional terminator (hairpin loop) downstream from the pelB-Fd-ΔpIII stop codon was added. Like pelB, ompA (22 aa; MKKTAIAIAVALAGFATVAQAA) directs export from the cytoplasm to the periplasm through the SEC pathway. The carboxy-terminal segment of pIII in pComb3H and pComb3X is shortened to aa 230–406 compared to aa 198–406 in pComb3, removing the carboxy-terminal fragment of domain N2 and a portion of the glycine-rich linker. Although both generations still allow sequential LC and Fd cloning via SacI/XbaI and XhoI/SpeI restriction sites, respectively, a key feature of the second-generation is the introduction of two asymmetric SfiI sites between ompA and $V_L$ and between $C_H1$ and $\Delta$pIII. With the Fab-encoding cassette flanked by asymmetric SfiI sites, directional cloning can be performed in one step. SfiI is a rare-cutting restriction enzyme that recognizes the 8-bp palindromic sequence 5′-GGCCNNNNNGGCC-3′, which is interrupted by a 5-bp degenerate sequence (N can be any nucleotide). It cuts within the degenerate sequence to generate a 5′ overhang (5′-GGCCN_NNN^NGGCC-3′). In pComb3H and pCom3X, the upstream SfiI site (labeled as "SfiI (a)" in Fig. 3B,C) is 5′-GGCCC_AGG^CGGCC-3′, whereas the downstream SfiI site (labeled as "SfiI (b)") is 5′-GGCCA_GGC^CGGCC-3′. Their different degenerate sequences and asymmetric 5′ overhangs allow an SfiI-cleaved expression cassette to be ligated in the correct orientation into the SfiI-cleaved phagemid. Importantly, unlike the 6-bp recognition sequences of SacI, XbaI, XhoI, and SpeI, the 8-bp recognition sequence of SfiI is absent from $V_H$ and $V_L$ genes. With the enzyme being commercially available in high quality and quantity (from, e.g., New England Biolabs, at a concentration of 20 U/µL, with 1 U defined as the amount of enzyme required to digest 1 µg plasmid in 1 h at 50°C), directional cloning by SfiI is the method of choice for the second-generation of pComb3 phagemids. This strategy is also compatible with cloning cDNAs that encode single-chain protein libraries of interest into pComb3H and pComb3X, including scFv, single domains, and zinc fingers. Here, ompA serves as the only signal peptide, and the protein of interest is either directly fused to $\Delta$pIII (pComb3H) or via a $(His)_6$ (HIS) and hemagglutinin (HA) tag (pComb3X). The HA tag (10 aa; YPYDVPDYAS) is an HA epitope against which commercial monoclonal antibodies of high affinity are available. These two tags in pComb3X, along with an amber (UAG on the mRNA; TAG on the cDNA) stop codon at the junction of HA tag and $\Delta$pIII (Fig. 3C), are the only differences to pComb3H. The amber stop codon enables readthrough in *E. coli* strains that harbor an amber suppressor tRNA, such as *supE44* (incorporation of glutamine) of XL1-Blue and ER2738, which are commonly used as host bacteria for phage display vectors of the pComb3 phagemid family. However, if an *E. coli* strain without amber suppressor tRNA (e.g., TOP10) is transformed with pComb3X or infected with pComb3X-based phage, translation of the encoded protein of interest terminates at the amber stop codon, thus aborting the expression of its fusion with $\Delta$pIII and producing a soluble protein equipped with a carboxy-terminal HIS tag for immobilized metal affinity chromatography (IMAC) purification followed by an HA tag for detection in ELISA and other immunoassays. As mentioned for pComb3, the $\Delta$pIII-encoding DNA of pComb3H and pComb3X can be removed by SpeI/NheI digestion followed by self-ligation. Note that this procedure also removes the HIS and HA tags in pComb3X. Maps and the complete nucleic acid sequences of the pComb3H and pComb3X phagemids can be found online at https://www.addgene.org/64133 and https://www.addgene.org/63890, respectively. Both phagemids are commercially available with a stuffer fragment between the asymmetric SfiI sites, to facilitate cloning.

## SPECIALIZED pComb3 DESCENDANTS pC3C AND pC3Csort

Two specialized phage display vectors derived from the second-generation of pComb3 phagemids are pC3C (Fig. 3D) and pC3Csort (Fig. 3E). Unlike pComb3H and pComb3X, which can be used as phage display and expression vectors for a variety of different dual-chain and single-chain proteins, pC3C (Hofer et al. 2007) and pC3Csort (Wilson et al. 2018) were developed specifically for phage display of Fab with human constant domains ($C_L$ and $C_H1$) and human or nonhuman variable domains ($V_L$ and $V_H$). As such, they have been mainly used for the generation and selection of human and chimeric rabbit/human Fab libraries from immune, naive, and synthetic human and rabbit antibody repertoires, respectively (Table 1). However, they also facilitate phage display of chimeric nonhuman/human Fab from other mammalian and avian antibody repertoires. With human $C_H1$ being common to all of these Fab, the SfiI (b) site was moved upstream to shorten the SfiI-flanked library insert from ~1.4 kb (pComb3H and pComb3X) to ~1.15 kb (pC3C and pC3Csort). Shortening the insert enables a two-step library assembly strategy in which $V_L$ ($V_\kappa$ and $V_\lambda$) and $V_H$ are amplified by reverse transcription polymerase chain reaction (RT-PCR) in a first step and then fused via $C_L$ ($C_\kappa$ and $C_\lambda$) by three-fragment overlap extension PCR in a second step, before SfiI digestion and directional cloning (Fig. 4). In contrast, the Fab-encoding cassettes of pComb3H and pComb3X require a three-step library assembly strategy, with two sequential two-fragment overlap extension PCRs to first fuse $V_L$ and $V_H$ with $C_L$ and $C_H1$, respectively, and then $V_L$–$C_L$ with $V_H$–$C_H1$. Reducing the required amplification steps from three to two saves time and can also potentially increase library diversity by reducing PCR biases. In addition, the SfiI (b) site was changed from 5′-GGCCA_GGC^CGGCC-3′ to 5′-GGCC_CCG^TCGGCC-3′ to remove a Dcm methylase site (CC^AGG) that interferes with SfiI cleavage efficiency. Accommodating the SfiI (b) site near the start of $C_H1$ required a single aa mutation (ASTKGPSVFPLAPSSKSTS[…] to ASTKGPSVFPLAP SAKSTS[…]) that does not affect Fab display or expression. Nonetheless, an upstream natural ApaI site (G_GGCC^C) (Fig. 2D) allows for the transfer of $V_H$ into dedicated Fab and IgG expression vectors with unmutated $C_H1$. Unlike pComb3X, pC3C and pC3Csort do not contain a HIS tag, and their HA tag is directly fused to ΔpIII without an intervening amber stop codon.

The difference between pC3C and pC3Csort is the addition of a sortase A (SORT) tag at the carboxyl terminus of the LC. The SORT tag used in pC3Csort (GGGGSLPETGG) contains the SORT recognition pentapeptide LPETG. SORT cleaves the TG amide bond in this motif and replaces the G with a glycine derivative of a small or large molecule of interest. For example, adding an excess of the

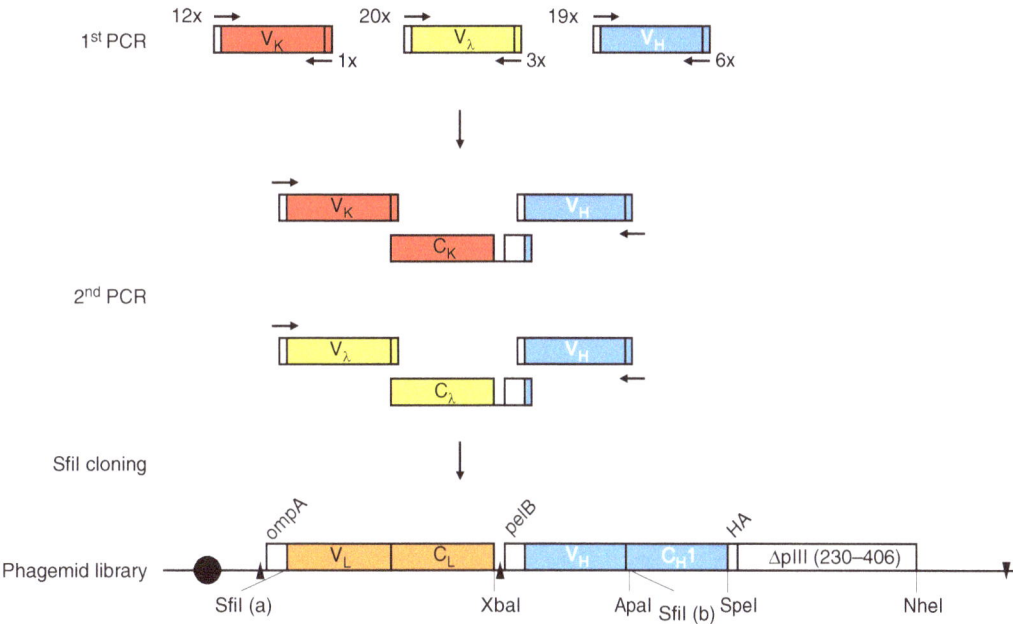

FIGURE 4. Assembly of Fab libraries in phagemid pC3C. Phagemid pC3C enables a two-step assembly strategy of human or chimeric nonhuman/human Fab libraries for phage display. Shown is the workflow for the generation of a human Fab library starting with a 1st PCR step in which cDNAs encoding $V_\kappa$, $V_\lambda$, and $V_H$ are amplified from reverse transcribed mRNA of an immune or naive human antibody repertoire. Diverse primer pairs (12 × 1 for $V_\kappa$, 20 × 3 for $V_\lambda$, and 19 × 6 for $V_H$) are used in 186 separate PCR reactions for each individual human antibody repertoire (Kwong et al. 2008). Subsequently, a second PCR step fuses the $V_\kappa$ and $V_H$ pools and the $V_\lambda$ and $V_H$ pools via a separately amplified invariable cDNA fragment that comprises $C_\kappa$ and $C_\lambda$, respectively, the downstream ribosome binding site, and pelB. The flanking primers in the second PCR step also introduce SfiI (a) and SfiI (b), which are cut and ligated into the correspondingly cut phagemid pC3C. Electroporation of the ligation mixture into Escherichia coli completes the phagemid library, which typically comprises $10^7$–$10^{10}$ independent transformants. As depicted in Figure 1, addition of helper phage to the host bacterial cells converts the phagemid-encoding Fab library to a phage-encoding and displaying Fab library. Using a different set of primers for the first PCR step, the same procedure can be used for the generation of chimeric nonhuman/human Fab libraries.

tetrapeptide GGGK that is biotinylated at the ε-amino group of its carboxy-terminal K to a pC3Csort-encoded Fab, adds a biotin tag to the Fab. This reaction is highly selective and efficient, and it has been applied to the site-specific biotinylation of Fab-displaying phage (Wilson et al. 2018). This strategy can be used for the enrichment of Fab-displaying phage over "bald" phage in selections on whole cells. Like pC3C, pC3Csort is compatible with human and chimeric nonhuman/human Fab libraries from immune, naive, and synthetic antibody repertoires. The map and complete nucleic acid sequence of the pC3C phagemid, which is commercially available from Kerafast, can be found online at https://www.kerafast.com/productgroup/724/pc3c-phage-display-plasmids. The accompanying plasmid pC3C-His, which replaces ΔpIII with a HIS tag to facilitate Fab-$(His)_6$ expression and purification (Kwong and Rader 2009), is also commercially available from Kerafast. Other plasmids compatible with pC3C and pC3Csort for *E. coli* expression of soluble Fab have been described (Stahl et al. 2010).

## CONCLUSIONS

Conceptualized in 1991, phage display vectors of the pComb3 phagemid family have been used widely for mining immunoglobulin and nonimmunoglobulin repertoires, and their design recapitulates key features of phagemid-based phage display. The pComb3 phagemid family comprises the original phage display vector pComb3, its descendants pComb3H, pComb3X, pC3C, and pC3Csort, and other derivatives not discussed in this review. Collectively, these phage display vectors are able to "comb" large libraries of naturally diverse and synthetically diversified proteins for distinctive molecular and functional properties. As such, they continue to serve as valuable tools for basic and applied research.

## REFERENCES

Andris-Widhopf J, Rader C, Steinberger P, Fuller R, Barbas CF III. 2000. Methods for the generation of chicken monoclonal antibody fragments by phage display. *J Immunol Methods* 242: 159–181. doi:10.1016/S0022-1759(00)00221-0

Baek H, Suk KH, Kim YH, Cha S. 2002. An improved helper phage system for efficient isolation of specific antibody molecules in phage display. *Nucl Acids Res* 30: e18. doi:10.1093/nar/30.5.e18

Barbas CF III, Kang AS, Lerner RA, Benkovic SJ. 1991. Assembly of combinatorial antibody libraries on phage surfaces: the gene III site. *Proc Natl Acad Sci* 88: 7978–7982. doi:10.1073/pnas.88.18.7978

Barbas CF III, Bain JD, Hoekstra DM, Lerner RA. 1992. Semisynthetic combinatorial antibody libraries: a chemical solution to the diversity problem. *Proc Natl Acad Sci* 89: 4457–4461. doi:10.1073/pnas.89.10.4457

Barbas CF III, Languino LR, Smith JW. 1993. High-affinity self-reactive human antibodies by design and selection: targeting the integrin ligand binding site. *Proc Natl Acad Sci* 90: 10003–10007. doi:10.1073/pnas.90.21.10003

Barbas CF III, Hu D, Dunlop N, Sawyer L, Cababa D, Hendry RM, Nara PL, Burton DR. 1994. In vitro evolution of a neutralizing human antibody to human immunodeficiency virus type 1 to enhance affinity and broaden strain cross-reactivity. *Proc Natl Acad Sci* 91: 3809–3813. doi:10.1073/pnas.91.9.3809

Baskar S, Suschak JM, Samija I, Srinivasan R, Childs RW, Pavletic SZ, Bishop MR, Rader C. 2009. A human monoclonal antibody drug and target discovery platform for B-cell chronic lymphocytic leukemia based on allogeneic hematopoietic stem cell transplantation and phage display. *Blood* 114: 4494–4502. doi:10.1182/blood-2009-05-222786

Burton DR, Barbas CF III, Persson MA, Koenig S, Chanock RM, Lerner RA. 1991. A large array of human monoclonal antibodies to type 1 human immunodeficiency virus from combinatorial libraries of asymptomatic seropositive individuals. *Proc Natl Acad Sci* 88: 10134–10137. doi:10.1073/pnas.88.22.10134

Cao G, Gao X, Zhan Y, Wang Q, Zhang Z, Dimitrov DS, Gong R. 2020. An engineered human IgG1 CH2 domain with decreased aggregation and nonspecific binding. *MAbs* 12: 1689027. doi:10.1080/19420862.2019.1689027

Cary SP, Lee J, Wagenknecht R, Silverman GJ. 2000. Characterization of superantigen-induced clonal deletion with a novel clan III-restricted avian monoclonal antibody: exploiting evolutionary distance to create antibodies specific for a conserved VH region surface. *J Immunol* 164: 4730–4741. doi:10.4049/jimmunol.164.9.4730

Chen W, Zhu Z, Feng Y, Xiao X, Dimitrov DS. 2008. Construction of a large phage-displayed human antibody domain library with a scaffold based on a newly identified highly soluble, stable heavy chain variable domain. *J Mol Biol* 382: 779–789. doi:10.1016/j.jmb.2008.07.054

Chung J, Rader C, Popkov M, Hur YM, Kim HK, Lee YJ, Barbas CF III. 2004. Integrin $α_{IIb}β_3$-specific synthetic human monoclonal antibodies and HCDR3 peptides that potently inhibit platelet aggregation. *FASEB J* 18: 361–363. doi:10.1096/fj.03-0586fje

Dinh Q, Weng NP, Kiso M, Ishida H, Hasegawa A, Marcus DM. 1996. High affinity antibodies against Lex and sialyl Lex from a phage display library. *J Immunol* 157: 732–738. doi:10.4049/jimmunol.157.2.732

Dreier B, Beerli RR, Segal DJ, Flippin JD, Barbas CF III. 2001. Development of zinc finger domains for recognition of the 5′-ANN-3′ family of DNA sequences and their use in the construction of artificial transcription factors. *J Biol Chem* 276: 29466–29478. doi:10.1074/jbc.M102604200

Dreier B, Fuller RP, Segal DJ, Lund CV, Blancafort P, Huber A, Koksch B, Barbas CF III. 2005. Development of zinc finger domains for recognition of the 5′-CNN-3′ family DNA sequences and their use in the construction of artificial transcription factors. *J Biol Chem* 280: 35588–35597. doi:10.1074/jbc.M506654200

Duchosal MA, Eming SA, Fischer P, Leturcq D, Barbas CF III, McConahey PJ, Caothien RH, Thornton GB, Dixon FJ, Burton DR. 1992. Immunization of hu-PBL-SCID mice and the rescue of human monoclonal Fab fragments through combinatorial libraries. *Nature* 355: 258–262. doi:10.1038/355258a0

Feng M, Bian H, Wu X, Fu T, Fu Y, Hong J, Fleming BD, Flajnik MF, Ho M. 2019. Construction and next-generation sequencing analysis of a large phage-displayed V$_{NAR}$ single-domain antibody library from six naive nurse sharks. *Antib Ther* 2: 1–11. doi:10.1093/abt/tby011

Glamann J, Burton DR, Parren PW, Ditzel HJ, Kent KA, Arnold C, Montefiori D, Hirsch VM. 1998. Simian immunodeficiency virus (SIV) envelope-specific Fabs with high-level homologous neutralizing activity: recovery from a long-term-nonprogressor SIV-infected macaque. *J Virol* 72: 585–592. doi:10.1128/JVI.72.1.585-592.1998

Goydel RS, Weber J, Peng H, Qi J, Soden J, Freeth J, Park H, Rader C. 2020. Affinity maturation, humanization, and co-crystallization of a rabbit anti-human ROR2 monoclonal antibody for therapeutic applications. *J Biol Chem* 295: 5995–6006. doi:10.1074/jbc.RA120.012791

Gram H, Marconi LA, Barbas CF III, Collet TA, Lerner RA, Kang AS. 1992. In vitro selection and affinity maturation of antibodies from a naive combinatorial immunoglobulin library. *Proc Natl Acad Sci* 89: 3576–3580. doi:10.1073/pnas.89.8.3576

Hofer T, Tangkeangsirisin W, Kennedy MG, Mage RG, Raiker SJ, Venkatesh K, Lee H, Giger RJ, Rader C. 2007. Chimeric rabbit/human Fab and IgG specific for members of the Nogo-66 receptor family selected for species cross-reactivity with an improved phage display vector. *J Immunol Methods* 318: 75–87. doi:10.1016/j.jim.2006.10.007

Hong J, Kwon HJ, Cachau R, Chen CZ, Butay KJ, Duan Z, Li D, Ren H, Liang T, Zhu J, et al. 2022. Dromedary camel nanobodies broadly neutralize SARS-CoV-2 variants. *Proc Natl Acad Sci* 119: e2201433119. doi:10.1073/pnas.2201433119

Hoogenboom HR, Griffiths AD, Johnson KS, Chiswell DJ, Hudson P, Winter G. 1991. Multi-subunit proteins on the surface of filamentous phage: methodologies for displaying antibody (Fab) heavy and light chains. *Nucl Acids Res* 19: 4133–4137. doi:10.1093/nar/19.15.4133

Kang AS, Barbas CF III, Janda KD, Benkovic SJ, Lerner RA. 1991. Linkage of recognition and replication functions by assembling combinatorial antibody Fab libraries along phage surfaces. *Proc Natl Acad Sci* 88: 4363–4366. doi:10.1073/pnas.88.10.4363

Kim HJ, McCoy MR, Majkova Z, Dechant JE, Gee SJ, Tabares-da Rosa S, Gonzalez-Sapienza GG, Hammock BD. 2012. Isolation of alpaca anti-hapten heavy chain single domain antibodies for development of sensitive immunoassay. *Anal Chem* 84: 1165–1171. doi:10.1021/ac2030255

Kuwata T, Katsumata Y, Takaki K, Miura T, Igarashi T. 2011. Isolation of potent neutralizing monoclonal antibodies from an SIV-Infected rhesus macaque by phage display. *AIDS Res Hum Retroviruses* 27: 487–500. doi:10.1089/aid.2010.0191

Kwong KY, Rader C. 2009. E. coli expression and purification of Fab antibody fragments. *Curr Protoc Protein Sci* Chapter 6, Unit 6 10. doi:10.1002/0471140864.ps0610s55

Kwong KY, Baskar S, Zhang H, Mackall CL, Rader C. 2008. Generation, affinity maturation, and characterization of a human anti-human NKG2D monoclonal antibody with dual antagonistic and agonistic activity. *J Mol Biol* 384: 1143–1156. doi:10.1016/j.jmb.2008.09.008

Lang IM, Barbas CF III, Schleef RR. 1996. Recombinant rabbit Fab with binding activity to type-1 plasminogen activator inhibitor derived from a phage-display library against human alpha-granules. *Gene* 172: 295–298. doi:10.1016/0378-1119(96)00021-2

Li W, Schafer A, Kulkarni SS, Liu X, Martinez DR, Chen C, Sun Z, Leist SR, Drelich A, Zhang L, et al. 2020. High potency of a bivalent human VH domain in SARS-CoV-2 animal models. *Cell* 183: 429.e416–441.e416. doi:10.1016/j.cell.2020.09.007

Mason NJ, Chester N, Xiong A, Rotolo A, Wu Y, Yoshimoto S, Glassman P, Gulendran G, Siegel DL. 2021. Development of a fully canine anti-canine CTLA4 monoclonal antibody for comparative translational research in dogs with spontaneous tumors. *MAbs* 13: 2004638. doi:10.1080/19420862.2021.2004638

McIntosh RS, Asghar MS, Watson PF, Kemp EH, Weetman AP. 1996. Cloning and analysis of IgG kappa and IgG lambda anti-thyroglobulin autoantibodies from a patient with Hashimoto's thyroiditis: evidence for in vivo antigen-driven repertoire selection. *J Immunol* 157: 927–935. doi:10.4049/jimmunol.157.2.927

Moreland NJ, Susanto P, Lim E, Tay MYF, Rajamanonmani R, Hanson BJ, Vasudevan SG. 2012. Phage display approaches for the isolation of monoclonal antibodies against dengue virus envelope domain III from human and mouse derived libraries. *Int J Mol Sci* 13: 2618–2635. doi:10.3390/ijms13032618

Nathan S, Rader C, Barbas CF III. 2005. Neutralization of *Burkholderia pseudomallei* protease by Fabs generated through phage display. *Biosci Biotechnol Biochem* 69: 2302–2311. doi:10.1271/bbb.69.2302

Norbury LJ, Basalaj K, Baska P, Kalinowska A, Zawistowska-Deniziak A, Yap HY, Wilkowski P, Wesolowska A, Wedrychowicz H. 2019. Construction of a novel phage display antibody library against *Fasciola hepatica*, and generation of a single-chain variable fragment specific for F. hepatica cathepsin L1. *Exp Parasitol* 198: 87–94. doi:10.1016/j.exppara.2019.02.001

Noren KA, Noren CJ. 2001. Construction of high-complexity combinatorial phage display peptide libraries. *Methods* 23: 169–178. doi:10.1006/meth.2000.1118

O'Brien PM, Aitken R, O'Neil BW, Campo MS. 1999. Generation of native bovine mAbs by phage display. *Proc Natl Acad Sci* 96: 640–645. doi:10.1073/pnas.96.2.640

Payne AS, Ishii K, Kacir S, Lin C, Li H, Hanakawa Y, Tsunoda K, Amagai M, Stanley JR, Siegel DL. 2005. Genetic and functional characterization of human pemphigus vulgaris monoclonal autoantibodies isolated by phage display. *J Clin Invest* 115: 888–899. doi:10.1172/JCI24185

Peng H, Nerreter T, Chang J, Qi J, Li X, Karunadharma P, Martinez GJ, Fallahi M, Soden J, Freeth J, et al. 2017. Mining naive rabbit antibody repertoires by phage display for monoclonal antibodies of therapeutic utility. *J Mol Biol* 429: 2954–2973. doi:10.1016/j.jmb.2017.08.003

Popkov M, Mage RG, Alexander CB, Thundivalappil S, Barbas CF III, Rader C. 2003. Rabbit immune repertoires as sources for therapeutic monoclonal antibodies: the impact of kappa allotype-correlated variation in cysteine content on antibody libraries selected by phage display. *J Mol Biol* 325: 325–335. doi:10.1016/S0022-2836(02)01232-9

Rader C, Cheresh DA, Barbas CF III. 1998. A phage display approach for rapid antibody humanization: designed combinatorial V gene libraries. *Proc Natl Acad Sci* 95: 8910–8915. doi:10.1073/pnas.95.15.8910

Rader C, Ritter G, Nathan S, Elia M, Gout I, Jungbluth AA, Cohen LS, Welt S, Old LJ, Barbas CF III. 2000. The rabbit antibody repertoire as a novel source for the generation of therapeutic human antibodies. *J Biol Chem* 275: 13668–13676. doi:10.1074/jbc.275.18.13668

Rader C, Popkov M, Neves JA, Barbas CF III. 2002. Integrin αvβ3 targeted therapy for Kaposi's sarcoma with an in vitro evolved antibody. *FASEB J* 16: 2000–2002. doi:10.1096/fj.02-0281fje

Rakonjac J, Model P. 1998. Roles of pIII in filamentous phage assembly. *J Mol Biol* 282: 25–41. doi:10.1006/jmbi.1998.2006

Rakonjac J, Bennett NJ, Spagnuolo J, Gagic D, Russel M. 2011. Filamentous bacteriophage: biology, phage display and nanotechnology applications. *Curr Issues Mol Biol* 13: 51–76. doi:10.21775/cimb.013.051

Roark JH, Bussel JB, Cines DB, Siegel DL. 2002. Genetic analysis of autoantibodies in idiopathic thrombocytopenic purpura reveals evidence of clonal expansion and somatic mutation. *Blood* 100: 1388–1398. doi:10.1182/blood.V100.4.1388.h81602001388_1388_1398

Roben P, Barbas SM, Sandoval L, Lecerf JM, Stollar BD, Solomon A, Silverman GJ. 1996. Repertoire cloning of lupus anti-DNA autoantibodies. *J Clin Invest* 98: 2827–2837. doi:10.1172/JCI119111

Rondot S, Koch J, Breitling F, Dübel S. 2001. A helper phage to improve single-chain antibody presentation in phage display. *Nat Biotechnol* 19: 75–78. doi:10.1038/83567

Russel M, Kidd S, Kelley MR. 1986. An improved filamentous helper phage for generating single-stranded plasmid DNA. *Gene* 45: 333–338. doi:10.1016/0378-1119(86)90032-6

Sambrook J, Fritschi EF, Maniatis T. 1989. *Molecular cloning: a laboratory manual*. Cold Spring Harbor Laboratory Press, Cold Spring Harbor, New York.

Samuelsson A, Chiodi F, Ohman P, Putkonen P, Norrby E, Persson MA. 1995. Chimeric macaque/human Fab molecules neutralize simian immunodeficiency virus. *Virology* 207: 495–502. doi:10.1006/viro.1995.1109

Schofield DJ, Glamann J, Emerson SU, Purcell RH. 2000. Identification by phage display and characterization of two neutralizing chimpanzee monoclonal antibodies to the hepatitis E virus capsid protein. *J Virol* **74**: 5548–5555. doi:10.1128/JVI.74.12.5548-5555.2000

Segal DJ, Dreier B, Beerli RR, Barbas CF III. 1999. Toward controlling gene expression at will: selection and design of zinc finger domains recognizing each of the 5′-GNN-3′ DNA target sequences. *Proc Natl Acad Sci* **96**: 2758–2763. doi:10.1073/pnas.96.6.2758

Short JM, Fernandez JM, Sorge JA, Huse WD. 1988. λ ZAP: a bacteriophage λ expression vector with in vivo excision properties. *Nucl Acids Res* **16**: 7583–7600. doi:10.1093/nar/16.15.7583

Siegel DL, Chang TY, Russell SL, Bunya VY. 1997. Isolation of cell surface-specific human monoclonal antibodies using phage display and magnetically-activated cell sorting: applications in immunohematology. *J Immunol Methods* **206**: 73–85. doi:10.1016/S0022-1759(97)00087-2

Smith GP. 1985. Filamentous fusion phage: novel expression vectors that display cloned antigens on the virion surface. *Science* **228**: 1315–1317. doi:10.1126/science.4001944

Smith GP, Petrenko VA. 1997. Phage display. *Chem Rev* **97**: 391–410. doi:10.1021/cr960065d

Stahl SJ, Watts NR, Rader C, DiMattia MA, Mage RG, Palmer I, Kaufman JD, Grimes JM, Stuart DI, Steven AC, et al. 2010. Generation and characterization of a chimeric rabbit/human Fab for co-crystallization of HIV-1 Rev. *J Mol Biol* **397**: 697–708. doi:10.1016/j.jmb.2010.01.061

Steinberger P, Kraft D, Valenta R. 1996. Construction of a combinatorial IgE library from an allergic patient. Isolation and characterization of human IgE Fabs with specificity for the major timothy grass pollen allergen, Phl p 5. *J Biol Chem* **271**: 10967–10972. doi:10.1074/jbc.271.18.10967

Steinberger P, Sutton JK, Rader C, Elia M, Barbas CF III. 2000. Generation and characterization of a recombinant human CCR5-specific antibody. A phage display approach for rabbit antibody humanization. *J Biol Chem* **275**: 36073–36078. doi:10.1074/jbc.M002765200

Steiner D, Forrer P, Stumpp MT, Pluckthun A. 2006. Signal sequences directing cotranslational translocation expand the range of proteins amenable to phage display. *Nat Biotechnol* **24**: 823–831. doi:10.1038/nbt1218

Tanaka F, Fuller R, Shim H, Lerner RA, Barbas CF III. 2004. Evolution of aldolase antibodies in vitro: correlation of catalytic activity and reaction-based selection. *J Mol Biol* **335**: 1007–1018. doi:10.1016/j.jmb.2003.11.014

Vieira J, Messing J. 1987. Production of single-stranded plasmid DNA. *Meth Enzymol* **153**: 3–11. doi:10.1016/0076-6879(87)53044-0

Williamson RA, Peretz D, Smorodinsky N, Bastidas R, Serban H, Mehlhorn I, DeArmond SJ, Prusiner SB, Burton DR. 1996. Circumventing tolerance to generate autologous monoclonal antibodies to the prion protein. *Proc Natl Acad Sci* **93**: 7279–7282. doi:10.1073/pnas.93.14.7279

Wilson HD, Li X, Peng H, Rader C. 2018. A Sortase A programmable phage display format for improved panning of Fab antibody libraries. *J Mol Biol* **430**: 4387–4400. doi:10.1016/j.jmb.2018.09.003

Yang WP, Green K, Pinz-Sweeney S, Briones AT, Burton DR, Barbas CF III. 1995. CDR walking mutagenesis for the affinity maturation of a potent human anti-HIV-1 antibody into the picomolar range. *J Mol Biol* **254**: 392–403. doi:10.1006/jmbi.1995.0626

Yang J, Baskar S, Kwong KY, Kennedy MG, Wiestner A, Rader C. 2011. Therapeutic potential and challenges of targeting receptor tyrosine kinase ROR1 with monoclonal antibodies in B-cell malignancies. *PLoS ONE* **6**: e21018. doi:10.1371/journal.pone.0021018

Yoo DK, Lee SR, Jung Y, Han H, Lee HK, Han J, Kim S, Chae J, Ryu T, Chung J. 2020. Machine learning–guided prediction of antigen-reactive in silico clonotypes based on changes in clonal abundance through biopanning. *Biomolecules* **10**: 421. doi:10.3390/biom10030421

Zhang MY, Shu Y, Phogat S, Xiao X, Cham F, Bouma P, Choudhary A, Feng YR, Sanz I, Rybak S, et al. 2003. Broadly cross-reactive HIV neutralizing human monoclonal antibody Fab selected by sequential antigen panning of a phage display library. *J Immunol Methods* **283**: 17–25. doi:10.1016/j.jim.2003.07.003

CHAPTER 5

# Generation and Selection of Phage Display Antibody Libraries in Fab Format

Christoph Rader[1]

*Department of Immunology and Microbiology, The Herbert Wertheim UF Scripps Institute for Biomedical Innovation & Technology, University of Florida, Jupiter, Florida 33458, USA*

Monoclonal antibodies (mAbs) have exceptional utility as research reagents and pharmaceuticals. As a complement to both traditional and contemporary single-B-cell cloning technologies, the mining of antibody libraries via display technologies—which mimic and simplify B cells by physically linking phenotype (protein) to genotype (protein-encoding DNA or RNA)—has become an important method for mAb discovery. Among these display technologies, phage display has been particularly successful for the generation of mAbs that bind to a wide variety of antigens with exceptional specificities and affinities. Rather than multivalent whole antibodies, phage display typically uses monovalent antibody fragments, such as "fragment antigen binding" (Fab), as the format of choice. The ~50-kDa Fab format consists of four immunoglobulin (Ig) domains on two polypeptide chains (light chain and shortened heavy chain), and exhibits its antigen binding site in a natural configuration found in bivalent IgG and other multivalent Ig molecules. The Fab fragment has a high melting temperature and a low tendency to aggregate, and can be readily converted to natural and nonnatural Ig formats without affecting antigen binding properties, which has made it a favored format for phage display for more than three decades. Here, I briefly summarize some of the approaches used for the generation and selection of phage display antibody libraries in Fab format, from human and nonhuman antibody repertoires.

## MONOCLONAL ANTIBODIES

Beyond their use as research reagents, monoclonal antibodies (mAbs) have found wide applications as diagnostic, preventative, and therapeutic biologics in human and veterinary medicine (Mullard 2021). The success of the antibody molecule as a pharmaceutical is, in one respect, grounded in its natural properties—specifically, its high affinity, specificity, and stability—along with its long circulatory half-life and its ability to bridge the adaptive and innate immune system through linking recognition and response. In another respect, its natural modularity has rendered it highly compatible with manipulation by protein engineering (Chiu et al. 2019).

The most common natural format of mAbs is the 150-kDa immunoglobulin G1 (IgG1) molecule (Fig. 1A), which consists of two identical 25-kDa light (L) chains and two identical 50-kDa heavy (H) chains (Saphire et al. 2001; Vidarsson et al. 2014). Each chain (light and heavy) contains a variable (V) domain ($V_L$ and $V_H$, respectively), followed by one constant domain in the light chain ($C_L$) and three constant domains in the heavy chain ($C_H1$, $C_H2$, and $C_H3$). These four polypeptide chains assemble in

[1]Correspondence: rader33458@gmail.com

© 2026 Cold Spring Harbor Laboratory Press
Cite this introduction as *Cold Spring Harb Protoc*; doi:10.1101/pdb.top107764

# Chapter 5

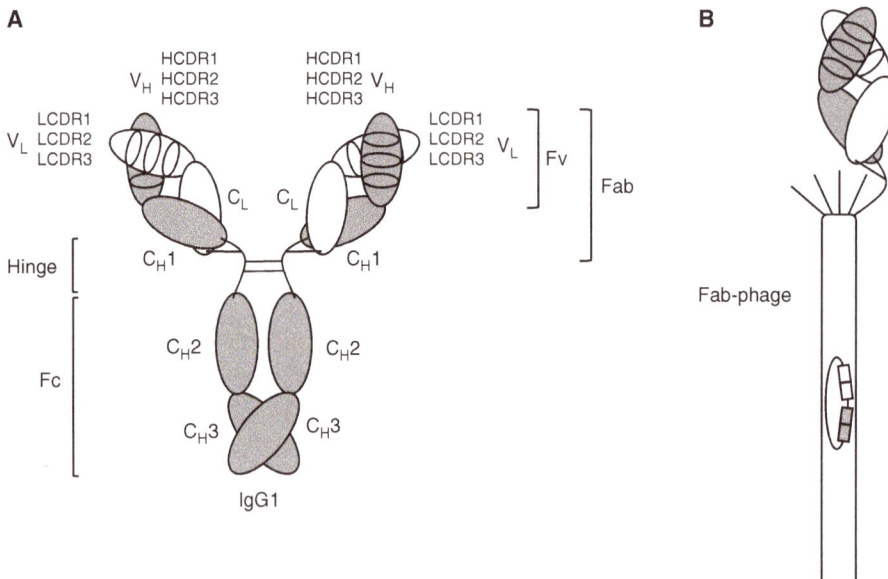

FIGURE 1. IgG1, Fab, and Fab-phage. (A) Depiction of an ~150-kDa IgG1 molecule with the two ~25-kDa light chains shown in white and the two ~50-kDa heavy chains shown in gray. Each of these chains (light and heavy) comprises an N-terminal variable domain ($V_L$ or $V_H$, respectively), each having three complementarity determining regions (CDRs), followed by one constant domain in the light chain ($C_L$) and three constant domains in the heavy chain ($C_H1$, $C_H2$, and $C_H3$). $V_L/V_H$ and $C_L/C_H1$ heterodimerize, and $C_H3/C_H3$ homodimerizes. The glycosylated $C_H2$ domain does not dimerize. The hinge region between $C_H1$ and $C_H2$ includes four interchain disulfide bridges: two between the heavy chains and two between heavy and light chains. The $V_L/V_H$ heterodimer is also known as "fragment variable" (Fv), and the combined $V_L/V_H$ and $C_L/C_H1$ heterodimer is known as the "fragment antigen binding" (Fab). The hinge region connects the two Fab arms of the IgG1 molecule to the "fragment crystallizable" (Fc). (B) Depiction of a Fab-displaying filamentous phage in phagemid-based phage display systems. Note the preserved C-terminal disulfide bridge between the light chain and heavy chain fragments. The Fab is fused to the N terminus of a C-terminal fragment of minor coat protein pIII and forms, together with four wild-type pIII copies, the spiky cap of the filamentous phage particle. The phagemid encapsulated by the filamentous phage particle encodes the displayed Fab, affording the physical linkage of genotype and phenotype.

a Y-shaped configuration that is stabilized by four disulfide bridges in the hinge region. The two arms of the Y are known as "fragment antigen binding" (Fab) fragments, and consist of a heterodimer of the light chain ($V_L$-$C_L$) and the N-terminal half ($V_H$-$C_H1$) of the heavy chain. The stem of the Y is known as the "fragment crystallizable" (Fc) fragment, and consists of a homodimer of the C-terminal half ($C_H2$-$C_H3$) of the heavy chains. The N-terminal portion of the Fab, the $V_L$ and $V_H$ heterodimer that is also known as the "fragment variable" (Fv) fragment, contains the antigen binding site, or paratope. The Fv is encoded by the $V_L$ and $J_L$ genes in the κ or λ light chain locus, and by the $V_H$, D, and $J_H$ genes in the heavy chain locus. In the human genome, the κ and λ light chain loci on chromosomes 2 and 22, respectively, comprise 44 functional $V_κ$, five functional $J_κ$, 37 functional $V_λ$, and seven functional $J_λ$ genes (Watson et al. 2015). The heavy chain locus on chromosome 14, on the other hand, comprises 44 functional $V_H$, 25 functional D, and six functional $J_H$ genes (Matsuda et al. 1998). During the initial stages of B-cell maturation in the bone marrow, $V_L$-$J_L$ and $V_H$-D-$J_H$ genes somatically rearrange to create functional light and heavy chain genes. These have hypervariable amino acid sequences, known as light and heavy chain complementarity determining region 3 (LCDR3 and HCDR3), encoded at the deliberately imprecise $V_L$-$J_L$ and $V_H$-D-$J_H$ fusion sites, respectively. Other regions with highly variable amino acid sequences are encoded by the $V_L$ (LCDR1 and LCDR2) and $V_H$ (HCDR1 and HCDR2) genes, and additional diversification in the variable domains is achieved by somatic hypermutation at later stages of B-cell maturation in the blood (Rajewsky 1996). The variable Ig domain comprises two antiparallel β sheets that interact through a hydrophobic core and are connected by a disulfide bridge. This configuration provides a highly stable scaffold that displays the CDRs as protruding β loops. In doing so, the variable Ig domain combines high amino acid sequence variability with high stability.

The six CDRs of the Fv fragment essentially define the paratope that binds an epitope on the antigen with typically high specificity and affinity. In addition, the presence of two Fab arms on each IgG affords an avidity gain through bivalent binding.

The antibody molecule is a unique protein that combines high variability in the primary structure with high stability of the secondary, tertiary, and quaternary structures. Taking advantage of the natural modular stability of the antibody molecule, antibody fragments can be used to engineer and evolve its affinity and specificity. The 50-kDa Fab fragment is particularly robust, with high solubility and dispersity, and a melting temperature ($T_m$) of ∼75°C (Garidel et al. 2020). It can be expressed in *Escherichia coli*, which is a requisite for phage display, is readily convertible to IgG1 and other natural and nonnatural Ig formats without affecting antigen binding properties, and is the modality of several therapeutic mAbs, including abciximab (Reopro) and ranibizumab (Lucentis) (Rader 2009).

## PHAGE DISPLAY ANTIBODY LIBRARIES

Like natural antibody repertoires, antibody libraries are collections of millions to billions of different antibodies that collectively cover a large antigen binding space (Lerner 2006). Advances in antibody discovery and engineering have resulted in the development of robust techniques and tools for the generation and selection of antibody libraries, in particular display technologies that physically link phenotype (protein) and genotype (protein-encoding DNA or RNA). As such, they mimic and simplify B cells, which feature a more complex physical and functional linkage of exposed antibody and concealed antibody-encoding DNA and RNA. This universal concept is common to both cell-based (phage display, virus display, bacterial display, yeast display, and mammalian cell display) (McCafferty and Schofield 2015) and cell-free (ribosome display and mRNA display) (Amstutz et al. 2001) display technologies.

For more than three decades, phage display has been extensively used for the mining of naive, immune, and synthetic antibody repertoires from humans and a variety of nonhuman species (Barbas et al. 1991; Clackson et al. 1991; Lerner 2006; Winter 2019). Indeed, numerous mAbs in research and medicine have been discovered de novo or evolved in vitro by phage display, including what has now become the most profitable drug in history, adalimumab (Humira), an anti-TNFα mAb for the treatment of inflammatory and autoimmune diseases (Rome and Kesselheim 2023).

The main ingredients of phage display (Smith 1985) are filamentous phage particles that are derived from filamentous phage that belong to the genus *Inovirus* and infect Gram-negative bacteria, including *E. coli*. Filamentous phage particles used for phage display are elongated protein tubes that (1) "display" a peptide or protein of interest on the outside, (2) harbor the DNA encoding the peptide or protein of interest on the inside, and (3) retain the ability to infect *E. coli*. Importantly, each particle is designed to only display and encode a single peptide or protein. Amino acid sequence diversification of the peptide or protein of interest is presented by billions to trillions of particles (with a typical titer of $10^{12}$/mL), which collectively constitute a library. Using the displayed peptide or protein, the library can be stringently selected for or against binding to a small or large molecule, a cell or tissue, or other natural or synthetic objects of interest. Because the DNA encoding the selected peptide or protein is selected concomitantly, and the linkage of phenotype and genotype is stable, selected pools of particles can be reamplified by infection of host bacterial cells and subjected to several rounds of selection. This process is also known as affinity selection or, in laboratory jargon, panning (Smith 2023).

The generation and selection of antibody libraries typically use phagemid-based phage display systems that require three components; namely, phagemid library, host bacterial cells, and helper phage. A detailed review of phagemid-based phage display at the conceptual, functional, and molecular level can be found in this collection (see Chapter 4, Overview: The pComb3 Phagemid Family of Phage Display Vectors [Rader 2024]). Briefly, phagemids are plasmids that contain the origin of replication and packaging signal of filamentous phage. After *E. coli* transformation, phagemids are replicated as double-stranded DNA (dsDNA), analogous to plasmids. When phagemid-transformed

*E. coli* are superinfected with helper phage, however, phagemids are replicated as single-stranded DNA (ssDNA), for packaging into filamentous phage particles. Helper phage, which are wild-type filamentous phage particles with a debilitated packaging signal, provide all filamentous phage proteins and ssDNA required to assemble infectious filamentous phage particles. Converting a phagemid library to a phage library thus requires the efficient transformation of host bacterial cells, typically done by electroporation, followed by their superinfection with helper phage. The reason that phagemid-based phage display is the method of choice for antibody libraries is the use of minor coat protein pIII as the physical linkage of antibody and filamentous phage particle. Because pIII—of which five copies are presented in the spiky cap of the filamentous phage particle—is required for *E. coli* infection, recombinant fusion of an antibody fragment (such as a Fab) interferes with the ability to propagate the phage library. The phagemid/helper phage partition solves this problem by using the phagemid to encode a recombinant fusion of Fab and the C-terminal fragment of pIII (ΔpIII), and the helper phage to encode wild-type pIII. Phagemid-transformed *E. coli* superinfected with helper phage produce infectious filamentous phage particles that display zero to one Fab-ΔpIII copies and four to five wild-type pIII copies and harbor the phagemid ssDNA encoding the Fab-ΔpIII recombinant fusion (Fig. 1B). The excess of wild-type pIII, however, comes at the price of an excess of "bald" phage, which have zero antibody-ΔpIII copies, but allows monovalent display of the Fab, enabling affinity-driven as opposed to avidity-driven selections. Monoclonality is another important feature of phagemid-based phage display. Only one phagemid clone is replicated and packaged in the host bacterial cell, and only one filamentous phage particle can infect it. This ensures that the filamentous phage particles pair cognate phenotype (a Fab) and genotype (a phagemid clone encoding that identical Fab). As such, a Fab library is presented as a large collection of filamentous phage particles that allow the effective propagation, selection, and identification of Fab.

Depending on the origin of the Fab library, it can cover a large antigen binding space that can be selected for particular properties, such as high affinity [high association rate constant ($k_{on}$), low dissociation rate constant ($k_{off}$), or both], high specificity, species cross-reactivity, pH dependency, and high stability. Filamentous phage particles are assembled in the periplasm of *E. coli*, the space between the inner and outer membranes, which provides an oxidizing environment needed for Ig domains to fold into disulfide bridge-stabilized β sandwiches.

As mentioned above, the use of phagemids as phage display vectors enabled the expansion of phage display from peptide to protein libraries. Phagemid pComb3 was among the first phage display vectors used for the generation and selection of antibody libraries in the 50-kDa Fab format (Barbas et al. 1991). Smaller antibody fragments, including the 25-kDa scFv format, which connects $V_L$ and $V_H$ through a polypeptide linker, and the 12.5-kDa single-variable domain format known as nanobody, are alternative phage display formats of antibody libraries. A review of the pComb3 phagemid family of phage display vectors, which includes pComb3 successors pComb3H, pComb3X, pC3C, and pC3Csort, is part of this collection (see Chapter 4, Overview: The pComb3 Phagemid Family of Phage Display Vectors [Rader 2024]). Collectively, pComb3 phagemids have been widely used for the de novo discovery of mAbs from natural (naive or immune) and synthetic antibody repertoires, as well as for the in vitro evolution of mAbs for affinity and specificity maturation and humanization.

The generation of an antibody library from an antibody repertoire encompasses the generation of a phagemid library that can be converted to a phage library suitable for various selection campaigns. A phagemid library and a phage library ideally fully capture and represent the amino acid sequence diversity in the antibody repertoire. In practice, however, this is difficult to achieve for natural antibody repertoires, as the original $V_L$ and $V_H$ pairs are cloned independently and recombined randomly (Beerli and Rader 2010). In antibody libraries that are derived from antibody repertoires of hyperimmunized animals, which are shaped by few $V_L$ and $V_H$ clonotypes, original or virtually original $V_L$ and $V_H$ pairs are restored and selectable. Naive antibody repertoires, on the other hand, can be diversified by random recombination of $V_L$ and $V_H$, creating new selectable specificities. Synthetic antibody repertoires are typically built on one or a few defined $V_L$ and $V_H$ scaffolds, with amino acid sequence diversification confined to the CDRs (Zhai et al. 2011; Nilvebrant and Sidhu 2018; Teixeira et al. 2021).

## GENERATION OF PHAGE DISPLAY ANTIBODY LIBRARIES IN FAB FORMAT

The first step in the generation of a phagemid library from a natural antibody repertoire is the preparation of total RNA from primary or secondary lymphoid organs (typically, bone marrow or spleen, respectively) or from peripheral blood mononuclear cells (PBMCs). The mRNA fraction is then reverse-transcribed to cDNA and subjected to PCR amplification using panels of sense and antisense primers that hybridize to the flanking sequences of $V_L$- and $V_H$-encoding cDNA. Suitable primer panels have been published for human antibodies, and also for antibodies from numerous other mammalian and avian species. To assemble Fab-encoding pComb3H and pComb3X cassettes (see Chapter 4, Overview: The pComb3 Phagemid Family of Phage Display Vectors [Rader 2024]), pools of $V_L$ and $V_H$ cDNAs are first fused to $C_L$ and $C_H 1$ cDNA, respectively, via an overlap extension PCR and then fused to each other in a second overlap extension PCR. Following digestion with restriction enzyme SfiI and ligation into SfiI-cut pComb3H or pComb3X, electroporation of *E. coli* creates the phagemid library, with the number of independent transformants defining the maximum number of Fab clones. The Fab cassette, which is flanked by asymmetric SfiI (GGCCN_NNN^NGGCC) sites [shown as SfiI (a) and SfiI (b) in Fig. 2] and is under the control of a *lacZ* promoter/operon, equips $V_L$-$C_L$ and $V_H$-$C_H 1$ with N-terminal signal peptides for periplasmic transport in the host bacterial cell and results in the fusion of the C terminus of $V_H$-$C_H 1$ with ΔpIII. The Fab-encoding pC3C and pC3Csort cassette is assembled by one three-fragment ($V_L$-$C_L$-$V_H$) rather than by two two-fragment ($V_L$-$C_L$ and $V_H$-$C_H 1$) overlap extension PCRs (Hofer et al. 2007; see also Chapter 4, Overview: The pComb3 Phagemid Family

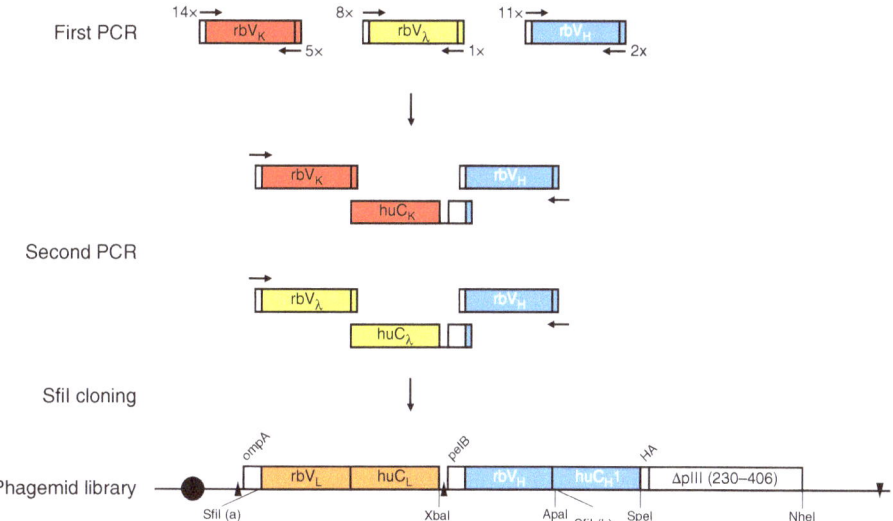

FIGURE 2. Assembly of chimeric rabbit/human "fragment antigen binding" (Fab) libraries in phagemid pC3C. Shown is a schematic of the workflow for the generation of a phagemid pC3C-based chimeric rabbit/human (rb/hu) Fab library, starting with a first PCR step in which rb$V_κ$-, rb$V_λ$-, and rb$V_H$-encoding cDNAs are amplified from reverse-transcribed mRNA of a naive or immune rabbit antibody repertoire. Diverse primer pairs (14 × 5 for rb$V_κ$, 8 × 1 for rb$V_λ$, and 11 × 2 for rb$V_H$) are used in 100 separate PCRs for each individual rabbit antibody repertoire. Subsequently, a second PCR step fuses the rb$V_κ$ and rb$V_H$ pools and the rb$V_λ$ and rb$V_H$ pools via a separately amplified invariable cDNA fragment that comprises hu$C_κ$ and hu$C_λ$, respectively; a downstream ribosome binding site; and pelB. The flanking primers in the second PCR step also introduce the asymmetric SfiI (a) (5'-GGCCC_AGG^CGGCC-3') and SfiI (b) (5'-GGCC_CCG^TCGGCC-3') restriction sites, which are cut and ligated into the correspondingly cut phagemid pC3C. Electroporation of the ligation mixture into *Escherichia coli* completes the phagemid library, which typically comprises $10^8$–$10^{10}$ independent transformants. Addition of helper phage to the host bacterial cells converts the phagemid-encoding Fab library to a phage-encoding and displaying Fab library. In pC3C, a lac promoter (black circle) drives bicistronic ompA-$V_L$-$C_L$ and pelB-$V_H$-$C_H 1$-HA-ΔpIII expression cassettes, with ompA and pelB serving as signal peptides, HA (hemagglutinin) as a detection tag, and ΔpIII, which is a C-terminal segment of the minor coat protein pIII of filamentous phage. Shine–Dalgarno sequences are indicated as black triangles, and a transcriptional terminator sequence is depicted as a reverse black triangle. Annotated features of pC3C are discussed in detail elsewhere (see Chapter 4, Overview: The pComb3 Phagemid Family of Phage Display Vectors [Rader 2024]).

of Phage Display Vectors [Rader 2024]). This is achieved by moving the SfiI (b) site upstream of $C_H1$. In this configuration, $C_H1$ is encoded by the vector rather than the insert. This design is compatible with human as well as chimeric nonhuman/human Fab formats.

Using bone marrow mononuclear cells (BMMCs) from healthy donors and PBMCs from patients, respectively, large naive and specialized immune human Fab-phage libraries based on pC3C have been generated and successfully selected (Kwong et al. 2008; Baskar et al. 2009). In addition, pC3C has been extensively used for the generation and selection of chimeric rabbit/human Fab-phage libraries to mine naive and immune rabbit antibody repertoires (Hofer et al. 2007; Peng et al. 2017; Weber et al. 2017). Thus far, three leads selected from pC3C-based chimeric rabbit/human Fab-phage libraries have been translated to clinical trials for cancer therapy (ClinicalTrials.gov IDs NCT02706392, NCT04441099, and NCT04877613). The workflow of building a chimeric rabbit/human Fab library in pC3C is shown in Figure 2. It is representative of the assembly of other Fab libraries that require different panels of sense and antisense primers for the PCR amplification of $V_\kappa$-, $V_\lambda$-, and $V_H$-encoding cDNAs but use a set of identical primers for the subsequent three-fragment overlap extension PCR.

Irrespective of their origin, the generation of antibody libraries for phagemid-based phage display requires diligence with respect to protocols and reagents. Five accompanying protocols (see Protocol 1: Generation of Antibody Libraries for Phage Display: Human Fab Format [Peng and Rader 2024a]; Protocol 2: Generation of Antibody Libraries for Phage Display: Chimeric Rabbit/Human Fab Format [Peng and Rader 2024b] (Fig. 2); Protocol 3: Generation of Antibody Libraries for Phage Display: Preparation of Electrocompetent *E. coli* [Peng and Rader 2024c]; Protocol 4: Generation of Antibody Libraries for Phage Display: Preparation of Helper Phage [Peng and Rader 2024d]; Protocol 5: Generation of Antibody Libraries for Phage Display: Library Reamplification [Peng and Rader 2024e]) describe several methods that are broadly applicable and adaptable to the generation of antibody libraries from human and nonhuman naive, immune, and synthetic antibody repertoires.

## SELECTION OF PHAGE DISPLAY ANTIBODY LIBRARIES IN FAB FORMAT

Once built, phage display antibody libraries can be subjected to selection using a variety of strategies tailored toward desired antigen binding properties (Ledsgaard et al. 2022). Antigens that are available as purified proteins are often immobilized on plastic surfaces, such as in wells of a 96-well ELISA plate (Fig. 3), followed by blocking with bovine serum albumin (BSA), incubation with the filamentous phage particles displaying the antibody library, and stringent washing. Remaining binders are eluted with trypsin or acidic buffer, added to host bacterial cells, and reamplified for another round of panning. As an alternative to immobilization on plastic surfaces, which can cause denaturation and epitope loss, antigens can be chemically or enzymatically conjugated to biotin and captured by streptavidin-coated magnetic beads. Filamentous phage particle binding to the beads via the antigen can then be isolated with a magnetic device. This method also allows selection of the antibody library using soluble rather than immobilized antigen if the biotinylated antigen and antibody library are mixed prior to adding streptavidin-coated magnetic beads. Gradually lowering the concentration of the soluble antigen over several rounds of panning favors the selection of high-affinity binders and is a common method in affinity maturation campaigns.

In cases where the antigen is not available as a purified protein but is targetable as an endogenously or ectopically expressed cell surface antigen, antibody libraries have been successfully selected on whole cells, typically mammalian cells (Alfaleh et al. 2017; Ledsgaard et al. 2022). Due to the presence of thousands of different cell surface antigens and the inherent stickiness of filamentous phage particles to mammalian cells, these panning campaigns are more challenging. However, they have been successfully applied to target-agnostic selections, enabling concerted antibody and antigen discovery. For example, we developed a modification of phagemid pC3C, named pC3Csort, which adds a C-terminal sortase A tag to the light chain of Fab (Wilson et al. 2018). This tag can be used for site-specifically introducing a biotin to filamentous phage particles that display Fab, while "bald"

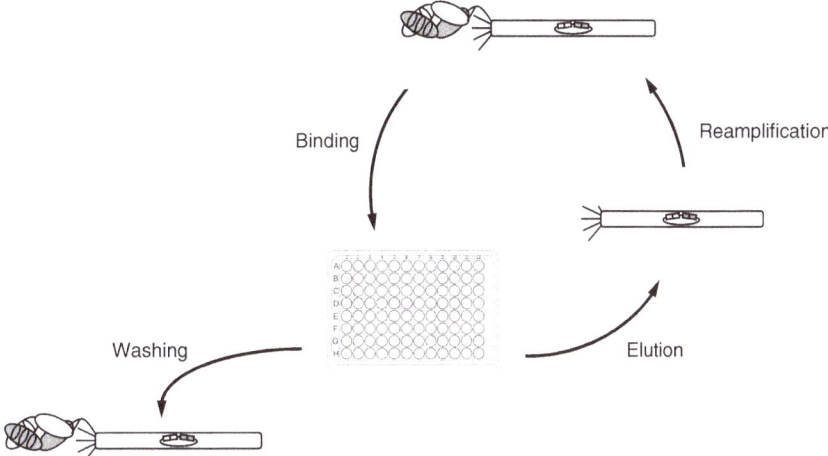

FIGURE 3. "Fragment antigen binding" (Fab)-phage library panning against immobilized antigen. Depicted is one round of panning that starts with adding a Fab-phage library to an antigen immobilized in one or more wells of a 96-well ELISA plate, for binding. Following stringent washing to remove nonspecific binders, specific binders are eluted using, for instance, trypsin (which removes the Fab phenotype but retains the Fab-encoding genotype) and then added to host bacterial cells for reamplification. Three to four rounds of panning are typically required to sufficiently enrich binders of high specificity and affinity from the Fab-phage library for identification by low-throughput screening. Screening methods using high-throughput sequencing only require one to two rounds of panning. Fab-phage libraries can also be selected against soluble antigen or on whole cells (see the text for details).

phage particles are not biotinylated. As such, specific binders can be efficiently separated from unspecific binders, facilitating target-defined and target-agnostic selections on whole cells (Wilson et al. 2018; Cyr et al. 2023).

One of the virtues of phage display is the ability to tailor the selection strategy toward unique binding properties that are so rare that they are difficult to find when mining antibody repertoires by screening rather than by selection campaigns. It allows identifying antibodies that can discriminate marginally varied amino acid sequences and conformations of highly homologous proteins by using one as bait and the other as decoy. Positive and negative selections can be done sequentially or simultaneously, and are also frequently applied to whole-cell panning (Alfaleh et al. 2017). When cross-reactivity to antigens (e.g., ortholog proteins from distinct species) is the objective, sequential positive selections can be applied, for instance, by alternating the bait over multiple rounds of panning. Elution of filamentous phage particles from immobilized or captured antigen can be tailored toward selecting, for instance, pH-, metal ion-, or small-molecule-dependent binding. In addition, selection campaigns can be designed to favor fast or slow association (high or low $k_{on}$ values) and fast or slow dissociation rate constants (high or low $k_{off}$ values) to diversify both thermodynamic and kinetic binding parameters of antibody–antigen interactions. Another selectable feature of broad utility is the stability of antibodies. In all, the ability of filamentous phage particles to tolerate substantial physical, chemical, and biological stress without losing integrity and infectivity has made highly stringent selection strategies possible.

To initially determine whether a selection campaign was successful, a phage ELISA designed to analyze the binding of phage pools collected after each round of panning is conducted. Subsequently, clones of interest are identified in a crude Fab ELISA and further analyzed by DNA fingerprinting and sequencing of the Fab-encoding phagemid cassette. Two accompanying protocols (see Protocol 6: Phage Display Selection of Antibody Libraries: Panning Procedures [Peng and Rader 2024f]; Protocol 7: Phage Display Selection of Antibody Libraries: Screening of Selected Binders [Peng and Rader 2024g]) provide step-by-step details for the selection of antibody libraries by phage display and their subsequent analysis.

Once Fabs of interest have been identified by this initial approach, they will need to be subjected to a set of rigorous analyses in order to determine their biophysical developability and biological suitability. These assays require pure Fabs of defined concentration. Their production is described in an accompanying protocol (see Protocol 8: Cloning, Expression, and Purification of Phage Display-Selected Fab for Biophysical and Biological Studies [Cyr et al. 2024]). Purified Fabs in milligram amounts enable advanced antigen binding analyses via, for instance, surface plasmon resonance, flow cytometry, X-ray crystallography, or cryogenic electron microscopy, which ultimately inform the success of phage display library generation and selection.

## CLOSING REMARKS

The mining of human and nonhuman antibody repertoires by phage display has delivered mAbs of broad utility in research and medicine. A key aspect is the conversion of diverse and diversified antibody repertoires to antibody libraries that are suitable for stringent selection strategies. This requires the phage display of a monovalent antibody fragment that robustly and reliably represents mAbs, such as Fab. Indeed, the Fab fragment has been a favored format for phage display antibody libraries since their inception in 1991. Several generations of the pComb3 phagemid family of phage display vectors have been tailored for generating and selecting antibody libraries in Fab format, as described in eight accompanying protocols.

## REFERENCES

Alfaleh MA, Jones ML, Howard CB, Mahler SM. 2017. Strategies for selecting membrane protein-specific antibodies using phage display with cell-based panning. *Antibodies* 6: 10. doi:10.3390/antib6030010

Amstutz P, Forrer P, Zahnd C, Plückthun A. 2001. In vitro display technologies: novel developments and applications. *Curr Opin Biotechnol* 12: 400–405. doi:10.1016/s0958-1669(00)00234-2

Barbas CF III, Kang AS, Lerner RA, Benkovic SJ. 1991. Assembly of combinatorial antibody libraries on phage surfaces: the gene III site. *Proc Natl Acad Sci* 88: 7978–7982. doi:10.1073/pnas.88.18.7978

Baskar S, Suschak JM, Samija I, Srinivasan R, Childs RW, Pavletic SZ, Bishop MR, Rader C. 2009. A human monoclonal antibody drug and target discovery platform for B-cell chronic lymphocytic leukemia based on allogeneic hematopoietic stem cell transplantation and phage display. *Blood* 114: 4494–4502. doi:10.1182/blood-2009-05-222786

Beerli RR, Rader C. 2010. Mining human antibody repertoires. *mAbs* 2: 365–378. doi:10.4161/mabs.12187

Chiu ML, Goulet DR, Teplyakov A, Gilliland GL. 2019. Antibody structure and function: the basis for engineering therapeutics. *Antibodies* 8: 55. doi:10.3390/antib8040055

Clackson T, Hoogenboom HR, Griffiths AD, Winter G. 1991. Making antibody fragments using phage display libraries. *Nature* 352: 624–628. doi:10.1038/352624a0

Cyr MG, Wilson HD, Spierling AL, Chang J, Peng H, Steinberger P, Rader C. 2023. Concerted antibody and antigen discovery by differential whole-cell phage display selections and multi-omic target deconvolution. *J Mol Biol* 435: 168085. doi:10.1016/j.jmb.2023.168085

Cyr MG, Peng H, Rader C. 2024. Cloning, expression, and purification of phage display-selected Fab for biophysical and biological studies. *Cold Spring Harb Protoc* doi:10.1101/pdb.prot108604

Garidel P, Eiperle A, Blech M, Seelig J. 2020. Thermal and chemical unfolding of a monoclonal IgG1 antibody: application of the multistate Zimm–Bragg theory. *Biophys J* 118: 1067–1075. doi:10.1016/j.bpj.2019.12.037

Hofer T, Tangkeangsirisin W, Kennedy MG, Mage RG, Raiker SJ, Venkatesh K, Lee H, Giger RJ, Rader C. 2007. Chimeric rabbit/human Fab and IgG specific for members of the Nogo-66 receptor family selected for species cross-reactivity with an improved phage display vector. *J Immunol Methods* 318: 75–87. doi:10.1016/j.jim.2006.10.007

Kwong KY, Baskar S, Zhang H, Mackall CL, Rader C. 2008. Generation, affinity maturation, and characterization of a human anti-human NKG2D monoclonal antibody with dual antagonistic and agonistic activity. *J Mol Biol* 384: 1143–1156. doi:10.1016/j.jmb.2008.09.008

Ledsgaard L, Ljungars A, Rimbault C, Sørensen CV, Tulika T, Wade J, Wouters Y, McCafferty J, Laustsen AH. 2022. Advances in antibody phage display technology. *Drug Discov Today* 27: 2151–2169. doi:10.1016/j.drudis.2022.05.002

Lerner RA. 2006. Manufacturing immunity to disease in a test tube: the magic bullet realized. *Angew Chem Int Ed Engl* 45: 8106–8125. doi:10.1002/anie.200603381

Matsuda F, Ishii K, Bourvagnet P, Kuma K, Hayashida T, Miyata T, Honjo T. 1998. The complete nucleotide sequence of the human immunoglobulin heavy chain variable region locus. *J Exp Med* 188: 2151–2162. doi:10.1084/jem.188.11.2151

McCafferty J, Schofield D. 2015. Identification of optimal protein binders through the use of large genetically encoded display libraries. *Curr Opin Chem Biol* 26: 16–24 doi:10.1016/j.cbpa.2015.01.003

Mullard A. 2021. FDA approves 100th monoclonal antibody product. *Nat Rev Drug Discov* 20: 491–495. doi:10.1038/d41573-021-00079-7

Nilvebrant J, Sidhu SS. 2018. Construction of synthetic antibody phage-display libraries. *Methods Mol Biol* 1701: 45–60. doi:10.1007/978-1-4939-7447-4_3

Peng H, Rader C. 2024a. Generation of antibody libraries for phage display: human Fab format. *Cold Spring Harb Protoc* doi:10.1101/pdb.prot108597

Peng H, Rader C. 2024b. Generation of antibody libraries for phage display: chimeric rabbit/human Fab format. *Cold Spring Harb Protoc* doi:10.1101/pdb.prot108598

Peng H, Rader C. 2024c. Generation of antibody libraries for phage display: preparation of electrocompetent *E. coli*. *Cold Spring Harb Protoc* doi:10.1101/pdb.prot108599

Peng H, Rader C. 2024d. Generation of antibody libraries for phage display: preparation of helper phage. *Cold Spring Harb Protoc* doi:10.1101/pdb.prot108600

Peng H, Rader C. 2024e. Generation of antibody libraries for phage display: library reamplification. *Cold Spring Harb Protoc* doi:10.1101/pdb.prot108601

Peng H, Rader C. 2024f. Phage display selection of antibody libraries: panning procedures. *Cold Spring Harb Protoc* doi:10.1101/pdb.prot108602

Peng H, Rader C. 2024g. Phage display selection of antibody libraries: screening of selected binders. *Cold Spring Harb Protoc* doi:10.1101/pdb.prot108603

Peng H, Nerreter T, Chang J, Qi J, Li X, Karunadharma P, Martinez GJ, Fallahi M, Soden J, Freeth J, et al. 2017. Mining naive rabbit antibody repertoires by phage display for monoclonal antibodies of therapeutic utility. *J Mol Biol* **429:** 2954–2973. doi:10.1016/j.jmb.2017.08.003

Rader C. 2009. Overview on concepts and applications of Fab antibody fragments. *Curr Protoc Protein Sci* **55:** 6.9.1–6.9.14. doi:10.1002/0471140864.ps0609s55

Rader C. 2024. The pComb3 phagemid family of phage display vectors. *Cold Spring Harb Protoc* doi:10.1101/pdb.over107756

Rajewsky K. 1996. Clonal selection and learning in the antibody system. *Nature* **381:** 751–758. doi:10.1038/381751a0

Rome BN, Kesselheim AS. 2023. Biosimilar competition for Humira is here: signs of hope despite early hiccups. *Arthritis Rheumatol* **75:** 1325–1327. doi:10.1002/art.42520

Saphire EO, Parren PWHI, Pantophlet R, Zwick MB, Morris GM, Rudd PM, Dwek RA, Stanfield RL, Burton DR, Wilson IA. 2001. Crystal structure of a neutralizing human IgG against HIV-1: a template for vaccine design. *Science* **293:** 1155–1159. doi:10.1126/science.1061692

Smith GP. 1985. Filamentous fusion phage: novel expression vectors that display cloned antigens on the virion surface. *Science* **228:** 1315–1317. doi:10.1126/science.4001944

Smith GP. 2023. Principles of affinity selection. *Cold Spring Harb Protoc* doi:10.1101/pdb.over107894

Teixeira AAR, Erasmus MF, D'Angelo S, Naranjo L, Ferrara F, Leal-Lopes C, Durrant O, Galmiche C, Morelli A, Scott-Tucker A, et al. 2021. Drug-like antibodies with high affinity, diversity and developability directly from next-generation antibody libraries. *mAbs* **13:** 1980942. doi:10.1080/19420862.2021.1980942

Vidarsson G, Dekkers G, Rispens T. 2014. IgG subclasses and allotypes: from structure to effector functions. *Front Immunol* **5:** 520. doi:10.3389/fimmu.2014.00520

Watson CT, Steinberg KM, Graves TA, Warren RL, Malig M, Schein J, Wilson RK, Holt RA, Eichler EE, Breden F. 2015. Sequencing of the human IG light chain loci from a hydatidiform mole BAC library reveals locus-specific signatures of genetic diversity. *Genes Immun* **16:** 24–34. doi:10.1038/gene.2014.56

Weber J, Peng H, Rader C. 2017. From rabbit antibody repertoires to rabbit monoclonal antibodies. *Exp Mol Med* **49:** e305. doi:10.1038/emm.2017.23

Wilson HD, Li X, Peng H, Rader C. 2018. A sortase A programmable phage display format for improved panning of Fab antibody libraries. *J Mol Biol* **430:** 4387–4400. doi:10.1016/j.jmb.2018.09.003

Winter G. 2019. Harnessing evolution to make medicines (Nobel lecture). *Angew Chem Int Ed Engl* **58:** 14438–14445. doi:10.1002/anie.201909343

Zhai W, Glanville J, Fuhrmann M, Mei L, Ni I, Sundar PD, Van Blarcom T, Abdiche Y, Lindquist K, Strohner R, et al. 2011. Synthetic antibodies designed on natural sequence landscapes. *J Mol Biol* **430:** 4387–4400. doi:10.1016/j.jmb.2011.07.018

# Protocol 1

# Generation of Antibody Libraries for Phage Display: Human Fab Format

Haiyong Peng and Christoph Rader[1]

*Department of Immunology and Microbiology, The Herbert Wertheim UF Scripps Institute for Biomedical Innovation & Technology, University of Florida, Jupiter, Florida 33458, USA*

Phage display is a powerful method for the de novo generation and affinity maturation of human monoclonal antibodies from naive, immune, and synthetic antibody repertoires. The pComb3 phagemid family of phage display vectors facilitates the selection of human monoclonal antibody libraries in the monovalent Fab format, which consists of human variable domains $V_H$ and $V_L$ ($V_\kappa$ or $V_\lambda$), fused to the human constant domains $C_H1$ of IgG1 and $C_L$ ($C_\kappa$ or $C_\lambda$), respectively. Here, we describe the use of a pComb3 derivative, phagemid pC3C, for the generation of human Fab libraries with randomly combined human variable domains ($V_H$, $V_\kappa$, and $V_\lambda$) of high sequence diversity, starting from the preparation of mononuclear cells from blood and bone marrow. Depending on the complexity of the parental antibody repertoire, the protocol can be scaled for yielding a library size of $10^8$–$10^{11}$ independent human Fab clones. As such, it can be used, for instance, for the generation of a large naive human Fab library from healthy individuals or for the generation of a specialized immune human Fab library from individuals with an endogenous antibody response of interest.

## MATERIALS

It is essential that you consult the appropriate Material Safety Data Sheets and your institution's Environmental Health and Safety Office for proper handling of equipment and hazardous materials used in this protocol.

RECIPES: Please see the end of this protocol for recipes indicated by <R>. Additional recipes can be found online at http://cshprotocols.cshlp.org/site/recipes.

### Reagents

ACK Lysing Buffer (Lonza 10-548E); store at room temperature
Agarose gel electrophoresis reagents:
  DNA gel loading dye, 6× (Thermo Fisher Scientific R0611)
  DNA ladders, 100-bp and 1-kb (Thermo Fisher Scientific SM0241 and SM0311).
  SYBR Safe DNA gel stain (Thermo Fisher Scientific S33102)
  TAE (Tris-acetate-EDTA) buffer (dilute from 50× TAE; Thermo Fisher Scientific B49)
    *For agarose-formaldehyde gel electrophoresis reagents, see Green and Sambrook (2022) (optional, see Step 15).*

UltraPure Agarose (Thermo Fisher Scientific 16500500)
BCP (1-bromo-3-chloro-propane; Molecular Research Center BP151); store at room temperature

---

[1]Correspondence: rader33458@gmail.com

© 2026 Cold Spring Harbor Laboratory Press
Cite this protocol as *Cold Spring Harb Protoc*; doi:10.1101/pdb.prot108597

Bovine serum albumin (BSA), 1% (w/v) in TBS
> To prepare, dissolve 0.5 g of BSA (MilliporeSigma A7030) in 50 mL of 1× TBS and sterilize by filtration through a 0.22-μm filter unit. Store at room temperature.

Carbenicillin (100 μg/μL)
> To prepare, dissolve 1 g of carbenicillin disodium (MilliporeSigma C1389) in 10 mL of highly pure water. Sterilize by filtration through a 0.22-μm filter. Store 1-mL aliquots in 1.5-mL microfuge tubes at −20°C.

Dry ice

Escherichia coli SS320 (nonamber suppressor strain; Biosearch Technologies 60512-2) or ER2738 (amber suppressor strain; Biosearch Technologies 60522-2) electrocompetent cells, with an efficiency of $\geq 4 \times 10^{10}$ or $\geq 2 \times 10^{10}$, respectively, of colony-forming units per microgram of pUC19 plasmid, and recovery medium (Biosearch Technologies 80026-1); store at −80°C
> As a cost-saving alternative to commercial sources, see Protocol 3: Generation of Antibody Libraries for Phage Display: Preparation of Electrocompetent E. coli (Peng and Rader 2024a), which describes the preparation of electrocompetent bacteria. SS320 is the result of mating male ($F^+$) XL1-Blue with female ($F^-$) MC1061 to achieve higher library transformation efficiency for electroporation (Sidhu et al. 2000). A preferred strain for library transformation that is not commercially available is SR320, a lytic phage-resistant variant of SS320 developed in Dr. Sachdev S. Sidhu's laboratory.

Ethanol (70% v/v, in water)
> Prepare fresh from pure ethanol (MilliporeSigma 459836-500ML) and RNase-free water (Thermo Fisher Scientific AM9937). Store at room temperature.

Glycerol, ultrapure (Thermo Fisher Scientific J64719.AP); store at room temperature

High-fidelity PCR reagents; store at −20°C:
  dNTP mix, 10 mM (2.5 mM of each dATP, dCTP, dGTP, and dTTP diluted in sterile water from 100 mM stock concentrations; Thermo Fisher Scientific 18427013)
  Platinum Taq DNA Polymerase, 5 units/μL, 10× PCR buffer ($-MgCl_2$), and 50 mM $MgCl_2$ (Thermo Fisher Scientific 15966005)

Human whole blood (50-mL, freshly drawn) or human bone marrow fluid and cells (20-mL, freshly aspirated) in 10-mL blood collection tubes with anticoagulant (e.g., BD Vacutainer Heparin Tubes [BD 367874] containing 158 USP units sodium heparin)
> See Step 1.
>
> For the generation of a specialized immune human Fab library, use blood and/or bone marrow from one or a few individuals with an endogenous antibody response of interest. For generation of a large naive human Fab library, use blood and/or bone marrow from six to 12 individuals with diverse backgrounds in terms of age, sex, and genetic ancestry.

Ice

Isopropanol (MilliporeSigma I9516); store at room temperature

Kanamycin (50 μg/μL)
> To prepare, dissolve 500 mg of kanamycin sulfate (MilliporeSigma 60615) in 10 mL of highly pure water. Sterilize by filtration through a 0.22-μm filter. Store 1-mL aliquots in 1.5-mL microfuge tubes at −20°C.

LB + 100 μg/mL carbenicillin plates <R>

LiCl, 7.5 M, RNase-free (Thermo Fisher Scientific AM9480); store at room temperature

Lymphocyte Separation Medium (Lonza 17-829E); store at room temperature

Oligonucleotides
> The names and DNA sequences of all primers used in this protocol are shown in Table 1. For each primer, prepare 20 μM working dilutions in sterile water.

PCR reagents:
  dNTP mix, 10 mM (2.5 mM of each dATP, dCTP, dGTP, and dTTP diluted in sterile water from 100 mM stock concentrations; Thermo Fisher Scientific 18427013); store at −20°C
  Taq DNA polymerase, 5 units/μL, 10× Taq buffer with $(NH_4)_2SO_4$, and 25 mM $MgCl_2$ (Thermo Fisher Scientific EP0405)

PEG-8000 (MilliporeSigma P2139); store at room temperature

TABLE 1. Primers for PCR amplification, PCR assembly, and DNA sequencing of huV$_L$ and huV$_H$[a,b]

**huV$_\kappa$ sense primers (12)**

| | |
|---|---|
| HUVκ-F1a | GCTACCGTGGCCCAGGCGGCCGACATCCAG**W**TGACCCAGTCTCC |
| HUVκ-F1b | GCTACCGTGGCCCAGGCGGCCGCCATCC**RGW**TGACCCAGTCTCC |
| HUVκ-F1c | GCTACCGTGGCCCAGGCGGCCGTCATCTGGATGACCCAGTCTCC |
| HUVκ-F2a | GCTACCGTGGCCCAGGCGGCCGATATTGTGATGACCCAGACTCC |
| HUVκ-F2b | GCTACCGTGGCCCAGGCGGCCGAT**R**TTGTGATGACTCAGTCTCC |
| HUVκ-F3a | GCTACCGTGGCCCAGGCGGCCGAAATTGTGTTGAC**R**CAGTCTCC |
| HUVκ-F3b | GCTACCGTGGCCCAGGCGGCCGAAATAGTGATGA**Y**GCAGTCTCC |
| HUVκ-F3c | GCTACCGTGGCCCAGGCGGCCGAAATTGTAATGACACAGTCTCC |
| HUVκ-F4 | GCTACCGTGGCCCAGGCGGCCGACATCGTGATGACCCAGTCTCC |
| HUVκ-F5 | GCTACCGTGGCCCAGGCGGCCGAAACGACACTCACGCAGTCTCC |
| HUVκ-F6a | GCTACCGTGGCCCAGGCGGCCGAAATTGTGCTGACTCAGTCTCC |
| HUVκ-F6b | GCTACCGTGGCCCAGGCGGCCGATGTTGTGATGACACAGTCTCC |

**huV$_\kappa$ antisense primer (1)**

| | |
|---|---|
| hucκ-r | GACAGATGGTGCAGCCACAGTTCG |

**huV$_\lambda$ sense primers (20)**

| | |
|---|---|
| HUVλ-F1a | GCTACCGTGGCCCAGGCGGCCCAGTCTGTGCTGACTCAGCCACC |
| HUVλ-F1b | GCTACCGTGGCCCAGGCGGCCCAGTCTGT**SS**TGACGCAGCCGCC |
| HUVλ-F1c | GCTACCGTGGCCCAGGCGGCCCAGTCTGTGTTGACGCAGCCGCC |
| HUVλ-F2a | GCTACCGTGGCCCAGGCGGCCCAGTCTGCCCTGACTCAGCCTCC |
| HUVλ-F2b | GCTACCGTGGCCCAGGCGGCCCAGTCTGCCCTGACTCAGCCTCG |
| HUVλ-F2c | GCTACCGTGGCCCAGGCGGCCCAGTCTGCCCTGACTCAGCCTGC |
| HUVλ-F3a | GCTACCGTGGCCCAGGCGGCCTCCTATG**W**GCTGACTCAGCCACC |
| HUVλ-F3b | GCTACCGTGGCCCAGGCGGCCTCCTATGAGCTGACTCAGCCACT |
| HUVλ-F3c | GCTACCGTGGCCCAGGCGGCCTCTTCTGAGCTGACTCAGGACCC |
| HUVλ-F3d | GCTACCGTGGCCCAGGCGGCCTCCTATGAGCTGACACAGC**YAY**C |
| HUVλ-F3e | GCTACCGTGGCCCAGGCGGCCTCCTATGAGCTGATGCAGCCAC |
| HUVλ-F4a | GCTACCGTGGCCCAGGCGGCCCTGCCTGTGCTGACTCAGCCCCCGT |
| HUVλ-F4b | GCTACCGTGGCCCAGGCGGCCCAGCCTGTGCTGACTCAATCATC |
| HUVλ-F4c | GCTACCGTGGCCCAGGCGGCCCAGCTTGTGCTGACTCAATCGCC |
| HUVλ-F5a9 | GCTACCGTGGCCCAGGCGGCCCAGCCTGTGCTGACTCAGCCA**Y**C |
| HUVλ-F5b | GCTACCGTGGCCCAGGCGGCCCAGGCTGTGCTGACTCAGCCG**K**C |
| HUVλ-F6 | GCTACCGTGGCCCAGGCGGCCAATTTTATGCTGACTCAGCCCCA |
| HUVλ-F7 | GCTACCGTGGCCCAGGCGGCCCAG**R**CTGTGGTGACTCAGGAGCC |
| HUVλ-F8 | GCTACCGTGGCCCAGGCGGCCCAGACTGTGGTGACCCAGGAGCC |
| HUVλ-F10 | GCTACCGTGGCCCAGGCGGCCCAGGCAGGGCTGACTCAGCCACC |

**huV$_\lambda$ antisense primers (3)**

| | |
|---|---|
| hujλ-r1 | GAGGGGGCAGCCTTGGGCTGACCTAGGACGGTGACCTTGGTCCCAG |
| hujλ-r23 | GAGGGGGCAGCCTTGGGCTGACCTAGGACGGTCAGCTTGGTCCCTC |
| hujλ-r7 | GAGGGGGCAGCCTTGGGCTGACCGAGGACGGTCAGCTGGGTGCCTC |

**huV$_H$ sense primers (19)**

| | |
|---|---|
| HUVH-F1a | GCTGCCCAACCAGCCATGGCCCAGGTGCAGCTGGTGCAGTCTGG |
| HUVH-F1b | GCTGCCCAACCAGCCATGGCCCAGGTYCAGCT**K**GTGCAGTCTGG |
| HUVH-F1c | GCTGCCCAACCAGCCATGGCCCAGGTCCAGCTGGTACAGTCTGG |
| HUVH-F1d | GCTGCCCAACCAGCCATGGCCCA**R**ATGCAGCTGGTGCAGTCTGG |
| HUVH-F1e | GCTGCCCAACCAGCCATGGCCCAGGT**S**CAGCTGGTGCA**R**TCTGG |
| HUVH-F2a | GCTGCCCAACCAGCCATGGCCCAG**R**TCACCTTGAAGGAGTCTGG |
| HUVH-F2b | GCTGCCCAACCAGCCATGGCCCAGGTCACCTTGAGGGAGTCTGG |
| HUVH-F3a | GCTGCCCAACCAGCCATGGCC**S**AGGTGCAGCTGGTGGAGTCTGG |
| HUVH-F3b | GCTGCCCAACCAGCCATGGCCGAGGTGCAGCTGTTGGAGTCTGG |
| HUVH-F3c | GCTGCCCAACCAGCCATGGCCGAGGTGCAGCTGGTGGAG**WCY**GG |
| HUVH-F3d | GCTGCCCAACCAGCCATGGCCGAAGTGCAGCTGGTGGAGTCTGG |
| HUVH-F3e | GCTGCCCAACCAGCCATGGCCCAGGTACAGCTGGTGGAGTCTGG |
| HUVH-F4a | GCTGCCCAACCAGCCATGGCCCAG**S**TGCAGCTGCAGGAGTCGGG |
| HUVH-F4b | GCTGCCCAACCAGCCATGGCCCAGGTGCAGCTACAGCAGTGGGG |
| HUVH-F4c | GCTGCCCAACCAGCCATGGCCCAGCTGCAGCTGCAGGAGTCCGG |
| HUVH-F4d | GCTGCCCAACCAGCCATGGCCCAGGTGCAGCTACAACAGTGGGG |
| HUVH-F5 | GCTGCCCAACCAGCCATGGCCGA**R**GTGCAGCTGGTGCAGTCTGG |
| HUVH-F6 | GCTGCCCAACCAGCCATGGCCCAGGTACAGCTGCAGCAGTCAGG |
| HUVH-F7 | GCTGCCCAACCAGCCATGGCCCAGGTGCAGCTGGTGCAATCTGG |

(continued)

TABLE 1. Continued

| | |
|---|---|
| **huV$_H$ antisense primers (6)** | |
| hujh-r1 | CGATGGGCCCTTGGTGGAGGCTGAGGAGACGGTGACCAGGGTGCCCTG |
| hujh-r2 | CGATGGGCCCTTGGTGGAGGCTGAGGAGACAGTGACCAGGGTGCCACG |
| hujh-r3 | CGATGGGCCCTTGGTGGAGGCTGAAGAGACGGTGACCATTGTCCCTTG |
| hujh-r45 | CGATGGGCCCTTGGTGGAGGCTGAGGAGACGGTGACCAGGGTYCCYTG |
| hujh-r6a | CGATGGGCCCTTGGTGGAGGCTGAGGAGACGGTGACCGTGGTCCCTTG |
| hujh-r6b | CGATGGGCCCTTGGTGGAGGCTGAGGAGACGGTGACCGTGGTCCCTTT |
| **Cκ-pelB and Cλ-pelB primers** | |
| HCK | CGAACTGTGGCTGCACCATCTGTC |
| HCL | GGTCAGCCCAAGGCTGCCCCCTC |
| pelb | GGCCATGGCTGGTTGGGCAGC |
| **Overlap extension PCR primers** | |
| 5′SFIHUVL | CGCTACCGTGGCCCAGGCGGCC |
| 3′sfivh | GAGGAGGAGGGCCGACGGGGCCAAGGGGAAGACCGATGGGCCCTTGGTGGAGGCTGA |
| **Sequencing primers** | |
| VLSEO | GATAACAATTGAATTCAGGAG |
| vhseq | TGAGTTCCACGACACCGT |

[a]Nucleotide ambiguity code: R = A or G, Y = C or T, S = C or G, and W = A or T.
[b]Sequences are shown 5′–3′.

Phagemid pC3C (Hofer et al. 2007), available from Kerafast (ENH067-FP); store at −20°C

Phosphate-buffered saline (PBS; pH 7.4, 10×; Thermo Fisher Scientific J62036.K2); store at room temperature

Plasmid pCκ (Hofer et al. 2007); store at −20°C

Plasmid pCλ (Kwong et al. 2008); store at −20°C

QIAGEN MinElute Gel Extraction Kit (QIAGEN 28606); store at room temperature

QIAGEN RNeasy MinElute Cleanup Kit (QIAGEN 74204); store at room temperature

QIAprep Spin Miniprep Kit (QIAGEN 27106); store at room temperature

Recovery Cell Culture Freezing Medium (Thermo Fisher Scientific 12648010); store at −20°C

RNA Storage Solution (1 mM sodium citrate, pH 6.5; Thermo Fisher Scientific AM7000); store at −20°C

SB (Super Broth) medium <R>

SfiI (20 units/μL; New England Biolabs R0123L) with 10× rCutSmart buffer (New England Biolabs B6004S); store at −20°C

*Formerly used reagent, SfiI (40 units/μL; Roche 11288059001), was discontinued.*

Sodium acetate, 3 M, pH 5.2, RNase-free (Thermo Fisher Scientific AM9740); store at room temperature

Sodium azide (NaN$_3$), 2% (w/v) in water

*To prepare, dissolve 2 g of NaN$_3$ (MilliporeSigma S2002) in 100 mL of highly pure water. Store at room temperature.*

Sodium chloride (NaCl; MilliporeSigma S9888); store at room temperature

SuperScript III First-Strand Synthesis System for RT-PCR (Thermo Fisher Scientific 18080051); store at −20°C [contains 50 μM oligo(dT), 10 mM dNTP, RNase-free water, 10× reverse transcriptase (RT) buffer, 25 mM MgCl$_2$, 100 mM DTT, 40 units/μL RNase OUT, 200 units/μL SuperScript III RT, and 2 units/μL *E. coli* RNase H]

T4 DNA Ligase (5 units/μL; Thermo Fisher Scientific EL0011; includes 10× T4 DNA Ligase buffer); store at −20°C.

*Formerly used reagent T4 DNA Ligase (5 units/μL; Roche 10909246103) was discontinued.*

Tetracycline (5 μg/μL)

Chapter 5

> *To prepare, dissolve 50 mg of tetracycline hydrochloride (MilliporeSigma T7660) in 10 mL of 75% (v/v) ethanol in water. Store 1-mL aliquots in 1.5-mL microfuge tubes at −20°C.*

TRI reagent (Molecular Research Center TR118); store at 4°C

Tris-buffered saline (TBS, 1×)

> *Prepare by diluting 10× TBS (Thermo Fisher Scientific J60764.K7) in highly pure water and sterilizing by filtration through a 0.22-µm filter unit. Store at room temperature.*

Trypan Blue 0.4% (Lonza 17-942E); store at room temperature

VCSM13 helper phage at $10^{11}$–$10^{12}$ plaque-forming units (pfu)/mL

> *Starting from a commercially available VCSM13 Interference-Resistance Helper Phage stock (∼1 × $10^{11}$ pfu/mL [Agilent 200251 or Thermo Fisher Scientific 50125059]; store at −80°C), prepare 500 mL of VCSM13 helper phage at $10^{11}$–$10^{12}$ pfu/mL, as described in Protocol 4: Generation of Antibody Libraries for Phage Display: Preparation of Helper Phage (Peng and Rader 2024b).*

Water, highly pure, sterile

> *The protocol uses highly pure water from a purification system (see "Equipment"). Use freshly. Several steps require sterile water, which is obtained by filtering highly pure water through a 0.22-µm filter unit. Store at room temperature.*

## Equipment

*A laboratory with standard molecular biology and microbiology equipment is required for this protocol.*

Agarose gel electrophoresis equipment (e.g., Owl horizontal electrophoresis systems [Thermo Fisher Scientific] with various tanks, casters, combs, and power supplies [EC-105] for wide gel [D-series] and mini gel [EasyCast B-series] analytical and preparative electrophoresis and blue-light transilluminator [e.g., VB-40 Visi-Blue Transilluminator VWR 95041-594])

Autoclave

Bunsen burner

Cell counter (e.g., ViCELL BLU Cell Viability Analyzer, Beckman Coulter C19196)

Centrifuge bottles (500-mL; polypropylene, wide mouth, sealing cap; certified for >12,000g and resistant to bleaching and autoclaving; e.g., Nalgene 3141-0500)

Centrifuge tubes (15- and 50-mL)

Conical tubes (50 mL)

Cryoboxes

Cryovials (2-mL)

Digital balance with 0.01-g readability

Electroporator (e.g., BTX ECM 630 Electroporation System, BTX 45-2051) and compatible electroporation cuvettes with 1-mm electrode gap (e.g., Cuvette Plus, BTX 45-0124)

Filtered pipette tips (10-, 20-, 100-, and 200-µL, and 1-mL)

Filters with Luer lock fitting (0.22-µm; e.g., Millex-GP Syringe Filters, Millipore SLGP033RS)

Filter units (0.22-µm; e.g., Millipore Stericup and Steriflip filter units)

Freezers (−20°C and −80°C)

Freezing container with isopropanol

Glass bottles (500-mL and 1-L, autoclaved)

Glass Erlenmeyer flasks (250-mL, and 1- and 2-L, autoclaved)

Heat blocks for 50°C, 65°C, and 85°C incubation

Hemocytometer or electronic counting device

Ice bucket

Incubator (37°C)

Laminar flow hood

Liquid nitrogen tank with cryobox storage racks

Magnetic hot plate stirrer

Magnetic stir bars

Microcentrifuge (e.g., Eppendorf 5425)

Microfuge tubes (1.5- and 2-mL)
Microwave oven
Paper towels
PCR tube strips (0.2-mL; e.g., Eppendorf 951010022)
Petri dishes (10-cm)
Pipettes (5-, 10-, and 25-mL)
Pipette controller
Refrigerated benchtop centrifuge with swinging bucket rotor (e.g., Beckman Coulter Avanti J-15R with JS-4.750 swinging bucket rotor)
Refrigerated microcentrifuge (e.g., Eppendorf 5425R)
Refrigerated floor centrifuge (e.g., Thermo Fisher Scientific Sorvall Lynx 4000) with fixed-angle rotor for six 500-mL centrifuge bottles (e.g., Thermo Fisher Scientific Fiberlite F12-6 × 500 LEX)
Razor blades
Refrigerator (4°C)
Round-bottom tubes with snap cap (14-mL)
Shaker at 37°C (e.g., Eppendorf New Brunswick Innova 40 Benchtop Incubator Shaker).

*Two separate shakers, one for phage-free conditions and one for phage, are required.*

Single-channel and multichannel micropipettes (1- to 1000-μL)
Syringes with Luer lock fitting (1-, 5-, and 10-mL)
Thermal cyclers (96-well; e.g., Eppendorf Mastercycler X50s 96-Well Silver Block Thermal Cycler 6311000010)
UV photometer (e.g., NanoDrop One$^C$ Microvolume UV-Vis Spectrophotometer, Thermo Fisher Scientific ND-ONEC-W)
Vortex mixer
Water baths (16°C and 37°C)
Water purification system (e.g., Hydro PicoPure UV Plus)

## METHOD

### Preparation of PBMCs from Human Blood or BMMCs from Human Bone Marrow

*The collection and processing of clinical specimens from human subjects must be approved by an Institutional Review Board (IRB).*

*Clinical specimens should be treated as potentially infectious, and appropriate safety precautions should be taken. Wear appropriate protection and handle with care.*

*For the generation of a specialized immune human Fab library, blood and/or bone marrow from one individual may be sufficient, whereas the generation of a large naive human Fab library is typically based on blood and/or bone marrow from six to 12 individuals with diverse backgrounds in terms of age, sex, and genetic ancestry. For each individual, carry out the following steps separately.*

1. To prepare peripheral blood mononuclear cells (PBMCs), start with 50 mL of freshly drawn human whole blood in five separate 10-mL blood collection tubes with anticoagulant; e.g., heparin (vacutainer tubes with green top). To prepare bone marrow mononuclear cells (BMMCs), start with 20 mL of freshly aspirated human bone marrow fluid and cells in two separate 10-mL blood collection tubes with anticoagulant.

   *Perform Steps 2–8 under sterile conditions in a laminar flow hood.*

2. For PBMCs, dilute 25 mL of blood with 25 mL of PBS in each of two separate 50-mL conical tubes. For BMMCs, dilute 20 mL of aspirate with 30 mL of PBS in a 50-mL conical tube.

3. For each mix, slowly layer 25 mL of the diluted blood/diluted aspirate onto 14 mL of Lymphocyte Separation Medium in separate 50-mL conical tubes. Then, centrifuge (no brake) at 800g for 20 min at room temperature to separate plasma/fluid in the upper phase, mononuclear cells

(lymphocytes and monocytes) in the interphase, polymorphonuclear cells (granulocytes) in the lower phase, and red cells (erythrocytes) at the bottom.

4. For each tube, carefully remove the upper phase without disturbing the interphase. As plasma with an endogenous antibody response of interest may be used for IgG purification, store it at −80°C if needed. Using a 5-mL pipette, carefully transfer the interphase to a separate clean 50-mL conical tube containing 25 mL of PBS. Bring volume to 50 mL with PBS and centrifuge at 300g for 10 min at 4°C.

    *As an optional step, remove remaining erythrocytes as follows: Resuspend the cell pellet in 2 mL of ACK Lysing Buffer, incubate for 2 min at room temperature, add PBS to 50 mL, and centrifuge at 300g for 10 min at 4°C.*

5. Remove and discard the supernatant. Resuspend and combine the cell pellets in a final volume of 50 mL of PBS.

6. Determine the number of viable cells by Trypan Blue staining using a cell counter.

    *Expect a yield of around $1 \times 10^6$–$2 \times 10^6$ PBMCs per milliliter from whole blood and $3 \times 10^6$–$4 \times 10^6$ BMMCs per milliliter of bone marrow aspirate, with a viability of >95%.*

7. For proceeding with fresh mononuclear cells, centrifuge a volume corresponding to $2.5 \times 10^7$ mononuclear cells at 300g for 10 min at 4°C. Proceed to Step 9 for total RNA preparation. For cell cryopreservation for later use, see Step 8.

8. For cryopreservation of PBMCs and BMMCs, centrifuge at 300g for 10 min at 4°C and resuspend the cell pellet in cold Recovery Cell Culture Freezing Medium to a concentration of $1 \times 10^7$ cells/mL. Transfer each 1 mL of the cell preparation to a 2-mL cryovial. After securely tightening the caps, immediately place the cryovials in a freezing container with isopropanol. Store the container overnight at −80°C before transferring the cryovials to a cryobox in a liquid nitrogen tank.

    *Protect hands and eyes when handling liquid nitrogen.*

    *Cryopreserved PBMCs and BMMCs may be stored for years in liquid nitrogen.*

## Preparation of Total RNA from PBMCs or BMMCs

*This protocol is written for $2.5 \times 10^7$ PBMCs or BMMCs and can be scaled up or down.*

*Keep all reagents RNase-free by changing gloves frequently and working with clean equipment in dust-free conditions.*

*TRI reagent contains a poison (phenol) and an irritant (guanidine thiocyanate). Wear appropriate protection and handle with care.*

9. Start with $2.5 \times 10^7$ mononuclear cells. Freshly prepared PBMCs or BMMCs (from Step 7) are preferred. For cryopreserved mononuclear cells (from Step 8), partially thaw five 2-mL cryovial freezing tubes, each containing $1 \times 10^7$ cells in 1 mL, in a 37°C water bath. Add 1 mL of PBS just before the cells are completely thawed. Determine the number of viable cells by Trypan Blue staining using a hemocytometer or an electronic counting device. Transfer $2.5 \times 10^7$ viable cells to a 15-mL centrifuge tube, add PBS to 15 mL, and centrifuge at 1500g for 10 min at 4°C.

10. Remove the supernatant, add 2.5 mL of TRI reagent (1 mL per $1 \times 10^7$ cells), resuspend the cell pellet by repetitive pipetting, and incubate for 5 min at room temperature.

11. Transfer 1.25 mL of the mix to each of two RNase-free 1.5-mL microfuge tubes, add 125 μL (0.1 volume) of BCP to each tube, vortex for 15 sec, incubate for 10 min at room temperature, and centrifuge at 12,000g for 15 min at 4°C.

12. For each of the two samples, transfer the upper colorless aqueous phase to a clean RNase-free microfuge tube (you can discard the lower red organic phase), and then add 625 μL of isopropanol, vortex for 15 sec, incubate for 10 min at room temperature, and centrifuge at 12,000g for 10 min at 4°C.

13. For each of the two samples, carefully decant and discard the supernatant without disturbing the white pellet. Then, add 1.25 mL of 70% (v/v) ethanol and centrifuge at 12,000g for 10 min at 4°C.

14. Carefully decant and discard the supernatant without disturbing the white pellet, and then air-dry in dust-free conditions for 10 min at room temperature. Then, dissolve each of the samples in 50 µL of RNA Storage Solution and then pool the two samples in a single clean RNase-free 1.5-mL microfuge tube.

    *As an optional step, further purify the RNA with the QIAGEN RNeasy MinElute Cleanup Kit following the manufacturer's protocol, or by precipitation with RNase-free 7.5 M LiCl, as described by Green and Sambrook (2020).*

15. Immediately remove a 2-µL aliquot and place the remaining sample on dry ice. Add 498 µL of RNase-free water to the 2-µL aliquot and measure the absorbance at 260 and 280 nm in a UV photometer.

    *Use the absorbance at 260 nm to calculate the RNA concentration. At 1-cm pathlength, an absorbance of 1 corresponds to 40 µg/mL RNA. The ratio of the absorbance at 260 and 280 nm is typically in the range of 1.6–1.9.*

    *The yield of total RNA from $2.5 \times 10^7$ mononuclear cells is expected to be ~100 µg.*

    *As an optional control step, RNA can be separated by electrophoresis under denaturing conditions in an agarose–formaldehyde gel (Green and Sambrook 2022). Two sharp bands of ~4.7 and ~1.9 kb corresponding to 28S and 18S ribosomal RNA should be visible in addition to an ~0.1- to 0.15-kb smear corresponding to 5S ribosomal RNA and transfer RNA (Green and Sambrook 2022).*

    *A possible reason for low RNA quantity or quality is RNase contamination, and this should be addressed by starting over with fresh reagents and clean disposables and equipment.*

16. Proceed to reverse transcription. If needed, store total RNA in RNA Storage Solution for several weeks at −80°C until ready to proceed. For long-term storage (months to years), add 0.1 volumes of RNase-free 3 M sodium acetate (pH 5.2) and 2.2 volumes of ethanol, vortex, and store at −80°C.

## Reverse Transcription of mRNA

*Process total RNA samples derived from different human specimens and subjects in parallel, and each in independent duplicates.*

17. In an RNase-free 1.5-mL microfuge tube, combine 30 µg of total RNA, 12 µL of 50 µM oligo(dT), and 12 µL of 10 mM dNTP, and bring the volume to 120 µL with RNase-free water. Incubate for 5 min at 65°C and chill for at least 1 min on ice. In a separate RNase-free 1.5-mL microfuge tube, prepare the reverse transcription (RT) reaction mixture by combining 24 µL of 10× RT buffer, 48 µL of 25 mM $MgCl_2$, 24 µL of 100 mM DTT, 12 µL of 40 units/µL RNase OUT, and 12 µL of 200 units/µL SuperScript III RT.

18. Add the prepared RT reaction mixture (~120 µL) to the prepared RNA/oligo(dT)/dNTP mixture (~120 µL), mix gently, and incubate for 50 min at 50°C. Terminate the reaction for 5 min at 85°C. Chill for at least 1 min on ice.

19. Collect the mix by brief centrifugation. Add 12 µL of 2 units/µL *E. coli* RNase H to each tube and incubate for 20 min at 37°C.

20. Collect the mix by brief centrifugation and store at −20°C until ready to proceed to PCR amplification.

    *The first-strand cDNA is stable for weeks at −20°C. For long-term storage (months to years), add 0.1 volumes of 3 M sodium acetate (pH 5.2) and 2.2 volumes of ethanol, vortex, and store at −80°C.*

## PCR Amplification of Human $V_\kappa$, $V_\lambda$, and $V_H$ cDNA, and Human $C_\kappa$ and $C_\lambda$ DNA

*Each sample will be subjected to 186 unique primer combinations in separate reactions (Fig. 1A). Take precautions to avoid the amplification of $huV_\kappa$, $huV_\lambda$, and $huV_H$ cDNA from contaminating sources in the laboratory, such as phagemids and phage, and be sure to include negative controls without first-strand cDNA. Consider using a PCR workstation.*

*If processing first-strand cDNA samples derived from different human specimens and subjects, do it in parallel in 0.2-mL PCR tubes.*

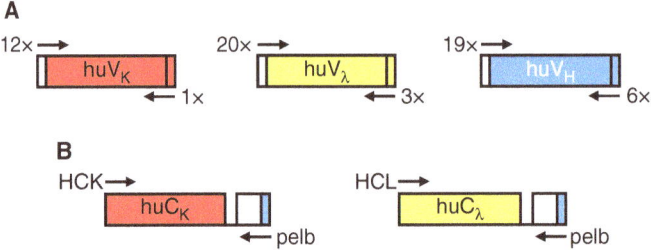

FIGURE 1. First PCR. (A) PCR amplification of cDNA encoding human variable domains. To amplify human variable domain cDNA by PCR from reverse-transcribed mRNA, 12, 60, and 114 unique primer combinations (Table 1) are used for huV$_\kappa$, huV$_\lambda$, and huV$_H$, respectively. (B) PCR amplification of human light chain constant domains. To amplify human constant domain DNA huC$_\kappa$ and huC$_\lambda$ by PCR from plasmids, sense primers HCK and HCL are separately combined with antisense primer pelb.

21. For human V$_\kappa$ (huV$_\kappa$) amplification, generate 12 primer combinations in separate 0.2-mL PCR tubes for each sample by combining 2 µL of the first-strand cDNA with 3 µL of 20 µM of one of the 12 huV$_\kappa$ sense primers and 3 µL of 20 µM of the huV$_\kappa$ antisense primer (Table 1).

22. For human V$_\lambda$ (huV$_\lambda$) amplification, generate 60 primer combinations in separate 0.2-mL PCR tubes for each sample by combining 2 µL of the first-strand cDNA with 3 µL of 20 µM of one of the 20 huV$_\lambda$ sense primers and 3 µL of 20 µM of one of the three huV$_\lambda$ antisense primers (Table 1).

23. For human V$_H$ (huV$_H$) amplification, generate 114 primer combinations in separate 0.2-mL PCR tubes for each sample by combining 2 µL of the first-strand cDNA with 3 µL of 20 µM of one of the 19 huV$_H$ sense primers and 3 µL of 20 µM of one of the six huV$_H$ antisense primers (Table 1).

24. Prepare a master mix of the PCR mixture sufficient for 186 reactions. In a 14-mL round-bottom tube, combine 2 mL of 10× Taq buffer with $(NH_4)_2SO_4$, 2 mL of 25 mM $MgCl_2$, 0.8 mL of 10 mM dNTP mix, 13.4 mL of sterile water, and 200 µL of 5 units/µL Taq DNA polymerase ($\sum$ 18.4 mL). Consider also the relevant controls.

    *Taq DNA polymerase has an error rate range from $10^{-a}$ to $10^{-t}$ depending on the PCR conditions. If higher fidelity is desired in the first PCR, use other thermostable DNA polymerases such as Platinum Taq DNA Polymerase, which is used for the second PCR (Steps 39–43).*

25. Add 92 µL of the prepared PCR mixture to each of the prepared first-strand cDNA/sense primer/antisense primer samples ($\sum$ 100 µL).

26. In a 96-well thermocycler, use these PCR parameters:

    2 min at 95°C;

    followed by 35 cycles of 30 sec at 95°C, 30 sec at 50°C, and 90 sec at 72°C;

    followed by 10 min at 72°C;

    followed by holding at room temperature.

27. Remove a 10-µL aliquot from each sample, add 2 µL of 6× DNA gel loading dye, and separate by electrophoresis on a 1% (w/v) agarose gel in TAE buffer using a 100-bp DNA ladder as reference.

    *The amplified huV$_\kappa$, huV$_\lambda$, and huV$_H$ cDNA should be visible as a bright band of ~400 bp. For both DNA analysis and preparation by agarose gel electrophoresis, use SYBR Safe stain and blue-light illumination rather than the hazardous combination of ethidium bromide and UV illumination.*

    *Not all 186 primer combinations are expected to work for each sample representing different human antibody repertoires. However, if the majority of ~400-bp bands are not or only weakly visible, a possible reason is impurity of the prepared total RNA. This can be addressed by adding the optional purification described in Step 14.*

28. Pool the remaining 90 µL of all amplified huV$_\kappa$ cDNA that originated from the same first-strand cDNA sample (12 × 90 µL = 1080 µL) and divide the pool into three 360-µL aliquots in separate

1.5-mL microfuge tubes. Then, add 0.1 volumes of 3 M sodium acetate (pH 5.2) and 2.2 volumes of ethanol to each, vortex, and store overnight at −20°C.

29. Pool the remaining 90 µL of all amplified huV$_\lambda$ cDNA that originated from the same first-strand cDNA sample (60 × 90 µL = 5400 µL) and divide the pool into fifteen 360-µL aliquots in separate 1.5-mL microfuge tubes. Then, add 0.1 volumes of 3 M sodium acetate (pH 5.2) and 2.2 volumes of ethanol to each, vortex, and store overnight at −20°C.

30. Pool the remaining 90 µL of all amplified huV$_H$ cDNA that originated from the same first-strand cDNA sample (114 × 90 µL = 10,260 µL) and divide the pool into twenty-eight 360-µL aliquots in 1.5-mL microfuge tubes. Then, add 0.1 volumes of 3 M sodium acetate (pH 5.2) and 2.2 volumes of ethanol to each, vortex, and store overnight at −20°C.

31. Proceed with two aliquots of the huV$_H$ cDNA, as they provide sufficient material for the following steps. Precipitate the huV$_H$ cDNA by centrifugation at 16,000g for 15 min at 4°C, decant and discard the supernatant, rinse the pellet with 1 mL of 70% (v/v) ethanol (room temperature), and air-dry for 10 min at room temperature. Dissolve each pellet in 100 µL of sterile water, pool the samples in a single tube, add 40 µL of 6× DNA gel loading dye, and separate by electrophoresis on a 1% (w/v) agarose gel in TAE buffer using a preparative comb and a 100-bp DNA ladder as reference.

32. Cut out the ∼400 bp band with a razor blade, dissect it further into smaller pieces, and transfer ∼0.3-g portions into 1.5-mL microfuge tubes. Purify huV$_H$ cDNA using reagents and protocols supplied by the QIAGEN MinElute Gel Extraction Kit. Elute huV$_H$ cDNA in 100 µL of sterile water and measure the concentration using a UV photometer.

    *Use the absorbance at 260 nm to calculate the DNA concentration. At 1-cm pathlength, an absorbance of 1 corresponds to 50 µg/mL DNA.*

33. Dilute the purified huV$_H$ cDNA with sterile water to a final concentration of 100 ng/µL and store at −20°C.

34. Repeat the procedure described in Steps 31–33 to purify huV$_\kappa$ and huV$_\lambda$ cDNA.

    *Purified huV$_H$, huV$_\kappa$, and huV$_\lambda$ cDNA may be stored for weeks at −20°C.*

35. To amplify the huC$_\kappa$-pelB DNA fragment, which is the middle fragment of the huV$_\kappa$/huC$_\kappa$/huV$_H$ cassette (Fig. 1B), prepare enough high-fidelity PCR master mix for 10 reactions. In a 1.5-mL microfuge tube, mix 10 µL of 100 ng/µL plasmid pCκ (Hofer et al. 2007) with 30 µL of 20 µM HCK (sense primer, Table 1), 30 µL of 20 µM pelb (antisense primer, Table 1), 100 µL of 10× PCR buffer (−MgCl$_2$), 40 µL of 10 mM dNTP mix, 40 µL of 50 mM MgCl$_2$, 740 µL of sterile water, and 10 µL of 5 units/µL Platinum Taq DNA Polymerase ($\sum$ 1 mL). Repeat for all 10 reactions.

36. In a 96-well thermocycler, run the ten 100-µL reactions in separate 0.2-mL PCR tubes using these PCR parameters:

    2 min at 94°C;

    followed by 20 cycles of 30 sec at 94°C, 30 sec at 55°C, and 30 sec at 72°C;

    followed by 10 min at 72°C;

    followed by holding at room temperature.

37. Pool all 10 reactions. Then, remove a 10-µL aliquot from the pool, add 2 µL of 6× DNA gel loading dye, and separate by electrophoresis on a 1% (w/v) agarose gel in TAE buffer using a 100-bp DNA ladder as reference. Precipitate and purify the amplified huC$_\kappa$-pelB DNA as described in Steps 30–33 for huV$_H$ cDNA, and dilute with sterile water to a final concentration of 100 ng/µL. Store at −20°C.

    *The amplified huC$_\kappa$-pelB DNA should be visible as a bright band of ∼400 bp. Cut out narrowly to avoid contamination of gel-purified huC$_\kappa$-pelB DNA with pCκ DNA.*

38. Repeat the procedure described in Steps 35–37 to amplify and purify the huC$_\lambda$-pelB DNA fragment, which is the middle fragment of the huV$_\lambda$/huC$_\lambda$/huV$_H$ cassette (Fig. 1B). Use

## Chapter 5

plasmid pCλ (Kwong et al. 2008) and sense primer HCL (Table 1) in place of pCκ and HCK, respectively.

### PCR Assembly of the huV$_L$/huC$_L$/huV$_H$ Cassettes

*Here, users will assemble the relevant cassettes using overlap extension PCR. Carry out several control PCRs with the individual fragments and the flanking primers first, to make sure that the reaction is not contaminated with an already assembled Fab expression cassette that would get amplified rapidly and diminish library complexity.*

39. For assembly of the huV$_κ$/huC$_κ$/huV$_H$ cassette by overlap extension PCR (Fig. 2A), prepare two high-fidelity PCR master mixes, each sufficient for 10 reactions.

    i. Master Mix I (without primers): In a 1.5-mL microfuge tube, mix 10 µL of 100 ng/µL huV$_H$ cDNA, 10 µL of 100 ng/µL huV$_κ$ cDNA, and 10 µL of 100 ng/µL huC$_κ$-pelB DNA. Add 50 µL of 10× PCR buffer (−MgCl$_2$), 20 µL of 10 mM dNTP mix, 20 µL of 50 mM MgCl$_2$, 375 µL of sterile water, and 5 µL of 5 units/µL Platinum Taq DNA Polymerase ($\sum$ 500 µL). Transfer 50 µL of the mix to 10 separate 0.2-mL tubes.

    ii. Master Mix II (with primers): In a 1.5-mL microfuge tube, mix 30 µL of 20 µM 5′SFIHUVL (sense primer, Table 1), 30 µL of 20 µM 3′sfivh (antisense primer, Table 1), 50 µL of 10× PCR buffer (−MgCl$_2$), 20 µL of 10 mM dNTP mix, 20 µL of 50 mM MgCl$_2$, 345 µL of sterile water, and 5 µL of 5 units/µL Platinum Taq DNA Polymerase ($\sum$ 500 µL). Keep on ice.

40. In a 96-well thermocycler, run ten 50-µL reactions of Master Mix I in 0.2-mL PCR tubes using these PCR parameters:

    2 min at 94°C;

    followed by 10 cycles of 30 sec at 94°C, 30 sec at 55°C, and 90 sec at 72°C;

    followed by 10 min at 72°C.

41. Cool to room temperature, add 50 µL of Master Mix II to each reaction, and run 20 additional cycles using the same PCR parameters in Step 40.

42. Pool all 10 reactions, remove a 10-µL aliquot, add 2 µL of 6× DNA gel loading dye, and separate by electrophoresis on a 1% (w/v) agarose gel in TAE buffer using 100-bp and 1-kb DNA ladders as reference.

    *The fused huV$_κ$/huC$_κ$/huV$_H$ cassette should be visible as a bright 1.2-kb band.*

43. Precipitate and purify the huV$_κ$/huC$_κ$/huV$_H$ DNA. To do this, first add 0.1 volumes of 3 M sodium acetate (pH 5.2) and 2.2 volumes of ethanol to the sample, vortex, and store overnight at −20°C. Then, precipitate and purify as described in Steps 31 and 32 for huV$_H$ cDNA. Expect a yield of at least 30 µg of DNA. Dilute with sterile water to a final concentration of 150 ng/µL and store at −20°C.

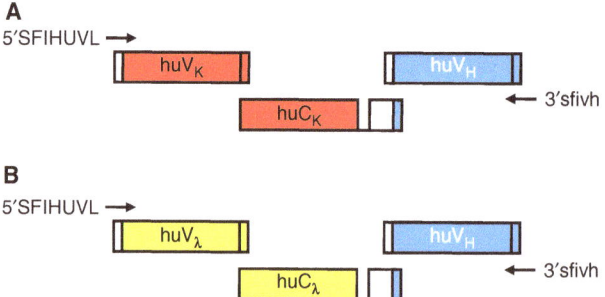

FIGURE 2. Second PCR. To assemble huV$_L$/huC$_L$/huV$_H$ by PCR, sense primer 5′SFIHUVL and antisense primer 3′sfivh are combined for the separate amplification of huV$_κ$/huC$_κ$/huV$_H$ (A) and huV$_λ$/huC$_λ$/huV$_H$ (B) by three-fragment overlap extension PCR.

44. Repeat the procedure described in Steps 39–42 to assemble and purify the huV$_\lambda$/huC$_\lambda$/huV$_H$ cassette (Fig. 2B). Use huV$_\lambda$ cDNA and huC$_\lambda$-pelB DNA in place of huV$_\kappa$ cDNA and huC$_\kappa$-pelB DNA, respectively.

    *Purified huV$_\kappa$/huC$_\kappa$/huV$_H$ and huV$_\lambda$/huC$_\lambda$/huV$_H$ cassettes may be stored for months at −20°C.*

## SfiI Digestion of huV$_L$/huC$_L$/huV$_H$ Cassettes and Phagemid pC3C

45. In two separate 1.5-mL microfuge tubes, digest the huV$_\kappa$/huC$_\kappa$/huV$_H$ and huV$_\lambda$/huC$_\lambda$/huV$_H$ cassettes by mixing 200 µL of 150 ng/µL (30 µg) DNA with 30 µL of 10× rCutSmart buffer, 50 µL of sterile water, and 20 µL of 20 units/µL SfiI. Incubate for 3 h at 50°C.

46. Cool to room temperature. Then, add 60 µL of 6× DNA gel loading dye and separate each sample by electrophoresis on a 1% (w/v) agarose gel in TAE buffer using a preparative comb and a 1-kb DNA ladder as reference.

47. Cut out the ∼1.2-kb band with a razor blade, dissect it further into smaller pieces, and transfer ∼0.3-g portions to separate 1.5-mL microfuge tubes.

48. Separately purify the SfiI-digested huV$_\kappa$/huC$_\kappa$/huV$_H$ and huV$_\lambda$/huC$_\lambda$/huV$_H$ cassettes using reagents and protocols supplied by the QIAGEN MinElute Gel Extraction Kit. Elute with sterile water so that you have 100 µL each of the SfiI-digested huV$_\kappa$/huC$_\kappa$/huV$_H$ and huV$_\lambda$/huC$_\lambda$/huV$_H$ cassettes.

49. Measure the DNA concentration of each 100-µL sample using a UV photometer.

    *Use the absorbance at 260 nm to calculate the DNA concentration. At 1-cm pathlength, an absorbance of 1 corresponds to 50 µg/mL DNA.*

50. Dilute with sterile water to a final concentration of 50 ng/µL and store at −20°C.

51. For SfiI digestion of phagemid pC3C, combine 75 µg of phagemid DNA with 45 µL of 10× rCutSmart buffer and 32 µL of 20 units/µL SfiI. Add sterile water to a total volume of 450 µL. Incubate for 3 h at 50°C.

52. Cool to room temperature. Then, add 90 µL of 6× DNA gel loading dye and separate by electrophoresis on a 1% (w/v) agarose gel in TAE buffer using a preparative comb and a 1-kb DNA ladder as reference.

53. Purify both the ∼3.5-kb band (vector) and the ∼1.2-kb band (insert) separately, as described for the SfiI-digested huV$_\kappa$/huC$_\kappa$/huV$_H$ and huV$_\lambda$/huC$_\lambda$/huV$_H$ expression cassettes in Steps 46–48. Dilute with sterile water to a final concentration of 100 ng/µL (vector) or 50 ng/µL (insert). Store at −20°C.

## Test Ligation and Transformation

54. In two separate 1.5-mL microfuge tubes, combine 9 µL of 100 ng/µL SfiI-digested pC3C vector (0.9 µg) with 12 µL of either 50 ng/µL SfiI-digested huV$_\kappa$/huC$_\kappa$/huV$_H$ or huV$_\lambda$/huC$_\lambda$/huV$_H$ cassette (0.6 µg). Then, add 10 µL of 10× T4 DNA Ligase buffer, 67 µL of sterile water, and 2 µL of 5 units/µL T4 DNA Ligase to each tube ($\sum$ 100 µL) of each mix. Prepare a test ligation mixture in parallel with SfiI-digested pC3C vector alone, as background control. Prepare another test ligation mixture in parallel by combining SfiI-digested pC3C vector with SfiI-digested pC3C insert, as religation control. Incubate for 24 h at 16°C.

55. Thaw electrocompetent bacteria for 10 min on ice. Cool the required number of cuvettes on ice.

    *For E. coli transformation, these instructions assume the use of a BTX ECM 630 electroporation system set for exponential decay at 1800 V, 25 µF, and 200 Ω when using compatible electroporation cuvettes with a 1-mm electrode gap. Each test ligation requires 50 µL of electrocompetent SS320, SR320, or ER2738 bacteria and one cuvette.*

56. Transfer 1 µL of each ligation mixture to a 1.5-mL microfuge tube and cool on ice.

57. Add 50 µL of the thawed electrocompetent bacteria to each 1-µL ligation mixture, mix by pipetting up and down once, and transfer immediately to a cuvette. Store for 1 min on ice.

58. Electroporate using the preset program (see note in Step 55).
59. Flush each cuvette immediately with a total of 3 mL (twice 1.5 mL) of recovery medium at room temperature.
60. Transfer to a 14-mL round-bottom tube with snap cap.
61. Shake at 250 rpm for 1 h at 37°C.
62. From each culture, plate 1 and 10 µL, each diluted in 100 µL of recovery medium, on separate LB + 100 µg/mL carbenicillin plates. Incubate overnight at 37°C.
63. Calculate the number of independent transformants that can be expected from one library ligation from the number of colonies × 30,000 (dilution factor) × 100 (fraction of test ligation that was transformed) × 20 (library ligation scale).

    *For example, 100 colonies on the 1-µL plate of a test ligation and transformation with the SfiI-digested cassettes predict $6 \times 10^9$ independent transformants per library ligation.*

    *The background control should give a much lower number of colonies, whereas the religation control typically gives a higher number of colonies.*

    *To proceed with library ligation, the number of predicted independent transformants per library ligation should be at least $1 \times 10^8$ with a background <10%. See Troubleshooting.*

64. To confirm that the library contains functional and diverse human Fab sequences, analyze 50 colonies from the test ligation with the SfiI-digested cassettes by Fab ELISA, as described in Protocol 7: Phage Display Selection of Antibody Libraries: Screening of Selected Binders (Peng and Rader 2024c). Expect >50% positive clones. Subsequently, analyze positive clones by DNA fingerprinting with AluI, which is also described in Protocol 7: Phage Display Selection of Antibody Libraries: Screening of Selected Binders (Peng and Rader 2024c). Each positive clone should give a unique DNA fingerprint.

## Phagemid and Phage Library Generation

65. For each library ligation (keep κ and λ separate), combine 180 µL of 100 ng/µL SfiI-digested pC3C vector (18 µg) with 240 µL of 50 ng/µL SfiI-digested huV$_κ$/huC$_κ$/huV$_H$ or huV$_λ$/huC$_λ$/huV$_H$ cassettes (12 µg), 200 µL of 10× T4 DNA Ligase buffer, 1.34 mL of sterile water, and 40 µL of 5 units/µL T4 DNA Ligase in a 14-mL round-bottom tube ($\sum$ 2 mL). Incubate for 24 h at 16°C.
66. Add 6 mL of QIAGEN buffer QG from the QIAGEN MinElute Gel Extraction Kit to the library ligation and incubate for 10 min at room temperature.
67. Add 2 mL of isopropanol, mix well, and divide each mix into three separate tubes. Proceed with DNA purification using reagents and protocols supplied by the QIAGEN MinElute Gel Extraction Kit. Elute DNA with 20 µL of sterile water, pool, and store on ice ($\sum$ 60 µL).
68. Thaw electrocompetent bacteria for 10 min on ice. Cool the required number of cuvettes on ice.

    *As stated above for the E. coli transformation of the test ligations, these instructions assume the use of a BTX ECM 630 electroporation system set for exponential decay at 1800 V, 25 µF, and 200 Ω when using compatible electroporation cuvettes with a 1-mm electrode gap.*

    *Each library ligation requires 1.5 mL (30 × 50 µL) of electrocompetent SS320, SR320, or ER2738 bacteria and 30 cuvettes.*

69. Transfer 2 µL of purified DNA from each library ligation to a 1.5-mL microfuge tube and cool on ice.
70. Add 50 µL of the thawed electrocompetent bacteria to the 2-µL sample, mix by pipetting up and down once, and transfer immediately to a cuvette. Incubate for 1 min on ice.
71. Electroporate using the preset program (see note in Step 68).
72. Flush the cuvette immediately with a total of 3 mL (twice 1.5 mL) of recovery medium at room temperature.

73. Pool recovered cultures (30 × 3 mL = 90 mL) and shake at 250 rpm for 1 h at 37°C in a 1-L Erlenmeyer flask.

    *Take precautions to avoid contaminating the* E. coli *culture with phage, including helper phage. Use a separate area and filtered pipette tips for pipetting and a phage-free shaker for growth. This is important, because the wild-type pIII protein of helper phage (unlike the C-terminal pIII protein domain encoded by the phagemid) renders* E. coli *immune to superinfection.*

74. Add 210 mL of SB, 120 μL of 100 μg/μL carbenicillin, and 600 μL of 5 μg/μL tetracycline. Shake at 250 rpm for 1 h at 37°C.

75. To determine the number of independent transformants, remove a 2-μL aliquot, add to 198 μL of SB medium in a 1.5-mL microfuge tube, mix, and plate 10 μL on an LB + 100 μg/mL carbenicillin plate. Incubate overnight at 37°C.

    *The next day, count the colonies to calculate the number of independent transformants, as described in Step 63.*

76. In the meantime, add 180 μL of 100 μg/μL carbenicillin to the remaining culture and continue shaking at 250 rpm for 1 h at 37°C.

77. Transfer 100 mL of the culture into three separate 2-L Erlenmeyer flasks. Add 5 mL of VCSM13 helper phage ($10^{11}$–$10^{12}$ pfu/mL) to each flask. Then, add 400 mL of SB, 400 μL of 100 μg/μL carbenicillin, and 800 μL of 5 μg/μL tetracycline ($\sum$ 500 mL) to each flask. Shake at 275 rpm for 90 min at 37°C.

78. Add 700 μL of 50 μg/μL kanamycin to each flask and continue shaking at 275 rpm overnight (12–16 h) at 30°C.

    *Using a lower temperature (i.e., 30°C instead of 37°C) in this step helps to increase library diversity by limiting the advantage of faster growth.*

79. From each overnight flask culture, transfer 200 mL to a 500-mL centrifuge bottle and centrifuge at 3000g for 15 min at 4°C. Save both the pellet and the supernatant.

80. For phagemid preparation, resuspend the pellet in 5 mL of QIAGEN buffer P1 from the QIAprep Spin Miniprep Kit and separately process four 250-μL samples using reagents and protocols supplied by the QIAprep Spin Miniprep Kit. Elute each with 80 μL of sterile water, pool, and store at −20°C.

    *This phagemid library stock serves as a backup from which the $huV_L$/$huC_L$/$huV_H$ cassettes can be retrieved by SfiI digestion.*

81. For phage precipitation, transfer the supernatant to a clean 500-mL centrifuge bottle with 8 g of PEG-8000 and 6 g of NaCl, and dissolve by shaking at 300 rpm for 5 min at 37°C.

82. Incubate for 30 min on ice. Centrifuge at 12,000g for 15 min at 4°C.

83. Discard the supernatant, drain the bottle by inverting on a paper towel for 10 min, and carefully wipe off any remaining liquid with a paper towel.

84. Resuspend the phage pellet in 2 mL of 1% (w/v) BSA in TBS by pipetting up and down along the side of the centrifuge tube.

85. Transfer the resuspended pellet to a 2-mL microfuge tube and pipette up and down with a 1-mL pipette tip.

86. Centrifuge at 16,000g for 5 min at 4°C.

87. Pass the supernatant through a 0.22-μm filter into a clean 2-mL microfuge tube.

88. In a 14-mL round-bottom tube, pool the phage preparations from the three overnight flask cultures of each library ligation (i.e., 3 × 2 mL = 6 mL). Store on ice for panning on the same day or add 0.01 volumes of 2% (w/v) $NaN_3$ for storage at 4°C. For long-term storage, add 1 volume of glycerol and store 1-mL aliquots in 1.5-mL microfuge tubes at −80°C.

    *Proceed to library selection as described in Protocol 6: Phage Display Selection of Antibody Libraries: Panning Procedures (Peng and Rader 2024d).*

*Only fresh phage, prepared on the same day, should be used for library selection, because the covalent linkage of Fab (phenotype) and phage (genotype) is susceptible to cleavage by contaminating proteases.*

*Fresh phage can be reamplified from the stored phage library as described in Protocol 5: Generation of Antibody Libraries for Phage Display: Library Reamplification (Peng and Rader 2024e).*

## TROUBLESHOOTING

*Problem (Step 63):* Low number of independent transformants in the test ligation.

*Solution:* The test ligation includes controls intended to help with narrowing down possible reasons for a low number of independent transformants. Low numbers of colonies on all plates indicate a problem with the SfiI-digested pC3C vector, or with one of the ligation or transformation reagents. If the religation of the SfiI-digested pC3C vector and insert gives a high number of colonies, whereas the ligation of the SfiI-digested pC3C vector and SfiI-digested $huV_\kappa/huC_\kappa/huV_H$ or $huV_\lambda/huC_\lambda/huV_H$ cassettes gives a low number of colonies, possible reasons are poor quality of the prepared DNA or the SfiI reagents. Address these issues accordingly by using fresh reagents to repeat the preparation of the SfiI-digested pC3C vector and SfiI-digested $huV_\kappa/huC_\kappa/huV_H$ or $huV_\lambda/huC_\lambda/huV_H$ cassettes.

## RECIPES

### LB + 100 µg/mL Carbenicillin Plates

Dissolve 25 g of LB (Lysogeny Broth) medium powder (MilliporeSigma L3522) in 1 L of highly pure water (e.g., that produced by Hydro PicoPure UV Plus) with 15 g of Bacto Agar (BD 214010). Autoclave for 15 min at 121°C and 15 psi. When cooled to ~50°C, add 1 mL of 100 µg/µL carbenicillin. Pour into 10-cm sterile Petri dishes (20–25 mL each), remove bubbles by quickly flaming the plates with a Bunsen burner, and allow to solidify at room temperature. Store for up to 3 mo at 4°C.

*Alternatively, prepoured LB + 100 µg/mL carbenicillin plates are commercially available (e.g., Teknova L1010).*

### SB (Super Broth) Medium

Mix 20 g of 3-(N-morpholino)propanesulfonic acid (MOPS; MilliporeSigma M1254), 60 g of Bacto Tryptone (Thermo Fisher Scientific 211705), and 40 g of Bacto Yeast Extract (Thermo Fisher Scientific 212750). Add highly pure water (e.g., that produced by Hydro PicoPure UV Plus) to a final volume of 1.9 L. Bring to pH 7.0 with 1 M sodium hydroxide (NaOH; Thermo Fisher Scientific A4782901). Bring volume to 2 L with highly pure water. Sterilize by autoclaving for 15 min at 121°C and 15 psi in two 1-L or four 500-mL glass bottles. Store for up to 2 yr at room temperature if unopened.

## ACKNOWLEDGMENTS

Work in our laboratory was supported by the National Cancer Institute, National Institutes of Health.

## REFERENCES

Green MR, Sambrook J. 2020. Precipitation of RNA with ethanol. *Cold Spring Harb Protoc* doi:10.1101/pdb.prot101717

Green MR, Sambrook J. 2022. Separation of RNA according to size: electrophoresis of RNA through agarose gels containing formaldehyde. *Cold Spring Harb Protoc* doi:10.1101/pdb.prot101758

Hofer T, Tangkeangsirisin W, Kennedy MG, Mage RG, Raiker SJ, Venkatesh K, Lee H, Giger RJ, Rader C. 2007. Chimeric rabbit/human Fab and IgG specific for members of the Nogo-66 receptor family selected for species cross-reactivity with an improved phage display vector. *J Immunol Methods* 318: 75–87. doi:10.1016/j.jim.2006.10.007

Kwong KY, Baskar S, Zhang H, Mackall CL, Rader C. 2008. Generation, affinity maturation, and characterization of a human anti-human NKG2D monoclonal antibody with dual antagonistic and agonistic activity. *J Mol Biol* 384: 1143–1156. doi:10.1016/j.jmb.2008.09.008

Peng H, Rader C. 2024a. Generation of antibody libraries for phage display: preparation of electrocompetent *E. coli*. *Cold Spring Harb Protoc* doi:10.1101/pdb.prot108599

Peng H, Rader C. 2024b. Generation of antibody libraries for phage display: preparation of helper phage. *Cold Spring Harb Protoc* doi:10.1101/pdb.prot108600

Peng H, Rader C. 2024c. Phage display selection of antibody libraries: screening of selected binders. *Cold Spring Harb Protoc* doi:10.1101/pdb.prot108603

Peng H, Rader C. 2024d. Phage display selection of antibody libraries: panning procedures. *Cold Spring Harb Protoc* doi:10.1101/pdb.prot108602

Peng H, Rader C. 2024e. Generation of antibody libraries for phage display: library reamplification. *Cold Spring Harb Protoc* doi:10.1101/pdb.prot108601

Sidhu SS, Lowman HB, Cunningham BC, Wells JA. 2000. Phage display for selection of novel binding peptides. *Methods Enzymol* 328: 333–363. doi:10.1016/s0076-6879(00)28406-1

Protocol 2

# Generation of Antibody Libraries for Phage Display: Chimeric Rabbit/Human Fab Format

Haiyong Peng and Christoph Rader[1]

*Department of Immunology and Microbiology, The Herbert Wertheim UF Scripps Institute for Biomedical Innovation & Technology, University of Florida, Jupiter, Florida 33458, USA*

Rabbit monoclonal antibodies are attractive reagents for research, and have also found use in diagnostic and therapeutic applications. This is owed to their high affinity and specificity, along with their ability to recognize epitopes conserved between mouse and human antigens. Phage display is a powerful method for the de novo generation, affinity maturation, and humanization of rabbit monoclonal antibodies from naive, immune, and synthetic antibody repertoires. Using phagemid family pComb3, a preferred phage display format is chimeric rabbit/human Fab, which consists of rabbit variable domains ($V_H$, $V_\kappa$, and $V_\lambda$) fused to human constant domains. The human constant domains, $C_H1$ of IgG1 and $C_L$ ($C_\kappa$ or $C_\lambda$), not only provide established purification and detection handles but also facilitate higher expression in *Escherichia coli* compared to the corresponding rabbit constant domains. Here, we describe the use of a pComb3 derivative, phagemid pC3C, for the generation of chimeric rabbit/human Fab libraries with randomly combined rabbit variable domains of high sequence diversity, starting from the preparation of total RNA from rabbit spleen and bone marrow. Depending on the complexity of the parental antibody repertoire, the protocol can be scaled for yielding a library size of $10^8$–$10^{11}$ independent chimeric rabbit/human Fab clones. As such, it can be used, for instance, for the generation of either specialized immune or large naive rabbit antibody libraries.

## MATERIALS

It is essential that you consult the appropriate Material Safety Data Sheets and your institution's Environmental Health and Safety Office for proper handling of equipment and hazardous materials used in this protocol.

RECIPES: Please see the end of this protocol for recipes indicated by <R>. Additional recipes can be found online at http://cshprotocols.cshlp.org/site/recipes.

### Reagents

Agarose gel electrophoresis reagents:
  DNA ladder, 100-bp and 1-kb (Thermo Fisher Scientific SM0241 and SM0311).
  DNA gel loading dye, 6× (Thermo Fisher Scientific R0611)
  SYBR Safe DNA gel stain (Thermo Fisher Scientific S33102)
  TAE (Tris-acetate-EDTA) buffer (dilute from 50× TAE; Thermo Fisher Scientific B49)
  UltraPure Agarose (Thermo Fisher Scientific 16500500)
    *For agarose–formaldehyde gel electrophoresis reagents, see Green and Sambrook (2022) (optional, see Step 7).*

---

[1]Correspondence: rader33458@gmail.com

© 2026 Cold Spring Harbor Laboratory Press
Cite this protocol as *Cold Spring Harb Protoc*; doi:10.1101/pdb.prot108598

BCP (1-bromo-3-chloro-propane (Molecular Research Center BP151); store at room temperature

Bovine serum albumin (BSA), 1% (w/v) in TBS

*To prepare, dissolve 0.5 g of BSA (MilliporeSigma A7030) in 50 mL of 1× TBS and sterilize by filtration through a 0.22-μm filter unit. Store at room temperature.*

Carbenicillin, 100 μg/μL

*To prepare, dissolve 1 g of carbenicillin disodium (MilliporeSigma C1389) in 10 mL of highly pure water. Sterilize by filtration through a 0.22-μm filter. Store 1-mL aliquots in 1.5-mL microfuge tubes at −20°C.*

Dry ice

*Escherichia coli* SS320 (nonamber suppressor strain; Biosearch Technologies 60512-2) or ER2738 (amber suppressor strain; Biosearch Technologies 60522-2) electrocompetent cells with an efficiency of $\geq 4 \times 10^{10}$ or $\geq 2 \times 10^{10}$, respectively, of colony-forming units per microgram of pUC19 plasmid, and recovery medium (Biosearch Technologies 80026-1); store at −80°C

*As a cost-saving alternative to commercial sources, see Protocol 3: Generation of Antibody Libraries for Phage Display: Preparation of Electrocompetent E. coli (Peng and Rader 2024a), which describes the preparation of electrocompetent bacteria. SS320 is the result of mating male ($F^+$) XL1-Blue with female ($F^-$) MC1061 to achieve higher library transformation efficiency for electroporation (Sidhu et al. 2000). A preferred strain for library transformation that is not commercially available is SR320, a lytic phage-resistant variant of SS320 developed in Dr. Sachdev S. Sidhu's laboratory.*

Ethanol (70%, v/v, in water)

*Prepare fresh from pure ethanol (MilliporeSigma 459836-500ML) and RNase-free water (Thermo Fisher Scientific AM9937). Store at room temperature.*

Glycerol, ultrapure (Thermo Fisher Scientific J64719.AP); store at room temperature

High-fidelity PCR reagents; store at −20°C:
  dNTP mix, 10 mM (2.5 mM of each dATP, dCTP, dGTP, and dTTP diluted in sterile water from 100 mM stock concentrations; Thermo Fisher Scientific 18427013).
  Platinum Taq DNA Polymerase, 5 units/μL, 10× PCR buffer (−$MgCl_2$), and 50 mM $MgCl_2$ (Thermo Fisher Scientific 15966005)

Ice

Isopropanol (MilliporeSigma I9516); store at room temperature

Kanamycin (50 μg/μL)

*To prepare, dissolve 500 mg of kanamycin sulfate (MilliporeSigma 60615) in 10 mL of highly pure water. Sterilize by filtration through a 0.22-μm filter. Store 1-mL aliquots in 1.5-mL microfuge tubes at −20°C.*

LB + 100 μg/mL carbenicillin plates <R>

Oligonucleotides

*The names and DNA sequences of all primers used in this protocol are shown in Table 1. For each primer, prepare 20 μM working dilutions in sterile water.*

PCR reagents; store at −20°C:
  dNTP mix, 10 mM (2.5 mM of each dATP, dCTP, dGTP, and dTTP diluted in sterile water from 100 mM stock concentrations; Thermo Fisher Scientific 18427013)
  Taq DNA polymerase, 5 units/μL, 10× Taq buffer with $(NH_4)_2SO_4$, and 25 mM $MgCl_2$ (Thermo Fisher Scientific EP0405)

PEG-8000 (MilliporeSigma P2139); store at room temperature

Phagemid pC3C (Hofer et al. 2007), available from Kerafast (ENH067-FP); store at −20°C

Plasmid pCκ (Hofer et al. 2007); store at −20°C

Plasmid pCλ (Kwong et al. 2008); store at −20°C

Qiagen MinElute Gel Extraction Kit (QIAGEN 28606); store at room temperature

QIAprep Spin Miniprep Kit (QIAGEN 27106); store at room temperature

RNA Storage Solution (1 mM sodium citrate, pH 6.5; Thermo Fisher Scientific AM7000); store at −20°C

SB (Super Broth) medium <R>

TABLE 1. Primers for PCR amplification, PCR assembly, and DNA sequencing of rbV$_L$ and rbV$_H$[a,b]

| | |
|---|---|
| **rbV$_\kappa$ sense primers (14)** | |
| RBVκ-F1 | GCTGGGGCCCAGGCGGCCGCCGTGCTGACCCAGACT |
| RBVκ-F2 | GCTGGGGCCCAGGCGGCCCAAGTGCTGACCCAGACT |
| RBVκ-F3 | GCTGGGGCCCAGGCGGCCCTTGTGATGACCCAGACT |
| RBVκ-F4 | GCTGGGGCCCAGGCGGCCGACCCTATGCTGACCCAGACT |
| RBVκ-F5 | GCTGGGGCCCAGGCGGCCGATGTCGTGATGACCCAGACT |
| RBVκ-F6 | GCTGGGGCCCAGGCGGCCGACCCTGTGCTGACCCAGACT |
| RBVκ-F7 | GCTGGGGCCCAGGCGGCCTATGTCATGATGACCCAGACT |
| RBVκ-Fa | GCTGGGGCCCAGGCGGCCGCCGTGATGACCCAGACT |
| RBVκ-Fb | GCTGGGGCCCAGGCGGCCCAAGGGCCAACCCAGACT |
| RBVκ-Fc | GCTGGGGCCCAGGCGGCCGTCGTGCTGACCCAGACT |
| RBVκ-Fd | GCTGGGGCCCAGGCGGCCATCAAAATGACCCAGACT |
| RBVκ-Fe | GCTGGGGCCCAGGCGGCCGACCCTGTGATGACCCAGACT |
| RBVκ-Ff | GCTGGGGCCCAGGCGGCCGATGGCGTGATGACCCAGACT |
| RBVκ-Fg | GCTGGGGCCCAGGCGGCCGACATTGTGCTGACCCAGACT |
| **rbV$_\kappa$ antisense primers (5)** | |
| rbvκ-r1 | AGATGGTGCAGCCACAGTTCGTTTGATTTCCAC**M**TTGGTGCC |
| rbvκ-r2 | AGATGGTGCAGCCACAGTTCGTTTGATCTCCA**S**CTTGGT**Y**CC |
| rbvκ-r3 | AGATGGTGCAGCCACAGTTCGTTTGAC**S**ACCACCTCGGTCCC |
| rbvκ-ra | AGATGGTGCAGCCACAGTTCGTAGGATCTCCAGCTCGGTCCC |
| rbvκ-rb | AGATGGTGCAGCCACAGTTCGTTCGACGACCACCTTGGTCCC |
| **rbV$_\lambda$ sense primers (8)** | |
| RBVλ-F1 | GCTGGGGCCCAGGCGGCCCAGCCTGTGCTGACTCAG |
| RBVλ-F2 | GCTGGGGCCCAGGCGGCCCAGTTTGTGCTGACTCAG |
| RBVλ-F3 | GCTGGGGCCCAGGCGGCCCAGCCTGCCCTCACTCAG |
| RBVλ-F4 | GCTGGGGCCCAGGCGGCCTCCTATGAGCTGACACAG |
| RBVλ-Fa | GCTGGGGCCCAGGCGGCCAGCGTTGTGTTCACGCAG |
| RBVλ-Fb | GCTGGGGCCCAGGCGGCCTCCCATGAGCTGACAAAG |
| RBVλ-Fc | GCTGGGGCCCAGGCGGCCTCCTTCGTGCTGACTCAG |
| RBVλ-Fd | GCTGGGGCCCAGGCGGCCCAGTTTGTGCTGAATCAA |
| **rbV$_\lambda$ antisense primer (1)** | |
| rbvλ-r1 | GAGGGGGCAGCCTTGGGCTGACCGCCTGTGACGGTCAGCTGGGTCCC |
| **rbV$_H$ sense primers (11)** | |
| RBVH-F1 | GCTGCCCAACCAGCCATGGCCCAGTCGGTGGAGGAGTCC**R**GG |
| RBVH-F2 | GCTGCCCAACCAGCCATGGCCCAGTCAGTGAAGGAGTCCGAG |
| RBVH-F3 | GCTGCCCAACCAGCCATGGCCCAGTCG**Y**TGGAGGAGTCCGGG |
| RBVH-F4 | GCTGCCCAACCAGCCATGGCCCAGGAGCAGCTGGAGGAGTCCGGG |
| RBVH-F5 | GCTGCCCAACCAGCCATGGCCCAGGAGCAGCTGAAGGAGTCCGG |
| RBVH-F6 | GCTGCCCAACCAGCCATGGCCCAG**R**AGCAGCTGGTGGAGTCCGG |
| RBVH-Fa | GCTGCCCAACCAGCCATGGCCCAGGAGCAGCAGAAGGAGTCCGGG |
| RBVH-Fb | GCTGCCCAACCAGCCATGGCCCAGTCGCTGGAGGAGTCCAGG |
| RBVH-Fc | GCTGCCCAACCAGCCATGGCCCAGTCGCTGGGGGAGTCCAGG |
| RBVH-Fd | GCTGCCCAACCAGCCATGGCCCAGACAGTGAAGGAGTCCGAG |
| RBVH-Fe | GCTGCCCAACCAGCCATGGCCCAGTCGCTGGAGGAATTCGGG |
| **rbV$_H$ antisense primers (2)** | |
| rbvh-r1 | CGATGGGCCCTTGGTGGAGGCTGA**R**GAGA**Y**GGTGACCAGGGTGCC |
| rbvh-r2 | CGATGGGCCCTTGGTGGAGGCTGAAGAGACGGTGACGAGGGTCCC |
| **Cκ-pelB and Cλ-pelB primers** | |
| HCK | CGAACTGTGGCTGCACCATCTGTC |
| HCL | GGTCAGCCCAAGGCTGCCCCCTC |
| pelb | GGCCATGGCTGGTTGGGCAGC |
| **Overlap extension PCR primers** | |
| 5′SFIVL | CTGCTGCTGGGGCCCAGGCGGCC |
| 3′sfivh | GAGGAGGAGGGCCGACGGGGCCAAGGGGAAGACCGATGGGCCCTTGGTGGAGGCTGA |
| **Sequencing primers** | |
| VLSEQ | GATAACAATTGAATTCAGGAG |
| vhseq | TGAGTTCCACGACACCGT |

[a]Nucleotide ambiguity code: **R** = A or G, **Y** = C or T, **S** = C or G, and **M** = A or C.
[b]Sequences are shown 5′–3′.

SfiI (20 units/μL; New England Biolabs R0123L) with 10× rCutSmart buffer (New England Biolabs B6004S); store at −20°C

*Formerly used reagent SfiI (40 units/μL; Roche 11288059001) was discontinued.*

Sodium acetate, 3 M, pH 5.2, RNase-free (Thermo Fisher Scientific AM9740); store at room temperature

Sodium azide (NaN$_3$), 2% (w/v) in water

*To prepare, dissolve 2 g of NaN$_3$ (MilliporeSigma S2002) in 100 mL of highly pure water. Store at room temperature.*

Sodium chloride (NaCl; MilliporeSigma S9888); store at room temperature

Spleen and bone marrow from one femur of naive or immunized rabbits, each in a 50-mL conical tube containing 10 mL of TRI reagent (Molecular Research Center TR118; store at 4°C), stored on dry ice or at −80°C

*See Step 1.*

*Users should get the proper ethical permissions from the relevant authorities before working with animals, and immunizing and harvesting organs.*

SuperScript III First-Strand Synthesis System for RT-PCR (Thermo Fisher Scientific 18080051); store at −20°C

*Contains 50 μM oligo(dT), 10 mM dNTP, RNase-free water, 10× reverse transcriptase (RT) buffer, 25 mM MgCl$_2$, 100 mM DTT, 40 units/μL RNase OUT, 200 units/μL SuperScript III RT, and 2 units/μL E. coli RNase H.*

T4 DNA Ligase (5 units/μL; Thermo Fisher Scientific EL0011; includes 10× T4 DNA Ligase buffer); store at −20°C

*Formerly used reagent T4 DNA Ligase (5 units/μL; Roche 10909246103) was discontinued.*

Tetracycline (5 μg/μL)

*To prepare, dissolve 50 mg of tetracycline hydrochloride (MilliporeSigma T7660) in 10 mL of 75% (v/v) ethanol. Store 1-mL aliquots in 1.5-mL microfuge tubes at −20°C.*

TRI reagent (Molecular Research Center TR118); store at 4°C

Tris-buffered saline (TBS, 1×)

*Prepare by diluting 10× TBS (Thermo Fisher Scientific J60764.K7) in highly pure water and sterilizing by filtration through a 0.22-μm filter unit. Store at room temperature.*

VCSM13 helper phage at $10^{11}$–$10^{12}$ plaque-forming units (pfu)/mL

*Starting from a commercially available VCSM13 Interference-Resistance Helper Phage stock (∼1 × $10^{11}$ pfu/mL [Agilent 200251 or Thermo Fisher Scientific 50125059]; store at −80°C), prepare 500 mL of VCSM13 helper phage at $10^{11}$–$10^{12}$ pfu/mL as described in Protocol 4: Generation of Antibody Libraries for Phage Display: Preparation of Helper Phage (Peng and Rader 2024b).*

Water, highly pure, sterile

*The protocol uses highly pure water from a purification system (see "Equipment"). Use freshly. Several steps require sterile water, which is obtained by filtering highly pure water through a 0.22-μm filter unit. Store at room temperature.*

## Equipment

*A laboratory with standard molecular biology and microbiology equipment is required for this protocol.*

Agarose gel electrophoresis equipment (e.g., Owl horizontal electrophoresis systems [Thermo Fisher Scientific] with various tanks, casters, combs, and power supplies [EC-105] for wide gel [D-series] and mini gel [EasyCast B-series] analytical and preparative electrophoresis and blue-light transilluminator [e.g., VB-40 Visi-Blue Transilluminator VWR 95041-594])

Autoclave

Bunsen burner

Centrifuge bottles (500-mL; polypropylene, wide mouth, sealing cap; certified for >12,000g and resistant to bleaching and autoclaving; e.g., Nalgene 3141-0500)

Centrifuge tubes (50-mL; polypropylene, sealing cap; certified for >20,000g; e.g., Nalgene Oak Ridge 3139-0050)
Conical tubes (50-mL)
Digital balance with 0.01-g readability
Electroporator (e.g., BTX ECM 630 Electroporation System, BTX 45-2051) and compatible electroporation cuvettes with 1-mm electrode gap (e.g., Cuvette Plus, BTX 45-0124)
Filtered pipette tips (10-, 20-, 100-, and 200-μL, and 1-mL)
Filters with Luer lock fitting (0.22-μm; e.g., Millex-GP Syringe Filters, Millipore SLGP033RS)
Filter units (0.22-μm; e.g., Millipore Stericup and Steriflip filter units)
Freezers (−20°C and −80°C)
Glass bottles (500-mL and 1-L, autoclaved)
Glass Erlenmeyer flasks (250-mL, and 1- and 2-L, autoclaved)
Heat blocks for 50°C, 65°C, and 85°C incubation
Homogenizer (e.g., PowerGen 125 [Thermo Fisher Scientific] or equivalent newer models)
Ice bucket
Incubator (37°C)
Magnetic hot plate stirrer
Magnetic stir bars
Microcentrifuge (e.g., Eppendorf 5425)
Microfuge tubes (1.5- and 2-mL)
Microwave oven
Paper towels
PCR tube strips (0.2-mL; e.g., Eppendorf 951010022)
Petri dishes (10-cm)
Pipettes (5-, 10-, and 25-mL)
Pipette controller
Refrigerated microcentrifuge (e.g., Eppendorf 5425R)
Refrigerated floor centrifuge (e.g., Thermo Fisher Scientific Sorvall Lynx 4000) with fixed-angle rotor for six 500-mL centrifuge bottles (e.g., Thermo Fisher Scientific Fiberlite F12-6 × 500 LEX)
Razor blades
Refrigerator (4°C)
Round-bottom tubes with snap cap (14-mL)
Shaker at 37°C (e.g., Eppendorf New Brunswick Innova 40 Benchtop Incubator Shaker).

*Two separate shakers, one for phage-free conditions and one for phage, are required.*

Single-channel and multichannel micro pipettes (1- to 1000-μL)
Syringes with Luer lock fitting (1-, 5-, and 10-mL)
Thermal cyclers (96-well; e.g., Eppendorf Mastercycler X50s 96-Well Silver Block Thermal Cycler 6311000010)
UV photometer (e.g., NanoDrop One[C] Microvolume UV-Vis Spectrophotometer, Thermo Fisher Scientific ND-ONEC-W)
Vortex mixer
Water baths (16°C and 37°C)
Water purification system (e.g., Hydro PicoPure UV Plus)

## METHOD

### Preparation of Total RNA from Rabbit Spleen and Bone Marrow

*Keep all reagents RNase-free by changing gloves frequently and working with clean equipment in dust-free conditions.*

*TRI reagent contains a poison (phenol) and an irritant (guanidine thiocyanate). Wear appropriate protection and handle with care.*

1. Thaw the spleen and bone marrow samples (which are in TRI reagent) for 15 min at 37°C in a water bath. Then, blend with a homogenizer at 50% output (1 min for bone marrow and 3 min for spleen), followed by incubation for 5 min at room temperature.

   *To generate an immune rabbit library, spleen and bone marrow from two immunized rabbits should be processed as four samples in parallel. For a naive rabbit antibody library, we recommend using 20 tissue samples from 10 naive rabbits of different strains from different farms, to achieve high diversity ($>10^{10}$). We describe below the steps for one sample.*

   *Rabbits (Oryctolagus cuniculus) have two $C_\kappa$ genes. The one that is more frequently found in rabbit antibodies with κ light chains encodes a cysteine residue at position 171 that forms an intrachain disulfide bond with a commonly found cysteine residue at position 80 in $V_\kappa$ (Weber et al. 2017). Because human $C_\kappa$ does not harbor a matching cysteine, preferred rabbit strains for generating chimeric rabbit/human Fab libraries have light chain repertoires enriched for $V_\kappa$ without cysteine 80 and $V_\lambda$. An example is the b9 allotype rabbit strain (Popkov et al. 2003).*

2. Add 20 mL of TRI reagent and centrifuge at 3000$g$ for 10 min at 4°C.

3. Transfer the supernatant (~30 mL) to an RNase-free 50-mL centrifuge tube, add 3 mL of BCP, vortex for 15 sec, incubate for 15 min at room temperature, and centrifuge at 17,500$g$ for 15 min at 4°C.

   *The mixture will separate into three phases.*

4. Carefully transfer the top colorless aqueous phase to a clean RNase-free 50-mL conical centrifuge tube. Precipitate RNA by adding 15 mL of isopropanol. Mix well and allow the mixture to stand for 10 min at room temperature. Centrifuge the sample at 17,500$g$ for 10 min at 4°C.

5. Carefully remove the supernatant without disturbing the white pellet, wash the RNA pellet with 30 mL of ice-cold 70% (v/v) ethanol. Vortex the tube briefly and centrifuge at 17,500$g$ for 10 min at 4°C.

6. Carefully remove the supernatant without disturbing the white pellet, and air-dry for 10 min at room temperature. Dissolve the pellet in 0.5 mL of RNA Storage Solution and transfer to an RNase-free 1.5-mL microfuge tube.

7. Immediately remove a 2-μL aliquot and store the remaining sample on dry ice. Add 498 μL of RNase-free water to the 2-μL aliquot and measure the absorbance at 260 and 280 nm in a UV photometer.

   *Use the absorbance at 260 nm to calculate the RNA concentration. At 1-cm pathlength, an absorbance of 1 corresponds to 40 μg/mL RNA. The ratio of the absorbance at 260 and 280 nm is typically in the range of 1.6–1.9.*

   *~2 mg of total RNA can be isolated from one sample, and only 0.1 mg is required to proceed.*

   *As an optional step, RNA can be separated by electrophoresis under denaturing conditions in an agarose–formaldehyde gel (Green and Sambrook 2022). Two sharp bands of ~4.7 and ~1.9 kb corresponding to 28S and 18S ribosomal RNA should be visible in addition to an ~0.1- to 0.15-kb smear corresponding to 5S ribosomal RNA and transfer RNA (Green and Sambrook 2022).*

   *A possible reason for low RNA quantity or quality is RNase contamination, and this should be addressed by starting over with fresh reagents and clean disposables and equipment.*

8. Proceed to reverse transcription. If needed, store total RNA at −80°C until ready to proceed. For long-term storage, add 0.1 volumes of RNase-free 3 M sodium acetate (pH 5.2) and 2.2 volumes of ethanol. Mix well and store at −80°C.

## Reverse Transcription of mRNA

*Process total RNA samples derived from rabbit spleen and bone marrow in parallel.*

9. In an RNase-free 1.5-mL microfuge tube, combine 30 μg of total RNA, 12 μL of 50 μM oligo(dT), and 12 μL of 10 mM dNTP, and bring the volume to 120 μL with RNase-free water. Incubate for 5 min at 65°C and chill for at least 1 min on ice.

Chapter 5

10. In a separate RNase-free 1.5-mL microfuge tube, prepare a reverse transcription (RT) reaction mixture by combining 24 µL of 10× RT buffer, 48 µL of 25 mM MgCl$_2$, 24 µL of 100 mM DTT, 12 µL of 40 units/µL RNase OUT, and 12 µL of 200 units/µL SuperScript III RT.

11. Add the prepared RT reaction mixture (~120 µL) to the prepared RNA/oligo(dT)/dNTP sample (~120 µL), mix gently, and incubate for 50 min at 50°C. Terminate the reaction for 5 min at 85°C. Chill for at least 1 min on ice.

12. Collect the reaction mix by brief centrifugation. Add 12 µL of 2 units/µL *E. coli* RNase H to each tube and incubate for 20 min at 37°C.

13. Collect the mix by brief centrifugation and store at −20°C.

*The first-strand cDNA is stable for weeks at −20°C. For long-term storage (months to years), add 0.1 volumes of 3 M sodium acetate (pH 5.2) and 2.2 volumes of ethanol, vortex, and store at −80°C.*

## PCR Amplification of Rabbit $V_\kappa$, $V_\lambda$, and $V_H$ cDNA, and Human $C_\kappa$ and $C_\lambda$ DNA

*Process first-strand cDNA samples derived from rabbit spleen and bone marrow in parallel in 0.2-mL PCR tubes. Each sample will be subjected to 100 unique primer combinations in separate reactions (Fig. 1A).*

*Take precautions to avoid the amplification of rbV$_\kappa$, rbV$_\lambda$, and rbV$_H$ cDNA from contaminating sources in the laboratory, such as phagemids and phage, and include negative controls without first-strand cDNA. Consider using a PCR workstation.*

14. For rabbit $V_\kappa$ (rbV$_\kappa$) amplification, generate 70 primer combinations in separate 0.2-mL PCR tubes for each sample by combining 2 µL of the first-strand cDNA with 3 µL of 20 µM of one of the 14 rbV$_\kappa$ sense primers and 3 µL of 20 µM of one of the five rbV$_\kappa$ antisense primers (Table 1).

15. For rabbit $V_\lambda$ (rbV$_\lambda$) amplification, generate eight primer combinations in separate 0.2-mL PCR tubes for each sample by combining 2 µL of the first-strand cDNA with 3 µL of 20 µM of one of the eight rbV$_\lambda$ sense primers and 3 µL of 20 µM of the rbV$_\lambda$ antisense primer (Table 1).

16. For rabbit $V_H$ (rbV$_H$) amplification, generate 22 primer combinations in separate 0.2-mL PCR tubes for each sample by combining 2 µL of the first-strand cDNA with 3 µL of 20 µM of one of the 11 rbV$_H$ sense primers and 3 µL of 20 µM of one of the two rbV$_H$ antisense primers (Table 1).

17. Prepare a master mix of the PCR mixture sufficient for 100 (number of combinations) × 2 (number of first-strand cDNA samples from spleen and bone marrow per rabbit) reactions. In a 14-mL round-bottom tube, combine 2 mL of 10× Taq buffer with (NH$_4$)$_2$SO$_4$, 2 mL of 25 mM MgCl$_2$, 0.8 mL of 10 mM dNTP mix, 13.4 mL of sterile water, and 200 µL of 5 units/µL Taq DNA polymerase ($\sum$ 18.4 mL). Consider also the relevant controls.

*Taq DNA polymerase has an error rate range from $10^{-4}$ to $10^{-5}$ depending on the PCR conditions. If higher fidelity is desired in the first PCR, use other thermostable DNA polymerases such as Platinum Taq DNA Polymerase, which is used for the second PCR (Steps 32–37).*

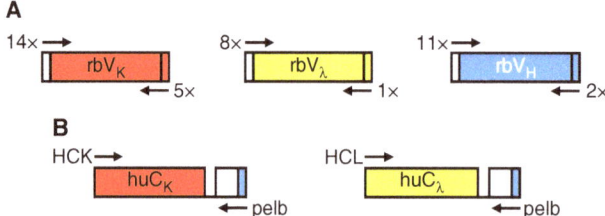

FIGURE 1. First PCR. (*A*) PCR amplification of cDNA-encoding rabbit variable domains. To amplify rabbit variable domain cDNA by PCR from reverse-transcribed mRNA, 70, eight, and 22 unique primer combinations (Table 1) are used for rbV$_\kappa$, rbV$_\lambda$, and rbV$_H$, respectively. (*B*) PCR amplification of human light chain constant domains. To amplify human constant domain DNA huC$_\kappa$ and huC$_\lambda$ by PCR from plasmids, sense primers HCK and HCL are separately combined with antisense primer pelb.

18. Add 92 μL of the prepared PCR mixture to each of the prepared first-strand cDNA/sense primer/antisense primer samples (∑ 100 μL).

19. In a 96-well thermocycler, use these PCR parameters:

    2 min at 95°C;

    followed by 35 cycles of 30 sec at 95°C, 30 sec at 50°C, and 90 sec at 72°C;

    followed by 10 min at 72°C;

    followed by holding at room temperature.

20. Remove a 10-μL aliquot from each sample, add 2 μL of 6× DNA gel loading dye, and separate by electrophoresis on a 1% (w/v) agarose gel in TAE buffer using a 100-bp DNA ladder as reference.

    *The amplified $rbV_\kappa$, $rbV_\lambda$, and $rbV_H$ cDNA should be visible as a bright band of ~400 bp.*

    *For both DNA analysis and preparation by agarose gel electrophoresis, use SYBR Safe stain and blue-light illumination rather than the hazardous combination of ethidium bromide and UV illumination.*

    *Not all 100 primer combinations are expected to work for each sample representing different rabbit antibody repertoires. However, if the majority of ~400-bp bands are not or only weakly visible, a possible reason is low RNA or first-strand cDNA quality, and this should be addressed by starting over with fresh reagents and clean disposables and equipment.*

21. Pool the remaining 90 μL of all amplified $rbV_\kappa$ cDNA that originated from the same first-strand cDNA sample (70 × 90 μL = 6300 μL) and divide the pool into seventeen 360-μL aliquots in separate 1.5-mL microfuge tubes. Then, add 0.1 volumes of 3 M sodium acetate (pH 5.2) and 2.2 volumes of ethanol to each, vortex, and store overnight at −20°C.

22. Pool the remaining 90 μL of all amplified $rbV_\lambda$ cDNA that originated from the same first-strand cDNA sample (8 × 90 μL = 720 μL) and divide the pool into two 360-μL aliquots in 1.5-mL microfuge tubes. Then, add 0.1 volumes of 3 M sodium acetate (pH 5.2) and 2.2 volumes of ethanol to each, vortex, and store overnight at −20°C.

23. Pool the remaining 90 μL of all amplified $rbV_H$ cDNA that originated from the same first-strand cDNA sample (22 × 90 μL = 1980 μL) and divide the pool into five 360-μL aliquots in separate 1.5-mL microfuge tubes. Then, add 0.1 volumes of 3 M sodium acetate (pH 5.2) and 2.2 volumes of ethanol to each, vortex, and store overnight at −20°C.

24. Proceed with two aliquots of the $rbV_H$ cDNA, as they provide sufficient material for the following steps. Precipitate $rbV_H$ cDNA by centrifugation at 16,000g for 15 min at 4°C, decant and discard the supernatant, rinse the pellet with 1 mL of 70% (v/v) ethanol (room temperature), and air-dry for 10 min at room temperature. Dissolve each pellet with 100 μL of sterile water, pool the samples, add 40 μL of 6× DNA gel loading dye, and separate by electrophoresis on a 1% (w/v) agarose gel in TAE buffer using a preparative comb and a 100-bp DNA ladder as reference.

25. Cut out the ~400-bp band with a razor blade, dissect it further into smaller pieces, and transfer ~0.3-g portions into 1.5-mL microfuge tubes. Purify $rbV_H$ cDNA using reagents and protocols supplied by the QIAGEN MinElute Gel Extraction Kit. Elute $rbV_H$ cDNA in 100 μL of sterile water and measure the concentration using a UV photometer.

    *Use the absorbance at 260 nm to calculate the DNA concentration. At 1-cm pathlength, an absorbance of 1 corresponds to 50 μg/mL DNA.*

26. Dilute the purified $rbV_H$ cDNA with sterile water to a final concentration of 100 ng/μL and store at −20°C.

27. Repeat the procedure described in Steps 24–26 to purify $rbV_\kappa$ and $rbV_\lambda$ cDNA.

    *Purified $rbV_H$, $rbV_\kappa$, and $rbV_\lambda$ cDNA may be stored for weeks at −20°C.*

28. To amplify the $huC_\kappa$-pelB DNA fragment, which is the middle fragment of the $rbV_\kappa/huC_\kappa/rbV_H$ cassette (Fig. 1B), prepare enough high-fidelity PCR master mix for 10 reactions. In a 1.5-mL microfuge tube, mix 10 μL of 100 ng/μL plasmid pCκ (Hofer et al. 2007) with 30 μL of 20 μM

Chapter 5

HCK (sense primer, Table 1), 30 µL of 20 µM pelb (antisense primer, Table 1), 100 µL of 10× PCR buffer (–MgCl$_2$), 40 µL of 10 mM dNTP mix, 40 µL of 50 mM MgCl$_2$, 740 µL of sterile water, and 10 µL of 5 units/µL Platinum Taq DNA Polymerase ($\sum$ 1 mL). Repeat for all 10 reactions.

29. In a 96-well thermocycler, run the ten 100-µL reactions in separate 0.2-mL PCR tubes using these PCR parameters:

    2 min at 94°C;

    followed by 20 cycles of 30 sec at 94°C, 30 sec at 55°C, and 30 sec at 72°C;

    followed by 10 min at 72°C;

    followed by holding at room temperature.

30. Pool all 10 reactions. Then, remove a 10-µL aliquot from the pool, add 2 µL of 6× DNA gel loading dye, and separate by electrophoresis on a 1% (w/v) agarose gel in TAE buffer using a 100-bp DNA ladder as reference. Precipitate and purify the amplified huC$_\kappa$-pelB DNA as described in Steps 24–26 for rbV$_H$ cDNA and dilute with sterile water to a final concentration of 100 ng/µL. Store at –20°C.

    *The amplified huC$_\kappa$-pelB DNA should be visible as a bright band of ~400 bp. Cut out narrowly to avoid contamination of gel-purified huC$_\kappa$-pelB DNA with pC$\kappa$ DNA.*

31. Repeat the procedure described in Steps 28–30 to amplify and purify a huC$_\lambda$-pelB DNA fragment, which is the middle fragment of the rbV$_\lambda$/huC$_\lambda$/rbV$_H$ cassette (Fig. 1B), using plasmid pC$\lambda$ (Kwong et al. 2008) and sense primer HCL (Table 1) in place of pC$\kappa$ and HCK, respectively.

## PCR Assembly of the rbV$_L$/huC$_L$/rbV$_H$ Cassettes

*Here, users will assemble the relevant cassettes using overlap extension PCR. Carry out several control PCRs with the individual fragments and the flanking primers first, to make sure that the reaction is not contaminated with an already assembled Fab expression cassette that would get amplified rapidly and diminish library complexity.*

32. For assembly of the rbV$_\kappa$/huC$_\kappa$/rbV$_H$ cassette by overlap extension PCR (Fig. 2A), prepare two high-fidelity PCR master mixes, each sufficient for 10 reactions.

    i. Master Mix I (without primers): In a 1.5-mL microfuge tube, mix 10 µL of 100 ng/µL rbV$_H$ cDNA, 10 µL of 100 ng/µL rbV$_\kappa$ cDNA, and 10 µL of 100 ng/µL huC$_\kappa$-pelB DNA. Add 50 µL of 10× PCR buffer (–MgCl$_2$), 20 µL of 10 mM dNTP mix, 20 µL of 50 mM MgCl$_2$, 375 µL of sterile water, and 5 µL of 5 units/µL Platinum Taq DNA Polymerase ($\sum$ 500 µL). Transfer 50 µL of the mix to 10 separate 0.2-mL tubes.

    ii. Master Mix II (with primers): In a 1.5-mL microfuge tube, mix 30 µL of 20 µM 5′SFIVL (sense primer, Table 1), 30 µL of 20 µM 3′sfivh (antisense primer, Table 1), 50 µL of 10× PCR buffer (–MgCl$_2$), 20 µL of 10 mM dNTP mix, 20 µL of 50 mM MgCl$_2$, 345 µL of sterile water, and 5 µL of 5 units/µL Platinum Taq DNA Polymerase ($\sum$ 500 µL). Keep on ice.

33. In a 96-well thermocycler, run the ten 50-µL reactions of Master Mix I in 0.2-mL PCR tubes using these PCR parameters:

    2 min at 94°C;

    followed by 10 cycles of 30 sec at 94°C, 30 sec at 55°C, and 90 sec at 72°C;

    followed by 10 min at 72°C.

34. Cool to room temperature, add 50 µL of Master Mix II to each reaction, and run 20 additional cycles using the same PCR parameters as in Step 33.

35. Pool all 10 reactions, remove a 10-µL aliquot, add 2 µL of 6× DNA gel loading dye, and separate by electrophoresis on a 1% (w/v) agarose gel in TAE buffer using 100-bp and 1-kb DNA ladders as reference.

    *The fused rbV$_\kappa$/huC$_\kappa$/rbV$_H$ cassette should be visible as a bright 1.2-kb band.*

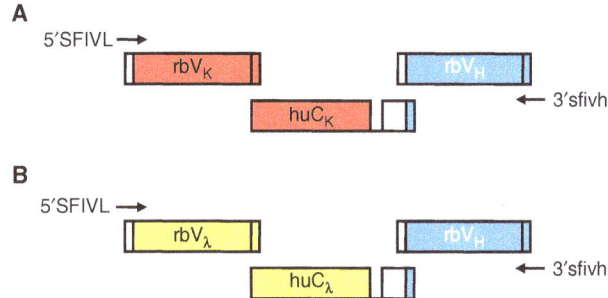

FIGURE 2. Second PCR. To assemble rbV$_L$/huC$_L$/rbV$_H$ by PCR, sense primer 5'SFIVL and antisense primer 3'sfivh are combined for the separate amplification of rbV$_\kappa$/huC$_\kappa$/rbV$_H$ (A) and rbV$_\lambda$/huC$_\lambda$/rbV$_H$ (B) by three-fragment overlap extension PCR.

36. Precipitate and purify the rbV$_\kappa$/huC$_\kappa$/rbV$_H$ DNA. To do this, first add 0.1 volumes of 3 M sodium acetate (pH 5.2) and 2.2 volumes of ethanol to the sample, vortex, and store overnight at −20°C. Then, precipitate and purify as described in Steps 24 and 25 for rbV$_H$ cDNA. Expect a yield of at least 30 µg of DNA. Dilute with sterile water to a final concentration of 150 ng/µL and store at −20°C.

37. Repeat the procedure described in Steps 32–36 to assemble and purify the rbV$_\lambda$/huC$_\lambda$/rbV$_H$ cassette (Fig. 2B). Use rbV$_\lambda$ cDNA and huC$_\lambda$-pelB DNA in place of rbV$_\kappa$ cDNA and huC$_\kappa$-pelB DNA, respectively.

   *Purified rbV$_\kappa$/huC$_\kappa$/rbV$_H$ and rbV$_\lambda$/huC$_\lambda$/rbV$_H$ cassettes may be stored for months at −20°C.*

## SfiI Digestion of rbV$_L$/huC$_L$/rbV$_H$ Cassettes and Phagemid pC3C

38. In two separate 1.5-mL microfuge tubes, digest the rbV$_\kappa$/huC$_\kappa$/rbV$_H$ and rbV$_\lambda$/huC$_\lambda$/rbV$_H$ cassettes by mixing 200 µL or 150 ng/µL (30 µg) DNA with 30 µL of 10× rCutSmart buffer, 50 µL of sterile water, and 20 µL of 20 units/µL SfiI. Incubate for 3 h at 50°C.

39. Cool to room temperature. Then, add 60 µL of 6× DNA gel loading dye and separate by electrophoresis on a 1% (w/v) agarose gel in TAE buffer using a preparative comb and a 1-kb DNA ladder as reference.

40. Cut out the ∼1.2-kb band with a razor blade, dissect it further into smaller pieces, and transfer ∼0.3-g portions to separate 1.5-mL microfuge tubes.

41. Separately purify the SfiI-digested rbV$_\kappa$/huC$_\kappa$/rbV$_H$ and rbV$_\lambda$/huC$_\lambda$/rbV$_H$ cassettes using reagents and protocols supplied by the QIAGEN MinElute Gel Extraction Kit. Elute with sterile water so that you have 100 µL each of the SfiI-digested rbV$_\kappa$/huC$_\kappa$/rbV$_H$ and rbV$_\lambda$/huC$_\lambda$/rbV$_H$ cassettes.

42. Measure the DNA concentration of each 100-µL sample using a UV photometer.

    *Use the absorbance at 260 nm to calculate the DNA concentration. At 1-cm pathlength, an absorbance of 1 corresponds to 50 µg/mL DNA.*

43. Dilute with sterile water to a final concentration of 50 ng/µL and store at −20°C.

44. For SfiI digestion of phagemid pC3C, combine 75 µg of phagemid DNA with 45 µL of 10× rCutSmart buffer, and 32 µL of 20 units/µL SfiI. Add sterile water to a total volume of 450 µL. Incubate for 3 h at 50°C.

45. Cool to room temperature. Then, add 90 µL of 6× DNA gel loading dye and separate by electrophoresis on a 1% (w/v) agarose gel in TAE buffer using a preparative comb and a 1-kb DNA ladder as reference.

Chapter 5

46. Purify both the ~3.5-kb band (vector) and the ~1.2-kb band (insert) separately, as described for the SfiI-digested rbV$_\kappa$/huC$_\kappa$/rbV$_H$ and rbV$_\lambda$/huC$_\lambda$/rbV$_H$ expression cassettes in Steps 40–42. Dilute with sterile water to a final concentration of 100 ng/µL (vector) or 50 ng/µL (insert). Store at −20°C.

## Test Ligation and Transformation

47. In two separate 1.5-mL microfuge tubes, combine 9 µL of 100 ng/µL SfiI-digested pC3C vector (0.9 µg) with 12 µL of 50 ng/µL SfiI-digested rbV$_\kappa$/huC$_\kappa$/rbV$_H$ or rbV$_\lambda$/huC$_\lambda$/rbV$_H$ cassette (0.6 µg). Then, add 10 µL of 10× T4 DNA Ligase buffer, 67 µL of sterile water, and 2 µL of 5 units/µL T4 DNA Ligase ($\sum$ 100 µL) to each mix. Prepare a test ligation mixture in parallel with SfiI-digested pC3C vector alone, as background control. Prepare another test ligation mixture in parallel by combining SfiI-digested pC3C vector with SfiI-digested pC3C insert, as religation control. Incubate for 24 h at 16°C.

48. Thaw electrocompetent bacteria for 10 min on ice. Cool the required number of cuvettes on ice.

    *For E. coli transformation, these instructions assume the use of a BTX ECM 630 electroporation system set for exponential decay at 1800 V, 25 µF, and 200 Ω when using compatible electroporation cuvettes with a 1-mm electrode gap.*

    *Each test ligation requires 50 µL of electrocompetent SS320, SR320, or ER2738 bacteria and one cuvette.*

49. Transfer 1 µL of each test ligation mixture to a 1.5-mL microfuge tube and cool on ice.

50. Add 50 µL of the thawed electrocompetent bacteria to each 1-µL test ligation mixture, mix by pipetting up and down once, and transfer immediately to a cuvette. Store for 1 min on ice.

51. Electroporate using the preset program (see note in Step 48).

52. Flush each cuvette immediately with a total of 3 mL (twice 1.5 mL) of recovery medium at room temperature.

53. Transfer to a 14-mL round-bottom tube with snap cap.

54. Shake at 250 rpm for 1 h at 37°C.

55. From each culture, plate 1 and 10 µL, each diluted in 100 µL of recovery medium, on LB + 100 µg/mL carbenicillin plates. Incubate overnight at 37°C.

56. Calculate the number of independent transformants that can be expected from one library ligation from the number of colonies × 30,000 (dilution factor) × 100 (fraction of test ligation that was transformed) × 20 (library ligation scale).

    *For example, 100 colonies on the 1-µL plate of a test ligation and transformation with the SfiI-digested cassettes predict $6 \times 10^9$ independent transformants per library ligation.*

    *The background control should give a much lower number of colonies, whereas the religation control typically gives a higher number of colonies.*

    *To proceed with library ligation, the number of predicted independent transformants per library ligation should be at least $1 \times 10^8$, with a background <10%. See Troubleshooting.*

57. To confirm that the library contains functional and diverse chimeric rabbit/human Fab sequences, analyze 50 colonies from the test ligation with the SfiI-digested cassettes by Fab ELISA, as described in Protocol 7: Phage Display Selection of Antibody Libraries: Screening of Selected Binders (Peng and Rader 2024c). Expect >50% positive clones. Subsequently, analyze positive clones by DNA fingerprinting with AluI, which is also described in Protocol 7: Phage Display Selection of Antibody Libraries: Screening of Selected Binders (Peng and Rader 2024c). Each positive clone should give a unique DNA fingerprint.

## Phagemid and Phage Library Generation

58. For each library ligation (keep κ and λ separate), combine 180 µL of 100 ng/µL SfiI-digested pC3C vector (18 µg) with 240 µL of 50 ng/µL SfiI-digested rbV$_\kappa$/huC$_\kappa$/rbV$_H$ or rbV$_\lambda$/huC$_\lambda$/rbV$_H$

cassette (12 μg), 200 μL of 10× T4 DNA Ligase buffer, 1.34 mL of sterile water, and 40 μL of 5 units/μL T4 DNA Ligase in a 14-mL round-bottom tube ($\sum$ 2 mL). Incubate for 24 h at 16°C.

59. Add 6 mL of QIAGEN buffer QG from the Qiagen MinElute Gel Extraction Kit to the library ligation and incubate for 10 min at room temperature.

60. Add 2 mL of isopropanol, mix well, and divide each mix into three separate tubes. Proceed with DNA purification using reagents and protocols supplied by the QIAGEN MinElute Gel Extraction Kit. Elute DNA with 20 μL of sterile water, pool, and store on ice ($\sum$ 60 μL).

61. Thaw electrocompetent bacteria for 10 min on ice. Cool the required number of cuvettes on ice.

    *As stated above for the E. coli transformation of the test ligations, these instructions assume the use of a BTX ECM 630 electroporation system set for exponential decay at 1800 V, 25 μF, and 200 Ω when using compatible electroporation cuvettes with a 1-mm electrode gap.*

    *Each library ligation requires 1.5 mL (30 × 50 μL) of electrocompetent SS320, SR320, or ER2738 bacteria and 30 cuvettes.*

62. Transfer 2 μL of purified DNA from each library ligation to a 1.5-mL microfuge tube and cool on ice.

63. Add 50 μL of the thawed electrocompetent bacteria to the 2-μL sample, mix by pipetting up and down once, and transfer immediately to a cuvette. Incubate for 1 min on ice.

64. Electroporate using the preset program (see note in Step 61).

65. Flush the cuvette immediately with a total of 3 mL (twice 1.5 mL) of recovery medium at room temperature.

66. Pool recovered cultures (30 × 3 mL = 90 mL) and shake at 250 rpm for 1 h at 37°C in a 1-L Erlenmeyer flask.

    *Take precautions to avoid contaminating the E. coli culture with phage, including helper phage. Use a separate area and filtered pipette tips for pipetting and a phage-free shaker for growth. This is important, because the wild-type pIII protein of helper phage (unlike the C-terminal pIII protein domain encoded by the phagemid) renders E. coli immune to superinfection.*

67. Add 210 mL of SB, 120 μL of 100 μg/μL carbenicillin, and 600 μL of 5 μg/μL tetracycline. Shake at 250 rpm for 1 h at 37°C.

68. To determine the number of independent transformants, remove a 2-μL aliquot, add to 198 μL of SB medium in a 1.5-mL microfuge tube, and plate 10 μL on an LB + 100 μg/mL carbenicillin plate. Incubate overnight at 37°C.

    *The next day, count the colonies to calculate the number of independent transformants, as described in Step 56.*

69. In the meantime, add 180 μL of 100 μg/μL carbenicillin to the remaining culture and continue shaking at 250 rpm for 1 h at 37°C.

70. Transfer 100 mL of the culture into three separate 2-L Erlenmeyer flasks. Add 5 mL of VCSM13 helper phage ($10^{11}$–$10^{12}$ pfu/mL) to each flask. Then, add 400 mL of SB, 400 μL of 100 μg/μL carbenicillin, and 800 μL of 5 μg/μL tetracycline ($\sum$ 500 mL) to each flask. Shake at 275 rpm for 90 min at 37°C.

71. Add 700 μL of 50 μg/μL kanamycin and continue shaking at 275 rpm overnight (12–16 h) at 30°C.

    *Using a lower temperature (i.e., 30°C instead of 37°C) in this step helps to increase library diversity by limiting the advantage of faster growth.*

72. From each overnight flask culture, transfer 200 mL to a 500-mL centrifuge bottle and centrifuge at 3000g for 15 min at 4°C. Save both the pellet and the supernatant.

73. For phagemid preparation, resuspend the pellet in 5 mL of QIAGEN buffer P1 from the QIAprep Spin Miniprep Kit and separately process four 250-μL samples using reagents and protocols

supplied by the QIAprep Spin Miniprep Kit. Elute each with 80 μL of sterile water, pool, and store at −20°C.

> *This phagemid library stock serves as a backup from which the $rbV_L/huC_L/rbV_H$ cassettes can be retrieved by SfiI digestion.*

74. For phage precipitation, transfer the supernatant to a clean 500-mL centrifuge bottle with 8 g of PEG-8000 and 6 g of NaCl and dissolve by shaking at 300 rpm for 5 min at 37°C.

75. Incubate for 30 min on ice. Centrifuge at 15,000$g$ for 15 min at 4°C.

76. Discard the supernatant, drain the bottle by inverting onto a paper towel for 10 min, and carefully wipe off any remaining liquid with a paper towel.

77. Resuspend the phage pellet in 2 mL of 1% (w/v) BSA in TBS by pipetting up and down along the side of the centrifuge tube.

78. Transfer the resuspended pellet to a 2-mL microfuge tube and pipette up and down with a 1-mL pipette tip.

79. Centrifuge at 16,000$g$ for 5 min at 4°C.

80. Pass the supernatant through a 0.22-μm filter into a clean 2-mL microfuge tube.

81. In a 14-mL round-bottom tube, pool the phage preparations from the three overnight flask cultures of each library ligation (i.e., 3 × 2 mL = 6 mL). Store on ice for panning on the same day or add 0.01 volumes of 2% (w/v) $NaN_3$ for storage at 4°C. For long-term storage, add 1 volume of glycerol and store 1-mL aliquots in 1.5-mL microfuge tubes at −80°C.

> *Proceed to library selection as described in Protocol 6: Phage Display Selection of Antibody Libraries: Panning Procedures (Peng and Rader 2024d).*

> *Only fresh phage prepared on the same day should be used for library selection because the covalent linkage of Fab (phenotype) and phage (genotype) is susceptible to cleavage by contaminating proteases.*

> *Fresh phage can be reamplified from the stored phage library as described in Protocol 5: Generation of Antibody Libraries for Phage Display: Library Ramplification (Peng and Rader 2024e).*

## TROUBLESHOOTING

*Problem (Step 56):* Low number of independent transformants in the test ligation.

*Solution:* The test ligation includes controls intended to help with narrowing down possible reasons for a low number of independent transformants. Low numbers of colonies on all plates indicate a problem with SfiI-digested pC3C vector, or with one of the ligation or transformation reagents. If the religation of SfiI-digested pC3C vector and insert gives a high number of colonies, whereas the ligation of the SfiI-digested pC3C vector and SfiI-digested $rbV_\kappa/huC_\kappa/rbV_H$ or $rbV_\lambda/huC_\lambda/rbV_H$ cassette gives a low number of colonies, possible reasons are poor quality of the prepared DNA or the SfiI reagents. Address these issues accordingly by using fresh reagents to repeat the preparation of the SfiI-digested pC3C vector and SfiI-digested $rbV_\kappa/huC_\kappa/rbV_H$ or $rbV_\lambda/huC_\lambda/rbV_H$ cassette.

## RECIPES

### LB + 100 µg/mL Carbenicillin Plates

Dissolve 25 g of LB (Lysogeny Broth) medium powder (MilliporeSigma L3522) in 1 L of highly pure water (e.g., that produced by Hydro PicoPure UV Plus) with 15 g of Bacto Agar (BD 214010). Autoclave for 15 min at 121°C and 15 psi. When cooled to ∼50°C, add 1 mL of 100 µg/µL carbenicillin. Pour into 10-cm sterile Petri dishes (20–25 mL each), remove bubbles by quickly flaming the plates with a Bunsen burner, and allow to solidify at room temperature. Store for up to 3 mo at 4°C.

*Alternatively, prepoured LB + 100 µg/mL carbenicillin plates are commercially available (e.g., Teknova L1010).*

### SB (Super Broth) Medium

Mix 20 g of 3-(N-morpholino)propanesulfonic acid (MOPS; MilliporeSigma M1254), 60 g of Bacto Tryptone (Thermo Fisher Scientific 211705), and 40 g of Bacto Yeast Extract (Thermo Fisher Scientific 212750). Add highly pure water (e.g., that produced by Hydro PicoPure UV Plus) to a final volume of 1.9 L. Bring to pH 7.0 with 1 M sodium hydroxide (NaOH; Thermo Fisher Scientific A4782901). Bring volume to 2 L with highly pure water. Sterilize by autoclaving for 15 min at 121°C and 15 psi in two 1-L or four 500-mL glass bottles. Store for up to 2 yr at room temperature if unopened.

## ACKNOWLEDGMENTS

Work in our laboratory was supported by the National Cancer Institute, National Institutes of Health.

## REFERENCES

Green MR, Sambrook J. 2022. Separation of RNA according to size: electrophoresis of RNA through agarose gels containing formaldehyde. *Cold Spring Harb Protoc* doi:10.1101/pdb.prot101758

Hofer T, Tangkeangsirisin W, Kennedy MG, Mage RG, Raiker SJ, Venkatesh K, Lee H, Giger RJ, Rader C. 2007. Chimeric rabbit/human Fab and IgG specific for members of the Nogo-66 receptor family selected for species cross-reactivity with an improved phage display vector. *J Immunol Methods* **318:** 75–87. doi:10.1016/j.jim.2006.10.007

Kwong KY, Baskar S, Zhang H, Mackall CL, Rader C. 2008. Generation, affinity maturation, and characterization of a human anti-human NKG2D monoclonal antibody with dual antagonistic and agonistic activity. *J Mol Biol* **384:** 1143–1156. doi:10.1016/j.jmb.2008.09.008

Peng H, Rader C. 2024a. Generation of antibody libraries for phage display: preparation of electrocompetent *E. coli*. *Cold Spring Harb Protoc* doi:10.1101/pdb.prot108599

Peng H, Rader C. 2024b. Generation of antibody libraries for phage display: preparation of helper phage. *Cold Spring Harb Protoc* doi:10.1101/pdb.prot108600

Peng H, Rader C. 2024c. Phage display selection of antibody libraries: screening of selected binders. *Cold Spring Harb Protoc* doi:10.1101/pdb.prot108603

Peng H, Rader C. 2024d. Phage display selection of antibody libraries: panning procedures. *Cold Spring Harb Protoc* doi:10.1101/pdb.prot108602

Peng H, Rader C. 2024e. Generation of antibody libraries for phage display: library reamplification. *Cold Spring Harb Protoc* doi:10.1101/pdb.prot108601

Popkov M, Mage RG, Alexander CB, Thundivalappil S, Barbas CF III, Rader C. 2003. Rabbit immune repertoires as sources for therapeutic monoclonal antibodies: the impact of kappa allotype-correlated variation in cysteine content on antibody libraries selected by phage display. *J Mol Biol* **325:** 325–335. doi:10.1016/s0022-2836(02)01232-9

Sidhu SS, Lowman HB, Cunningham BC, Wells JA. 2000. Phage display for selection of novel binding peptides. *Methods Enzymol* **328:** 333–363. doi:10.1016/s0076-6879(00)28406-1

Weber J, Peng H, Rader C. 2017. From rabbit antibody repertoires to rabbit monoclonal antibodies. *Exp Mol Med* **49:** e305. doi:10.1038/emm.2017.23

## Protocol 3

# Generation of Antibody Libraries for Phage Display: Preparation of Electrocompetent *E. coli*

Haiyong Peng and Christoph Rader[1]

*Department of Immunology and Microbiology, The Herbert Wertheim UF Scripps Institute for Biomedical Innovation & Technology, University of Florida, Jupiter, Florida 33458, USA*

The size of an antibody library, that is, the phage display-selectable diversity, is restricted mainly by its transformation into the host bacterial cells. Electroporation is the most efficient method for transforming *Escherichia coli* with plasmids, including phagemids. Here, we describe the preparation of electrocompetent *E. coli* for the generation of phagemid-encoded antibody libraries encompassing $10^9$–$10^{11}$ independent transformants. To become electrocompetent, the bacterial suspension has to have high resistance, i.e., low ionic strength, which is achieved by gradually and gently transferring bacteria grown to mid-log phase to 10% (v/v) glycerol in highly pure water. The electrocompetent *E. coli* must be F plasmid-harboring bacteria, referred to as $F^+$ or male, in order to express F pili and be susceptible to infection by filamentous phage during library generation. In addition, it is necessary to apply antibiotic (e.g., tetracycline) pressure to retain the F plasmid, as it tends to segregate from bacteria. This protocol also includes assays for analyzing the prepared electrocompetent *E. coli* for competency, and evaluating potential contamination with helper phage, phagemid and phagemid-derived filamentous phage, and lytic phage.

## MATERIALS

It is essential that you consult the appropriate Material Safety Data Sheets and your institution's Environmental Health and Safety Office for proper handling of equipment and hazardous materials used in this protocol.

RECIPES: Please see the end of this protocol for recipes indicated by <R>. Additional recipes can be found online at http://cshprotocols.cshlp.org/site/recipes.

### Reagents

Carbenicillin (100 µg/µL)

*To prepare, dissolve 1 g of carbenicillin disodium (MilliporeSigma C1389) in 10 mL of highly pure water. Sterilize by filtration through a 0.22-µm filter. Store 1-mL aliquots in 1.5-mL microfuge tubes at −20°C.*

*Escherichia coli* strain of choice, such as SS320 (nonamber suppressor strain; Biosearch Technologies 60512-2) or ER2738 (amber suppressor strain; Biosearch Technologies 60522-2).

*SS320 is the result of mating male ($F^+$) XL1-Blue with female ($F^-$) MC1061, to achieve higher library transformation efficiency for electroporation (Sidhu et al. 2000). A preferred strain for library transformation that is not commercially available is SR320, a lytic phage-resistant variant of SS320 developed in the laboratory of Dr. Sachdev S. Sidhu.*

---

[1]Correspondence: rader33458@gmail.com

© 2026 Cold Spring Harbor Laboratory Press
Cite this protocol as *Cold Spring Harb Protoc*; doi:10.1101/pdb.prot108599

Glycerol, 10% (v/v) and 20% (v/v) in highly pure sterile water (Thermo Fisher Scientific J64719.AP); store at room temperature and chill on ice before use

Ice

Kanamycin (50 µg/µL)

*To prepare, dissolve 500 mg of kanamycin sulfate (MilliporeSigma 60615) in 10 mL of highly pure water. Sterilize by filtration through a 0.22-µm filter. Store 1-mL aliquots in 1.5-mL microfuge tubes at −20°C.*

LB plates <R>

*Alternatively, prepoured LB plates are commercially available (e.g., Teknova L1100).*

LB + 100 µg/mL carbenicillin plates <R>

LB + 30 µg/mL kanamycin plates <R>

LB top agar <R>

Liquid nitrogen

Plasmid pUC19 plasmid (0.5 µg/µL; Thermo Fisher Scientific SD0061); store at −20°C

Recovery medium (Biosearch Technologies 80026-1); store at 4°C

SB (Super Broth) medium <R>

Tetracycline (5 µg/µL)

*To prepare, dissolve 50 mg of tetracycline hydrochloride (MilliporeSigma T7660) in 10 mL of 75% (v/v) ethanol. Store 1-mL aliquots in 1.5-mL microfuge tubes at −20°C.*

Water, highly pure, sterile

*The protocol uses highly pure water from a purification system (see "Equipment"). Use freshly. Several steps require sterile water, which is obtained by filtering highly pure water through a 0.22-µm filter unit. Store at room temperature.*

2× YT medium <R>

2× YT plates <R>

## Equipment

*A laboratory with standard molecular biology and microbiology equipment is required for this protocol.*

Autoclave

Bunsen burner

Centrifuge bottles (500-mL; polypropylene, wide mouth, sealing cap; certified for >12,000$g$ and resistant to bleaching and autoclaving; e.g., Nalgene 3141-0500); chill on ice before use

*Keep free of phage.*

Conical tubes (50-mL); chill on ice before use

Dewar flask

Digital balance with 0.01-g readability

Electroporator (e.g., BTX ECM 630 Electroporation System, BTX 45-2051) and compatible electroporation cuvettes with 1-mm electrode gap (e.g., Cuvette Plus, BTX 45-0124)

Filtered pipette tips (10-, 20-, 100-, and 200-µL, and 1-mL)

Filters with Luer lock fitting (0.22-µm; e.g., Millex-GP Syringe Filters, Millipore SLGP033RS)

Filter units (0.22-µm; e.g., Millipore Stericup and Steriflip filter units)

Freezers (−20°C and −80°C)

Glass bottles (100- and 500 mL, and 1-L, autoclaved)

Glass Erlenmeyer flasks (250-mL and 2-L, autoclaved)

Ice bucket

Incubator (37°C)

Inoculating loops

Magnetic hot plate stirrer

Magnetic stir bars

Microfuge tubes (1.5- and 2-mL)

Microwave oven
Orbital shaking platform
Paper towels
Petri dishes (10-cm)
Pipettes (5-, 10-, 25-, and 50-mL)
Pipette controller
Plate spreaders
Razor blades
Refrigerated benchtop centrifuge with swinging bucket rotor (e.g., Beckman Coulter Avanti J-15R with JS-4.750 swinging bucket rotor)
Refrigerated floor centrifuge (e.g., Thermo Fisher Scientific Sorvall Lynx 4000) with fixed-angle rotor for six 500-mL centrifuge bottles (e.g., Thermo Fisher Scientific Fiberlite F12-6 × 500 LEX)
Refrigerator (4°C)
Round-bottom tubes with snap cap (14-mL)
Shaker at 37°C (e.g., Eppendorf New Brunswick Innova 40 Benchtop Incubator Shaker).
  *Keep free of phage.*

Single-channel micropipettes (1- to 1000-μL)
Spectrophotometer (e.g., NanoDrop One$^C$ Microvolume UV-Vis Spectrophotometer, Thermo Fisher Scientific ND-ONEC-W)
Syringes with Luer lock fitting (1-, 5-, and 10-mL)
Vacuum line
Water purification system (e.g., Hydro PicoPure UV Plus)

## METHOD

*Take precautions to avoid contaminating the* E. coli *culture with phage, including helper phage. This is important, because the wild-type pIII protein of helper phage (unlike the C-terminal pIII protein domain encoded by the phagemid) renders* E. coli *immune to superinfection. Use a separate area and filtered pipette tips for pipetting, and a phage-free shaker for growth. Use pipettors, Erlenmeyer flasks, and centrifuge bottles that have never been in contact with phage. Bleach the bench top, centrifuge, and rotor thoroughly. Use highly pure water to wash and rinse all glassware and plasticware that come in touch with the bacterial cultures and suspensions.*

*Chill centrifuge bottles, conical tubes, microcentrifuge tubes, pipettes, and pipette tips.*

1. Thaw cells of *E. coli* strain SR320, SS320, or ER2738 and streak on a 2× YT plate (without antibiotics). Grow overnight at 37°C.
2. Inoculate 50 mL of SB (without antibiotics) in a 250-mL glass Erlenmeyer flask with a single colony from the plate from Step 1 and grow at 250 rpm overnight at 37°C.
3. The next day, prepare six cultures in separate 2-L glass Erlenmeyer flasks. In each of them, inoculate 450 mL of SB containing 10 μg/mL tetracycline (i.e., add 900 μL of 5 μg/μL tetracycline) with 1 mL of the overnight culture. Grow the cultures at 250 rpm and 37°C to an optical density (OD) at 600 nm of 0.5–0.8.
   *Start checking the OD after 1.5 h. The bacterial doubling time is ∼ 20 min.*
4. Once the appropriate OD is reached, chill the cultures as well as six 500-mL centrifuge bottles for 15 min on ice.
   *From here on, everything should be kept on ice, and steps should proceed as rapidly as possible.*
5. Transfer each of the chilled cultures to one of the chilled 500-mL centrifuge bottles and centrifuge at 3000$g$ for 15 min at 4°C.
6. Aspirate the supernatant using a vacuum source, add 25 mL of chilled 10% (v/v) glycerol to the pellet, and rotate on an orbital shaking platform at 200 rpm for 15 min on ice.
   *Use ice buckets.*

7. Refill the centrifuge bottles to 450 mL with chilled 10% (v/v) glycerol and centrifuge at 3000g for 15 min at 4°C.
8. Repeat Steps 6 and 7.
9. Thoroughly but carefully aspirate the supernatant using a vacuum source, add 25 mL of chilled 10% (v/v) glycerol to the pellet, and place on an orbital shaker at 200 rpm for 15 min on ice.
10. Transfer the 25-mL suspension to a chilled 50-mL conical tube, and add 10 mL of chilled 20% (v/v) glycerol to the bottom, making sure not to disturb the 25-mL suspension on the top. Then, centrifuge at 3000g for 15 min at 4°C.
11. Thoroughly but carefully aspirate the supernatant using a vacuum source. Then, add 400 µL of chilled 20% (v/v) glycerol to the pellet and resuspend very gently using a chilled blue (i.e., 1-mL) pipette tip. Combine all six samples in one chilled 50-mL conical tube and mix by swirling gently.
12. Use a snipped-off yellow pipette tip to immediately transfer 50-µL aliquots into approximately 75 chilled 1.5-mL microcentrifuge tubes. Then, immediately flash-freeze in liquid nitrogen in a Dewar flask and store at −80°C.

    *The next day, test the prepared electrocompetent* E. coli *for competency and potential contamination with (1) helper phage, (2) phagemid and phagemid-derived filamentous phage, and (3) lytic phage, by following the steps described in "Competency Test" and "Contamination Test," respectively.*

## Competency Test

*Warm plates to 37°C in an incubator before plating cells.*

13. Thaw electrocompetent bacteria from Step 12 for 10 min on ice. Cool a cuvette on ice.

    *For* E. coli *transformation, these instructions assume the use of a BTX ECM 630 electroporation system set for exponential decay at 1800 V, 25 µF, and 200 Ω. The competency test requires one 50-µL aliquot of the prepared electrocompetent* E. coli *and one 1-mm cuvette.*

14. Place 1 µL of a 10 pg/µL dilution of plasmid pUC19 in a 1.5-mL microfuge tube and cool on ice.
15. Add 50 µL of the thawed electrocompetent *E. coli* to the plasmid, mix by pipetting up and down once, and transfer immediately to the cuvette. Store for 1 min on ice.
16. Electroporate using the preset program (see note in Step 13).
17. Flush the cuvette immediately with a total of 3 mL (twice 1.5 mL) of recovery medium at room temperature.
18. Transfer the mix to a 14-mL round-bottom tube with snap cap and shake at 250 rpm for 1 h at 37°C.
19. Add 7 mL of SB medium to the mix. Take 1 µL of this mix, dilute it in 100 µL of SB, and plate the dilution on an LB + 100 µg/mL carbenicillin plate. Incubate overnight at 37°C.
20. Calculate the competency in colony-forming units (cfu) per microgram of plasmid as follows: number of colonies $\times 10^4$ (dilution factor) $\times 10^4$ (fraction of 1 µg of plasmid that was transformed).

    *For example, 200 colonies on the 1-µL plate indicate a competency of $2 \times 10^{10}$ cfu/µg ($200 \times 10^4 \times 10^4 = 2 \times 10^{10}$), which is a typical result.*

## Contamination Tests

*Warm plates to 37°C in an incubator before plating cells.*

21. Thaw two 50-µL aliquots of the prepared electrocompetent *E. coli* (from Step 12) for 10 min on ice. To test for a potential contamination with helper phage, plate 25 µL of the thawed bacteria directly onto an LB + 30 µg/mL kanamycin plate. Incubate overnight at 37°C. If colonies grow, discard and remake the electrocompetent *E. coli* batch.

22. To test for a potential contamination with phagemid and phagemid-derived filamentous phage, thaw an aliquot of the prepared electrocompetent *E. coli* (from Step 12) for 10 min on ice. Then, plate 25 µL of the thawed bacteria directly onto an LB + 100 µg/mL carbenicillin plate. Incubate overnight at 37°C. If colonies grow, discard and remake electrocompetent *E. coli* batch.

23. To test for a potential contamination with lytic phage, thaw an aliquot of the prepared electrocompetent *E. coli* (from Step 12) for 10 min on ice. Then, mix 25 µL of the thawed bacteria with liquefied LB top agar (melted in a microwave oven and cooled to <50°C right before use) and pour onto an LB plate (no antibiotics). Incubate overnight at 37°C. If plaques form, discard and remake the electrocompetent *E. coli* batch.

   *Larger plaques indicate lytic phage contamination, whereas smaller plaque-like structures are a result of slower growth due to helper phage contamination.*

## RECIPES

### LB Plates

Dissolve 25 g of LB (Lysogeny Broth) medium powder (MilliporeSigma L3522) in 1 L of highly pure water with 15 g of Bacto Agar (BD 214010). Autoclave for 15 min at 121°C and 15 psi. When cooled to ~50°C, pour into 10-cm sterile Petri dishes (20–25 mL each), remove bubbles by flaming the plates with a Bunsen burner, and allow to solidify at room temperature.
Store for up to 3 mo at 4°C.

### LB + 100 µg/mL Carbenicillin Plates

Dissolve 25 g of LB (Lysogeny Broth) medium powder (MilliporeSigma L3522) in 1 L of highly pure water (e.g., that produced by Hydro PicoPure UV Plus) with 15 g of Bacto Agar (BD 214010). Autoclave for 15 min at 121°C and 15 psi. When cooled to ~50°C, add 1 mL of 100 µg/µL carbenicillin. Pour into 10-cm sterile Petri dishes (20–25 mL each), remove bubbles by quickly flaming the plates with a Bunsen burner, and allow to solidify at room temperature. Store for up to 3 mo at 4°C.

   *Alternatively, prepoured LB + 100 µg/mL carbenicillin plates are commercially available (e.g., Teknova L1010).*

### LB + 30 µg/mL Kanamycin Plates

Dissolve 25 g of LB (Lysogeny Broth) medium powder (MilliporeSigma L3522) in 1 L of highly pure water (e.g., that produced with a Hydro PicoPure UV Plus system) with 15 g of Bacto Agar (BD 214010). Autoclave for 15 min at 121°C and 15 psi. When cooled to ~50°C, add 600 µL of 50 µg/µL kanamycin. Pour into 10-cm sterile Petri dishes (20–25 mL each), remove bubbles by flaming the plates with a Bunsen burner, and allow to solidify at room temperature. Store for up to 3 mo at 4°C. Warm to 37°C in an incubator before plating.

### LB Top Agar

Dissolve 1.25 g of LB (Lysogeny Broth) medium powder (MilliporeSigma L3522) in 50 mL of highly pure water (e.g., that produced with a Hydro PicoPure UV Plus system) with 0.35 g of Bacto Agar (BD 214010). Autoclave for 15 min at 121°C and 15 psi. When cooled, store at 4°C. Store for up to 3 mo if unopened. Melt in a microwave oven before use.

### SB (Super Broth) Medium

Mix 20 g of 3-(N-morpholino)propanesulfonic acid (MOPS; MilliporeSigma M1254), 60 g of Bacto Tryptone (Thermo Fisher Scientific 211705), and 40 g of Bacto Yeast Extract (Thermo Fisher Scientific 212750). Add highly pure water (e.g., that produced by Hydro PicoPure UV Plus) to a final volume of 1.9 L. Bring to pH 7.0 with 1 M sodium hydroxide (NaOH; Thermo Fisher Scientific A4782901). Bring volume to 2 L with highly pure water. Sterilize by autoclaving for 15 min at 121°C and 15 psi in two 1-L or four 500-mL glass bottles. Store for up to 2 yr at room temperature if unopened.

### 2× YT Plates

Add 7.5 g of Bacto Agar (BD 214010) to 500 mL of 2× YT medium <R>. Autoclave for 15 min at 121°C and 15 psi. Pour into 10-cm sterile Petri dishes (20–25 mL each), remove bubbles by flaming the plates with a Bunsen burner, and allow to solidify at room temperature. Store for up to 3 mo at 4°C.

### 2× YT Medium

Measure ∼900 mL of distilled $H_2O$. Add 16 g of Bacto Tryptone, 10 g of Bacto yeast extract, and 5 g of NaCl. Mix to dissolve. Adjust pH to 7.0 with 5 N NaOH. Adjust to 1 L with distilled $H_2O$. Sterilize by autoclaving.

## ACKNOWLEDGMENTS

Work in our laboratory was supported by the National Cancer Institute, National Institutes of Health.

## REFERENCE

Sidhu SS, Lowman HB, Cunningham BC, Wells JA. 2000. Phage display for selection of novel binding peptides. *Methods Enzymol* **328:** 333–363. doi:10.1016/s0076-6879(00)28406-1

## Protocol 4

# Generation of Antibody Libraries for Phage Display: Preparation of Helper Phage

Haiyong Peng and Christoph Rader[1]

*Department of Immunology and Microbiology, The Herbert Wertheim UF Scripps Institute for Biomedical Innovation & Technology, University of Florida, Jupiter, Florida 33458, USA*

The generation and selection of antibody libraries by phagemid-based phage display requires three components; namely, phagemid library, host bacterial cells, and helper phage. The use of helper phage is necessary for the selection of phagemid libraries by phage display because it provides all genes needed for production of infectious phage particles. Here, we describe the generation of high-titer helper phage preparations suitable for phagemid-based phage display. The approach is based on helper phage VCSM13, which includes a gene for kanamycin resistance and a mutated packaging signal that, in the presence of a phagemid with an unmutated packaging signal, favors the production of infectious phage particles with phagemid phenotype and genotype.

## MATERIALS

It is essential that you consult the appropriate Material Safety Data Sheets and your institution's Environmental Health and Safety Office for proper handling of equipment and hazardous materials used in this protocol.

RECIPES: Please see the end of this protocol for recipes indicated by <R>. Additional recipes can be found online at http://cshprotocols.cshlp.org/site/recipes.

### Reagents

*Escherichia coli* ER2738 (amber suppressor strain; Biosearch Technologies 60522-2) electrocompetent cells

Kanamycin, 50 µg/µL

   To prepare, dissolve 500 mg of kanamycin sulfate (MilliporeSigma 60615) in 10 mL of highly pure water. Sterilize by filtration through a 0.22-µm filter. Store 1-mL aliquots in 1.5-mL microfuge tubes at −20°C.

LB plates <R>

   Alternatively, prepoured LB plates are commercially available (e.g., Teknova L1100).

SB (Super Broth) medium <R>

Tetracycline, 5 µg/µL

   To prepare, dissolve 50 mg of tetracycline hydrochloride (MilliporeSigma T7660) in 10 mL of 75% (v/v) ethanol in water. Store 1-mL aliquots in 1.5-mL microfuge tubes at −20°C.

---

[1]Correspondence: rader33458@gmail.com

© 2026 Cold Spring Harbor Laboratory Press

Cite this protocol as *Cold Spring Harb Protoc*; doi:10.1101/pdb.prot108600

VCSM13 Interference-Resistance Helper Phage stock (~1 × 10$^{11}$ pfu/mL [Agilent 200251 or Thermo Fisher Scientific 50125059]); store at −80°C

Water, highly pure, sterile

*The protocol uses highly pure water from a purification system (see "Equipment"). Use freshly. Several steps require sterile water, which is obtained by filtering highly pure water through a 0.22-µm filter unit. Store at room temperature.*

## Equipment

*A laboratory with standard molecular biology and microbiology equipment is required for this protocol.*

Autoclave
Bunsen burner
Centrifuge tubes (50-mL)
Digital balance with 0.01-g readability
Filtered pipette tips (10-, 20-, 100-, and 200-µL, and 1-mL)
Filters with Luer lock fitting (0.22-µm; e.g., Millex-GP Syringe Filters, Millipore SLGP033RS)
Freezers (−20°C and −80°C)
Glass bottles (500-mL and 1-L, autoclaved)
Glass Erlenmeyer flasks (2-L, autoclaved)
Incubator (37°C)
Microfuge tubes (1.5- and 2-mL)
Microwave oven
Petri dishes (10-cm)
Pipettes (5-, 10-, and 25-mL)
Pipette controller
Refrigerated benchtop centrifuge with swinging bucket rotor (e.g., Beckman Coulter Avanti J-15R with JS-4.750 swinging bucket rotor)
Refrigerator (4°C)
Round-bottom tubes with snap cap (14-mL)
Shaker at 37°C (e.g., Eppendorf New Brunswick Innova 40 Benchtop Incubator Shaker)

*Two separate shakers, one for phage-free conditions and one for phage, are required*

Single-channel and multichannel micro pipettes (1- to 1000-µL)
Water bath (70°C)
Water purification system (e.g., Hydro PicoPure UV Plus)

## METHOD

1. Inoculate 4 µL of ER2738 (from a single electrocompetent bacterial aliquot) into 2 mL of SB medium in a 50-mL centrifuge tube. Add 4 µL of 5 µg/µL tetracycline and shake at 275 rpm for ~2 h at 37°C.

    *The aliquot of electrocompetent bacteria used can be thawed from and frozen at −80°C multiple times.*

2. Prewarm three LB agar plates at 37°C.

3. Prepare 10$^{-6}$, 10$^{-7}$, and 10$^{-8}$ dilutions of commercial helper phage VCSM13 in 1 mL of SB medium in 1.5-mL microfuge tubes.

4. Mix 1 µL of each of these dilutions and 50 µL of the ER2738 culture from Step 1 in three separate 14-mL round-bottom tubes with snap cap and incubate for 15 min at room temperature. In the meantime, liquefy LB top agar in a microwave oven and cool it to <50°C.

5. Add 3 mL of the cooled—but still liquefied—LB top agar to each tube, mix, and pour onto the plain prewarmed LB agar plates from Step 2. Incubate overnight at 37°C.

   *Plaques are bacterial colonies that grow slower due to helper phage infection.*

   *Plates with plaques can be stored for up to 1 wk at 4°C.*

6. The next day, in a 50-mL centrifuge tube, inoculate 10 mL of SB prewarmed to 37°C with 20 μL of ER2738 (from a single electrocompetent bacterial aliquot). Then, add 20 μL of 5 μg/μL tetracycline and shake at 275 rpm for ∼2 h at 37°C.

7. Select a plate from the ones from Step 5 from which a single plaque can be picked. Then, use a 20-μL pipette tip to transfer a single plaque to the culture and shake at 275 rpm for ∼2 h at 37°C.

8. Transfer the infected culture to a 2-L Erlenmeyer flask containing 500 mL of SB prewarmed to 37°C. Add 1 mL of 5 μg/μL tetracycline and 700 μL of 50 μg/μL kanamycin, and shake at 250 rpm overnight (12–16 h) at 37°C.

9. Transfer the culture to 10 separate 50-mL centrifuge tubes and centrifuge at 2500$g$ for 15 min at room temperature.

10. Transfer the supernatants to clean 50-mL centrifuge tubes and discard the pellets. Incubate the supernatants in a water bath for 20 min at 70°C.

11. Centrifuge at 2500$g$ for 15 min at room temperature, transfer the supernatants to clean 50-mL centrifuge tubes, and discard the pellets. Store at 4°C.

12. To determine the titer of the helper phage preparation, repeat Steps 1–5. Count the number of plaques on the plates. Calculate the titer of the helper phage preparation in plaque-forming units per milliliter as follows: number of plaques (e.g., 50) × dilution factor (e.g., $1 \times 10^7$) × 1000.

    *Expect the titer to be in a range from $10^{11}$ to $10^{12}$ pfu/mL.*

    *Although the titer will drop over time, helper phage preparations can be stored for several months at 4°C.*

## RECIPES

### LB Plates

Dissolve 25 g of LB (Lysogeny Broth) medium powder (MilliporeSigma L3522) in 1 L of highly pure water with 15 g of Bacto Agar (BD 214010). Autoclave for 15 min at 121°C and 15 psi. When cooled to ∼50°C, pour into 10-cm sterile Petri dishes (20–25 mL each), remove bubbles by flaming the plates with a Bunsen burner, and allow to solidify at room temperature.
Store for up to 3 mo at 4°C.

### SB (Super Broth) Medium

Mix 20 g of 3-(N-morpholino)propanesulfonic acid (MOPS; MilliporeSigma M1254), 60 g of Bacto Tryptone (Thermo Fisher Scientific 211705), and 40 g of Bacto Yeast Extract (Thermo Fisher Scientific 212750). Add highly pure water (e.g., that produced by Hydro PicoPure UV Plus) to a final volume of 1.9 L. Bring to pH 7.0 with 1 M sodium hydroxide (NaOH; Thermo Fisher Scientific A4782901). Bring volume to 2 L with highly pure water. Sterilize by autoclaving for 15 min at 121°C and 15 psi in two 1-L or four 500-mL glass bottles. Store for up to 2 yr at room temperature if unopened.

## ACKNOWLEDGMENTS

Work in our laboratory was supported by the National Cancer Institute, National Institutes of Health.

Protocol 5

# Generation of Antibody Libraries for Phage Display: Library Reamplification

Haiyong Peng and Christoph Rader[1]

*Department of Immunology and Microbiology, The Herbert Wertheim UF Scripps Institute for Biomedical Innovation & Technology, University of Florida, Jupiter, Florida 33458, USA*

Phage display of Fab libraries enables the de novo discovery and in vitro evolution of monoclonal antibodies. Fab libraries are collections of millions to billions of different antibodies that collectively cover a large antigen or epitope binding space. To preserve the diversity of the Fab library for repeated selection campaigns, it is recommended to use the original phage from the Fab library generation rather than reamplified phage, if practically possible. This is because reamplification will bias the Fab library for clones that are expressed at higher rates. Fab-phage, however, should only be used if they have been prepared on the same day, to avoid proteolytic cleavage of the physical linkage of phenotype (phage-displayed Fab protein) and genotype (phage-encapsulated Fab DNA). Thus, in practice, reamplification of a Fab-phage library cannot usually be avoided. Here, we describe the steps for the reamplification of an original Fab-phage library prior to its selection. The protocol can also be used to reamplify Fab-phage from the third or later panning rounds when enriched clones are unlikely to be lost by reamplification biases.

## MATERIALS

It is essential that you consult the appropriate Material Safety Data Sheets and your institution's Environmental Health and Safety Office for proper handling of equipment and hazardous materials used in this protocol.

RECIPES: Please see the end of this protocol for recipes indicated by <R>. Additional recipes can be found online at http://cshprotocols.cshlp.org/site/recipes.

### Reagents

Bovine serum albumin (BSA), 1% (w/v) in TBS

*To prepare, dissolve 0.5 g of BSA (MilliporeSigma A7030) in 50 mL of 1× TBS and sterilize by filtration through a 0.22-µm filter unit. Store for up to 3 mo at room temperature if unopened.*

Carbenicillin, 100 µg/µL

*To prepare, dissolve 1 g of carbenicillin disodium (MilliporeSigma C1389) in 10 mL of highly pure water. Sterilize by filtration through a 0.22-µm filter. Store 1-mL aliquots in 1.5-mL microfuge tubes at −20°C.*

*Escherichia coli* ER2738 (amber suppressor strain; Biosearch Technologies 60522-2) electrocompetent cells; store at −80°C

Ice

---

[1]Correspondence: rader33458@gmail.com

© 2026 Cold Spring Harbor Laboratory Press
Cite this protocol as *Cold Spring Harb Protoc*; doi:10.1101/pdb.prot108601

Kanamycin, 50 µg/µL
> To prepare, dissolve 500 mg of kanamycin sulfate (MilliporeSigma 60615) in 10 mL of highly pure water. Sterilize by filtration through a 0.22-µm filter. Store 1-mL aliquots in 1.5-mL microfuge tubes at −20°C.

LB + 100 µg/mL carbenicillin plates <R>

Original Fab-phage library (see, i.e., Protocol 1: Generation of Antibody Libraries for Phage Display: Human Fab Format [Peng and Rader 2024a] or Protocol 2: Generation of Antibody Libraries for Phage Display: Chimeric Rabbit/Human Fab Format [Peng and Rader 2024b])

PEG-8000 (MilliporeSigma P2139); store at room temperature

SB (Super Broth) medium <R>

Sodium chloride (NaCl; MilliporeSigma S9888); store at room temperature

Tetracycline, 5 µg/µL
> To prepare, dissolve 50 mg of tetracycline hydrochloride (MilliporeSigma T7660) in 10 mL of 75% (v/v) ethanol. Store 1-mL aliquots in 1.5-mL microfuge tubes at −20°C.

Tris-buffered saline (TBS), 1×
> Prepare by diluting 10× TBS (Thermo Fisher Scientific J60764.K7) in highly pure water and sterilizing by filtration through a 0.22-µm filter unit. Store for up to 3 mo at room temperature if unopened.

VCSM13 helper phage at $10^{11}$–$10^{12}$ plaque-forming units (pfu)/mL
> Starting from a commercially available VCSM13 Interference-Resistance Helper Phage stock (∼1 × $10^{11}$ pfu/mL [Agilent 200251 or Thermo Fisher Scientific 50125059]; store at −80°C), prepare 500 mL of VCSM13 helper phage at $10^{11}$–$10^{12}$ pfu/mL, as described in Protocol 4: Generation of Antibody Libraries for Phage Display: Preparation of Helper Phage (Peng and Rader 2024c).

Water, highly pure, sterile
> The protocol uses highly pure water from a purification system (see "Equipment"). Use freshly. Several steps require sterile water, which is obtained by filtering highly pure water through a 0.22-µm filter unit. Store at room temperature.

## Equipment

*A laboratory with standard molecular biology and microbiology equipment is required for this protocol.*

Autoclave
Bunsen burner
Centrifuge bottles (500-mL; polypropylene, wide mouth, sealing cap; certified for >12,000g and resistant to bleaching and autoclaving; e.g., Nalgene 3141-0500)
Centrifuge tubes (50-mL)
Digital balance with 0.01-g readability
Filtered pipette tips (10-, 20-, 100-, and 200-µL, and 1-mL)
Filters with Luer lock fitting (0.22-µm; e.g., Millex-GP Syringe Filters, Millipore SLGP033RS)
Filter units (0.22-µm; e.g., Millipore Stericup and Steriflip filter units)
Freezers (−20°C and −80°C)
Glass bottles (500-mL and-1 L, autoclaved)
Glass Erlenmeyer flasks (250-mL, autoclaved)
Ice bucket
Incubator (37°C)
Magnetic hot plate stirrer
Magnetic stir bars
Microfuge tubes (1.5- and 2-mL)
Petri dishes (10-cm)
Pipettes (5-, 10-, and 25-mL)
Pipette controller
Refrigerated benchtop centrifuge with swinging bucket rotor (e.g., Beckman Coulter Avanti J-15R with JS-4.750 swinging bucket rotor)

Refrigerated floor centrifuge (e.g., Thermo Fisher Scientific Sorvall Lynx 4000) with fixed-angle rotor for six 500-mL centrifuge bottles (e.g., Thermo Fisher Scientific Fiberlite F12-6 × 500 LEX)
Refrigerated microcentrifuge (e.g., Eppendorf 5425R)
Refrigerator (4°C)
Round-bottom tubes with snap cap (14-mL)
Shaker at 37°C (e.g., Eppendorf New Brunswick Innova 40 Benchtop Incubator Shaker).
*Two separate shakers, one for phage-free conditions and one for phage, are required.*

Single-channel pipettes (1–1000 μL)
Syringes with Luer lock fitting (1-, 5-, and 10-mL)
UV photometer (e.g., NanoDrop One$^C$ Microvolume UV-Vis Spectrophotometer, Thermo Fisher Scientific ND-ONEC-W)
Water purification system (e.g., Hydro PicoPure UV Plus)

## METHOD

1. Inoculate 50 μL of ER2738 (from a single electrocompetent ER2738 aliquot) into 50 mL of SB medium in a 250-mL Erlenmeyer flask. Add 100 μL of 5 μg/μL tetracycline and shake at 250 rpm for ∼1.5–2.5 h at 37°C to an $OD_{600}$ of ∼1.

   *Take precautions to avoid contaminating the E. coli culture with phage, including helper phage. Use a separate area and filtered pipette tips for pipetting and a phage-free shaker for growth. This is important, because the wild-type pIII protein of helper phage (unlike the C-terminal pIII protein domain encoded by the phagemid) renders E. coli immune to superinfection.*

2. Add 10 μL of a stored original Fab-phage library to the ER2738 culture and incubate without shaking for 15 min at room temperature.

3. Add 20 μL of 100 μg/μL carbenicillin, remove a 1-μL aliquot for use in Step 4, and shake the remaining culture at 275 rpm for 1 h at 37°C.

4. In the meantime, to determine the number of independently reamplified phage, add the 1-μL aliquot from Step 3 to 10 mL of SB medium in a 14-mL round-bottom tube. Mix and plate 10 μL on an LB + 100 μg/mL carbenicillin plate. Incubate overnight at 37°C. The next day, calculate the number of independently reamplified phage as follows: number of colonies × 50,000 (culture volume) × 1000 (dilution factor); e.g., $500 \times 50,000 \times 1000 = 2.5 \times 10^{10}$.

   *This number should be five to 10 times higher than the size (i.e., the number of independent transformants) of the original Fab-phage library.*

5. After the 1-h incubation from Step 3, add 30 μL of 100 μg/μL carbenicillin to the culture from Step 3 and continue shaking at 275 rpm for 1 h at 37°C.

6. Add 2 mL of VCSM13 helper phage ($10^{11}$–$10^{12}$ pfu/mL) and transfer the entire mix to a 500-mL centrifuge bottle. Add 148 mL of SB, 150 μL of 100 μg/μL carbenicillin, and 300 μL of 5 μg/μL tetracycline (Σ 200 mL).

7. Incubate the culture with shaking at 275 rpm for 90 min at 37°C. Then, add 280 μL of 50 μg/μL kanamycin and continue shaking at 275 rpm overnight (12–16 h) at 30°C.

   *Using a lower temperature (i.e., 30°C instead of 37°C) in this step helps to maintain library diversity by limiting the advantage of faster growth.*

8. The next day, transfer the culture to a 500-mL centrifuge bottle and centrifuge at 3000g for 15 min at 4°C.

9. For phage precipitation, transfer the supernatant to a clean 500-mL centrifuge bottle containing 8 g of PEG-8000 and 6 g of NaCl and dissolve by shaking at 300 rpm for 5 min at 37°C.

10. Incubate for 30 min on ice. Centrifuge at 12,000g for 15 min at 4°C.

11. Discard the supernatant, drain the bottle by inverting on a paper towel for 10 min, and carefully wipe off any remaining liquid with a paper towel.
12. Resuspend the phage pellet in 2 mL of 1% (w/v) BSA in TBS by pipetting up and down along the side of the centrifuge tube.
13. Transfer the pellet to a 2-mL microfuge tube and pipette up and down with a 1-mL pipette tip.
14. Centrifuge at 16,000$g$ for 5 min at 4°C.
15. Pass the supernatant through a 0.22-μm filter into a clean 2-mL microfuge tube. Store on ice for panning on the same day.

    *Only fresh phage prepared on the same day should be used for library selection because the covalent linkage of Fab (phenotype) and phage (genotype) is susceptible to cleavage by contaminating proteases.*

    *Proceed to library selection as described in Protocol 6: Phage Display Selection of Antibody Libraries: Panning Procedures (Peng and Rader 2024d).*

## RECIPES

### LB + 100 μg/mL Carbenicillin Plates

Dissolve 25 g of LB (Lysogeny Broth) medium powder (MilliporeSigma L3522) in 1 L of highly pure water (e.g., that produced by Hydro PicoPure UV Plus) with 15 g of Bacto Agar (BD 214010). Autoclave for 15 min at 121°C and 15 psi. When cooled to ~50°C, add 1 mL of 100 μg/μL carbenicillin. Pour into 10-cm sterile Petri dishes (20–25 mL each), remove bubbles by quickly flaming the plates with a Bunsen burner, and allow to solidify at room temperature. Store for up to 3 mo at 4°C.

*Alternatively, prepoured LB + 100 μg/mL carbenicillin plates are commercially available (e.g., Teknova L1010).*

### SB (Super Broth) Medium

Mix 20 g of 3-(N-morpholino)propanesulfonic acid (MOPS; MilliporeSigma M1254), 60 g of Bacto Tryptone (Thermo Fisher Scientific 211705), and 40 g of Bacto Yeast Extract (Thermo Fisher Scientific 212750). Add highly pure water (e.g., that produced by Hydro PicoPure UV Plus) to a final volume of 1.9 L. Bring to pH 7.0 with 1 M sodium hydroxide (NaOH; Thermo Fisher Scientific A4782901). Bring volume to 2 L with highly pure water. Sterilize by autoclaving for 15 min at 121°C and 15 psi in two 1-L or four 500-mL glass bottles. Store for up to 2 yr at room temperature if unopened.

## ACKNOWLEDGMENTS

Work in our laboratory was supported by the National Cancer Institute, National Institutes of Health.

## REFERENCES

Peng H, Rader C. 2024a. Generation of antibody libraries for phage display: human Fab format. *Cold Spring Harb Protoc* doi:10.1101/pdb.prot108597

Peng H, Rader C. 2024b. Generation of antibody libraries for phage display: chimeric rabbit/human Fab format. *Cold Spring Harb Protoc* doi:10.1101/pdb.prot108598

Peng H, Rader C. 2024c. Generation of antibody libraries for phage display: preparation of helper phage. *Cold Spring Harb Protoc* doi:10.1101/pdb.prot108600

Peng H, Rader C. 2024d. Phage display selection of antibody libraries: panning procedures. *Cold Spring Harb Protoc* doi:10.1101/pdb.prot108602

Protocol 6

# Phage Display Selection of Antibody Libraries: Panning Procedures

Haiyong Peng and Christoph Rader[1]

*Department of Immunology and Microbiology, The Herbert Wertheim UF Scripps Institute for Biomedical Innovation & Technology, University of Florida, Jupiter, Florida 33458, USA*

The mining of naive, immune, and synthetic antibody repertoires by phage display has been widely applied to the de novo generation and in vitro evolution of monoclonal antibodies from multiple species. Once built, phage display antibody libraries can be selected by a variety of different strategies tailored toward the desired antigen binding properties. Here, we describe the selection of antibody libraries generated in a phage display vector of the pComb3 phagemid family. The approach includes panning procedures for immobilized antigens, biotinylated antigens in solution, and cell surface antigens. Although the typical format of these antibody libraries is human Fab or chimeric nonhuman/human Fab, the basic selection strategies provided in this protocol are compatible with a variety of formats.

## MATERIALS

It is essential that you consult the appropriate Material Safety Data Sheets and your institution's Environmental Health and Safety Office for proper handling of equipment and hazardous materials used in this protocol.

RECIPES: Please see the end of this protocol for recipes indicated by <R>. Additional recipes can be found online at http://cshprotocols.cshlp.org/site/recipes.

### Reagents

Bovine serum albumin (BSA), 1% (w/v) in TBS

*To prepare, dissolve 0.5 g of BSA (MilliporeSigma A7030) in 50 mL of 1× TBS and sterilize by filtration through a 0.22-µm filter unit. Store at room temperature.*

Bovine serum albumin (BSA) 3% (w/v) in TBS

*To prepare, dissolve 1.5 g of BSA (MilliporeSigma A7030) in 50 mL of 1× TBS, sterilize by filtration through a 0.22-µm filter unit, and store at room temperature.*

Carbenicillin, 100 µg/µL

*To prepare, dissolve 1 g of carbenicillin disodium (MilliporeSigma C1389) in 10 mL of highly pure water. Sterilize by filtration through a 0.22-µm filter. Store 1-mL aliquots in 1.5-mL microfuge tubes at −20°C.*

*Escherichia coli* ER2738 (amber suppressor strain; Biosearch Technologies 60522-2) electrocompetent cells; store at −80°C (alternatively, see Protocol 3: Generation of Antibody Libraries for Phage Display: Preparation of Electrocompetent *E. coli* [Peng and Rader 2024a])

---

[1]Correspondence: rader33458@gmail.com

© 2026 Cold Spring Harbor Laboratory Press
Cite this protocol as *Cold Spring Harb Protoc*; doi:10.1101/pdb.prot108602

Ice

Kanamycin, 50 µg/µL

*To prepare, dissolve 500 mg of kanamycin sulfate (MilliporeSigma 60615) in 10 mL of highly pure water. Sterilize by filtration through a 0.22-µm filter. Store 1-mL aliquots in 1.5-mL microfuge tubes at −20°C.*

LB + 100 µg/mL carbenicillin plates <R>

Original or reamplified Fab-phage library (see Protocol 1: Generation of Antibody Libraries for Phage Display: Human Fab Format [Peng and Rader 2024b], Protocol 2: Generation of Antibody Libraries for Phage Display: Chimeric Rabbit/Human Fab Format [Peng and Rader 2024c], or Protocol 5: Generation of Antibody Libraries for Phage Display: Library Reamplification [Peng and Rader 2024d])

PEG-8000 (MilliporeSigma P2139); store at room temperature.

Reagents needed only if the library will be panned against a biotinylated antigen (see "Panning Against Biotinylated Antigen in Solution"):

BiotinTag Micro Biotinylation Kit (MilliporeSigma BTAG); store at 4°C.

Dynabeads MyOne Strepatavidin C1 (Thermo Fisher Scientific 65001); store at 4°C.

Purified biotinylated antigen of interest (25–100 µg) in TBS; store at −80°C.

Reagents needed only if the library will be panned against an immobilized antigen (see "Panning Against Immobilized Antigen"):

Purified antigen of interest (0.5–5 µg) in carrier-free buffer; store at −80°C

Reagents needed only if the library will be panned against whole cells (see "Panning Against Eukaryotic Cells in Solution"):

BSA, 1% (w/v) in PBS

*To prepare, dissolve 0.5 g of BSA (MilliporeSigma A7030) in 50 mL of 1× PBS. Sterilize by filtration through a 0.22-µm filter unit, and store for up to 3 mo at room temperature if unopened.*

Eukaryotic cells ($1 \times 10^7$ to $5 \times 10^7$); nonadherent or suspended adherent eukaryotic cells originating from cultured cell lines or from primary cells

*The collection and processing of clinical specimens from human subjects must be approved by an Institutional Review Board (IRB). Clinical specimens should be treated as potentially infectious, and appropriate safety precautions should be taken. Wear appropriate protection and handle with care.*

High-affinity rat anti-HA IgG1 3F10 (Roche Life Science Products available from, e.g., MilliporeSigma 11867423001); store at 4°C

Human serum (e.g., MilliporeSigma H6914) or fetal bovine serum (FBS; e.g., Thermo Fisher Scientific Gibco FBS); store at −20°C.

Phosphate-buffered saline (PBS, 1×)

*Prepare by diluting 10× PBS (Thermo Fisher Scientific J61196.AP) in highly pure water. Sterilize by filtration through a 0.22-µm filter unit, and store for up to 3 mo at room temperature if unopened.*

Sodium azide ($NaN_3$), 2% (w/v)

*Dissolve 0.2 g of $NaN_3$ (MilliporeSigma S8032) in 10 mL of highly pure water. Store at room temperature.*

Trypsin, 10 mg/mL in PBS

*Dissolve 50 mg of Gibco Trypsin 1:250 (Thermo Fisher Scientific 27250018) in 5 mL of 1× PBS and sterilize by filtration through a 0.22 µm filter unit. Use freshly.*

SB (Super Broth) medium <R>

Sodium chloride (NaCl; MilliporeSigma S9888); store at room temperature

Tetracycline, 5 µg/µL

*To prepare, dissolve 50 mg of tetracycline hydrochloride (MilliporeSigma T7660) in 10 mL of 75% (v/v) ethanol. Store 1-mL aliquots in 1.5-mL microfuge tubes at −20°C.*

Tris-buffered saline (TBS, 1×)

*Prepare by diluting 10× TBS (Thermo Fisher Scientific J60764.K7) in highly pure water and sterilizing by filtration through a 0.22-µm filter unit. Store for up to 3 mo at room temperature if unopened.*

Trypsin, 10 mg/mL in TBS
> *To prepare, dissolve 50 mg of Gibco Trypsin 1:250 (Thermo Fisher Scientific 27250018) in 5 mL of 1× TBS. Sterilize by filtration through a 0.22 μm filter unit. Use freshly.*

Tween 20, 0.05% (v/v) in TBS
> *To prepare, dilute 25 μL of Tween 20 (MilliporeSigma P1379) in 50 mL of 1× TBS. Sterilize by filtration through a 0.22 μm filter unit, and store for up to 3 mo at room temperature if unopened.*

VCSM13 helper phage at $10^{11}$–$10^{12}$ plaque-forming units (pfu)/mL.
> *Starting from a commercially available VCSM13 Interference-Resistance Helper Phage stock (~$1 \times 10^{11}$ pfu/mL [Agilent 200251 or Thermo Fisher Scientific 50125059]; store at −80°C), prepare 500 mL of VCSM13 helper phage at $10^{11}$–$10^{12}$ pfu/mL as described in Protocol 4: Generation of Antibody Libraries for Phage Display: Preparation of Helper Phage (Peng and Rader 2024e).*

Water, highly pure, sterile
> *The protocol uses highly pure water from a purification system (see "Equipment"). Use freshly. Several steps require sterile water, which is obtained by filtering highly pure water through a 0.22-μm filter unit. Store at room temperature.*

## Equipment

*A laboratory with standard molecular biology and microbiology equipment is required for this protocol.*

Autoclave
Bunsen burner
Centrifuge bottles (500-mL; polypropylene, wide mouth, sealing cap; certified for >12,000g and resistant to bleaching and autoclaving; e.g., Nalgene 3141-0500)
Centrifuge tubes (50-mL)
Conical centrifuge tubes (15-mL)
Digital balance with 0.01-g readability
Equipment needed only if the library will be panned against a biotinylated antigen (see "Panning Against Biotinylated Antigen in Solution"):
    Dynabeads MPC-S magnetic particle concentrator (Thermo Fisher Scientific A13346)
    Minirotator (e.g., Glas-Col 099A MR1512)
    Screw-cap tubes (2-mL)
Equipment needed only if the library will be panned against an immobilized antigen (see "Panning Against Immobilized Antigen"):
    ELISA plate, 96-well, half-area (Costar 3690; Thermo Fisher Scientific 07-200-37)
    Plate sealer (SealPlate Sealing Film; Excel Scientific 100-SEAL-PLT)
Equipment needed only if the library will be panned against whole cells (see "Panning Against Eukaryotic Cells in Solution"):
    ELISA plate, 96-well, half-area (Costar 3690; Thermo Fisher Scientific 07-200-37)
    Laminar flow hood
    Liquid nitrogen tank with cryobox storage racks
    Plate sealer (SealPlate Sealing Film; Excel Scientific 100-SEAL-PLT)
Filtered pipette tips (10-, 20-, 100-, and 200-μL, and 1-mL)
Filters with Luer lock fitting (0.22-μm; e.g., Millex-GP Syringe Filters, Millipore SLGP033RS)
Filter units (0.22-μm; e.g., Millipore Stericup and Steriflip filter units)
Freezers (−20°C and −80°C)
Glass bottles (500-mL and 1-L, autoclaved)
Glass Erlenmeyer flasks (250-mL, and 1- and 2-L, autoclaved)
Ice bucket
Incubator (37°C)
Magnetic hot plate stirrer
Magnetic stir bars

Chapter 5

Microfuge tubes (1.5- and 2-mL)
Paper towels
Petri dishes (10-cm)
Pipettes (5-, 10-, and 25-mL)
Pipette controller
Refrigerated benchtop centrifuge with swinging bucket rotor (e.g., Beckman Coulter Avanti J-15R with JS-4.750 swinging bucket rotor)
Refrigerated floor centrifuge (e.g., Thermo Fisher Scientific Sorvall Lynx 4000) with fixed-angle rotor for six 500-mL centrifuge bottles (e.g., Thermo Fisher Scientific Fiberlite F12-6 × 500 LEX)
Refrigerated microcentrifuge (e.g., Eppendorf 5425R)
Refrigerator (4°C)
Round-bottom tubes with snap cap (14-mL)
Shaker at 37°C (e.g., Eppendorf New Brunswick Innova 40 Benchtop Incubator Shaker).

*Two separate shakers, one for phage-free conditions and one for phage, are required.*

Single-channel pipettes (1- to 1000-µL)
Syringes with Luer lock fitting (1-, 5-, and 10-mL)
Water purification system (e.g., Hydro PicoPure UV Plus)

## METHOD

*Three different options are described below, depending on the type of antigen the library is panned against. If purified stable antigen is available, it can be used immobilized (see "Panning Against Immobilized Antigen") or, following biotinylation, in solution (see "Panning Against Biotinylated Antigen in Solution"). In the third option, "Panning Against Eukaryotic Cells in Solution," we describe a basic protocol for panning the library against nonadherent or suspended adherent human cells, which may be the option of choice if the antigen is unknown or purified stable antigen is unavailable. Using different libraries or different antigens, up to four panning experiments for the first two options, and up to two panning experiments for the third option, can be carried out in parallel. Figure 1 depicts the steps involved in each panning round, which takes 24 h to complete.*

### Panning Against Immobilized Antigen

1. Coat one to two wells of a 96-well half-area ELISA plate with 0.1–1 µg of antigen in 25 µL of TBS, cover the wells with plate sealer, and incubate for 1 h at 37°C.

    *In the first panning round, coat two wells for each library to be selected. One well is sufficient in each of all subsequent panning rounds.*

2. Shake out the coating solution, add 150 µL of 3% (w/v) BSA in TBS to block each coated well, cover with plate sealer, and incubate for 1 h at 37°C.

    *To save time, carry out Steps 1 and 2 in parallel to harvesting fresh Fab-phage in the relevant library generation protocols; i.e., in Protocol 1: Generation of Antibody Libraries for Phage Display: Human Fab Format (Peng and Rader 2024b) or in Protocol 2: Generation of Antibody Libraries for Phage Display: Chimeric Rabbit/Human Fab Format (Peng and Rader 2024c) when starting with a freshly prepared original human Fab-phage library or chimeric rabbit/human Fab-phage library, respectively, or in Protocol 5: Generation of Antibody Libraries for Phage Display: Library Reamplification (Peng and Rader 2024d) when starting with a freshly prepared reamplified Fab-phage library. Likewise, in subsequent panning rounds, carry out Steps 1 and 2 in parallel to Steps 14–20 in this protocol.*

3. Shake out the blocking solution, add 50 µL of the freshly prepared original or reamplified phage preparation in 1% (w/v) BSA in TBS (see "Materials") to each blocked well, cover with plate sealer, and incubate for 2 h at 37°C.

4. In the meantime, inoculate 4 µL of ER2738 (from a single electrocompetent ER2738 aliquot) into 2 mL of SB medium in a 50-mL centrifuge tube. Add 4 µL of 5 µg/µL tetracycline and

shake at 275 rpm for ∼2 h at 37°C. Grow one culture for each parallel panning experiment and one additional culture for determining the input titer.

*The E. coli aliquot can be thawed and frozen at 80°C multiple times.*

*Take precautions to avoid contaminating the E. coli culture with phage, including helper phage. Use a separate area and filtered pipette tips for pipetting and a phage-free shaker for growth. This is important, because the wild-type pIII protein of helper phage (unlike the C-terminal pIII protein domain encoded by the phagemid) renders E. coli immune to superinfection.*

5. Shake out the phage solution, add 150 μL of 0.05% (v/v) Tween 20 in TBS to each prepared well, pipette vigorously up and down five times, and incubate for 5 min at room temperature (a multichannel pipette may be used to wash two or more wells at the same time). Shake out the washing solution and repeat this washing step four additional times (if first panning round), five to 10 times (if second and third panning round), and 10–15 times (if fourth and any subsequent panning rounds).

6. Shake out the washing solution, add 50 μL of freshly prepared 10 mg/mL trypsin in TBS to each prepared well, cover with plate sealer, and incubate for 30 min at 37°C.

7. Pipette vigorously up and down 10 times and transfer the combined trypsin solution (50 μL/well; i.e., 2 × 50 μL in the first panning round, and 1 × 50 μL in all subsequent panning rounds) to the prepared 2 mL of ER2738 culture (from Step 4). Incubate for 15 min at room temperature.

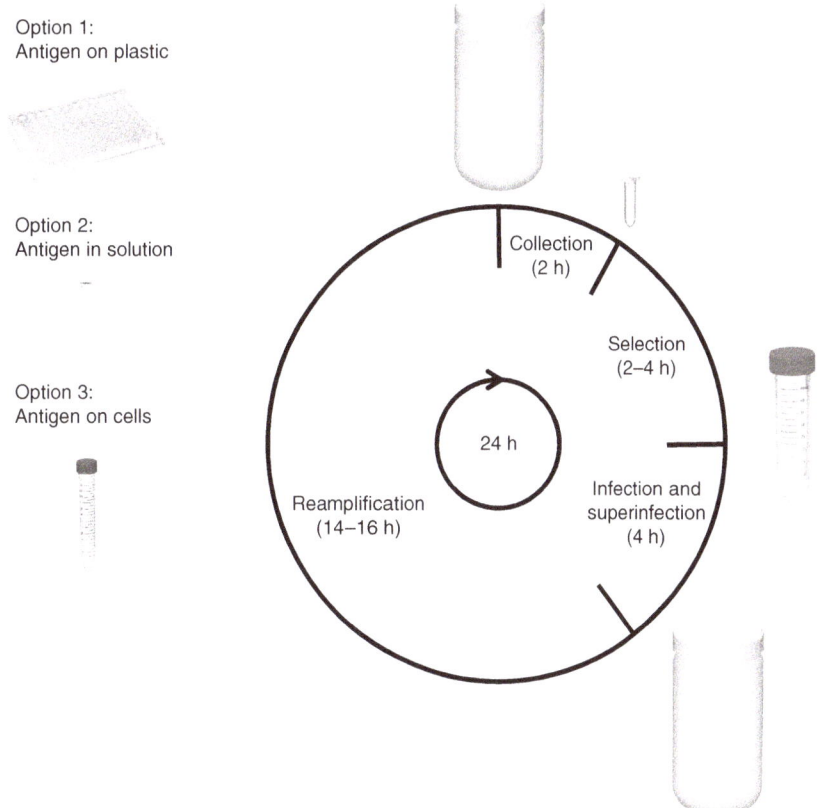

FIGURE 1. Panning rounds described in this protocol. Each panning round consists of a 24-h cycle of phage collection, phage selection (with one of the options shown on the *left*), phage infection followed by helper phage superinfection, and phage reamplification. Plasticware used for collection (a 500-mL centrifuge tube and a 2-mL microfuge tube), selection (a 96-well half-area ELISA plate, a 2-mL screw cap tube, or a 15-mL conical centrifuge tube depending on option 1, 2, or 3, respectively, selected for the antigen), and infection, superinfection, and reamplification (50- and 500-mL centrifuge tubes) is shown.

8. To each phage-infected 2-mL ER2738 culture, add 6 mL of SB, 3.2 µL of 100 µg/µL carbenicillin, and 12 µL of 5 µg/µL tetracycline. Shake at 250 rpm for 1 h at 37°C.

9. For determining the output titer, remove a 2-µL aliquot, add to 198 µL of SB medium in a 1.5-mL microfuge tube, and plate 100 µL of the mix on an LB + 100 µg/mL carbenicillin plate. Incubate overnight at 37°C and count the colonies.

   *The number of colonies (e.g., 50) ×8000 (volume of output solution, in microliters) determines the output number for each panning round (e.g., $4 \times 10^5$). Output numbers can range anywhere from $1 \times 10^4$ to $1 \times 10^7$. They are typically low in the second panning round and sometimes increase by a factor of 10–100 from the second to third or third to fourth panning round, indicating a successful library selection. For low-output titers, see Troubleshooting.*

10. For determining the input titer, prepare a $10^{-8}$ dilution of the phage preparation by serial dilutions in SB medium. Transfer 150 µL of the ER2738 culture (from Step 4) into a 1.5-mL microfuge tube, add 1 µL of the $10^{-8}$ dilution, incubate for 15 min at room temperature, and plate the entire volume on an LB + 100 µg/mL carbenicillin plate. Incubate overnight at 37°C and count the colonies.

    *The number of colonies (e.g., 50) $\times 1 \times 10^8$ (dilution factor) ×50 or 100 (volume of input phage preparation, i.e., 50 µL/well for one to two wells) determines the input number for each panning round (e.g., $5 \times 10^{11}$). Input numbers vary less than output numbers, with a typical range from $1 \times 10^{11}$ to $1 \times 10^{12}$. For low-input titers, see Troubleshooting.*

11. Add 4.8 µL of 100 µg/µL carbenicillin to the culture in the 50-mL centrifuge tube (from Step 8) and continue shaking at 250 rpm for 1 h at 37°C.

12. Add 1 mL of VCSM13 helper phage ($10^{11}$–$10^{12}$ pfu/mL) and transfer the mix into a 500-mL centrifuge bottle. Add 91 mL of SB, 92 µL of 100 µg/µL carbenicillin, and 184 µL of 5 µg/µL tetracycline (Σ 100 mL). Shake at 275 rpm for 90 min at 37°C.

13. Add 140 µL of 50 µg/µL kanamycin and continue shaking at 275 rpm overnight (14–16 h) at 30°C.

    *Using a lower temperature (i.e., 30°C instead of 37°C) in this step helps to maintain clonal diversity by limiting the advantage of faster growth. If there is no overnight growth, see Troubleshooting.*

14. Centrifuge at 3000g for 15 min at 4°C. For phage precipitation, transfer the supernatant to a clean 500-mL centrifuge bottle containing 4 g of PEG-8000 and 3 g of NaCl and dissolve by shaking at 300 rpm for 5 min at 37°C.

15. Incubate for 30 min on ice. Centrifuge at 12,000g for 15 min at 4°C.

16. Discard the supernatant, drain the bottle by inverting on a paper towel for 10 min, and carefully wipe off any remaining liquid with a paper towel.

17. Resuspend the phage pellet in 2 mL of 1% (w/v) BSA in TBS by pipetting up and down along the side of the centrifuge tube.

18. Transfer to a 2-mL microfuge tube, and pipette up and down with a 1-mL pipette tip.

19. Centrifuge at 16,000g for 5 min at 4°C.

20. Pass the supernatant through a 0.22-µm filter into a clean 2-mL microfuge tube. Store on ice and continue with the next panning round (i.e., go back to Step 1) until the selected number of rounds has been completed.

    *Typically, four panning rounds are sufficient to select a panel of human Fabs that bind with high affinity and specificity. If the panning experiment needs to be interrupted, add 0.01 volumes of 2% (w/v) sodium azide to the phage preparation, store at 4°C, and resume the panning experiment by first reamplifying phage, as described in Protocol 5: Generation of Antibody Libraries for Phage Display: Library Reamplification (Peng and Rader 2024d).*

    *Only fresh phage prepared on the same day should be used for library selection because the covalent linkage of Fab (phenotype) and phage (genotype) is susceptible to cleavage by contaminating proteases.*

21. Proceed to analyze library selection.

    *Before proceeding with the analysis of library selection, as described in Protocol 7: Phage Display Selection of Antibody Libraries: Screening of Selected Binders (Peng and Rader 2024f), phage preparations and output plates from each panning round can be stored for up to 1 wk at 4°C.*

## Panning Against Biotinylated Antigen in Solution

*The steps below describe the use of biotinylated antigen to bind Fab-phage in solution and then capture and isolate the complex by streptavidin-coated magnetic beads. To avoid a loss of Fab-phage binders present in low copy number in the first panning round, it is important to not exceed the capturing capacity of the beads. Alternatively, biotinylated antigen can be first coated on the beads and then used to capture Fab-phage binders. In either case, ensure the absence or removal of free biotin.*

22. Transfer 1 mL of the freshly prepared phage preparation in 1% (w/v) BSA in TBS to a 2-mL screw cap tube. Add 5 µL of 1 µg/µL biotinylated antigen in TBS. Using a minirotator, rotate the mixture at 10 rpm for 1 h at 37°C.

    *Recommended biotinylation reagents and protocols are supplied by the BiotinTag Micro Biotinylation Kit, which is based on an N-hydroxysuccinimide derivative of biotin that reacts with primary amino groups provided by the ε-amino group of lysine or the N terminus in the protein or peptide antigens. Note that this modification may remove natural epitopes or add artificial epitopes.*

    *To select for high affinity, gradually lower the concentration of the biotinylated antigen in solution from one panning round to the next. Ultimately, use subnanomolar concentrations to select Fab-phage binders with subnanomolar affinities.*

23. In the meantime, inoculate 4 µL of ER2738 (from a single electrocompetent ER2738 aliquot, which can be thawed and frozen at −80°C multiple times) into 2 mL of SB medium in a 50-mL centrifuge tube. Add 4 µL of 5 µg/µL tetracycline and shake at 275 rpm for ∼2 h at 37°C. Grow one culture for each parallel panning experiment and one additional culture for determining the input titer.

    *Take precautions to avoid contaminating the E. coli culture with phage, including helper phage. Use a separate area and filtered pipette tips for pipetting and a phage-free shaker for growth. This is important, because the wild-type pIII protein of helper phage (unlike the C-terminal pIII protein domain encoded by the phagemid) renders E. coli immune to superinfection.*

24. Wash 50 µL of Dynabeads MyOne Strepatavidin C1 with 1 mL of TBS according to the manufacturer's protocol and add to the incubation mixture. Rotate at 10 rpm for 30 min at 37°C.

25. Place the sample on a Dynabeads MPC-S magnetic particle concentrator for 2 min. Then, remove the supernatant with a pipette while keeping the sample on the magnet.

26. Remove the sample from the magnet and resuspend the beads in 1 mL of 0.05% (v/v) Tween 20 in TBS. Rotate at 10 rpm for 5 min at 37°C. Place the sample on the magnet for 2 min. Remove the supernatant with a pipette while keeping the sample on the magnet. Repeat this washing step four additional times (if first panning round), five to 10 times (if second and third panning round), and 10–15 times (if fourth and all subsequent panning rounds).

27. After removing the last supernatant, resuspend the beads in 50 µL of freshly prepared 10 mg/mL trypsin in TBS and incubate for 30 min at 37°C.

28. Add the prepared 2 mL of ER2738 culture (from Step 23) directly to the trypsinized beads. Incubate for 15 min at room temperature. Continue as described in Steps 8–21, using the additional culture from Step 23 for determining the input titer as described in Step 10.

## Panning Against Eukaryotic Cells in Solution

*Below we describe a basic protocol for the selection of Fab-phage libraries against whole cells. The protocol uses a simple strategy that is based on positive selection only. Depending on the origin of the Fab-phage library, the antigen of interest, and the available cells, it may be advisable to include consecutive or simultaneous negative and positive selection with antigen-negative (or antigen-masked) (Popkov et al. 2004) and antigen-positive (Siegel*

Chapter 5

*2002; Siva et al. 2008) cells. These advanced protocols may avoid the selection of both phage that stick to cells nonspecifically and phage that bind to common cell surface antigens.*

*A variant of pC3C, named pC3Csort (described in Chapter 4, Overview: The pComb3 Phagemid Family of Phage Display Vectors [Rader 2024]), has been used to overcome challenges of whole-cell panning by applying a Fab biotinylation and selection (FBC) strategy (Wilson et al. 2018), which can be combined with high-throughput sequencing for highly efficient target-defined or target-agnostic panning campaigns (Cyr et al. 2023).*

29. Using 500 ng of high-affinity rat anti-HA IgG1 3F10 as immobilized antigen in each of two wells (Step 1), carry out one panning round as described in Steps 2–20.

    *Panning against immobilized high-affinity rat anti-HA IgG1 3F10 eliminates phage that does not display functional human Fab with the hemagglutinin (HA) tag. The following phagemids of the pComb3 family are compatible with this strategy, as they encode an HA tag at the C terminus of the heavy chain fragment: pComb3X, pC3C, and pC3Csort (see Chapter 4, Overview: The pComb3 Phagemid Family of Phage Display Vectors [Rader 2024]). Including this polishing step before and during panning against human cells was found to be critical for successful selection of Fab-phage libraries on whole cells (Baskar et al. 2009).*

30. Inoculate 4 µL of ER2738 (from a single electrocompetent ER2738 aliquot, which can be thawed and frozen at −80°C multiple times) into 2 mL of SB medium in a 50-mL centrifuge tube. Add 4 µL of 5 µg/µL tetracycline and shake at 275 rpm for ∼2 h at 37°C. Grow two cultures for each parallel panning experiment and one additional culture for determining the input titer.

    *Take precautions to avoid contaminating the E. coli culture with phage, including helper phage. Use a separate area and filtered pipette tips for pipetting and a phage-free shaker for growth. This is important, because the wild-type pIII protein of helper phage (unlike the C-terminal pIII protein domain encoded by the phagemid) renders E. coli immune to superinfection.*

31. Mix 0.5 mL of the phage preparation with 0.5 mL of 5% (v/v) human serum or FBS in PBS and 0.5 mL of 1% (w/v) BSA in PBS in a 15-mL conical centrifuge tube. Add 15 µL (0.01 volumes) of 2% (w/v) $NaN_3$ and incubate for 30 min at room temperature.

    *In this blocking step, use serum that the human cells have been exposed to, e.g., human serum for primary human cells and FBS for mammalian cell lines.*

    *Sodium azide is added to prevent internalization of phage.*

32. Collect $1 \times 10^7$ to $5 \times 10^7$ nonadherent or suspended adherent eukaryotic cells (see "Materials") in 1.5 mL of 5% (v/v) human serum or FBS in PBS or other medium, add to the 1.5-mL phage preparation from Step 31, and incubate for 30 min at room temperature with gentle swirling every 5 min to keep cells suspended.

33. Add 12 mL of PBS to the mix, centrifuge at 300g for 10 min at room temperature, and decant and discard the supernatant. Repeat this washing step twice more with 15 mL of PBS.

34. Resuspend the cells in 0.6 mL of freshly prepared 10 mg/mL trypsin in PBS and shake at 250 rpm for 30 min at 37°C.

35. Pipette vigorously up and down 10 times and transfer half of the trypsin solution to each of the two prepared 2-mL ER2738 cultures from Step 30. Incubate for 15 min at room temperature. Continue as described in Steps 8–21, using the additional culture from Step 30 for determining the input titer as described in Step 10.

## TROUBLESHOOTING

*Problem (Steps 9 and 10):* No colonies on both input and output plates.
*Solution:* Finding no colonies on both input and output plates is likely caused by contamination of the ER2738 cultures with helper phage. Repeat the panning round with fresh reagents and clean equipment for growing ER2738 cultures in phage-free conditions.

*Problem (Step 9):* No colonies on output plates.

*Solution:* Finding colonies on input but no colonies on output plates is likely caused by a loss of Fab-phage binders due to too stringent selection conditions. Increase the amount of antigen used for panning and reduce the number of washing steps. If the problem persists, consider switching to a different panning option.

*Problem (Step 13):* No overnight growth.

*Solution:* Finding no overnight growth is typically accompanied by finding no colonies on output plates, and this is likely caused by a loss of Fab-phage binders due to too stringent selection conditions. Increase the amount of antigen used for panning and reduce the number of washing steps. If the problem persists, consider switching to a different panning option.

# RECIPES

## LB + 100 µg/mL Carbenicillin Plates

Dissolve 25 g of LB (Lysogeny Broth) medium powder (MilliporeSigma L3522) in 1 L of highly pure water (e.g., that produced by Hydro PicoPure UV Plus) with 15 g of Bacto Agar (BD 214010). Autoclave for 15 min at 121°C and 15 psi. When cooled to ∼50°C, add 1 mL of 100 µg/µL carbenicillin. Pour into 10-cm sterile Petri dishes (20–25 mL each), remove bubbles by quickly flaming the plates with a Bunsen burner, and allow to solidify at room temperature. Store for up to 3 mo at 4°C.

*Alternatively, prepoured LB + 100 µg/mL carbenicillin plates are commercially available (e.g., Teknova L1010).*

## SB (Super Broth) Medium

Mix 20 g of 3-(N-morpholino)propanesulfonic acid (MOPS; MilliporeSigma M1254), 60 g of Bacto Tryptone (Thermo Fisher Scientific 211705), and 40 g of Bacto Yeast Extract (Thermo Fisher Scientific 212750). Add highly pure water (e.g., that produced by Hydro PicoPure UV Plus) to a final volume of 1.9 L. Bring to pH 7.0 with 1 M sodium hydroxide (NaOH; Thermo Fisher Scientific A4782901). Bring volume to 2 L with highly pure water. Sterilize by autoclaving for 15 min at 121°C and 15 psi in two 1-L or four 500-mL glass bottles. Store for up to 2 yr at room temperature if unopened.

# ACKNOWLEDGMENTS

Work in our laboratory was supported by the National Cancer Institute, National Institutes of Health.

# REFERENCES

Baskar S, Suschak JM, Samija I, Srinivasan R, Childs RW, Pavletic SZ, Bishop MR, Rader C. 2009. A human monoclonal antibody drug and target discovery platform for B-cell chronic lymphocytic leukemia based on allogeneic hematopoietic stem cell transplantation and phage display. *Blood* 114: 4494–4502. doi:10.1182/blood-2009-05-222786

Cyr MG, Wilson HD, Spierling AL, Chang J, Peng H, Steinberger P, Rader C. 2023. Concerted antibody and antigen discovery by differential whole-cell phage display selections and multi-omic target deconvolution. *J Mol Biol* 435: 168085. doi:10.1016/j.jmb.2023.168085

Peng H, Rader C. 2024a. Generation of antibody libraries for phage display: preparation of electrocompetent E. coli. *Cold Spring Harb Protoc* doi:10.1101/pdb.prot108599

Peng H, Rader C. 2024b. Generation of antibody libraries for phage display: human Fab format. *Cold Spring Harb Protoc* doi:10.1101/pdb.prot108597

Peng H, Rader C. 2024c. Generation of antibody libraries for phage display: chimeric rabbit/human Fab format. *Cold Spring Harb Protoc* doi:10.1101/pdb.prot108598

Peng H, Rader C. 2024d. Generation of antibody libraries for phage display: library reamplification. *Cold Spring Harb Protoc* doi:10.1101/pdb.prot108601

Peng H, Rader C. 2024e. Generation of antibody libraries for phage display: preparation of helper phage. *Cold Spring Harb Protoc* doi:10.1101/pdb.prot108600

Peng H, Rader C. 2024f. Phage display selection of antibody libraries: screening of selected binders. *Cold Spring Harb Protoc* doi:10.1101/pdb.prot108603

Popkov M, Rader C, Barbas CF III. 2004. Isolation of human prostate cancer cell reactive antibodies using phage display technology. *J Immunol Methods* **291:** 137–151. doi: 10.1016/j.jim.2004.05.004

Rader C. 2024. The pComb3 phagemid family of phage display vectors. *Cold Spring Harb Protoc* doi: 10.1101/pdb.over107756

Siegel DL. 2002. Selecting antibodies to cell-surface antigens using magnetic sorting techniques. *Methods Mol Biol* **178:** 219–226. doi: 10.1385/1-59259-240-6:219

Siva AC, Kirkland RE, Lin B, Maruyama T, McWhirter J, Yantiri-Wernimont F, Bowdish KS, Xin H. 2008. Selection of anti-cancer antibodies from combinatorial libraries by whole-cell panning and stringent subtraction with human blood cells. *J Immunol Methods* **330:** 109–119. doi: 10.1016/j.jim.2007.11.008

Wilson HD, Li X, Peng H, Rader C. 2018. A Sortase a programmable phage display format for improved panning of Fab antibody libraries. *J Mol Biol* **430:** 4387–4400. doi:10.1016/j.jmb.2018.09.003

Protocol 7

# Phage Display Selection of Antibody Libraries: Screening of Selected Binders

Haiyong Peng and Christoph Rader[1]

*Department of Immunology and Microbiology, The Herbert Wertheim UF Scripps Institute for Biomedical Innovation & Technology, University of Florida, Jupiter, Florida 33458, USA*

Phage display selection of antibody libraries is a powerful method for generating and evolving monoclonal antibodies. The pComb3 phagemid family of phage display vectors facilitates the mining of antibody libraries in Fab format from human and nonhuman antibody repertoires. Here, we describe the screening for monoclonal Fab binders after selection of a polyclonal pool of Fab binders to an antigen of interest, with the goal of identifying and sequencing monoclonal antibodies that bind the antigen with high affinity and specificity. The screening cascade involves a phage ELISA, followed by a crude Fab ELISA and DNA fingerprinting and sequencing. The protocol outlines phage and crude Fab ELISAs using purified antigen immobilized on microplates, native antigen expressed on eukaryotic cells, or both.

## MATERIALS

It is essential that you consult the appropriate Material Safety Data Sheets and your institution's Environmental Health and Safety Office for proper handling of equipment and hazardous materials used in this protocol.

RECIPES: Please see the end of this protocol for recipes indicated by <R>. Additional recipes can be found online at http://cshprotocols.cshlp.org/site/recipes.

### Reagents

Agarose gel electrophoresis reagents:
    DNA gel loading dye, 6× (Thermo Fisher Scientific R0611)
    DNA ladder, 100-bp (Thermo Fisher Scientific SM0241).
    SYBR Safe DNA gel stain (Thermo Fisher Scientific S33102)
    TAE (Tris-acetate-EDTA) buffer (dilute from 50× TAE; Thermo Fisher Scientific B49)
    UltraPure Agarose (Thermo Fisher Scientific 16500500)
AluI (10 units/µL; Thermo Fisher Scientific ER0011) with 10× Tango buffer; store at −20°C.
Autoinduction medium (optional; see Steps 16 and 27) (e.g., Novagen Overnight Express Instant TB Medium, MilliporeSigma 71491)
Bovine serum albumin (BSA), 1% (w/v) in TBS
    *To prepare, dissolve 0.5 g of BSA (MilliporeSigma A7030) in 50 mL of 1× TBS and sterilize by filtration through a 0.22-µm filter unit. Store for up to 3 mo at room temperature if unopened.*

---

[1]Correspondence: rader33458@gmail.com

© 2026 Cold Spring Harbor Laboratory Press
Cite this protocol as *Cold Spring Harb Protoc*; doi:10.1101/pdb.prot108603

Bovine serum albumin (BSA) 3% (w/v) in TBS

*To prepare, dissolve 1.5 g of BSA (MilliporeSigma A7030) in 50 mL of 1× TBS, sterilize by filtration through a 0.22-μm filter unit and store for up to 3 mo at room temperature if unopened.*

Carbenicillin, 100 μg/μL

*To prepare, dissolve 1 g of carbenicillin disodium (MilliporeSigma C1389) in 10 mL of highly pure water. Sterilize by filtration through a 0.22-μm filter. Store 1-mL aliquots in 1.5-mL microfuge tubes for up to 1 yr at −20°C.*

Eukaryotic cells ($10^7$–$10^8$); nonadherent or suspended adherent eukaryotic cells originating from cultured cell lines or from primary cells and expressing the antigen of interest on their surface (optional; Steps 9–15 and 27–36)

*The collection and processing of clinical specimens from human subjects must be approved by an Institutional Review Board (IRB). Clinical specimens should be treated as potentially infectious, and appropriate safety precautions should be taken. Wear appropriate protection and handle with care.*

Goat antihuman Fab pAbs conjugated to HRP (optional; Steps 26 and 36); e.g., Peroxidase AffiniPure F(ab′)$_2$ Fragment Goat Anti-Human IgG, F(ab′)$_2$ fragment-specific (Jackson ImmunoResearch 109-036-097); store at 4°C

Goat antihuman κ light chain polyclonal antibodies (pAbs) (SouthernBiotech 2060-01); store at 4°C

Goat antihuman λ light chain pAbs (SouthernBiotech 2070-01); store at 4°C

High-affinity rat anti-HA IgG1 3F10 (Roche Life Science Products available from, e.g., MilliporeSigma 11867423001); store at 4°C

Horse radish peroxidase (HRP) substrate solution; e.g., ABTS [2,2′-azino-bis(3-ethylbenzthiazoline-6-sulfonic acid] One Component HRP Microwell Substrate (Surmodics IVD ABTS-0100-01); store at 4°C

Ice

Isopropyl-β-D-thiogalactoside (IPTG) 0.5 M

*To prepare, dissolve 6 g of isopropyl-β-D-thiogalactoside (IPTG [Gold Biotechnology I2481C25]; store at −20°C) in 50 mL of water. Sterilize by filtration through a 0.22-μm filter unit. Store 1-mL aliquots in 1.5-mL microfuge tubes for up to 1 yr at −20°C.*

LB + 100 μg/mL carbenicillin plates <R>

Mouse anti-M13 mAb A5B3 conjugated to horse radish peroxidase (HRP) (Thermo Fisher Scientific MA5-36125); store at 4°C (short term) or −20°C (long term)

Oligonucleotides

*The names and DNA sequences of all primers used in this protocol are shown in Table 1. For each primer, prepare 20 μM working dilutions in sterile water.*

PCR reagents; store at −20°C:
dNTP mix, 10 mM (2.5 mM of each dATP, dCTP, dGTP, and dTTP diluted in sterile water from 100 mM stock concentrations; Thermo Fisher Scientific 18427013)
Taq DNA polymerase, 5 units/μL, 10× Taq buffer with $(NH_4)_2SO_4$, and 25 mM $MgCl_2$ (Thermo Fisher Scientific EP0405]

Phosphate-buffered saline (PBS, 1×)

*Dilute 10× PBS (Thermo Fisher Scientific J61196.AP) in highly pure water, sterilize by filtration through a 0.22-μm filter unit, and store for up to 3 mo at room temperature if unopened.*

Purified antigen of interest (5–50 μg) in carrier-free buffer; store at −80°C

QIAprep Spin Miniprep Kit (QIAGEN 27106); store at room temperature

SB (Super Broth) medium <R>

Selected Fab-phage library; phage preparations and output plates from each panning round (see Protocol 6: Phage Display Selection of Antibody Libraries: Panning Procedures [Peng and Rader 2024a])

Tris-buffered saline (TBS, 1×)

*Prepare by diluting 10× TBS (Thermo Fisher Scientific J60764.K7) in highly pure water and sterilizing by filtration through a 0.22-μm filter unit. Store for up to 3 mo at room temperature if unopened.*

TABLE 1. Primers for DNA fingerprinting of $V_L$-$C_L$-$V_H$ and DNA sequencing of $V_L$ and $V_H$[a]

**Fingerprinting primers**
| | |
|---|---|
| 5′SFIHUVL | CGCTACCGTGGCCCAGGCGGCC (use for hu$V_L$) |
| 5′SFIVL | CTGCTGCTGGGGCCCAGGCGGCC (use for rb$V_L$) |
| 3′sfivh | GAGGAGGAGGGCCGACGGGGCCAAGGGGAAGACCGATGGGCCCTTGGTGGAGGCTGA |

**Sequencing primers**
| | |
|---|---|
| VLSEQ | GATAACAATTGAATTCAGGAG |
| vhseq | TGAGTTCCACGACACCGT |

[a]Sequences are shown 5′–3′.

Tween 20, 0.05% (v/v) in TBS
*To prepare, dilute 25 µL of Tween 20 (MilliporeSigma P1379) in 50 mL of 1× TBS. Sterilize by filtration through a 0.22 µm filter unit and store for up to 3 mo at room temperature if unopened.*

Water, highly pure, sterile
*The protocol uses highly pure water from a purification system (see "Equipment"). Use freshly. Several steps require sterile water, which is obtained by filtering highly pure water through a 0.22-µm filter unit. Store at room temperature.*

## Equipment

*A laboratory with standard molecular biology and microbiology equipment is required for this protocol.*

Agarose gel electrophoresis equipment (e.g., Owl horizontal electrophoresis systems [Thermo Fisher Scientific] with various tanks, casters, combs, and power supplies [EC-105] for wide gel [D-series] analytical electrophoresis and blue-light transilluminator [e.g., VB-40 Visi-Blue Transilluminator, VWR 95041-594])

Autoclave

Bunsen burner

Centrifuge tubes (50-mL)

Digital balance with 0.01-g readability

Filtered pipette tips (10-, 20-, 100-, and 200-µL, and 1-mL)

Filters with Luer lock fitting (0.22-µm; e.g., Millex-GP Syringe Filters, Millipore SLGP033RS)

Filter units (0.22-µm; e.g., Millipore Stericup and Steriflip filter units)

Freezers (−20°C and −80°C)

Glass bottles (500-mL and 1-L, autoclaved)

Ice bucket

Incubator (37°C)

Magnetic hot plate stirrer

Magnetic stir bars

Microfuge tubes (1.5- and 2-mL)

Microplate, polypropylene, 96-well, 2-mL deep (optional; Steps 16 and 27) (e.g., PlateOne Deep 96-Well 2-mL Polypropylene Plate, USA Scientific 1896-2110)

Microplate, polystyrene, 96-well, clear, flat-bottom, half-area (Costar 3690, Thermo Fisher Scientific 07-200-37)

Microplate, polystyrene, 96-well, clear, V-bottom, tissue culture-treated (Costar 3894, Thermo Fisher Scientific 07-200-96)

Microplate reader (e.g., Tecan Spark Multimode Microplate Reader)

Microwave oven

Orbital shaker (e.g., VWR Standard 1000 Orbital Shaker 89032-088)

PCR tube strips (0.2-mL; e.g., Eppendorf 951010022)

Petri dishes (10-cm)

Pipettes (5-, 10-, and 25-mL)

Pipette controller

Plate sealer (SealPlate Sealing Film, Excel Scientific 100-SEAL-PLT)

Refrigerated benchtop centrifuge with swinging bucket rotor (e.g., Beckman Coulter Avanti J-15R with JS-4.750 swinging bucket rotor) and compatible microplate carriers

Refrigerator (4°C)

Round-bottom tubes with snap cap (14-mL)

Shaker at 37°C (e.g., Eppendorf New Brunswick Innova 40 Benchtop Incubator Shaker).

*Two separate shakers, one for phage-free conditions and one for phage, are required.*

Single-channel and multichannel micropipettes (1- to 1000-µL)

Syringes with Luer lock fitting (1-, 5-, and 10-mL)

Thermal cyclers (96-well; e.g., Eppendorf Mastercycler X50s 96-Well Silver Block Thermal Cycler 6311000010)

Water purification system (e.g., Hydro PicoPure UV Plus)

## METHOD

*Following Fab-phage library selection (see Protocol 6: Phage Display Selection of Antibody Libraries: Panning Procedures [Peng and Rader 2024a]), an enrichment of binders over multiple rounds of selection is detected by analyzing the phage preparations of each panning round by a polyclonal phage ELISA (see "Analysis of Library Selection by Phage ELISA"). If purified stable antigen is available, it is immobilized on a microplate for the phage ELISA (see "Phage ELISA with Immobilized Antigen"). Alternatively, carrying out the phage ELISA with eukaryotic cells in solution (see "Phage ELISA with Eukaryotic Cells in Solution") is the option of choice if the antigen is unknown or purified stable antigen is unavailable. After detecting an enrichment of binders in the phage ELISA, a subsequent monoclonal crude (i.e., unpurified) Fab ELISA is done to analyze individual clones (see "Analysis of Library Selection by Crude Fab ELISA"). Again, if the crude Fab ELISA reveals individual positive clones, DNA fingerprinting and sequencing (see "DNA Fingerprinting and Sequencing") is used to identify unique and redundant binders, which can then be cloned, expressed, and purified in different antibody formats tailored for specific downstream analyses and applications of interest (see Protocol 8: Cloning, Expression, and Purification of Phage Display-Selected Fab for Biophysical and Biological Studies [Cyr et al. 2024]).*

### Analysis of Library Selection by Phage ELISA

*As a first assessment of whether or not a Fab-phage library selection (see Protocol 6: Phage Display Selection of Antibody Libraries: Panning Procedures [Peng and Rader 2024a]) was successful in terms of enrichment of binders, a phage ELISA is carried out to analyze the binding of each phage preparation acquired over the course of the panning experiment. Below, we describe the steps to do this with either purified antigen (Steps 1–8) or with eukaryotic cells in solution (Steps 9–15).*

#### Phage ELISA with Immobilized Antigen

1. For each phage preparation to test (e.g., four panning rounds yield four phage preparations), coat two wells in row A of a 96-well half-area microplate with 0.1–1 µg of antigen in 25 µL of TBS, cover the wells with plate sealer, and incubate for 1 h at 37°C or overnight at 4°C.

    *The amount of antigen used for immobilization depends on its molecular weight, on the number of epitopes per antigen, and on epitope preservation following immobilization.*

2. Shake out the coating solution, add 150 µL of 3% (w/v) BSA in TBS to each coated well and to a corresponding empty well in row B for blocking, cover with a plate sealer, and incubate for 1 h at 37°C.

3. In the meantime, dilute 50 µL of each phage preparation of the selected Fab-phage library (see "Materials") with 200 µL of 1% (w/v) BSA in TBS, mix, and store on ice.

4. Shake out the blocking solution from Step 2, add 50 µL of the diluted phage preparations from Step 3 to two wells in row A and two wells in row B, cover with a plate sealer, and incubate for 2 h at 37°C.

5. Shake out the phage solution. Use a multichannel pipette to wash each well 10 times each with 150 µL of 0.05% (v/v) Tween 20 in TBS.

6. Shake out the last washing solution, add 50 µL of a 1:1000 dilution of mouse anti-M13 mAb A5B3 conjugated to HRP in 1% (w/v) BSA in TBS to each well, cover with a plate sealer, and incubate for 1 h at 37°C.

7. Shake out the detecting antibody solution. Use a multichannel pipette to wash each well 10 times each with 150 µL of TBS.

8. Shake out the last washing solution, add 50 µL of HRP substrate solution to each well, and incubate for 5 min to 1 h at room temperature until the green color is evident. Determine the absorbance at 405 nm with a microplate reader.

    *A successful panning experiment typically reveals strong row A antigen (but only weak row B background) reactivity in phage preparations from the third and/or fourth panning round. If successful, proceed to "Analysis of Library Selection by Crude Fab ELISA." If unsuccessful, see Troubleshooting.*

### Phage ELISA with Eukaryotic Cells in Solution

9. On the basis of input titers determined during library selection (see Protocol 6: Phage Display Selection of Antibody Libraries: Panning Procedures [Peng and Rader 2024a]), dilute phage preparations of the selected Fab-phage library (see "Materials") to ~$10^{11}$ phage in 75 µL of PBS and store on ice.

10. Wash the cells twice with PBS and then prepare a suspension of $1 \times 10^7$ eukaryotic cells in 2 mL of PBS. Then, add 100 µL ($5 \times 10^5$ eukaryotic cells) to eight row A and eight row B wells of a 96-well clear V-bottom tissue culture-treated microplate. For each phage preparation to be tested (e.g., four panning rounds yield four phage preparations), prepare two wells in row A and two wells in row B.

    *If available, include antigen-positive (rows A and B) and antigen-negative (rows C and D) eukaryotic cells to assess background binding.*

11. Centrifuge the plate at 300g for 5 min at room temperature, decant and discard the supernatant, and resuspend the cells in (1) 75 µL of the prepared phage preparation from Step 9 (for row A) or (2) 75 µL of PBS (for row B). Incubate the covered plate with gentle orbital shaking (e.g., 50 rpm using a VWR Standard 1000 Orbital Shaker) for 1 h at room temperature.

12. Centrifuge the plate at 300g for 5 min at room temperature, decant and discard the supernatant, and resuspend the cells in 100 µL of PBS with a multichannel pipette. Repeat this washing step once more.

13. Decant and discard the final washing solution and resuspend the cells in 100 µL of a 1:1000 dilution of mouse anti-M13 mAb A5B3 conjugated to HRP in PBS. Incubate the covered plate with gentle orbital shaking for 30 min at room temperature.

14. Centrifuge the plate at 300g for 5 min at room temperature, decant and discard the supernatant, and resuspend the cells in 100 µL of PBS with a multichannel pipette. Repeat this washing step once more.

15. Decant and discard the final washing solution and resuspend the cells in 50 µL of HRP substrate solution. Incubate the covered plate for 5 min to 1 h at room temperature until the green color is evident. Determine the absorbance at 405 nm with a microplate reader.

    *A successful panning experiment typically reveals strong row A antigen (but only weak row B background) reactivity in phage preparations from the third and/or fourth panning round. If successful, proceed to "Analysis of Library Selection by Crude Fab ELISA." If unsuccessful, see Troubleshooting.*

Chapter 5

## Analysis of Library Selection by Crude Fab ELISA

*If the phage ELISA revealed reactive phage preparations from the third, fourth, or any subsequent panning round, crude Fab (unpurified Fab-ΔpIII fusion protein and its proteolytic fragments) prepared from colonies of the output plate that corresponds to the most reactive phage preparation should be analyzed next, and is described below. Depending on the availability of the antigen, and similar to the phage ELISA, the crude Fab ELISA can be carried out with immobilized antigen, with eukaryotic cells in solution, or with both. Both options are listed below.*

### Crude Fab ELISA with Immobilized Antigen

*The crude Fab ELISA with immobilized antigen is designed to (1) identify clones with antigen (but not background) reactivity and (2) compare the expression level of positive (and, possibly, negative) clones through capturing with the goat antihuman κ or λ light chain pAbs. If the phage ELISA revealed strong antigen reactivity, typically ≥90% clones in the crude Fab ELISA are positive.*

16. Prepare thirty-two 14-mL round-bottom tubes with snap cap, each containing 3 mL of SB medium supplemented with 100 µg/mL carbenicillin, and inoculate 31 of them with a single colony from the output plate(s) that correspond(s) to the reactive phage preparation(s) of the selected Fab-phage library (see "Materials"). Use the remaining tube as negative control for both culture growth and ELISA. Shake at 300 rpm for 8 h at 37°C.

    *As an alternative to SB medium and subsequent IPTG induction, use autoinduction medium (Studier 2005) for higher yields. For higher throughput, use a 2-mL deep 96-well plate containing 0.4 mL of autoinduction medium supplemented with 100 µg/mL carbenicillin in each well.*

17. Draw a numbered square grid for 32 colonies on the back of a prewarmed LB + 100 µg/mL carbenicillin plate and transfer a 2-µL aliquot of each culture from Step 16 to an identifiable square. Incubate overnight at 37°C.

    *This backup plate will be used in Step 37 for phagemid preparation.*

18. Add 12 µL of 0.5 M IPTG to each culture (to a final concentration of 2 mM) from Step 16. Continue shaking at 300 rpm overnight (12–16 h) at 30°C.

19. Centrifuge the overnight cultures at 3000g for 15 min at 4°C. Transfer 1 mL of each supernatant to separate 1.5-mL microfuge tubes and discard the remaining supernatant and pellet. Store on ice.

20. Use two 96-well half-area microplates. On the first plate, coat each of 32 wells with 0.1–1 µg of antigen in 25 µL of TBS, and each of another 32 wells with 200 ng of goat antihuman κ light chain pAbs in 25 µL of TBS, and leave the remaining 32 wells empty. On the second plate, coat each of 32 wells with 0.1–1 µg of antigen in 25 µL of TBS, and each of another 32 wells with 200 ng of goat antihuman λ light chain pAbs in 25 µL of TBS, and leave the remaining 32 wells empty. Cover each plate with a plate sealer, and incubate for 1 h at 37°C or overnight at 4°C.

    *The amount of antigen used for immobilization depends on its molecular weight, on the number of epitopes per antigen, and on epitope preservation following immobilization.*

21. Shake out the coating solution, add 150 µL of 3% (w/v) BSA in TBS to each coated or empty well for blocking, cover with a plate sealer, and incubate for 1 h at 37°C.

22. Shake out the blocking solution, and add 50 µL of the supernatant prepared in Step 19 to each well so that each of the 32 samples is incubated with BSA-blocked antigen (on both plates), BSA-blocked goat antihuman κ light chain pAbs (on the first plate), BSA-blocked goat anti-λ light chain pAbs (on the second plate), and BSA alone (on both plates). Cover with a plate sealer and incubate for 2 h at 37°C.

23. Shake out the supernatant. Use a multichannel pipette to wash each well 10 times each with 150 µL of TBS.

24. Shake out the last washing solution, add 50 µL of a 1:1000 dilution of rat anti-HA mAb 3F10 conjugated to HRP in 1% (w/v) BSA in TBS to each well, cover with a plate sealer, and incubate for 1 h at 37°C.

25. Shake out the detecting antibody solution. Use a multichannel pipette to wash each well 10 times each with 150 μL of TBS.

26. Shake out the last washing solution, add 50 μL of HRP substrate solution to each well, and incubate for 5 min to 1 h at room temperature until the green color is evident. Determine the absorbance at 405 nm with a microplate reader.

   *The detecting antibody rat anti-HA mAb 3F10 conjugated to HRP binds to the hemagglutinin (HA) decapeptide tag that is located between $huC_H1$ and $\Delta pIII$ in pComb3 family phagemids pComb3X, pC3C, and pC3Csort (see Chapter 4, Overview: The pComb3 Phagemid Family of Phage Display Vectors [Rader 2024]). Alternatively, and for pComb3H, which does not have an HA tag, use goat antihuman Fab pAbs conjugated to HRP. Although the Fab-$\Delta pIII$ fusion protein is assembled in the Escherichia coli periplasm, detectable quantities of Fab-$\Delta pIII$ fusion protein and its proteolytic fragments are found in the supernatant due to outer membrane leakage and cell death. The inclusion of immobilized goat antihuman $\kappa$ and $\lambda$ light chain pAbs in addition to immobilized antigen of interest serves as a positive control confirming Fab and HA expression. In addition, it relates the specific antigen reactivity of the unpurified Fab-$\Delta pIII$ fusion protein to its concentration in the supernatant.*

   *If positive clones have been identified, proceed to "DNA Fingerprinting and Sequencing." Otherwise, see Troubleshooting.*

## Crude Fab ELISA with Eukaryotic Cells in Solution

*The crude Fab ELISA with eukaryotic cells in solution is designed to identify clones with antigen (but not background) reactivity. If the phage ELISA revealed strong antigen reactivity, typically $\geq$90% clones in the crude Fab ELISA are positive.*

27. Prepare thirty-two 14-mL round-bottom tubes with snap cap, each containing 3 mL of SB medium supplemented with 100 μg/mL carbenicillin, and inoculate 31 of them with a single colony from the output plate(s) that correspond(s) to the reactive phage preparation(s) of the selected Fab-phage library (see "Materials"). Use the remaining tube as negative control for both culture growth and ELISA. Shake at 300 rpm for 8 h at 37°C.

   *As an alternative to SB medium and subsequent IPTG induction, use autoinduction medium (Studier 2005) for higher yields. For higher throughput, use a 2-mL deep 96-well plate containing 0.4 mL of autoinduction medium supplemented with 100 μg/mL carbenicillin in each well.*

28. Draw a numbered square grid for 32 colonies on the back of a prewarmed LB + 100 μg/mL carbenicillin plate and transfer a 2-μL aliquot of each culture from Step 16 to an identifiable square. Incubate overnight at 37°C.

    *This backup plate will be used in Step 37 for phagemid preparation.*

29. Add 12 μL of 0.5 M IPTG to each culture (to a final concentration of 2 mM). Continue shaking at 300 rpm overnight (12–16 h) at 30°C.

30. Centrifuge the overnight cultures at 3,000g for 15 min at 4°C. Transfer 1 mL of each supernatant to separate 1.5-mL microfuge tubes and discard the remaining supernatant and pellet. Store on ice.

31. Wash the cells twice with PBS and prepare a suspension of $2 \times 10^7$ eukaryotic cells in 4 mL of PBS. Then, add 100 μL ($5 \times 10^5$ eukaryotic cells) to each of 32 wells of a 96-well clear V-bottom tissue culture-treated microplate.

    *If available, include antigen-positive and antigen-negative eukaryotic cells to assess background binding.*

32. Centrifuge the plate at 300g for 5 min at room temperature, decant and discard the supernatant, and resuspend the cells in 75 μL of the supernatant prepared in Step 30. Incubate the covered plate with gentle orbital shaking (e.g., 50 rpm using a VWR Standard 1000 Orbital Shaker) for 1 h at room temperature.

33. Centrifuge the plate at 300g for 5 min at room temperature, decant and discard the supernatant, and resuspend the cells in 100 μL of PBS with a multichannel pipette. Repeat this washing step once more.

34. Decant and discard the final washing solution and resuspend the cells in 100 µL of a 1:1000 dilution of rat anti-HA mAb 3F10 conjugated to HRP in PBS. Incubate the covered plate for 30 min at room temperature.

35. Centrifuge the plate at 300g for 5 min at room temperature, decant and discard the supernatant, and resuspend the cells in 100 µL of PBS with a multichannel pipette. Repeat this washing step once more.

36. Decant and discard the final washing solution and resuspend the cells in 50 µL of HRP substrate solution. Incubate the covered plate for 5 min to 1 h at room temperature until the green color is evident. Determine the absorbance at 405 nm with a microplate reader.

   *The detecting antibody rat anti-HA mAb 3F10 conjugated to HRP binds to the hemagglutinin (HA) decapeptide tag that is located between $huC_H1$ and $\Delta pIII$ in pComb3 family phagemids pComb3X, pC3C, and pC3Csort (see Chapter 4, Overview: The pComb3 Phagemid Family of Phage Display Vectors [Rader 2024]). Alternatively, and for pComb3H, which does not have an HA tag, use goat antihuman Fab pAbs conjugated to HRP. Although the Fab-$\Delta pIII$ fusion protein is assembled in the E. coli periplasm, detectable quantities of Fab-$\Delta pIII$ fusion protein and its proteolytic fragments are found in the supernatant due to outer membrane leakage and cell death.*

   *If positive clones have been identified, proceed to "DNA Fingerprinting and Sequencing." Otherwise, see Troubleshooting.*

## DNA Fingerprinting and Sequencing

*If the crude Fab ELISA revealed positive clones, users need to prepare phagemid and then analyze the clones by DNA fingerprinting and sequencing. The steps below are based on the analysis of 20 positive clones and describe the PCR amplification of the $V_L/C_L/V_H$ segment of the Fab-encoding cassette of the phagemid, its digestion by restriction enzyme AluI, and the preparation for DNA sequencing of the $V_L$- and $V_H$-encoding segments. AluI is a frequently cutting restriction enzyme with the recognition sequence AG^CT. It can identify clones with different sequences prior to DNA sequencing. Unique clones selected from naive antibody libraries typically reveal multiple AluI fragments of varied sizes when analyzed on a dense agarose gel. We refer to the unique pattern of these AluI fragments as a fingerprint.*

37. Prepare twenty 14-mL round-bottom tubes with snap cap, each containing 2 mL of SB medium supplemented with 100 µg/mL carbenicillin, and inoculate each with a colony from the backup plate prepared in Steps 17 and 28 that corresponds to a positive clone. Shake at 300 rpm overnight (12–16 h) at 37°C.

38. Centrifuge at 3000g for 15 min at 4°C. Discard the supernatant and resuspend the pellet in 250 µL of QIAGEN buffer P1 from the QIAprep Spin Miniprep Kit. Continue the phagemid preparation using reagents and protocols supplied by the QIAprep Spin Miniprep Kit. Elute in 20 µL of water and store phagemid preparations at −20°C.

39. For PCR amplification of the $V_L/C_L/V_H$ segment, transfer 1 µL of each sample into separate 0.2-mL PCR tubes. Then, prepare a PCR master mix sufficient for 20 reactions as follows: In a 1.5-mL microfuge tube, combine 75 µL of 10× PCR buffer, 75 µL of 25 mM $MgCl_2$, 60 µL of 10 mM dNTP mix, 22.5 µL of 20 µM sense primer (C-5′SFIHUVL for $huV_L/huC_L/huV_H$ or C-5′SFIVL for $rbV_L/huC_L/rbV_H$), 22.5 µL of 20 µM c-3′sfivh (antisense primer), 466 µL of water, and 4 µL of 5 units/µL Taq DNA polymerase (Σ 725 µL). Add 29 µL of the PCR master mix to each prepared 1-µL sample.

   *See Protocol 1: Generation of Antibody Libraries for Phage Display: Human Fab Format (Peng and Rader 2024b) and Protocol 2: Generation of Antibody Libraries for Phage Display: Chimeric Rabbit/Human Fab Format (Peng and Rader 2024c) for the generation of Fab-phage libraries with SfiI-flanked human ($huV_L/huC_L/huV_H$) and chimeric rabbit/human ($rbV_L/huC_L/rbV_H$) cassettes, respectively.*

40. In a 96-well thermocycler, use these PCR parameters:

    2 min at 95°C;

    followed by 30 cycles of 30 sec at 95°C, 30 sec at 52°C, and 90 sec at 72°C;

    followed by 10 min at 72°C;

    followed by holding at room temperature.

41. Transfer a 10-µL aliquot from each sample to a separate 1.5-mL microfuge tube.

42. Prepare an AluI master mix sufficient for 20 reactions as follows: In a 1.5-mL microfuge tube, combine 90 µL of 10× Tango buffer, 495 µL of water, and 15 µL of 10 units/µL AluI (Σ 600 µL). Then, add 20 µL of the AluI master mix to each prepared 10-µL sample and incubate for 2 h at 37°C.

43. Add 3 µL of 6× gel loading dye solution to each sample and separate by electrophoresis on a 4% (w/v) agarose gel in TAE buffer using a 100-bp DNA ladder as reference.

    *Expect to identify distinct repeated AluI fingerprints. Each unique AluI fingerprint indicates a different DNA sequence. However, identical AluI fingerprints may still have slightly different DNA and amino acid sequences that are only detectable by DNA sequencing. See Troubleshooting.*

44. Contract a Sanger sequencing service to further analyze the phagemid preparations. Use sense primer VLSEQ (a sense primer located upstream of ompA in pComb3H, pComb3X, pC3C, and pC3Csort) and vhseq [an antisense primer located in $C_H1$ of these phage display vectors downstream from the SfiI (b) site in pC3C and pC3Csort] (see Chapter 4, Overview: The pComb3 Phagemid Family of Phage Display Vectors [Rader 2024]) to determine the DNA and deduced amino acid sequences of light and heavy chain variable domains.

    *In general, a small panel of selected human Fab that reveals strong antigen reactivity and reasonable expression levels in the crude Fab ELISA as well as a unique amino acid sequence should be pursued. See Troubleshooting.*

    *As described in Protocol 8: Cloning, Expression, and Purification of Phage Display-Selected Fab for Biophysical and Biological Studies (Cyr et al. 2023), following recombinant removal of ΔpIII, selected human or chimeric nonhuman/human Fab can be expressed and purified in milligram quantities, facilitating a variety of downstream applications. Depending on the application, promising candidates can subsequently be converted to a variety of alternative antibody formats, including monovalent scFv or bivalent scFv–Fc and IgG.*

## TROUBLESHOOTING

*Problem (Steps 8 and 15):* No or only weak specific antigen reactivity (i.e., row A signals lower than twofold above row B background) in phage ELISA.

*Solution:*

- Suboptimal selection conditions. Carry out two additional panning rounds.
- Low affinity of the selected Fab-phage. Consider alternative panning procedures as described in Protocol 6: Phage Display Selection of Antibody Libraries: Panning Procedures (Peng and Rader 2024a).

*Problem (Steps 26 and 36):* Low percentage of positive clones, or no or only weak specific antigen reactivity (i.e., signals lower than twofold above background) in crude Fab ELISA.

*Solution:*

- Suboptimal selection conditions. Carry out two additional panning rounds as described in Protocol 6: Phage Display Selection of Antibody Libraries: Panning Procedures (Peng and Rader 2024a).
- Low expression of the Fab-ΔpIII fusion protein. Repeat with purified Fab as described in Protocol 8: Cloning, Expression, and Purification of Phage Display-Selected Fab for Biophysical and Biological Studies (Cyr et al. 2024) or after conversion to IgG.
- Weak affinity of the selected Fab. Consider more stringent or alternative panning procedures as described in Protocol 6: Phage Display Selection of Antibody Libraries: Panning Procedures (Peng and Rader 2024a).

*Problem (Steps 43 and 44):* All positive clones have an identical DNA fingerprint and sequence.
*Solution:* Too stringent selection conditions. Analyze positive clones from the previous panning round or consider less stringent panning procedures by, e.g., increasing the amount of antigen and decreasing the number of washing steps as described in Protocol 6: Phage Display Selection of Antibody Libraries: Panning Procedures (Peng and Rader 2024a).

*Problem (Steps 43 and 44):* No repeated DNA fingerprint and sequence among positive clones.
*Solution:* Suboptimal selection conditions. Carry out two additional panning rounds applying more stringency or consider alternative panning procedures as described in Protocol 6: Phage Display Selection of Antibody Libraries: Panning Procedures (Peng and Rader 2024a).

## RECIPES

### LB + 100 μg/mL Carbenicillin Plates

Dissolve 25 g of LB (Lysogeny Broth) medium powder (MilliporeSigma L3522) in 1 L of highly pure water (e.g., that produced by Hydro PicoPure UV Plus) with 15 g of Bacto Agar (BD 214010). Autoclave for 15 min at 121°C and 15 psi. When cooled to ∼50°C, add 1 mL of 100 μg/μL carbenicillin. Pour into 10-cm sterile Petri dishes (20–25 mL each), remove bubbles by quickly flaming the plates with a Bunsen burner, and allow to solidify at room temperature. Store for up to 3 mo at 4°C.

*Alternatively, prepoured LB + 100 μg/mL carbenicillin plates are commercially available (e.g., Teknova L1010).*

### SB (Super Broth) Medium

Mix 20 g of 3-(N-morpholino)propanesulfonic acid (MOPS; MilliporeSigma M1254), 60 g of Bacto Tryptone (Thermo Fisher Scientific 211705), and 40 g of Bacto Yeast Extract (Thermo Fisher Scientific 212750). Add highly pure water (e.g., that produced by Hydro PicoPure UV Plus) to a final volume of 1.9 L. Bring to pH 7.0 with 1 M sodium hydroxide (NaOH; Thermo Fisher Scientific A4782901). Bring volume to 2 L with highly pure water. Sterilize by autoclaving for 15 min at 121°C and 15 psi in two 1-L or four 500-mL glass bottles. Store for up to 2 yr at room temperature if unopened.

## ACKNOWLEDGMENTS

Work in our laboratory was supported by the National Cancer Institute, National Institutes of Health.

## REFERENCES

Cyr MG, Peng H, Rader C. 2024. Cloning, expression, and purification of phage display-selected Fab for biophysical and biological studies. *Cold Spring Harb Protoc* doi:10.1101/pdb.prot108604

Peng H, Rader C. 2024a. Phage display selection of antibody libraries: panning procedures. *Cold Spring Harb Protoc* doi:10.1101/pdb.prot108602

Peng H, Rader C. 2024b. Generation of antibody libraries for phage display: human Fab format. *Cold Spring Harb Protoc* doi:10.1101/pdb.prot108597

Peng H, Rader C. 2024c. Generation of antibody libraries for phage display: chimeric rabbit/human Fab format. *Cold Spring Harb Protoc* doi:10.1101/pdb.prot108598

Rader C. 2024. The pComb3 phagemid family of phage display vectors. *Cold Spring Harb Protoc* doi:10.1101/pdb.prot107756

Studier FW. 2005. Protein production by auto-induction in high density shaking cultures. *Protein Expr Purif* 41: 207–234. doi:10.1016/j.pep.2005.01.016

Protocol 8

# Cloning, Expression, and Purification of Phage Display-Selected Fab for Biophysical and Biological Studies

Matthew G. Cyr, Haiyong Peng, and Christoph Rader[1]

*Department of Immunology and Microbiology, The Herbert Wertheim UF Scripps Institute for Biomedical Innovation & Technology, University of Florida, Jupiter, Florida 33458, USA*

The antigen-binding fragment (Fab) is the ∼50-kDa monovalent arm of an antibody molecule. In the laboratory, the Fab can be produced via either enzymatic digestion or recombinant expression, and its use facilitates the accurate assessment of affinity and specificity of monoclonal antibodies. The high melting temperature of the Fab, together with its low tendency to aggregate and ready conversion to natural and nonnatural immunoglobulin (Ig) formats (without affecting antigen binding properties), have made it a preferred format for phage display, as well as a tool for accurate assessment of affinity, specificity, and developability of monoclonal antibodies. Here, we outline a strategy to clone, express, and purify human or chimeric nonhuman/human Fabs that have previously been selected by phage display. Fabs purified using this approach, which results in milligram amounts, enable a variety of downstream biophysical and biological assays that ultimately inform the success of phage display library generation and selection.

## MATERIALS

It is essential that you consult the appropriate Material Safety Data Sheets and your institution's Environmental Health and Safety Office for proper handling of equipment and hazardous materials used in this protocol.

RECIPES: Please see the end of this protocol for recipes indicated by <R>. Additional recipes can be found online at http://cshprotocols.cshlp.org/site/recipes.

### Reagents

2-Mercaptoethanol (MilliporeSigma M63689); store at room temperature
Agarose gel electrophoresis reagents:
    DNA ladder, 100-bp (Thermo Fisher Scientific SM0241)
    DNA gel loading dye, 6× (Thermo Fisher Scientific R0611)
    SYBR Safe DNA gel stain (Thermo Fisher Scientific S33102)
    TAE (Tris-acetate-EDTA) buffer (dilute from 50× TAE, Thermo Fisher Scientific B49)
    UltraPure Agarose (Thermo Fisher Scientific 16500500)
Amicon Ultra-15 Centrifugal Filters (15-mL sample volume, 30-kDa MWCO; MilliporeSigma UFC903008)

---

[1]Correspondence: rader33458@gmail.com

© 2026 Cold Spring Harbor Laboratory Press
Cite this protocol as *Cold Spring Harb Protoc*; doi:10.1101/pdb.prot108604

Amicon Ultra-4 Centrifugal Filters (4-mL sample volume, 30-kDa MWCO; MilliporeSigma UFC803008)

BamHI HF, 20 units/µL (New England Biolabs R3136S) with 10× rCutSmart buffer (New England Biolabs B6004S) (optional; see Step 2); store at −20°C

Carbenicillin, 100 µg/µL

*To prepare, dissolve 1 g of carbenicillin disodium (MilliporeSigma C1389) in 10 mL of highly pure water. Sterilize by filtration through a 0.22-µm filter. Store 1-mL aliquots in 1.5-mL microfuge tubes at −20°C.*

Chloramphenicol, 100 µg/µL

*To prepare, dissolve 1 g of chloramphenicol (MilliporeSigma C0378) in 10 mL of pure ethanol (MilliporeSigma 459836-500ML). Store 1-mL aliquots in 1.5-mL microfuge tubes at −20°C.*

Dulbecco's phosphate-buffered saline (DPBS), without calcium, without magnesium, pH 7.0–7.3 (Thermo Fisher Scientific J67802.K2); store at room temperature

Ethanol, 20% (v/v) in water

*Prepare fresh from pure ethanol (MilliporeSigma 459836-500ML) and highly pure water. Store at room temperature.*

Glycerol, 50% (v/v) in water

*Mix 50 mL of glycerol, ultrapure (Thermo Fisher Scientific J64719.AP; store at room temperature) with 50 mL of highly pure water. Sterilize by autoclaving for 15 min at 121°C and 15 psi. Store at room temperature.*

Ice

Imidazole, 5 mM in DPBS

*Dissolve 170.2 mg of imidazole (MilliporeSigma I2399) in 500 mL of DPBS. Sterilize by filtration through a 0.22-µm filter unit. Store at room temperature.*

Imidazole, 250 mM in DPBS

*Dissolve 8.51 g of imidazole (MilliporeSigma I2399) in 500 mL of DPBS. Sterilize by filtration through a 0.22-µm filter unit. Store at room temperature.*

Isopropyl-β-D-thiogalactoside (IPTG), 1 M

*To prepare, dissolve 6 g of isopropyl-β-D-thiogalactoside (IPTG [Gold Biotechnology I2481C25]; store at −20°C) in 25 mL of water. Sterilize by filtration through a 0.22-µm filter unit. Store 1-mL aliquots in 1.5-mL microfuge tubes at −20°C.*

LB + 100 µg/mL carbenicillin plates <R>

LB + 100 µg/mL carbenicillin + 25 µg/mL chloramphenicol plates <R>

Liquid nitrogen

NdeI, 20 units/µL (New England Biolabs R0111S) with 10× rCutSmart buffer (New England Biolabs B6004S) (optional; see Step 2); store at −20°C

Oligonucleotides specific for the T7 promoter (5′-TAATACGACTCACTATAGGG-3′) and the T7 terminator (5′-CTAGTTATTGCTCAGCGGT-3′) for DNA sequencing

*For each primer, prepare 20 µM working dilutions in sterile water.*

One Shot TOP10 Chemically Competent *Escherichia coli* (twenty-one 50-µL aliquots; store at −80°C) including SOC Medium (6 mL; store at 4°C) (Thermo Fisher Scientific C404003)

PageRuler Prestained Protein Ladder, 10- to 180-kDa (Thermo Fisher Scientific 26616); store at −20°C

PageBlue Protein Staining Solution (Thermo Fisher Scientific 24620); store at room temperature

PCR reagents (optional; Step 2); store at −20°C:

dNTP mix, 10 mM (2.5 mM of each dATP, dCTP, dGTP, and dTTP diluted in sterile water from 100 mM stock concentrations; Thermo Fisher Scientific 18427013)

Taq DNA polymerase, 5 units/µL, 10× Taq buffer with $(NH_4)_2SO_4$, and 25 mM $MgCl_2$ (Thermo Fisher Scientific EP0405)

pET-11a plasmid (MilliporeSigma 69436); store at −20°C

*For plasmid map and sequence, see https://www.sigmaaldrich.com/US/en/product/mm/69436.*

Phagemid preparations from Fab-phage library selected over several panning rounds (i.e., see Protocol 7: Phage Display Selection of Antibody Libraries: Screening of Selected Binders [Peng and Rader 2024])

QIAprep Spin Miniprep Kit (QIAGEN 27106); store at room temperature

QIAquick Gel Extraction Kit (QIAGEN 28704); store at room temperature

Rosetta 2(DE3) Singles Competent Cells (twenty-two 50-µL aliquots; store at −80°C) including SOC Medium (4 × 2 mL; store at 4°C) (MilliporeSigma 71400)

SB (Super Broth) medium <R>

SDS-PAGE gel, NuPAGE 4%–12% Bis-Tris, 1.0–1.5 mM, Mini Protein Gels, 10 wells (Thermo Fisher Scientific NP0321BOX); store at 4°C

NuPAGE MES SDS Buffer Kit (for Bis-Tris Gels; Thermo Fisher Scientific NP0060), including 500 mL of 20× NuPAGE MES SDS Running Buffer and 10 mL of 4× NuPAGE LDS Sample Buffer; store at 4°C

SfiI, 20 units/µL (New England Biolabs R0123L) with 10× rCutSmart buffer (New England Biolabs B6004S); store at −20°C

T4 DNA Ligase (5 units/µL; Thermo Fisher Scientific EL0011; includes 10× T4 DNA Ligase buffer); store at −20°C

Water, highly pure, sterile

*The protocol uses highly pure water from a purification system (see "Equipment"). Use freshly. Several steps require sterile water, which is obtained by filtering highly pure water through a 0.22-µm filter unit. Store at room temperature.*

## Equipment

*A laboratory with standard molecular biology and microbiology equipment is required for this protocol.*

Agarose gel electrophoresis equipment (e.g., Owl horizontal electrophoresis systems [Thermo Fisher Scientific] with various tanks, casters, combs, and power supplies [EC-105] for mini gel [EasyCast B-series] preparative electrophoresis and blue-light transilluminator [e.g., VB-40 Visi-Blue Transilluminator VWR 95041-594])

Autoclave

Bunsen burner

Centrifuge bottles, 500-mL (polypropylene, wide-mouth, sealing cap; certified for >12,000g and resistant to bleaching and autoclaving; e.g., Nalgene 3141-0500)

Centrifuge tubes (50-mL)

Dewar flask

Digital balance with 0.01-g readability

Filtered pipette tips (10-, 20-, 100-, and 200-µL, and 1-mL)

Filters with Luer lock fitting (0.22-µm; e.g., Millex-GP Syringe Filters, Millipore SLGP033RS)

Filter units (0.22-µm; e.g., Millipore Stericup and Steriflip filter units)

Freezers (−20°C and −80°C)

Gel staining trays

Glass bottles (500-mL and 1-L, autoclaved)

Glass Erlenmeyer flasks, 2-L (autoclaved)

Heat blocks for 42°C, 50°C, and 95°C incubation

His tag purification column compatible with a medium pressure liquid chromatography system (e.g., complete His Tag Purification Column, MilliporeSigma 6781543001; store in 20% [v/v] ethanol at 4°C)

Ice bucket

Incubator (37°C)

Inoculating loops (sterile)

Laboratory film (i.e., Parafilm)

Magnetic hot plate stirrer

Magnetic stir bars

Medium pressure liquid chromatography system (e.g., an ÄKTA chromatography system from Cytiva or an NGC chromatography system from Bio-Rad) with 150-mL superloop (optional; see Steps 19–21)

Microfuge tubes (1.5- and 2-mL)

Microwave oven

$OD_{600}$ DiluPhotometer (Implen OD600-GO)

Orbital shaker (e.g., VWR Standard 1000 Orbital Shaker 89032-088)

PD-10 Desalting Columns (Cytiva 17085101), disposable (optional; see Step 24); store at room temperature

PCR tube strips (0.2-mL; e.g., Eppendorf 951010022)

Peristaltic Pump P-1 with tubing and connectors (Cytiva 18111091) (optional; Steps 19–21)

Petri dishes (10-cm)

Pipettes (5-, 10-, and 25-mL)

Pipette controller

Refrigerated benchtop centrifuge with swinging bucket rotor (e.g., Beckman Coulter Avanti J-15R with JS-4.750 swinging bucket rotor) and compatible microplate carriers

Refrigerated floor centrifuge (e.g., Thermo Fisher Scientific Sorvall Lynx 4000) with fixed-angle rotor for six 500-mL centrifuge bottles (e.g., Thermo Fisher Scientific Fiberlite F12-6 × 500 LEX)

Refrigerator (4°C)

Round-bottom tubes with snap cap (14-mL)

SDS-PAGE equipment compatible with NuPAGE gels and reagents (e.g., Invitrogen Mini Gel Tank [Thermo Fisher Scientific A25977] and Invitrogen PowerEase Touch 350W Power Supply [Thermo Fisher Scientific PSC350M])

Shaker at 37°C (e.g., Eppendorf New Brunswick Innova 40 Benchtop Incubator Shaker)

Single-channel micropipettes (1- to 1000-μL)

Syringes with Luer lock fitting (1-, 5-, and 10-mL)

Thermal cyclers (96-well; e.g., Eppendorf Mastercycler X50s 96-Well Silver Block Thermal Cycler 6311000010)

UV photometer (e.g., NanoDrop One[C] Microvolume UV-Vis Spectrophotometer; Thermo Fisher Scientific ND-ONEC-W)

Water purification system (e.g., Hydro PicoPure UV Plus)

## METHOD

*This protocol outlines a strategy to produce human or chimeric nonhuman/human Fabs following their selection by phage display. This involves cloning, followed by Fab expression and purification. The steps below follow this organization.*

*For Fab production in E. coli, the pET-11a plasmid is a preferred vector, which enables T7 promoter-driven and IPTG-inducible expression of a protein of interest when transformed into E. coli strain DE3 and its variants. A pET-11a plasmid with a Fab expression cassette compatible with pC3C and pC3Csort was published (Stahl et al. 2010), and similar constructs with Fab expression cassettes compatible with pComb3H and pComb3X (see Chapter 4, Overview: The pComb3 Phagemid Family of Phage Display Vectors [Rader 2024]) can be easily cloned as described in this protocol.*

*With ompA for the light chain and pelB for the heavy chain fragment, the Fabs will be initially secreted through the inner membrane into the E. coli periplasm, which is a potential source for purification. Large amounts of Fab, however, leak through the outer membrane into the medium, which allow for facile purification from the supernatant after centrifugation.*

*The Fab expression cassettes include a hexahistidine (His$_6$) tag, encoded at the C terminus of the heavy chain fragment. This allows for universal Fab purification by immobilized metal affinity chromatography (IMAC), avoiding harsh elution conditions and ensuring the removal of free light chain.*

*The cloning procedure takes 3 d to go from phagemid to pET-11a clones, and the subsequent expression and purification takes another 4 d.*

## Fab Cloning

*In Protocol 7: Phage Display Selection of Antibody Libraries: Screening of Selected Binders (Peng and Rader 2024), the initial analysis of output colonies after Fab-phage library selection by crude Fab ELISA, DNA fingerprinting, and DNA sequencing is described. Start with the phagemid preparation of positive clones of interest generated as part of that or similar protocols.*

1. Quantify the DNA concentration of the phagemid preparation by measuring the absorbance at 260 nm and multiplying the value by the conversion factor of 50 ng/µL (concentration of pure dsDNA with an $A_{260} = 1.0$) to get the concentration in nanograms per microliter. Dilute to 100 ng/µL in sterile water to streamline further use.

   *After the next step, store the remaining phagemid preparation at −20°C.*

2. Prepare a master mix in a microfuge tube to digest 1 µg of phagemid using SfiI to excise the Fab-encoding cassette (ompA/V$_L$/C$_L$/pelB/V$_H$/C$_H$1 in pComb3H and pComb3X; ompA/V$_L$.C$_L$/pelB/V$_H$ in pC3C and pC3Csort) (see Chapter 4, Overview: The pComb3 Phagemid Family of Phage Display Vectors [Rader 2024]), keeping all components on ice. To do this, mix the following (volumes given per reaction, add in order): 5 µL of 10× rCutSmart buffer, 34 µL of sterile water, and 1 µL of 20 units/µL SfiI. Mix gently and aliquot 40 µL into a microfuge tube. Add 10 µL (1 µg) of phagemid preparation, mix gently, and incubate for 3 h at 50°C.

   *SfiI requires two sites for activity, such as the two asymmetric SfiI sites on pComb3H, pComb3X, pC3C, and pC3Csort (see Chapter 4, Overview: The pComb3 Phagemid Family of Phage Display Vectors [Rader 2024]). It cannot be used for cutting a single site.*

   *If a pET-11a plasmid is available that already has a Fab-His$_6$ insert (with asymmetric SfiI sites flanking ompA/V$_L$/C$_L$/pelB/V$_H$/C$_H$1 in pComb3H and pComb3X or ompA/V$_L$/C$_L$/pelB/V$_H$ in pC3C and pC3Csort) (see Chapter 4, Overview: The pComb3 Phagemid Family of Phage Display Vectors [Rader 2024]), digest it in an identical manner. If the commercially available pET-11a plasmid without Fab insert is the starting point, PCR-amplify the entire Fab cassette with primers that add NdeI and BamHI restriction sites (and an additional 3–6 bases past the cleavage site for maximal efficiency) and a His$_6$ tag, and digest both the pET-11a plasmid and the PCR product with these enzymes to clone into the T7 promoter open reading frame (ORF). Use a Fab insert that is not cleaved by NdeI and BamHI.*

3. Separate the cleavage products by agarose gel electrophoresis (1%–1.5% gel), and excise the bands corresponding to the SfiI-flanked Fab-encoding cassette (∼1.4 kb for ompA/V$_L$/C$_L$/pelB/V$_H$/C$_H$1 and ∼1.1 kb for ompA/V$_L$/C$_L$/pelB/V$_H$). Dissolve the gel slices by adding 3 volumes of Buffer QG (QIAquick Gel Extraction Kit); i.e., 300 µL of Buffer QG for each 100-mg gel. Proceed with the supplier's protocol and elute with 30–50 µL of sterile water.

   *Purify the SfiI-cut or NdeI/BamHI-cut pET-11a plasmid analogously.*

4. Quantify the DNA of both the vector and the insert, as described in Step 1.

5. Prepare a ligation reaction in 0.2-mL PCR tubes on ice, using a 3:1 molar ratio of insert:vector, with 50 ng of vector (given a 1.1- or 1.4-kb insert and a 5.6-kb vector, this requires 30–40 ng of insert), 1 µL of 10× T4 DNA Ligase buffer, and 0.5 µL of 5 units/µL T4 DNA Ligase, and bring to 10 µL with sterile water. Incubate for 15 min at room temperature, and then transfer the reaction to ice.

   *Enzyme and buffer are both sensitive to changes in temperature. Remove the enzyme from the freezer only when needed and keep it on ice. Aliquot the buffer to avoid repeat freeze–thaw cycles.*

6. Transform the ligation reaction into One Shot TOP10 Chemically Competent *E. coli* according to the supplier's protocol. Briefly, thaw a single 50-µL aliquot of competent bacteria on ice per reaction, add 1 µL of cold ligation reaction, mix gently by pipetting, and incubate for 30 min on

ice. Heat-shock for 30 sec at 42°C followed by a 2-min incubation on ice. Then, add 500 μL of SOC medium (room temperature) and incubate while shaking at 250 rpm for 1 h at 37°C. Plate 50 μL onto prewarmed LB + 100 μg/mL carbenicillin plates and incubate overnight at 37°C. Make sure to use controls at this stage to verify digestion, ligation, and transformation efficiency (i.e., undigested vector alone, digested vector alone, and insert alone).

> *Competent bacteria are sensitive to temperature changes. Keep on ice until it is time for the heat shock.*
>
> *A larger volume of the ligation reaction can be transformed as long as it does not exceed 10% of the transformation volume.*
>
> *It is important to let the mixture of ligation reaction and competent bacteria sit for 30 min on ice prior to heat shock without agitation during or after this incubation.*
>
> *Given the asymmetric SfiI sites, there should be no self-ligation of the digested vector.*

7. The next day (day 2), pick colonies (one to three per ligation reaction) to inoculate 3–5 mL of LB supplemented with 100 μg/mL carbenicillin in separate 14-mL round-bottom tubes with snap caps. Seal the plates with laboratory film and store at 4°C as backup in case additional clones are needed. Incubate the tubes while shaking at 250 rpm overnight at 37°C.

8. The next day (day 3), remove 100 μL of the overnight culture and mix with 100 μL of 50% (v/v) glycerol to make backup stocks of these clones. Store at −80°C.

9. Centrifuge the remaining overnight culture at 3000g for 15 min at 4°C, discard the supernatant, resuspend the pellet in 250 μL of QIAGEN buffer P1 from the QIAprep Spin Miniprep Kit, and process using reagents and protocols supplied by the QIAprep Spin Miniprep Kit. Elute with 50 μL of sterile water and store at −20°C.

10. Confirm correct vector and insert sizes by restriction digestion (e.g., NdeI/BamHI and, separately, SfiI) followed by DNA sequencing using oligonucleotides specific for the T7 promoter and the T7 terminator (see "Materials").

> *With a functional Fab cassette confirmed, proceed to section "Fab Expression."*

## Fab Expression

*This section takes 3 d, with day 1 for Step 11, day 2 for Step 12, and day 3 for Step 14.*

11. Transform the Fab-encoding pET-11a plasmid (5–20 ng total) into Rosetta 2(DE3) Singles Competent Cells according to the supplier's protocol and analogously to Step 6. Plate onto prewarmed LB + 100 μg/mL carbenicillin + 25 μg/mL chloramphenicol plates and incubate overnight at 37°C.

> *The Rosetta 2(DE3) strain is a derivative of E. coli strain BL21 and is designed for the expression of eukaryotic proteins with codons rarely used in E. coli. Rosetta 2 supplies tRNAs for the seven rare codons AUA, AGG, AGA, CUA, CCC, GGA, and CGG on a plasmid, pRARE, with a chloramphenicol resistance marker. It also includes a chromosomally encoded T7 RNA polymerase under an IPTG-inducible lacUV5 promoter, the result of an integration of λDE3 phage.*

12. The next day, inoculate one to three colonies into separate starter cultures of 5-mL SB medium supplemented with 100 μg/mL carbenicillin and 25 μg/mL chloramphenicol in 14-mL round-bottom tubes with snap caps. Incubate while shaking at 250 rpm overnight at 37°C.

13. The next day, transfer the culture with the most growth into a 2-L Erlenmeyer flask containing 500 mL of SB medium supplemented with 100 μg/mL carbenicillin and 25 μg/mL chloramphenicol. Incubate while shaking at 250 rpm for 2–4 h at 37°C. Monitor the optical density at 600 nm ($OD_{600}$) of the culture, using an $OD_{600}$ DiluPhotometer and sterile SB medium as a blank.

> *Ensure the flask is not sealed tightly, to allow for air exchange.*

14. Once the $OD_{600}$ reaches 0.6–0.8, add 0.5 mL of 1 M IPTG (1 mM final concentration) and continue to incubate while shaking at 250 rpm overnight at 37°C.

    *The E. coli growth rates can vary depending on the sequence of the Fab being expressed, so it is important to closely monitor each culture individually if multiple Fab clones are being expressed.*

    *There is no need to lower the temperature to 30°C, as Fab is more efficiently secreted from the periplasm to the medium at 37°C.*

## Fab Purification

*There are several options for purifying recombinant human Fab using prepacked chromatography columns, such as those specific for the light chain constant domain (HiTrap KappaSelect or HiTrap LambdaFabSelect from Cytiva), the heavy chain constant domain 1 (CaptureSelect CH1-XL from Thermo Fisher Scientific), or the $His_6$ tag (cOmplete His Tag Purification Column from MilliporeSigma or HisTrap HP from Cytiva). The following protocol is optimized for a 1-mL cOmplete His Tag Purification Columns attached to either a medium pressure liquid chromatography system with a 150-mL superloop or a peristaltic pump. Refer to the supplier's recommendations for loading, washing, elution, and storage buffer specifications.*

15. Split the culture into two separate 500-mL centrifuge bottles (~250 mL each), ensure the volume is balanced in each, and centrifuge at 12,000g for 15 min at 4°C.

    *Keep supernatant and purified Fab fractions on ice from this point forward.*

16. As soon as centrifugation is complete, decant the supernatant from both bottles into a single 500-mL 0.22-μm filter unit.

17. Wash the 1-mL cOmplete His Tag Purification Column (stored in 20% [v/v] ethanol) with 10 mL of sterile water, followed by flushing with 10 mL of 5 mM imidazole in DPBS (washing buffer).

    *The suggested imidazole concentrations in the washing buffer (5 mM), the supernatant (5 mM), and the elution buffer (250 mM) are dependent on the chemical composition of the immobilized metal affinity chromatography (IMAC) matrix used, the $His_6$-tagged protein, and the expression system. The NaCl concentration of DPBS is 140 mM NaCl. The addition of NaCl up to 500 mM can help reduce nonspecific binding and increase the purity of the final product. See the supplier's protocol and troubleshooting for the cOmplete His Tag Purification Column.*

18. Dissolve 170.2 mg of imidazole (to a final concentration of 5 mM) in the 500 mL of supernatant.

19. Save a 1-mL aliquot on ice and load the remaining mix onto the cOmplete His Tag Purification Column at 1 mL/min. Save the flowthrough on ice.

    *Transfer the prepared ~500-mL supernatant to a 150-mL superloop (in four portions) when using a medium pressure liquid chromatography system, or load prepared supernatant directly when using a peristaltic pump. Keep fluids and column refrigerated during the loading, washing, and elution process.*

20. Wash column with 20 mL of 5 mM imidazole in DPBS. Save the wash on ice.

21. Elute with 10 mL of 250 mM imidazole in DPBS, collecting 1-mL fractions.

22. Measure the absorbance of the fractions at 280 nm ($A_{280}$) to determine their protein concentration. The average $A_{280}$ value for a Fab at 1 mg/mL is 1.35; thus, divide the $A_{280}$ by 1.35 to determine the concentration in milligrams per milliliter.

    *Depending on the quality and concentration of imidazole in the elution buffer, the background $A_{280}$ may be as high as 0.4–0.5, as older or impure imidazole is known to absorb at this wavelength. In that case, the yield may not be accurately determined at this step, and may instead be done in Step 26, after removal of imidazole.*

23. Analyze protein-containing fractions, supernatant, and flowthrough, and wash by nonreducing SDS-PAGE. To do this, mix 9–18 μL of each protein-containing fraction with 3–6 μL of 4× NuPAGE LDS Sample Buffer and heat for 10 min at 95°C. Load onto a NuPAGE 4%–12% Bis-Tris Mini Protein Gel along with a PageRuler Prestained Protein Ladder (10–180-kDa) and run at

150 V for 1–2 h in 1× NuPAGE MES SDS Running Buffer. Stain the gel for 1 h with PageBlue Protein Staining Solution and destain for at least 3 h in water using an orbital shaker.

*The nonreduced Fab should be visible as an ~50-kDa band.*

24. Pool the fractions containing Fab. Then, buffer-exchange into DPBS or buffer of choice using either of the following two options:
    - Use disposable PD-10 Desalting Column, according to the supplier's directions.
    - Using Amicon Ultra-15 Centrifugal Filters (for 15- or 4-mL sample volume, 30-kDa MWCO), concentrate the pooled fractions, and dilute in new buffer until the imidazole concentration is <1 mM (i.e., >250-fold dilution).

25. Using Amicon Ultra-15 Centrifugal Filters (for 15- or 4-mL sample volume, 30-kDa MWCO), concentrate to 1–2 mg/mL.

26. Measure $A_{280}$ as described in Step 22.

27. Assess the purity of Fabs by running nonreducing and reducing SDS-PAGE as described in Step 23. Per lane, load 5–10 μg of Fab in 1× NuPAGE LDS Sample Buffer with and without 5% (v/v) 2-mercaptoethanol for reducing and nonreducing conditions, respectively.

    *The reduced Fab should be visible as two ~25-kDa bands that may overlap. The nonreduced Fab should again be visible as an ~50-kDa band.*

28. Prepare 100-μL aliquots of Fab in 0.2-mL PCR tubes, flash-freeze in liquid nitrogen in a Dewar flask, and store at −80°C.

    *To avoid denaturing and aggregating the purified protein, freeze and thaw as quickly as possible and avoid repeated freeze–thaw cycles. Minimize the air interface of frozen aliquots by using small tubes.*

    *Proceed to downstream biophysical and biological assays, such as affinity, specificity, developability, and activity measurements.*

## RECIPES

### LB + 100 μg/mL Carbenicillin Plates

Dissolve 25 g of LB (Lysogeny Broth) medium powder (MilliporeSigma L3522) in 1 L of highly pure water (e.g., that produced by Hydro PicoPure UV Plus) with 15 g of Bacto Agar (BD 214010). Autoclave for 15 min at 121°C and 15 psi. When cooled to ~50°C, add 1 mL of 100 μg/μL carbenicillin. Pour into 10-cm sterile Petri dishes (20–25 mL each), remove bubbles by quickly flaming the plates with a Bunsen burner, and allow to solidify at room temperature. Store for up to 3 mo at 4°C.

*Alternatively, prepoured LB + 100 μg/mL carbenicillin plates are commercially available (e.g., Teknova L1010).*

### LB + 100 μg/mL Carbenicillin + 25 μg/mL Chloramphenicol Plates

Dissolve 25 g of LB (Lysogeny Broth) medium powder (MilliporeSigma L3522) in 1 L of highly pure water (e.g., that produced by the Hydro PicoPure UV Plus system) with 15 g of Bacto Agar (BD 214010). Autoclave for 15 min at 121°C and 15 psi. When cooled to ~50°C, add 1 mL of 100 μg/μL carbenicillin and 250 μL of 100 μg/μL chloramphenicol. Pour into 10-cm sterile Petri dishes (20–25 mL each), remove bubbles by quickly flaming the plates with a Bunsen burner, and allow to solidify at room temperature. Store for up to 3 mo at 4°C.

*SB (Super Broth) Medium*

Mix 20 g of 3-(N-morpholino)propanesulfonic acid (MOPS; MilliporeSigma M1254), 60 g of Bacto Tryptone (Thermo Fisher Scientific 211705), and 40 g of Bacto Yeast Extract (Thermo Fisher Scientific 212750). Add highly pure water (e.g., that produced by Hydro PicoPure UV Plus) to a final volume of 1.9 L. Bring to pH 7.0 with 1 M sodium hydroxide (NaOH; Thermo Fisher Scientific A4782901). Bring volume to 2 L with highly pure water. Sterilize by autoclaving for 15 min at 121°C and 15 psi in two 1-L or four 500-mL glass bottles. Store for up to 2 yr at room temperature if unopened.

## ACKNOWLEDGMENTS

Work in our laboratory was supported by the National Cancer Institute, National Institutes of Health.

## REFERENCES

Peng H, Rader C. 2024. Phage display selection of antibody libraries: screening of selected binders. *Cold Spring Harb Protoc* doi:10.1101/pdb.prot108603

Rader C. 2024. The pComb3 phagemid family of phage display vectors. *Cold Spring Harb Protoc* doi:10.1101/pdb.prot107756

Stahl SJ, Watts NR, Rader C, DiMattia MA, Mage RG, Palmer I, Kaufman JD, Grimes JM, Stuart DI, Steven AC, et al. 2010. Generation and characterization of a chimeric rabbit/human Fab for co-crystallization of HIV-1 Rev. *J Mol Biol* **397:** 697–708. doi:10.1016/j.jmb.2010.01.061

CHAPTER 6

# Generation of Chicken Antibody Libraries and Selection of Antigen Binders

Hyunji Yang,[1] Jisu Chae,[1] Hyori Kim,[2] Jinsung Noh,[3] and Junho Chung[1,4]

[1]Cancer Research Institute, Seoul National University College of Medicine, Seoul 03080, South Korea; [2]Convergence Medicine Research Center, Asan Medical Center, Seoul 05505, South Korea; [3]Bio-MAX Institute, Seoul National University, Seoul 08826, South Korea

Chicken antibodies have been widely used for research and diagnostic purposes. Chicken antibodies are often cross-reactive to epitopes shared by humans, nonhuman primates, and other mammals, and can be tested in many mouse disease models, which provides an advantage for their preclinical study and evaluation. In addition, the variable region of chicken antibodies has unique structural characteristics, including noncanonical cysteine residues in the heavy chain complementarity-determining region (CDR)3 and a long heavy chain CDR3, which together with a short light chain CDR enable the formation of unconventional antibody paratopes. As chickens have single functional copies of the $V_H$ and $J_H$ genes, and the somatic gene conversion process usually involves $D_H$ genes, all functional VDJ gene fragments can be obtained from the B-cell repertoire using a single PCR primer set, without any primer bias. As for the light chain, chickens only have a $V_\lambda$ light chain, composed of a single $V_\lambda$ and $J_\lambda$ gene pair. Therefore, the chicken light chain repertoire can also be accurately amplified using a single primer set. This unbiased reconstitution of the chicken B-cell repertoire provides a great advantage not only in the construction of phage display libraries but also for the in silico selection of antigen binders from a virtual B-cell receptor repertoire. Here, we introduce the use of chicken antibodies in research, diagnostic, and therapeutic fields. In addition, the chromosomal organization of chicken immunoglobulin genes and its diversification mechanisms for shaping the antibody repertoire are also discussed.

## INTRODUCTION

Chicken antibodies have been widely used for research and immunodiagnostic purposes, and have also been tested for their therapeutic potential. The wide application of chicken antibodies in these fields has been nicely reviewed in several papers (Zhang et al. 2010; Spillner et al. 2012; Lee et al. 2017b, 2021; Thirumalai et al. 2019). The significant phylogenetic distance between chickens and mammals provides a great advantage for the use of chickens as a host animal for immunization, especially when the target antigen is conserved among mammals (Gassmann et al. 1990; Lee et al. 2017b). This is relevant because protein antigens often share a high degree of identity among mammalian species, whose self-tolerance mechanism against the homologous proteins leads to low immunogenicity and restricts their value as host animals (Matsushita et al. 1998). Chickens can produce antibodies against highly conserved mammalian antigens, especially against novel human functional epitopes that are masked in mice due to sequence conservation. Furthermore, chicken antibodies are often cross-

[4]Correspondence: jjhchung@snu.ac.kr; junhochung@me.com

© 2026 Cold Spring Harbor Laboratory Press
Cite this introduction as *Cold Spring Harb Protoc*; doi:10.1101/pdb.top108210

reactive to the corresponding mouse ortholog. Cross-reactivity to mouse and primate orthologs is important for preclinical development of therapeutic antibodies, which often relies on murine models (Gjetting et al. 2019). In addition, chicken antibodies possess distinctive structural characteristics compared to their mammalian counterparts: noncanonical cysteine residues in heavy chain complementarity-determining region (CDR)3, long heavy chain CDR3 (Wu et al. 2012), and short light chain CDR1 (Matsushita et al. 1998; Shih et al. 2012; Conroy et al. 2014). Another advantage is that different physiological features in chickens might play a role in the development of unique antibodies. For example, chickens have a body temperature of ∼42°C–43°C, higher than those of mice or rabbits (Zhang et al. 2010). Hydrogen bonds are exothermic and more stable at low temperatures (Reverberi and Reverberi 2007). Conversely, the strength of the hydrophobic bond increases with temperature. For these reasons, chickens have been widely used as an alternative host animal to mammals for antibody development.

Chickens express only three classes of Ig heavy chain (IgH) genes: immunoglobulin M (IgM), IgY, and IgA (Sun et al. 2012). No bird species has been found to possess an IgD-encoding gene (Zhao et al. 2000; Lundqvist et al. 2001). The IgM in birds has a structure very similar to that of the mammalian IgM (Dahan et al. 1983; Magor et al. 1998; Choi et al. 2010). In chickens, IgY is the main class of serum antibody, with concentrations ranging from 5 to 15 mg/mL in the serum of laying hens, compared to the lower concentration of IgM (1–3 mg/mL) and IgA (0.3–0.5 mg/mL) (Rose et al. 1974; Kowalczyk et al. 1985; Spillner et al. 2012). IgY is the avian homolog of human G (IgG) in mammals, and the evolutionary ancestor of IgG and IgE (Parvari et al. 1988; Taylor et al. 2008, 2009; Lee et al. 2017b). Structurally, IgY consists of four heavy chain constant (CH) domains but contains no hinge region (Warr et al. 1995).

In this topic introduction, we first further discuss some of the advantages of chicken antibodies for research, diagnostic, and therapeutic purposes. We then discuss the chromosomal organization of chicken immunoglobulin genes and the mechanisms used to create the antibody repertoire, including somatic gene conversion, which provides unique advantages for engineering chicken antibodies. We then briefly highlight some current biomedical applications of chicken antibodies, focusing on those that have been used in clinical trials. Last, we briefly introduce four accompanying protocols that describe the preparation of chicken antibody libraries and selection methodologies for antigen binders via phage display technology.

## ADVANTAGES OF CHICKEN ANTIBODIES FOR BIOTECHNOLOGICAL USES

Polyclonal chicken antibodies can be isolated from egg yolk without euthanizing the immunized chickens, which is made possible due to the transfer of antibodies from the maternal side to the egg yolk. Among the isotypes expressed by chickens, IgY dominates the polyclonal antibodies extracted from the egg yolk (Hamal et al. 2006; Lee et al. 2017b). On average, a hen can lay an egg per day, each containing up to 100 mg of IgY, ∼10% of which is expected to be specific to the immunogen (Gassmann et al. 1990; Pauly et al. 2009; Lee et al. 2017b). High yield, acceptable specificity, and lack of need for the euthanizing of the host animal make egg yolk–derived polyclonal IgY the most popular format of chicken antibodies against an epitope of interest among commercially used ones.

IgY is more resistant to proteolysis than its mammalian counterparts, and retains 40% of its activity after incubation with trypsin or chymotrypsin for 8 h (Hatta et al. 1993), which might be attributed to its additional carbohydrate side-chains (four vs. two) (Gilgunn et al. 2016); higher hydrophobic nature, with an isoelectric point between 5.76 and 7.6 due to the longer Fc portion of the molecule (Pereira et al. 2019); and higher stiffness of the molecule due to the lack of a hinge region (Spillner et al. 2012). Similar to IgG, IgY is involved in opsonization, complement system activation, and most effector functions in chickens (Janeway 2001). IgY, however, is unable to bind to or activate the mammalian complement system (Larsson et al. 1992), Fc receptors (Schmidt et al. 1993; Lee et al. 2017b), rheumatoid factors (Larsson et al. 1991), and the erythrocyte agglutinogens A and B

(Gutiérrez Calzado et al. 2003). This excellent stability and its inertness in the human system prompted the prophylactic use of intranasal polyclonal IgY derived from eggs of hens immunized with viral antigens. In a double-blind, randomized, placebo-controlled phase 1 study, anti-SARS-CoV-2 RBD IgY administered intranasally for 14 d in 48 healthy adults demonstrated an excellent safety and tolerability profile, with no evidence of systemic absorption (Frumkin et al. 2022).

## DIVERSITY OF THE CHICKEN IG REPERTOIRE

The diversity of the chicken Ig repertoire is generated by three biological mechanisms: germline gene rearrangement, somatic gene conversion, and somatic hypermutation. Similar to mammals, and upon B-cell development, chickens (*Gallus gallus domestica*) generate Ig-coding regions in their genome through an irreversible gene rearrangement process. During gene rearrangement, variable ($V$), diversity ($D$), and joining ($J$) gene segments of the heavy chain, and the $V$ and $J$ gene segments of the light chain are recombined to generate the variable domain of an antibody. In the chicken IgH locus on chromosome 31, only a single functional $V_H$ segment, four $D_H$ segments, and a single $J_H$ segment have been identified (Reynaud et al. 1989, 1991; Lefranc and Lefranc 2001). Therefore, V(D)J recombination in chickens can generate a very limited diversity of IgH chains. To increase such diversity, chicken Ig genes use a unique mechanism of somatic gene conversion. There are more than 90 pseudo-$V_H$ genes upstream of $D_H$ (Mansikka 1992; Lefranc and Lefranc 2001). Analysis of the functional chicken IgH genes revealed that somatic gene conversion plays a major role in the generation of $V_H$ diversity in chickens. As a nonreciprocal process, somatic gene conversion can use the upstream pseudo-$V_H$ segments as donor sequences to repeatedly modify the functional $V_H$ that has been recombined with the D and J segments. Most hyperconversion events occur at the $D_H$ segments for the $V_H$, and are responsible for CDR3 diversification (Reynaud et al. 1989; Lee et al. 2017b). This mechanism ensures that all expressed chicken $V_H$ have a single leader sequence (from the single functional $V_H$) and largely diverse coding regions (McCormack et al. 1991; Mansikka 1992). In mammals, the $\alpha$ gene, encoding for the IgA constant region, is located in the most 3′ region of the IgH locus. However, in chickens, the $\alpha$ gene is positioned in the middle of the IgH locus and is inverted, with a different transcriptional orientation from that of $\mu$ and $\upsilon$ (Home et al. 1992; Zhao et al. 2000). Despite the differences, the bird $\alpha$ gene is still expressed through a class switch recombination process (Sun et al. 2012). Although two versions of the Ig light chain (IgL) gene ($\lambda$ and $\kappa$) are found in mammals and reptiles, only the $\lambda$ gene is found in chickens, on chromosome 15 (Reynaud et al. 1987; Magor et al. 1994). In this $\lambda$ locus, there is only a single functional $V_\lambda$ gene, and there are approximately 25 pseudo-$V_\lambda$ genes present upstream of a single pair of $J_\lambda$ and $C_\lambda$ (Lefranc and Lefranc 2001; Sun et al. 2012). Hyperconversion events in the $V_\lambda$ are restricted only to the $V_\lambda$ segment and not $J_\lambda$ (Thompson and Neiman 1987; Parvari et al. 1990; Lee et al. 2017b).

In this context, somatic gene conversion is also the major mechanism responsible for chicken IgL diversity (McCormack and Thompson 1990; McCormack et al. 1991; Sun et al. 2012). The combination of heavy and light chains can also generate a highly diverse antibody library in chickens, potentially producing a repertoire of approximately $3 \times 10^9$ combinations (Davison 2008; Ratcliffe 2008). Upon antigenic stimulation, the variable domain can further acquire somatic point mutations by an activation-induced cytidine deaminase (AID) mechanism (Lee et al. 2017b).

A great advantage of using the chicken immunoglobulin repertoire is the high level of identity between the single functional $V_H$ and $V_\lambda$ genes and their pseudogenes (Wu et al. 2012), and the mechanism of gene conversion saving the 5′ and 3′ region of functional $V_H$ and $V_\lambda$ genes. Gene conversion was not observed at the 5′ region of framework 1 on $V_H$ and $V_\lambda$ or at the 3′ region of framework 4 derived from $J_H$ or $J_\lambda$ during the process of diversification (McCormack et al. 1991). This allows the use of one or two primer sets for the relatively unbiased amplification of $V_H$ and $V_\lambda$ genes from the B-cell repertoire, individually. In contrast, the unbiased amplifications of mammalian $V_H$ and $V_\lambda$ genes require the use of numerous primer combinations. Indeed, an in silico chicken $V_H$ gene

Chapter 6

repertoire was created using a single PCR primer set from chicken B-cell populations obtained through an immunization process (Kim et al. 2019). Afterward, the $V_H$ genes were clustered into components based on sequence similarity. By monitoring the increase in the number of $V_H$ genes from somatic hypermutation or somatic gene conversion, as well as the summated frequency of $V_H$ genes in each component from the immunization process, $V_H$ gene components formed in response to the immunized antigen were successfully retrieved from the in silico generated B-cell repertoires.

## ANTIBODY ENGINEERING

The greater degree of sequence similarity among the chicken $V_H$ and $V_\lambda$ genes compared to those in other host animals provides unique opportunities for humanization and optimization. The accumulated knowledge regarding $V_H$ or $V_\lambda$ framework sequence engineering, either for humanization or enhancing physicochemical properties, can be more easily applied to other chicken antibodies due to their sequence similarity. More interestingly, most of the chicken $V_H$ or $V_\lambda$ framework residues were found to be interchangeable with other amino acids without significantly affecting their affinity (Shin et al. 2018). In addition, as several framework residues can be mutated to cysteine without multimer formation through intermolecular disulfide bond or loss of antigen reactivity, these mutated chicken antibodies can be used for site-specific conjugation with cytotoxic drugs, dyes, or chemicals (Yoon et al. 2016). This provides unique opportunities for the development of antibody–drug conjugates, as the Fc region of the antibody could be preserved for its interaction with Fc receptors and complement proteins. In addition to the framework residues, the CDRs can also be mutated to enhance their characteristics, particularly for antigen binding. For example, the HCDR3 residues of the chicken antibody have been mutated to achieve higher specificity and affinity (Hu et al. 2012; Gjetting et al. 2019).

## BIOMEDICAL APPLICATIONS

Two chicken monoclonal antibodies have been tested in clinical trials. Sym021, an anti-PD1 humanized chicken antibody, was tested in a phase 1 clinical trial (Lee et al. 2017a). For development, antibodies with unique cognate $V_H/V_\lambda$ CDRs representing 46 clonal families were grafted in silico. After CDR grafting, 34 clonotypes were successfully humanized, with a success rate of 74%. In surface plasmon resonance analysis, Sym021 bound PD1 with a monovalent affinity of 30 pM. Recently, a phase 2 clinical trial (NCT05338931) for humanized chicken anti-CD19 single-chain variable fragment (scFv)-based chimeric antigen receptor (CAR) T-cell therapy was initiated (WO2019125070A1) (Kim et al. 2021).

## GENERATION OF CHICKEN ANTIBODY USING PHAGE DISPLAY LIBRARY TECHNOLOGY

The chicken immune system has been studied for decades, and it is now easy to produce and obtain antibodies through the immunization of chickens. Chicken antibodies offer many advantages over mammalian antibodies due to an increased phylogenetic distance between birds and mammals. The chicken immune system can recognize epitopes on the mammalian proteins more readily and often detect epitopes that differ from those detected by mammals. Furthermore, animal care costs are lower for chickens than those of rabbits, and the antibodies can be purified in large amounts from egg yolk, making laying hens highly efficient producers of polyclonal antibodies (Glick 1979; Carlander et al. 1999).

The effective isolation of specific antibodies from immunological repertoires requires an antigen-driven diverse library for a specific antigen and an efficient selection procedure, such as bio-panning and phage ELISA. Through the course of immunization, antibody titers should be carefully monitored; for instance, via ELISA. Once the titer reaches an appropriate level, mRNA can be prepared from the spleen, bursa of Fabricius, and bone marrow to construct a cDNA library (Bäck et al. 1973; Jeurissen et al. 1988). In the accompanying protocol, we provide a step-by-step description of this procedure (see Protocol 1: Chicken Immunization Followed by RNA Extraction and cDNA Synthesis for Antibody Library Preparation [Yang et al. 2024a]).

Once the cDNA is prepared, a phage display chicken scFv library can be generated, using primers specific for the chicken immunoglobulin variable regions, a phage display vector, and a helper phage. The multiple primer sets used in the accompanying protocol have been newly designed to cover a single functional immunoglobulin heavy chain and $\lambda$ chain genes. They also individually amplify 82% and 86% of pseudo-$V_H$ and $V_\lambda$ genes of the chicken. For details, see Protocol 2: Generation of a Phage Display Chicken Single-Chain Variable Fragment Library (Yang et al. 2024b) and Protocol 3: Preparation of VCSM13 Helper Phage for Display Library Reamplification and Bio-Panning (Yang et al. 2024c).

Antigen-binding clones can be enriched from the phage display chicken scFv library through panning. Additionally, high-throughput sequencing (HTS) analysis can then be used to determine the pattern of enrichment of scFv clones throughout the panning. The scFv clones from the last round of panning or chemically synthesized scFv clones based on nucleotide sequence in HTS analysis should be subjected to phage ELISA to confirm antigen reactivity. Moreover, it is possible to train a machine learning algorithm to derive in silico antigen binding clones from a repertoire of HTS sequences. This provides a method for identifying the possible antigen binders, which allows for the characterization of various antibody libraries inaccessible by conventional methods (Yoo et al. 2020). For details of this process, see Protocol 4: Selection of Antigen Binders from a Chicken Single-Chain Variable Fragment Library (Yang et al. 2024d).

## CONCLUDING REMARKS

In conclusion, chicken antibodies exhibit unique advantages for the generation of antibodies for research, diagnostic, and therapeutic use. There is only a single functional $V_H$ segment, four $D_H$ segments, and a single $J_H$ segment that have been identified in chicken IgH genes, as well as a single functional $V_\lambda$ gene in chicken Ig$\lambda$ genes. Although each $V_H$ and $V_\lambda$ gene has 90 and 25 pseudogenes, respectively, gene conversion was not observed at the 5′ region of framework 1 on $V_H$ and $V_\lambda$ during the process of Ig diversification. Moreover, the chicken immunoglobulin repertoire shows a high level of identity between the single functional $V_H$ and $V_\lambda$ genes and their pseudogenes. This allows the use of specific primer sets for the unbiased amplification of $V_H$ and $V_\lambda$ genes from the B-cell repertoire, respectively. Phage display technology and HTS analysis can be effectively used to select and identify a diverse panel of chicken antibodies to a variety of antigens of interest.

## REFERENCES

Bäck O, Bäck R, Hemmingsson EJ, Lidén S, Linna TJ. 1973. Migration of bone marrow cells to the bursa of Fabricius and the spleen in the chicken. Scand J Immunol 2: 357–366. doi:10.1111/j.1365-3083.1973.tb02044.x

Carlander D, Stålberg J, Larsson A. 1999. Chicken antibodies: a clinical chemistry perspective. Ups J Med Sci 104: 179–189. doi:10.3109/03009739909178961

Choi JW, Kim JK, Seo HW, Cho BW, Song G, Han JY. 2010. Molecular cloning and comparative analysis of immunoglobulin heavy chain genes from Phasianus colchicus, Meleagris gallopavo, and Coturnix japonica. Vet Immunol Immunopathol 136: 248–256. doi:10.1016/j.vetimm.2010.03.014

Conroy PJ, Law RH, Gilgunn S, Hearty S, Caradoc-Davies TT, Lloyd G, O'Kennedy RJ, Whisstock JC. 2014. Reconciling the structural attributes of avian antibodies. J Biol Chem 289: 15384–15392. doi:10.1074/jbc.M114.562470

Dahan A, Reynaud CA, Weill JC. 1983. Nucleotide sequence of the constant region of a chicken mu heavy chain immunoglobulin mRNA. Nucleic Acids Res 11: 5381–5389. doi:10.1093/nar/11.16.5381

Davison F. 2008. The importance of the avian immune system and its unique features. In Avian immunology (ed. Davison F et al.), pp. 1–11. Academic Press, London, UK.

Frumkin LR, Lucas M, Scribner CL, Ortega-Heinly N, Rogers J, Yin G, Hallam TJ, Yam A, Bedard K, Begley R, et al. 2022. Egg-derived anti-SARS-CoV-2 immunoglobulin Y (IgY) with broad variant activity as intranasal prophylaxis against COVID-19. medRxiv doi:10.1101/2022.01.07.22268914.

Gassmann M, Thömmes P, Weiser T, Hubscher U. 1990. Efficient production of chicken egg yolk antibodies against a conserved mammalian protein. *FASEB J* 4: 2528–2532. doi:10.1096/fasebj.4.8.1970792

Gilgunn S, Millán Martín S, Wormald MR, Zapatero-Rodríguez J, Conroy PJ, O'Kennedy RJ, Rudd PM, Saldova R. 2016. Comprehensive N-glycan profiling of avian immunoglobulin Y. *PLoS ONE* 11: e0159859. doi:10.1371/journal.pone.0159859

Gjetting T, Gad M, Fröhlich C, Lindsted T, Melander MC, Bhatia VK, Grandal MM, Dietrich N, Uhlenbrock F, Galler GR, et al. 2019. Sym021, a promising anti-PD1 clinical candidate antibody derived from a new chicken antibody discovery platform. *MAbs* 11: 666–680. doi:10.1080/19420862.2019.1596514

Glick B. 1979. The avian immune system. *Avian Dis* 23: 282–289. doi:10.2307/1589557

Gutiérrez Calzado E, Cruz Mario E, Samón Chávez T, Vázquez Luna E, Corona Ochoa Z, Schade R. 2003. Extraction of a monospecific Coombs-reagent from chicken eggs. *ALTEX* 20: 21–25.

Hamal KR, Burgess SC, Pevzner IY, Erf GF. 2006. Maternal antibody transfer from dams to their egg yolks, egg whites, and chicks in meat lines of chickens. *Poult Sci* 85: 1364–1372. doi:10.1093/ps/85.8.1364

Hatta H, Tsuda K, Akachi S, Kim M, Yamamoto T, Ebina T. 1993. Oral passive immunization effect of anti-human rotavirus IgY and its behavior against proteolytic enzymes. *Biosci Biotechnol Biochem* 57: 1077–1081. doi:10.1271/bbb.57.1077

Home WA, Ford JE, Gibson DM. 1992. L chain isotype regulation in horse. I. Characterization of Ig$\lambda$ genes. *J Immunol* 149: 3927–3936. doi:10.4049/jimmunol.149.12.3927

Hu Z-Q, Liu J-L, Li H-P, Xing S, Xue S, Zhang J-B, Wang J-H, Nölke G, Liao Y-C. 2012. Generation of a highly reactive chicken-derived single-chain variable fragment against *Fusarium verticillioides* by phage display. *Int J Mol Sci* 13: 7038–7056. doi:10.3390/ijms13067038

Janeway C. 2001. *Immunobiology 5: the immune system in health and disease*. Garland Pub., New York.

Jeurissen SH, Janse EM, Ekino S, Nieuwenhuis P, Koch G, De Boer GF. 1988. Monoclonal antibodies as probes for defining cellular subsets in the bone marrow, thymus, bursa of Fabricius, and spleen of the chicken. *Vet Immunol Immunopathol* 19: 225–238. doi:10.1016/0165-2427(88)90110-9

Kim S, Lee H, Noh J, Lee Y, Han H, Yoo DK, Kim H, Kwon S, Chung J. 2019. Efficient selection of antibodies reactive to homologous epitopes on human and mouse hepatocyte growth factors by next-generation sequencing-based analysis of the B cell repertoire. *Int J Mol Sci* 20: 417. doi:10.3390/ijms20020417

Kim K-H, Patel RP, Lee YG, Kim S, Choi J-H, Kim S-M, Kim G-B, Lee J-H, Lee H-J, Park J-H, et al. 2021. Abstract LB030: a novel anti-CD19 chimeric antigen receptor T cell product targeting a membrane-proximal domain of CD19. *Cancer Res* 81: LB030. doi:10.1158/1538-7445.AM2021-LB030

Kowalczyk K, Daiss J, Halpern J, Roth TF. 1985. Quantitation of maternal-fetal IgG transport in the chicken. *Immunology* 54: 755–762.

Larsson A, Karlsson-Parra A, Sjöquist J. 1991. Use of chicken antibodies in enzyme immunoassays to avoid interference by rheumatoid factors. *Clin Chem* 37: 411–414. doi:10.1093/clinchem/37.3.411

Larsson A, Wejåker PE, Forsberg PO, Lindahl T. 1992. Chicken antibodies: a tool to avoid interference by complement activation in ELISA. *J Immunol Methods* 156: 79–83. doi:10.1016/0022-1759(92)90013-J

Lee HK, Jin J, Kim SI, Kang MJ, Yi EC, Kim JE, Park JB, Kim H, Chung J. 2017a. A point mutation in the heavy chain complementarity-determining region 3 (HCDR3) significantly enhances the specificity of an anti-ROS1 antibody. *Biochem Biophys Res Commun* 493: 325–331. doi:10.1016/j.bbrc.2017.09.023

Lee W, Syed Atif A, Tan SC, Leow CH. 2017b. Insights into the chicken IgY with emphasis on the generation and applications of chicken recombinant monoclonal antibodies. *J Immunol Methods* 447: 71–85. doi:10.1016/j.jim.2017.05.001

Lee L, Samardzic K, Wallach M, Frumkin LR, Mochly-Rosen D. 2021. Immunoglobulin Y for potential diagnostic and therapeutic applications in infectious diseases. *Front Immunol* 12: 696003. doi:10.3389/fimmu.2021.696003

Lefranc MP, Lefranc G. 2001. *The immunoglobulin FactsBook*. Academic Press, London, UK.

Lundqvist ML, Middleton DL, Hazard S, Warr GW. 2001. The immunoglobulin heavy chain locus of the duck. Genomic organization and expression of D, J, and C region genes. *J Biol Chem* 276: 46729–46736. doi:10.1074/jbc.M106221200

Magor KE, Higgins DA, Middleton DL, Warr GW. 1994. cDNA sequence and organization of the immunoglobulin light chain gene of the duck, *Anas platyrhynchos*. *Dev Comp Immunol* 18: 523–531. doi:10.1016/S0145-305X(06)80006-6

Magor KE, Warr GW, Bando Y, Middleton DL, Higgins DA. 1998. Secretory immune system of the duck (*Anas platyrhynchos*). Identification and expression of the genes encoding IgA and IgM heavy chains. *Eur J Immunol* 28: 1063–1068. doi:10.1002/(SICI)1521-4141(199803)28:03<1063::AID-IMMU1063>3.0.CO;2-O

Mansikka A. 1992. Chicken IgA H chains. Implications concerning the evolution of H chain genes. *J Immunol* 149: 855–861. doi:10.4049/jimmunol.149.3.855

Matsushita K, Horiuchi H, Furusawa S, Horiuchi M, Shinagawa M, Matsuda H. 1998. Chicken monoclonal antibodies against synthetic bovine prion protein peptide. *J Vet Med Sci* 60: 777–779. doi:10.1292/jvms.60.777

McCormack WT, Thompson CB. 1990. Chicken IgL variable region gene conversions display pseudogene donor preference and 5′ to 3′ polarity. *Genes Dev* 4: 548–558. doi:10.1101/gad.4.4.548

McCormack WT, Tjoelker LW, Thompson CB. 1991. Avian B-cell development: generation of an immunoglobulin repertoire by gene conversion. *Annu Rev Immunol* 9: 219–241. doi:10.1146/annurev.iy.09.040191.001251

Parvari R, Avivi A, Lentner F, Ziv E, Tel-Or S, Burstein Y, Schechter I. 1988. Chicken immunoglobulin gamma-heavy chains: limited VH gene repertoire, combinatorial diversification by D gene segments and evolution of the heavy chain locus. *EMBO J* 7: 739–744. doi:10.1002/j.1460-2075.1988.tb02870.x

Parvari R, Ziv E, Lantner F, Heller D, Schechter I. 1990. Somatic diversification of chicken immunoglobulin light chains by point mutations. *Proc Natl Acad Sci* 87: 3072–3076. doi:10.1073/pnas.87.8.3072

Pauly D, Dorner M, Zhang X, Hlinak A, Dorner B, Schade R. 2009. Monitoring of laying capacity, immunoglobulin Y concentration, and antibody titer development in chickens immunized with ricin and botulinum toxins over a two-year period. *Poult Sci* 88: 281–290. doi:10.3382/ps.2008-00323

Pereira EPV, van Tilburg MF, Florean E, Guedes MIF. 2019. Egg yolk antibodies (IgY) and their applications in human and veterinary health: a review. *Int Immunopharmacol* 73: 293–303. doi:10.1016/j.intimp.2019.05.015

Ratcliffe MJH. 2008. B cells, the bursa of fabricius and the generation of antibody repertoires. In *Avian immunology* (ed. Davison F et al.), pp. 67–89. Academic Press, London, UK.

Reverberi R, Reverberi L. 2007. Factors affecting the antigen-antibody reaction. *Blood Transfus* 5: 227–240.

Reynaud CA, Anquez V, Grimal H, Weill JC. 1987. A hyperconversion mechanism generates the chicken light chain preimmune repertoire. *Cell* 48: 379–388. doi:10.1016/0092-8674(87)90189-9

Reynaud CA, Dahan A, Anquez V, Weill JC. 1989. Somatic hyperconversion diversifies the single $V_H$ gene of the chicken with a high incidence in the D region. *Cell* 59: 171–183. doi:10.1016/0092-8674(89)90879-9

Reynaud CA, Anquez V, Weill JC. 1991. The chicken D locus and its contribution to the immunoglobulin heavy chain repertoire. *Eur J Immunol* 21: 2661–2670. doi:10.1002/eji.1830211104

Rose ME, Orlans E, Buttress N. 1974. Immunoglobulin classes in the hen's egg: their segregation in yolk and white. *Eur J Immunol* 4: 521–523. doi:10.1002/eji.1830040715

Schmidt P, Erhard MH, Schams D, Hafner A, Folger S, Lösch U. 1993. Chicken egg antibodies for immunohistochemical labeling of growth hormone and prolactin in bovine pituitary gland. *J Histochem Cytochem* 41: 1441–1446. doi:10.1177/41.9.8354884

Shih HH, Tu C, Cao W, Klein A, Ramsey R, Fennell BJ, Lambert M, Ní Shúilleabháin D, Autin B, Kouranova E, et al. 2012. An ultra-specific avian antibody to phosphorylated tau protein reveals a unique mechanism for phosphoepitope recognition. *J Biol Chem* 287: 44425–44434. doi:10.1074/jbc.M112.415935

Shin JW, Kim SI, Yoon A, Jin J, Park HB, Kim H, Chung J. 2018. The versatility of framework regions of chicken $V_H$ and $V_L$ to mutations. *Immune Netw* 18: e3. doi:10.4110/in.2018.18.e3

Spillner E, Braren I, Greunke K, Seismann H, Blank S, du Plessis D. 2012. Avian IgY antibodies and their recombinant equivalents in research, diagnostics and therapy. *Biologicals* 40: 313–322. doi:10.1016/j.biologicals.2012.05.003

Sun Y, Liu Z, Ren L, Wei Z, Wang P, Li N, Zhao Y. 2012. Immunoglobulin genes and diversity: what we have learned from domestic animals. *J Anim Sci Biotechnol* **3**: 18. doi:10.1186/2049-1891-3-18

Taylor AI, Gould HJ, Sutton BJ, Calvert RA. 2008. Avian IgY binds to a monocyte receptor with IgG-like kinetics despite an IgE-like structure. *J Biol Chem* **283**: 16384–16390. doi:10.1074/jbc.M801321200

Taylor AI, Fabiane SM, Sutton BJ, Calvert RA. 2009. The crystal structure of an avian IgY-Fc fragment reveals conservation with both mammalian IgG and IgE. *Biochemistry* **48**: 558–562. doi:10.1021/bi8019993

Thirumalai D, Visaga Ambi S, Vieira-Pires RS, Xiaoying Z, Sekaran S, Krishnan U. 2019. Chicken egg yolk antibody (IgY) as diagnostics and therapeutics in parasitic infections—a review. *Int J Biol Macromol* **136**: 755–763. doi:10.1016/j.ijbiomac.2019.06.118

Thompson CB, Neiman PE. 1987. Somatic diversification of the chicken immunoglobulin light chain gene is limited to the rearranged variable gene segment. *Cell* **48**: 369–378. doi:10.1016/0092-8674(87)90188-7

Warr GW, Magor KE, Higgins DA. 1995. IgY: clues to the origins of modern antibodies. *Immunol Today* **16**: 392–398. doi:10.1016/0167-5699(95)80008-5

Wu L, Oficjalska K, Lambert M, Fennell BJ, Darmanin-Sheehan A, Ní Shúilleabháin D, Autin B, Cummins E, Tchistiakova L, Bloom L, et al. 2012. Fundamental characteristics of the immunoglobulin $V_H$ repertoire of chickens in comparison with those of humans, mice, and camelids. *J Immunol* **188**: 322–333. doi:10.4049/jimmunol.1102466

Yang H, Chae J, Kim H, Noh J, Chung J. 2024a. Chicken immunization followed by RNA extraction and cDNA synthesis for antibody library preparation. *Cold Spring Harb Protoc* doi:10.1101/pdb.prot108568

Yang H, Chae J, Kim H, Noh J, Chung J. 2024b. Generation of a phage display chicken single-chain variable fragment library. *Cold Spring Harb Protoc* doi:10.1101/pdb.prot108213

Yang H, Chae J, Kim H, Noh J, Chung J. 2024c. Preparation of VCSM13 helper phage for display library reamplification and bio-panning. *Cold Spring Harb Protoc* doi:10.1101/pdb.prot108569

Yang H, Chae J, Kim H, Noh J, Chung J. 2024d. Selection of antigen binders from a chicken single-chain variable fragment library. *Cold Spring Harb Protoc* doi:10.1101/pdb.prot108211

Yoo DK, Lee SR, Jung Y, Han H, Lee HK, Han J, Kim S, Chae J, Ryu T, Chung J. 2020. Machine learning-guided prediction of antigen-reactive in silico clonotypes based on changes in clonal abundance through bio-panning. *Biomolecules* **10**: 421.

Yoon A, Shin JW, Kim S, Kim H, Chung J. 2016. Chicken scFvs with an artificial cysteine for site-directed conjugation. *PLoS ONE* **11**: e0146907. doi:10.1371/journal.pone.0146907

Zhang X, Chen H, Tian Z, Chen S, Schade R. 2010. Chicken monoclonal IgY antibody: a novel antibody development strategy. *Avian Biol Res* **3**: 97–106. doi:10.3184/175815510X12823014530963

Zhao Y, Rabbani H, Shimizu A, Hammarström L. 2000. Mapping of the chicken immunoglobulin heavy-chain constant region gene locus reveals an inverted α gene upstream of a condensed υ gene. *Immunology* **101**: 348–353. doi:10.1046/j.1365-2567.2000.00106.x

Protocol 1

# Chicken Immunization Followed by RNA Extraction and cDNA Synthesis for Antibody Library Preparation

Hyunji Yang,[1] Jisu Chae,[1] Hyori Kim,[2] Jinsung Noh,[3] and Junho Chung[1,4]

[1]Cancer Research Institute, Seoul National University College of Medicine, Seoul 03080, South Korea; [2]Convergence Medicine Research Center, Asan Medical Center, Seoul 05505, South Korea; [3]Bio-MAX Institute, Seoul National University, Seoul 08826, South Korea

Effective isolation of specific antibodies from immunological repertoires requires the generation of a diverse library against a specific antigen of interest, as well as efficient selection procedures, such as biopanning and phage ELISA. Key to this is the generation of a good immune response in the host, followed by preparation of high-quality RNA and cDNA from which a library can be constructed by the amplification and cloning of immunoglobulin heavy and light chain genes. The first step in the construction of such an "immune library" is a successful course of immunization. Detection of a strong serum antibody titer will theoretically then result in a pool of extracted RNA that is enriched for transcripts of genes encoding the antibody of interest. Chicken antibodies have been widely used for research and diagnostic purposes, largely because of both their cross-reactivity to epitopes shared by humans, mice, primates, and other mammals, and their simple characteristics, with chickens featuring single functional copies of $V_H/J_H$ and $V_\lambda/J_\lambda$ gene pairs. In chickens, antibodies against an antigen of interest can be detected in the serum as soon as 5–7 d after immunization. Once the antibody titer reaches an appropriate level in the serum, the spleen, bursa of Fabricius, and bone marrow are then harvested, and antibody libraries can be prepared from extracted RNA. Here, we describe a protocol for chicken immunization with an antigen of interest, followed by RNA extraction from the relevant tissues and cDNA synthesis, which users can use for antibody library construction.

## MATERIALS

It is essential that you consult the appropriate Material Safety Data Sheets and your institution's Environmental Health and Safety Office for proper handling of equipment and hazardous materials used in this protocol.

### Reagents

1-bromo-3-chloropropane (BCP; Sigma-Aldrich B9673)
   *Store protected from light at 4°C.*

Absolute ethanol
   *Store at −20°C.*

BSA (bovine serum albumin; Millipore 82-100-6), 3% (w/v) in 1× PBS
BSA or another suitable control antigen

---

[4]Correspondence: jjhchung@snu.ac.kr

© 2026 Cold Spring Harbor Laboratory Press
Cite this protocol as *Cold Spring Harb Protoc*; doi:10.1101/pdb.prot108568

Chickens (see Step 1 for requirements)
DEPC-treated water (Invitrogen AM9915G)
Ethanol (Sigma-Aldrich 1.00971), 75% (v/v) in DEPC-treated water
Freund's incomplete adjuvant (Sigma-Aldrich F5506)
Horseradish peroxidase substrate (e.g., ABTS solution; Thermo Fisher 002024) or another appropriate substrate, depending on the used conjugates
Immunizing antigen

*Use a storage buffer acceptable for use in animals, such as PBS.*

*The antigen can be a recombinant protein or native protein. For one chicken, at least 5 μg of protein is typically needed per immunization; however, the conditions can be varied depending on individual needs.*

Isopropanol (EMSURE 1.09634.1011)
LiCl precipitation solution, 7.5 M (Invitrogen AM9480)
PBS (Gibco A1286301)
PBS-Tween tablets (Calbiochem 524653-1EACN)
Rabbit anti-chicken IgG H + L (Jackson ImmunoResearch 303-036-003 or 303-056-003) or goat anti-chicken IgY H + L (Thermo Fisher A-11039)
RNAlater (Thermo Fisher AM7024)
Sodium acetate, 3 M, pH 5.2, RNase-free (Sigma-Aldrich 567422)
Sodium bicarbonate ($NaHCO_3$; Sigma-Aldrich S6014), 0.1 M in distilled water, pH 8.6
SUPERSCRIPT IV first strand system (Invitrogen 18091050)

*Store at −20°C.*

*The kit contains all components needed for cDNA synthesis.*

TRI reagent (contains phenol and guanidine thiocyanate; Thermo Fisher 15596018)

## Equipment

BD Vacutainer EDTA tubes (BD 367861)
Benchtop refrigerated centrifuge (Eppendorf EP-5810R)
Conical tubes, 15-mL and 50-mL (SPL 50015 and 50050, respectively)
ELISA plates (e.g., 96-well flat-bottom assay plates with high-binding surface, half area or standard area; Corning Costar 3690 or 3590)
GentleMACS M-tubes (Miltenyi Biotec 130-096-335)
GentleMACS tissue dissociator (Miltenyi Biotec 130-093-235)
MicroAmp PCR caps (Axygen AX.PCR-02CP-C)
MicroAmp PCR tubes (Axygen AX.PCR-0208-C)
Microcentrifuge tubes, 2-mL (Sigma-Aldrich BR780546)
Multiskan microplate photometer (Thermo Fisher 51119000)
NanoDrop microvolume UV-Vis spectrophotometer (Thermo Fisher A38189)
T100 PCR thermal cycler (Bio-Rad 1861096) or similar

## METHOD

*First, users will immunize at least three chickens and determine the titer of serum antibody at regular intervals during injections. Once the titer reaches an appropriate level, users will harvest relevant organs (bone marrow, bursa of Fabricius, and spleen), and then extract total RNA and perform first-strand cDNA synthesis using an oligo(dT) primer and reverse transcriptase following a standard procedure, with the ultimate goal of preparing an antibody library.*

*Users should get the proper ethical permissions from the relevant authorities before working with animals and immunizing and harvesting organs.*

Chapter 6

## Immunization and Titering Immune Serum by ELISA

1. Immunize at least three chickens according to the immunization schedule below:

   | Species | Breed | Age | Sex | Immunogen quantity | Injection interval | Bleeding |
   |---------|-------|-----|-----|--------------------|--------------------|----------|
   | Chicken | White leghorn | 3–6 mo | Either (but preferably female; see note) | 5 µg/immunization | 2–3 wk | 1 wk postinjection |

   *Female chickens are usually preferred because large amounts (up to 100 mg) of antibodies can be extracted from the eggs of immunized animals (Amro et al. 2018; Redwan et al. 2021).*

   *Collect preimmunization blood from the chicken. It can be used as a control to determine the serum titer for a specific antigen.*

   *Although these are standard guidelines developed for library construction, the conditions can be varied depending on individual needs.*

2. Bleed each chicken separately at 1 wk postinjection, according to facility regulations. Typically, 1–2 mL of blood is obtained from a single bleed. Collect the blood into BD Vacutainer EDTA tubes and transfer the blood to a 15-mL conical tube.

   *A regular bleeding schedule is important for following the progress of the antibody titer. Knowing whether the titer is increasing over time may help determine whether alterations in the immunization protocol are needed.*

   *The bleeding does not need to be after 1 wk precisely; however, a 1- to 2-wk interval after immunization might help to determine the titer of serum antibody (Matsuda et al. 1976; Muzyka et al. 2017).*

3. Centrifuge the sample at 3000$g$ for 15 min at 25°C (room temperature) in a benchtop centrifuge, to separate the serum.

4. Transfer the serum to a 2-mL tube and store at −80°C until use.

5. Prepare antigen-coated ELISA plates by incubating separate wells with 25 µL of the designated antigen, and the control antigen at a concentration between 0.1 and 1 µg/well in 0.1 M $NaHCO_3$, and prepare the same numbers of empty wells for the BSA (wells coated only with the blocking buffer; will be done the next day; see note for Step 6). Incubate overnight (>16 h) at 4°C.

   *Be sure to include a control antigen, different from the immunized one (but with the same tag, if applicable), to check whether the titer is specific for the immunized antigen.*

   *Consider enough wells for the immunized antigen, the control antigen, and BSA blocker (to check the background signal) for fivefold serial dilutions each of chicken serum. In addition, consider wells for serum samples acquired from preimmunization to final immunization, bleeding after the latest injection, to test for gradual changes in serum antibody titer with time. All assays should be performed at least in duplicate.*

6. Wash the coated ELISA plate (coated with the immunized antigen and control antigen) with distilled water or 0.05% (v/v) PBS-Tween solution and blot dry by firmly slapping the plate on clean paper towels. Block the ELISA plate by filling each well with 150 µL of 3% (w/v) BSA in 1× PBS. Incubate for 1 h at 37°C.

   *For checking the background signal with BSA, the same sets of uncoated ELISA plate are also blocked by applying with 150 µL of 3% (w/v) BSA in 1× PBS and incubating for 1 h at 37°C.*

7. Prepare fivefold serial dilutions of the serum samples in 3% (w/v) BSA in 1× PBS, beginning with a 1:500 dilution and ending with 1:62,500 dilution.

   *Each serum dilution is added at 25 µL per well (see Step 8). Therefore, be sure to prepare enough volume to add to all the wells of interest (including those with the immunized antigen, the control antigen, and BSA). Each serum dilution should be prepared enough to do the ELISA in duplicate form or more.*

   *All sera are serially diluted fivefold from 1:500 to 1:62,500 and tested from preimmunization to last immunization. It is usually tested in duplicate form or more.*

8. Wash the ELISA plate with water or 0.05% (v/v) PBS-Tween solution and blot dry. Apply 25 µL of each serum dilution to the three sets of designated wells (i.e., the immunized antigen, control antigen, and BSA), and incubate for 1 h at 37°C.

9. Wash the ELISA plate 10 times with water or three times with 0.05% (v/v) PBS-Tween solution and blot dry. Dilute rabbit anti-chicken IgG H + L antibody in 3% (w/v) BSA in 1× PBS at a ratio of 1:5000 and apply 25 µL to each well. Incubate for 1 h at 37°C.

10. Wash the plate 10 times with water or three times with 0.05% (v/v) PBS-Tween solution and blot dry. Apply 50 µL of substrate solution (e.g., ABTS solution) to all three sets of wells. Using a microplate reader, measure the optical density at 405 nm ($OD_{405}$) at two or three appropriate time points.

    *We do the first reading at 15 min at room temperature after adding the substrate. The second reading is usually done at 30 min at 37°C after the first reading.*

11. Plot the results of absorbance for the titration to compare the response to unrelated antigen/BSA wells and/or consecutive bleeds.

    *The "titer" refers to the highest dilution at which antigen-specific binding is detectable above background binding to a control antigen and BSA via ELISA assay. For the purpose of phage display, determining whether the titer is relatively high or low is more important than obtaining a precise titer (see Fig. 1 for an example).*

## Organ Harvesting, Followed by RNA Purification with Optional LiCl Precipitation

*Below we describe the harvesting of chicken spleen, bursa of Fabricius, and bone marrow, and the preparation of total RNA. These tissues will be enriched with B cells encoding antibodies reactive to the immunogen of interest if successful immunization was achieved, as evidenced by the ELISA assay described in the previous section. RNA samples prepared from these tissues can be used for cDNA synthesis, which can be followed by amplification of the $V_H$ and $V_\lambda$ genes through PCR with gene-specific primers to prepare an antibody library. If no or only weak amplification of these genes is obtained, a dramatic improvement can be achieved by purifying the total RNA by lithium chloride precipitation, and we also describe such steps below (see Step 22).*

*Follow standard procedures for working with RNA (wear gloves and use RNase-free materials).*

12. Harvest the spleen, bursa of Fabricius, and bone marrow from three chickens according to facility regulations. After removal, immediately transfer the tissues to separate 50-mL tubes (one per tissue type, three tubes in total), each containing 10 mL of RNAlater solution.

    *Harvesting should only be done by trained personnel.*

13. Transfer the tissues separately to a GentleMACS M-tube containing 10 mL of TRI reagent. Homogenize the samples using the GentleMACS dissociator in an "RNA program" mode.

14. Add 20 mL of TRI reagent to each sample and centrifuge at 2500g (3500 rpm in a Beckman GPR tabletop centrifuge) for 10 min at 4°C.

15. Transfer the supernatants to separate 50-mL centrifuge tubes and discard the pellets. Add 3 mL of BCP to each supernatant, vortex for 15 sec, and incubate for 15 min at room temperature.

16. Centrifuge at 17,500g (12,000 rpm in a Beckman JA-20 rotor) for 15 min at 4°C.

17. Transfer each upper, colorless aqueous phase to a separate fresh 50-mL centrifuge tube. Add 15 mL of isopropanol, vortex for 15 sec, and incubate for 10 min at room temperature.

18. Centrifuge at 17,500g for 10 min at 4°C.

19. Remove each supernatant carefully and discard. Add 30 mL of 75% (v/v) ethanol to each pellet, but do not resuspend. Centrifuge at 17,500g for 10 min at 4°C.

20. Remove each supernatant carefully and discard. Air-dry the pellets briefly at room temperature (do not dry under vacuum), dissolve each in 500 µL of DEPC-treated water, and store in a microcentrifuge tube at −80°C.

    *For storage periods exceeding a few weeks, precipitate the isolated total RNA by adding 0.1 vol of RNase-free 3 M sodium acetate (pH 5.2), and 2 vol of ethanol. Then, place the tube at −80°C.*

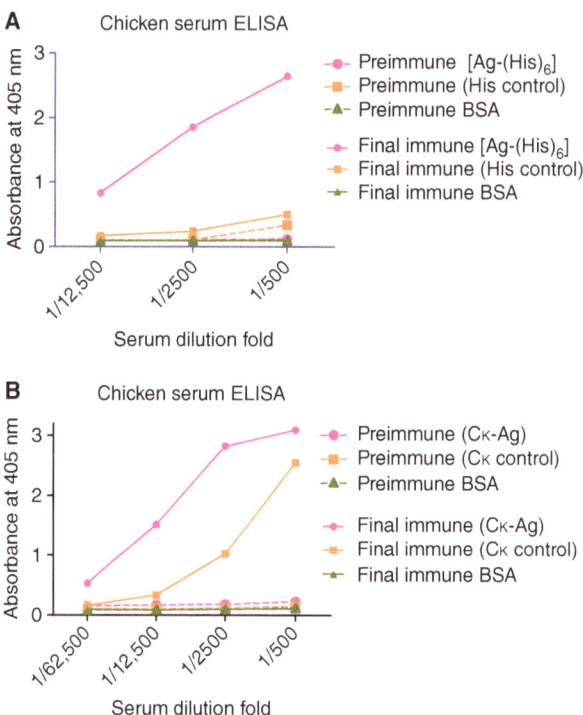

FIGURE 1. Titration of serum antibodies. (A) In this example, an adult female chicken was immunized with an antigen of interest-(His)$_6$ conjugate over a period of 2 mo, receiving injections of 5 µg for the first immunization, and 2.5 µg for the second and third (final) immunizations at 2-wk intervals. The serum extracted from preimmunization and after the final immunization were titered against the original immunizing conjugate [Ag-(His)$_6$], a His-tagged control protein (His control), and bovine serum albumin (BSA), as described in this protocol. The secondary antibody was a goat anti-chicken IgY (heavy chain + light chain) (Thermo Fisher A-11039) conjugated to horseradish peroxidase. The animal in this example has an antibody titer against the antigen of interest between 1:500 and 1:12,500, determined by comparing [Ag-(His)$_6$] binding, His-tagged control protein binding, and BSA binding. (B) In this example, an adult female chicken was immunized with an antigen of interest-(human Ig κ constant [Cκ]) conjugate over a period of 2 mo, receiving injections of 5 µg four times for immunization, with a time interval of 2 wk in between injections. The serum extracted from preimmunization and after the final immunization were titered against the original immunizing antigen with a different conjugate (human Igκ constant, Cκ-Ag), a Cκ-tagged control protein (Cκ control), and BSA, according to the protocol described here. The secondary antibody was a goat anti-chicken IgY (heavy chain + light chain) (Thermo Fisher A-11039) conjugated to horseradish peroxidase. The animal in this example has an antibody titer against the antigen of interest between 1:500 and 1:62,500, determined by comparing Cκ-Ag binding, Cκ-tagged control protein binding, and BSA binding.

21. Determine the optical density at 230, 260, and 280 nm for the RNA samples. Calculate the ratios of 260/280 and 260/230 to determine RNA purity (typically in the ranges of 1.6–1.9 and 2.0–2.2, respectively) and check the RNA concentration (40 ng/µL RNA gives an $OD_{260} = 1$).

    *Alternatively, just use a NanoDrop spectrophotometer.*

    *If additional purification of total RNA is needed, proceed to Step 22. If not, proceed to cDNA synthesis (Step 30).*

22. Transfer 300 µL of the isolated total RNA to a 1.5-mL microcentrifuge tube. Add 32 µL of RNase-free 7.5 M LiCl, vortex, and incubate for 2 h on ice.

23. Centrifuge at full speed in a microcentrifuge for 30 min at 4°C.

24. Remove the supernatant carefully and discard. Dissolve the pellet in 300 µL of DEPC-treated water, add 32 µL of RNase-free 7.5 M LiCl, vortex, and incubate for 2 h on ice.

25. Remove the supernatant carefully and discard. Dissolve the pellet in 300 µL of DEPC-treated water and add 30 µL of RNase-free 3 M sodium acetate (pH 5.2) and 600 µL of absolute ethanol. Vortex and incubate for 30 min at −20°C.

26. Centrifuge at full speed in a microcentrifuge for 30 min at 4°C. Remove the supernatant carefully and discard. Add 500 µL of 70% (v/v) ethanol in DEPC-treated water and centrifuge at full speed in a microcentrifuge for 10 min at 4°C.

27. Remove the supernatant carefully and discard. Air-dry the pellet briefly at room temperature and dissolve in 100 µL of DEPC-treated water. Store at −80°C.

    *For storage periods exceeding a few weeks, precipitate the isolated total RNA by adding 0.1 vol of RNase-free 3 M sodium acetate (pH 5.2) and 2 vol of ethanol. Then, store at −80°C*

28. Determine the concentration of the purified total RNA as described in Step 21.

29. Proceed to cDNA synthesis.

## cDNA Synthesis

30. Mix 1 µg of isolated total RNA prepared from each of the three organs (i.e., spleen, bursa of Fabricius and bone marrow) in separate 0.2-mL tubes with 1 µL of 50 µM oligo(dT)$_{20}$, 1 µL of 10 mM dNTP mix (10 mM each, included in SUPERSCRIPT IV first-strand system), and 10 µL of DEPC-treated water. Incubate each RNA mixture for 5 min at 65°C using a T100 PCR thermal cycler and then place them for at least 1 min on ice. Centrifuge briefly.

31. Per each cDNA reaction planned, prepare the following reaction mixture in a 0.2-mL tube, using the reagents included in the SUPERSCRIPT IV first strand system:

    | | |
    |---|---|
    | 5× SSIV buffer | 4 µL |
    | 100 mM DTT | 1 µL |
    | Ribonuclease inhibitor | 1 µL |
    | SUPERSCRIPT IV reverse transcriptase | 1 µL |

32. Add the above reaction mixture to each RNA mixture. Incubate for 10 min at 50°C and inactivate the reaction by incubating for 10 min at 80°C. The synthesized first-strand cDNA is now ready to be used for library construction.

    *The first-strand cDNA is stable when stored at −20°C.*

33. Proceed to downstream applications, including, for instance, generation of a phage display chicken single-chain variable fragment library, as described in Protocol 2: Generation of a Phage Display Chicken Single-Chain Variable Fragment Library (Yang et al. 2024).

## REFERENCES

Amro WA, Al-Qaisi W, Al-Razem F. 2018. Production and purification of IgY antibodies from chicken egg yolk. *J Genet Eng Biotechnol* 16: 99–103. doi:10.1016/j.jgeb.2017.10.003

Matsuda H, Baba T, Bito Y. 1976. The augmentation of antibody responses by preliminary intrabursal priming in the chicken. *Immunology* 31: 119–124.

Muzyka D, Lillehoj H, Rula O, Stegniy B. 2017. Immune response and distribution of antigen in chickens after infection LPAIV (H4N6). *Online J Public Health Inform* 9: 608274. doi:10.5210/ojphi.v9i1.7690

Redwan EM, Aljadawi AA, Uversky VN. 2021. Simple and efficient protocol for immunoglobulin Y purification from chicken egg yolk. *Poult Sci* 100: 100956. doi:10.1016/j.psj.2020.12.053

Yang H, Chae J, Kim H, Noh J, Chung J. 2024. Generation of a phage display chicken single-chain variable fragment library. *Cold Spring Harb Protoc* doi:10.1101/pdb.prot108213

# Protocol 2

# Generation of a Phage Display Chicken Single-Chain Variable Fragment Library

Hyunji Yang,[1] Jisu Chae,[1] Hyori Kim,[2] Jinsung Noh,[3] and Junho Chung[1,4]

[1]*Cancer Research Institute, Seoul National University College of Medicine, Seoul 03080, South Korea;*
[2]*Convergence Medicine Research Center, Asan Medical Center, Seoul 05505, South Korea;* [3]*Bio-MAX Institute, Seoul National University, Seoul 08826, South Korea*

Phage-displayed antibody fragment libraries can be constructed using essentially any species that is easily immunized, as long as the immunoglobulin variable region gene sequences are known. This protocol describes the procedures for the generation of a phage-displayed chicken single-chain variable fragment (scFv) library after immunization with a target antigen. Briefly, the rearranged heavy chain variable region ($V_H$) genes and the $\lambda$ light chain variable region ($V_\lambda$) genes are amplified separately and are linked through two separate PCR steps to give the final scFv genes. The genes are then cloned into pComb3XSS to generate the phage display chicken scFv library, which can then be used for test and final library ligations.

## MATERIALS

It is essential that you consult the appropriate Material Safety Data Sheets and your institution's Environmental Health and Safety Office for proper handling of equipment and hazardous materials used in this protocol.

RECIPES: Please see the end of this protocol for recipes indicated by <R>. Additional recipes can be found online at http://cshprotocols.cshlp.org/site/recipes.

### Reagents

Agarose (Thermo Fisher 15510-027)
BSA (Sigma-Aldrich A7906), 1% (w/v) in TBS, filter-sterilized
Carbenicillin (Goldbio C-103-100), 100 mg/mL in water, filter-sterilized
cDNA prepared from immunized chickens (see Protocol 1: Chicken Immunization Followed by RNA Extraction and cDNA Synthesis for Antibody Library Preparation [Yang et al. 2024a])
Chicken scFv primers, 100 μM (as listed below)
  Overlap extension primers
    CSC-F (forward): 5′-GAGGAGGAGGAGGAGGAGGTGGCCCAGGCGGCCCTGACTCAG-3′
    CSC-F2 (forward): 5′-GAGGAGGAGGAGGAGGAGGTGGCCCAGGCGGCCCAGGCAGCA-3′
    CSC-B (reverse): 5′-GAGGAGGAGGAGGAGGAGCTGGCCGGCCTGGCCACTAGTGGAGG-3′
  $V_H$ primers
    CSCVHo-FL (forward), long linker: 5′-GGTCAGTCCTCTAGATCTTCCGGCGGTGGTGGCAGCTCCGGTGGTGGCGGTTCCGCCGTGACGTTGGACGAG-3′

[4]Correspondence: jjhchung@snu.ac.kr

CSCVHo-FL2 (forward), long linker: 5′-GGTCAGTCCTCTAGATCTTCCGGCGGTGGTGGC
AGCTCCGGTGGTGGCGGTTCCGCCGTGACGTTGGATGAGTC-3′

CSCVHo-FL3 (forward), long linker: 5′-GGTCAGTCCTCTAGATCTTCCGGCGGTGGTGGC
AGCTCCGGTGGTGGCGGTTCCGCCGTGACATTGGACGAGTC-3′

CSCG-B (reverse): 5′-CTGGCCGGCCTGGCCACTAGTGGAGGAGACGATGACTTCGGTCC-3′

$V_\lambda$ primers

CSCVL (forward): 5′-GTGGCCCAGGCGGCCCTGACTCAGCCGTCCTCGGTGTC-3′

CSCVL2 (forward): 5′-GTG GCCCAGGCGGCCCTGACTCAGCCGTCCTCGG-3′

CSCVL3 (forward): 5′-GTGCCCAGGCGGCCCAGGCAGCACTGACTCAGC-3′

CLJo-B (reverse): 5′-GGAAGATCTAGAGGACTGACCTAGGACGGTCAGGGTTGTCC-3′

DNA loading buffer (6×) <R>

DNA molecular weight markers, 100-bp or 1-kb (Thermo Fisher 15628050 or 10787026)

Electrocompetent *Escherichia coli* cells, ER2738 (NEB E4104) or XL1-Blue (Stratagene 200228)

Ethidium bromide solution (20,000×, Sigma-Aldrich E1510)

Kanamycin (Goldbio K-120-25), 50 mg/mL in distilled water, filter-sterilized

KAPA HiFi HotStart PCR kit (Roche 07958897001) (250-μL × 50-μL reactions)

*All kit contents are stable at −20°C. Bring all solutions to 4°C and make sure the contents have thawed completely before use.*

LB + 100 μg/mL carbenicillin plates <R>

LB agar <R>

LB-kanamycin plates <R>

LB top agar <R>, liquefied (see Step 27)

pComb3XSS vector (Creative Biogen VPT4013)

PEG-8000 (polyethylene glycerol; Sigma-Aldrich 25322-68-3)

Plasmid DNA Miniprep Kit (QIAGEN 27106X4)

QIAquick Gel Extraction Kit (QIAGEN 28706X4)

SB medium <R>

SfiI restriction enzyme (10 units/μL; Thermo Fisher ER1821) and 10× restriction enzyme buffer G (supplied with SfiI restriction enzyme)

SOC medium <R>

Sodium acetate (Sigma-Aldrich S2889), 3 M in distilled water, pH ∼5.2–6

Sodium azide (Sigma-Aldrich 769320), 2% (w/v) in distilled water

Sodium chloride (Sigma-Aldrich S9888)

T4 DNA ligase, supplied with 5× reaction buffer (1 unit/μL; Thermo Fisher 15224-025)

TAE buffer (50×) <R>

TBS (50 mM Tris-HCl at pH 7.5, 150 mM NaCl), autoclaved (Sigma-Aldrich T6664)

Tetracycline (Sigma-Aldrich T3258), 5 mg/mL stock in ethanol

UltraPure DNase/RNase-free distilled water (Invitrogen 10977015)

VCSM13 helper phage, prepared according to Protocol 3: Preparation of VCSM13 Helper Phage for Display Library Reamplification and Bio-Panning (Yang et al. 2024b)

## Equipment

Avanti J-E centrifuge (Beckman Coulter 369003)

Benchtop orbital shaker (Thermo Fisher SHKE4000)

Conical tubes, 50-mL (SPL 50050)

Gene Pulser apparatus with pulse controller (Bio-Rad 165-2100)

Gene Pulser/MicroPulser cuvettes (Bio-Rad 1652086)

MicroAmp PCR caps (Axygen AX.PCR-02CP-C)

MicroAmp PCR tubes (Axygen AX.PCR-0208-C)

Microcentrifuge tubes, 1.5-mL (Axygen MCT-150-C)

Chapter 6

Microcentrifuge tubes, 2-mL (Sigma-Aldrich BR780546)
Mini Gel Doc for UV and blue light (Accuris Instruments E5001-SDB-E)
NanoDrop spectrophotometer (Thermo Fisher 8353-30-0010)
Polypropylene bottles, 500-mL (Beckman Coulter 355607)
Polypropylene pop-top tubes, 14-mL (SPL 40014)
Refrigerated microcentrifuge (Sigma-Aldrich 75772441)
Syringe filters, 0.2-μm (Pall 4612)
T100 PCR thermal cycler (Bio-Rad 1861096)

## METHOD

*Here, users will first amplify heavy chain variable region ($V_H$) and λ light chain variable region ($V_\lambda$) genes from cDNA generated from RNA obtained from chickens immunized with an antigen of interest (see Protocol 1: Chicken Immunization Followed by RNA Extraction and cDNA Synthesis for Antibody Library Preparation [Yang et al. 2024a]). The amplified genes are then subjected to restriction enzyme digestion, followed by cloning into the pComb3XSS vector, to generate the phage display chicken single-chain variable fragment (scFv) library (Zalles et al. 2020). In this protocol, we describe a chicken scFv library that uses a long linker (GQSSRSSGGGGSSGGGS), which tends to form scFv monomers displayed on the phage surface (Holliger et al. 1993; McGuinness et al. 1996). The primers listed in this protocol are designed to be compatible with the pComb3XSS vector (Andris-Widhopf et al. 2000; Chapter 4, Overview: The pComb3 Phagemid Family of Phage Display Vectors [Rader 2023]), and cover all functional genes and pseudogenes of chicken immunoglobulin (Maizels 2005). Figures 1–3 show a schematic overview of the steps that are used to generate the PCR products.*

### Amplification of $V_H$ and $V_\lambda$ Genes

1. To amplify the rearranged $V_H$ and $V_\lambda$ genes, first prepare the relevant forward primer mixtures for each gene. To do this, mix the 100 μM stock of each forward primer described below in equal ratio as follows:
   a. $V_H$ primer combination (forward): CSCVHo-FL, CSCVHo-FL2, and CSCVHo-FL3.
   b. $V_\lambda$ primer combination (forward): CSCVL, CSCVL2, and CSCVL3.

2. Set up two separate 100-μL PCR reactions, one for $V_H$ genes and one for $V_\lambda$ genes, by mixing the following reagents (see Figs. 1, 2), and using the primer combinations listed above:

| Quantity | Item |
| --- | --- |
| 1 μg | cDNA (see note below) |
| 0.75 μL from 100 μM stock mixture | Forward primer (5′ primer) mixture |
| 0.75 μL from 100 μM stock | Reverse primer (3′ primer, see note below) |
| 20 μL | 5× KAPA HiFi buffer |
| 3 μL | 10 mM KAPA dNTP mix |
| 2 μL | KAPA HiFi HotStart DNA polymerase (1 unit/μL) |
| Add water to a final volume of 100 μL | |

*Users can make separate libraries, each derived from each of the three source organs (the spleen, bursa of Fabricius, and bone marrow), or they can pool all the cDNA obtained from those tissues and make a single library.*

*For the $V_H$ forward primer combination, use primer CSCG-B as reverse.*

*For the $V_\lambda$ forward primer combination, use primer CLJo-B as reverse.*

3. Place the PCR tubes in the PCR machine and use the following program:
   Denaturation for 5 min at 95°C.
   Twenty-five cycles of 30 sec at 98°C, 30 sec at 65°C, and 60 sec at 72°C.
   Extension for 10 min at 72°C.

# Generation of Chicken Antibody Libraries

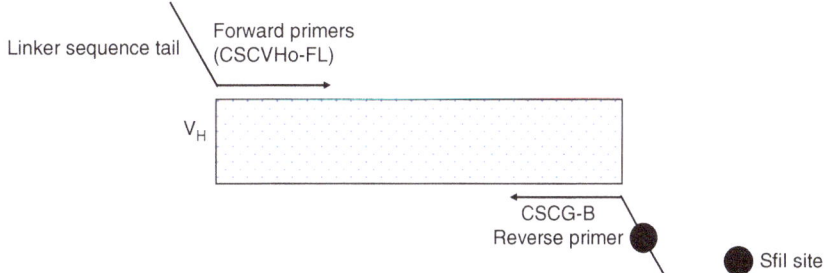

FIGURE 1. First-round PCR amplification of chicken heavy chain variable region ($V_H$) sequences for the construction of a single-chain variable fragment (scFv) library. The CSCVHo-FL primers are paired with the CSCG-B reverse primer to amplify $V_H$ segments from chicken cDNA (dotted box). The forward primer has a sequence tail that corresponds to the linker sequence that is then used in an overlap extension PCR (see Fig. 3). The reverse primer has a sequence tail containing an SfiI site; this tail is recognized by the reverse extension primer used in the second round of PCR.

4. Mix each PCR reaction with 20 µL of 6× DNA loading buffer and run the reactions on a 2% (w/v) agarose gel in 1× TAE electrophoresis running buffer (prepared by diluting the 50× stock with deionized water).

5. Cut out the DNA bands while visualizing with a Mini Gel Doc in blue-light mode. Place each DNA band in a separate tube for purification.

   *The expected size for $V_H$ is ~450 bp and that for $V_\lambda$ is 350 bp. If amplification is unsuccessful, see Troubleshooting.*

6. Purify the DNA using the QIAquick Gel Extraction Kit or similar, following the manufacturer's instructions.

7. Quantify PCR products using a NanoDrop spectrophotometer at 260 nm (1 O.D. unit = 50 µg/mL).

   *At least 2 µg of $V_H$ and $V_\lambda$ gene products are needed for the second round of PCR (overlap extension). If yield is too low, repeat the first round of PCR and combine the products.*

## Assembling the scFv ($V_\lambda$–Linker–$V_H$) Gene

*Using the purified $V_H$ and $V_\lambda$ gene fragments (Step 7), scFv genes can be generated by overlap extension PCR. The primers in the first round of PCR include sequences that serve as the overlap arm (see Fig. 3). The linker sequence between two variable heavy and light chain genes is 5'-GQSSRSSGGGGSSGGGGS-3'.*

8. To generate the overlapped $V_H$ and $V_\lambda$ genes, first prepare the relevant CSC-F and CSC-F2 primer mixture. To do this, mix an aliquot of the 100 µM stock of each CSC-F and CSC-F2 primer in a

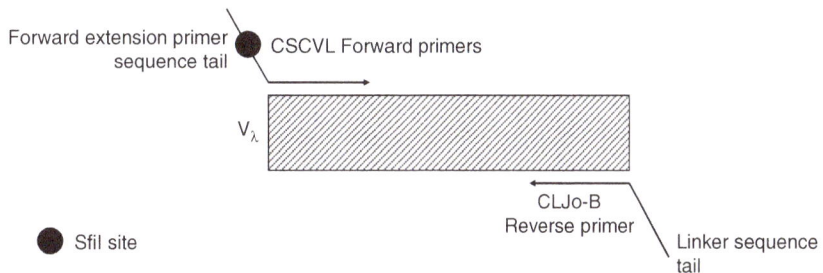

FIGURE 2. Amplification of chicken λ light chain variable region ($V_\lambda$) sequences for the construction of a single-chain variable fragment (scFv) library. The CSCVL forward primers are combined with the CLJo-B reverse primer to amplify $V_\lambda$ gene segments from chicken cDNA (slashed box). CSCVL has a 5' sequence tail that contains an SfiI site and is recognized by the forward extension primer in the second round of PCR. The reverse primer has a linker sequence tail that is used in the overlap extension reaction (see Fig. 3).

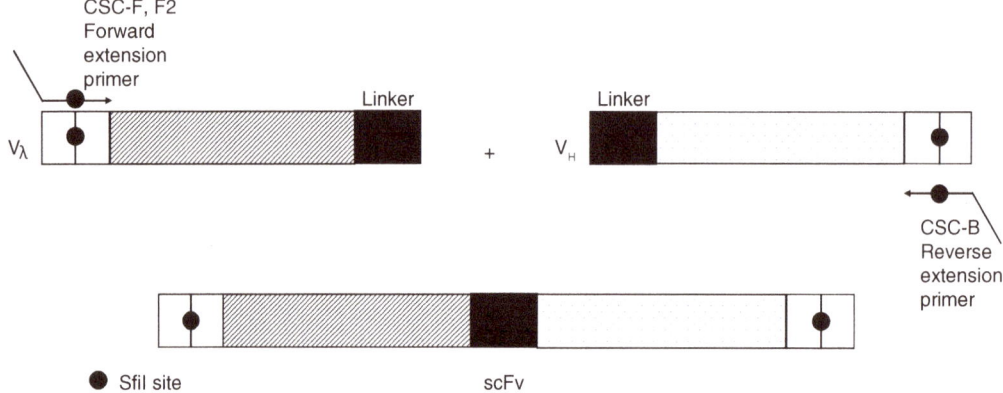

FIGURE 3. Overlap extension PCR to combine the chicken λ light chain variable region ($V_\lambda$) and heavy chain variable region ($V_H$) fragments for the construction of single-chain variable fragment (scFv) libraries. The forward and reverse extension primers used in this second round of PCR (CSC-F, F2, and CSC-B) recognize the sequence tails that were added in the first round of PCR (see Figs. 1, 2).

single tube in an equal ratio. Then, assemble 10 separate 100-µL PCR reactions, each with the following components:

| Quantity | Item |
| --- | --- |
| 100 ng | Purified $V_H$ product |
| 100 ng | Purified $V_\lambda$ product |
| 0.75 µL from 100 µM stock mixture | CSC-F, F2 primer mixture |
| 0.75 µL from 100 µM stock | CSC-B primer |
| 20 µL | 5× KAPA HiFi buffer |
| 3 µL | 10 mM KAPA dNTP mix |
| 2 µL | KAPA HiFi HotStart DNA polymerase (1 unit/µL) |
| Add water to a final volume of 100 µL | |

9. Place the PCR tubes in a thermocycler and use the following program:
   Denaturation for 5 min at 95°C.
   Twenty-five cycles of 30 sec at 98°C, 30 sec at 65°C, and 90 sec at 72°C.
   Extension for 10 min at 72°C.

10. Mix each PCR reaction with 20 µL of 6× DNA loading buffer and then run the ∼120-µL reactions on a 1.5% (w/v) agarose gel in 1× TAE electrophoresis running buffer. Cut out 750-bp bands, pool them, and purify the pooled DNA using an appropriate gel extraction kit, as in Step 6.
    *Run the entire DNA solution mixed with 6× DNA loading dye using several wells in the agarose gel.*

11. Quantitate yields by reading the O.D. at 260 nm (1 O.D. unit = 50 µg/mL) using a NanoDrop spectrophotometer.
    *At least 10 µg of scFv gene fragments is needed to proceed further. If yields are too low, repeat PCR and combine the products.*

## Restriction Digestion of the scFv Gene Fragments and the pComb3XSS Vector

12. Prepare the restriction digestion solution for the scFv gene fragments and the pComb3XSS vector individually as listed below, and incubate the reactions for 5 h at 50°C.

| | |
|---|---|
| 10 µg | Purified scFv gene (from Step 11) |
| 360 units | SfiI (36 units/µg of DNA) |
| 20 µL | 10× buffer G |
| Add water to a final volume of 200 µL | |

| | | |
|---|---|---|
| 20 µg | pComb3XSS (contains the stuffer fragment between the two SfiI cloning sites) | |
| 120 units | SfiI (6 units/µg of DNA) | |
| 20 µL | 10× buffer G | |
| Add water to a final volume of 200 µL | | |

13. Run the digested scFv gene fragments and digested vector separately on a 1.5% (w/v) gel and a 1% (w/v) gel, respectively.

    *Run the gel long enough to separate linearized vector DNA (size ~5000 bp; cut only at a single SfiI restriction site) and uncut—and possibly supercoiled—vector DNA from the desired double-cut vector DNA (size ~3400 bp). The band of the supercoiled vector DNA can be seen near 3400 bp, slightly above the desired double-cut vector DNA.*

14. Cut the bands corresponding to the digested scFv gene fragments (~750 bp), digested vector (~3400 bp), and stuffer segment (1600 bp) from the gel.

15. Quantify the purified scFv gene fragments, digested vector, and stuffer fragment by measuring the O.D. at 260 nm.

16. Store for 1 wk at 4°C or for long-term storage at −20°C before ligation. It is recommended, however, to proceed to ligation immediately after restriction digestion and purification.

## TEST LIGATION

*Perform small-scale ligations to assess the suitability of the vector and inserts for high-efficiency ligation and transformation. The ligation efficiency of the vector DNA can be tested by ligating it with the gel-purified stuffer fragment that is generated during the SfiI digestion of the vector DNA. These ligations should be done in parallel, and both should include the same amount of vector DNA.*

*In test ligations, set up three sets of ligation reactions. In the first set, the digested vector is ligated with scFv gene fragments. In the second set, the digested vector DNA is ligated with the stuffer segment. When the efficiency of this ligation reaction is >10-fold superior to that of the ligation reaction with digested PCR product (the first set), the restriction digestion of the scFv gene fragments should be repeated, as the efficiency of restriction digestion on PCR product is often found to be low. In the third set, the digested vector is ligated without any insert, which represents the degree of contamination in vector DNA sample with uncut and single-cut vector DNA.*

17. Set up three sets of reactions as described below, and incubate them overnight at 16°C.

| Ligation mixture set 1: small-scale test ligation | |
|---|---|
| 140 ng | pComb3XSS, SfiI-digested and purified |
| 70 ng | scFv gene fragment, SfiI-digested and purified |
| 4 µL | 5× T4 ligase buffer |
| 1 µL | T4 DNA ligase |
| Add water to a volume of 20 µL | |

| Ligation mixture set 2: control insert | |
|---|---|
| 140 ng | pComb3XSS, SfiI-digested and purified |
| 140 ng | Stuffer fragment, SfiI-digested and purified |
| 4 µL | 5× T4 ligase buffer |
| 1 µL | T4 DNA ligase |
| Add water to a volume of 20 µL | |

Chapter 6

| Ligation mixture set 3: test for vector self-ligation | |
|---|---|
| 140 ng | pComb3XSS, SfiI-digested and purified |
| 4 µL | 5× T4 ligase buffer |
| 1 µL | T4 DNA ligase |
| Add water to a volume of 20 µL | |

18. Warm an appropriate amount of SOC medium (at least 3 mL per sample) to room temperature. Label one 14-mL polypropylene tube for each sample and aliquot 2 mL of SOC into each tube. Chill one electroporation cuvette on ice for each sample.

19. Set the Gene Pulser II apparatus to 25 µF capacitance, 2.5 kV, and 200 Ω on the pulse controller unit.

20. Gently thaw aliquots of electrocompetent cells (50-µL aliquot per sample) on ice. Prepare the appropriate amount of cell samples to perform the test ligations, competency test (see Step 21), and "contamination test" (see Step 27).

21. After the cells are thawed, dispense 50 µL of cells and 1.0 µL of the corresponding 20-µL ligation mix (mixtures 1, 2, and 3 from Step 17). In addition, perform a "competency test": Add 10 ng of uncut pComb3XSS vector into a chilled 1.5-mL tube and dispense 50 µL of cells into it. Incubate for 1 min on ice.

    *The competency of cells used for the transformation of small-scale ligations, the so-called "competency test," is crucial and should be at least $3 \times 10^9$ colony-forming units (cfu)/µg of supercoiled control plasmid DNA (e.g., uncut pComb3XSS vector).*

22. Transfer each DNA–cell mixture to a separate prechilled electroporation cuvette and shake the mixture to the bottom.

23. Place each cuvette into the sliding cuvette chamber and apply one pulse at the settings listed above (Step 19). This should result in a pulse of 12.5 kV/cm with a time constant of 4–5 msec.

    *DNA containing too much salt will make the sample too conductive, resulting in arcs at high voltage. If arcing occurs, see Troubleshooting.*

24. Immediately add 1 mL of SOC medium to each cuvette and gently resuspend the cells.

    *A delay of just 1 min before adding the SOC can cause a threefold decrease in transformation efficiency.*

25. Transfer each mixture of cells and medium to a separate, labeled polypropylene tube containing an additional 2 mL of SOC. Incubate the cultures while shaking at 225–250 rpm for 1 h at 37°C.

26. Dilute each culture $10^4$–$10^6$ times with SOC and then plate 100 µL of each on separate LB + carbenicillin plates. Incubate the plates overnight at 37°C.

    *Add 10 µL of culture medium into 990 µL of fresh SOC medium to dilute the culture $10^2$ times and repeat it to get the culture diluted $10^4$–$10^6$ times.*

    *For accurate results, plate cells in triplicate.*

    *Do not dilute more than 1/100 at a time. For the competency test, since the corresponding dilution factor is not sufficient, dilute it to $10^7$ to accurately measure the number of colonies mentioned in Step 21.*

27. Check for potential contamination of electrocompetent *Escherichia coli* as follows, to ensure the *E. coli* stock is not contaminated with other DNA, helper phage, or lytic phage:

    i. Plate 25 µL of the commercial electrocompetent *E. coli* cells directly onto an LB agar + carbenicillin plate. Incubate overnight at 37°C.

       *No colonies should grow.*

    ii. To test contamination with helper phage, plate 25 µL of the prepared electrocompetent *E. coli* cells directly onto an LB agar + kanamycin plate. Incubate overnight at 37°C.

        *No colonies should grow.*

# Generation of Chicken Antibody Libraries

iii. To test contamination with lytic phage (and secondary test for contamination with helper phage), mix 25 μL of the electrocompetent *E. coli* cells with 3 mL of liquefied LB top agar (<50°C) and pour onto a plain LB agar plate. Incubate overnight at 37°C.

*No plaque should be formed.*

28. The next day, count the colonies on each plate (from Step 26) and calculate the total number of transformants using the formula below. Average the numbers from the three plates. The top formula is for the "competency test," and the bottom one is for the test ligations (mixtures 1, 2, and 3).

$$(\text{\# of colonies}) \times \left(\frac{\text{culture vol. (μL)}}{\text{plating vol. (μL)}}\right) \times \left(\frac{1}{\text{plasmid DNA (10 ng)}}\right) \times \left(\frac{10^3 \text{ng}}{1 \text{ μg}}\right) \times (\text{dilution fold})$$

$$= \left(\frac{\text{colony-forming unit (cfu)}}{\text{μg of plasmid}}\right)$$

$$(\text{\# of colonies}) \times \left(\frac{\text{culture vol. (μL)}}{\text{plating vol. (μL)}}\right) \times \left(\frac{\text{total ligation vol. (μL)}}{\text{ligation vol. transformed (μL)}}\right) \times (\text{dilution fold})$$

$$= (\text{total transformants})$$

29. Calculate the number of "mixture 1" transformants per microgram of vector DNA. If the number of colonies in this test ligation mixture plate does not exceed $1 \times 10^7$, do not proceed to the "real" library ligation.

30. At the same time, count the colonies obtained from ligation mixtures 2 and 3 for an indication of vector quality and ligation efficiency.

*A good vector DNA preparation should yield at least $10^8$ cfu/μg of vector DNA and should have <10% (ideally <5%) background ligation (calculated colony-forming units per microgram of vector DNA in ligation mixture 3). If the background is >10% or the ligation efficiency is low, see Troubleshooting. Similarly, if there is too much difference between mixtures 1 and 2, see Troubleshooting.*

## Library Ligation

*A major factor determining the quality of an antibody library is its complexity; i.e., the number of different antibodies in the library. The greater the complexity of the library, the more likely one is to select antibodies of the required affinity and/or specificity against the antigen of interest. There is no way to determine the absolute complexity of an antibody library, but the complexity cannot be higher than the number of independent transformants after library ligation and transformation. Thus, the number of independent transformants minus background (as estimated from the test ligations) is used to describe the complexity of an antibody library. This number is sometimes referred to as the library size. For antibody libraries derived from immune animals, reasonable library sizes are in the range of $10^7$–$10^8$ independent transformants. Depending on the outcome of the test ligations, one to 10 library ligations of the format given in the section below are necessary to achieve this number.*

31. For a single library ligation, combine the following reagents:

| | |
|---|---|
| 1.4 μg | pComb3XSS, SfiI-digested and purified (from Step 15) |
| 700 ng | scFv gene fragment, SfiI-digested and purified (from Step 15) |
| 40 μL | 5× T4 ligase buffer |
| 10 μL | T4 DNA ligase |
| Add water to a volume of 200 μL | |

*Ideally, the final library size should be around $10^8$, but it should be at least $5 \times 10^7$ total transformants. Determine the number of single library ligations needed to achieve this size.*

32. Incubate each 200-μL ligation mixture overnight at 16°C. Afterward, ethanol-precipitate the DNA using standard NaOAc–ethanol precipitation methods (see Green and Sambrook 2016).

*The ethanol-precipitated ligation reaction can be stored for months at −20°C.*

33. Dissolve each pellet in 15 µL of distilled water by pipetting and brief heating at 37°C, followed by gentle vortexing.
34. Place the ligated library samples and a corresponding number of cuvettes for 10 min on ice. At the same time, thaw 500 µL of electrocompetent *E. coli* cells on ice for a single library ligation. If more than four transformations are to be performed, thaw the electrocompetent *E. coli* cells with time intervals (i.e., in a staggered way) to avoid leaving each of them on ice for >20 min.
35. Add the electrocompetent *E. coli* cells to the ligated library sample, mix by pipetting up and down once, and transfer the mix to a cuvette. Incubate for 1 min on ice.
36. Electroporate at 2.5 kV, 25 µF, and 200 Ω. Expect τ to be ~4.0 msec. Flush the cuvette immediately with 1 mL and then twice with 2 mL of SOC medium at room temperature and combine the ~5-mL mix in a 50-mL polypropylene tube. Shake at 250 rpm for 1 h at 37°C.
37. Add 10 mL of prewarmed (37°C) SOC medium and 3 µL of 100 mg/mL carbenicillin (when XL1-Blue is used, add also 30 µL of 5 mg/mL tetracycline). Then, to titer the transformed bacteria, dilute 2 µL of the culture in 200 µL of SOC medium, and plate 100 and 10 µL of this 1:100 dilution on separate LB agar + carbenicillin plates. Incubate the plates overnight at 37°C.

    *The next day, calculate the total number of transformants by counting the number of colonies, multiplying by the culture volume, and dividing by the plating volume. Use the following formula:*

    $$(\text{\# of colonies}) \times \left(\frac{\text{culture vol. (15 mL)}}{\text{plating vol. (10 or 100 µL)}}\right) \times \left(\frac{1000 \text{ µL}}{1 \text{ mL}}\right) \times (\text{dilution fold})$$
    $$= \text{total transformants of library ligation}$$

38. Shake the remaining culture at 250 rpm for 1 h at 37°C, add 4.5 µL of 100 mg/mL carbenicillin, and shake at 250 rpm for an additional hour at 37°C.
39. Transfer the remaining culture to a 500-mL polypropylene centrifuge bottle and add 183 mL of prewarmed (37°C) SOC medium and 92.5 µL of 100 mg/mL carbenicillin (when XL1-Blue is used, add also 370 µL of 5 mg/mL tetracycline). Shake the ~200-mL culture at 250 rpm for 1.5–2 h at 37°C.
40. Add 280 µL of 50 mg/mL kanamycin and continue shaking at 250 rpm overnight at 37°C.
41. Spin at 3000*g* (e.g., 4000 rpm in a Beckman JA-10 rotor) for 15 min at 4°C. For phage precipitation, transfer the supernatant to a clean 500-mL centrifuge bottle, and add 8 g of PEG-8000 (to 4% [w/v]) and 6 g of sodium chloride (to 3% [w/v]). Dissolve the solids by shaking at 250 rpm for 10 min at 37°C. Store for 30 min on ice.
42. Save the bacterial pellet for subsequent phagemid DNA preparations, which can be done, for example, using the Plasmid DNA Miniprep Kit (QIAGEN 27106X4).
43. Spin the sample from Step 41 at 15,000*g* (e.g., 9000 rpm in a Beckman JA-10 rotor) for 15 min at 4°C. Discard the supernatant, drain the bottle by inverting on a paper towel for at least 10 min, and wipe off the remaining liquid from the upper part of the centrifuge bottle with a paper towel.
44. Resuspend the phage pellet in 2 mL of 1% (w/v) BSA in TBS by pipetting up and down along the side of the centrifuge bottle. Transfer the suspension to a 2-mL microcentrifuge tube. Resuspend further by pipetting up and down with a 1-mL pipet tip. Spin at full speed in a microcentrifuge for 5 min at 4°C, and pass the supernatant through a 0.2-µm filter into a 2-mL microcentrifuge tube.

    *This fresh phage preparation can now be used for panning. For panning, see Protocol 4: Selection of Antigen Binders from Chicken Single-Chain Variable Fragment Library (Yang et al. 2024c).*

    *Users can add sodium azide to 0.02% (w/v) for storage at 4°C. Stored phage preparations should be reamplified prior to panning. For library reamplification, proceed to "Library Reamplification."*

    *There is no way to determine the absolute complexity of an antibody library. High-throughput sequencing (HTS) analysis of the phagemid DNA prepared from the pellet, however, can be used to get an approxi-*

*mation of the diversity of the library. As a reference, for the naive human scFv panning library, sized in the range of $10^8$ to $10^9$ independent transformants, we prepared phagemid DNA (see Step 42) to analyze its diversity through HTS. In comparison to immune animals, our reference human data are sufficient to cover the library's diversity, as reasonable library sizes for antibody libraries derived from immune animals typically range from $10^7$ to $10^8$ independent transformants. We used the Illumina NovaSeq 6000 250 PE sequencer for HTS and obtained data on human immunoglobulin (Ig) heavy and light chain sequences separately from the scFv library due to limitations in the maximum sequencing length (maximum ~450 bp). On average, we obtained at least $3 \times 10^6$ reads per type of phagemid (e.g., the phagemid prepared from the real library ligations or the first round of panning procedure). After further processing the raw data, we identified around $1 \times 10^5$ unique human Ig heavy chain sequences. In other words, HTS can cover 1/100 to 1/1000 of the library diversity, providing a means to assess the quality of the chicken library or estimate its genuine diversity. However, there is no standard percentage of unique sequences among total sequences in a chicken, as it can vary based on factors such as the performance of the sequencer, the assigned throughput for each HTS preparation made by different phagemids, and the quality of raw data. Therefore, data obtained from HTS are recommended for the referenced observation of the library in terms of the sequence quality and diversity.*

## Library Reamplification

*Phage preparations should ideally be used for bio-panning only if they have been prepared on the same day, because proteases present in trace levels cleave the displayed antibodies. Furthermore, the rate of production of antibody-displaying phage is influenced by variations in antibody sequence, such that phages that display different antibodies are produced at different rates. Thus, it can be assumed that reamplification of an existing antibody library reduces its complexity, and repeated reamplifications should be avoided. However, library reamplification is sometimes necessary, so we describe the steps for this below. For reamplification, we recommend the use of only original phage preparations that were directly obtained from library ligation and transformation.*

45. Inoculate 50 mL of SB medium in a 250-mL Erlenmeyer flask with 50 µL of a competent *E. coli* preparation (when XL1-Blue is used, add 100 µL of 5 mg/mL tetracycline).

46. Shake at 250 rpm for 1.5–2.5 h at 37°C, until the optical density (OD) at 600 nm is ~0.8.
    *Be careful to avoid contaminating the culture with phage.*

47. Add 10 µL of the phage library preparation (from Step 44) to the culture and incubate for 15 min at room temperature.

48. Add 10 µL of 100 mg/mL carbenicillin. To titer the phage-infected bacteria (the resulting number should be well above the library size), dilute the culture medium 1000-fold and plate 100 µL of the dilution on LB + carbenicillin plates. Test potential contamination with phagemids, helper phage, and lytic phage using electrocompetent cells, as described in Step 27. For accurate results, plate cells in triplicate. Incubate the plates overnight at 37°C. Calculate the number of transformants using the equation below:

$$(\# \text{ of colonies}) \times \left( \frac{\text{culture vol. } (50 \text{ mL})}{\text{plating vol. } (100 \text{ µL})} \right) \times \left( \frac{1000 \text{ µL}}{1 \text{ mL}} \right) \times \left( \frac{\text{total vol. of phage } (x \text{ µL})}{\text{infected vol. of phage } (10 \text{ µL})} \right)$$
$$\times (\text{dilution fold}) = \text{total transformants}$$

49. Transfer the infected 50-mL culture to a 500-mL polypropylene centrifuge bottle and shake at 300 rpm for 1 h at 37°C. Add 15 µL of 100 mg/mL carbenicillin and shake at 300 rpm for an additional hour at 37°C.

50. Add 2 mL of VCSM13 helper phage ($10^{12}$–$10^{13}$ pfu/mL) (see Protocol 3: Preparation of VCSM13 Helper Phage for Display Library Reamplification and Bio-Panning [Yang et al. 2024b]). Add 148 mL of prewarmed (37°C) SB medium and 75 µL of 100 mg/mL carbenicillin (when XL1-Blue is used, add 300 µL of 5 mg/mL tetracycline instead of the carbenicillin). Shake the ~200-mL culture at 300 rpm for 1.5–2 h at 37°C.

51. Add 280 µL of 50 mg/mL kanamycin and continue shaking at 300 rpm overnight at 37°C.

52. Spin at 3000g (e.g., 4000 rpm in a Beckman JA-10 rotor) for 15 min at 4°C. Save the bacterial pellet for phagemid DNA preparations, which can be done using, for example, the Plasmid DNA Miniprep Kit (QIAGEN 27106X4). Store the phagemid DNA at −20°C. For phage precipitation, transfer the supernatant to a clean 500-mL centrifuge bottle, and add 8 g of PEG-8000 (to 4% [w/v]) and 6 g of sodium chloride (to 3% [w/v]). Dissolve the solids by shaking at 300 rpm for 5 min at 37°C. Store for 30 min on ice.

53. Spin at 15,000g (e.g., 9000 rpm in a Beckman JA-10 rotor) for 15 min at 4°C. Discard the supernatant, drain the bottle by inverting on a paper towel for at least 10 min, and wipe off the remaining liquid from the upper part of the centrifuge bottle with a paper towel.

54. Resuspend the phage pellet in 2 mL of 1% (w/v) BSA in TBS by pipetting up and down along the side of the centrifuge bottle. Transfer the suspension to a 2-mL microcentrifuge tube. Resuspend further by pipetting up and down using a 1-mL pipet tip, spin at full speed in a microcentrifuge for 5 min at 4°C, and pass the supernatant through a 0.2-μm filter into a sterile 2-mL microcentrifuge tube.

*This fresh phage preparation can now be used for panning. For panning, see Protocol 4: Selection of Antigen Binders from a Chicken Single-Chain Variable Fragment Library (Yang et al. 2024c).*

*Sodium azide can be added to 0.02% (w/v) for storage at 4°C. Only freshly prepared phage (prepared the same day) should be used for panning.*

## TROUBLESHOOTING

*Problem (Step 5):* Unsuccessful amplification of $V_H$ and $V_\lambda$ genes.
*Solution:* The sizes of $V_H$ and $V_\lambda$ gene fragments are ∼400 and 350 bp, respectively. Increase the volume of cDNA up to 5 μg. Alternatively, increase the purity of the RNA sample with lithium chloride precipitation (see Protocol 1: Chicken Immunization Followed by RNA Extraction and cDNA Synthesis for Antibody Library Preparation [Yang et al. 2024a]).

*Problem (Step 23):* Arcing in electroporation.
*Solution:* Decrease the amount of DNA, and/or increase the volume of cells to 60–70 μL.

*Problem (Step 30):* Too much difference in the number of colony-forming units between ligation mixtures 1 and 2 (>10-fold).
*Solution:* Perform the restriction digestion of scFv gene fragments again.

*Problem (Step 30):* Background in transformation is >10% or the ligation efficiency is low.
*Solution:* Digest new vector and repeat the ligations.

## RECIPES

### DNA Loading Buffer (6×)

30% (v/v) glycerol
0.25% (w/v) bromophenol blue
0.25% (w/v) xylene cyanol FF

Store at 4°C.

### LB + 100 µg/mL Carbenicillin Plates

1. In a 1-L bottle or flask, combine 32 g of LB agar (Thermo Fisher 22700041) with 1 L of water. Stir and autoclave for 15 min at 121°C at 15 psi.
2. When cooled to 42°C–45°C, add carbenicillin to a final concentration of 100 µg/mL. Pour into Petri dishes and allow to solidify.

Store for up to 1 mo at 4°C.

### LB Agar

Agar (20 g/L)
NaCl (10 g/L; Sigma-Aldrich S9625)
Tryptone (10 g/L; BD 211705)
Yeast extract (5 g/L; BD 212750)

Add $H_2O$ to a final volume of 1 L. Adjust the pH to 7.0 with 5 N NaOH. Autoclave. Pour into Petri dishes (~25 mL per 100-mm plate).

### LB-Kanamycin Agar Plates

Kanamycin, filter-sterilized (50 mg/mL stock)
LB agar <R>
Autoclave 1 L of LB agar. Cool to 55°C. Add 1 mL of the kanamycin stock (i.e., 50 mg). Pour into Petri dishes (~25 mL per 100-mm plate).

### LB Top Agar

1. Add 0.35 g of Bacto agar (Sigma-Aldrich A5306) and 1.25 g of LB medium (Sigma-Aldrich L3397) to 50 mL of distilled water in a glass bottle.
2. Autoclave for 20 min at 120°C. Then, aliquot 10 mL per 50-mL tube and store at 4°C.
3. Melt in a microwave and place in a water bath to keep it at 50°C until use.

### SB Medium

30 g tryptone
20 g yeast extract
10 g MOPS (3-[N-morpholino]-propanesulfonic acid)

Bring to 1 L total volume with $H_2O$, stir until dissolved, and titrate to pH 7. Sterilize by autoclaving at 15 psi on liquid cycle for 20 min at 121°C.

## SOC Medium

Per liter: To 950 mL of deionized H$_2$O, add:

| | |
|---|---|
| Tryptone | 20 g |
| Yeast extract | 5 g |
| NaCl | 0.5 g |

SOC medium is identical to SOB medium, except that it contains 20 mM glucose. To prepare SOB medium, combine the above ingredients and shake until the solutes have dissolved. Add 10 mL of a 250 mM solution of KCl. (This solution is made by dissolving 1.86 g of KCl in 100 mL of deionized H$_2$O.) Adjust the pH of the medium to 7.0 with 5 N NaOH (~0.2 mL). Adjust the volume of the solution to 1 L with deionized H$_2$O. Sterilize by autoclaving for 20 min at 15 psi (1.05 kg/cm$^2$) on liquid cycle. Just before use, add 5 mL of a sterile solution of 2 M MgCl$_2$. (This solution is made by dissolving 19 g of MgCl$_2$ in 90 mL of deionized H$_2$O. Adjust the volume of the solution to 100 mL with deionized H$_2$O and sterilize by autoclaving for 20 min at 15 psi [1.05 kg/cm$^2$] on liquid cycle.)

After the SOB medium has been autoclaved, allow it to cool to 60°C or less. Add 20 mL of a sterile 1 M solution of glucose. (This solution is made by dissolving 18 g of glucose in 90 mL of deionized H$_2$O. After the sugar has dissolved, adjust the volume of the solution to 100 mL with deionized H$_2$O and sterilize by passing it through a 0.22-μm filter.)

## TAE Buffer (50×)

| Reagent | Quantity (for 1 L) |
|---|---|
| Tris (hydroxymethyl) aminomethane | 242 g |
| Acetic acid | 57.1 mL |
| Na$_2$EDTA | 7.43 g |

Dissolve in distilled H$_2$O and make volume up to 1 L.

## REFERENCES

Andris-Widhopf J, Rader C, Steinberger P, Fuller R, Barbas CF III. 2000. Methods for the generation of chicken monoclonal antibody fragments by phage display. *J Immunol Methods* **242**: 159–181. doi:10.1016/S0022-1759(00)00221-0

Green MR, Sambrook J. 2016. Precipitation of DNA with ethanol. *Cold Spring Harb Protoc* doi: 10.1101/pdb.prot093377

Holliger P, Prospero T, Winter G. 1993. "diabodies": small bivalent and bispecific antibody fragments. *Proc Natl Acad Sci* **90**: 6444–6448. doi:10.1073/pnas.90.14.6444

Maizels N. 2005. Immunoglobulin gene diversification. *Annu Rev Genet* **39**: 23–46. doi:10.1146/annurev.genet.39.073003.110544

McGuinness BT, Walter G, FitzGerald K, Schuler P, Mahoney W, Duncan AR, Hoogenboom HR. 1996. Phage diabody repertoires for selection of large numbers of bispecific antibody fragments. *Nat Biotechnol* **14**: 1149–1154. doi:10.1038/nbt0996-1149

Rader C. 2023. The pComb3 phagemid family of phage display vectors. *Cold Spring Harb Protoc* doi: 10.1101/pdb.over107756

Yang H, Chae J, Kim H, Noh J, Chung J. 2024a. Chicken immunization followed by RNA extraction and cDNA synthesis for antibody library preparation. *Cold Spring Harb Protoc* doi: 10.1101/pdb.prot108568

Yang H, Chae J, Kim H, Noh J, Chung J. 2024b. Preparation of VCSM13 helper phage for display library reamplification and bio-panning. *Cold Spring Harb Protoc* doi: 10.1101/pdb.prot108569

Yang H, Chae J, Kim H, Noh J, Chung J. 2024c. Selection of antigen binders from a chicken single-chain variable fragment library. *Cold Spring Harb Protoc* doi: 10.1101/pdb.prot108211

Zalles M, Smith N, Saunders D, Saran T, Thomas L, Gulej R, Lerner M, Fung K-M, Chung J, Hwang K, et al. 2020. Assessment of an scFv antibody fragment against ELTD1 in a G55 glioblastoma xenograft model. *Transl Oncol* **13**: 100737. doi:10.1016/j.tranon.2019.12.009

Protocol 3

# Preparation of VCSM13 Helper Phage for Display Library Reamplification and Bio-Panning

Hyunji Yang,[1] Jisu Chae,[1] Hyori Kim,[2] Jinsung Noh,[3] and Junho Chung[1,4]

[1]*Cancer Research Institute, Seoul National University College of Medicine, Seoul 03080, South Korea;* [2]*Convergence Medicine Research Center, Asan Medical Center, Seoul 05505, South Korea;* [3]*Bio-MAX Institute, Seoul National University, Seoul 08826, South Korea*

Phage-displayed antibody libraries can be constructed using any species that is easily immunized. The pComb3XSS phagemid vector is commonly used for library cloning and phage display. This phagemid encodes the origin of replication of the filamentous bacteriophage f1 but lacks all the genes required for replication and assembly of phage particles. The replication and the assembly of phage from these phagemids thus requires a "helper" phage that provides the genes essential for those steps during library production and bio-panning. One of those helper phages is VCSM13. In this protocol, we describe the preparation of VCSM13 helper phage. Users should prepare VCSM13 helper phage for library reamplification and for bio-panning.

## MATERIALS

It is essential that you consult the appropriate Material Safety Data Sheets and your institution's Environmental Health and Safety Office for proper handling of equipment and hazardous materials used in this protocol.

RECIPES: Please see the end of this protocol for recipes indicated by <R>. Additional recipes can be found online at http://cshprotocols.cshlp.org/site/recipes.

### Reagents

*Escherichia coli*, strain ER2738 (NEB E4104)
Kanamycin, 50 mg/mL in water, filter-sterilized
LB agar <R>, prewarmed to 37°C
LB top agar <R>, liquefied
SB medium <R>
VCSM13 helper phage (Stratagene 200251)

### Equipment

Benchtop orbital shaker (Thermo Fisher SHKE4000)
Benchtop refrigerated centrifuge (Eppendorf EP-5810R)
Conical tubes, 50-mL (SPL 50050)

---

[4]Correspondence: jjhchung@snu.ac.kr

© 2026 Cold Spring Harbor Laboratory Press
Cite this protocol as *Cold Spring Harb Protoc*; doi:10.1101/pdb.prot108569

Erlenmeyer flasks, 2-L (Corning CL431255)
Spectrophotometer
Water bath (Sigma-Aldrich Z690619)

## METHOD

*A mutation in the VCSM13 origin of replication makes production of the helper phage less efficient than that of pComb3X phage. Thus, addition of helper phage VCSM13 to cells that have been transformed with pComb3X phagemid will ultimately produce a mixed phage population that predominantly contains pComb3X phage.*

*VCSM13 contains a gene conferring kanamycin resistance, and kanamycin can thus be used to select for E. coli infected with VCSM13.*

*It takes 3 d to get the VCSM13 phage; therefore, users should plan accordingly to prepare it prior to library reamplification (see Protocol 2: Generation of a Phage Display Chicken Single-Chain Variable Fragment Library [Yang et al. 2024a]) or bio-panning (see Protocol 4: Selection of Antigen Binders from a Chicken Single-Chain Variable Fragment Library [Yang et al. 2024b]).*

1. Inoculate 2 mL of SB medium with 2 µL of recently thawed *E. coli* ER2738 and shake at 250 rpm for 1 h at 37°C, until it reaches an $OD_{600}$ of ~0.8.

2. Prepare 1 mL of $10^{-6}$, $10^{-7}$, and $10^{-8}$ dilutions of the commercially obtained VCSM13 preparation (usually in the range of $1 \times 10^{11}$ plaque-forming units [pfu]/mL) in SB medium.

    *Add 10 µL of culture medium into 990 µL of freshly aliquoted SB medium to dilute the culture $10^2$ times and repeat it to get the culture diluted $10^6$–$10^8$ times. Do not dilute more than 1/100 at a time.*

3. Add 1 µL of each of the dilutions from Step 2 to three separate 50-µL aliquots of the ER2738 culture and incubate for 15 min at room temperature.

4. Mix each of the dilutions with 3 mL of liquefied LB top agar (<50°C) and pour onto separate plain LB agar plates that have been prewarmed to 37°C. Incubate overnight at 37°C.

    *The aim is to obtain a plate from which single VCSM13 plaques (i.e., E. coli colonies that are growing slower because of infection with VCSM13) can be easily picked.*

5. The next day, repeat Step 1 and then inoculate 10 mL of prewarmed (37°C) SB medium with 10 µL of *E. coli* ER2738 culture in a 50-mL conical tube. Allow growth at 250 rpm for 1 h at 37°C.

6. Use a pipet tip to transfer a single VCSM13 plaque from one of the plates from Step 4 to the culture from Step 5 and incubate at 250 rpm for 2 h at 37°C.

7. Transfer the infected 10-mL culture to a 2-L Erlenmeyer flask containing 500 mL of prewarmed (37°C) SB. Add 700 µL of 50 mg/mL kanamycin to a final concentration of 70 µg/mL, and continue shaking at 250 rpm overnight at 37°C.

8. The next day, transfer the culture to 10 separate 50-mL conical tubes and spin at 2500*g* (e.g., 3500 rpm in a Benchtop refrigerated centrifuge) for 15 min.

9. Transfer each of the supernatants to fresh separate 50-mL conical tubes and incubate in a water bath for 20 min at 70°C.

10. Spin all tubes at 2500*g* for 15 min.

11. Transfer the supernatants to separate fresh 50-mL conical tubes and store at 4°C.

    *The supernatants can be stored for months at 4°C.*

12. Determine the titer of the VCSM13 preparation as follows:

    i. Inoculate 2 mL of SB medium with 2 µL of freshly thawed ER2738, and allow growth at 250 rpm for 1 h at 37°C.

    ii. Prepare 1 mL of $10^{-7}$, $10^{-8}$, and $10^{-9}$ dilutions of the VCSM13 preparation in SB medium. Use 1 µL of each of these dilutions to infect three separate 50-µL aliquots of the ER2738 culture and incubate for 15 min at room temperature.

iii. Add 3 mL of liquefied LB top agar (<50°C) to each mix and pour onto separate plain LB agar plates. Incubate overnight at 37°C.

iv. Determine the titer of the VCSM13 preparation from the number of plaques. For example, 50 plaques on the plate derived from the $10^{-8}$ dilution corresponds to a titer of $5 \times 10^{12}$ pfu/mL, per the following equation:

$$\text{Phage titer} \left(\frac{\text{plaque forming unit, pfu}}{\text{mL}}\right) = \frac{\text{No. of plaque}}{\text{innoculation vol. } (1\ \mu L)} \times \frac{10^3\ \mu L}{1\ \text{mL}} \times \text{dilution fold } (10^n).$$

*Expect the titer to be in the range of $10^{12}$–$10^{13}$ pfu/mL. Although this titer will decrease over time, VCSM13 preparations are stable for months at 4°C.*

13. Use the VCSM13 preparation for your desired application; for instance, for library reamplification (see Protocol 2: Generation of a Phage Display Chicken Single-Chain Variable Fragment Library [Yang et al. 2024a] or bio-panning (see Protocol 4: Selection of Antigen Binders from a Chicken Single-Chain Variable Fragment Library [Yang et al. 2024b]).

# RECIPES

## LB Agar

Agar (20 g/L)
NaCl (10 g/L; Sigma-Aldrich S9625)
Tryptone (10 g/L; BD 211705)
Yeast extract (5 g/L; BD 212750)

Add $H_2O$ to a final volume of 1 L. Adjust the pH to 7.0 with 5 N NaOH. Autoclave. Pour into Petri dishes (∼25 mL per 100-mm plate).

## LB Top Agar

1. Add 0.35 g of Bacto agar (Sigma-Aldrich A5306) and 1.25 g of LB medium (Sigma-Aldrich L3397) to 50 mL of distilled water in a glass bottle.
2. Autoclave for 20 min at 120°C. Then, aliquot 10 mL per 50-mL tube and store at 4°C.
3. Melt in a microwave and place in a water bath to keep it at 50°C until use.

## SB Medium

30 g tryptone
20 g yeast extract
10 g MOPS (3-[N-morpholino]-propanesulfonic acid)
Bring to 1 L total volume with $H_2O$, stir until dissolved, and titrate to pH 7. Sterilize by autoclaving at 15 psi on liquid cycle for 20 min at 121°C.

# REFERENCES

Yang H, Chae J, Kim H, Noh J, Chung J. 2024a. Generation of a phage display chicken single-chain variable fragment library. *Cold Spring Harb Protoc* doi:10.1101/pdb.prot108213

Yang H, Chae J, Kim H, Noh J, Chung J. 2024b. Selection of antigen binders from chicken scFv library. *Cold Spring Harb Protoc* doi:10.1101/pdb.prot108211

## Protocol 4

# Selection of Antigen Binders from a Chicken Single-Chain Variable Fragment Library

Hyunji Yang,[1] Jisu Chae,[1] Hyori Kim,[2] Jinsung Noh,[3] and Junho Chung[1,4]

[1]*Cancer Research Institute, Seoul National University College of Medicine, Seoul 03080, South Korea;*
[2]*Convergence Medicine Research Center, Asan Medical Center, Seoul 05505, South Korea;* [3]*Bio-MAX Institute, Seoul National University, Seoul 08826, South Korea*

Antibody production against an antigen of interest is highly efficient in chickens, and the use of chicken antibody libraries in phage display can result in high-affinity single-chain variable fragments (scFvs) for multiple applications. After library preparation from an animal immunized with the antigen of interest, the next step involves the identification of antigen binders. Here, we describe a process for the screening of a phage display chicken library using a technique called bio-panning. It consists of several rounds of binding scFv-displaying phage to antigens, followed by washing, elution, and reamplification. We also describe the steps for assessing clone pools obtained after bio-panning via an ELISA-based procedure known as "phage ELISA" to identify single clones. Last, we provide the steps for using high-throughput sequencing to analyze the pool of selected clones.

## MATERIALS

It is essential that you consult the appropriate Material Safety Data Sheets and your institution's Environmental Health and Safety Office for proper handling of equipment and hazardous materials used in this protocol.

RECIPES: Please see the end of this protocol for recipes indicated by <R>. Additional recipes can be found online at http://cshprotocols.cshlp.org/site/recipes.

### Reagents

Acid elution buffer (0.1 M citrate in water, pH adjusted to 3.1 using 1 N HCl)
Agarose (Thermo Fisher 15510-027)
AMPure XP (Beckman Coulter A63881)
Ammonium sulfate (Sigma-Aldrich A4418), 3 M in water
Antigen of interest

*It could be recombinant protein or native protein. Binding capacity of Dynabeads M-270 Epoxy varies depending on the ligand used (e.g., antibodies, peptides, or proteins), but it is typically 5–10 µg of IgG per milligram of beads. Use >3 µg of pure ligand per $10^7$ Dynabeads M-270 Epoxy for coupling. Avoid buffers with amino or sulfogroups (e.g., Tris). Sugars or stabilizers can inhibit binding and should be removed from the ligand columns prior to coupling.*

BSA (Sigma-Aldrich A7906), 3% (w/v) in PBS, filter-sterilized

[4]Correspondence: jjhchung@snu.ac.kr

© 2026 Cold Spring Harbor Laboratory Press
Cite this protocol as *Cold Spring Harb Protoc*; doi:10.1101/pdb.prot108211

TABLE 1. Chicken scFv HTS primers

| Type | Name | Sequence |
|---|---|---|
| $V_H$ primer | CSCVHo-FL_NGS (forward), long linker | 5′-ACACTCTTTCCCTACACGACGCTCTTCCGATCTCTCCGGTGGTGGCGGTTC-3′ |
| $V_H$ primer | CSCG-B_NGS (reverse) | 5′-TGACTGGAGTTCAGACGTGTGCTCTTCCGATCTCCATGGTGATGGTGATGGTGC-3′ |
| $V_\lambda$ primer | CSCVL_NGS (forward) | 5′-ACA CTC TTT CCC TAC ACG ACG CTC TTC CGA TCT CAC TGG CTG GTT TCG CTA CC-3′ |
| $V_\lambda$ primer | CLJo-B_NGS (reverse), long linker | 5′-TGACTGGAGTTCAGACGTGTGCTCTTCCGATCTGCCGGAAGATCTAGAGGACTGACC-3′ |

(scFv) Single-chain variable fragment; (HTS) high-throughput sequencing; ($V_H$) heavy chain variable region; ($V_\lambda$) λ light chain variable region.

BSA or another suitable antigen for control (Sigma-Aldrich A7906)
Carbenicillin (Goldbio C-103-100), 100 mg/mL in water, filter-sterilized
DEPC-treated water (Invitrogen AM9915G)
DNA loading buffer (6×) <R>
DNA molecular weight markers, 100-bp or 1-kb (Thermo Fisher 15628050 or 10787026)
Dynabeads M-270 Epoxy (Invitrogen 14301)
*Escherichia coli* cells, ER2738 (NEB E4104) or XL1-Blue (Stratagene 200228)
Ethanol (EMSURE, 1.00983.1011), 70% (v/v) in water
Ethidium bromide solution (Sigma-Aldrich E1510)
HRP substrate (e.g., ABTS solution, Thermo Fisher 002024) or another appropriate substrate, depending on the conjugate used
HRP-conjugated anti-HA, clone 3F10 from rat, high-affinity (Roche Molecular Biochemicals 12013819001)
  *Use at 1:2000, diluted in 3% (w/v) BSA in PBS.*
HRP-conjugated anti-M13 monoclonal antibody from mouse (Sino Biological 11973-MM05T-H)
  *Use at 1:4000 diluted in 3% (w/v) BSA in PBS.*
Kanamycin (Goldbio K-120-25), 50 mg/mL in water, filter-sterilized
KAPA HiFi HotStart PCR kit (Roche 07958897001) (250-μL × 50-μL reactions)
LB + 100 μg/mL carbenicillin plates <R>
LB top agar <R>
NaHCO$_3$, 0.1 M in water, pH 8.6
Neutralization solution (2 M Tris base [Sigma-Aldrich 648310-M] in water, pH adjusted to 9.1 using 1 N NaOH)
Primers
  Chicken scFv HTS primers (Table 1)
  Indexing primers (Table 2)
  Sanger sequencing primers
    Ompseq, $V_\lambda$ (5′-AAGACAGCTATCGCGATTGCAG-3′)
    HRML-F, $V_H$ (5′-GGTCAGTCCTCTAGATCTTCC-3′)
Output titering plates (see Step 15)
PBS (phosphate-buffered saline) (Sigma-Aldrich P5493)
PEG-8000 (polyethylene glycol 8000; Sigma-Aldrich 25322-68-3)
Phage library preparation, fresh, as described in Protocol 2: Generation of a Phage Display Chicken Single-Chain Variable Fragment Library (Yang et al. 2024a)
Plasmid DNA Miniprep Kit (QIAGEN 27106X4)
SB medium <R>
Skim milk (BD 232100), 5% (w/v) in PBS
Sodium azide (Sigma-Aldrich 769320), 2% (w/v) in water
Sodium chloride (Sigma-Aldrich S9888)
Sodium phosphate (Sigma-Aldrich 255793 and 229903), 0.1 M in water, pH 7.4
TAE buffer (50×) <R>, for making 1× running buffer with deionized water.

Chapter 6

TABLE 2. Forward and reverse index list

| Name | Forward index | Sequence |
|---|---|---|
| F1 | TATAGCCT | 5′-AATGATACGGCGACCACCGAGATCTACACTATAGCCTACACTCTTTCCCTACACGACGCTCTTCCGATCT-3′ |
| F2 | ATAGAGGC | 5′-AATGATACGGCGACCACCGAGATCTACACATAGAGGCACACTCTTTCCCTACACGACGCTCTTCCGATCT-3′ |
| F3 | CCTATCCT | 5′-AATGATACGGCGACCACCGAGATCTACACCCTATCCTACACTCTTTCCCTACACGACGCTCTTCCGATCT-3′ |
| F4 | GGCTCTGA | 5′-AATGATACGGCGACCACCGAGATCTACACGGCTCTGAACACTCTTTCCCTACACGACGCTCTTCCGATCT-3′ |
| F5 | AGGCGAAG | 5′-AATGATACGGCGACCACCGAGATCTACACAGGCGAAGACACTCTTTCCCTACACGACGCTCTTCCGATCT-3′ |
| F6 | TAATCTTA | 5′-AATGATACGGCGACCACCGAGATCTACACTAATCTTAACACTCTTTCCCTACACGACGCTCTTCCGATCT-3′ |
| F7 | CAGGACGT | 5′-AATGATACGGCGACCACCGAGATCTACACCAGGACGTACACTCTTTCCCTACACGACGCTCTTCCGATCT-3′ |
| F8 | GTACTGAC | 5′-AATGATACGGCGACCACCGAGATCTACACGTACTGACACACTCTTTCCCTACACGACGCTCTTCCGATCT-3′ |
| F9 | CGAGGTCT | 5′-AATGATACGGCGACCACCGAGATCTACACCGAGGTCTACACTCTTTCCCTACACGACGCTCTTCCGATCT-3′ |
| F10 | TTGTTCAA | 5′-AATGATACGGCGACCACCGAGATCTACACTTGTTCAAACACTCTTTCCCTACACGACGCTCTTCCGATCT-3′ |
| F11 | AACTTGCA | 5′-AATGATACGGCGACCACCGAGATCTACACAACTTGCAACACTCTTTCCCTACACGACGCTCTTCCGATCT-3′ |
| F12 | CTAAGATC | 5′-AATGATACGGCGACCACCGAGATCTACACCTAAGATCACACTCTTTCCCTACACGACGCTCTTCCGATCT-3′ |
| Name | Reverse index | Sequence |
| R1 | ATTACTCG | 5′-CAAGCAGAAGACGGCATACGAGATCGAGTAATGTGACTGGAGTTCAGACGTGTGCTCTTCCGATCT-3′ |
| R2 | TCCGGAGA | 5′-CAAGCAGAAGACGGCATACGAGATTCTCCGGAGTGACTGGAGTTCAGACGTGTGCTCTTCCGATCT-3′ |
| R3 | CGCTCATT | 5′-CAAGCAGAAGACGGCATACGAGATAATGAGCGGTGACTGGAGTTCAGACGTGTGCTCTTCCGATCT-3′ |
| R4 | GAGATTCC | 5′-CAAGCAGAAGACGGCATACGAGATGGAATCTCGTGACTGGAGTTCAGACGTGTGCTCTTCCGATCT-3′ |
| R5 | ATTCAGAA | 5′-CAAGCAGAAGACGGCATACGAGATTTCTGAATGTGACTGGAGTTCAGACGTGTGCTCTTCCGATCT-3′ |
| R6 | GAATTCGT | 5′-CAAGCAGAAGACGGCATACGAGATACGAATTCGTGACTGGAGTTCAGACGTGTGCTCTTCCGATCT-3′ |
| R7 | CTGAAGCT | 5′-CAAGCAGAAGACGGCATACGAGATAGCTTCAGGTGACTGGAGTTCAGACGTGTGCTCTTCCGATCT-3′ |
| R8 | TAATGCGC | 5′-CAAGCAGAAGACGGCATACGAGATGCGCATTAGTGACTGGAGTTCAGACGTGTGCTCTTCCGATCT-3′ |

This list contains a sequence of index adapters arranged in the Illumina flow cell to enforce the recommended pairing strategy.

Tetracycline (Sigma-Aldrich T3258), 5 mg/mL stock in ethanol

Tween 20 (polyoxyethylene sorbitan monolaurate) (Sigma-Aldrich P1379), 0.05% (v/v) in PBS, filter-sterilized

VCSM13 helper phage (see Protocol 3: Preparation of VCSM13 Helper Phage for Display Library Reamplification and Bio-Panning [Yang et al. 2024b])

## Equipment

Avanti J-E centrifuge (Beckman Coulter 369003)
Benchtop orbital shaker (Thermo Fisher SHKE4000)
Conical tubes, 50-mL (SPL 50050)
Deep-well plate, 2-mL (Corning P-2ML-SQ-C)
ELISA plates (e.g., 96-well flat-bottom assay plates with high-binding surface, half area or standard area, Corning Costar 3690 or 3590)
ELISA reader for 96-well plates (Thermo Fisher 51119000)
Illumina NovaSeq 6000 System (Illumina)
Illumina NovaSeq 6000 SP Reagent Kit v1.5 (500 cycles) (Illumina 20028402)
Magnetic particle concentrator (Thermo Fisher A13346)
MicroAmp PCR caps (Axygen AX.PCR-02CP-C)
MicroAmp PCR tubes (Axygen AX.PCR-0208-C)
Microtubes, 1.5-mL (Axygen MCT-150-C) and 2-mL (Sigma-Aldrich BR780546)
Multichannel pipettor (optional)
Pasteur pipet (Corning CLS7095B5X)
Polypropylene centrifuge bottles, 500-mL (Beckman Coulter 355607)
QIAgen Plasmid DNA Miniprep Kit (QIAGEN 27106X4)
QIAquick Gel Extraction Kit (QIAGEN 28706X4)
Refrigerated microcentrifuge (Sigma-Aldrich 75772441)
Rotating mixer (Sigma-Aldrich BMSR5010)

SPRIPlate 96 Super Magnet Plate (Beckman A32782)
Syringe filter, 0.2-µm (Pall 4612)
T100 PCR thermal cycler (Bio-Rad 1861096)
UV-Vis Spectrophotometer (Agilent Cary 60 UV-Vis)

## METHOD

*In this protocol, we first describe the panning of a chicken single-chain variable fragment (scFv)-displaying phage library on antigens immobilized to magnetic beads, which involves binding phage to the beads, multiple washing steps, elution by low pH, and reamplification (Fig. 1). In every round, binding clones are selected and amplified. These binding clones will then dominate the phage library after three or four rounds. The input of each round is usually in the range of $10^{12}$ phage particles, and the output is usually in the range of $10^5$–$10^8$ phage, depending on the number of washing steps and the degree of enrichment occurring at a given round. A 10-fold to 100-fold increase in output after the third or fourth round is typical. After several rounds of panning, the resultant phage pool can be tested in phage ELISA to detect antigen binders, and the steps for this are also provided.*

*To investigate the enrichment pattern of positive clones (possible antigen binders), users will also prepare high-throughput sequencing (HTS) libraries using phagemid DNA obtained in every round of panning. Users will thus be able to obtain, via HTS, the heavy chain ($V_H$) and light chain sequences of binders that have binding affinity against the antigen of interest.*

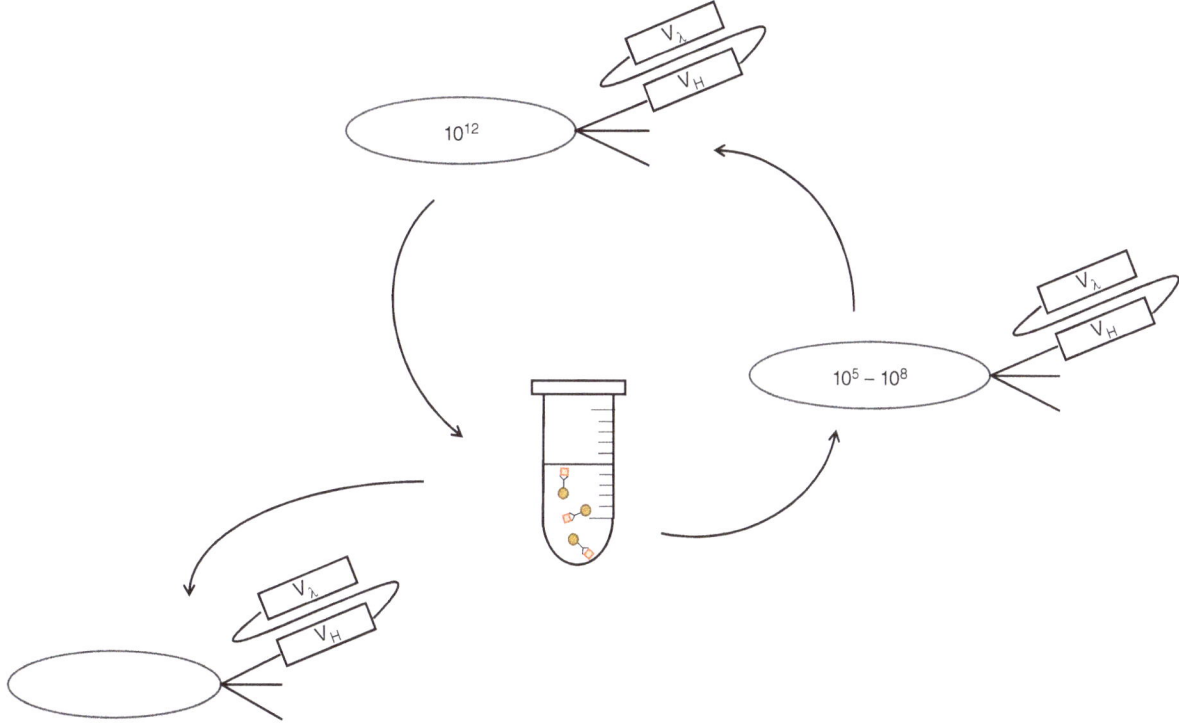

FIGURE 1. Library panning on immobilized antigens. Panning involves several rounds of binding phage to an antigen immobilized to magnetic beads, a defined number of washing steps, elution by low pH, and reamplification. During each round, specific binding clones are selected for and amplified. Shown is a schematic representation of the bio-panning process using phage display technology. The oval represents the filamentous phage, and the three lines behind it depict the carboxy-terminal domain of the minor coat protein (coat protein III), where antibody fragments are expressed. (*Top*) Before conducting bio-panning, the ideal number of phage is around $10^{12}$. In the tube, the magnetic epoxy beads conjugated with various antigens of interest are shown, with magnetic beads in brown and antigens in red. (*Right*) Following several rounds of bio-panning, the number of rescued phage typically ranges from $10^5$ to $10^8$. A specific antigen binder—a possible antibody clone—is shown at the *left*.

Chapter 6

Bio-Panning

1. Calculate the volume of beads to be used (see note below). Mix the beads well and transfer the suitable volume of resuspended beads to a 1.5-mL microcentrifuge tube.

   *We recommend $1 \times 10^7$ beads for the first round of a single library, and a minimum of $5 \times 10^6$ beads may be used for the rounds that follow.*

   *To prevent the loss of beads during the washing step, maintain the bead volume >50 μL and place the microcentrifuge tube on the MPC for at least 4 min before removing the supernatant.*

2. Wash the beads by adding 0.1 M sodium phosphate (pH 7.4) for a final concentration of $1 \times 10^9$ beads/mL, vortexing for 30 sec, and rotating for an additional 10 min at room temperature. Incubate for 4 min on a magnetic particle concentrator (MPC) to capture the beads and then discard the supernatant.

3. Repeat the wash in Step 2 one additional time.

4. Determine the desired number of panning rounds (see note below). Prepare the target antigen and add the solution to the prewashed beads in a microcentrifuge tube. Then, add an equal volume of 3 M ammonium sulfate for a final concentration of 1.5 M ammonium sulfate. Incubate for 16–24 h at 37°C with rotation.

   *Usually, users do four to five rounds of panning for a single library; however, it depends on the input/output titer and its enrichment pattern during the panning. The binding capacity of the beads is 3 μg of protein per $1 \times 10^7$ beads, according to the manufacturer.*

5. Place the microcentrifuge tube on the MPC and remove the supernatant. Wash the coated beads with 500 μL of 3% (w/v) BSA in PBS for a total of four times. For each washing step, resuspend the beads, place on the MPC for 4 min, and remove the supernatant. After the four washes, resuspend the beads in 500 μL of 3% (w/v) BSA in PBS, rotate for 1 h at room temperature, and place on the MPC for 4 min to then remove the supernatant.

6. Perform an additional wash by incubating the beads in 100 μL of 1× PBS with 0.5% (v/v) Tween 20 for 10 min on a rotator, followed by placing the tube on the MPC for 4 min. Aspirate and discard the supernatant, and resuspend the beads in PBS with 0.02% (w/v) sodium azide to make a solution of $5 \times 10^6$ beads/50 μL. Aliquot the 50-μL bead aliquot in a microcentrifuge tube and store at 4°C until use.

   *Use the beads within a week.*

   *Decide how many aliquots users should prepare prior to doing the bio-panning (see Step 1).*

7. To start bio-panning, resuspend the aliquoted beads (for a single library, use $1 \times 10^7$ beads for the first round and $5 \times 10^6$ beads thereafter) by pipetting or vortexing. Place the microcentrifuge tube on the MPC for 4 min and wash once with 100 μL of 3% (w/v) BSA in PBS.

8. Add 500 μL of the phage library to the beads and incubate on a rotator for 2 h at room temperature.

9. In the meantime, inoculate 3 μL of *Escherichia coli* from a frozen stock in 15 mL of SB medium in a 50-mL conical tube (if XL1-Blue is used, also add 4 μL of 5 mg/mL tetracycline). Shake the culture at 250 rpm for 1.5–2.5 h at 37°C, until the optical density (OD) at 600 nm ($OD_{600}$) reaches 0.7–1.

   *A bacterial culture should be freshly prepared for each round of panning. The total volume of inoculated culture should be adjusted depending on the number of libraries (2 mL/library). An extra 125 μL of culture of E. coli per library is needed for input titration and contamination tests.*

10. Place the microcentrifuge tube from Step 8 on the MPC and aspirate and discard the phage supernatant.

11. Add 500 μL of 0.05% (v/v) Tween 20 in PBS to the beads and pipet vigorously up and down once. Place the tube in the MPC and discard the wash solution.

    *For the second and third rounds of panning, repeat this washing step five times. For the fourth round, the number of washing steps and the concentration of Tween 20 may be adjusted depending on the output titer.*

12. After removing the final wash solution, resuspend the beads with 60 μL of elution buffer. For the following rounds, use 30 μL. Incubate for 5 min on a rotator and place on the MPC. Transfer the eluate to a new 50-mL conical tube and neutralize the pH with 2 M Tris; use 1.5 μL of 2 M Tris per 30 μL of elution buffer.

13. When the $OD_{600}$ of the *E. coli* culture from Step 9 reaches 0.7–1, add 2 mL of the *E. coli* culture into the 50-mL conical tube containing the eluate (from Step 12), shake gently, and incubate for 15 min at room temperature.

14. Add 6 mL of SB medium and 1.6 μL of 100 mg/mL carbenicillin (when XL1-Blue is used, also add 12 μL of 5 M/mL tetracycline), and then proceed to output titration.

15. For output titration, dilute 2 μL of the sample in 200 μL of SB medium, and plate 100 and 10 μL of the diluted sample on separate LB agar + carbenicillin plates. Shake the remaining culture at 250 rpm for 1 h at 37°C, add 2.4 μL of 100 M/mL carbenicillin, and shake for an additional hour at 250 rpm and 37°C.

16. In the meantime, proceed with input titration by infecting 50 μL of the *E. coli* culture at $OD_{600}$ of 0.7–1 (from Step 9) with 1 μL of a $10^{-8}$ diluted phage supernatant used for bio-panning from Step 8 (see note below). Shake gently, incubate for 15 min at room temperature, and plate on an LB agar + carbenicillin plate.

    *Do not dilute the phage supernatant more than 1/100 at a time. For a 1:100 dilution, add 10 μl of phage supernatant to 990 μL of fresh SB medium. Repeat until you obtain the $10^{-8}$ dilution.*

17. Test the electrocompetent *E. coli* cells (from Step 9) for phage contamination as follows:

    i. Plate 25 μL of the prepared electrocompetent *E. coli* cells directly onto an LB agar + carbenicillin plate. Incubate overnight at 37°C.
        *No colonies should grow.*

    ii. To test contamination with helper phage, plate 25 μL of the prepared electrocompetent *E. coli* cells directly onto an LB agar + kanamycin plate. Incubate overnight at 37°C.
        *No colonies should grow.*

    iii. To test contamination with lytic phage (and secondary test for contamination with helper phage), mix 25 μL of the prepared electrocompetent *E. coli* cells with 3 mL of liquefied LB top agar (<50°C) and pour onto a plain LB agar plate. Incubate overnight at 37°C.
        *No plaque should be formed.*
        *If there are any colonies on the plate, see Troubleshooting.*

18. Incubate the input (see Step 16) and output (see Step 15) plates overnight at 37°C. The next day, calculate the output and input titer by multiplying the number of colonies by the culture volume and dividing by the plating volume. For output titration, multiply by the dilution fold as well.

$$(\text{\# of colonies}) \times \left(\frac{\text{culture vol. } (50\,\mu L)}{\text{plating vol. } (50\,\mu L)}\right) \times \left(\frac{\text{total vol. of phage } (x\,\mu L)}{\text{infected vol. of phage } (1\,\mu L)}\right) \times (\text{dilution fold})$$
$$= \text{Total transformants of input titration}$$

$$(\text{\# of colonies}) \times \left(\frac{\text{culture vol. } (8000\,\mu L)}{\text{plating vol. } (10 \text{ or } 100\,\mu L)}\right) \times (\text{dilution fold})$$
$$= \text{Total transformants of output titration}$$

19. Add 1 mL of VCSM13 helper phage ($10^{12}$ to $10^{13}$ pfu) to the 8-mL culture (from Step 15) and transfer to a 500-mL polypropylene centrifuge bottle. Add 91 mL of prewarmed (37°C) SB medium and 46 μL of 100 mg/mL carbenicillin (when XL1-Blue is used, also add 184 μL of 5 mg/mL tetracycline). Shake the ~100-mL culture at 300 rpm for 1.5–2 h at 37°C.

20. Add 140 µL of 50 mg/mL kanamycin to the culture from Step 19 and continue shaking at 300 rpm overnight at 37°C.

21. Centrifuge the 100-mL culture at 3000g (e.g., 4000 rpm in a Beckman JA-10 rotor) for 15 min at 4°C. Save the bacterial pellet for phagemid DNA preparation (which can be done, for example, with Plasmid DNA Miniprep Kit (QIAGEN 27106X4; see note below). For phage precipitation, transfer the supernatant to a clean 500-mL polypropylene centrifuge bottle and add 4 g of PEG-8000 (to 4% [w/v]) and 3 g of sodium chloride (to 3% [w/v]). Shake at 300 rpm for 5–10 min at 37°C, until the solids dissolve. Then, incubate for 30 min on ice.

    *Store the phagemid DNA obtained at each panning round at −20°C. It can be used for library reamplification, as described in Protocol 2: Generation of a Phage Display Chicken Single-Chain Variable Fragment Library (Yang et al. 2024a), and "High-Throughput Sequencing Analysis" described below.*

22. Centrifuge the PEG precipitation at 15,000g (e.g., 9000 rpm in a Beckman JA-10 rotor) for 15 min at 4°C. Discard the supernatant, drain the centrifuge bottle by inverting on a paper towel for at least 10 min, and wipe off any remaining liquid from the upper part of the bottle with a paper towel.

23. Resuspend the phage pellet in 2 mL of 1% (w/v) BSA in PBS by pipetting up and down along the side of the centrifuge bottle. Transfer to a 2-mL microcentrifuge tube and resuspend further by pipetting up and down using a 1-mL pipet tip. Spin at full speed in a microcentrifuge for 5 min at 4°C, and pass the supernatant through a 0.2-µm syringe filter into a sterile 2-mL microcentrifuge tube. Add sodium azide to final concentration of 0.02% (w/v) for storage at 4°C.

    *Phage preparations can be used immediately for the next round of bio-panning or stored for months at 4°C. If they are stored, reamplification would be required before bio-panning, as described in Protocol 2: Generation of a Phage Display Chicken Single-Chain Variable Fragment Library (Yang et al. 2024a). Only freshly prepared phage (prepared on the same day) should be used for bio-panning.*

## Small-Scale Preparation of (PEG-Precipitated) Antibody Fragment-Displaying Phage

*This section describes the preparation of antibody-displaying phage derived from single clones to select the binders via phage ELISA. It is a slightly modified and downscaled version of the procedure for the generation of PEG-precipitated phage for panning. This workflow facilitates the generation of PEG-precipitated phage from many single clones. The volume and concentration of PEG-precipitated phage obtained are usually sufficient for testing by ELISA and flow cytometry to identify clones exhibiting relevant binding characteristics.*

24. Inoculate 500 µL of SB containing 15 µg/mL carbenicillin with a single colony from an output titer plate from Step 15. Repeat for the number of colonies you want to test in a 2-mL deep well.

    *Typically, 96–192 individual colonies from output titer plates from the last one or two rounds of a panning experiment are used. If the phage pools had been previously analyzed by ELISA, as described later in this protocol, the results will reveal which output titer plates should be used for the analysis of single clones.*

25. Shake the cultures at 300 rpm for 1.5 h at 37°C until they reach an $OD_{600}$ of 0.7–1.

    *Increase the incubation time for larger cultures; for example, a 100-mL culture should be grown for 8 h before proceeding to the next step.*

26. Add 100 µL of helper phage to the cultures. The titer of the helper phage should be at least $5 \times 10^9$ pfu/mL (see Protocol 3: Preparation of VCSM13 Helper Phage for Display Library Reamplification and Bio-Panning[(Yang et al. 2024b)]).

27. Shake the cultures at 300 rpm for another 2 h at 37°C.

28. Add 50 mg/mL kanamycin to each culture, to a final concentration of 70 µg/mL.

29. Shake the cultures at 300 rpm for at least 10 h (typically overnight) at 37°C.

30. Centrifuge the cultures at 2800g (3500 rpm in a tabletop centrifuge) for 15 min. Transfer 400 µL of each culture supernatant to a separate microcentrifuge tube. Add 100 µL of 5× PEG/NaCl to each tube, mix, and precipitate phage by incubating for 30 min on ice.

*The culture supernatant can be used directly, without the need for PEG precipitation, for the selection of antigen binders via phage ELISA. The supernatant, however, is generally not suitable for use in flow cytometry assays without it.*

*The bacterial pellet can be used to prepare phagemid using a standard plasmid minipreparation method (e.g., QIAprep Spin Miniprep Kit, QIAGEN). It can be stored for 1 wk at −20°C; however, it is recommended to prepare the phagemid freshly as soon as possible. Such bacterial pellets are important when identifying clones with meaningful signal-to-background ratio in phage ELISA. See Step 45.*

31. Spin the tubes in a microcentrifuge at 14,000 rpm and 4°C. Carefully aspirate and discard the supernatant with a Pasteur pipet. Spin the tubes for another 20 sec and aspirate and discard the remaining supernatant.

32. Resuspend each phage pellet in 50 µL of PBS + 1% (w/v) BSA + 0.02% (w/v) sodium azide. Store them at 4°C until use.

    *The PEG-precipitated phage can be used in binding assays for several weeks, and is good for reinfection for up to 1 yr.*

## Phage ELISA

*After several rounds of panning, the resultant phage pool can be tested in an ELISA to evaluate the success of the panning experiment. Such "phage ELISA" has several advantages over the often tedious analysis of single clones via flow cytometry or conventional ELISA. It is more useful and convenient to screen out the possible binders in a single experiment.*

*Both suspended and PEG-precipitated phage from single clones can be tested in ELISA according to the procedure described below. Although we recommend the use of fresh phage preparations for panning, PEG-precipitated phage stored at 4°C can be used for the phage ELISA for up to 1 wk. Phage preparations that have been stored for longer can be reamplified as described in Protocol 2: Generation of a Phage Display Chicken Single-Chain Variable Fragment Library (Yang et al. 2024a).*

33. Dilute the antigens (the antigen of interest, as well as a control antigen) in 0.1 M $NaHCO_3$ (pH 8.6). Prepare 30 µL of antigen solution containing 0.05–0.5 µg of antigen for each well to be coated. Furthermore, to check the background binding of phage, plan to use another set of wells blocked but not antigen-coated (blocked-only well; see Step 36). Plan also to use blank wells (see Step 37).

    *The volume and amounts of antigen given above are for half-area assay plates (e.g., Corning Costar 3690). If standard-size plates are used, double the volumes throughout the protocol. Prepare the appropriate antigen-coated wells according to the numbers of single clone that you want to test via phage ELISA.*

    *If you want to test a pool of phage generated while doing bio-panning, coat the well of half-area assay plates with the antigen according to the number of panning rounds.*

34. Add 25 µL of antigen solution (antigen of interest or control) to each of the corresponding ELISA plate wells required and incubate overnight at 4°C, or for at least 1 h at 37°C.

35. Shake out the antigen solution and wash twice. To wash, hold the plate under running deionized tap water for 20 sec and then shake the excess liquid out.

36. Block antigen-coated and blocked-only wells by adding 150 µL of 5% (w/v) skim milk in PBS. Incubate for 1 h at 37°C and shake out the blocking solution from the wells.

37. Dilute the phage solution from Step 32 threefold in 5% (w/v) skim milk in PBS. Prepare enough so that you have 55 µL of the dilution per well. Add 50 µL of the diluted phage to (1) antigen-coated and blocked wells, (2) blocked-only wells, and (3) blank wells, and incubate for 1–2 h at 37°C.

    *In "Small-Scale Preparation of (PEG-Precipitated) Antibody Fragment-Displaying Phage," we describe the preparation of antibody-displaying phage derived from single clones.*

    *Blank wells are needed to check the successful amplification of phages.*

38. Dilute HRP-conjugated anti-M13 antibodies or HRP-conjugated anti-HA antibodies with 5% (w/v) skim milk in PBS (see Materials for dilution factors). Prepare 55 µL per well.

    *Anti-HA conjugates can be used only with phages displaying scFv tagged with HA peptide, such as for pComb3X-based phage display.*

## Chapter 6

39. Wash the plate from Step 37 10 times with deionized tap water, as described in Step 35.

40. Add 50 µL of diluted HRP-conjugated anti-M13 or anti-HA antibody to each well, and incubate for 1 h at 37°C.

41. Prepare the substrate (e.g., ABTS solution for HRP-conjugated antibodies) as recommended by the supplier. Prepare 55 µL per well.

42. Wash the plate 10 times with deionized tap water, as described in Step 35.

43. Add 50 µL of substrate solution to each well.

44. Incubate at room temperature and use an ELISA plate reader to read the absorbance at the wavelength defined by the manufacturer at two to three time points. Typically, the first reading is done at 15 min after adding the substrate. The second reading is typically done 30 min at 37°C after the first reading.

45. Identify the clones with meaningful signal-to-background ratio and retrieve the corresponding bacterial pellet (from Step 30) for phagemid preparation.

    *For the purpose of monitoring a successful phage display selection, determining whether the signal is relatively high or low is more important than obtaining a precise titer. The ELISA signal from the antigen of interest should increase with each panning round, but the signal from the control antigen and blocking-only should remain low throughout the panning. An increasing signal from the control antigen might indicate nonspecific sticky clones or plastic binders. At the end of panning, the ELISA reading obtained with the antigen of interest is usually three or more times the reading with the control antigen and blocking-only signal.*

46. Analyze and compare the scFv sequences by Sanger sequencing using the Ompseq and HRML-F primers.

### High-Throughput Sequencing Analysis of the Chicken scFv Library for In Silico Selection of Antigen Binders

*Fast and reliable analysis of the panned antibody fragment library, at a polyclonal as well as at a cloned level, is imperative to quickly assess whether the panning experiment was successful, and to select antibody fragment clones that have the desired specificity and affinity. As described in the previous sections, one way of evaluating this is using ELISA-based procedures. After several rounds of panning, the resultant phage pools can be tested by "phage ELISA." If the selected phage pools show specific binding to the antigen of interest, single clones are analyzed. Phagemid DNA is prepared from selected clones and used for DNA sequence analysis, as described above. In addition, high-throughput sequencing (HTS) analysis can then be used to determine the complexity of the chicken scFv library, whose composition changes throughout the panning process, with certain clones being enriched and others being depleted (Yang et al. 2017; Kim et al. 2021; Lee et al. 2021). Below, we describe the steps for using HTS. Analysis of the library composition dynamics can result in the in silico identification of additional binder candidates.*

47. Prepare phagemid DNA from the bacterial cell pellet from Step 21. Then, for each clone, amplify the heavy chain variable region ($V_H$) and $\lambda$ light chain variable region ($V_\lambda$) gene fragments via PCR (see Figs. 2, 3). The primers contain an Illumina adapter, and sequences of either the vector backbone or the scFv linker. For the amplification, prepare the following PCR mixtures, using the primer combinations listed in the note below:

| Volume | Reagent |
| --- | --- |
| 1 µL | Phagemid DNA (~0.1 µg) |
| 0.75 µL of 10 µM stock | Forward primer |
| 0.75 µL of 10 µM stock | Reverse primer |
| 5 µL | 5× KAPA HiFi buffer |
| 0.75 µL | 10 mM KAPA dNTP mix |
| 0.5 µL | KAPA HiFi HotStart DNA polymerase (1 unit/µL) |
| Add water to a final volume of 25 µL | |

*Because of limitations in sequencing length in the Illumina sequencing platform, the PCR products should be separated into $V_H$ and $V_\lambda$.*

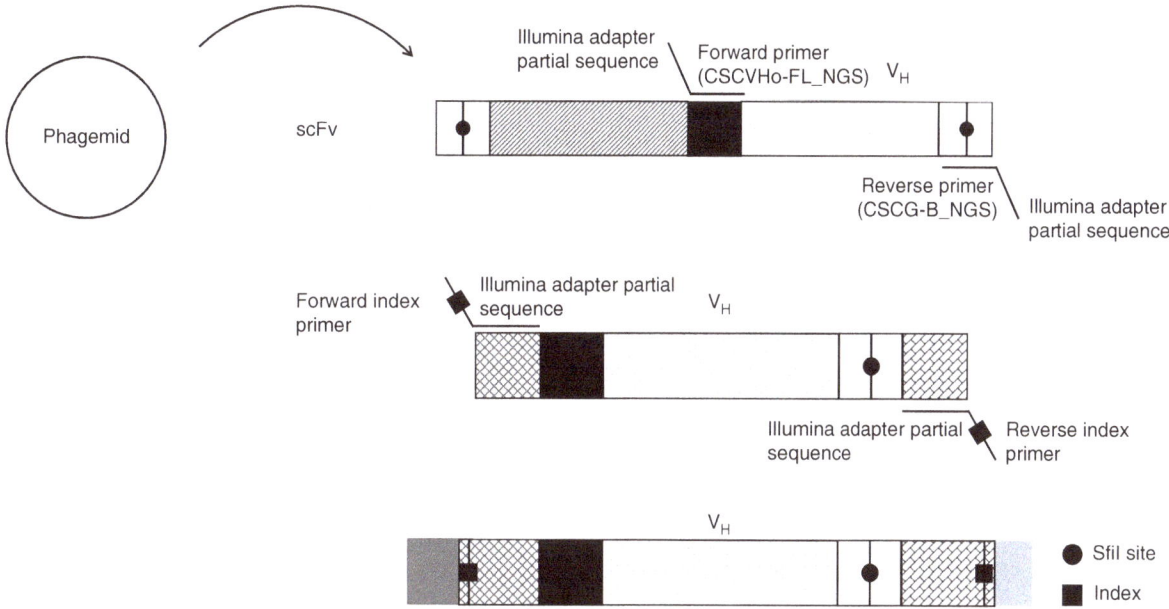

FIGURE 2. First-round PCR amplification of chicken heavy chain variable region ($V_H$) sequences for high-throughput sequencing (HTS) analysis. The primer CSCVHo-FL_NGS is paired with the CSCG-B_NGS reverse primer to amplify $V_H$ segments from phagemid obtained from every round of bio-panning. The forward and reverse primers have a sequence tail that corresponds to the partial sequences of an Illumina adapter used in HTS analysis. These tails are recognized by the index primer used in the second-round PCR. Forward and reverse index primers used in the second-round PCR allow identification of individual $V_H$ sequences obtained during the different bio-panning rounds.

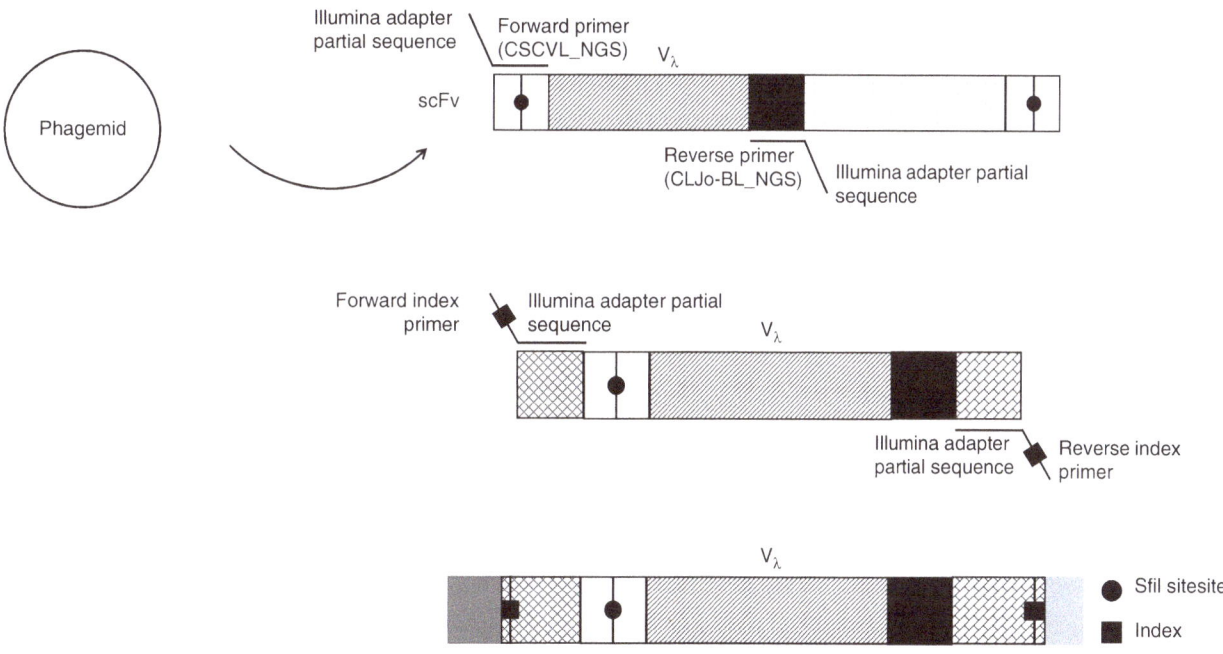

FIGURE 3. The first-round PCR amplification of chicken λ light chain variable region ($V_\lambda$) sequences for high-throughput sequencing (HTS) analysis. The primer CSCVL_NGS is paired with the CLJo-BL_NGS reverse primer to amplify $V_\lambda$ segments from phagemid obtained from every round of bio-panning. The forward and reverse primers have a sequence tail that corresponds to the partial sequences of an Illumina adapter used in HTS analysis. These tails are recognized by the index primer used in the second-round PCR. Forward and reverse index primers used in the second-round PCR allow identification of individual $V_\lambda$ sequences obtained during the different bio-panning rounds.

Chapter 6

*For $V_H$, use a CSCVHo-FL_NGS and CSCG-B_NGS primer combination. For $V_\lambda$, use a CSCVL_NGS and CLJo-B_NGS primer combination.*

48. Perform the PCR under the following conditions:

    | | |
    |---|---|
    | 5 min at 95°C | |
    | 10 cycles of | 30 sec at 98°C |
    | | 30 sec at 60°C |
    | | 60 sec at 72°C |
    | Followed by 10 min at 72°C | |

49. Load an aliquot of the PCR samples on a 2% (w/v) agarose gel in 1 × TAE and run the gel for an appropriate amount of time and at an appropriate voltage. Make sure to run the gel with a suitable molecular weight marker. The expected sizes of amplified $V_H$ and $V_\lambda$ are ∼400 and ∼350 bp, respectively.

50. Purify the PCR products using the AMPure XP purification kit, following the manufacturer's instructions. The volume of AMPure bead solution should be equivalent to that of the PCR sample to be purified. Mix the sample with magnetic beads by pipetting, and incubate the mixture for 5 min at room temperature for maximum recovery. Place the PCR tubes on a SPRIPlate 96 Super Magnet Plate for 2 min to separate beads from the solution, and remove the solution.

51. Dispense 200 µL of 70% (v/v) ethanol to each tube, incubate for 30 sec at room temperature, and discard the ethanol. Repeat this washing step one additional time. Air-dry or aspirate out the ethanol.

52. Add 40 µL of DEPC-treated water to each tube and resuspend the beads. Incubate for 5 min at room temperature. Place the PCR tubes on SPRIPlate 96 Super Magnet Plate for 2 min to separate the beads from the solution. Transfer each supernatant to a new tube.

53. Prepare the following PCR mixtures, for indexing, with separate reactions for the different purified products:

    | Volume | Reagent |
    |---|---|
    | 15 µL | Purified $V_H$ or $V_\lambda$ product |
    | 0.75 µL of 10 µM stock | Forward index primer (Table 2) |
    | 0.75 µL of 10 µM stock | Reverse index primer (Table 2) |
    | 5 µL | 5× KAPA HiFi buffer |
    | 0.75 µL | 10 mM KAPA dNTP mix |
    | 0.5 µL | KAPA HiFi HotStart DNA polymerase (1 unit/µL) |
    | Add water to a final volume of 25 L | |

    *The indexing PCR aims to incorporate, into each sample, a unique index to identify the source, i.e., different panning rounds or various target antigens, of the sequences obtained in the HTS analysis. Hence, the purified PCR products from Step 52 are tagged with their own indexes at both ends, where the combination of the forward and reverse indexes are different. The primers in the first-round PCR (Step 47) provide overhang sequences that are also present in the indexing PCR primers, to enable their binding and to generate final products with indexes (see Table 2).*

54. Perform the indexing PCR under the following conditions:

    | | |
    |---|---|
    | 3 min at 95°C | |
    | 10 cycles of | 30 sec at 98°C |
    | | 30 sec at 60°C |
    | | 1 min at 72°C |
    | Followed by 5 min at 72°C | |

55. Run the reactions on an 1.5% (w/v) agarose gel, cut out the correct-sized bands, and purify the DNA by resin binding (e.g., using QIAquick Gel Extraction Kit, QIAGEN). Elute the DNA in 35 μL of distilled water.

    *The expected sizes of amplified $V_H$ and $V_\lambda$ with the indexes are ∼450 and ∼400 bp, respectively.*

56. Purify the DNA once again with the AMPure XP purification kit, as described in Step 50.

57. Perform HTS with the Illumina NovaSeq 6000 sequencing system using the purified library from Step 56.

58. Analyze the enrichment pattern of individual sequences following published protocols (Yang et al. 2017; Noh et al. 2019) to determine possible binders. The source code for analyzing HTS raw data is available at https://doi.org/10.5281/zenodo.7844471, which was developed in a previous article (Kim et al. 2021).

    *Two sets of sequencing data are required to use the code in the repository link above: HTS raw data from phagemid DNA and sequence information of positive scFv clones with confirmed binding activity from phage ELISA.*

    *Additionally, a machine-learning algorithm can be used to facilitate the in silico selection of possible antigen–binder pipelines (Yoo et al. 2020). The binder candidate may be obtained by chemical gene synthesis or other alternatives like high-throughput position-encoded cloning aided with HTS analysis.*

## TROUBLESHOOTING

*Problem (Step 17):* There are colonies or plaques in the test plates assessing contamination.
*Solution:* Discard all of the input/output titration plates and perform the following round of bio-panning with freshly thawed *E. coli* cells (ER2738 or XL1-Blue).

## RECIPES

### DNA Loading Buffer (6×)

30% (v/v) glycerol
0.25% (w/v) bromophenol blue
0.25% (w/v) xylene cyanol FF

Store at 4°C.

### LB + 100 μg/mL Carbenicillin Plates

1. In a 1-L bottle or flask, combine 32 g of LB agar (Thermo Fisher 22700041) with 1 L of water. Stir and autoclave for 15 min at 121°C at 15 psi.
2. When cooled to 42°C–45°C, add carbenicillin to a final concentration of 100 μg/mL. Pour into Petri dishes and allow to solidify.

    Store for up to 1 mo at 4°C.

### LB Top Agar

1. Add 0.35 g of Bacto agar (Sigma-Aldrich A5306) and 1.25 g of LB medium (Sigma-Aldrich L3397) to 50 mL of distilled water in a glass bottle.
2. Autoclave for 20 min at 120°C. Then, aliquot 10 mL per 50-mL tube and store at 4°C.
3. Melt in a microwave and place in a water bath to keep it at 50°C until use.

### SB Medium

30 g tryptone
20 g yeast extract
10 g MOPS (3-[N-morpholino]-propanesulfonic acid)
Bring to 1 L total volume with $H_2O$, stir until dissolved, and titrate to pH 7. Sterilize by autoclaving at 15 psi on liquid cycle for 20 min at 121°C.

### TAE Buffer (50×)

| Reagent | Quantity (for 1 L) |
| --- | --- |
| Tris (hydroxymethyl) aminomethane | 242 g |
| Acetic acid | 57.1 mL |
| $Na_2EDTA$ | 7.43 g |

Dissolve in distilled $H_2O$ and make volume up to 1 L.

## REFERENCES

Kim SI, Noh J, Kim S, Choi Y, Yoo DK, Lee Y, Lee H, Jung J, Kang CK, Song K-H, et al. 2021. Stereotypic neutralizing $V_H$ antibodies against SARS-CoV-2 spike protein receptor binding domain in patients with COVID-19 and healthy individuals. *Sci Transl Med* 13: eabd6990. doi:10.1126/scitranslmed.abd6990

Lee Y, Yoo DK, Noh J, Ju S, Lee E, Lee H, Kwon S, Chung J. 2021. Amplification of a minimally biased antibody repertoire for in vitro display using a universal primer-based amplification method. *J Immunol Methods* 496: 113089. doi:10.1016/j.jim.2021.113089

Noh J, Kim O, Jung Y, Han H, Kim JE, Kim S, Lee S, Park J, Jung RH, Kim SI, et al. 2019. High-throughput retrieval of physical DNA for NGS-identifiable clones in phage display library. *MAbs* 11: 532–545. doi:10.1080/19420862.2019.1571878

Yang W, Yoon A, Lee S, Kim S, Han J, Chung J. 2017. Next-generation sequencing enables the discovery of more diverse positive clones from a phage-displayed antibody library. *Exp Mol Med* 49: e308. doi:10.1038/emm.2017.22

Yang H, Chae J, Kim H, Noh J, Chung J. 2024a. Generation of a phage display chicken single-chain variable fragment library. *Cold Spring Harb Protoc* doi:10.1101/pdb.prot108213

Yang H, Chae J, Kim H, Noh J, Chung J. 2024b. Preparation of VCSM13 helper phage for display library reamplification and bio-panning. *Cold Spring Harb Protoc* doi:10.1101/pdb.prot108569

Yoo DK, Lee SR, Jung Y, Han H, Lee HK, Han J, Kim S, Chae J, Ryu T, Chung J. 2020. Machine learning-guided prediction of antigen-reactive in silico clonotypes based on changes in clonal abundance through bio-panning. *Biomolecules* 10: 421. doi: 10.3390/biom10030421

CHAPTER 7

# Structural Survey of Antigen Recognition by Synthetic Human Antibodies

Maryna Gorelik,[1] Shane Miersch,[1] and Sachdev S. Sidhu[2]

*School of Pharmacy, University of Waterloo, Waterloo, Ontario N2G 1C5, Canada*

Synthetic antibody libraries have been used extensively to isolate and optimize antibodies. To generate these libraries, the immunological diversity and the antibody framework(s) that supports it outside of the binding regions are carefully designed/chosen to ensure favorable functional and biophysical properties. In particular, minimalist, single-framework synthetic libraries pioneered by our group have yielded a vast trove of antibodies to a broad array of antigens. Here, we review their systematic and iterative development to provide insights into the design principles that make them a powerful tool for drug discovery. In addition, the ongoing accumulation of crystal structures of antigen-binding fragment (Fab)–antigen complexes generated with synthetic antibodies enables a deepening understanding of the structural determinants of antigen recognition and usage of immunoglobulin sequence diversity, which can assist in developing new strategies for antibody and library optimization. Toward this, we also survey here the structural landscape of a comprehensive and unbiased set of 50 distinct complexes derived from these libraries and compare it to a similar set of natural antibodies with the goal of better understanding how each achieves molecular recognition and whether opportunities exist for iterative improvement of synthetic libraries. From this survey, we conclude that despite the minimalist strategies used for design of these synthetic antibody libraries, the overall structural interaction landscapes are highly similar to natural repertoires. We also found, however, some key differences that can help guide the iterative design of new synthetic libraries via the introduction of positionally tailored diversity.

## INTRODUCTION

Monoclonal antibodies (Abs) are not only invaluable tools in biological research but are also increasingly used as therapeutic agents (Alfaleh et al. 2020). Abs can be generated through traditional hybridoma technology, a variety of single B-cell cloning technologies that rely on animal immunization (Lee et al. 2014; Pedrioli and Oxenius 2021), or through in vitro display technologies (i.e., phage, yeast, mRNA, bacterial, etc.) (Valldorf et al. 2022) that isolate Ab fragments from combinatorial libraries using in vitro selection techniques on purified proteins, virus-like particles, or even antigen-expressing cells (Nixon et al. 2019). For in vitro methods, the starting immune repertoires —whether natural naive (McCafferty et al. 1990; Marks et al. 1991) or immunized (Shen et al. 2007), or synthetic (Barbas et al. 1992) or semi-synthetic—play a crucial role in the performance of the library and the characteristics of the Abs that can be isolated from them. In particular, Ab libraries with paratopes constructed entirely from synthetic diversity have proven to be versatile tools for testing hypotheses regarding sequence diversity and function (Fellouse et al. 2004, 2006), for analyzing

---

[1]These authors contributed equally to this work.
[2]Correspondence: sachdev.sidhu@uwaterloo.ca

© 2026 Cold Spring Harbor Laboratory Press
Cite this overview as *Cold Spring Harb Protoc*; doi:10.1101/pdb.over107759

biomolecular interactions (Lee et al. 2004a,b; Bond et al. 2005), and for the selection and optimization of therapeutic Abs (Nelson et al. 2018; Pavlovic et al. 2018; Nabhan et al. 2023).

Because synthetic Ab libraries are completely naive, they are not subject to restrictions imposed by self-tolerance, which culls autoreactive B-cell clones from immunorepertoires to mitigate the risk of autoimmunity. Protein drug targets that are highly conserved across species can pose challenges in eliciting an immune response when used as antigens for hybridoma technologies that rely on immunization, because they are not recognized as foreign (Zhou et al. 2009). Synthetic libraries, on the other hand, suffer from no such constraints. From a more fundamental point of view, the ability to precisely control the design of synthetic diversity also offers a straightforward approach for facile characterization and continual, iterative improvement of synthetic library designs based on empirical observations.

Despite the enormous potential of synthetic Ab libraries, the field developed more slowly than phage display applications that use natural repertoires. For the most part, this was because the development of precisely designed synthetic repertoires is considerably more difficult than the cloning and exploitation of natural repertoires. In the latter case, the construction of libraries from natural sources requires sequence knowledge of immunoglobulin genes and an understanding of the fundamental principles of Ab structure and function to efficiently transfer the in vivo repertoire to an in vitro display system using molecular biology techniques (Vaughan et al. 1996; Lai and Lim 2023). On the other hand, synthetic repertoires are generated by chemical synthesis of degenerate oligonucleotides, and the introduction of this combinatorial diversity into Ab complementarity-determining regions (CDRs) requires both molecular techniques to capture this diversity and also extensive detailed knowledge of Ab structure and function and of the positional diversity that can generate effective Abs.

To contextualize this decades-long effort, we provide here an overview of the design principles and considerations that have been used to generate high-quality synthetic Ab libraries and discuss the advantages of the single-framework approach, also highlighting the use of this framework in approved therapeutic Abs. The systematic assessment of diversity and concurrent evolution is traced from the earliest minimalist libraries generated on a single framework to now, representing nearly 30 years of development. During this period, a growing number of crystal structures of synthetic antigen-binding fragments (Fabs) in complex with their target antigen have been published that provide exquisite details of how they achieve molecular recognition. Close inspection of these details and a comparison to those of natural Ab complexes can, in turn, be used to obtain insights and develop new design strategies to further improve synthetic repertoires (Burkovitz and Ofran 2016; Moreno et al. 2022). Toward this aim, we also surveyed the structural landscape of both natural and synthetic antibodies by analyzing the Fab–antigen interfaces of 50 distinct complexes to characterize the paratopes of each antibody. By comparing the two sets, we sought to contrast key features of both, including paratope size, framework, residue and CDR usage, and CDR length, sequence, and conformational diversity, to determine whether synthetic antibodies, designed in part using minimalist strategies (single-framework, fixed CDRs, restricted diversity), effectively recapitulate the usage of natural immune repertoires and eliminate or propagate development liabilities that can be found in natural antibodies. Similarities and differences between the two are discussed to understand how diversity is used. We also discuss preconceptions in design, limitations of this survey, and ways that synthetic diversity can be refined and suggest hypotheses that can inform and improve the design of next-generation synthetic Ab libraries.

## SYNTHETIC ANTIBODY LIBRARIES

### Principles of Library Design

Synthetic Ab libraries are constructed by introducing designed combinatorial diversity (using chemically synthesized, degenerate oligonucleotides) into the CDRs that are involved specifically in antigen recognition (Fellouse and Sidhu 2006). Thus, a major consideration in devising a synthetic library is the nature and valency of the Ab fragment that is used as the phage-displayed framework into which synthetic diversity is added. Most commonly, the entire Fab, the single-chain variable fragment (scFv) consisting of the light- and heavy-chain variable domains ($V_L$ and $V_H$), or the $V_H$ domain alone is displayed in either mono- or bivalent formats, which can have critical influence over the properties of the Abs selected (Lee et al. 2004b; Sidhu et al. 2004). Beyond the basic molecular architecture of the displayed Ab, the framework (the portion of the Ab that is not subjected to diversification but rather serves as a stable scaffold for displaying diverse paratopes) is a key decision point in the type of library to be made or used. Many successful libraries have used a single framework consisting of a highly stable $V_H$/$V_L$ pair (Silacci et al. 2005; Yang et al. 2009; Persson et al. 2013), whereas others have relied on multiple $V_H$ and $V_L$ frameworks in diverse combinations (Knappik et al. 2000; Rothe et al. 2008; Prassler et al. 2011; Tiller et al. 2013). The single-framework option simplifies library design and analysis but is thought to limit the diversity presented for selection. On the other hand, supposed diversity gains of libraries built with multiple $V_H$/$V_L$ pairs may be negated by unfavorable framework combinations. Although frameworks generally provide structural stability for CDRs, different frameworks can exhibit differences in $V_H$/$V_L$ packing, tilt angles, and interface area (Teplyakov et al. 2016), which can influence solubility, production yields, immunogenicity, and in vivo stability, to render some Abs to be poorly expressing, nonfunctional, aggregating, or difficult to engineer (Tiller et al. 2013). Alternatively, reducing library complexity by limiting the number of frameworks can facilitate the use of optimized frameworks and downstream engineering.

Insofar as the diversity of synthetic libraries is most often now introduced into the CDRs in a tailored fashion using mutagenic degenerate (Lee et al. 2004a,b; Sidhu et al. 2004; Fellouse et al. 2005) or trinucleotide (Virnekas et al. 1994; Knappik et al. 2000; Rothe et al. 2008; Persson et al. 2013) oligonucleotides, there is opportunity to devise innovative design strategies with the fundamental aim of generating a library that can yield high-affinity and specific Abs targeting a variety of epitopes on virtually any antigen. Perhaps the most widely adopted design philosophy has been to emulate natural repertoires, and this has been increasingly informed by attempts to capture the diversity of the human immune repertoire using high-throughput sequencing technologies (Glanville et al. 2009; Briney et al. 2019; Marks and Deane 2020). In these efforts, it has proven critical to balance the understanding of the diversity of naive repertoires found in the peripheral B-cell compartment with knowledge of the diversity of functional, mature Abs represented by the structural survey of natural Abs in complex with their antigens.

An appreciation of the liabilities that may impede development, even for natural Abs (Jain et al. 2017; Raybould et al. 2019), has spurred alternative design strategies to reduce the presence of residues that can adversely affect Ab function and properties (e.g., poor solubility, glycosylation sites, reactive cysteines, residues prone to oxidation or isomerization, etc.) and increase the likelihood of selecting Abs with favorable biophysical properties (high expression, solubility, and stability).

Finally, perhaps the most crucial property of a synthetic library is the choice of CDR residues targeted for diversification and the allowable positional diversity at each residue. Although a sufficient number of both diversified CDR positions and amino acids available at each position is required to randomize a library, theoretical diversity can be lost during library construction when it far exceeds the cellular transformation efficiency and, consequently, limits the diversity that can be captured during library construction. Thus, judicious and careful design of diversification strategies is recommended to focus diversity on positions that frequently participate in binding, and limit the presence of residues that can compromise folding or binding, or introduce developability liabilities. In short, successful designs should incorporate just enough diversity in the right positions and of the right kinds to enable the selection of stable Abs that recognize cognate antigens with high affinity and specificity and that are easily developable.

## Trastuzumab Single-Framework Libraries

Based on the principles discussed above, numerous synthetic libraries have been developed and used with great success to develop Abs with high affinity and specificity against diverse proteins (Knappik et al. 2000; Silacci et al. 2005; Rothe et al. 2008; Yang et al. 2009; Shi et al. 2010; Persson et al. 2013; Tiller et al. 2013). Although a comprehensive survey of all synthetic Abs and the libraries from which they originated is beyond the scope of this overview, we have focused instead on a library design that was originally developed at Genentech using minimalist design principles (Fellouse et al. 2004, 2005, 2006, 2007; Sidhu et al. 2004; Persson et al. 2013), which, over the past two decades, has been used to develop specific, high-affinity Abs to a multitude of targets for a variety of diagnostic and therapeutic applications by our group and others (Paduch et al. 2013; Hornsby et al. 2015; Miersch et al. 2015, 2021). This library design is based on a single, optimized framework, which greatly simplifies the selection, analysis, and engineering of developable Abs suited for therapeutic purposes.

Synthetic Ab libraries at Genentech were developed with a single human framework derived from the consensus sequences of the most abundant human subclasses; namely, $V_H$ subgroup III and $V_L$ κ subgroup I (Kabat et al. 1991). This framework was originally used for the humanization of murine Abs (Carter et al. 1992; Presta et al. 1993, 1997; Werther et al. 1996), and of the 15 Abs that Genentech has had approved as therapeutics (including those developed in collaboration with others) (Lu et al. 2020; Wang et al. 2021), nine are based on this framework (Table 1). Specifically, the anti-ErbB2 Ab humanized 4D5 or trastuzumab (Carter et al. 1992) was chosen as the library scaffold because the trastuzumab Fab is well expressed in bacterial and mammalian cells and had been displayed previously on phage (Garrard and Henner 1993). Furthermore, the $V_H$ subgroup III Fab possesses a Protein-A–binding site that can facilitate engineering of the Fab on phage and purification from bacteria (Starovasnik et al. 1999), and high-resolution crystal structures of the Fab, alone (Eigenbrot et al. 1993) or in complex with antigen (Cho et al. 2003), were available to provide structural details to inform engineering efforts.

Using this framework, multiple libraries were developed that differed in the portion of Ab displayed for selection, the positions diversified, and the nature of diversification at different positions, aimed at exploring the basic principles of natural immune repertoire evolution and iteratively improving library design (Table 2). The earliest iterations of the library explored mono- and bivalent scFv versions of the 4D5 framework, introducing diversity to solvent-accessible positions using degenerate oligonucleotides that emulated the diversity observed in natural repertoires (Sidhu et al.

TABLE 1. Genentech's approved antibodies in 4D5 framework

| Ab name | DrugBank accession | Target(s) | Indication[a] | PDB ID |
|---|---|---|---|---|
| Atezolizumab | DB11595 | PD-L1 | Non-small-cell lung cancer, small-cell lung cancer, hepatocellular carcinoma, melanoma, alveolar soft part sarcoma | 5X8L |
| Bevacizumab | DB00112 | VEGF | MCC (metastatic colorectal cancer), nonsquamous non-small-cell lung cancer, metastatic breast cancer glioblastoma, metastatic colorectal cancer, nonsquamous non-small-cell lung cancer | 6BFT |
| Efalizumab[b] | DB00095 | LFA-1 | Adult patients with moderate to severe chronic plaque psoriasis | 3EOA |
| Emicizumab | DB13923 | Factor IX/X | Prophylaxis to prevent/reduce the frequency of bleeding episodes of adult and pediatric patients with hemophilia A | NA |
| Mosunetuzumab[c] (CD20 arm) | DB15434 | CD20/CD3 | Adult patients with relapsed or refractory follicular lymphoma (FL) who have received at least two prior systemic therapies | NA |
| Ocrelizumab | DB11988 | CD20 | Adult patients with relapsing or primary progressive forms of multiple sclerosis | NA |
| Omalizumab | DB00043 | IgE | Moderate to severe persistent allergic asthma that is not controlled by inhaled steroids | 5HYS |
| Pertuzumab | DB06366 | HER2 | HER2-positive metastatic breast cancer | 1S78 |
| Ranibizumab | DB01270 | VEGF | Neovascular (wet) age-related macular degeneration (AMD), macular edema following retinal vein occlusion, diabetic macular edema, diabetic retinopathy, and myopic choroidal neovascularization | NA |
| Trastuzumab | DB | HER2 | HER2-positive metastatic breast cancer, gastric or gastroesophageal junction adenocarcinoma | 1N8Z |

(PDB) Protein Data Bank.
[a]Indications obtained from DrugBank entries.
[b]Voluntarily withdrawn from the market after three confirmed and one suspected case of progressive multifocal leukoencephalopathy (PML) were spontaneously reported.
[c]Granted accelerated approval on December 22, 2022.

TABLE 2. An overview of minimalist libraries based on the 4D5 framework

| Library | Valency | Framework | L3 | H1 | H2 | H3 | Observed maximum $K_D$ (nM) | Reference(s) |
|---|---|---|---|---|---|---|---|---|
| Lib-scFv | Mono | | — | Degenerate | Degenerate | DVK | 100 | Sidhu et al. 2004 |
| Lib-scFv$_2$ | Bi | | — | Degenerate | Degenerate | DVK | >100 | Lee et al. 2004b |
| Lib 1-Fab | Mono | | — | Degenerate | Degenerate | DVK | 200 | Lee et al. 2004b |
| Lib 1-Fab'-zip | Bi | | — | Degenerate | Degenerate | DVK | 41 | Lee et al. 2004b |
| Lib 2-Fab | Mono | | — | Degenerate | Degenerate | DVK | 0.6 | Lee et al. 2004a |
| Lib 2-Fab | Mono | | — | Degenerate | Degenerate | NVT | NA | Lee et al. 2004a |
| Lib 2-Fab | Mono | | — | Degenerate | Degenerate | NNS | NA | Lee et al. 2004a |
| Fab'-zip | Bi | | — | Degenerate tetranomial | Degenerate tetranomial | Degenerate tetranomial[a] | >10,000 | Fellouse et al. 2004 |
| Library A | Bi | 4D5 | — | YS | YS | YS[a] | >200 | Fellouse et al. 2005 |
| Library B | Bi | | YS | YS | YS | YS[a] | 60 | Fellouse et al. 2005, 2006 |
| Library C | Bi | | YS[a,b] | YS[b] | YS[b] | Degenerate tetranomial[a,b] | 1.6 | |
| Library D | Bi | | Tailored Y/S/G+19/20[a,b] | YS[b] | YS[b] | Tailored Y/S/G+19/20[a,b] | 4.4 | Fellouse et al. 2007 |
| YS | Bi | | YS | YS | YS | YS[a] | ND | Birtalan et al. 2008 |
| Y/S/R | Bi | | YS | YS | YS | Trinomial Y/S/R[a] | 3.1 | |
| Y/S/G | Bi | | YS | YS | YS | Trinomial Y/S/G[a] | 0.3 | |
| Y/S/G/R | Bi | | YS | YS | YS | Tetranomial Y/S/G/R[a] | 4.9 | |
| Library F | Bi | | Tailored Y/S/G+19/20[a,b] | YS[b] | YS[b] | Tailored Y/S/G+19/20[a,b] | 0.7 | Persson et al. 2013 |

(Fab) Antigen-binding fragment.
[a]Introduction of length diversity.
[b]The restricted diversification of nonparatope residues.

2004). A similar approach was taken in the subsequent iteration, which instead used mono- and bivalent Fab formats, exploring different CDR-H3 degeneracy and modest length diversity. These libraries yielded Abs with high affinities in the subnanomolar range, showing strong proof of concept that a single-framework library with restricted diversity could achieve the affinities observed from natural sources (Lee et al. 2004a).

To further explore restrictions with bivalent Fab display, tetranomial diversity was introduced into the CDRs of heavy chains (Fellouse et al. 2004), with some length diversity in CDR-H3. Although these results revealed that specific binding clones could be successfully obtained from restricted diversity libraries, naive clones exhibited affinities only in the micromolar range. However, affinity maturation of naive clones with tetranomial diversity in solvent-exposed positions of the light-chain CDRs resulted in specific clones with low nanomolar affinities whose paratopes were dominated by Tyr, suggesting the importance of this residue in mediating recognition (Fellouse et al. 2006). Natural Abs are known to exhibit strong biases for particular amino acids within the CDR loops, including Tyr and Ser residues, with Tyr sidechains contributing to a disproportionate number of antigen contacts relative to other residues (Mian et al. 1991; Padlan 1994; Davies and Cohen 1996).

Based on these findings, extreme restriction of diversity was explored by randomizing similar solvent-exposed positions in the heavy-chain CDRs with only Tyr and Ser. This confirmed that specific clones could be obtained with affinities in the low nanomolar range (Fellouse et al. 2005). From this foundation, binary code diversity was expanded into CDR-L3, which resulted in further improvements in affinity (Fellouse et al. 2005). Further expansion of diversity into nonparatope residues that were not solvent-exposed but could provide structural diversity for CDRs, together with the use of tetranomial diversity in CDR-H3, enabled the isolation of clones with single-digit nanomolar affinities, comparable to the affinities obtained from natural sources (Fellouse et al. 2006). Systematic addition of tailored diversity and additional refinement to CDRs L3 and H3 resulted in subsequent iterations of the library from which specific, high (single-digit to subnanomolar) affinities could be routinely obtained (Fellouse et al. 2007; Birtalan et al. 2008; Persson et al. 2013; Hanna et al. 2020; Nilvebrant et al. 2021).

## STRUCTURAL ANALYSIS OF SYNTHETIC ANTIBODIES

The current, state-of-the-art synthetic Ab libraries (Fellouse et al. 2004, 2005, 2006; Persson et al. 2013) have been designed on two basic assumptions, which are directly informed by minimalist design principles that reflect the dominant functional usage of natural immune repertoires rather than the full diversity of human repertoires: first, that these libraries would yield Abs that interact with the antigen in a manner similar to that of natural Abs, preferentially using the same CDR positions and the same types of amino acids that are used to engage the antigen in a manner that is conformationally similar to that observed in nature, and second, that by limiting or eliminating altogether residues in natural repertoires that can impede therapeutic development, the developability of Abs isolated from the library is considered and ensured at the design stage.

Now, after more than two decades of successful development of Abs from synthetic libraries and the elucidation of many high-resolution Ab–antigen structures, we can examine whether this is the case and validate whether the design principles assumed are generally effective or require further optimization. Importantly, we can now explore whether there are unexpected interactions formed by synthetic Abs, positional preferences that can be exploited, conformational limitations imposed by the use of a single framework, or design elements that can be dispensed with if observed to be underused when designing the next generation of synthetic libraries. In this section, we review these questions by structural comparison of the published structures of 50 synthetic and 50 natural Ab–antigen complexes. In doing so, we provide a broad structural survey of the observed interactions and conformations formed by single-framework synthetic Abs, compare these to those formed by natural Abs, and illustrate the iterative design principles that can help to optimize future libraries.

## Data Set of Structures of Synthetic and Natural Ab–Antigen Complexes

To obtain a general overview of the structural details of antigen recognition by synthetic and natural Abs, two data sets were compiled, each containing approximately 50 structures for one of the classes (synthetic or natural) (Table 3). To ensure a meaningful and consistent analysis of the Ab–antigen interactions, the structures had to meet the following criteria: (1) X-ray crystal structures with resolution of <3.5 Å, (2) Ab–antigen interface of >500 Å$^2$ on each side, (3) Abs in Fab format bound to protein antigen, and (4) Abs containing a κ $V_L$. Although this latter restriction limits the scope of the study to only Abs bearing the κ $V_L$, it offers an opportunity to probe theoretical questions of relevance to the field of Ab structure and function and provides practical insight into the principles of library design. To eliminate potential bias, whereas the 50 structures analyzed for each data set corresponded to 50 unique Abs that used κ $V_L$, they otherwise sampled a variety of frameworks and possessed no significant similarity in CDR sequences.

To compile the synthetic Ab set, the Protein Data Bank (PDB) was manually searched to identify structures that met the above criteria and were derived from trastuzumab-based synthetic libraries designed by our group and others that used similar strategies for library construction (Paduch et al. 2013; Hornsby et al. 2015; Miersch et al. 2015, 2021). The resulting data set contained 50 unique Abs bound to 37 different antigens, including diverse proteins from humans, yeast, bacteria, and viruses (Fig. 1A; Table 3).

For the assembly of a complementary natural Ab data set, the Structural Antibody Database (SAbDab) was searched to compile a list of structures of human Abs in complex with protein antigens, restricting to those that, like trastuzumab, contained the γ-1κ framework but otherwise allowing for a diversity of heavy-chain frameworks, as shown in Table 3 (far right). The list was then manually curated to select structures with natural Abs isolated from human B cells, eliminating all Abs generated by synthetic methods, humanized Abs, and Abs with unclear origins. The final set contained 50 unique Abs bound to 36 unique antigens, among which a maximum of two Abs was allowed for each antigen to ensure a diverse set of interactions (Fig. 1B; Table 3). For example, although there were numerous structures corresponding to Abs in complex with the receptor-binding domain of SARS-CoV-2, only two of these Abs were chosen for analysis. In summary, eight of the 50 natural Abs possessed the same $V_H$ subgroup III and $V_L$ κ subgroup I frameworks as trastuzumab, whereas 10 $V_H$ III and 12 $V_L$ κ I were partnered with other frameworks, and all other frameworks were represented at least once, with $V_H$ subgroup III being the most abundant, followed by subgroup I. Similarly, all other $V_L$ subgroups were represented at least once in the set of natural Abs, with $V_L$ subgroups I and III being equally abundant.

## Analysis of Ab Residues Involved in Antigen Interaction

To characterize the Ab residues involved in antigen interaction, we determined the structural paratope for each Ab. Although the structural paratope is often defined as the solvent-accessible surface area (SASA) that is buried upon complex formation with the antigen (Richards 1977; Reis et al. 2022), a definition of an interatomic distance between Ab and antigen residues of <4.0 Å is also used to identify interacting residues (McConkey et al. 2003; Akbar et al. 2021) between the Ab and the antigen. Although the latter provides more detailed information regarding Ab–antigen interactions in the interface, it can exclude elements of the molecular surface that are nonetheless important for understanding how the paratope contributes to Ab properties. Given the ease of analyzing SASA, the fact that it captures a more comprehensive view of the paratope, and its widespread use by others (Chen et al. 2013; Mitternacht 2016; Hebditch and Warwicker 2019; Myung et al. 2023), this method, expressed mathematically below, was used exclusively:

$$\text{buried surface area} = \text{SASA}_{Ag} + \text{SASA}_{Fab} - \text{SASA}_{Ag-Fab}.$$

Each structure was thus analyzed using the GETAREA webserver (https://curie.utmb.edu/getarea.html) (von Freyberg et al. 1993), which further enables calculation of SASA per residue. Calculation

Chapter 7

TABLE 3. List of PDB structures selected for the comparison of synthetic and natural Abs

| Synthetic Abs | | $K_D{}^a$ nM | Natural Abs | | | Framework HC/LC |
|---|---|---|---|---|---|---|
| PDB | Antigen | | nM | Antigen | | |
| 1TZH | Human vascular endothelial growth factor (VEGF) | 1.8 | 4YDK | HIV-1 envelope CD4 binding site | | IGHV3/IGKV1 |
| 2QR0 | | 7.8 | 6MTO | HIV-1 envelope membrane-proximal external region (MPER) | | IGHV1/IGKV3 |
| 2FJG | | 1.5 | 4YWG | HIV-1 envelope V1–V2 region | | IGHV4/IGKV3 |
| 2FJH | | 12 | 7LM8 (2)$^b$ | SARS-CoV-2 spike glycoprotein RBD | | IGHV1/IGKV1 |
| | | | | | | IGHV5/IGKV1 |
| 3BDY | | 26 | 5GMQ | MERS-CoV spike glycoprotein RBD | | IGHV1/IGKV3 |
| 3PNW | Tudor domain containing protein 3 (TDR3) Tudor domain | 1 | 7KQG | Influenza B hemagglutinin (HA) | | IGHV3/IGKV1 |
| 4ZFG | Angiopoietin-2 (ANG2) | 5/5 | 4HG4 | Influenza A hemagglutinin (HA) globular domain | | IGHV1/IGKV1 |
| 7RTH | α-Lysozyme nanobody | 21 | 5IBL | | | IGHV5/IGKV3 |
| 3R1G | β-Secretase 1 (BACE1) | 1.3 | 3SDY | Influenza A hemagglutinin (HA) stem | | IGHV1/IGKV1 |
| 4JQI | β-Arrestin 1 (ARRB1) | NR | 6PZF | Influenza A neuraminidase (NA) | | IGHV3/IGKV1 |
| 3G6J | Complement component C3b | 1.2 | 6Q20 | | | IGHV3/IGKV1 |
| 1ZA3 | Death receptor 5 (DR5) | 34 | 6WO4 | Hepatitis C envelope glycoprotein E2 | | IGHV1/IGKV1 |
| 2H9G | | 2 | 6WO5 | | | IGHV1/IGKV2 |
| 3P0Y | Epidermal growth factor receptor (EGFR) | 1.9/0.4 | 5VIG | Zika virus envelope protein DIII | | IGHV3/IGKV1 |
| 3GRW | Fibroblast growth factor receptor 3 (FGFR3) | <1 | 6DFI | Zika virus envelope protein DIII | | IGHV4/IGKV3 |
| 6O39 | Frizzled class receptor 5 (FZD5) | 1.7/0.3 | 5VIC | Dengue 1 virus envelope protein DIII | | IGHV3/IGKV1 |
| 6O3A | Frizzled class receptor (FZD7) | 41/5.2 | 6OE4 | RSV virus prefusion F0 glycoprotein | | IGHV4/IGKV1 |
| 6O3B | | 57/4.5 | 6APB | RSV virus postfusion F0 glycoprotein | | IGHV2/IGKV1 |
| 2HFG | BLyS receptor 3 (BR3) | 0.6 | 6BL1 | RSV virus G peptide | | IGHV4/IGKV1 |
| 3N85 | Epidermal growth factor receptor (HER2) | <1 | 6UVO | | | IGHV4/IGKV3 |
| 3K2U | Hepatocyte growth factor activator (HGFA) | <1 | 6N8D | GII.4 2002 norovirus capsid P domain | | IGHV3/IGKV3 |
| 4IOF | Insulin-degrading enzyme (IDE) | 4 | 6N81 | | | IGHV3/IGKV1 |
| 4DKE | Interleukin-34 (IL-34) | 0.1$^c$ | 7L7R (2)$^b$ | Crimean–Congo hemorrhagic fever CCHFV Gc prefusion monomer | | IGHV3/IGKV1 |
| | | | | | | IGHV4/IGKV1 |
| 4DKF | | NR | 3CSY | Ebola glycoprotein | | IGHV3/IGKV4 |
| 5BO1 | Jagged canonical Notch ligand 1 (JAG1) | 0.8 | 6oz9 | | | IGHV4/IGKV3 |
| 7KEO | K29-linked di-ubiquitin (Ub) | 3.1 | 7S8H (2)$^b$ | Lassa virus glycoprotein 2 | | IGHV3/IGKV1 |
| | | | | | | IGHV4/IGKV3 |
| 3DVG | K63-linked di-ubiquitin (Ub) | 8.7 | 6B9J | Vaccinia virus D8 protein | | IGHV3/IGKV3 |
| 4K94 | Tyrosine protein kinase KIT domain D4D5 | 0.6 | 7LSE | Tick-borne encephalitis virus (TBEV) envelope domain III (EDIII) | | IGHV3/IGKV3 |
| 3SOB | Lipoprotein receptor-related protein 6 (LRP6) | NR | 5VOB | Human cytomegalovirus (HCMV) pentamer | | IGHV3/IGKV3 |
| 2QQK | Neuropilin-2 (NRP2) | 0.2 | 5Y11 | Severe fever with thrombocytopenia syndrome virus (SFTSV) glycoprotein N | | IGHV5/IGKV1 |
| 3L95 | Notch 1-negative regulatory region (NNR) | 2.5 | 5O14 | Factor H binding protein fHbp (vaccine against Men B) | | IGHV5/IGKV1 |
| 4I18 | Prolactin receptor (PRL) | 5.6 | 6XZW | | | IGHV3/IGKV3 |
| 6CW2 | Saccharomyces cerevisiae Ada2/Gcn5 | NR | 5O1R | Neisserial heparin binding antigen NHBA (4CMenB vaccine) | | IGHV4/IGKV3 |
| 6CW3 | | NR | 5D1Q | Staphylococcus aureus IsdB NEAT2 domain | | IGHV1/IGKV1 |
| 5EII | S. cerevisiae anti-silencing factor 1 (ASF1) | NR | 5D1Z | | | IGHV4/IGKV1 |
| 5UCB | | NR | 4IDJ | S. aureus α-hemolysin | | IGHV3/IGKV3 |
| 5UEA | | NR | 7CE2 | Tetanus toxin (TeNT) | | IGHV4/IGKV3 |
| 4XTR | S. cerevisiae Get3-Pep12 | <1 | 6PHB | Plasmodium falciparum vaccine Pfs25 | | IGHV4/IGKV1 |

(continued)

TABLE 3. Continued

| Synthetic Abs | | $K_D$[a] | | Natural Abs | | Framework |
| --- | --- | --- | --- | --- | --- | --- |
| PDB | Antigen | nM | | Antigen | nM | HC/LC |
| 3PGF | *Escherichia coli* maltose binding protein (MBP) | 5/67 | 6PHC | *Plasmodium falciparum* C-terminal domain of CSP (ctCSP) | | IGHV3/IGKV2 |
| 5BJZ | | 1.5/20 | 7RXP | *Plasmodium vivax* reticulocyte binding protein 2b (PvRBP2b) | | IGHV3/IGKV2 |
| 5BK1 | | 1/152 | 6WOZ | | | IGHV3/IGKV2 |
| 5BK2 | | 0.9 | 6WQO | | | IGHV4/IGKV1 |
| 5C8J | *E. coli* YidC | <1 | 6R2S | *Plasmodium vivax* Duffy binding protein (PvDBP) | | IGHV4/IGKV1 |
| 7MDJ | *Streptomyces lividans* KcsA | 3.7 | 6BFQ | Granulocyte–macrophage colony-stimulating factor (GM-CSF) | | IGHV4/IGKV1 |
| 4NNP | *S. aureus* MntC | 114/189 | 6RLO | CD9 large extracellular loop | | IGHV3/IGKV4 |
| 6CBV | *E. coli* BRIL | 0.3 | 1IQD | Human factor VIII | | IGHV1/IGKV3 |
| 7KLH | SARS-CoV-2 spike glycoprotein RBD | <1 | 5XMH | Human IgG1 Fc | | IGHV1/IGKV3 |
| 5EU7 | HIV-1 Integrase | NR | | | | |
| 5W2B | Ebola nucleoprotein | 4.1 | | | | |
| 6XMI | *Salmonella* phage 22 terminase | 72 | | | | |

(Abs) Antibodies.
[a]Literature $K_D$ values are provided for synthetic Abs to their target antigen where available. In most cases, $K_D$ values were not reported for natural Abs to their target antigen and thus are omitted.
[b]"(2)" indicates that two different Abs within the same PDB file were analyzed.
[c]Refers to the affinity matured Fab variant 1.1.

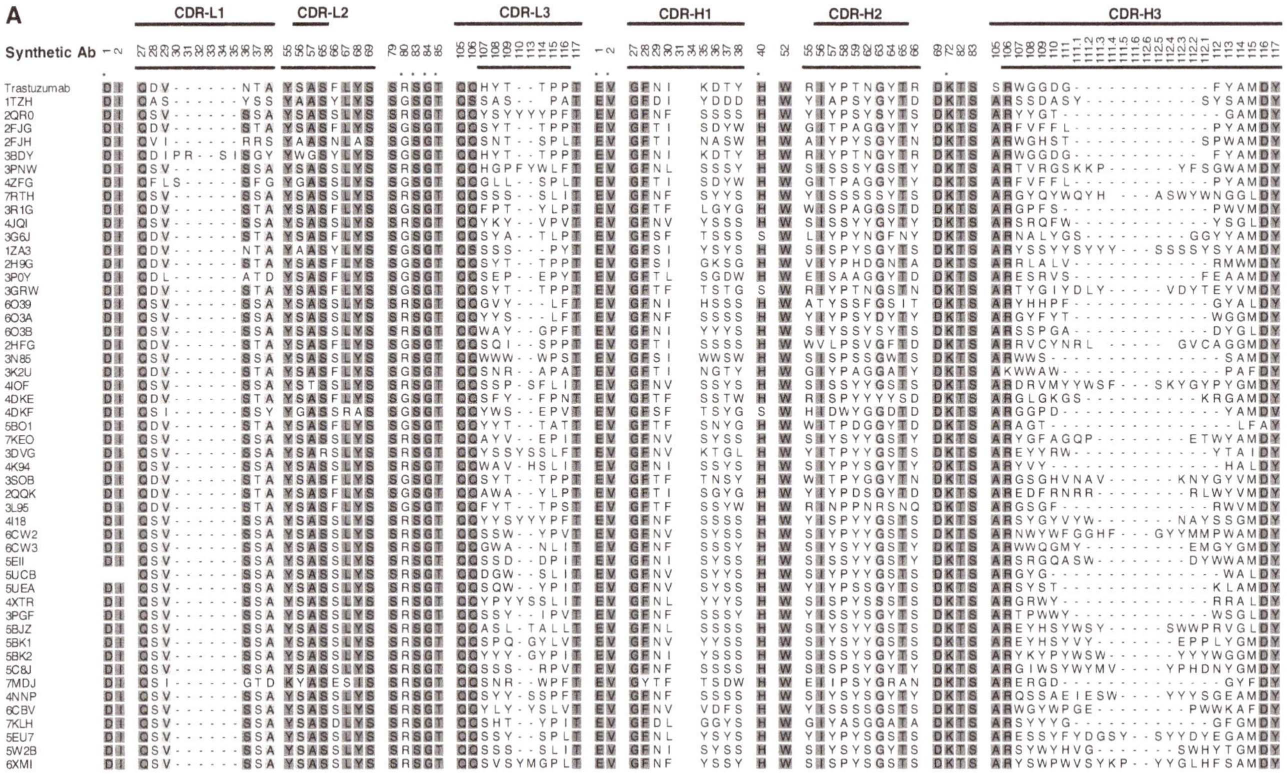

# Chapter 7

FIGURE 1. Alignment of $V_H$ and $V_L$ sequences for the antibody (Ab) data sets. Alignments are shown for synthetic Abs (A) and natural Abs (B). Only positions identified to be important for antigen recognition based on a minimum 0.5% contribution in the paratope in synthetic and/or natural Abs are shown. Positions with ≥90% or 70%–90% conservation across the 50 structures chosen per data set are shaded dark or light gray, respectively. Complementarity-determining regions (CDRs) are indicated, according to the existing IMGT definition, by bold lines *above* the position numbers and numbered according to the IMGT nomenclature (Lefranc et al. 2003). Bold lines *below* the position numbers indicate alternate CDRs, defined based on the analysis in Figure 2 as continuous stretches of amino acids containing positions that contribute to antigen interaction based on a minimum 0.5% contribution in the paratope in >5% of structures analyzed from the synthetic and/or natural antibody sets. Asterisks indicate residues found outside of CDR definitions that nevertheless contribute to antigen interaction in >5% of structures analyzed.

of the SASA at the residue level is a validated analytical method for determining SASA (Fraczkiewicz and Braun 1998), which has been used and referenced extensively as a viable means for performing these calculations (Xu and Zhang 2009; Al Mughram et al. 2021; Sraphet and Javadi 2022) and is superior to solvent-excluded surface-based methods (Cai et al. 2011).

In cases of multiple instances of Fab–antigen complexes within the asymmetric unit, a representative complex was chosen for analysis, and any crystallographic contacts between the complexes were ignored. The calculation was first performed on each complete PDB file, which calculated SASA for each residue in the Fab in the presence of the antigen. The PDB file was then edited to remove coordinates that corresponded to residues in the antigen and was analyzed again in the same way to calculate SASA in the absence of antigen. The structural paratope was then calculated as the change in SASA of each Fab residue upon complexation with the antigen.

This analysis revealed that the synthetic and natural Ab sets matched well in terms of the size range of structural paratopes. For the synthetic Abs, the sizes ranged from 555 to 1422 Å$^2$, with an average of 913 Å$^2$, whereas for the natural Abs, the sizes ranged from 536 to 1388 Å$^2$, with an average of 833 Å$^2$. Figure 2 summarizes the relative contribution of each position in the Ab to the structural paratope averaged over 50 structures each for synthetic and natural Abs. Positions are not shown if they did not contribute to antigen interaction or if they were only present in a small number of structures due to CDR length variability. Although CDR definitions have been devised based on structural and sequence analysis of Abs and Ab complexes, it has been acknowledged that these definitions (Kabat, Chothia,

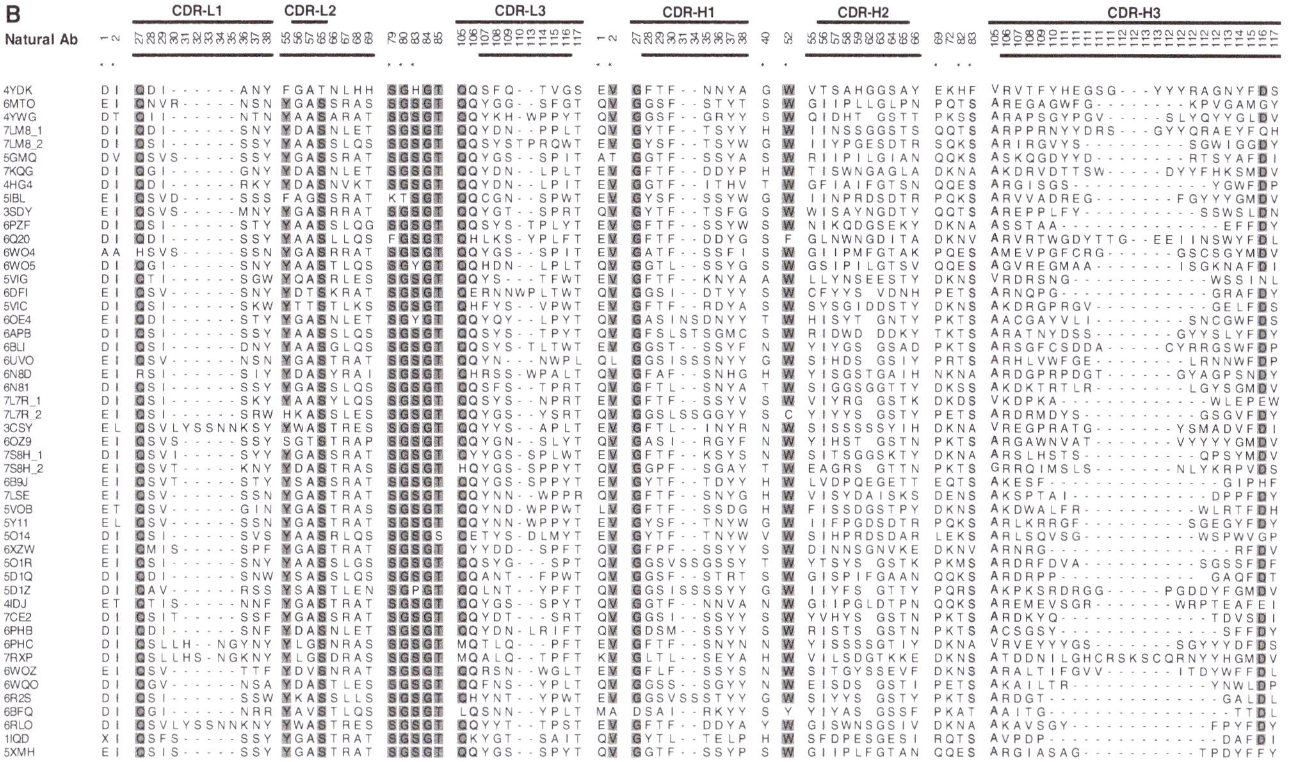

FIGURE 1. Continued.

IMGT) oversimplify, do not always accurately capture interface residues, and should be taken as an approximation of the paratope (Sela-Culang et al. 2013; Dondelinger et al. 2018). Thus, to more accurately compare the loop lengths and positions found within the paratope for the purposes of devising strategies for engineering functional Abs, the CDRs were defined here as continuous stretches containing amino acids that are found in the paratope, according to the SASA analysis, in at least some of the structures. Notably, similar approaches have been used by others to circumvent disparities between CDRs and paratopes (Kunik et al. 2012; Stave and Lindpaintner 2013), and ultimately, our analyses selected the same residues in the natural and synthetic Ab sets (see Fig. 3).

For each residue in the structural paratope, its contribution to antigen recognition was measured with two parameters. First, we calculated the percentage of structures where a residue at position X contributed to the interaction interface (Fig. 2, contributing residues %, black bars). To avoid the noise introduced by very minor interactions, a residue was only counted to contribute to antigen interaction if its relative contribution to the structural paratope was $\geq 0.5\%$. For example, if 25 out of 50 structures contained a residue at position X with $\geq 0.5\%$ relative contribution to the structural paratope, the contributing residue percentage value at position X was calculated to be 50%. Second, we calculated the relative contribution to the structural paratope at position X averaged over the 50 structures (Fig. 2, average contribution %, white bars). For example, if at position X, a residue contributed 5% to the structural paratope in Structure 1 and 3% to the structural paratope in Structure 2, then the average contribution percentage value at position X was calculated to be 4% (this example is for two structures, but analysis was performed for all 50 structures). Overall, we saw good correlation between these two measures in Figure 2, insofar as the more frequently a residue was observed to contribute to the paratope, the greater its average contribution, with correlation coefficients of $R > 0.85$ both for $V_H$ and $V_L$, across natural and synthetic Abs. Although these measures reflect the general importance of a given position in the structural paratope, they nevertheless provide different information. The former describes which residues tend to be used and how frequently in different structures, whereas the latter measures the contribution each residue tends to make to the overall paratope within a structure.

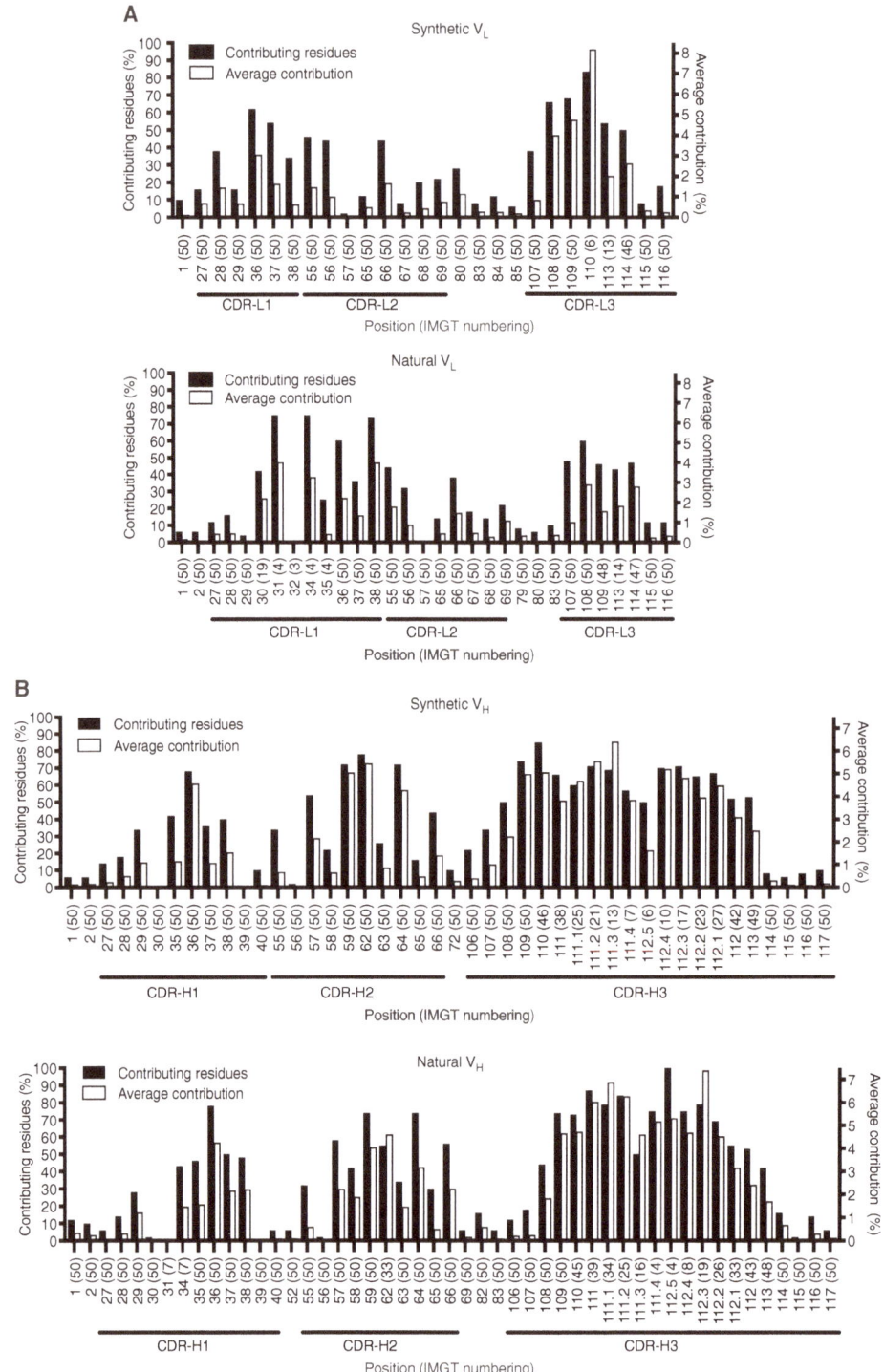

FIGURE 2. Relative contributions of antibody (Ab) residues to the structural paratope for the analyzed synthetic and natural Abs. Data are plotted for light-chain variable ($V_L$) domains (A) and heavy-chain variable ($V_H$) domains (B). At each position (x-axis), the black bars (left y-axis) show the percentage of contributing residues (contributing residues %), and the white bars (right y-axis) show average relative contribution to binding interface (average contribution %). The contributing residue percentage was calculated by dividing the number of structures with ≥0.5% relative contribution to the structural paratope at position X by the total number of structures available for position X. Average contribution percentage was obtained by calculating relative contribution to the structural paratope at position X for each structure and then averaging these values. Note that the relative contribution values for each position are not normally distributed but follow a strong positive skew distribution, with a significant proportion of values being 0 and standard deviations exceeding the average values. Thus, the average values shown are not meant to represent the most observed values for a given position but rather provide some measure of comparison between synthetic and natural data sets. Only positions with the contributing residues ≥5% and that have residues in ≥5% of structures are shown. Complementarity-determining regions (CDRs) are indicated and are defined as the continuous stretches of amino acids containing positions that contribute to the structural paratope in ≥5% of structures analyzed. The total number of structures that contain a residue at a given position is indicated in parentheses. Positions are numbered according to the IMGT nomenclature (Lefranc et al. 2003).

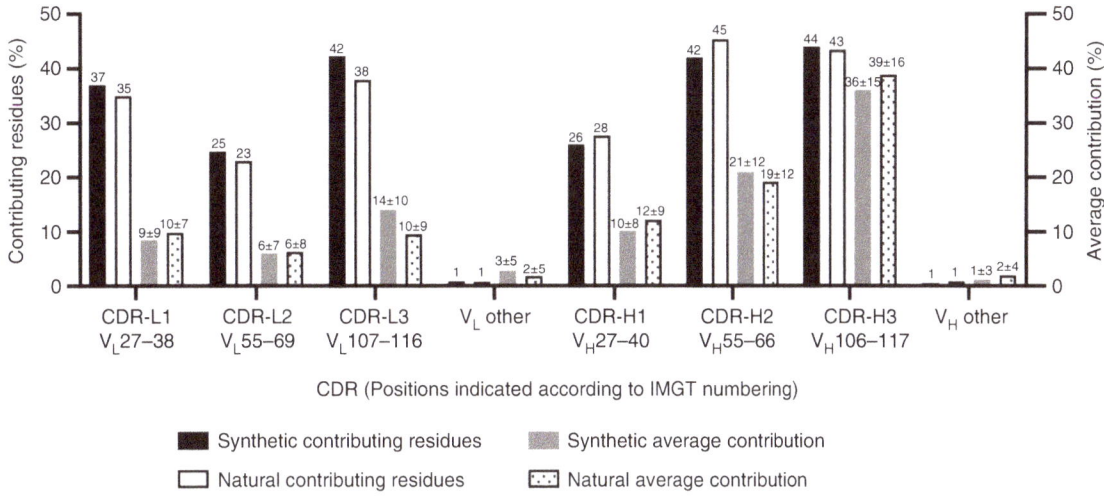

FIGURE 3. Relative contributions of complementarity-determining regions (CDRs) to the structural paratope for the analyzed synthetic and natural antibodies (Abs). The CDRs were defined based on the analysis described in Figure 2, and the positions assigned to each CDR are indicated on the x-axis. For each CDR (x-axis), the black bars (left y-axis) show the percentage of contributing residues (contributing residues %), and the white bars (right y-axis) show average relative contribution to the structural paratope (average contribution %). The contributing residue percentage was obtained by calculating the total number of residues with ≥0.5% relative contribution to the structural paratope within a given CDR in the 50 structures and dividing it by total number of residues within the same CDR in the 50 structures (values are shown *above* the bars for clarity). The average contribution percentage was obtained by calculating relative contribution of a given CDR to the structural paratope in each structure and then taking an average over 50 structures. The values of averages ± standard deviation are shown *above* the bars. Note that large standard deviations are reflective of strong positive skew of the distribution, where the majority of values are lower than the average and even include some 0 values. The same analysis was performed for all positions outside the CDRs, which were grouped as light-chain variable ($V_L$) other and heavy-chain variable ($V_H$) other for $V_L$ and $V_H$ domains, respectively.

The analyses described above allowed us to directly compare the structural paratopes of the curated synthetic and natural Ab data sets (Fig. 2). Remarkably, we found that, at the level of individual positions, the interaction landscapes were very similar. More specifically, the degree to which each position tends to contribute to the structural paratope in the synthetic Abs closely mirrors natural Abs, with some exceptions. In the $V_L$ domain (Fig. 2A), CDRs L2 and L3 are especially similar between the two data sets, with positions 55, 56, and 66 in CDR-L2 and positions 107–114 in CDR-L3 dominating antigen interactions. CDR-L1 interactions are less similar, with natural Abs tending to have longer CDR-L1 loops and position 28 playing a more important role in synthetic Abs, whereas the opposite is true for position 38. This is a likely a consequence of synthetic library design that eschewed diversification in L1 (Persson et al. 2013) to focus it on the more frequently used CDR-L3. In the $V_H$ domain (Fig. 2B), positions 35–38 in CDR-H1; positions 57, 59, 62, 64, and 66 in CDR-H2; and positions 108–113 in CDR-H3 dominate antigen interactions in both data sets. However, positions 58 and 65 in CDR-H2 play a more prominent role in natural Abs, which is likely a consequence of the diversity designs used in the synthetic libraries.

Consistent with the overall similarity at the level of individual positions, a more comprehensive look at the relative contributions of individual CDRs toward antigen interaction (Fig. 3) reveals a similar picture for the synthetic and natural Abs. Based on the number of residues contributing to the structural paratope versus the total number of CDR residues over the 50 structures (i.e., the contributing residue percentage), the highest proportion of residues used is found in CDRs H3, H2, and L3, in which nearly half of the their CDR residues participate in the paratope, whereas only roughly one-quarter of residues in CDRs H1 and L2 are used, and usage is closely mirrored between synthetic and natural Abs. In considering the average contribution percentage of the CDR to the overall paratope, CDR-H3 is, as expected, dominant, contributing 35%–40% of the overall paratope, followed by CDR-H2, which uses roughly half that. CDRs H1, L3, L2, and L1 constitute an average of only 5%–15% of the overall CDRs, again with no substantial differences in usage between natural and synthetic Ab sets. Given the drastically different origins of the synthetic and natural Abs, the level of conservation in CDR usage observed between the two data sets is unexpected. Detailed examination of small differences that are observed will be helpful to inform the design of future naive synthetic libraries and libraries for affinity maturation.

To determine whether additional opportunities for novel library design strategies exist, we explored whether residues outside the conventionally defined CDRs are involved in antigen recognition. Figure 2 shows that in both synthetic and natural data sets, there are few positions outside the CDR boundaries that contribute to antigen recognition. However, these are worth mentioning, as they present opportunities to mediate antigen recognition through residues outside the CDRs that are normally diversified. The position outside of the traditional CDR definitions with the most significant paratope contribution is $V_L$ position 80, which lies between CDRs L2 and L3 and corresponds to the recently identified CDR-L4 (Kelow et al. 2020). We observed this, however, only in the synthetic Abs, which tend to have an Arg residue at this position (Figs. 1 and 2). In natural Abs, on the other hand, position 80 is often a Gly residue, and although it is sometimes involved in antigen recognition, it makes contributions to the paratope much less frequently, presumably due to its lack of a sidechain. In addition to residue 80, other residues in this loop can be found in the paratopes of both natural and synthetic Abs. In the heavy chain, residues 82 and 83 of the H4 loop contribute to some natural paratopes, whereas synthetic Abs do not make use of this loop (Fig. 2B). Interestingly, both heavy and light chains make use of residues at the base of the ascending D strand, distal to the DE loop, and the two N-terminal positions of both the $V_H$ and $V_L$ domains are observed to be included in paratopes in both sets of Abs, albeit in rare instances. Each of these observations suggest additional opportunities for diversification outside of defined CDRs in future libraries.

## Core Structural Paratope

For a more detailed look at the most crucial positions involved in antigen recognition, we defined a core structural paratope and examined which amino acids tend to contribute to antigen recognition at these positions (Fig. 4). The core positions were defined as those that are present in the paratope of at least 60% of structures and contribute to antigen recognition (relative contribution to binding interface ≥0.5% of structural paratope area) in at least 20% of structures. This stringent definition excludes positions that are not present in a majority of analyzed paratopes and eliminates those that make only minor contributions in few structures. The use of change in SASA as a means of defining a core paratope has been previously validated by determining and demonstrating strong correlation with the confidence level of protein–protein interactions for surface atoms (Peng et al. 2014). Thus, these criteria were used to define the core structural paratope in synthetic Abs, and the same positions were analyzed within the natural Abs for comparison.

Comparison of the types of amino acids involved in antigen recognition at different positions between synthetic and natural Ab data sets revealed similarities and differences (Fig. 4A). In both sets, Ser and Tyr residues contribute to antigen interaction at most positions, which is especially true for positions outside CDR-H3. However, Trp residues, which often contribute to antigen interaction in synthetic paratopes, are much rarer in the examined natural paratopes. This is especially evident upon comparison of CDR-H3 loops and is an important consideration given prior observations that Trp, although capable of mediating high-affinity interactions (Birtalan et al. 2010), can be detrimental to specificity (Birtalan et al. 2010; Kelly et al. 2018) and may also be omitted from libraries without compromising affinity (Kelly et al. 2018).

Mapping of the core paratope onto the Fab structure revealed that the examined synthetic and natural Abs are very similar in terms of the relative contributions of individual paratope residues to antigen recognition (Fig. 4B). However, synthetic Abs relied more heavily on aromatic residues (predominantly Tyr and Trp) for antigen recognition, whereas, in natural Abs, greater positional diversity was observed, and aromatic residues played a lesser role.

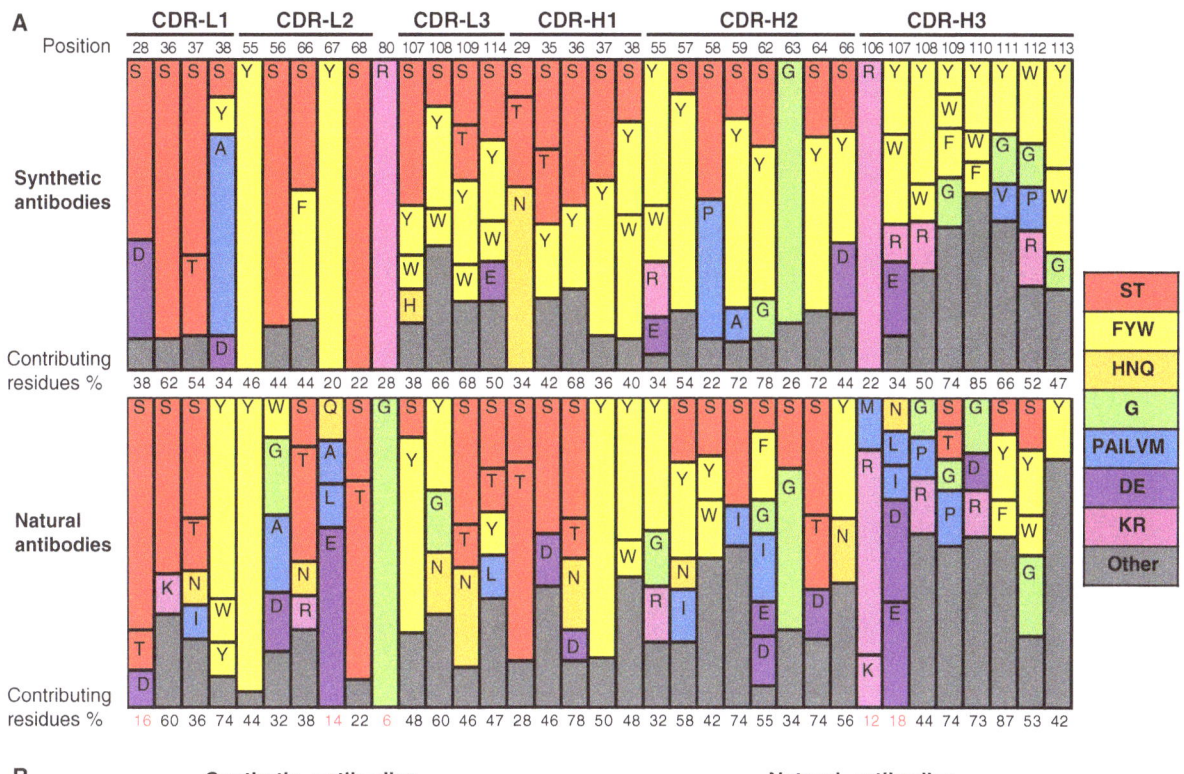

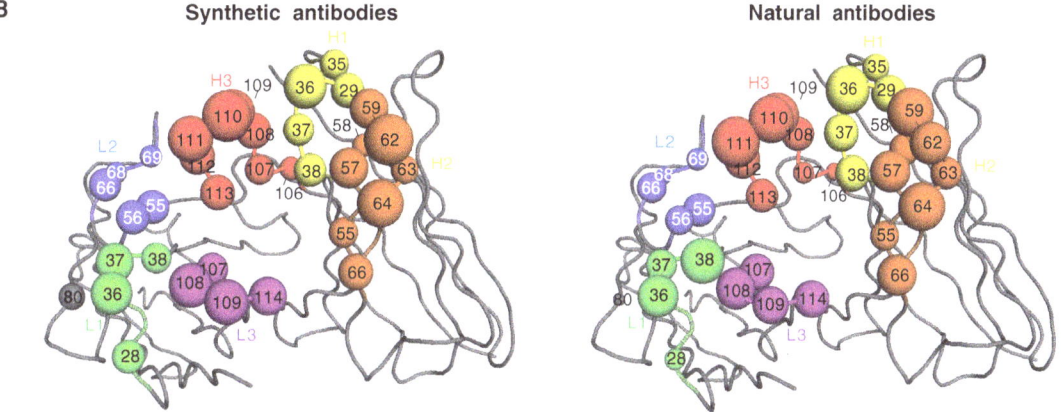

FIGURE 4. Core structural paratope of the analyzed synthetic antibodies (Abs). The core structural paratope of the analyzed synthetic Abs was defined as positions that are present in ≥60% of the selected structures and contribute to antigen recognition in ≥20% of the structures. The same positions in natural Abs are shown for comparison. (*A*) For each position in the core structural paratope of synthetic Abs, the amino acid distributions are shown for synthetic Abs (*top*) and natural Abs (*bottom*). Amino acids that are present at ≥10% frequency are shown separately (colored bars), whereas amino acids present at <10% frequency are grouped together as "other" (gray bars). The length of each bar is proportional to the frequency, and the total length of the bars at each position is equal to 100%. (*B*) The core structural paratope of synthetic Abs mapped onto the structure of the trastuzumab antigen-binding fragment (Fab) (Protein Data Bank entry 1N8Z). The size of each sphere is proportional to the frequency at which residues at that position contribute to antigen recognition in synthetic Abs (*left*) and natural Abs (*right*).

Considering that the evaluated synthetic Abs were derived from libraries built with the trastuzumab framework and were diversified at specific positions with tailored ratios of amino acid subsets, it is not surprising to see differences in amino acids engaged in antigen recognition at some positions. At the same time, it is remarkable that other positions tend to use the same amino acids for antigen engagement within both data sets, suggesting some selection pressure for these residues. Both the differences and similarities highlighted by our analyses may help inform the design of future synthetic

Ab libraries with desirable properties. For example, shifting the diversification schemes away from aromatic residues toward more polar amino acids observed within natural Abs may result in libraries with more hydrophilic paratopes, contributing to a higher yield of Abs with good developability properties. Furthermore, fixing certain positions toward a particular amino acid that tends to prevail in antigen recognition within natural Abs may produce libraries with more focused diversification schemes and with a better chance of generating functional Abs.

## Length and Composition of CDRs L3 and H3

To further compare the synthetic and natural Ab data sets, we focused on CDRs L3 and H3 (Fig. 5). In both natural and synthetic Abs, these CDRs are the most variable in terms of length and amino acid composition (Persson et al. 2013), and this observation holds true in both the natural and synthetic Ab data sets. Overall, the CDR-3 length distributions exhibited similar patterns between the synthetic and natural Ab data sets (Fig. 5A). In the case of CDR-L3, six-residue loops dominated both data sets, accounting for 66% and 62% of synthetic and natural Abs, respectively. The lengths of CDR-H3 loops were more variable, with loops ranging from eight to 22 residues being present in both data sets. However, there were biases for some loop lengths in each data set, with 12-residue loops and 14- or 17-residue loops being highly prevalent in synthetic or natural Abs, respectively. Notably, most natural and synthetic CDR-L1 loops contained five residues, but a substantial number of natural CDR-L1 loops contained six residues and some contained nine, 10, or 11 residues. Interestingly, this

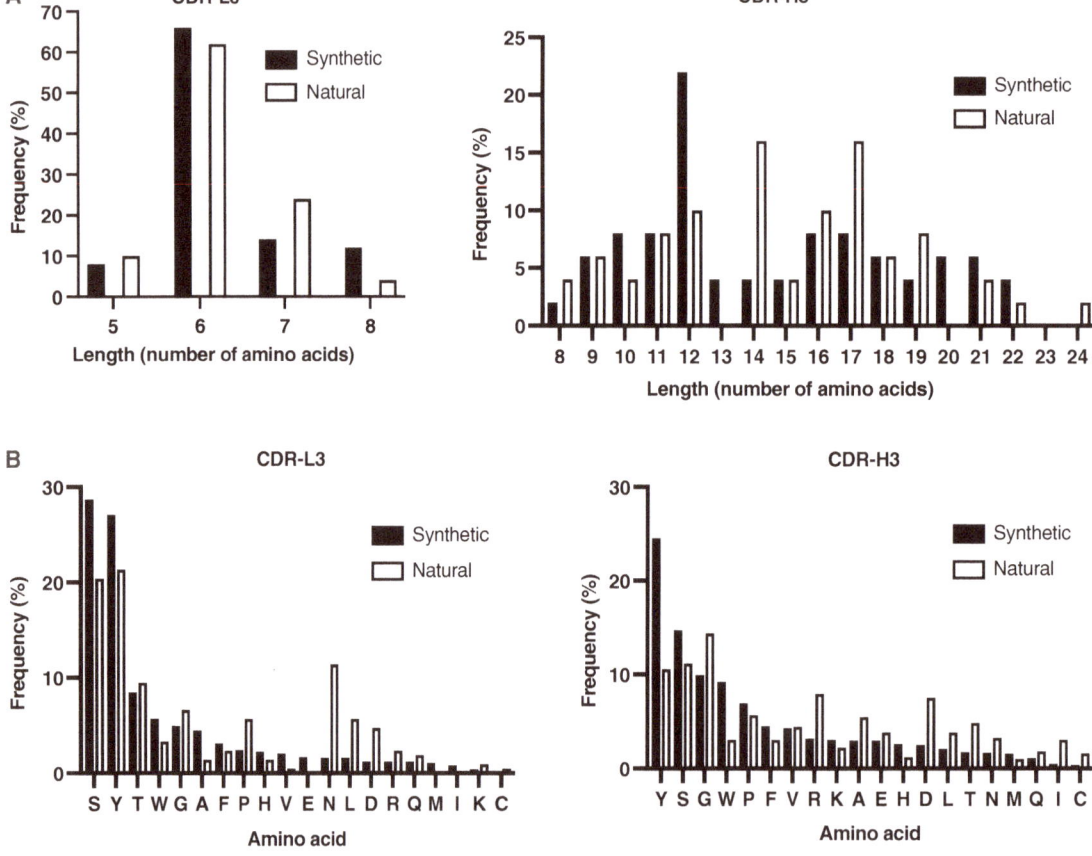

FIGURE 5. Analysis of complementarity-determining regions (CDRs) L3 and H3. (A) Distribution of CDR-L3 lengths (positions 107–116, *left*) and CDR-H3 lengths (positions 106–117, *right*) for the selected synthetic and natural antibodies (Abs). (B) Amino acid composition of CDR-L3 (positions 107–114, *left*) and CDR-H3 (positions 107–113, *right*) for the selected synthetic and natural Abs. The percent abundance within the CDR is shown for each amino acid, and the amino acids are ordered from the most to least abundant within the synthetic Abs.

difference arises from fixing both the length and sequence of CDR-L1 in synthetic libraries. In contrast, CDR-L2 displayed highly similar length diversity in both Ab sets despite being fixed in synthetic libraries, suggesting that length diversity, as designed, is dispensable here.

Consistent with the above analysis of structural paratope residues, comparison of the amino acid composition within CDR-3 loops of natural and synthetic Abs revealed both similarities and differences (Fig. 5B). Within CDRs L1 and L2, the amino acid frequencies are very similar at most positions despite being fixed within the synthetic libraries, suggesting strong natural selection of these residues, with the exception of residues 56 and 68 in CDR-L2, which display significant diversity in the paratopes of natural Abs. Within CDR-L3, Ser and Tyr are the most abundant amino acids in both synthetic and natural data sets, and Asn is third most prevalent in natural Abs but is rare in synthetic Abs. Interestingly, despite the prevalence of Asn residues in the CDR-L3 loops of natural Abs, there are no glycosylation sites in the CDR-L3 loops of any of the 50 Abs in our data set (Fig. 1). In CDR-H3 loops, Gly and Arg are abundant and are observed with similar frequencies between the synthetic and natural data sets, though Gly is prevalent in the other CDRs of natural Abs. The aromatic amino acids Tyr and Trp dominate CDR-H3 structural paratope positions in the synthetic Abs (Fig. 4). Conversely, the negatively charged amino acid Asp is fairly common within the CDR-H3 loops of natural Abs but is absent in the synthetic Abs, and overall, Asp appears to make a greater contribution to the paratopes of natural Abs relative to their synthetic counterparts. Notably, Cys residues were found to be least frequent in CDRs L3 and H3 in both data sets. This suggests that during evolution of natural Abs, there is a strong selective pressure against unpaired Cys residues that could form spurious disulfides, and in synthetic Abs, Cys residues are excluded by design.

## Canonical Conformations of CDR Loops

As Ab structures began to accumulate, the structural analysis of CDRs revealed that they adopt a small number of main-chain conformations that were classified as canonical structures (Chothia and Lesk 1987). As more structures have accumulated, the original classification has been updated, expanded, and refined (Martin and Thornton 1996; Al-Lazikani et al. 1997; North et al. 2011; Kelow et al. 2022), and web tools now facilitate CDR analysis and assignment to "standard conformations" (Adolf-Bryfogle et al. 2015).

The different approaches to generating synthetic libraries, whether on a single framework or on a multitude of frameworks, raises an interesting question regarding conformational representation. Although it has been argued intuitively that single-framework libraries "lose the structural diversity present across the different frameworks of natural Abs" (Rothe et al. 2008) and that their "structural diversity does not approach that of other naive libraries" (Knappik et al. 2000), these statements have not been supported with evidence.

To explore this question, the canonical cluster assignments were obtained for each of the CDRs in the data sets and analyzed for comparison of the diversity of conformations represented in the natural and synthetic Abs using PyIgClassify2 (Adolf-Bryfogle et al. 2015; Kelow et al. 2022). Plots of the assigned conformations versus the number of times they were observed in each data set were determined for each of CDRs (Fig. 6). In CDRs L1, L2, and H1, clustering was similarly dominated by a single conformation in both data sets, suggesting that conformational diversity in these CDRs is limited despite the variety of the frameworks, and that CDRs L1 and L2 are fixed for length and sequence in most analyzed synthetic Abs. Notable differences were observed in CDR-L1, which in addition to the dominant L1-11-1 conformation, had a preference for L1-12-1 or L1-11-* in natural or synthetic Abs, respectively. In CDR-H2, while the H2-10-1 conformation was the most represented in both sets, the natural set also exhibited similar representation of H2-9-1 and H2-10-1 and lesser but significant representation from H2-9-*, H2-10-*, and H2-10-2, suggesting some conformational limitations in CDR-H2 of synthetic Abs. Remarkably, CDR-L3 in both the synthetic and natural sets is dominated by the L3-9-cis7-1 conformation, with >40% of the Abs in each set assigned as such. The remaining diversity was dispersed throughout a wide variety of clusters, both overlapping and unique for each set. In contrast, the most observed CDR-H3 clusters in the natural and synthetic sets were H3-18-* or H3-13-*, respectively, indicating different loop length preferences. Beyond that, no

Chapter 7

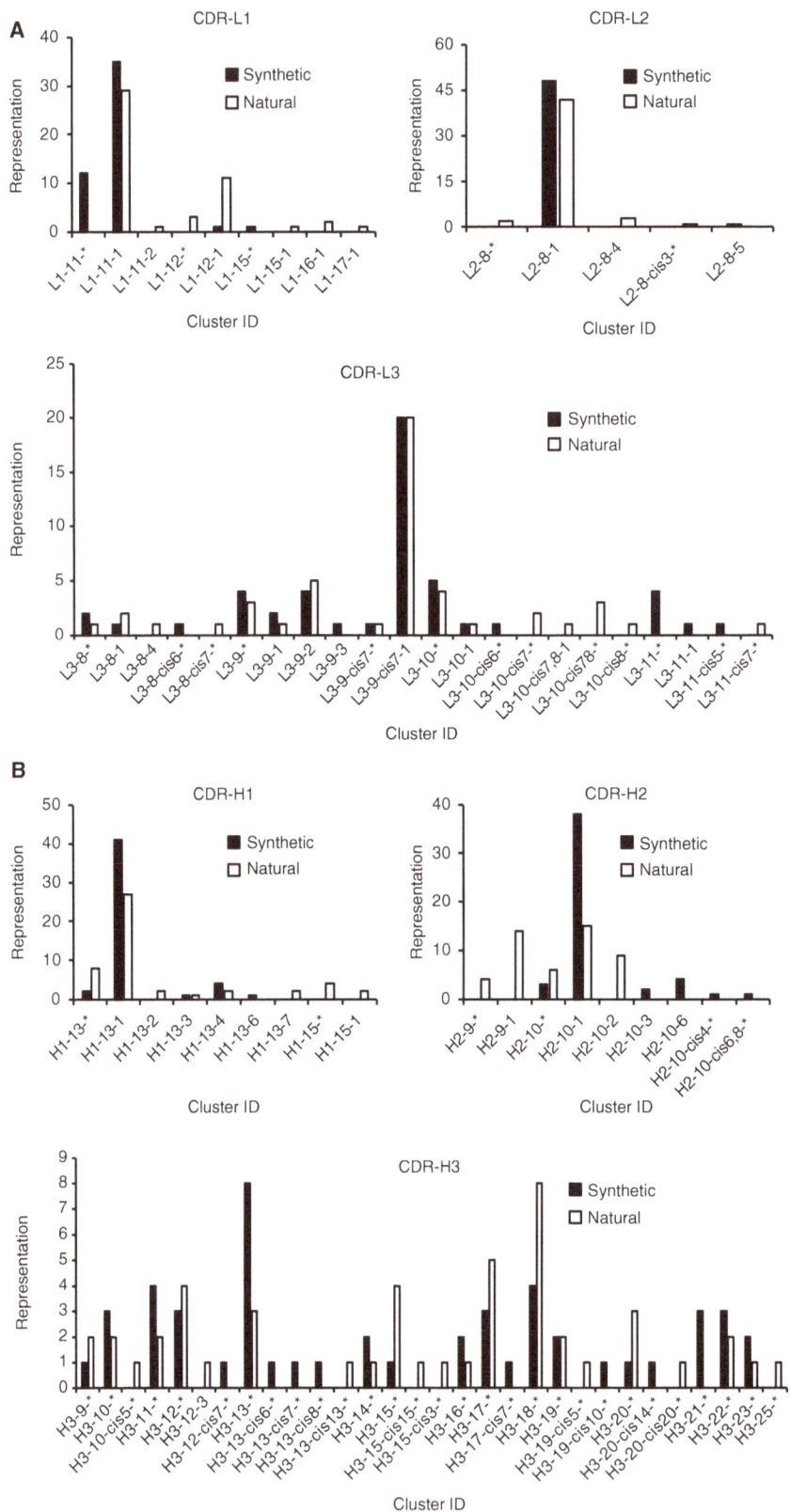

FIGURE 6. Classification of antibody (Ab) complementarity-determining regions (CDRs) into canonical clusters. Canonical cluster assignments for all six CDRs were determined from the Protein Data Bank files for each of the Ab:Ag complexes using PyIgClassify 2 (Kelow et al. 2022). The number of times a cluster ID (x-axis) was observed in the structure set was tallied (y-axis) and plotted for each assignment for both the synthetic and natural Ab sets.

other cluster was dominant, and each set displayed diversity of conformational clusters with overlapping and distinct clusters in each.

Contrary to intuitive assertions (Knappik et al. 2000; Rothe et al. 2008), the results of our quantitative survey of canonical structures show that Abs derived from a single-framework library are not conformationally constrained relative to the 13 different framework pairs represented in the natural Ab set. Recognizing that these synthetic libraries have generated Abs to thousands of structurally different antigens, these observations are in accord with empirical success. Overall, these findings suggest that single-framework libraries appear to possess no immunological blind spots that limit the breadth of antigens that can be targeted with high affinity and specificity.

## Limitations

Although our survey provides the most extensive structural comparison of natural and synthetic Abs to date, inherent biases in the data sets chosen for analysis limit the broader interpretation of the results. First, the exclusive use of the κ-light-chain isotype in the synthetic libraries chosen for analysis skews the data and precludes application of the findings to Abs containing the λ light chain. Several groups have noted differences in the biophysical, sequence, and structural properties of κ- and λ-light-chain Abs (DeKosky et al. 2016; Townsend et al. 2016; Raybould et al. 2019), and thus there is no expectation that the observations and conclusions made here would extrapolate beyond other κ Abs. The incorporation of λ frameworks into other synthetic libraries (Knappik et al. 2000; Rothe et al. 2008; Prassler et al. 2011), however, creates a similar opportunity to structurally compare natural versus synthetic λ-light-chain Abs to determine whether biophysical and structural differences observed in nature are maintained or altered based on the synthetic diversity introduced. Second, though the set of κ natural Abs sourced for this study is small relative to the available κ Ab structures found in the PDB, the broad diversity of CDR-3 loop lengths, sequences, and conformations observed in the set analyzed suggests that any unintentional biases introduced by structure complex selection do not appreciably skew the data obtained for the natural set of Abs. Last, by focusing strictly on a structural approach to analyzing the success of library design, this study is biased toward functional, high-affinity Abs. However, this provides little insight into the physicochemical behavior of the Abs obtained. Early iterations of the library provided tight, specific binding clones that were, however, dominated by Tyr in the paratope. Likely, such a clone would be poorly developable, and thus alternate measures of library success that are not captured by this methodology would also be needed to provide a more comprehensive assessment of library quality.

## CONCLUSIONS AND PERSPECTIVES

To date, natural Abs have dominated the therapeutic and structural landscape, with most Ab therapeutics and structures published being derived from natural rather than synthetic sources. Inarguably, the lessons learned from the sequencing of natural Ab repertoires have informed the design of synthetic libraries, and most synthetic approaches tend to emulate natural Ab repertoires to varying degrees (Knappik et al. 2000; Sidhu et al. 2004; Zhai et al. 2011). Early designs focused on emulating Ab sequences from genomic sources (Johnson and Wu 2000; Sidhu et al. 2004), presumably with varying degrees of functional and structural adaptation. More recently, herculean efforts to sequence the peripheral $IgM^+$ B-cell compartment (containing a mix of naive, memory, and plasma cells) (Glanville et al. 2009) continue to inform Ab library design and reflect both naive unadapted and functionally adapted repertoires. These approaches have undoubtedly achieved the first aim of a synthetic library—the development of synthetic repertoires that closely resemble natural repertoires—and enable the generation of specific and high-affinity Abs targeting highly diverse antigens.

Insofar as natural Abs can also exhibit developability liabilities (Jain et al. 2017; Raybould et al. 2019), design strategies aimed at optimizing synthetic libraries for therapeutic applications must also consider which elements of the natural repertoire are to be emulated and which are not. Thus, from a structural perspective, although it is important to determine whether the diverse landscape of Ab interactions in which natural Abs participate is reflected in the paratopes of synthetic Abs, this, however, reflects only the functional side of the balance between function and developability that ensures that a functional Ab is of sufficient quality for therapeutic applications. As synthetic libraries are increasingly used for therapeutic development, design strategies must also ensure that the properties possessed by the Abs are amenable to development; in other words, that they can be produced in high yield and in soluble and monodisperse form, and that they possess low nonspecific or self-binding that could otherwise compromise manufacturing or good pharmacokinetic behavior.

With these issues in mind, we described here a detailed structural analysis of 50 Fab–antigen complexes with synthetic Abs derived from the minimalist, single-framework libraries described here. By conducting the same analysis on a set of 50 natural Fab–antigen complexes that included Abs with diverse frameworks found in human Ab repertoires, the objective was to assess a variety of interface measures that would enable comparison of natural and synthetic paratopes. By comparing the participation of specific positions, amino acid types, and CDRs, we aimed to obtain insights that could be used to devise strategies for the design of new iterations of already highly optimized synthetic libraries.

Overall, the structural interaction landscapes were remarkably similar between the synthetic and natural Ab data sets, given the highly restricted nature of these minimalist libraries. Natural and synthetic Abs used similar positions with similar trends in frequency observed (Fig. 2) and made similar use of and contribution to the paratopes (Fig. 3). Differences and similarities in the observed positional diversity between natural and synthetic Abs suggest opportunities for library refinement and strategies for optimization to enhance or reduce "naturalness" where appropriate (Figs. 4 and 5). An unexpected result of these analyses, however, is the largely equivalent use of conformational diversity in the two Ab sets, suggesting that the single-framework approach to synthetic library generation does not constrain conformational diversity, as has previously been asserted (Fig. 6).

In conclusion, our structural analyses suggest that Abs derived from synthetic Ab libraries with a trastuzumab framework interact with antigens in a manner similar to natural Abs but with some key distinctions. We conclude that, in many aspects, synthetic Ab libraries successfully recapitulate much of the structural landscape of natural Ab–antigen interactions, both in the way that CDRs are used in paratopes and in the conformational diversity exhibited by CDRs. The similarities and differences that were observed between the two sets, particularly with regards to residue usage, provide a rational path to library optimization, but as has been emphasized previously, there is a trade-off between specificity, affinity, solubility, and stability, and thus the priorities of any design strategy must be carefully considered. Structural analysis nevertheless provides a critical reflection on the quality of synthetic Ab libraries, and the insights obtained here provide ample suggestions on ways to further improve both the functional performance and developability of the Abs they generate.

## REFERENCES

Adolf-Bryfogle J, Xu Q, North B, Lehmann A, Dunbrack RL. 2015. PyIgClassify: a database of antibody CDR structural classifications. *Nucleic Acids Res* **43**: D432–D438. doi:10.1093/nar/gku1106

Akbar R, Robert PA, Pavlović M, Jeliazkov JR, Snapkov I, Slabodkin A, Weber CR, Scheffer L, Miho E, Haff IH, et al. 2021. A compact vocabulary of paratope-epitope interactions enables predictability of antibody-antigen binding. *Cell Rep* **34**: 108856. doi:10.1016/j.celrep.2021.108856

Alfaleh MA, Alsaab HO, Mahmoud AB, Alkayyal AA, Jones ML, Mahler SM, Hashem AM. 2020. Phage display derived monoclonal antibodies: from bench to bedside. *Front Immunol* **11**: 567223. doi:10.3389/fimmu.2020.01986

Al-Lazikani B, Lesk AM, Chothia C. 1997. Standard conformations for the canonical structures of immunoglobulins. *J Mol Biol* **273**: 927–948. doi:10.1006/jmbi.1997.1354

Al Mughram MH, Herrington NB, Catalano C, Kellogg GE. 2021. Systematized analysis of secondary structure dependence of key structural features of residues in soluble and membrane-bound proteins. *J Struct Biol* **5**: 100055. doi:10.1016/j.yjsbx.2021.100055

Barbas CF, Bain JD, Hoekstra DM, Lerner RA. 1992. Semisynthetic combinatorial antibody libraries: a chemical solution to the diversity problem. *Proc Natl Acad Sci* **89**: 4457–4461. doi:10.1073/pnas.89.10.4457

Birtalan S, Zhang Y, Fellouse FA, Shao L, Schaefer G, Sidhu SS. 2008. The intrinsic contributions of tyrosine, serine, glycine and arginine to the affinity and specificity of antibodies. *J Mol Biol* **377**: 1518–1528. doi:10.1016/j.jmb.2008.01.093

Birtalan S, Fisher RD, Sidhu SS. 2010. The functional capacity of the natural amino acids for molecular recognition. *Mol Biosyst* **6**: 1186–1194. doi:10.1039/b927393j

Bond CJ, Wiesmann C, Marsters JC, Sidhu SS. 2005. A structure-based database of antibody variable domain diversity. *J Mol Biol* **348**: 699–709. doi:10.1016/j.jmb.2005.02.063

Briney B, Inderbitzin A, Joyce C, Burton DR. 2019. Commonality despite exceptional diversity in the baseline human antibody repertoire. *Nature* **566**: 393–397. doi:10.1038/s41586-019-0879-y

Burkovitz A, Ofran Y. 2016. Understanding differences between synthetic and natural antibodies can help improve antibody engineering. *MAbs* **8:** 278–287. doi:10.1080/19420862.2015.1123365

Cai Q, Ye X, Wang J, Luo R. 2011. On-the-fly numerical surface integration for finite-difference Poisson-Boltzmann methods. *J Chem Theory Comput* **7:** 3608–3619. doi:10.1021/ct200389p

Carter P, Presta L, Gorman CM, Ridgway JBB, Henner D, Wong WLT, Rowland AM, Kotts C, Carver ME, Shepard HM. 1992. Humanization of an anti-p185HER2 antibody for human cancer therapy. *Proc Natl Acad Sci* **89:** 4285–4289. doi:10.1073/pnas.89.10.4285

Chen J, Sawyer N, Regan L. 2013. Protein–protein interactions: General trends in the relationship between binding affinity and interfacial buried surface area. *Protein Sci* **22:** 510–515. doi:10.1002/pro.2230

Cho HS, Mason K, Ramyar KX, Stanley AM, Gabelli SB, Denney DW, Leahy DJ. 2003. Structure of the extracellular region of HER2 alone and in complex with the Herceptin Fab. *Nature* **421:** 756–760. doi:10.1038/nature01392

Chothia C, Lesk AM. 1987. Canonical structures for the hypervariable regions of immunoglobulins. *J Mol Biol* **196:** 901–917. doi:10.1016/0022-2836(87)90412-8

Davies DR, Cohen GH. 1996. Interactions of protein antigens with antibodies. *Proc Natl Acad Sci* **93:** 7–12. doi:10.1073/pnas.93.1.7

DeKosky BJ, Lungu OI, Park D, Johnson EL, Charab W, Chrysostomou C, Kuroda D, Ellington AD, Ippolito GC, Gray JJ, et al. 2016. Large-scale sequence and structural comparisons of human naive and antigen-experienced antibody repertoires. *Proc Natl Acad Sci* **113:** E2636–E2645. doi:10.1073/pnas.1525510113

Dondelinger M, Filée P, Sauvage E, Quinting B, Muyldermans S, Galleni M, Vandevenne MS. 2018. Understanding the significance and implications of antibody numbering and antigen-binding surface/residue definition. *Front Immunol* **9:** 412684. doi:10.3389/fimmu.2018.02278

Eigenbrot C, Randal M, Presta L, Carter P, Kossiakoff AA. 1993. X-ray structures of the antigen-binding domains from three variants of humanized anti-p185HER2 antibody 4D5 and comparison with molecular modeling. *J Mol Biol* **229:** 969–995. doi:10.1006/jmbi.1993.1099

Fellouse FA, Sidhu SS. 2006. Making antibodies in bacteria. In *Making and using antibodies* (ed. Howard GC, Kaser MR), pp. 171–194. CRC Press, Boca Raton, FL.

Fellouse FA, Wiesmann C, Sidhu SS. 2004. Synthetic antibodies from a four-amino-acid code: a dominant role for tyrosine in antigen recognition. *Proc Natl Acad Sci* **101:** 12467–12472. doi:10.1073/pnas.0401786101

Fellouse FA, Li B, Compaan DM, Peden AA, Hymowitz SG, Sidhu SS. 2005. Molecular recognition by a binary code. *J Mol Biol* **348:** 1153–1162. doi:10.1016/j.jmb.2005.03.041

Fellouse FA, Barthelemy PA, Kelley RF, Sidhu SS. 2006. Tyrosine plays a dominant functional role in the paratope of a synthetic antibody derived from a four amino acid code. *J Mol Biol* **357:** 100–114. doi:10.1016/j.jmb.2005.11.092

Fellouse FA, Esaki K, Birtalan S, Raptis D, Cancasci VJ, Koide A, Jhurani P, Vasser M, Wiesmann C, Kossiakoff AA, et al. 2007. High-throughput generation of synthetic antibodies from highly functional minimalist phage-displayed libraries. *J Mol Biol* **373:** 924–940. doi:10.1016/j.jmb.2007.08.005

Fraczkiewicz R, Braun W. 1998. Exact and efficient analytical calculation of the accessible surface areas and their gradients for macromolecules. *J Comput Chem* **19:** 319–333. doi:10.1002/(SICI)1096-987X(199802)19:3<319::AID-JCC6>3.0.CO;2-W

Garrard LJ, Henner DJ. 1993. Selection of an anti-IGF-1 Fab from a Fab phage library created by mutagenesis of multiple CDR loops. *Gene* **128:** 103–109. doi:10.1016/0378-1119(93)90160-5

Glanville J, Zhai W, Berka J, Telman D, Huerta G, Mehta GR, Ni I, Mei L, Sundar PD, Day GMR, et al. 2009. Precise determination of the diversity of a combinatorial antibody library gives insight into the human immunoglobulin repertoire. *Proc Natl Acad Sci* **106:** 20216–20221. doi:10.1073/pnas.0909775106

Hanna R, Cardarelli L, Patel N, Blazer LL, Adams JJ, Sidhu SS. 2020. A phage-displayed single-chain Fab library optimized for rapid production of single-chain IgGs. *Protein Sci* **29:** 2075–2084. doi:10.1002/pro.3931

Hebditch M, Warwicker J. 2019. Web-based display of protein surface and pH-dependent properties for assessing the developability of biotherapeutics. *Sci Rep* **9:** 1969. doi:10.1038/s41598-018-36950-8

Jain T, Sun T, Durand S, Hall A, Houston NR, Nett JH, Sharkey B, Bobrowicz B, Caffry I, Yu Y, et al. 2017. Biophysical properties of the clinical-stage antibody landscape. *Proc Natl Acad Sci* **114:** 944–949. doi:10.1073/pnas.1616408114

Johnson G, Wu TT. 2000. Kabat database and its applications: 30 years after the first variability plot. *Nucleic Acids Res* **28:** 214–218. doi:10.1093/nar/28.1.214

Kabat EA, Wu TT, Foeller C, Perry HM, Gottesman KS. 1991. *Sequences of Proteins of Immunological Interest*. [Publisher], [Publisher City], [Publisher State/Country]. https://books.google.com/books/about/Sequences_of_Proteins_of_Immunological_I.html?id=3jMvZYW2ZtwC

Kelly RL, Le D, Zhao J, Wittrup KD. 2018. Reduction of nonspecificity motifs in synthetic antibody libraries. *J Mol Biol* **430:** 119–130. doi:10.1016/j.jmb.2017.11.008

Kelow SP, Adolf-Bryfogle J, Dunbrack RL. 2020. Hiding in plain sight: structure and sequence analysis reveals the importance of the antibody DE loop for antibody-antigen binding. *MAbs* **12:** 1840005. doi:10.1080/19420862.2020.1840005

Kelow S, Faezov B, Xu Q, Parker M, Adolf-Bryfogle J, Dunbrack RL. 2022. A penultimate classification of canonical antibody CDR conformations. bioRxiv doi:10.1101/2022.10.12.511988

Knappik A, Ge L, Honegger A, Pack P, Fischer M, Wellnhofer G, Hoess A, Wölle J, Plückthun A, Virnekäs B. 2000. Fully synthetic human combinatorial antibody libraries (HuCAL) based on modular consensus frameworks and CDRs randomized with trinucleotides. *J Mol Biol* **296:** 57–86. doi:10.1006/jmbi.1999.3444

Kunik V, Peters B, Ofran Y. 2012. Structural consensus among antibodies defines the antigen binding site. *PLOS Comput Biol* **8:** e1002388. doi:10.1371/journal.pcbi.1002388

Lai JY, Lim TS. 2023. Construction of naïve and immune human Fab phage display library. *Methods Mol Biol* **2702:** 39–58. doi:10.1007/978-1-0716-3381-6_3

Lee CV, Liang WC, Dennis MS, Eigenbrot C, Sidhu SS, Fuh G. 2004a. High-affinity human antibodies from phage-displayed synthetic Fab libraries with a single framework scaffold. *J Mol Biol* **340:** 1073–1093. doi:10.1016/j.jmb.2004.05.051

Lee CV, Sidhu SS, Fuh G. 2004b. Bivalent antibody phage display mimics natural immunoglobulin. *J Immunol Methods* **284:** 119–132. doi:10.1016/j.jim.2003.11.001

Lee EC, Liang Q, Ali H, Bayliss L, Beasley A, Bloomfield-Gerdes T, Bonoli L, Brown R, Campbell J, Carpenter A, et al. 2014. Complete humanization of the mouse immunoglobulin loci enables efficient therapeutic antibody discovery. *Nat Biotechnol* **32:** 356–363. doi:10.1038/nbt.2825

Lefranc MP, Pommié C, Ruiz M, Giudicelli V, Foulquier E, Truong L, Thouvenin-Contet V, Lefranc G. 2003. IMGT unique numbering for immunoglobulin and T cell receptor variable domains and Ig superfamily V-like domains. *Dev Comp Immunol* **27:** 55–77. doi:10.1016/S0145-305X(02)00039-3

Lu RM, Hwang YC, Liu IJ, Lee CC, Tsai HZ, Li HJ, Wu HC. 2020. Development of therapeutic antibodies for the treatment of diseases. *J Biomed Sci* **27:** 1–30. doi:10.1186/s12929-019-0592-z

Marks C, Deane CM. 2020. How repertoire data are changing antibody science. *J Biol Chem* **295:** 9823–9837. doi:10.1074/jbc.REV120.010181

Marks JD, Hoogenboom HR, Bonnert TP, McCafferty J, Griffiths AD, Winter G. 1991. By-passing immunization. Human antibodies from V-gene libraries displayed on phage. *J Mol Biol* **222:** 581–597. doi:10.1016/0022-2836(91)90498-U

Martin ACR, Thornton JM. 1996. Structural families in loops of homologous proteins: automatic classification, modelling and application to antibodies. *J Mol Biol* **263:** 800–815. doi:10.1006/jmbi.1996.0617

McCafferty J, Griffiths AD, Winter G, Chiswell DJ. 1990. Phage antibodies: filamentous phage displaying antibody variable domains. *Nature* **348:** 552–554. doi:10.1038/348552a0

McConkey BJ, Sobolev V, Edelman M. 2003. Discrimination of native protein structures using atom–atom contact scoring. *Proc Natl Acad Sci* **100:** 3215–3220. doi:10.1073/pnas.0535768100

Mian IS, Bradwell AR, Olson AJ. 1991. Structure, function and properties of antibody binding sites. *J Mol Biol* **217:** 133–151. doi:10.1016/0022-2836(91)90617-F

Mitternacht S. 2016. FreeSASA: an open source C library for solvent accessible surface area calculations. *F1000Research* **5:** 189. doi:10.12688/f1000research.7931.1

Moreno E, Valdés-Tresanco MS, Molina-Zapata A, Sánchez-Ramos O. 2022. Structure-based design and construction of a synthetic phage display nanobody library. *BMC Res Notes* **15**: 124. doi:10.1186/s13104-022-06001-7

Myung Y, Pires DEV, Ascher DB. 2023. Understanding the complementarity and plasticity of antibody–antigen interfaces. *Bioinformatics* **39**: btad392. doi:10.1093/bioinformatics/btad392

Nabhan AN, Webster JD, Adams JJ, Blazer L, Everrett C, Eidenschenk C, Arlantico A, Fleming I, Brightbill HD, Wolters PJ, et al. 2023. Targeted alveolar regeneration with Frizzled-specific agonists. *Cell* **186**: 2995–3012.e15. doi:10.1016/j.cell.2023.05.022

Nelson B, Adams J, Kuglstatter A, Li Z, Harris SF, Liu Y, Bohini S, Ma H, Klumpp K, Gao J, et al. 2018. Structure-guided combinatorial engineering facilitates affinity and specificity optimization of anti-CD81 antibodies. *J Mol Biol* **430**: 2139–2152. doi:10.1016/j.jmb.2018.05.018

Nilvebrant J, Ereño-Orbea J, Gorelik M, Julian MC, Tessier PM, Julien JP, Sidhu SS. 2021. Systematic engineering of optimized autonomous heavy-chain variable domains. *J Mol Biol* **433**: 167241. doi:10.1016/j.jmb.2021.167241

Nixon AML, Duque A, Yelle N, McLaughlin M, Davoudi S, Pedley NM, Haynes J, Brown KR, Pan J, Hart T, et al. 2019. A rapid in vitro methodology for simultaneous target discovery and antibody generation against functional cell subpopulations. *Sci Rep* **9**: 842. doi:10.1038/s41598-018-37186-2

North B, Lehmann A, Dunbrack RL. 2011. A new clustering of antibody CDR loop conformations. *J Mol Biol* **406**: 228–256. doi:10.1016/j.jmb.2010.10.030

Padlan EA. 1994. Anatomy of the antibody molecule. *Mol Immunol* **31**: 169–217. doi:10.1016/0161-5890(94)90001-9

Pavlovic Z, Adams JJ, Blazer LL, Gakhal AK, Jarvik N, Steinhart Z, Robitaille M, Mascall K, Pan J, Angers S, et al. 2018. A synthetic anti-Frizzled antibody engineered for broadened specificity exhibits enhanced anti-tumor properties. *MAbs* **10**: 1157–1167. doi:10.1080/19420862.2018.1515565

Pedrioli A, Oxenius A. 2021. Single B cell technologies for monoclonal antibody discovery. *Trends Immunol* **42**: 1143–1158. doi:10.1016/j.it.2021.10.008

Peng HP, Lee KH, Jian JW, Yang AS. 2014. Origins of specificity and affinity in antibody-protein interactions. *Proc Natl Acad Sci* **111**: E2656–E2665. doi:10.1073/pnas.1401131111

Persson H, Ye W, Wernimont A, Adams JJ, Koide A, Koide S, Lam R, Sidhu SS. 2013. CDR-H3 diversity is not required for antigen recognition by synthetic antibodies. *J Mol Biol* **425**: 803–811. doi:10.1016/j.jmb.2012.11.037

Prassler J, Thiel S, Pracht C, Polzer A, Peters S, Bauer M, Nörenberg S, Stark Y, Kölln J, Popp A, et al. 2011. HuCAL PLATINUM, a synthetic Fab library optimized for sequence diversity and superior performance in mammalian expression systems. *J Mol Biol* **413**: 261–278. doi:10.1016/j.jmb.2011.08.012

Presta LG, Lahr SJ, Shields RL, Porter JP, Gorman CM, Fendly BM, Jardieu PM. 1993. Humanization of an antibody directed against IgE. *J Immunol* **151**: 2623–2632. doi:10.4049/jimmunol.151.5.2623

Presta LG, Chen H, O'Connor SJ, Chisholm V, Meng YG, Krummen L, Winkler M, Ferrara N. 1997. Humanization of an anti-vascular endothelial growth factor monoclonal antibody for the therapy of solid tumors and other disorders. *Cancer Res* **57**: 4593–4599.

Raybould MIJ, Marks C, Krawczyk K, Taddese B, Nowak J, Lewis AP, Bujotzek A, Shi J, Deane CM. 2019. Five computational developability guidelines for therapeutic antibody profiling. *Proc Natl Acad Sci* **116**: 4025–4030. doi:10.1073/pnas.1810576116

Reis PBPS, Barletta GP, Gagliardi L, Fortuna S, Soler MA, Rocchia W. 2022. Antibody-antigen binding interface analysis in the big data era. *Front Mol Biosci* **9**: 945808. doi:10.3389/fmolb.2022.945808

Richards FM. 1977. Areas, volumes, packing and protein structure. *Annu Rev Biophys Bioeng* **6**: 151–176. doi:10.1146/annurev.bb.06.060177.001055

Rothe C, Urlinger S, Löhning C, Prassler J, Stark Y, Jäger U, Hubner B, Bardroff M, Pradel I, Boss M, et al. 2008. The human combinatorial antibody library HuCAL GOLD combines diversification of all six CDRs according to the natural immune system with a novel display method for efficient selection of high-affinity antibodies. *J Mol Biol* **376**: 1182–1200. doi:10.1016/j.jmb.2007.12.018

Sela-Culang I, Kunik V, Ofran Y. 2013. The structural basis of antibody-antigen recognition. *Front Immunol* **4**: 64858. doi:10.3389/fimmu.2013.00302

Shen Y, Yang X, Dong N, Xie X, Bai X, Shi Y. 2007. Generation and selection of immunized Fab phage display library against human B cell lymphoma. *Cell Res* **17**: 650–660. doi:10.1038/cr.2007.57

Shi L, Wheeler JC, Sweet RW, Lu J, Luo J, Tornetta M, Whitaker B, Reddy R, Brittingham R, Borozdina L, et al. 2010. De novo selection of high-affinity antibodies from synthetic fab libraries displayed on phage as pIX fusion proteins. *J Mol Biol* **397**: 385–396. doi:10.1016/j.jmb.2010.01.034

Sidhu SS, Li B, Chen Y, Fellouse FA, Eigenbrot C, Fuh G. 2004. Phage-displayed antibody libraries of synthetic heavy chain complementarity determining regions. *J Mol Biol* **338**: 299–310. doi:10.1016/j.jmb.2004.02.050

Silacci M, Brack S, Schirru G, Mårlind J, Ettorre A, Merlo A, Viti F, Neri D. 2005. Design, construction, and characterization of a large synthetic human antibody phage display library. *Proteomics* **5**: 2340–2350. doi:10.1002/pmic.200401273

Sraphet S, Javadi B. 2022. Application of hierarchical clustering to analyze solvent-accessible surface area patterns in *Amycolatopsis* lipases. *Biology* **11**: 652. doi:10.3390/biology11050652

Starovasnik MA, O'Connell MP, Fairbrother WJ, Kelley RF. 1999. Antibody variable region binding by Staphylococcal protein A: thermodynamic analysis and location of the Fv binding site on E-domain. *Protein Sci* **8**: 1423–1431.

Stave JW, Lindpaintner K. 2013. Antibody and antigen contact residues define epitope and paratope size and structure. *J Immunol* **191**: 1428–1435. doi:10.4049/jimmunol.1203198

Teplyakov A, Obmolova G, Malia TJ, Luo J, Muzammil S, Sweet R, Almagro JC, Gilliland GL. 2016. Structural diversity in a human antibody germline library. *MAbs* **8**: 1045–1063. doi:10.1080/19420862.2016.1190060

Tiller T, Schuster I, Deppe D, Siegers K, Strohner R, Herrmann T, Berenguer M, Poujol D, Stehle J, Stark Y, et al. 2013. A fully synthetic human Fab antibody library based on fixed VH/VL framework pairings with favorable biophysical properties. *MAbs* **5**: 445. doi:10.4161/mabs.24218

Townsend CL, Laffy JMJ, Wu YCB, O'Hare JS, Martin V, Kipling D, Fraternali F, Dunn-Walters DK. 2016. Significant differences in physicochemical properties of human immunoglobulin κ and λ CDR3 regions. *Front Immunol* **7**: 388. doi:10.3389/fimmu.2016.00388

Valldorf B, Hinz SC, Russo G, Pekar L, Mohr L, Klemm J, Doerner A, Krah S, Hust M, Zielonka S. 2022. Antibody display technologies: selecting the cream of the crop. *Biol Chem* **403**: 455–477. doi:10.1515/hsz-2020-0377

Vaughan TJ, Williams AJ, Pritchard K, Osbourn JK, Pope AR, Earnshaw JC, McCafferty J, Hodits RA, Wilton J, Johnson KS. 1996. Human antibodies with sub-nanomolar affinities isolated from a large non-immunized phage display library. *Nat Biotechnol* **14**: 309–314. doi:10.1038/nbt0396-309

Virnekas B, Ge L, Plukthun A, Schneider KC, Wellnhofer G, Moroney SE. 1994. Trinucleotide phosphoramidites: ideal reagents for the synthesis of mixed oligonucleotides for random mutagenesis. *Nucleic Acids Res* **22**: 5600–5607. doi:10.1093/nar/22.25.5600

Wang SS, Yan Y, Ho K. 2021. US FDA-approved therapeutic antibodies with high-concentration formulation: summaries and perspectives. *Antib Ther* **4**: 262–272. doi:10.1093/abt/tbab027

Werther WA, Gonzalez TN, O'Connor SJ, McCabe S, Chan B, Hotaling T, Champe M, Fox JA, Jardieu PM, Berman PW, et al. 1996. Humanization of an anti-lymphocyte function-associated antigen (LFA)-1 monoclonal antibody and reengineering of the humanized antibody for binding to rhesus LFA-1. *J Immunol* **157**: 4986–4995. doi:10.4049/jimmunol.157.11.4986

Xu D, Zhang Y. 2009. Generating triangulated macromolecular surfaces by Euclidean Distance Transform. *PLoS ONE* **4**: e8140. doi:10.1371/journal.pone.0008140

Yang HY, Kang KJ, Chung JE, Shim H. 2009. Construction of a large synthetic human scFv library with six diversified CDRs and high functional diversity. *Mol Cells* **27**: 225–235. doi:10.1007/s10059-009-0028-9

Zhai W, Glanville J, Fuhrmann M, Mei L, Ni J, Sundar PD, Van Blarcom T, Abdiche Y, Lindquist K, Strohner R, et al. 2011. Synthetic antibodies designed on natural sequence landscapes. *J Mol Biol* **412**: 55–71. doi:10.1016/j.jmb.2011.07.018

Zhou H, Wang Y, Wang W, Jia J, Li Y, Wang Q, Wu Y, Tang J. 2009. Generation of monoclonal antibodies against highly conserved antigens. *PLoS One* **4**: e6087. doi:10.1371/journal.pone.0006087

CHAPTER 8

# Beyond Natural Immune Repertoires: Synthetic Antibodies

Gianluca Veggiani[1,2] and Sachdev S. Sidhu[1,3]

[1]The Anvil Institute, Kitchener, Ontario N2G 1H6, Canada; [2]Department of Pathobiological Sciences, School of Veterinary Medicine, Louisiana State University, Baton Rouge, Louisiana 70803, USA

Synthetic antibody libraries, in which the antigen-binding sites are precisely designed, offer unparalleled precision in antibody engineering, exceeding the potential of natural immune repertoires and constituting a novel generation of research tools and therapeutics. Recent advances in artificial intelligence–driven technologies and their integration into synthetic antibody discovery campaigns hold the promise to further streamline and effectively develop antibodies. Here, we provide an overview of synthetic antibodies. Our associated protocol describes how to develop highly diverse and functional synthetic antibody phage display libraries.

## BACKGROUND

Over the last three decades, monoclonal antibodies (mAbs) have evolved from research tools to powerful therapeutics for many diseases. Monoclonal antibodies have traditionally been developed through the generation of hybridomas obtained from the fusion of primary human splenocytes with immortalized myeloma cell lines (Olsson and Kaplan 1980; Reichert et al. 2005) or via the Epstein–Barr virus-mediated immortalization of primary human lymphocytes (Kozbor et al. 1982). Although these methods are effective, they are expensive, produce insufficient quantities of mAbs, and cannot be used to isolate mAbs against self-antigens (Beerli and Rader 2010). Furthermore, these approaches are time-consuming, and thus are unsuitable for the rapid discovery and evaluation of mAbs required in the case of pandemic outbreaks (Kelley 2020).

In vitro selection methods, notably phage display, allow the rapid isolation of mAbs for targets, such as self-antigens and toxins, that are unsuitable for the hybridoma technology. Additionally, such technologies circumvent the need for animal immunizations, are more amenable to automation, and enable more precise control over antibody selection parameters (Bradbury et al. 2011; Valldorf et al. 2022). Consequently, in vitro antibody display methods are now routinely used to develop mAbs for many diverse applications, and the robustness and simplicity of phage display make this technology the most widely used platform for the discovery of therapeutic antibody candidates.

Although early phage-displayed antibody libraries were developed by amplifying the V(D)J-rearranged immunoglobulin repertoire from B-cell cDNA of immunized animals, subsequent advancements in the technology allowed the construction of large naive libraries obtained from the natural antibody repertoire of hosts that were not exposed to antigens (Ferrara et al. 2022). Both of these methods allowed the development of large phage-displayed antibody libraries that enabled the isolation of mAbs currently used in the clinic (Osbourn et al. 2005; Li et al. 2008), but they were limited to

[3]Correspondence: sachdev.sidhu@uwaterloo.ca

© 2026 Cold Spring Harbor Laboratory Press
Cite this introduction as *Cold Spring Harb Protoc*; doi:10.1101/pdb.top107768

the diversity encoded by natural immune systems and often generated mAbs with suboptimal biophysical and pharmacological properties.

Today, the vast knowledge of antibody structure and function facilitates the design of synthetic antibody libraries in which diversity is generated through precisely designed synthetic DNA. The majority of synthetic antibody libraries are constructed by introducing diversity in the complementarity-determining regions (CDRs), within optimized frameworks, by using degenerate combinations of mono- or trinucleotide units to create highly diverse antigen-binding sites (Shim 2015). Given that only a limited number of residues in the antibody framework can establish interactions with antigens (Kunik et al. 2012a,b), such a strategy maximizes the chemical diversity introduced in the CDRs while minimizing the risk of immunogenicity and ensuring high stability and protein production (Tiller et al. 2013; Douillard et al. 2019). Additionally, the well-defined nature of synthetic antibodies makes them amenable to further library designs, such as in the case of affinity maturation, thereby reducing the time and the optimization required to generate therapeutic antibody candidates (Tiller et al. 2017). Finally, computational and artificial intelligence–based methods are increasingly being incorporated into phage display-mediated synthetic antibody discovery and engineering pipelines, and could greatly simplify the development and lower the cost of biologics (Robert et al. 2022; Kim et al. 2023).

Our work has shown that a single stable framework can support a large array of antibody functions for diverse antigens, and that synthetic antibody libraries, whose diversity is not genetically constrained, can enable antigen recognition via mechanisms that are distinct from those of natural antibodies (Persson et al. 2013). Furthermore, we have shown that even simplified synthetic antibody libraries in which only a subset of the six CDRs was diversified, or in which the chemical diversity of each randomized position was reduced only to tyrosine and serine, are highly functional (Persson et al. 2013).

In our associated protocol, we describe principles and methods for the construction of a highly diverse synthetic phage-displayed fragment antigen-binding (Fab) library that has enabled the isolation of numerous high-affinity antibodies (see Protocol 1: Generation and Selection of Synthetic Human Antibody Libraries via Phage Display [Veggiani and Sidhu 2023]) (Miersch et al. 2017; Pavlovic et al. 2018; Chen et al. 2019; Tao et al. 2019; Gallo et al. 2021; Mishra et al. 2022; Philpott et al. 2022). These Fabs are based on a single antibody framework derived from the highly stable anti-HER2 antibody 4D5 (Albanell and Baselga 1999). Genetic diversity is introduced in the library via oligonucleotide-directed mutagenesis using an optimized version of the Kunkel method (Liu et al. 2020) (see Fig. 1A in Protocol 1: Generation and Selection of Synthetic Human Antibody Libraries via Phage Display [Veggiani and Sidhu 2023]). A large phage-displayed synthetic antibody library containing more than 10 billion unique Fabs can be obtained by introducing sequence diversity into the CDRs using precisely designed synthetic mutagenic oligonucleotides. Following electroporation in an *Escherichia coli* host, the genetic library is converted into a phage-displayed synthetic Fab library that can be used for isolating antibodies against virtually any target of interest.

In summary, synthetic antibody libraries hold promise for the development of antibodies with properties beyond those of natural antibodies, allowing the rapid discovery and development of antibodies capable of targeting virtually any antigen.

## REFERENCES

Albanell J, Baselga J. 1999. Trastuzumab, a humanized anti-HER2 monoclonal antibody, for the treatment of breast cancer. *Drugs Today (Barc)* 35: 931–946.

Beerli RR, Rader C. 2010. Mining human antibody repertoires. *mAbs* 2: 365–378. doi:10.4161/mabs.12187

Bradbury ARM, Sidhu S, Dübel S, McCafferty J. 2011. Beyond natural antibodies: the power of in vitro display technologies. *Nat Biotechnol* 29: 245–254. doi:10.1038/nbt.1791

Chen G, Karauzum H, Long H, Carranza D, Holtsberg FW, Howell KA, Abaandou L, Zhang B, Jarvik N, Ye W, et al. 2019. Potent neutralization of staphylococcal enterotoxin B in vivo by antibodies that block binding to the T-cell receptor. *J Mol Biol* 431: 4354–4367. doi:10.1016/j.jmb.2019.03.017

Douillard P, Freissmuth M, Antoine G, Thiele M, Fleischanderl D, Matthiessen P, Voelkel D, Kerschbaumer RJ, Scheiflinger F, Sabarth N. 2019. Optimization of an antibody light chain framework enhances expression, biophysical properties and pharmacokinetics. *Antibodies (Basel)* 8: E46. doi:10.3390/antib8030046

Ferrara F, Erasmus MF, D'Angelo S, Leal-Lopes C, Teixeira AA, Choudhary A, Honnen W, Calianese D, Huang D, Peng L, et al. 2022. A pandemic-enabled comparison of discovery platforms demonstrates a naïve antibody library can match the best immune-sourced antibodies. *Nat Commun* 13: 462. doi:10.1038/s41467-021-27799-z

Gallo E, Kelil A, Haughey M, Cazares-Olivera M, Yates BP, Zhang M, Wang N-Y, Blazer L, Carderelli L, Adams JJ, et al. 2021. Inhibition of cancer cell adhesion, migration and proliferation by a bispecific antibody that targets two distinct epitopes on αv integrins. *J Mol Biol* **433:** 167090. doi:10.1016/j.jmb.2021.167090

Kelley B. 2020. Developing therapeutic monoclonal antibodies at pandemic pace. *Nat Biotechnol* **38:** 540–545. doi:10.1038/s41587-020-0512-5

Kim J, McFee M, Fang Q, Abdin O, Kim PM. 2023. Computational and artificial intelligence-based methods for antibody development. *Trends Pharmacol Sci* **44:** 175–189. doi:10.1016/j.tips.2022.12.005

Kozbor D, Lagarde AE, Roder JC. 1982. Human hybridomas constructed with antigen-specific Epstein–Barr virus-transformed cell lines. *Proc Natl Acad Sci* **79:** 6651–6655. doi:10.1073/pnas.79.21.6651

Kunik V, Ashkenazi S, Ofran Y. 2012a. Paratome: an online tool for systematic identification of antigen-binding regions in antibodies based on sequence or structure. *Nucl Acids Res* **40:** W521–W524. doi:10.1093/nar/gks480

Kunik V, Peters B, Ofran Y. 2012b. Structural consensus among antibodies defines the antigen binding site. *PLoS Comput Biol* **8:** e1002388. doi:10.1371/journal.pcbi.1002388

Li S, Kussie P, Ferguson KM. 2008. Structural basis for EGF receptor inhibition by the therapeutic antibody IMC-11F8. *Structure* **16:** 216–227. doi:10.1016/j.str.2007.11.009

Liu B, Long S, Liu J. 2020. Improving the mutagenesis efficiency of the Kunkel method by codon optimization and annealing temperature adjustment. *New Biotechnol* **56:** 46–53. doi:10.1016/j.nbt.2019.11.004

Miersch S, Maruthachalam BV, Geyer CR, Sidhu SS. 2017. Structure-directed and tailored diversity synthetic antibody libraries yield novel anti-EGFR antagonists. *ACS Chem Biol* **12:** 1381–1389. doi:10.1021/acschembio.6b00990

Mishra N, Teyra J, Boytz R, Miersch S, Merritt TN, Cardarelli L, Gorelik M, Mihalic F, Jemth P, Davey RA, et al. 2022. Development of monoclonal antibodies to detect for SARS-CoV-2 proteins. *J Mol Biol* **434:** 167583. doi:10.1016/j.jmb.2022.167583

Olsson L, Kaplan HS. 1980. Human–human hybridomas producing monoclonal antibodies of predefined antigenic specificity. *Proc Natl Acad Sci* **77:** 5429–5431. doi:10.1073/pnas.77.9.5429

Osbourn J, Groves M, Vaughan T. 2005. From rodent reagents to human therapeutics using antibody guided selection. *Methods* **36:** 61–68. doi:10.1016/j.ymeth.2005.01.006

Pavlovic Z, Adams JJ, Blazer LL, Gakhal AK, Jarvik N, Steinhart Z, Robitaille M, Mascall K, Pan J, Angers S, et al. 2018. A synthetic anti-Frizzled antibody engineered for broadened specificity exhibits enhanced antitumor properties. *MAbs* **10:** 1157–1167. doi:10.1080/19420862.2018.1515565

Persson H, Ye W, Wernimont A, Adams JJ, Koide A, Koide S, Lam R, Sidhu SS. 2013. CDR-H3 diversity is not required for antigen recognition by synthetic antibodies. *J Mol Biol* **425:** 803–811. doi:10.1016/j.jmb.2012.11.037

Philpott DN, Gomis S, Wang H, Atwal R, Kelil A, Sack T, Morningstar B, Burnie C, Sargent EH, Angers S, et al. 2022. Rapid on-cell selection of high-performance human antibodies. *ACS Cent Sci* **8:** 102–109. doi:10.1021/acscentsci.1c01205

Reichert JM, Rosensweig CJ, Faden LB, Dewitz MC. 2005. Monoclonal antibody successes in the clinic. *Nat Biotechnol* **23:** 1073–1078. doi:10.1038/nbt0905-1073

Robert PA, Akbar R, Frank R, Pavlović M, Widrich M, Snapkov I, Slabodkin A, Chernigovskaya M, Scheffer L, Smorodina E, et al. 2022. Unconstrained generation of synthetic antibody–antigen structures to guide machine learning methodology for antibody specificity prediction. *Nat Comput Sci* **2:** 845–865. doi:10.1038/s43588-022-00372-4

Shim H. 2015. Synthetic approach to the generation of antibody diversity. *BMB Rep* **48:** 489–494. doi:10.5483/BMBRep.2015.48.9.120

Tao Y, Mis M, Blazer L, Ustav M Jr, Steinhart Z, Chidiac R, Kubarakos E, O'Brien S, Wang X, Jarvik N, et al. 2019. Tailored tetravalent antibodies potently and specifically activate Wnt/Frizzled pathways in cells, organoids and mice. *eLife* **8:** e46134. doi:10.7554/eLife.46134

Tiller T, Schuster I, Deppe D, Siegers K, Strohner R, Herrmann T, Berenguer M, Poujol D, Stehle J, Stark Y, et al. 2013. A fully synthetic human Fab antibody library based on fixed VH/VL framework pairings with favorable biophysical properties. *MAbs* **5:** 445–470. doi:10.4161/mabs.24218

Tiller KE, Chowdhury R, Li T, Ludwig SD, Sen S, Maranas CD, Tessier PM. 2017. Facile affinity maturation of antibody variable domains using natural diversity mutagenesis. *Front Immunol* **8:** 986. doi:10.3389/fimmu.2017.00986

Valldorf B, Hinz SC, Russo G, Pekar L, Mohr L, Klemm J, Doerner A, Krah S, Hust M, Zielonka S. 2022. Antibody display technologies: selecting the cream of the crop. *Biol Chem* **403:** 455–477. doi:10.1515/hsz-2020-0377

Veggiani G, Sidhu SS. 2023. Generation and selection of synthetic human antibody libraries via phage display. *Cold Spring Harb Protoc* doi:10.1101/pdb.prot108347

## Protocol 1

# Generation and Selection of Synthetic Human Antibody Libraries via Phage Display

Gianluca Veggiani[1,2] and Sachdev S. Sidhu[1,3]

[1]*The Anvil Institute, Kitchener, Ontario N2G 1H6, Canada;* [2]*Department of Pathobiological Sciences, School of Veterinary Medicine, Louisiana State University, Baton Rouge, Louisiana 70803, USA*

Synthetic antibody libraries enable the development of antibodies that can recognize virtually any antigen, with affinity and specificity profiles that are superior to those of natural antibodies. By using highly stable and optimized frameworks, synthetic antibody libraries can be rapidly generated by precisely designing synthetic DNA, allowing absolute control over the position and chemical diversity introduced while expanding the sequence space for antigen recognition. Here, we describe a detailed protocol for the generation of highly diverse synthetic antibody phage display libraries based on a single framework, with diversity genetically incorporated by using finely designed mutagenic oligonucleotides. This general method enables the facile construction of large antibody libraries with precisely tunable features, resulting in the rapid development of recombinant antibodies for virtually any antigen.

## MATERIALS

It is essential that you consult the appropriate Material Safety Data Sheets and your institution's Environmental Health and Safety Office for proper handling of equipment and hazardous materials used in this protocol.

RECIPES: Please see the end of this protocol for recipes indicated by <R>. Additional recipes can be found online at http://cshprotocols.cshlp.org/site/recipes.

### Reagents

1-kb DNA Ladder (FroggaBio DM010-R500)
2× YT medium <R>
2× YT/Carb/Cam medium <R>
2× YT/Carb/Kan medium <R>
2× YT/Carb/Kan/Uridine medium <R>
2× YT/Tet medium <R>
10 mM adenosine 5′-triphosphate (ATP) (New England Biolabs P0756S)
10 mM deoxynucleoside triphosphate (dNTP) mix (Life Technologies 18427013)
Antigen solution of interest

*Antigen targets for selection campaigns can be purchased from commercial suppliers or produced in the laboratory using standard recombinant DNA technology methods.*

[3]Correspondence: sachdev.sidhu@uwaterloo.ca

© 2026 Cold Spring Harbor Laboratory Press
Cite this protocol as *Cold Spring Harb Protoc*; doi:10.1101/pdb.prot108347

Anti-M13 antibody HRP-conjugated (SinoBiological 11973-MM05T-H)
Bovine serum albumin (BSA) (Millipore-Sigma A7030-500G)
Carbenicillin (Carb; 100 mg/mL) <R>
Carrier protein (e.g., GST or MBP, lacking the target fusion) or streptavidin (optional; see Step 50)
Dithiothreitol (DTT; 100 mM) (Millipore-Sigma D9779-250MG)
*Escherichia coli* CJ236 electrocompetent cells (Lucigen 60701-2)
*E. coli* OmniMAX 2 T1$^R$ cells (ThermoFisher C854003)
*E. coli* SS320 cells (Lucigen 60512-2) preinfected with M13KO7 helper phage (Sidhu et al. 2000)
Exonuclease I (ExoI) (Millipore-Sigma GEE70073X)
Hydrochloric acid (HCl; Millipore-Sigma H1758-500ML; 100 mM HCl in sterile $H_2O$ for irrigation)

*Filter-sterilize.*

Kanamycin (Kan; 25 mg/mL) <R>
KPL TMB Microwell Peroxidase Substrate System (Mandel KP-50-76-03)
LB agar plates <R>

*Prepare LB-Carb plates containing 100 µg/mL carbenicillin, LB-Kan plates containing 25 µg/mL kanamycin, and LB-Tet plates containing 10 µg/mL tetracycline.*

M13KO7 helper phage (Life Technologies 18311019)
MLB buffer <R>
MP buffer <R>
Mutagenic oligonucleotide

*Oligonucleotides can be provided by multiple suppliers and custom-designed to meet the desired requirements. Mutagenic oligonucleotides are designed following standard primer design guidelines.*

PBT buffer (pH 7.4) <R>
PCR primers for Sanger sequencing (see Step 85)
PEG (20%)/2.5 M NaCl <R>
Phagemid vector
Phosphate-buffered saline (PBS; pH 7.4) <R>
Phosphate-buffered saline (PBS; pH 7.4) <R> containing 0.5% (w/v) BSA
Phosphoric acid ($H_3PO_4$; Millipore-Sigma 04107-1L; 1 M in Milli Q-purified $H_2O$)
PT buffer <R>
QIAquick Gel Extraction kit (QIAGEN 28706X4)
QIAprep Spin Miniprep kit (QIAGEN 27106)
Shrimp alkaline phosphatase (SAP) (Millipore-Sigma GEE70092X)
SOC medium <R>
Sterile $H_2O$ for irrigation (B. Braun Medical Inc. 971-R5000-01)
SYBR Safe (Life Technologies S33102)
T4 DNA ligase (New England Biolabs M0202L)
T4 polynucleotide kinase (PNK; New England Biolabs M0201L)
T7 polymerase (New England Biolabs M0274L)
TAE <R>
Taq polymerase (New England Biolabs M0267L)
Tetracycline (10 mg/mL) <R>
TM buffer (10×) <R>
ThermoPol buffer (New England Biolabs B9004S)
Tris(hydroxymethyl)aminomethane (Millipore-Sigma 252859-500G; 1 M, pH 11.0 in sterile $H_2O$ for irrigation [B. Braun Medical Inc. 971-R5000-01], filter-sterilized)
Ultrapure agarose (ThermoFisher 16500500)
Ultrapure glycerol (ThermoFisher 15514011) (optional; see Steps 49 and 57)

Chapter 8

## Equipment

0.2-cm gap electroporation cuvettes (BTX Harvard Apparatus 45-0125)
96-well microtubes (VWR 89005-582)
Adhesive sealing film (VWR 732-0077)
Baffled flask, 250-mL
Baffled flask, 2-L
BTX ECM-630 electroporation system (BTX)
Centrifuge
Centrifugation bottle, 1-L
Falcon tubes, 50-mL, 15-mL
Filter pipette tips
Gel electrophoresis apparatus
Shaking Incubator (New Brunswick Scientific, Innova 44)
Static Incubator (Fisher Scientific, Isotemp)
MaxiSorp 96-well plates (Fisher Scientific 12-565-135)
MaxiSorp 384-well plates (Fisher Scientific 12-565-347)
Microcentrifuge
Microcentrifuge tubes, 1.5-mL
Mini tube 96-well system (Axygen Scientific 89005-582)
Multichannel pipette
PCR tubes
Plastic spatulas (VWR 80081-190)
Spectrophotometer

## METHOD

*See the Discussion for additional information on the procedure including phagemid design and library construction.*

### Preparation of dU-Single-Stranded DNA (ssDNA) Template

*Highly pure dU-double-stranded DNA (dsDNA) phagemid template is crucial for successful library construction. The following method is adapted and modified from the QIAGEN QIAprep spin M13 kit protocol. We recommend using 20 µg of dU-dsDNA template for the construction of one library containing $10^{10}$ unique Fab clones. To maximize the likelihood of displaying Fabs harboring the desired mutations, we recommend introducing stop codons at sites to be diversified in the library (stop template). The outlined procedure is sufficient to obtain ~20 µg of stop template. If a lower stop template amount is obtained, we suggest scaling up.*

1. Chemically transform 500 ng of the phagemid vector encoding the desired stop template to be mutated into competent *E. coli* CJ236 (or an analogous *dut−/ung−* bacterial host) using the protocol described by Sambrook and Russell (2006). Using plastic spatulas, plate transformed bacteria on LB-Carb plates to select for the phagemid vector and grow bacteria overnight at 37°C.

2. Using a pipette tip, pick a single colony of *E. coli* CJ236 harboring the phagemid vector and inoculate 1 mL of 2× YT/Carb/Cam medium. Incubate the culture for 30 min at 37°C at 200 rpm. Add 1 µL of M13KO7 helper phage ($10^{10}$ pfu/mL) to the culture and allow the infection to proceed for 2 h at 37°C at 200 rpm.

    *Carbenicillin (carb) is used to select for the presence of phagemid vector, whereas chloramphenicol (cam) selects for the F' episome in* E. coli *CJ236 cells.*

    *To avoid pipette contamination, always use filter tips when handling phage.*

3. Add 25 µg/mL kanamycin to the culture media to select for bacteria coinfected with the M13KO7 helper phage. Incubate the culture for 6 h at 37°C at 200 rpm.

4. Transfer the culture to a baffled 250 mL flask containing 30 mL 2× YT/Carb/Kan/uridine medium and incubate for 20 h at 37°C at 200 rpm.

5. Transfer the culture to a 50-mL Falcon tube. Pellet the bacteria by centrifuging at 11,000$g$ for 10 min at 4°C.

6. Collect the phage-containing supernatant and place it in a new 50 mL Falcon tube. Add 5 mL PEG/NaCl solution to precipitate phage. Mix tube by inversion and incubate on ice for 20 min.

7. Precipitate phage by centrifugation at 24,800$g$ for 30 min at 4°C. Decant the supernatant and further precipitate phage at 24,800$g$ for another 3 min at 4°C.

8. Carefully remove residual supernatant and resuspend the precipitated phage in 500 µL sterile PBS pH 7.4. Transfer the phage to a 1.5-mL microcentrifuge tube and centrifuge it at 24,800$g$ for 5 min at 4°C to remove bacterial cell debris. Transfer the phage-containing supernatant to a new 1.5-mL microcentrifuge tube.

9. Add 7 µL MP buffer to the phage-containing solution and incubate the mixture for 5 min at room temperature.

10. Apply the sample to a QIAprep spin column from the QIAprep Spin Miniprep kit, and centrifuge at 11,000$g$ for 30 sec at 25°C. Discard the flowthrough. Phage particles will be bound to the column.

11. Add 700 µL MLB buffer to the QIAprep spin column, and centrifuge at 11,000 $g$ for 30 sec at 25°C. Discard the flowthrough.

12. Add 700 µL MLB buffer and incubate the column for 1 min at 25°C. Centrifuge the column at 11,000$g$ for 30 sec at 25°C. Discard the flowthrough. The extracted DNA will be bound to the column's matrix.

13. Add 700 µL PE buffer and centrifuge the column at 11,000$g$ for 30 sec at 25°C. Discard the flowthrough.

14. Remove residual proteins and salt by adding 700 µL PE buffer. Centrifuge the column at 11,000$g$ for 30 sec at 25°C. Discard flowthrough.

15. Remove excess ethanol (contained in the PE buffer) by centrifuging the column at 11,000$g$ for 30 sec at 25°C.

16. Transfer the column into a new 1.5-mL microcentrifuge tube. Add 50 µL sterile H$_2$O for irrigation to the center of the column. Incubate the column for 10 min at 37°C. Centrifuge the column at 24,800$g$ for 1 min at 25°C and save the eluate (containing purified dU-ssDNA).

17. Add 50 µL sterile H$_2$O to the middle of the column and incubate for 10 min at 37°C. Centrifuge the column at 24,800$g$ for 1 min at 25°C. Save the eluate and combine it with the previously eluted dU-ssDNA.

18. Analyze the purity of the dU-ssDNA by loading 5 µL onto a 1% agarose gel containing the SYBR Safe DNA gel stain alongside 1-kb DNA Ladder using TAE buffer and standard procedures. The obtained dU-ssDNA should give the typical supercoiled, open-circular, and linear banding pattern of plasmid DNA and lack any smears. For efficient library cloning, dU-ssDNA with a ≥90% purity level should be used.

    See Troubleshooting.

19. Spectrophotometrically determine the dU-ssDNA concentration. An absorbance value of 1.0 at 260 nm indicates a concentration of 33 ng/µL dU-ssDNA. This procedure typically yields dU-ssDNA concentrations in the 250–500 ng/µL range.

20. Aliquot 20 µg of dU-ssDNA into 1.5-mL microcentrifuge tubes. Store samples at −20°C. Avoid repeated freeze and thaw cycles.

Chapter 8

## In Vitro Synthesis of Covalently Closed Circular (CCC)-dsDNA

*A three-step procedure is used for the synthesis of CCC-dsDNA. Following phosphorylation, mutagenic oligonucleotides are annealed to a dU-ssDNA template. The 5'-phosphorylated mutagenic oligonucleotides are extended through the activity of T7 DNA polymerase and ligated to form CCC-dsDNA using T4 DNA ligase. It is possible to obtain phage-displayed libraries containing more than 10 billion unique synthetic antibodies by using 20 μg of highly pure dU-ssDNA.*

21. Phosphorylate each mutagenic oligonucleotide by assembling the following reaction in a PCR tube:

    0.6 μg mutagenic oligonucleotide
    2 μL 10× TM buffer
    2 μL 10 mM ATP
    1 μL 100 mM DTT
    2 μL T4 polynucleotide kinase (20 units)

    Add sterile irrigation $H_2O$ to bring the final reaction volume to 20 μL.

    Incubate the reaction mixture for 1 h at 37°C. Use of phosphorylated mutagenic oligonucleotides immediately after phosphorylation is recommended. However, 5'-phosphorylated oligonucleotides can be stored for ~1 mo at −20°C without significantly affecting their performance.

    *For the construction of synthetic Fab libraries in which complementarity-determining region (CDR) length variations are desired, it is possible to use pools of mutagenic oligonucleotides. See Discussion for more information.*

22. Anneal phosphorylated mutagenic oligonucleotides to 20 μg purified dU-ssDNA. In case multiple CDRs are to be diversified, two or more mutagenic oligonucleotides can be used simultaneously. Ensure that no sequences within the mutagenic oligonucleotides overlap with each other:

    20 μg dU-ssDNA
    25 μL 10× TM buffer
    20 μL phosphorylated oligonucleotide (or oligonucleotide pools)

    Add sterile irrigation $H_2O$ to bring the final reaction volume to 250 μL.

    *If multiple mutagenic oligonucleotides are used simultaneously, we recommend using 20 μL of each oligonucleotide to maximize binding to the template dU-ssDNA. The indicated dU-ssDNA quantities provide an oligonucleotide:template molar ratio of 3:1, assuming that the oligonucleotide:template length ratio is 1:100.*

23. Incubate the reaction mixture for 3 min at 90°C, for 3 min at 50°C, and for 5 min at 25°C. Annealing of mutagenic oligonucleotides can be performed on a thermocycler or on a dry block heater.

24. Immediately after annealing of the mutagenic oligonucleotides to the dU-ssDNA, perform in vitro synthesis of CCC-dsDNA by assembling the following reaction:

    250 μL annealed oligonucleotide/dU-ssDNA mixture
    10 μL 10 mM ATP
    10 μL 10 mM dNTP mix
    15 μL 100 mM DTT
    5 μL T4 DNA ligase (2000 units)
    3 μL T7 DNA polymerase (30 units)

    Incubate the reaction mixture overnight at 25°C.

    *Prior to library construction, it is advisable to test the efficiency of the mutagenesis reaction using 1/20 of the amounts and reaction volumes described above. The reaction efficiency can be evaluated by agarose gel electrophoresis immediately after in vitro synthesis of the CCC-dsDNA heteroduplex. A successful reaction is shown in Figure 1C and will result in the complete conversion of the dU-ssDNA to CCC-dsDNA characterized by lower electrophoretic mobility.*

    *See Troubleshooting.*

25. Purify and desalt the CCC-dsDNA using the QIAquick gel extraction kit. Add 1 mL of buffer QG to the reaction mixture containing the synthesized CCC-dsDNA and apply the sample to two

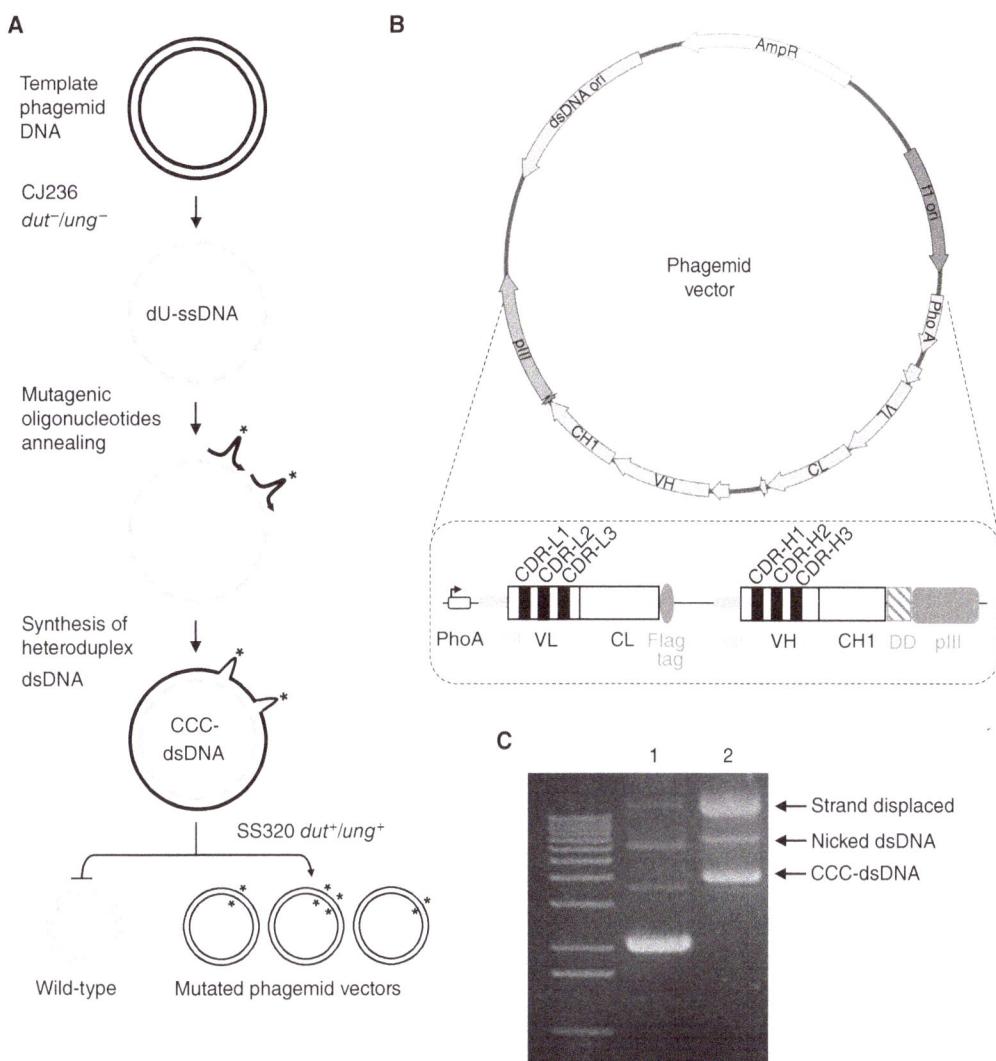

FIGURE 1. Design of a phagemid vector for Fab display and library construction workflow. (A) The dU-ssDNA is prepared from phage particles produced using *Escherichia coli* CJ236 *dut−/ung−* cells (or other *dut−/ung−* strains) that enable the preparation of uracil-containing templates. Mutations (asterisks) are introduced through the annealing of oligonucleotides (arrows) to the dU-ssDNA template (dashed, gray circles). Using T7 DNA polymerase and T4 DNA ligase, the mutagenic oligonucleotides are extended and ligated to synthesize covalently closed circular, double-stranded DNA (CCC-dsDNA, black circles). The resulting CCC-dsDNA is then transformed into competent *E. coli* SS320 *dut+/ung+* cells (or other *dut+/ung+* strains), where dU-ssDNA is inactivated, and mismatched regions are repaired to either wild-type or mutant sequence. (B) The presented phagemid vector designed for the construction of synthetic Fab libraries contains both a bacterial (dsDNA ori) and a filamentous phage (f1 ori) origin of replication. In addition to a selectable marker that confers resistance to ampicillin and carbenicillin (Amp$^R$), the vector contains a bicistronic Fab expression cassette, the details of which are shown in the inset. The bicistronic expression of both variable light (VL) and variable heavy (VH) chains are controlled by a *PhoA* promoter, and both chains contain a stII leader peptide for secretion of expressed polypeptides into the bacterial periplasm. The first open reading frame encodes the variable (VL) and the constant domain (CL) of the light chain in frame with a Flag tag to allow simple detection of displayed Fab protein on the phage surface by ELISA. The second open reading frame encodes the variable (VH) and the first constant domain (CH1) of the heavy chain fused to a dimerization domain (DD) required for bivalent Fab display (Lee et al. 2004). Furthermore, the heavy chain is fused to the carboxy-terminal domain of a truncated pIII minor coat protein. Following expression and secretion in the bacterial periplasm, the light and heavy chains can appropriately pair to form a functional Fab anchored to the phage inner membrane via fusion to pIII protein. Superinfection of *E. coli* cells harboring the phagemid vector with M13 helper phage results in assembly of Fabs displayed on the surfaces of phage particles. (C) Agarose gel electrophoresis of in vitro synthesized CCC-dsDNA. Upon enzymatic extension with mutagenic oligonucleotides, dU-ssDNA template (lane *1*) is converted to CCC-dsDNA (lane *2*). The band with the lowest electrophoretic mobility (strand displaced) is the result of the activity of the T7 DNA polymerase, which like other DNA polymerases, can catalyze strand displacement synthesis on nicked, duplex DNA templates (Lechner et al. 1983). Unlike CCC-dsDNA, the strand displaced product provides a low mutation frequency (~20%) and can be transformed in competent *E. coli* SS320 cells, albeit with lower efficiency than CCC-dsDNA. The intermediate band (nicked dsDNA) derives from correctly extended but unligated dsDNA. (1C, Reproduced from Tonikian et al. 2007.)

QIAquick spin columns. Centrifuge the columns at 11,000g for 1 min at 25°C. Discard the flowthrough.

*DNA binding to QIAquick spin columns is efficient only at pH ≤ 7.5, under which the QG buffer color is yellow. If the reaction mixture turns orange or violet upon addition of buffer QC, adjust the pH of the solution by adding 10 μL 3 M sodium acetate pH 5.0 before applying the sample to the columns.*

26. Add 750 μL PE buffer to each column and centrifuge at 24,800g for 30 sec at 25°C. Discard the flowthrough.

27. Add 750 μL PE buffer to each column and incubate for 1 min at 25°C. Centrifuge columns at 24,800g for 30 sec at 25°C. Discard the flowthrough.

    *Incubation after the addition of PE buffer ensures greater salt removal from the DNA-containing solution, therefore resulting in greater quality of the plasmid solution and preventing potential arcing during electroporation.*

28. Transfer the columns into two new 1.5-mL microcentrifuge tubes and remove excess ethanol (contained in the PE buffer) by centrifugation at 24,800g for 1 min at 25°C.

29. Transfer the columns into two new 1.5-mL microcentrifuge tubes and add 30 μL sterile $H_2O$ for irrigation to the center of each membrane. Incubate the columns for 10 min at 37°C and elute DNA by centrifuging the columns at 24,800g for 1 min at 25°C.

30. Combine the eluants from the two columns and determine the concentration of the CCC-dsDNA heteroduplex by measuring absorbance at 260 nm ($A_{260}$ = 1.0 for 50 ng/μL CCC-dsDNA). Eluted DNA can be used immediately for electroporation of *E. coli*, or it can be frozen at −20°C for later use. Avoid repeated freeze and thaw cycles.

31. Analyze the efficiency of the mutagenesis reaction by loading 5 μL CCC-dsDNA alongside the dU-ssDNA template and 1-kb DNA Ladder onto a 1% agarose gel using standard procedures. A successful reaction results in the complete conversion of dU-ssDNA template into CCC-dsDNA (Fig. 1C).

    *See Troubleshooting.*

## Conversion of CCC-dsDNA into a Phage-Displayed Fab Library

*The final step required to construct highly diverse phage-displayed antibody libraries is the transformation of the CCC-dsDNA heteroduplex into an E. coli dut+/ung+ host containing the F′ episome to enable M13 bacteriophage infection and propagation. See the Discussion for additional information. The use of E. coli SS320 electrocompetent cells preinfected with M13KO7 simplifies the transformation procedure by bypassing the need for superinfection of bacteria with helper phage.*

32. Prewarm 25 mL SOC medium at 37°C (20 mL in a 250-mL baffled flask and 5 mL in a 15-mL tube).

33. Chill purified CCC-dsDNA heteroduplex (20 μg are sufficient to construct one phage-displayed library) and a 0.2-cm gap electroporation cuvette on ice.

34. Thaw a 350 μL aliquot of electrocompetent *E. coli* SS320 cells (∼$3 \times 10^{11}$ cfu/mL) on ice. Add the purified CCC-dsDNA to the cells and mix by pipetting several times. Avoid introducing bubbles and exceeding the maximum volume capacity of the cuvette.

35. Transfer the mixture to the chilled electroporation cuvette and wipe the sides of the cuvette with a paper tissue. Electroporate according to the manufacturer's instructions. For the BTX ECM-630 electroporation system (BTX) we use the following settings: 2.5 kV field strength, 125 Ω resistance, and 50 μF capacitance.

36. Immediately after electroporation, rescue bacterial cells by adding 1 mL prewarmed SOC medium to the cuvette. Transfer the cells and SOC to the 250-mL baffled flask containing 20 mL of prewarmed SOC medium. Rinse the cuvette with the remaining prewarmed 4 mL of SOC medium to ensure the collection of all transformed cells.

*The temperature and time gap for bacterial rescue are crucial for successful library construction. Always keep electrocompetent E. coli SS320 cells, CCC-dsDNA and electroporation cuvettes on ice before transformation. Quickly adding prewarmed SOC medium to electroporated cells favors cell recovery from electric shock, ensuring a greater cell survival rate (~50% cells survive after electroporation).*

37. Incubate cells for 30 min at 37°C with 200 rpm shaking.

38. To determine library diversity, transfer 10 µL from the 250-mL baffled flask and perform eight 10-fold serial dilutions into 90 µL 2× YT medium in a 96-well plate. Plate 5 µL of each dilution, using a multichannel pipette, onto an LB-Carb plate (selecting phagemid-containing cells), an LB-Kan plate (to determine the titer of M13KO7 infected cells), and an LB-Tet plate (to determine the total cell concentration). Incubate plates overnight at 37°C.

    *The titer from the LB-Tet and LB-Kan plates should be approximately the same. The titer of phagemid-containing cells on LB-Carb plates is expected to be 10-fold lower. As an alternative, it is possible to estimate phage titer spectrophotometrically ($OD_{268} = 1.0$ for a solution containing $5 \times 10^{12}$ phage/mL).*

    *See Troubleshooting.*

39. Immediately after 30 min of incubation, transfer the culture containing electroporated E. coli SS320 cells into a 2-L baffled flask containing 500 mL 2× YT/Carb/Kan medium. Incubate the culture for 20 hours at 37°C with 200 rpm shaking.

40. Following amplification of phage, transfer the culture to a 1-L centrifuge bottle and centrifuge bacteria at 11,000g for 10 min at 4°C.

41. Inoculate 25 mL 2× YT/Tet medium with a single colony of E. coli OmniMAX 2 T1$^R$ cells from a fresh LB-Tet plate. Incubate the culture for ~6 h at 37°C with 200 rpm shaking.

    *Tetracycline is used to select the F' episome in OmniMAX 2 T1$^R$ cells.*

42. Transfer the supernatant from Step 40 to a fresh 1-L bottle containing 120 mL PEG/NaCl solution prechilled on ice. To precipitate the phage, mix by inversion and incubate on ice for 20 min.

43. Precipitate the phage by centrifugation at 11,950g for 30 min at 4°C. Decant the supernatant and centrifuge at 11,950g for additional 3 min at 4°C. Remove the remaining supernatant with a pipette.

44. Gently resuspend the phage in 20 mL prechilled filter-sterilized PBT buffer. Transfer the solution into a 50-mL Falcon tube and remove bacterial cell debris by centrifugation at 24,800g for 10 min at 4°C.

45. Transfer the supernatant to a clean new tube containing 5 mL ice-cold PEG/NaCl solution. Mix the tube by inversion and incubate it on ice for 20 min.

46. Centrifuge the phage at 24,800g for 30 min at 4°C. Decant the supernatant and centrifuge at 24,800g for additional 3 min at 4°C. Remove the remaining supernatant with a pipette. Resuspend the phage pellet in 5 mL filter-sterilized PBT buffer.

47. Determine the concentration of the phage library by infecting mid-log phase ($OD_{600} = 0.6-0.8$) E. coli OmniMAX 2 T1$^R$ cells from Step 41. Dilute 10 µL of the phage library (from Step 46) into 90 µL 2× YT medium and perform 12 10-fold serial dilutions. Transfer 90 µL of mid-log phase E. coli OmniMAX 2 T1$^R$ cells into 12 wells of a 96-well plate and infect cells by adding 10 µL of a given phage dilution to each well. Seal the plate with a gas-permeable adhesive sealing film and incubate the plate for 30 min at 37°C with 200 rpm shaking.

48. Using a multichannel pipette, spot 5 µL of each dilution in a grid-like pattern onto an LB-Carb plate (selecting phagemid-containing cells), an LB-Kan plate, and an LB-Tet plate. Incubate the plates overnight at 37°C and then, for each dilution, count the number of colonies that grew onto the LB-Carb plate. To avoid counting cell doublets we suggest counting the number of single colonies present at each dilution. For determining the phage titer use the following formula:

$$(25 \text{ mL}) * [(\text{number of colonies}/5 \text{ µL}) * (\text{dilution factor})] * 1000.$$

Chapter 8

*Here 25 mL is the culture volume used to rescue bacteria following electroporation (Step 36) and 5 µL is the volume spotted onto the LB agar plates. The dilution factor refers to the factor by which the bacterial culture from Step 36 is diluted to give colonies.*

*The expected phage concentration is $10^{12}$–$10^{13}$ pfu/mL.*

49. Use the obtained library immediately for selection experiments. Alternatively, to store the library, add glycerol to a final concentration of 40% and store at −80°C. We recommend dividing the library into 1-mL aliquots for avoiding freeze and thaw cycles. Frozen libraries can be stored indefinitely at −80°C.

## Selection of High-Affinity Synthetic Fabs

*Here, we describe the most common method of selecting antibodies, which relies on incubating the phage-displayed Fab library with the antigen immobilized on an immunoplate. See the Discussion for more information. We recommend fusing the target of interest to a carrier protein, such as glutathione S-transferase (GST) or maltose-binding protein (MBP), to circumvent difficulties in expression and purification of the target, as well as to maximize optimal orientation for antigen presentation to the antibody library (Feng et al. 2011), and prevent target unfolding. Alternatively, to avoid target unfolding due to nonspecific hydrophobic interactions with the matrix of immunoplates, it is possible to immobilize site-specifically biotinylated targets (Fairhead and Howarth 2015) onto streptavidin-coated immunoplates.*

50. Coat MaxiSorp 96-well plate wells with 100 µL of antigen solution at 5–10 µg/mL in PBS pH 7.4 and incubate overnight at 4°C or for 2 h at 37°C. The number of wells to be coated is dependent on the diversity of the library. Typically, for libraries containing 10 billion unique Fabs, we use eight wells. In the case of targets fused to carrier proteins or immobilized onto the plate via the streptavidin-biotin system, coat an equal number of wells containing 10 µg/mL of carrier protein (e.g., GST or MBP, lacking the target fusion) or streptavidin. These wells will be used to deplete antibodies that bind to the carrier protein or streptavidin from the phage pool.

51. Remove the coating solution and block wells by adding 300 µL/well of PBS pH 7.4 containing 0.5% (w/v) BSA. Incubate the immunoplate for 1 h at 25°C with gentle shaking.

52. Remove the blocking solution and wash wells three times with 300 µL/well of PT buffer.

53. (Optional) When using targets fused to carrier proteins or immobilized to plates via streptavidin, perform a negative selection step by applying 100 µL of library phage solution in PBT buffer to wells coated only with streptavidin or the carrier protein. Following incubation for 1 h at 25°C with gentle shaking, transfer phage pools depleted of Fabs that recognize streptavidin or the carrier protein to wells containing the desired target antigen (Step 54). If performing this negative selection, wash wells as in Step 55 and elute phage particles as described in Step 56.

54. Add 100 µL of library phage solution in PBT buffer to each of the coated wells. Incubate for 1 h at 25°C with gentle shaking.

    *The number of phage particles in each well should exceed the library diversity by 1000-fold. Thus, for a library with a diversity of $1 \times 10^{10}$, $1 \times 10^{13}$ phage should be used.*

55. Remove the phage solution and wash wells with 300 µL/well PT buffer. The stringency of the washing step can be modified by adjusting the number of washing steps performed, the duration of each step (e.g., by incubating wells for 5 min with PT buffer before proceeding to the next wash step), and the concentration of Tween-20 used in the buffer. In a typical selection experiment, we incrementally increase the number of washing steps with each round of selection, starting from four wash steps in round 1 to ≥20 wash steps in round 5.

56. To elute bound phage, add 100 µL/well of 100 mM HCl. Incubate for 5 min at 25°C with gentle shaking. Transfer the HCl solution to a 1.5-mL microcentrifuge tube and immediately neutralize the pH of the solution by adding 1/10 of the HCl volume of 1 M Tris pH 11.

57. Use half of the eluted phage solution to infect 10 volumes of mid-log phase OmniMAX 2 T1$^R$ cells ($OD_{600}$ = 0.6–0.8) growing in 2× YT/Tet medium as in Steps 41 and 47. Incubate culture for

30 min at 37°C with 200 rpm shaking. Store the remaining half of the eluted phage for up to 1 wk at 4°C or for longer periods of time at −20°C following the addition of 30% glycerol to phage. In case of bacterial cell contamination, storing the eluted phage allows reinfection and amplification of the phage pool without the need to reperform the selection.

58. Following the 30 min incubation, determine the number of eluted phages by diluting 10 µL of phage-infected OmniMAX 2 T1$^R$ cells into 90 µL 2× YT medium and performing eight 10-fold serial dilutions as in Step 48. Plate 5 µL of each dilution onto an LB-Carb plate. Incubate plates overnight at 37°C. Determine the enrichment ratio by dividing the number of counted colonies of phage eluted from target-immobilized wells by the number of colonies obtained from phage eluted from wells coated only with streptavidin or carrier proteins.

59. To the remainder of the OmniMAX 2 T1$^R$ cells infected with eluted phage in Step 57, add M13KO7 helper phage to a final concentration of $10^{10}$ pfu/mL. Incubate for 1 h at 37°C with shaking at 200 rpm.

60. Transfer the superinfected *E. coli* OmniMAX 2 T1$^R$ cells to a 250-mL baffled flask containing 25 volumes of 2× YT/Carb/Kan and incubate overnight at 37°C with 200 rpm shaking.

61. Transfer the culture to a 50-mL Falcon tube and pellet bacterial cells by centrifugation for 10 min at 11,000$g$ and 4°C.

62. Transfer the supernatant to a new 50-mL Falcon tube containing 5 mL ice-cold PEG/NaCl. Mix the tube by inversion and incubate on ice for 20 min.

63. Precipitate phage by centrifuging for 30 min at 24,800$g$ and 4°C. Decant the supernatant and centrifuge for 3 min at 24,800$g$ and 4°C. Remove the remaining supernatant with a pipette. Resuspend the phage pellet in 1 mL of filter-sterilized PBT buffer.

64. Transfer the phage to a 1.5-mL microcentrifuge tube and centrifuge the tube for 5 min at 24,800$g$ and 4°C to remove bacterial cell debris. Transfer the phage-containing supernatant to a new 1.5-mL microcentrifuge tube and determine the phage concentration spectrophotometrically as described in Step 38.

65. Repeat the selection cycle (Steps 54–64) until the enrichment ratio has plateaued. Typically, significant enrichment is observed from round 3 and selections beyond round 5 are seldom necessary.

66. Pick individual clones for analysis of Fab identity via DNA sequencing and analysis of binding specificity via phage enzyme-linked immunosorbent assay (ELISA).

## Binding Analysis of Selected Fabs by Phage ELISA

*Following the selection process, the resulting phage pool contains target-binding clones, but likely also contains nonbinding clones as well as nonspecific binders. Binding specificity can be easily assessed by direct-binding phage ELISA using the culture medium in which phage particles displaying unique Fabs are secreted (Fig. 2). The procedure described below is optimized for the use of 384-well MaxiSorp immunoplates.*

67. Inoculate 96 mini tubes of a Mini tube 96-well system (Axygen Scientific 89005-582) with 400 µL 2× TY/Carb medium containing M13KO7 helper phage to a final concentration of $10^{10}$ pfu/mL with single colonies harboring selected phagemid from Step 58. Incubate overnight at 37°C with 200 rpm shaking.

   *We recommend using 96-well mini tubes as they ensure good aeration and optimal growth of bacterial cultures. However, 96-deep-well plates can be used as an alternative.*

68. Coat MaxiSorp immunoplate wells with 25 µL of antigen solution, carrier proteins or streptavidin at 2 µg/mL in PBS pH 7.4. Incubate overnight at 4°C or for 2 h at 37°C.

69. Remove the coating solution and block each well by adding 60 µL PBS pH 7.4 containing 0.5% (w/v) BSA. Incubate the immunoplate for 1 h at 25°C with gentle shaking.

Chapter 8

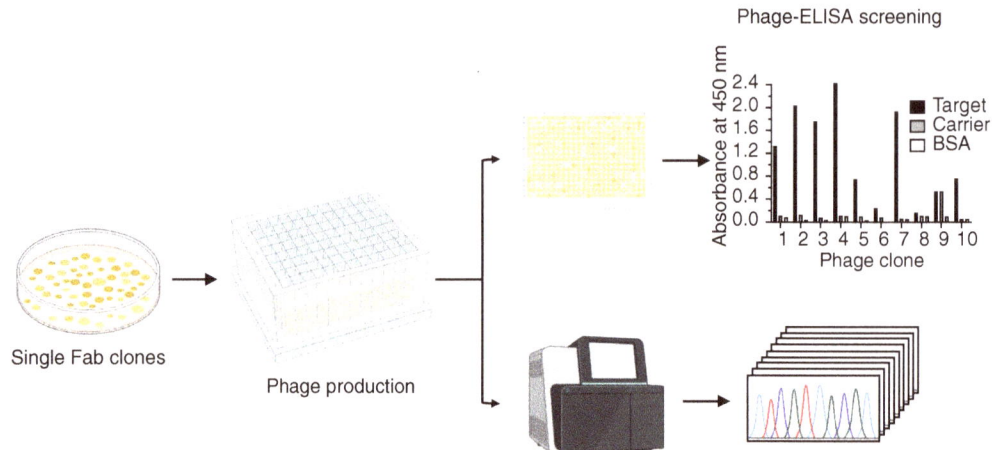

FIGURE 2. Workflow for identification of target-specific antibodies. Following selection, phage pools likely contain target-specific Fabs and also nonbinding or unspecific clones. To identify target-specific Fabs, following the production of phage particles displaying a unique Fab clone, monoclonal phage ELISA is performed. Binding specificity is determined by comparing the absorbance measured in wells coated with the target antigen to that of wells coated with the carrier protein only or BSA (used as a negative control). Typically, clones that display 10 times greater binding to target antigen than the carrier protein are considered specific. Selected target-specific clones are sequenced by Sanger DNA sequencing to determine the identity of each Fab.

70. Remove the blocking solution and wash each well four times with 90 μL PT buffer.

71. Centrifuge the 96-well microtubes for 10 min at 4°C at 2750g to remove bacteria, and transfer phage-containing supernatant to new 96-well microtubes.

72. Dilute the phage-containing supernatant threefold with PBT buffer.

73. Transfer 25 μL of each diluted phage supernatant to immunoplate wells from Step 70 and incubate for 30 min at 25°C with gentle shaking. In addition to binding to the target antigen and carrier proteins or streptavidin, binding should also be assayed using wells blocked with BSA only (negative control).

74. Remove the phage solution and wash each well four times with 90 μL PT buffer.

75. Add 25 μL horseradish peroxidase-conjugated anti-M13 antibody diluted 1/3000 in PBT buffer to each well. Incubate the immunoplate for 30 min at 25°C with gentle shaking.

76. Remove the anti-M13 HRP-conjugated antibody solution and wash each well six times with 90 μL PT buffer.

77. Add 25 μL of freshly prepared TMB substrate to each well. Incubate the plate for 5–10 min at 25°C with gentle shaking and allow color to develop.

78. Quench the colorimetric reaction by adding 25 μL 1 M $H_3PO_4$ to each well and read the immunoplate spectrophotometrically at 450 nm.

79. To determine binding specificity, subtract absorbance values obtained from wells blocked with BSA only and compare the measured absorbance intensity of wells coated with the target antigen to that of wells coated only with the carrier protein or streptavidin.

## DNA Sequencing of Target-Specific Phage Clones

*After direct binding analysis, Fab identity can be deduced by sequencing of the phagemid DNA encapsulated in the phage virions. Although next-generation sequencing is emerging as a more robust and comprehensive method to identify target-specific antibody clones (Yang et al. 2017; Juds et al. 2020), Sanger DNA sequencing is still widely used.*

*Here, we describe the method to perform Sanger DNA sequencing in a high-throughput 96-well format by using the same phage-containing supernatant used for screening of target-specific Fab clones.*

80. Assemble a PCR reaction mixture as follows:
    2.5 µL 10× ThermoPol buffer
    0.25 µL 10 mM dNTP mix
    0.25 µL each PCR primer
    2 µL phage supernatant from Step 71
    0.5 µL Taq polymerase (2.5 units)

    Add sterile irrigation $H_2O$ to bring the final reaction volume to 25 µL.
    PCR primers can be designed to amplify the phagemid DNA fragment to be sequenced. The reaction mixture can be scaled up and dispensed in a 96-well PCR plate for high-throughput sequencing.

    *We recommend designing primers annealing at least 100 bp upstream and downstream from the VH- and VL-encoding sequences. Since The VH and VL size is ~650 bp we suggest amplifying and sequencing each chain separately.*

81. Amplify the DNA fragment using the following PCR program: 5 min at 95°C, 30 cycles of amplification (30 sec at 95°C, 45 sec at 55°C, 1 min at 68°C), 5 min at 68°C, and storage at 4°C.

82. Assess the presence of PCR amplicons by loading 5 µL of each reaction onto a 1% agarose gel using standard procedures.

83. Assemble a clean-up reaction mixture by mixing:
    0.5 µL SAP (2.5 units)
    0.3 µL ExoI (3 units)

    Add sterile irrigation $H_2O$ to bring the final reaction volume to 2 µL.
    Dispense 2 µL of clean-up reaction mixture into each well of the 96-well PCR plate containing PCR amplicons.

    *ExoI is needed to degrade the residual PCR primers, whereas SAP is required to dephosphorylate the remaining dNTPs. Both enzymes are added directly to the PCR reaction after thermal cycling, without changing buffer.*

84. Incubate the clean-up reactions for 20 min at 37°C and inactivate the enzymes for 15 min at 80°C.

85. Analyze the PCR products that have been treated with ExoI and SAP by Sanger DNA sequencing.

## TROUBLESHOOTING

*Problem (Step 18):* dU-ssDNA bands appear as a smear on the agarose gel.
*Solution:* Apply an additional wash (Step 14) to the purification column, as the high-salt PE buffer composition will remove residual nuclease activity. To further increase the purity, it is possible to perform an ethanol precipitation of the obtained dU-ssDNA using the protocol described by Green and Sambrook (2016).

*Problem (Steps 24 and 31):* The dU-ssDNA is not entirely converted to CCC-dsDNA.
*Solution:* We suggest optimizing the reaction by varying the oligonucleotide:template molar ratio and/or testing different oligonucleotide annealing temperatures. Conditions that will enable complete conversion of dU-ssDNA but result in large amounts of strand displaced DNA relative to CCC-dsDNA (Fig. 1C) might still be electroporated as strand displaced products can be transformed in *E. coli* hosts, albeit with lower efficiency. In the case of complete conversion of dU-ssDNA but low

CCC-dsDNA yields, it is advisable to increase the number of in vitro CCC-dsDNA synthesis reactions performed.

*Problem (Step 38):* A low number of transformants is observed.

*Solution:* The described procedure allows the development of phage display libraries containing $10^{10}$ unique antibody clones. If the number of transformants is low, we suggest repeating the in vitro CCC-dsDNA synthesis reactions (Steps 21–24) and electroporating the obtained CCC-dsDNA immediately after its purification (Steps 25–31). To increase the electroporation efficiency, we suggest including an additional wash during CCC-dsDNA purification (Step 14), as contaminant salts and proteins can lower electroporation performance. In addition, we suggest diluting the CCC-dsDNA concentration and performing multiple electroporations in parallel as the number of transformants typically decreases with increasing CCC-dsDNA concentration.

## DISCUSSION

### Phagemid Design

A phagemid for displaying the library template as a bivalent Fab (Fig. 1B) is a specialized vector that contains a double-stranded DNA origin of replication (dsDNA ori) for replication in *E. coli*, and an ssDNA filamentous phage origin of replication (f1 ori) that enables ssDNA to be packaged into phage particles. Additionally, the presented phagemid vector contains a β-lactamase gene (*Bla*) conferring resistance to ampicillin and carbenicillin, but other selectable markers can be used.

The Fab is displayed on the surface of the phage particle via fusion of the heavy chain to the carboxy-terminal domain of the truncated M13 bacteriophage minor coat protein 3 (pIII) (Fig. 1B). Bicistronic Fab expression is controlled by the alkaline phosphatase promoter (*PhoA*) as such promoter is well-suited for expression of Fab proteins in *E. coli* (Luo et al. 2019), but alternatively, other promoters can be used. Following translation, the light chain and the heavy chain are secreted into the bacterial periplasm via the stII secretion peptide, where they can assemble in the oxidizing environment of the periplasm to form the heterodimeric Fab protein. However, other leader peptides have been used to direct secretion of antibodies into the periplasm (Thie et al. 2008).

The phagemid presented here allows the display of bivalent Fabs via the insertion of a dimerization domain (DD), consisting of the IgG1 core hinge region, placed between the first constant Ig domain (CH1) and the pIII protein (Fig. 1C; Persson et al. 2013). It should be noted that the bivalent display of Fabs might result in libraries with increased avidity that could pose concerns for applications requiring stringent selections solely relying on the intrinsic affinity of Fabs. However, bivalently displayed Fabs have the advantage of efficiently mimicking the binding mode of natural IgGs, being readily reformatted as full-length antibodies and allowing the isolation of binders with weak affinity that cannot be recovered using monovalent Fab libraries (Lee et al. 2004). Furthermore, the DD used in the presented phagemid enables the formation of stable VH:VL heterodimers that results in a controlled avidity effect, thereby avoiding the artifacts typical of extreme avidity as observed in unstable and prone to oligomerize scFv libraries (Lee et al. 2004).

Display of Fab heterodimers on the surface of phage particles is achieved by coinfecting phagemid-harboring *E. coli* cells with helper phage, such as M13KO7, which provides all the additional coat proteins necessary for the assembly of virions.

### Library Construction

The library here described takes advantage of the highly stable and well-characterized framework derived from the human anti-HER2 antibody 4D5 (Albanell and Baselga 1999), which contains variable domains from the predominant $V_H 3$ and $V\kappa 1$ subgroups (Cox et al. 1994; Suzuki et al. 1995). First, mutagenic oligonucleotides are used to introduce, via oligonucleotide-directed mutagenesis (Huang et al. 2012), stop codons at sites to be diversified in the library. In this way, the presence of stop codons eliminates wild-type Fab display, ensuring the display of diversified Fabs only. The stop codon–containing template can then be used to introduce genetic diversity into the CDRs via oligonucleotide-directed mutagenesis. Uracil-containing ssDNA (dU-ssDNA) stop codon–containing template (purified from an *E. coli dut−/ung−* strain) is annealed to mutagenic oligonucleotides designed to replace stop codons with desired degenerate codons. Phosphorylated mutagenic oligonucleotides prime the T7 DNA polymerase–mediated synthesis of a complementary DNA strand, which is ligated by the activity of T4 DNA ligase, thus forming a covalently closed circular dsDNA (CCC-dsDNA) heteroduplex. Upon transformation of the CCC-dsDNA into an *E. coli dut+/ung+* host, the uracil-containing DNA strand is preferentially degraded (Warner et al. 1981), resulting in efficient replication and propagation of the randomized strand.

The mutagenic oligonucleotides are designed with complementarity to the sequences immediately preceding and following the introduced stop codons. The oligonucleotide sequence between the two complementary regions can be of any desired sequence and length, therefore enabling facile and precise control over the introduced library diversity. Furthermore, provided the absence of overlap, mutations of multiple regions farther apart in the template dU-ssDNA can be accomplished in a single reaction by annealing multiple mutagenic oligonucleotides. In this way, CDR diversity can be introduced not only by the encoded genetic diversity of mutagenic oligonucleotides but also by varying the length of CDRs using pools of degenerate oligonucleotides.

## CCC-dsDNA Conversion

In this protocol, we use the *E. coli* SS320 strain preinfected with M13KO7 helper phage that enables high-efficiency electroporation and phage production (Sidhu et al. 2000). Methods and procedures for preparing M13KO7 helper phage and electrocompetent *E. coli* SS320 cells have been previously described (Tonikian et al. 2007). Upon electroporation of the in vitro synthesized CCC-dsDNA, each *E. coli* SS320 cell will be able to produce phage particles displaying a unique Fab without the need for further helper phage infection.

## Selection of High-Affinity Synthetic Fabs

Phage-displayed synthetic antibody libraries can be used to isolate high-affinity antibodies for virtually any target of interest. There are many strategies amenable for the successful selection of highly specific and tightly binding antibodies (Zhang et al. 2004; Jara-Acevedo et al. 2016; Chen et al. 2020; Ministro et al. 2020; Kelil et al. 2021). However, the antibody isolation strategies should be carefully designed to retain natural protein conformation and maximize target accessibility, as well as to ensure antibody function in particular applications (e.g., correctly folded targets are needed to isolate antibodies functional in flow-cytometry experiments, whereas wholly or partly denatured antigens might be preferable for isolating antibodies that perform well in western blot assays) (Uhlen et al. 2016). Here, we describe the most commonly used method, which relies on incubating the phage-displayed Fab library with the antigen immobilized on an immunoplate.

# RECIPES

### 2× YT/Carb/Kan/Uridine Medium

2× YT medium <R>
100 µg/mL carbenicillin from a stock solution of Carbenicillin (Carb; 100 mg/mL) <R>
25 µg/mL kanamycin from a stock solution of Kanamycin (Kan; 25 mg/mL) <R>
0.25 µg/mL uridine from a stock solution of Uridine (0.25 mg/mL) <R>

Freshly prepare.

## Chapter 8

### 2× YT/Carb/Cam Medium

2× YT medium <R>
100 µg/mL carbenicillin from a stock solution of Carbenicillin (Carb; 100 mg/mL) <R>
10 µg/mL chloramphenicol from a stock solution of Chloramphenicol (Cam; 10 mg/mL) <R>

Freshly prepare.

### 2× YT/Carb/Kan Medium

2× YT medium <R>
100 µg/mL carbenicillin from a stock solution of Carbenicillin (Carb; 100 mg/mL) <R>
25 µg/mL kanamycin from a stock solution of Kanamycin (Kan; 25 mg/mL) <R>

Freshly prepare.

### 2× YT/Tet Medium

2× YT medium <R>
10 µg/mL tetracycline a stock solution of Tetracycline (10 mg/mL) <R>

Freshly prepare.

### 2× YT Medium

Measure ∼900 mL of distilled H$_2$O. Add 16 g of Bacto Tryptone, 10 g of Bacto yeast extract, and 5 g of NaCl. Mix to dissolve. Adjust pH to 7.0 with 5 N NaOH. Adjust to 1 L with distilled H$_2$O. Sterilize by autoclaving.

### 40% PEG 8000

| Reagent | Amount to add (for 50 mL) | Final concentration |
|---|---|---|
| PEG 8000 (Sigma-Aldrich 89510-250G-F) | 20 g | 40% (w/v) |
| ddH$_2$O (sterile) | Add to 50 mL | |

Transfer the PEG 8000 to a 50-mL conical centrifuge tube and add ddH$_2$O to the 50-mL line. Mix vigorously to suspend the PEG 8000 in the liquid and displace air in the powder. Continue mixing (e.g., on a rotator) and adding ddH$_2$O to 50 mL until the PEG 8000 is completely dissolved. To rapidly dissolve the PEG 8000, the suspension can be briefly microwaved for no more than 15 sec with the unscrewed lid placed loosely on top of the conical tube. (Do not attempt to microwave the tube with the lid screwed on.) To prevent gradual loss of activity due to polyether peroxidation and chain shortening, store for up to 6 mo at 4°C in the dark or freeze 10-mL aliquots at −20°C for long-term storage.

### Chloramphenicol (Cam; 10 mg/mL)

Prepare 10 mg/mL chloramphenicol (Millipore-Sigma C3175-100MG) in 2-propanol (Millipore-Sigma I9516-1L).

Store for up to 4 mo at −20°C.

### Carbenicillin (Carb; 100 mg/mL)

Prepare 100 mg/mL carbenicillin disodium salt (Millipore-Sigma C1389-5G) in milli-Q-purified H$_2$O, filter sterilize.

Store indefinitely at −20°C.

### EDTA

EDTA (ethylenediamenetetraacetic acid)
NaOH

To prepare EDTA at 0.5 M (pH 8.0): Add 186.1 g of disodium EDTA·2H$_2$O to 800 mL of H$_2$O. Stir vigorously on a magnetic stirrer. Adjust the pH to 8.0 with NaOH (~20 g of NaOH pellets). Dispense into aliquots and sterilize by autoclaving. The disodium salt of EDTA will not go into solution until the pH of the solution is adjusted to ~8.0 by the addition of NaOH.

### Kanamycin (Kan; 25 mg/mL)

Prepare 25 mg/mL kanamycin sulfate (Millipore-Sigma K4000-25G) in milli-Q-purified H$_2$O, filter-sterilize.

Store indefinitely at −20°C.

### LB Agar Plates

7.5 g agar (Millipore-Sigma 05039-500G)
16 g of Bacto Tryptone (Millipore-Sigma T7293-1KG)
10 g of yeast extract (Millipore-Sigma 70161-2.5KG)
5 g of NaCl (Millipore-Sigma S3014-5KG)

Mix to dissolve and adjust pH to 7.0 with 5 N NaOH. Adjust to 1 L with distilled H$_2$O. Sterilize by autoclaving. Add an appropriate antibiotic to the desired concentration before pouring plates.

Store plates for up to 1 mo at 4°C.

### MP Buffer

Dissolve 3.3 g citric acid monohydrate (Millipore-Sigma C1909-25G) at room temperature in 2 mL sterile H$_2$O for irrigation (B. Braun Medical Inc. 971-R5000-01). Filter-sterilize. Although the solution can be stored at −20°C, we recommend preparing immediately before its use for optimal results.

### MLB Buffer

1 M sodium perchlorate (Millipore-Sigma 410241-100G)
30% (v/v) 2-propanol (Millipore-Sigma I9516-1L)

Store the solution indefinitely at −20°C.

### PEG (20%)/2.5 M NaCl

| Reagent | Amount to add (for 50 mL) | Final concentration |
|---|---|---|
| 40% PEG 8000 <R> | 25 mL | 20% |
| NaCl (5 M) | 25 mL | 2.5 M |

Store for up to 6 mo at 4°C.

### PBT Buffer (pH 7.4)

Phosphate-buffered saline (PBS; pH 7.4) <R>
0.5% (w/v) Bovine serum albumin (BSA) (Millipore-Sigma A7030-500G)
0.05% (v/v) Tween-20 (Millipore-Sigma P1379-250ML)

Filter-sterilize. The solution can be stored for 1 mo at 4°C. However, to avoid any source of contamination, we recommend preparing it immediately before use.

### Phosphate-Buffered Saline (PBS; pH 7.4)

8 mM $Na_2HPO_4$ (Millipore-Sigma S9763-100G)
1.5 mM $KH_2PO_4$ (Millipore-Sigma P5655-100G)
3 mM KCl (Millipore-Sigma P9541-500G)
137 mM NaCl (Millipore-Sigma S3014-5KG)

Adjust pH to 7.4 using HCl and autoclave. Store the solution indefinitely at room temperature. If any precipitate forms in the buffer during storage, it should be redissolved by warming up the buffer before use.

### PT Buffer

Phosphate-buffered saline (PBS; pH 7.4) <R>
0.05% (v/v) Tween-20 (Millipore-Sigma P1379-250ML)

Store the solution protected from light indefinitely at room temperature.

### SOC Medium

Per liter: To 950 mL of deionized $H_2O$, add:

| | |
|---|---|
| Tryptone | 20 g |
| Yeast extract | 5 g |
| NaCl | 0.5 g |

SOC medium is identical to SOB medium, except that it contains 20 mM glucose. To prepare SOB medium, combine the above ingredients and shake until the solutes have dissolved. Add 10 mL of a 250 mM solution of KCl. (This solution is made by dissolving 1.86 g of KCl in 100 mL of deionized $H_2O$.) Adjust the pH of the medium to 7.0 with 5 N NaOH (∼0.2 mL). Adjust the volume of the solution to 1 L with deionized $H_2O$. Sterilize by autoclaving for 20 min at 15 psi (1.05 kg/cm$^2$) on liquid cycle. Just before use, add 5 mL of a sterile solution of 2 M $MgCl_2$. (This solution is made by dissolving 19 g of $MgCl_2$ in 90 mL of deionized $H_2O$. Adjust the volume of the solution to 100 mL with deionized $H_2O$ and sterilize by autoclaving for 20 min at 15 psi [1.05 kg/cm$^2$] on liquid cycle.)

After the SOB medium has been autoclaved, allow it to cool to 60°C or less. Add 20 mL of a sterile 1 M solution of glucose. (This solution is made by dissolving 18 g of glucose in 90 mL of deionized $H_2O$. After the sugar has dissolved, adjust the volume of the solution to 100 mL with deionized $H_2O$ and sterilize by passing it through a 0.22-μm filter.)

### TM Buffer (10×)

0.1 M $MgCl_2$ (Millipore-Sigma 1374248-1G)
0.5 M Tris-HCl pH 7.5
Filter-sterilize.

Store the solution for up to 1 mo at 4°C.

### TAE

Prepare a 50× stock solution in 1 L of H$_2$O:
  242 g of Tris base
  57.1 mL of acetic acid (glacial)
  100 mL of 0.5 M EDTA (pH 8.0)

The 1× working solution is 40 mM Tris-acetate/1 mM EDTA.

### Tetracycline (10 mg/mL)

Prepare 10 mg/mL tetracycline hydrochloride (Millipore-Sigma T7660-25G) in ethanol (Millipore-Sigma 493511-1L). Protect the solution from light and store it indefinitely at −20°C.

### Uridine (0.25 mg/mL)

Prepare 0.25 mg/mL uridine (Millipore-Sigma U3750-1G) in sterile H$_2$O for irrigation (B. Braun Medical Inc. 971-R5000-01).

Filter-sterilize and store for 1 mo at −20°C.

## REFERENCES

Albanell J, Baselga J. 1999. Trastuzumab, a humanized anti-HER2 monoclonal antibody, for the treatment of breast cancer. *Drugs Today (Barc)* **35:** 931–946.

Chen L, Zhu C, Guo H, Li R, Zhang L, Xing Z, Song Y, Zhang Z, Wang F, Liu X, et al. 2020. Epitope-directed antibody selection by site-specific photocrosslinking. *Sci Adv* **6:** eaaz7825. doi:10.1126/sciadv.aaz7825

Cox JP, Tomlinson IM, Winter G. 1994. A directory of human germ-line Vκ segments reveals a strong bias in their usage. *Eur J Immunol* **24:** 827–836. doi:10.1002/eji.1830240409

Fairhead M, Howarth M. 2015. Site-specific biotinylation of purified proteins using BirA. *Methods Mol Biol* **1266:** 171–184. doi:10.1007/978-1-4939-2272-7_12

Feng B, Luo Y, Ge F, Wang L, Huang L, Dai Y. 2011. Site-oriented immobilization of fusion antigen directed by an affinity ligand, and its validation in an immunoassay. *Surf Interface Anal* **43:** 1304–1310. doi:10.1002/sia.3712

Green MR, Sambrook J. 2016. Precipitation of DNA with ethanol. *Cold Spring Harb Protoc* doi:10.1101/pdb.prot093377

Huang R, Fang P, Kay BK. 2012. Improvements to the Kunkel mutagenesis protocol for constructing primary and secondary phage-display libraries. *Methods* **58:** 10–17. doi:10.1016/j.ymeth.2012.08.008

Jara-Acevedo R, Diez P, Gonzalez-Gonzalez M, Degano RM, Ibarrola N, Gongora R, Orfao A, Fuentes M. 2016. Methods for selecting phage display antibody libraries. *Curr Pharm Des* **22:** 6490–6499. doi:10.2174/1381612822666161007153127

Juds C, Schmidt J, Weller MG, Lange T, Beck U, Conrad T, Börner HG. 2020. Combining phage display and next-generation sequencing for materials sciences: a case study on probing polypropylene surfaces. *J Am Chem Soc* **142:** 10624–10628. doi:10.1021/jacs.0c03482

Kelil A, Gallo E, Banerjee S, Adams JJ, Sidhu SS. 2021. CellectSeq: in silico discovery of antibodies targeting integral membrane proteins combining in situ selections and next-generation sequencing. *Commun Biol* **4:** 561. doi:10.1038/s42003-021-02066-5

Lechner RL, Engler MJ, Richardson CC. 1983. Characterization of strand displacement synthesis catalyzed by bacteriophage T7 DNA polymerase. *J Biol Chem* **258:** 11174–11184. doi:10.1016/S0021-9258(17)44401-2

Lee CV, Sidhu SS, Fuh G. 2004. Bivalent antibody phage display mimics natural immunoglobulin. *J Immunol Methods* **284:** 119–132. doi:10.1016/j.jim.2003.11.001

Luo M, Zhao M, Cagliero C, Jiang H, Xie Y, Zhu J, Yang H, Zhang M, Zheng Y, Yuan Y, et al. 2019. A general platform for efficient extracellular expression and purification of Fab from *Escherichia coli*. *Appl Microbiol Biotechnol* **103:** 3341–3353. doi:10.1007/s00253-019-09745-8

Ministro J, Manuel AM, Goncalves J. 2020. Therapeutic antibody engineering and selection strategies. *Adv Biochem Eng Biotechnol* **171:** 55–86. doi:10.1007/10_2019_116

Persson H, Ye W, Wernimont A, Adams JJ, Koide A, Koide S, Lam R, Sidhu SS. 2013. CDR-H3 diversity is not required for antigen recognition by synthetic antibodies. *J Mol Biol* **425:** 803–811. doi:10.1016/j.jmb.2012.11.037

Sambrook J, Russell DW. 2006. Preparation and transformation of competent *E. coli* using calcium chloride. *Cold Spring Harb Protoc* doi:10.1101/pdb.prot3932

Sidhu SS, Lowman HB, Cunningham BC, Wells JA. 2000. Phage display for selection of novel binding peptides. *Methods Enzymol* **328:** 333–363. doi:10.1016/S0076-6879(00)28406-1

Suzuki I, Pfister L, Glas A, Nottenburg C, Milner EC. 1995. Representation of rearranged VH gene segments in the human adult antibody repertoire. *J Immunol* **154:** 3902–3911. doi:10.4049/jimmunol.154.8.3902

Thie H, Schirrmann T, Paschke M, Dübel S, Hust M. 2008. SRP and Sec pathway leader peptides for antibody phage display and antibody fragment production in E. coli. *New Biotechnol* **25:** 49–54. doi:10.1016/j.nbt.2008.01.001

Tonikian R, Zhang Y, Boone C, Sidhu SS. 2007. Identifying specificity profiles for peptide recognition modules from phage-displayed peptide libraries. *Nat Protoc* **2:** 1368–1386. doi:10.1038/nprot.2007.151

Uhlen M, Bandrowski A, Carr S, Edwards A, Ellenberg J, Lundberg E, Rimm DL, Rodriguez H, Hiltke T, Snyder M, et al. 2016. A proposal for validation of antibodies. *Nat Methods* **13:** 823–827. doi:10.1038/nmeth.3995

Warner HR, Duncan BK, Garrett C, Neuhard J. 1981. Synthesis and metabolism of uracil-containing deoxyribonucleic acid in *Escherichia coli*. *J Bacteriol* **145:** 687–695. doi:10.1128/jb.145.2.687-695.1981

Yang W, Yoon A, Lee S, Kim S, Han J, Chung J. 2017. Next-generation sequencing enables the discovery of more diverse positive clones from a phage-displayed antibody library. *Exp Mol Med* **49:** e308. doi:10.1038/emm.2017.22

Zhang Y, Pool C, Sadler K, Yan H, Edl J, Wang X, Boyd JG, Tam JP. 2004. Selection of active ScFv to G-protein-coupled receptor CCR5 using surface antigen-mimicking peptides. *Biochemistry* **43:** 12575–12584. doi:10.1021/bi0492152

CHAPTER 9

# Considerations for Using Phage Display Technology in Therapeutic Antibody Drug Discovery

Mary Ann Pohl[1,3] and Juan C. Almagro[2]

[1]Ailux Biologics, Somerville, Massachusetts 02143, USA; [2]GlobalBio, Inc., Cambridge, Massachusetts 02138, USA

Phage display is a versatile and effective platform for the identification and engineering of biologic-based therapeutics. Using standard molecular biology laboratory techniques, one can create a highly diverse and functional antibody phage-displayed library, and rapidly identify antibody fragments that bind to a target of interest with exquisite specificity and high affinity. Here, we discuss key aspects for the development of an antibody discovery strategy to harness the power of phage display technology to obtain molecules that can successfully be developed into therapeutics, including target validation, antibody design goals, and considerations for preparing and executing phage panning campaigns. Careful design and implementation of discovery campaigns—regardless of the target—provides the best chance of identifying desirable antibody fragments for further therapeutic development, so these principles can be applied to any new discovery project.

## INTRODUCTION

Between the time George Smith first performed the experiments to display a peptide on the surface of an M13 bacteriophage (Smith 1985) and when he was then awarded the Nobel Prize in Chemistry along with Sir Gregory Winter in 2018 for using phage display to engineer antibody-based drugs (Smith 2019), this technology became a powerful approach for the identification of peptides and antibodies with specific binding and functional characteristics. With the explosion of the development of biologic-based therapeutics in the last few decades (Lawrence 2005; Scolnik 2009; Nelson et al. 2010; Ecker et al. 2015), phage display has replaced or complemented traditional hybridoma fusion technology as a method for antibody discovery, in both academic and pharmaceutical laboratories, and the success of this approach is demonstrated by the several dozens of phage-derived antibodies in clinical development. As of 2022, 14 antibodies have been approved by the U.S. Food and Drug Administration (FDA) and/or The European Medicines Agency (EMA) for the treatment of diverse diseases (for reviews, see Alfaleh et al. 2020; André et al. 2022).

Although phage display allows for the rapid identification of panels of antibody binders to a specific target, there are several caveats to this methodology that must be considered, as well as elements of the lead identification workflow before and after phage panning that must be considered to increase the chances of success. This becomes especially important as an increasing number of academic researchers are now looking to identify therapeutics as part of their translational research programs or in collaboration with biotechnology/pharmaceutical companies. Indeed, there is growing support to

[3]Correspondence: maryann.pohl@xtalpi.com

© 2026 Cold Spring Harbor Laboratory Press
Cite this introduction as *Cold Spring Harb Protoc*; doi:10.1101/pdb.top107757

encourage academic drug development (Frearson and Wyatt 2010; Everett 2015), but researchers in those settings are not necessarily trained in the process of designing and implementing an antibody discovery strategy that will yield molecules with the desirable profile to become therapeutics.

Here, we discuss some of the major considerations for developing an antibody discovery campaign with phage display that includes phage library selection, along with panning and screening approaches. These considerations typically include evaluating target characteristics, establishing antibody design goals, and creating a lead identification workflow. In addition, and as a part of this collection, we provide protocols to generate a semisynthetic library that can be used as a source of specific and developable antibodies for therapeutic development (Protocol 1: Semisynthetic Phage Display Library Construction: Design and Synthesis of Diversified Single-Chain Variable Fragments and Generation of Primary Libraries [Almagro and Pohl 2024a], Protocol 2: Semisynthetic Phage Display Library Construction: Generation of Filtered Libraries [Almagro and Pohl 2024b], and Protocol 3: Semisynthetic Phage Display Library Construction: Generation of Single-Chain Variable Fragment Secondary Libraries [Almagro and Pohl 2024c]).

## OVERVIEW OF THE ANTIBODY DISCOVERY PROCESS

Before beginning efforts to identify a therapeutic antibody, it is important to understand the general process and workflow that must be in place for efficient selection and characterization of molecules with the desired binding and/or functional properties. Figure 1 outlines this process; each of these steps could be discussed separately in dedicated reviews, but here we simply briefly discuss each of them and focus on the use of phage display for antibody discovery. We also focus on the early stages of the process, prior to preclinical development and Investigational New Drug (IND) filing.

The first step in the antibody discovery process is to develop a strategy to isolate molecules with the desired functional profile that can eventually be developed into biotherapeutics. As it is often said in science, "one day of planning experiments at your desk is worth two days at the bench." Thus, it is important to create a plan to mitigate potential risks and maximize chances for success. To begin, a molecule must be identified and validated as a relevant antigen that, when targeted with antibodies, elicits a therapeutic effect in vivo. In short, it needs to be determined whether a given molecule (1) is associated with disease, (*outlegends*f8*2) exhibits adequate expression on the cell surface or in the body fluids, and (3) is druggable by an antibody. In other words, an antibody drug is anticipated to be able to access the target in vivo and mediate a therapeutic benefit. Target identification and validation

| Target identification and validation | Feasibility assessment | Antibody design goals |
|---|---|---|
| • Primary data demonstrating a correlation of disease outcome and presence or function of the target<br>• For newly identified targets, tool antibodies can be generated for further elucidation of target biology and validation of disease relevance | • Tissue expression profile (potential for antigen sink and off-target effects)<br>• Protein localization, structure, and post-translational modifications<br>• Related proteins–homologs in humans and other species<br>• Target pathway (known protein–protein interactions and signaling cascades) | • Desired therapeutic format (mAb, CAR-T, ADC, bispecific, etc.)<br>• Binding specificity, selectivity, and affinity<br>• Cross-reactivity (NHP, mouse homologs)<br>• In vitro activity in biochemical or cell-based assay(s)<br>• Efficacy in disease-relevant animal model<br>• Favorable pharmacodynamics (safety profile)<br>• Favorable pharmacokinetics (sufficient half-life)<br>• Biophysical properties that allow for manufacturability and stability |

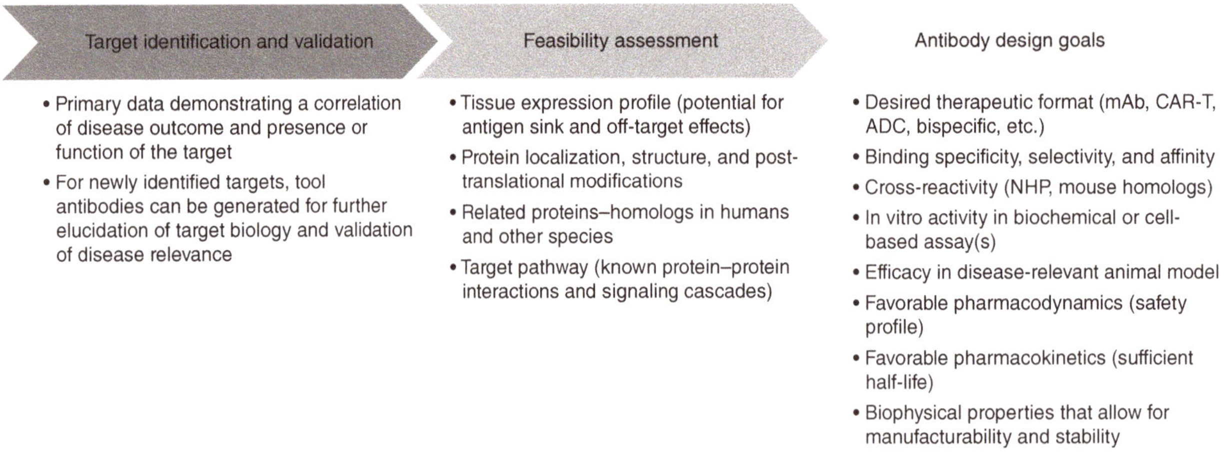

FIGURE 1. Development of an antibody discovery strategy. Listed are the major stages of preparation for the antibody discovery and development process, with factors for consideration at each stage to develop an antibody discovery strategy. See the text for details.

often arise from data generated in basic research laboratories. The research program may suggest, for example, that such a molecule is overexpressed in certain tumors or autoimmune disease (for reviews, see Carter et al. 2004; Hughes et al. 2011). Based on that preliminary information, a panel of proof-of-concept-specific antibodies to this molecule are generated and used for expression profiling in tumor versus normal cells, evaluating tissue specificity, or determining whether target engagement by an antibody interferes with protein–protein actions that result in alterations in signaling cascades to mediate a specific phenotype. In this case, the antibody discovery process may proceed without having fully elucidated the target biology, because the antibodies themselves will help answer these questions and thus validate the molecule as a druggable target.

The next step in developing the discovery strategy is the target feasibility assessment (Fig. 1). This process includes analysis of characteristics of the target, such as its class (e.g., soluble, membrane-bound protein, or a multipass transmembrane protein), structure, and post-translational modifications. For example, if the target is a multipass transmembrane protein, it is important to use a fragment of the protein in its native conformation as selector in the discovery campaign—and not unstructured linear peptide fragments—to identify antibodies that recognize the proper conformation of protein. Furthermore, glycosylation can significantly affect antibody binding site availability; therefore, it is prudent to determine the number of putative N-glycosylation sites on the target, which can be predicted in silico through computational tools such as PROSITE (Sigrist et al. 2013). If the target is a glycoprotein and is produced recombinantly for antibody discovery, it is important to do so in mammalian cells as opposed to in bacteria, to ensure glycosylation occurs and the molecule is in a conformation as close as possible to that of the functional native protein.

It is also important to determine the target's closest homologs in humans as well as orthologs in relevant species, mainly in rodents and nonhuman primates (NHPs), that could eventually be used for in vivo efficacy and safety studies. Knowing the level of similarity between a human target and that of its rodent and/or NHP orthologs will determine the likelihood that any antibodies identified will be cross-reactive to these species. Cross-reactivity between humans and relevant species such as rodents and NHPs used in efficacy and safety studies is often desired, so that no surrogate antibodies are needed. Moreover, if a given target is from a family of proteins with a high level of similarity among them, it may be necessary to incorporate counterscreens into the antibody screening funnel to ensure the selectivity of the antibody, and thus avoid unwanted toxicity due to targeting closely related molecules (Kushwaha et al. 2014). Alternatively, it may be wise to focus antibody generation efforts on domains of the protein that are most divergent from homologs, to improve the chances of identifying target-selective mAbs.

Once the aforementioned information on the target is obtained, a set of additional success criteria should be defined up front for the antibody discovery and optimization campaigns. These criteria include but are not limited to affinity, isotype, and potency in cell-based assays and/or in vitro functional assays, in addition to efficacy, safety, pharmacokinetics, pharmacodynamics, and developability. For instance, if the intended mechanism of action (MOA) is to disrupt a ligand–receptor interaction, then an affinity higher than the ligand's affinity to the receptor may be required. In some instances, such affinity should be in the low picomolar range (Tiwari et al. 2017). If the goal, however, requires merely binding to a membrane-bound protein, an affinity in the low nanomolar range may be sufficient. If the MOA is to block a ligand–receptor interaction, and no effector function such as antibody-dependent cellular cytotoxicity (ADCC) is desired, a human isotype such as immunoglobulin (Ig)G4 or IgG2 should be considered. On the other hand, if the MOA is supposed to destroy a tumoral cell via Fc engagement, ADCC is necessary, and thus IgG1 should be the isotype of choice.

In terms of efficacy, specific benchmarks can be set, such as, for instance, equal or greater reduction in tumor volume compared to existing therapeutics to that target in a mouse xenograft model. Depending on the intended product profile of the therapeutic, some of these criteria are more important than others. For example, if the end product is a chimeric T-cell antigen receptor (CAR-T) cell, the antibody fragment will be expressed on the patient's own cells, so manufacturability as a soluble recombinant protein is not a major consideration (Huang et al. 2020). Conversely, if the final product is an IgG, manufacturability traits such as high expression yield from the production cells, solubility in the formulation buffer, minimal or no aggregation, and short- and long-term stability are critical for development of the antibody-based drug. Alternatively, the goal may be to develop a molecule that requires less frequent dosing than current therapies. In this case, pharmacokinetics of the antibody is a top priority. Importantly, each of these antibody criteria must be as well defined as possible prior to initiation of the antibody discovery campaign and tailored to the specific antibody development program. It is essential to precisely outline as many antibody design goals as possible at the beginning of the project, as they will dictate the critical path for lead identification and optimization efforts, preclinical and clinical development costs and timelines, and, ultimately, probability of success.

The predetermined criteria defined in the antibody design goals allows for creation of a lead identification screening funnel that will be applied to any antibodies derived from a hybridoma fusion, sequence recovery from B-cell sorting, or, in our case, an antibody phage-displayed selection campaign (Fig. 2). After deciding on the selection strategy (i.e., target presentation, concentration, rounds of panning, and elution conditions, among others), the first step in most screening funnels is a primary binding assay, using either the soluble recombinant protein or cells expressing the target. From there, primary hits can be screened (or counterscreened) for cross-reactivity with orthologs from relevant species in efficacy and/or safety studies as well as with related proteins to assess selectivity. Secondary screenings can include binding to cells endogenously expressing the target of interest or evaluating binding in target knockout cells, to confirm specificity in a more relevant biological context. Triaged hits can then be produced as IgGs with the desired isotype and screened for functions such as cell-based blockage of a ligand–receptor interaction, or via a cell-based neutralization assay. It is important that in parallel to the assessment of the in vitro biological activity of those hits that have demonstrated the desired binding and functional profiles, a preliminary assessment of the developability profile of the molecules is performed, including monomeric content after purification, identity via intact mass spectrometry, thermal stability, and solubility.

For those hits that have demonstrated the desired functional and developability profiles, an in vivo efficacy study can be considered. Based on the target biology, a disease-relevant animal model is selected, and parameters of the study can be established. If sufficient potency in vivo is observed as compared to either a reference molecule (if known) or a pre-established criterium (if no benchmark is available), lead candidate molecules can move into the optimization stage. At this stage, the number of hits has likely decreased from the hundreds down to a handful. Lead optimization is a field of its own, generally including a more thorough biophysical assessment of molecules, and improvement of properties such as affinity, cross-reactivity, and selectivity; expression in the manufacture platform

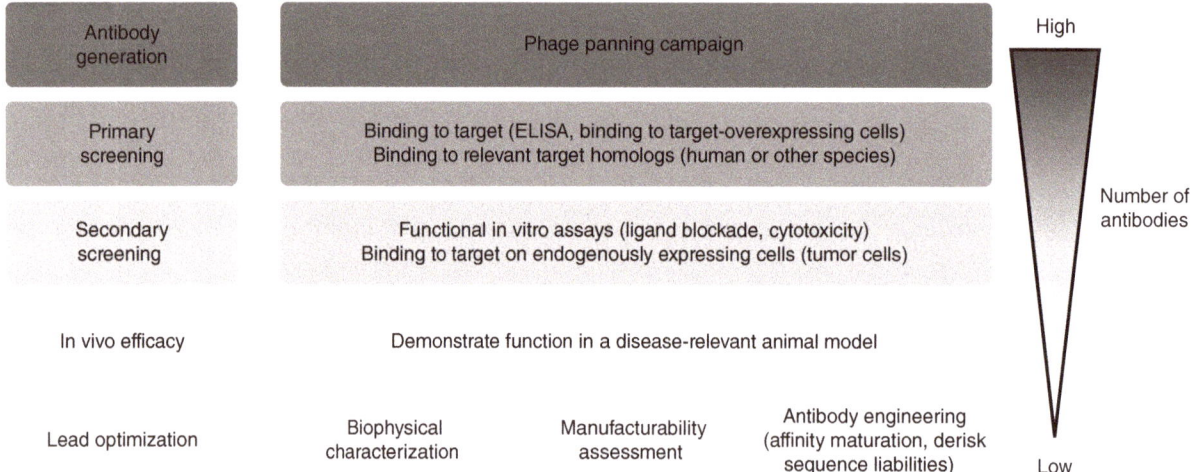

FIGURE 2. General screening funnel for antibody lead identification. During a phage display-based antibody discovery campaign, polyclonal pools of binders to the target are enriched during phage panning, and then single antibody binders are identified in the primary screen. These then move to secondary screening, with select hits moving into in vivo testing and biophysical optimization and engineering. At each step, the number of antibodies is triaged, starting with hundreds or thousands of antibodies down to a few.

(commonly CHO cells); stability under stress conditions; and concentration in formulation buffers. These activities usually fall outside of an academic research setting or discovery laboratory, but rather take place within specific, specialized teams in biotechnology or pharmaceutical companies.

With a clearly defined antibody discovery strategy and screening funnel, preparation of reagents and assays can begin. An essential step for the identification of a lead antibody is antigen and assay quality control (QC), which is often overlooked in academic laboratories. All recombinant proteins used for panning or screening should be assessed for purity by SDS-PAGE analysis and size exclusion chromatography. A good rule of thumb is >90% purity and visible lack of aggregation. Further QC may include testing proteins for binding to their ligands by enzyme-linked immunosorbent assay (ELISA) or kinetic methods (surface plasmon resonance, biolayer interferometry), which confirms that the antigen is in the native and relevant conformation. Additional QC can also include ELISA or kinetic assessment with a reference antibody against the target. If performing phage selection using whole cells as antigen, it is important to assess antigen density and relative expression levels of the target on the cell surface by flow cytometry.

## PHAGE DISPLAY LIBRARY CONSTRUCTION AND SELECTION FOR THERAPEUTIC MONOCLONAL ANTIBODY IDENTIFICATION

Like all antibody generation methods, there are advantages and disadvantages to various types of phage display libraries and their use in a discovery campaign. There are three major types of phage display libraries, which are outlined in Table 1. Fully naive libraries are those in which the natural immune repertoire is captured, usually from healthy donors. With such a library, the repertoire has not been skewed toward one particular target, other than what the donor has naturally been exposed to in the course of their lifetime. Here, the general approach to maximize diversity is to increase the number of donors used in the library construction as well as to use primers for antibody gene amplification that fully capture the variable gene rearrangement repertoire. If careful measures are taken to amplify these gene fragments, and multiple ligations/electroporation reactions are performed, these libraries can routinely reach a very high diversity, commonly on the order of $10^9$–$10^{11}$ clones of unique antibody sequences (Almagro et al. 2019). It should be noted that a naive library is not reliant on the development of any specific immune response, and thus has the advantage that it can be used for antibody selections with a wide array of targets. This makes such a library especially attractive for therapeutic antibody discovery, as the same library can be panned repeatedly against various antigens of interest, saving significant time and cost as a laboratory moves through a portfolio of discovery programs. Well-known examples of such libraries are those developed by the Xoma Corporation, which were generated in both Fab and single-chain variable fragment (scFv) format from 30 healthy human donors (Schwimmer et al. 2013). These libraries have been panned using

TABLE 1. Types of phage display libraries and their defining characteristics

| Characteristic | Type of library | | |
|---|---|---|---|
| | Naive | Immune | Synthetic/semisynthetic |
| Repertoire | Large, from healthy human donors | Small, from immunized source | Artificial or combination of artificial and naturally introduced diversity in CDRs |
| Diversity (no. of unique mAb clones) | $10^9$–$10^{11}$, depending on number of donors | $10^6$–$10^8$ | Usually equivalent to fully naive |
| Affinity | Can be low and require maturation | Usually higher due to prior in vivo affinity maturation | Can be low and require maturation |
| Humanization | Not required | Required | Not required |
| Application | Target-agnostic | Target-specific | Target-agnostic |
| Developability | May encounter developability issues due to inherent sequence liabilities | Humanization may be required | Framework selection may be limiting |

(mAb) Monoclonal antibody, (CDRs) complementarity-determining regions.

hundreds of targets, generating several therapeutic antibodies for clinical development, and have been licensed for use at several biotechnology companies since their inception.

Despite the versatility of naive libraries, antibodies from a naive library often require additional engineering, as the repertoire of antibodies did not naturally evolve before the library was generated to produce therapeutic antibodies, and thus construction of the library does not take into account important biophysical features for monoclonal antibody (mAb) development, such as thermostability, solubility at the (high) concentrations needed in therapy, and high expression yield in manufacturing platforms. Thus, naive library-derived antibodies may need to undergo affinity maturation to increase affinity or other efforts to increase stability, because this normally occurs in vivo during the B-cell maturation process, and antibodies in a naive repertoire were not all subjected to this process (Mesin et al. 2016).

Alternatively, immune libraries can be an attractive type of phage library for discovery (Table 1). These libraries have shown great value in infectious diseases such as AIDS or COVID-19 (Ubah and Palliyil 2017; Sokullu et al. 2021), where individuals are naturally infected and survive the infection by developing an immune response against the pathogen. However, in cancer or autoimmune diseases where the immunization of humans is not possible, the library is constructed using material from animals, such as mice, chickens, or llamas, immunized with the target of interest, (Lim and Chan 2016). The antibody repertoire in the immune libraries is much more focused on antibodies specific for the target used in the immunization protocol, which results in antibodies that have a higher affinity for the original immunogen, as affinity maturation occurred in vivo. Despite higher affinities, in the case where the antibodies are obtained from a nonhuman source, immune library-derived antibodies require additional downstream engineering efforts to minimize potential immunogenicity issues, a process known as antibody humanization. Pharmaceutical/biotechnology companies often perform these efforts in antibody engineering or lead optimization teams. However, the emergence of several transgenic mouse models expressing fully human immunoglobulin repertoires, such as Alloy, Omni-Mouse, AlivaMab, and VelocImmune mice, has allowed for the generation of human antibodies directly derived from a rodent immunization campaign, and many of these models have reached the market or are available for licensing (Murphy et al. 2014; Brüggemann et al. 2015).

A third option is the use of synthetic or semisynthetic libraries (Table 1). There is variation in the design here, ranging from the entire antibody repertoire produced artificially (fully synthetic) to a semisynthetic library where natural diversity is only introduced at particular regions of the antibody fragment; for example, at the complementarity-determining regions (CDRs) (Valadon et al. 2019; Davydova 2022). Fully synthetic libraries have predefined designed scaffold(s), with variation in certain positions of the CDRs introduced artificially by PCR or cloning of synthetic DNA fragments. To avoid introduction of stop codons in the randomized fragments that lead to nonfunctional clones, synthetic libraries such as the HuCAL Library series used trinucleotide mutagenesis (TRIM) technology to generate diversity (Knappik et al. 2000; Rothe et al. 2008; Prassler et al. 2011). Depending on the desired application of the library, the extent of the diversity either can be controlled to cast a wide net by introducing high diversity in the CDRs or can be carefully introduced at specific residues of the antigen-binding site. Synthetic libraries can also be designed with the final therapeutic product in mind, such that a particular scaffold that is known to be stable and that features desirable biophysical properties can be chosen (Tiller et al. 2013; Azevedo Reis Teixeira et al. 2021).

Semisynthetic libraries are a hybrid approach to generating a versatile library for antibody discovery. This type of library takes advantage of the control over scaffold selection and variation of select residues that is afforded by a synthetic library. However, it also introduces natural diversity in the CDRs; for instance, in the third CDR of the heavy chain (HCDR3), which is commonly (but not always) the most important region of the binding site for antigen interaction and is difficult to design due to the high diversity and flexibility of this region (Almagro et al. 2014; D'Angelo et al. 2018). Examples of semisynthetic libraries, which are described in detail below and for which protocols for construction are included in this collection, are ALTHEA Gold Plus Libraries (see Protocol 1: Semisynthetic Phage Display Library Construction: Design and Synthesis of Diversified Single-Chain Variable Fragments and Generation of Primary Libraries [Almagro and Pohl 2024a], Protocol 2:

Semisynthetic Phage Display Library Construction: Generation of Filtered Libraries [Almagro and Pohl 2024b], and Protocol 3: Semisynthetic Phage Display Library Construction: Generation of Single-Chain Variable Fragment Secondary Libraries [Almagro and Pohl 2024c]).

Like naive libraries, synthetic and semisynthetic libraries can be valuable assets for a laboratory focused on antibody discovery, as they are target-agnostic and can be used to isolate antibodies for virtually any target of interest. Moreover, these types of libraries allow for the identification of antibody binders without dependence on a lengthy immunization protocol and a robust immune response. However, although it has been reported that high-affinity binders can be identified from synthetic (Ferrara et al. 2022) and semisynthetic libraries (Guzmán-Bringas et al. 2023), some antibodies will require additional engineering to achieve the desired affinity.

Besides the source of the antibody repertoire to be used in a naive, synthetic, or semisynthetic library, it is also important to select features of the library that are conducive to the discovery workflow of interest and desired output. This includes the format of the antibody fragment that will best fit the therapeutic modality in development. For example, if the desired final molecule will be a conventional IgG, a Fab phage display library may be preferable, as it is closer to the final format of the therapeutic. Fabs are commonly more stable than scFvs, as they are stabilized by the $C_H1:C_L$ interface and a C-terminal disulfide bridge in addition to the $V_H:V_L$ interface of the scFvs. However, Fab libraries can be more difficult to construct and may not be as well expressed in *Escherichia coli* as a smaller scFv fragment (Omar and Lim 2018). Efforts have been made, however, to optimize scFv construction (Zhu and Dimitrov 2009).

Other features of the library must also be carefully selected. For scFv libraries, selection of the proper format, either $V_H$–linker–$V_L$ or $V_L$–linker–$V_H$, and the nature of the linker (usually a flexible glycine–serine repeat) between the $V_H$ and $V_L$ fragments can impact the outcome of the selection process. Longer linkers allow for the scFv to fold back on itself, promoting monomer formation, whereas shorter linkers lead to dimerization between $V_H$ and $V_L$ of two scFvs, thus forming diabodies (Arndt et al. 1998). Diabodies could result in apparent affinity increases due to multiple antibody fragments interacting with the target, often referred to as avidity effect (Rudnick and Adams 2009), which departs from the true affinity of the IgG that is the most common final therapeutic format of antibodies. Some other standard features of a phagemid vector include the presence of an amber stop codon to allow for soluble antibody production free of the pIII fragment directly from the phagemid vector, as well as a tag for detection. Oftentimes, 6X-His, myc, HA, or V5 tags are chosen, but depending on what tags are commonly found on the antigens you will be panning against, you may want to avoid one or more of these. For instance, the His tag is very useful for scFv purification via nickel or cobalt affinity columns as well as in detection assays using anti-His antibodies. However, many commercially available antigens have His tags, and thus a His-tagged scFv may be a liability to developing assays with anti-His detection antibodies.

## A CASE STUDY IN LIBRARY CONSTRUCTION FOR THERAPEUTIC MONOCLONAL ANTIBODY DISCOVERY: THE ALTHEA GOLD PLUS LIBRARIES

ALTHEA Gold Plus Libraries are an upgrade of ALTHEA Gold Libraries, described in detail by Valadon et al. (2019). A comparison of ALTHEA Gold Libraries with state-of-the-art antibody phage-displayed libraries described in the literature over the last 10 yr has been discussed by Almagro et al. (2019). Among others, these libraries include naive libraries implemented by Xoma (Schwimmer et al. 2013) and synthetic libraries such as HuCAL Platinum (Prassler et al. 2011) and Ylanthia (Tiller et al. 2013) implemented by Morphosis, which are used by large pharmaceutical and biotechnology companies as their antibody discovery platforms.

ALTHEA Gold Libraries consist of two scFv sublibraries built with synthetically generated, well-known human *IGHV* and *IGKV* germline genes, combined with natural human HCDR3/JH (H3J) fragments obtained from peripheral blood mononuclear cells from a large pool of healthy human donors. One synthetic *IGHV* gene (*IGHV3-23*) provides a universal variable domain of heavy chain ($V_H$) scaffold, paired with two variable light ($V_L$) scaffolds obtained from the *IGKV3-20* and *IGKV4-01* genes that furnish two different topographies at the antigen-binding site, and hence the potential to bind distinct epitopes on the target.

All CDRs in these libraries are synthetically diversified, with the exception of the HCDR3. A strategic approach was implemented for synthetic $V_H$ and $V_L$ generation, whereby scaffolds are mutated at specific positions identified as being in contact with antigens in the known antigen–antibody complex structures that were available at the time of the library design (more than 2000 antibody structures are available at the Protein Data Bank, https://www.rcsb.org/). The diversification regime consists of high-usage amino acids found at those positions in corresponding human germline genes, and more than 60,000 antibody sequences compiled at the National Center for Biotechnology Information (https://www.ncbi.nlm.nih.gov/) and the international ImMunoGeneTics information system (https://www.imgt.org).

In addition to the unique design of the ALTHEA Gold Libraries, the functionality, stability, and diversity of the libraries are improved throughout a three-step construction process (Fig. 3). In the first step, fully synthetic primary libraries (PLs) are generated by combining the diversified scaffolds with a set of 90 synthetic neutral H3J germline gene fragments. These neutral H3J fragments serve as a "placeholder" to avoid structural constraints that a single H3J fragment could impart during subsequent steps in the library construction process and result in limited sequence diversity in the final libraries.

The second construction step consists of selecting or filtering PLs, based on the natural capacity of the *Staphylococcus aureus* protein A or of the *Peptostreptococcus magnus* protein L to bind human $V_H$ or $V_L$ scaffolds, respectively. Protein A binds the framework 3 of human *IGVH3* gene family members (Graille et al. 2000), whereas protein L binds the framework 1 of human *IGKV1*, *IGKV3*, and *IGKV4* gene families (Nilson et al. 1992; Graille et al 2002). These properties of protein A and protein L have been extensively used to select for well-folded and stable antibody fragments after incubation under denaturing or destabilizing conditions (Jespers et al. 2004; Hussack et al. 2011). However, the use of protein A or L has been limited to enriching the final antibody libraries and/or selecting for variants after mutagenesis to improve stability. In the accompanying protocols, we use protein L to generate

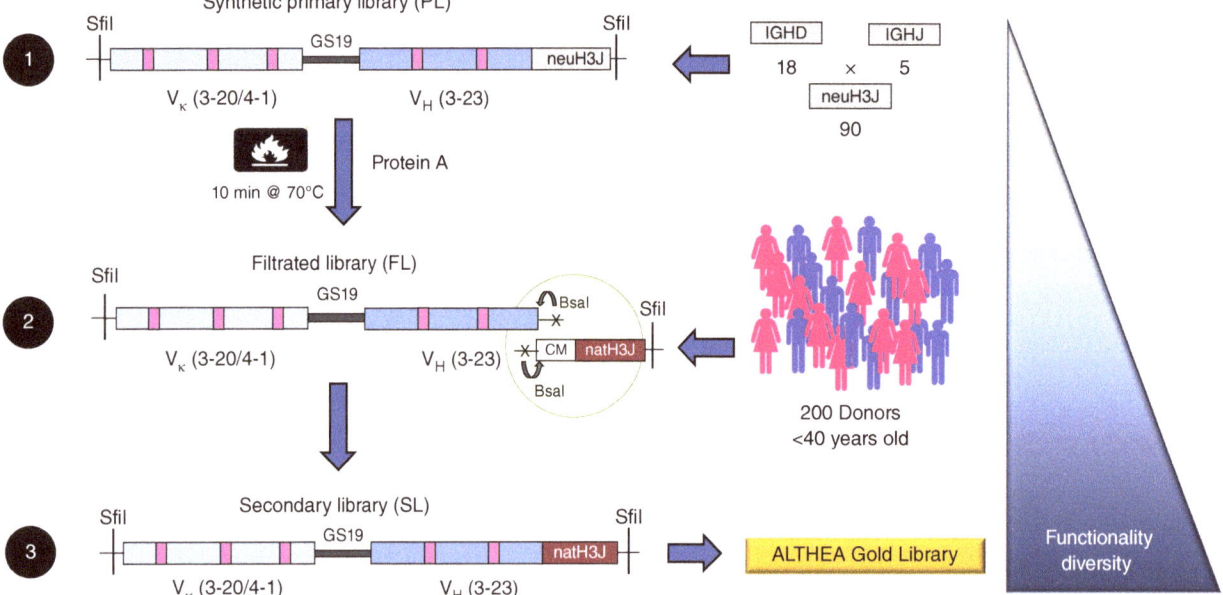

FIGURE 3. Three-step construction process to generate highly diverse and functional semisynthetic libraries. Figure reproduced from Valadon et al. (2019). See the text for details.

filtrated libraries (FLs) as an intermediary step to enrich the PLs with well-folded antibody fragments, taking place after harsh incubation conditions and before replacing the neutral H3J fragments with natural H3J fragments (see Protocol 1: Semisynthetic Phage Display Library Construction: Design and Synthesis of Diversified Single-Chain Variable Fragments and Generation of Primary Libraries [Almagro and Pohl 2024a]; Protocol 2: Semisynthetic Phage Display Library Construction: Generation of Filtered Libraries [Almagro and Pohl 2024b]; Protocol 3: Semisynthetic Phage Display Library Construction: Generation of Single-Chain Variable Fragment Secondary Libraries [Almagro and Pohl 2024c]). In the filtration process, the well-folded antibody fragments fused to phage particles bind to protein L, whereas truncated fragments resulting from premature stop codons or nonfunctional ones resulting from insertions and/or deletions are removed by centrifugation. These nonfunctional antibody variants erode the repertoire of functional antibody fragments in the library, leading to poor performance, so it is important that they are removed.

In the third and final step, the resulting stable synthetic antibody fragments of the FLs are combined with natural H3J fragments to generate secondary libraries. Because most of the antibody diversity is concentrated in the HCDR3, the natural H3J fragments restore the diversity lost during the filtration process. Moreover, because the natural H3J fragments have been under selective pressures to produce functional antibodies in vivo, the combination of highly stable synthetic antibody fragments from the FLs and a repertoire of natural H3J fragments results in highly diverse, stable, and functional libraries ready for therapeutic antibody discovery.

Validation of ALTHEA Gold Libraries with seven targets yielded specific antibodies in all cases. Further characterization of the isolated antibodies indicates $K_D$ values as human IgG1 molecules in the single-digit and subnanomolar range. Thermal stability ($T_m$) of all tested Fabs was between 75°C and 80°C, demonstrating that ALTHEA Gold Libraries are a valuable source of specific, high-affinity, and highly stable antibodies. Several target-specific antibodies isolated from these libraries have been characterized in detail (Pedraza-Escalona et al. 2021; Dao et al. 2022a,b), with one mAb even moving to Phase 1 clinical trials (NCT06017258 and NCT06128044).

The upgrade to ALTHEA Gold Plus Libraries from ALTHEA Gold Libraries results from adding two additional $V_L$ scaffolds (IGKV1-39 and IGKV3-11), for a total of four scFv semisynthetic libraries. These new scaffolds increase the structural diversity of the libraries, with the potential to generate a more diverse set of antibodies. Recently, application and validation of the four synthetic $V_L$ scaffolds used in ALTHEA Gold Plus Libraries were published by Mendoza-Salazar et al. (2022). In this work, the four synthetic $V_L$ libraries were used as counterparts of an immune $V_H$ repertoire obtained from a patient with coronavirus disease 2019 (COVID-19) infected with the severe acute respiratory syndrome coronavirus 2 (SARS-CoV-2) B.1.617.2 (Delta) variant. Using standard molecular cloning techniques, the variable light chain diversity of the ALTHEA libraries was combined with the $V_H$ repertoire from the COVID-19 convalescent patient. After three rounds of selection with SARS-CoV-2 wild-type (WT) receptor-binding domain, 34 unique scFv molecules were obtained. One of the antibodies, called IgG-A7, recognized the SARS-CoV-2 receptor-binding domain from Wuhan-Hu-1 (WT), Delta, and B.1.1.529 (Omicron) strains with high affinity. IgG-A7 also neutralizes the three strains of SARS-CoV-2 in plaque reduction neutralization tests and protects 100% of K18-hACE2 transgenic mice expressing human angiotensin-converting enzyme 2 (hACE2) infected with WT SARS-CoV-2 at a dose of 5 mg/kg (González-González et al. 2022). The developability profile of IgG-A7 matches that of marketed therapeutic antibodies, thus demonstrating the potential of the synthetic $V_L$ scaffolds used in ALTHEA Gold Plus Libraries to generate high-affinity and functional antibodies with potential as leads for development of antibody-based drugs.

The three accompanying protocols for implementing ALTHEA Gold Plus Libraries include Protocol 1: Semisynthetic Phage Display Library Construction: Design and Synthesis of Diversified Single-Chain Variable Fragments and Generation of Primary Libraries (Almagro and Pohl 2024a), Protocol 2: Semisynthetic Phage Display Library Construction: Generation of Filtered Libraries (Almagro and Pohl 2024b), and Protocol 3: Semisynthetic Phage Display Library Construction: Generation of Single-Chain Variable Fragment Secondary Libraries (Almagro and Pohl 2024c), and below we discuss general considerations to increase the chances of success during the implementation of the protocols.

# Chapter 9

## DEVELOPING A PANNING STRATEGY

After following the elements discussed above, a library that fits the needs of your discovery strategies can be generated, and a discovery campaign can be started. The first step in this process is to develop a strategy based on the characteristics of the target and available reagents. For instance, the nature of the target will have an important impact on the antibody design goals. If the target or a relevant fragment of the target can be produced as a high-quality recombinant protein that passes QC analysis, this provides a straightforward approach for the identification of binders. The simplest method of phage panning with soluble protein is a solid phase panning, whereby the target antigen is immobilized on a polystyrene surface, on either a microwell plate or immunotube (Kontermann 2001; Russo et al. 2018). This simple panning approach, however, has the disadvantage that protein adherence to the plastic may cause minor structural changes, resulting in the protein not having the same conformation as in solution. This may affect the conformation or availability of functional epitopes important for obtaining antibodies with the desired functional profile (Carmen and Jermutus 2002). In addition, in solid phase panning, the amount of antigen attached to the polystyrene cannot be controlled as well as in solution, as only a fraction of the protein added to the well/tube will adhere to the plate and saturate the available surface.

Alternatively, panning can be performed in solution, using the soluble target with an affinity tag for capture/immobilization, often biotin (Wenzel et al. 2020). In this panning modality, the antigen–biotin conjugate can be isolated using magnetic nanobeads coated with streptavidin, and the non-bound phage in solution can be separated from the bound phage. The twofold advantage of this process is that (1) the concentration of the protein captured by the beads (and the number of beads) can be controlled, which is especially important when trying to select for higher-affinity antibodies, and (2) the target is in its native conformation in solution, thus avoiding conformational changes that may occur in solid phase panning. One potential issue with this panning method, however, is that because the antigen requires labeling, the biotin conjugation could result in blocking desired epitopes. This can generally be avoided, though, by introducing a site-specific biotinylation tag on the antigen expression construct, such as an AviTag, so that the degree of labeling is 1:1 and the biotinylation occurs in a region that does not compromise the integrity of functional epitopes (Gräslund et al. 2017). Some commercial vendors offer Avi-tagged proteins for this purpose.

Some of the most challenging targets to isolate antibodies against include MHC complexes loaded with specific peptides to generate TCR-mimic antibodies, membrane-bound proteins, and ion channels. In the case of MHC:peptide complexes, there are challenges in generating appropriate antigens for selection and confirming that antibodies identified from the panning campaign are specific for the MHC in complex with the particular peptide of interest. One approach here is the use of biotinylated peptide–MHC antigens, and established methods for producing such complexes in bacteria have been published (Denkberg et al. 2000). Two case studies of identification of highly specific antibodies to peptide–MHC complexes by phage display are described by Dao et al. (2022a,b).

Membrane proteins, on the other hand, are difficult to express in soluble form, and the exposed exterior surface of the proteins in some cases are peptides that conformationally depend on the protein embedded in the membrane, and thus are difficult to produce as recombinant functional protein fragments. Ion channels are similarly notoriously difficult targets to generate antibodies against (Wilkinson et al. 2015), as often the goal is to isolate an antibody that blocks function due to occlusion of the channel pore by the antibody. This often leaves little "real estate" for an antibody to bind to, and is reliant on the pore complex to be present in its native, multimeric complex, ideally in one conformation (pore open or closed). In this case, whole cells or membrane fractions can be used for panning, or rather virus-like particles (VLPs), nanodisks, or amphipols loaded with the ion channel of interest. Amphipols consist of an amphiphilic polymer designed to serve as a "belt" to constrain the oligomer into its native membrane-bound conformation, resulting in a soluble antigen that can be used for panning (Hou et al. 2020). Similarly, nanodiscs are disc-shaped complexes consisting of a lipid bilayer surrounded by amphipathic protein that can also constrain a membrane protein suitable for phage selections (Pavlidou et al. 2013). Specificity can be an issue, however, as both the extracellular and intracellular portions of the ion channel are exposed in the phage panning mixture. For a review of antibodies generated against ion channels, see Haustrate et al. (2019).

Another approach to isolate antibodies against difficult targets using phage display is cell-based panning (Fahr and Frenzel 2018). In this panning modality, the phage library is directly incubated with cells expressing the target of interest. The advantage of this method is that the target of interest is in its native conformation and functional configuration. One of the drawbacks of cell-based panning, however, is that the expression level of the target may not be sufficient, and so it may be difficult to isolate binders to the protein. Another major difficulty with cell-based panning is specificity, as there are many other proteins than the target of interest on the cell surface that will compete for binding to the antibodies in the library, and thus can dampen the enrichment of specific binders to the target. Thus, it is important to first deplete the phage library of nonspecific binders and binders specific for proteins other than the target of interest with a cell line that does not express the target of interest. This step, called subtraction or negative selection, is ideally performed with the cell line used to generate the recombinant cell line; i.e., the cell line that has not been transfected with the target. Alternatively, one can alternate between soluble antigen and cells expressing the target to selectively isolate the antibodies binding both and remove phages that bind to irrelevant targets on the cell surface.

## OPERATIONAL CONSIDERATIONS BEFORE STARTING A PHAGE PANNING CAMPAIGN

Once a phage display library has been constructed, it is important to store the library under conditions to maintain its diversity. A library can be stored either as purified virions, or as a frozen bacterial stock that requires culture growth and infection with helper phage for amplification of virions displaying antibodies. If a library is stored as phage stock, it should be mixed with glycerol at 50% (v/v) and stored at −80°C, and should not be subjected to multiple freeze–thaw cycles. It should be noted that this is not to protect the phage virions themselves, which are quite robust and stable (Branston et al. 2013), but rather to minimize irreversible unfolding of the antibody fragments expressed on the phage surface and/or degradation by traces of proteolytic enzymes present derived from the originating bacterial culture. Both processes will lead to significant losses in diversity of the library when panning, but are not reflected in the titer of the phage library, as the phage particles are still infectious regardless of the presence of a displayed antibody. To avoid losses in library diversity due to antibody fragments unfolding and degradation, it is common practice to avoid large phage stock preparations and use freshly amplified libraries from bacterial stock as input for the first round of panning when possible. Of course, library preparation every time a new campaign is started is time- and resource-consuming, and thus a proper balance must be struck between having a phage stock ready and preparation right before a discovery campaign, depending on the demands of a given laboratory. It is important to note that it is never ideal to amplify a library from an existing phage stock, as this practice introduces amplification bias of certain clones as an unintended consequence of growth advantage of some clones (particularly phage with defective antibody fragments or without the fusion proteins displayed altogether) over others that may express valuable antibodies, which can adversely affect library diversity (Matochko et al. 2012).

Phage display panning campaigns require relatively standard equipment and reagents that are usually found in laboratories running routine experiments in molecular biology and biochemistry. With that said, there are several steps when preparing to efficiently execute rounds of phage panning. This is akin to the "mise-en-place" philosophy of chefs, which translates to "everything in its place." The idea is to have all ingredients on hand and prepared before you start cooking or, in this case, phage panning. It is important to ensure that all instruments are in working order, e.g., centrifuges, bacterial shaking incubators, spectrophotometer, among others, and that all reagents, including high-quality antigen, helper phage stock, culture media, and buffers, are ready and in sufficient supply. The laboratory bench and pipettes should be decontaminated with 10% (v/v) bleach, especially if multiple panning campaigns are to be performed in parallel, and sterile plastic disposable material is preferred over reusables.

Time management of panning rounds is another important consideration. Depending on the protocol, a single round of panning can be completed in 1 d, or it can be broken up into 2 d. In the 1-d scenario, the panning is completed, and eluted phage is amplified in a bacterial culture overnight, to be harvested fresh the next day for the subsequent round of panning. On this schedule, a panning campaign can be completed in 1 wk. However, one runs the risk of having to start over from the beginning (Round 1) if anything happens during later rounds of panning to the eluate or amplified eluate, as there is no "backup" stock. Alternatively, in the 2-d method, a bacterial culture is infected with eluted phage and spread onto solid medium for overnight growth. Scrapings from the agar plate can then be used the next day to grow up a log phase bacterial culture for helper phage infection and phage production. The advantage here is threefold: (1) Plating the eluate onto solid media allows for more slow-growing clones to "catch up" to those that would have a growth advantage in liquid medium, potentially increasing diversity of the eluted phage pool; (2) one can in parallel titer the output of the round of panning or just see the density of the bacteria, and thus assess how the panning is going; and (3) glycerol stock of the plate scrapings can be stored to preserve each round of panning. The bacterial glycerol stocks can be used to reamplify phage from that round for repeating panning, panning on a different antigen, or isolation of single colonies from that round for monoclonal screening.

It is also crucial to avoid contamination while panning, especially with helper phages. Due to the infectious properties of helper phages, phagemids, and F-pilus-expressing *E. coli* (Boeke et al. 1982), if the bacterial culture to be used for eluate amplification first becomes infected with helper phages, it deems those bacterial cells impervious to infection with phagemid virus, which can result in a loss of a panning round. Also, if performing more than one campaign in parallel, it is important to avoid cross-contamination. The best way to avoid contamination is to follow good laboratory practice, including the use of filter pipette tips and disposable plastic cultureware, and disinfecting surfaces frequently with freshly prepared 10% (v/v) bleach.

## TECHNICAL CONSIDERATIONS TO ISOLATE ANTIBODIES OF INTEREST

Regardless of the panning approach used, i.e., solid phase, solution panning, or cell-based selection, it is important to incorporate the overall antibody design goals into the panning schema. For example, if cross-reactivity to mouse or NHP orthologs of the target is required, these targets can be incorporated into alternating rounds of selection to enrich for binders that recognize shared epitopes between species. In contrast, if selectivity is required, and there are specific isoforms, paralogs, or related proteins in the target protein family, a depletion or subtraction step can be added at the beginning of each panning round, to remove cross-reactive binders with the unwanted targets from the phage library.

Besides the target selection, additional measures can be taken to isolate antibodies of interest. For example, if antibodies with higher affinity are desired, the target concentration can be decreased successively over each round of panning, usually fivefold to 10-fold per round. The easiest way to accomplish this is in a solution panning, where the target concentration can be tightly controlled. As the phage pool is enriched for binders to the target over each round of panning, lowering the antigen concentration creates competition in solution, with the highest-affinity antibodies retaining binding and outcompeting lower-affinity binders for antigen engagement.

Another way to select for higher-affinity antibodies is to increase washing stringency over successive panning rounds; for example, 10 washes in Round 1, 15 washes in Round 2, and 20 washes in Round 3. Adding detergent such as Tween to the wash buffer can help decrease nonspecific interactions between phage and the target, thus helping to isolate antibodies of interest. The washing stringency can also be increased by having longer incubation times in the wash buffer. This will select for antibodies with a slower off rate, and thus remove binders with a lower dissociation constant. In phage-based affinity maturation campaigns, wash times will often be extended to 2 h and even overnight to select for the highest-affinity clones (Thie et al. 2009). Additionally, one can add an excess of nonbiotinylated target during the overnight incubation, to outcompete the lower-affinity binders bound to the target. Finally, it should be considered that the elution method can affect the specificity of clones isolated. Instead of a general, nonspecific elution method using extreme pH such as Tris-glycine pH 2.2 or triethylamine pH 10, which will indiscriminately disrupt antigen–antibody interactions, a more focused competitive elution can be used. In this elution strategy, a nonbiotinylated target or a known functional antibody can be added to the phage pool to compete for phage-displayed antibodies. With this approach, only target or epitope-specific clones will be eluted from the phage–bead complexes (Duan and Siegumfeldt 2010).

Although affinity is important for many therapeutic antibody programs, in some cases it is not as critical as obtaining antibodies binding the right epitope to achieve the desired function. Indeed, selecting a diverse panel of antibodies is often more important than obtaining a few high-affinity binders. The kinetics of a functional antibody can be improved via an affinity maturation campaign, but a high-affinity antibody cannot be engineered into a functional antibody it if does not bind the right epitope. Thus, the initial goal of a discovery campaign should be focused on identifying antibodies that recognize as many epitopes as possible, to increase the chances of targeting the right functional epitope, so it can be used for proof-of-concept studies or as substrate for further optimization. If this is the case, panning selection conditions can be tweaked accordingly. To isolate a more diverse pool of antibodies, the selection conditions in the first round of panning are most important, because these are the clones that will be carried throughout the rest of the selection rounds. If conditions are too stringent in the first round, diversity will be significantly compromised. To avoid this pitfall, a higher concentration of antigen and a minimal number of washes can be used in Round 1, to avoid discarding lower-affinity but potentially functional binders. Also, fewer rounds of panning can be performed to avoid any clonal selection/outgrowth in later panning rounds. Keep in mind, however, that the diversity of the antibodies isolated from any campaign is limited by the source library, which is why careful construction of highly diverse and functional libraries is critical for a successful discovery campaign.

## CONCLUSION

There are many variables in a therapeutic antibody discovery campaign that must be considered before and during the lead identification process to increase the chances of success. With that said, once streamlined standard protocols, processes, and reagents are established in an antibody discovery laboratory (e.g., protein/antibody production and QC capabilities, a validated naive phage display library or immune library construction protocol, and high-throughput screening methods), campaigns can run efficiently for rapid identification of antibody candidates for further characterization. The most crucial portions of this workflow are (1) the antibody phage display library as the source of binders, as it will have the biggest impact on sequence diversity and quality of recovered antibodies; (2) the quality of the reagents, in particular the targets, as they will determine whether one is selecting for the right conformation/function; and (3) validated assays that will guide the decision on the best clones to move forward in the discovery and optimization campaigns. Moreover, the basic steps during the selection process, such as binding, washing, elution, and phage amplification, generally remain consistent among discovery campaigns, but minor changes in the selection protocol can significantly affect the results of the campaign, so careful recordkeeping of experimental variations is critical. Using information from previous antibody discovery efforts can also help to move a program forward. Ultimately, however, the success of a campaign is often determined empirically. It is best to try both proven and new creative panning approaches to deliver a diverse panel of antibody sequences for further testing, as hundreds of antibodies will eventually be narrowed down to a few leads for full preclinical development efforts.

In summary, the following points will pave the way for a successful phage-based lead identification approach. First, establish antibody design goals. Second, library source and method of construction have the biggest effect on sequence diversity and utility, so carefully select or build a library that will best fit the antibody design goals of the project. Third, perform rigorous quality control assessment of all reagents. Finally, design and implement library selection strategies and screening methods that maximize the chances of enrichment for antibodies with the desired target binding and functional properties.

## COMPETING INTEREST STATEMENT

M.A.P. is an employee of Xtalpi, Inc. J.C.A. is founder and CEO of GlobalBio, Inc., and has commercial interest in ALTHEA Gold Plus Libraries.

## REFERENCES

Alfaleh MA, Alsaab HO, Mahmoud AB, Alkayyal AA, Jones ML, Mahler SM, Hashem AM. 2020. Phage display derived monoclonal antibodies: from bench to bedside. *Front Immunol* 11: 1986. doi:10.3389/fimmu.2020.01986

Almagro JC, Pohl MA. 2024a. Semisynthetic phage display library construction: design and synthesis of diversified single-chain variable fragments (scFvs) and generation of primary libraries. *Cold Spring Harb Protoc* doi:10.1101/pdb.prot108614

Almagro JC, Pohl MA. 2024b. Semisynthetic phage display library construction: generation of filtered libraries. *Cold Spring Harb Protoc* doi:10.1101/pdb.prot108615

Almagro JC, Pohl MA. 2024c. Semisynthetic phage display library construction: generation of single-chain variable fragment secondary libraries. *Cold Spring Harb Protoc* doi:10.1101/pdb.prot108616

Almagro JC, Teplyakov A, Luo J, Sweet RW, Kodangattil S, Hernandez-Guzman F, Gilliland GL. 2014. Second antibody modeling assessment (AMA-II). *Proteins* 82: 1553–1562. doi:10.1002/prot.24567

Almagro JC, Pedraza-Escalona M, Arrieta HI, Pérez-Tapia SM. 2019. Phage display libraries for antibody therapeutic discovery and development. *Antibodies (Basel)* 8: 44. doi:10.3390/antib8030044

André AS, Moutinho I, Dias JNR, Aires-da-Silva F. 2022. In vivo phage display: a promising selection strategy for the improvement of antibody targeting and drug delivery properties. *Front Microbiol* 13: 962124. doi:10.3389/fmicb.2022.962124

Arndt KM, Müller KM, Plückthun A. 1998. Factors influencing the dimer to monomer transition of an antibody single-chain Fv fragment. *Biochemistry* 37: 12918–12926. doi:10.1021/bi9810407

Azevedo Reis Teixeira A, Erasmus MF, D'Angelo S, Naranjo L, Ferrara F, Leal-Lopes C, Durrant O, Galmiche C, Morelli A, Scott-Tucker A, et al. 2021. Drug-like antibodies with high affinity, diversity and developability directly from next-generation antibody libraries. *MAbs* 13: 1980942. doi:10.1080/19420862.2021.1980942

Boeke JD, Model P, Zinder ND. 1982. Effects of bacteriophage f1 gene III protein on the host cell membrane. *Mol Gen Genet* 186: 185–192. doi:10.1007/BF00331849

Branston SD, Stanley EC, Ward JM, Keshavarz-Moore E. 2013. Determination of the survival of bacteriophage M13 from chemical and physical challenges to assist in its sustainable bioprocessing. *Biotechnol Bioproc E* 18: 560–566. doi:10.1007/s12257-012-0776-9

Brüggemann M, Osborn MJ, Ma B, Hayre J, Avis S, Lundstrom B, Buelow R. 2015. Human antibody production in transgenic animals. *Arch Immunol Ther Exp (Warsz)* 63: 101–108. doi:10.1007/s00005-014-0322-x

Carmen S, Jermutus L. 2002. Concepts in antibody phage display. *Brief Funct Genomic Proteomic* 1: 189–203. doi:10.1093/bfgp/1.2.189

Carter P, Smith L, Ryan M. 2004. Identification and validation of cell surface antigens for antibody targeting in oncology. *Endocr Relat Cancer* 11: 659–687. doi:10.1677/erc.1.00766

D'Angelo S, Ferrara F, Naranjo L, Erasmus MF, Hraber P, Bradbury ARM. 2018. Many routes to an antibody heavy-chain CDR3: necessary, yet insufficient, for specific binding. *Front Immunol* 9: 395. doi:10.3389/fimmu.2018.00395

Dao T, Mun SS, Molvi Z, Korontsvit T, Klatt MG, Khan AG, Nyakatura EK, Pohl MA, White TE, Balderes PJ, et al. 2022a. A TCR mimic monoclonal antibody reactive with the "public" phospho-neoantigen pIRS2/HLA-A*02:01 complex. *JCI Insight* 7: e151624. doi:10.1172/jci.insight.151624

Dao T, Mun S, Korontsvit T, Khan AG, Pohl MA, White T, Klatt MG, Andrew D, Lorenz IC, Scheinberg DA. 2022b. A TCR mimic monoclonal antibody for the HPV-16 E7-epitope p11-19/HLA-A*02:01 complex. *PLoS ONE* 17: e0265534. doi:10.1371/journal.pone.0265534

Davydova EK. 2022. Protein engineering: advances in phage display for basic science and medical research. *Biochemistry (Mosc)* 87: S146–S110. doi:10.1134/S0006297922140127

Denkberg G, Cohen CJ, Segal D, Kirkin AF, Reiter Y. 2000. Recombinant human single-chain MHC-peptide complexes made from *E. coli* by in vitro refolding: functional single-chain MHC-peptide complexes and tetramers with tumor associated antigens. *Eur J Immunol* 30: 3522–3532. doi:10.1002/1521-4141(200012)30:12<3522::AID-IMMU3522>3.0.CO;2-D

Duan Z, Siegumfeldt H. 2010. An efficient method for isolating antibody fragments against small peptides by antibody phage display. *Comb Chem High Throughput Screen* 13: 818–828. doi:10.2174/138620710792927376

Ecker DM, Jones SD, Levine HL. 2015. The therapeutic monoclonal antibody market. *MAbs* 7: 9–14. doi:10.4161/19420862.2015.989042

Everett JR. 2015. Academic drug discovery: current status and prospects. *Expert Opin Drug Discov* 10: 937–944. doi:10.1517/17460441.2015.1059816

Fahr W, Frenzel A. 2018. Phage display and selections on cells. *Methods Mol Biol* 1701: 321–330. doi:10.1007/978-1-4939-7447-4_17

Ferrara F, Erasmus MF, D'Angelo S, Leal-Lopes C, Teixeira AA, Choudhary A, Honnen W, Calianese D, Huang D, Peng L, et al. 2022. A pandemic-enabled comparison of discovery platforms demonstrates a naïve antibody library can match the best immune-sourced antibodies. *Nat Commun* 13: 462. doi:10.1038/s41467-021-27799-z

Frearson J, Wyatt P. 2010. Drug discovery in academia- the third way? *Expert Opin Drug Discov* 5: 909–919. doi:10.1517/17460441.2010.506508

González-González E, Carballo-Uicab G, Salinas-Trujano J, Cortés-Paniagua MI, Vázquez-Leyva S, Vallejo-Castillo L, Mendoza-Salazar I, Gómez-Castellano K, Pérez-Tapia SM, Almagro JC. 2022. In vitro and in vivo characterization of a broadly neutralizing anti-SARS-CoV-2 antibody isolated from a semi-immune phage display library. *Antibodies* 11: 57. doi:10.3390/antib11030057

Graille M, Stura EA, Corper AL, Sutton BJ, Taussig MJ, Charbonnier J-B, Silverman GJ. 2000. Crystal structure of a *Staphylococcus aureus* protein A domain complexed with the Fab fragment of a human IgM antibody: structural basis for recognition of B-cell receptors and superantigen activity. *Proc Natl Acad Sci* 97: 5399–5404. doi:10.1073/pnas.97.10.5399

Graille M, Harrison S, Crump MP, Findlow SC, Housden NG, Muller BH, Battail-Poirot N, Sibai G, Sutton BJ, Taussig MJ, et al. 2002. Evidence for plasticity and structural mimicry at the immunoglobulin light chain-protein L interface. *J Biol Chem* 277: 47500–47506. doi:10.1074/jbc.M206105200

Gräslund S, Savitsky P, Müller-Knapp S. 2017. In vivo biotinylation of antigens in *E. coli*. *Methods Mol Biol* 1586: 337–344. doi:10.1007/978-1-4939-6887-9_22

Guzmán-Bringas OU, Gómez-Castellano KM, González-González E, Salinas-Trujano J, Vázquez-Leyva S, Vallejo-Castillo L, Pérez-Tapia SM, Almagro JC. 2023. Discovery and optimization of neutralizing SARS-CoV-2 antibodies using ALTHEA Gold Plus Libraries™. *Int J Mol Sci* 24: 4609. doi:10.3390/ijms24054609

Haustrate A, Hantute-Ghesquier A, Prevarskaya N, Lehen'kyi V. 2019. Monoclonal antibodies targeting ion channels and their therapeutic potential. *Front Pharmacol* 10: 606. doi:10.3389/fphar.2019.00606

Hou X, Outhwaite IR, Pedi L, Long SB. 2020. Cryo-EM structure of the calcium release-activated calcium channel Orai in an open conformation. *Elife* 9: e62772. doi:10.7554/eLife.62772

Huang R, Li X, He Y, Zhu W, Gao L, Liu Y, Gao L, Wen Q, Zhong JF, Zhang C, et al. 2020. Recent advances in CAR-T cell engineering. *J Hematol Oncol* 13: 86. doi:10.1186/s13045-020-00910-5

Hughes JP, Rees S, Kalindjian SB, Philpott KL. 2011. Principles of early drug discovery. *Br J Pharmacol* 162: 1239–1249. doi:10.1111/j.1476-5381.2010.01127.x

Hussack G, Hirama T, Ding W, Mackenzie R, Tanha J. 2011. Engineered single-domain antibodies with high protease resistance and thermal stability. *PLoS ONE* 6: e28218. doi:10.1371/journal.pone.0028218

Jespers L, Schon O, Famm K, Winter G. 2004. Aggregation-resistant domain antibodies selected on phage by heat denaturation. *Nat Biotechnol* 22: 1161–1165. doi:10.1038/nbt1000

Knappik A, Ge L, Honegger A, Pack P, Fischer M, Wellnhofer G, Hoess A, Wölle J, Plückthun A, Virnekäs B. 2000. Fully synthetic human combinatorial antibody libraries (HuCAL) based on modular consensus frameworks and CDRs randomized with trinucleotides. *J Mol Biol* 296: 57–86. doi:10.1006/jmbi.1999.3444

Kontermann RE. 2001. Immunotube selections. In *Antibody engineering* (ed. Kontermann R, Dübel S), pp. 137–148. Springer Lab Manuals. Springer, Berlin, Heidelberg.

Kushwaha R, Schäfermeyer KR, Downie AB. 2014. A protocol for phage display and affinity selection using recombinant protein baits. *J Vis Exp* 84: e50685. doi:10.3791/50685

Lawrence S. 2005. Biotech drug market steadily expands. *Nat Biotechnol* 23: 1466. doi:10.1038/nbt1205-1466

Lim TS, Chan SK. 2016. Immune antibody libraries: manipulating the diverse immune repertoire for antibody discovery. *Curr Pharm Des* 22: 6480–6489. doi:10.2174/1381612822666160923111924

Matochko WL, Chu K, Jin B, Lee SW, Whitesides GM, Derda R. 2012. Deep sequencing analysis of phage libraries using Illumina platform. *Methods* 58: 47–55. doi:10.1016/j.ymeth.2012.07.006

Mendoza-Salazar I, Gómez-Castellano KM, González-González E, Gamboa-Suasnavart R, Rodríguez-Luna SD, Santiago-Casas G, Cortés-Paniagua MI, Pérez-Tapia SM, Almagro JC. 2022. Anti-SARS-CoV-2 Omicron antibodies isolated from a SARS-CoV-2 Delta semi-immune phage display library. *Antibodies (Basel)* 11: 13. doi:10.3390/antib11010013

Mesin L, Ersching J, Victora GD. 2016. Germinal center B cell dynamics. *Immunity* 45: 471–482. doi:10.1016/j.immuni.2016.09.001

Murphy AJ, Macdonald LE, Stevens S, Karow M, Dore AT, Poburksy K, Huang TT, Poueymirou WT, Esau L, Meola M, et al. 2014. Mice with megabase humanization of their immunoglobulin genes generate antibodies as efficiently as normal mice. *Proc Natl Acad Sci* 111: 5153–5158. doi:10.1073/pnas.1324022111

Nelson AL, Dhimolea E, Reichert JM. 2010. Development trends for human monoclonal antibody therapeutics. *Nat Rev Drug Discov* 9: 767–774. doi:10.1038/nrd3229

Nilson BH, Solomon A, Björck L, Åkerström B. 1992. Protein L from *Peptostreptococcus magnus* binds to the κ light chain variable domain. *J Biol Chem* 267: 2234–2239.

Omar N, Lim TS. 2018. Construction of naive and immune human Fab phage-display library. *Methods Mol Biol* 1701: 25–44. doi:10.1007/978-1-4939-7447-4_2

Pavlidou M, Hänel K, Möckel L, Willbold D. 2013. Nanodiscs allow phage display selection for ligands to non-linear epitopes on membrane proteins. *PLoS ONE* 8: e72272. doi:10.1371/journal.pone.0072272

Pedraza-Escalona M, Guzmán-Bringas O, Arrieta-Oliva HI, Gómez-Castellano K, Salinas-Trujano J, Torres-Flores J, Muñoz-Herrera JC, Camacho-Sandoval R, Contreras-Pineda P, Chacón-Salinas R, et al. 2021. Isolation and characterization of high affinity and highly stable anti-Chikungunya virus antibodies using ALTHEA Gold Libraries™. *BMC Infect Dis* 21: 1121. doi:10.1186/s12879-021-06717-0

Prassler J, Thiel S, Pracht C, Polzer A, Peters S, Bauer M, Nörenberg S, Stark Y, Kölln J, Popp A, et al. 2011. HuCAL PLATINUM, a synthetic Fab library optimized for sequence diversity and superior performance in mammalian expression systems. *J Mol Biol* 413: 261–278. doi:10.1016/j.jmb.2011.08.012

Rothe C, Urlinger S, Löhning C, Prassler J, Stark Y, Jäger U, Hubner B, Bardroff M, Pradel I, Boss M, et al. 2008. The human combinatorial antibody library HuCAL GOLD combines diversification of all six CDRs according to the natural immune system with a novel display method for efficient selection of high-affinity antibodies. *J Mol Biol* 376: 1182–1200. doi:10.1016/j.jmb.2007.12.018

Rudnick SI, Adams GP. 2009. Affinity and avidity in antibody-based tumor targeting. *Cancer Biother Radiopharm* 24: 155–161. doi:10.1089/cbr.2009.0627

Russo G, Meier D, Helmsing S, Wenzel E, Oberle F, Frenzel A, Hust M. 2018. Parallelized antibody selection in microtiter plates. *Methods Mol Biol* 1701: 273–284. doi:10.1007/978-1-4939-7447-4_14

Schwimmer LJ, Huang B, Giang H, Cotter RL, Chemla-Vogel DS, Dy FV, Tam EM, Zhang F, Toy P, Bohmann DJ, et al. 2013. Discovery of diverse and functional antibodies from large human repertoire antibody libraries. *J Immunol Methods* 391: 60–71. doi:10.1016/j.jim.2013.02.010

Scolnik PA. 2009. Mabs: a business perspective. *MAbs* 1: 179–184. doi:10.4161/mabs.1.2.7736

Sigrist CJ, de Castro E, Cerutti L, Cuche BA, Hulo N, Bridge A, Bougueleret L, Xenarios I. 2013. New and continuing developments at PROSITE. *Nucleic Acids Res* 41: D344–D347. doi:10.1093/nar/gks1067

Smith GP. 1985. Filamentous fusion phage: novel expression vectors that display cloned antigens on the virion surface. *Science* 228: 1315–1317. doi:10.1126/science.4001944

Smith GP. 2019. Phage display: simple evolution in a Petri dish (Nobel lecture). *Angew Chem Int Ed Engl* 58: 14428–14437. doi:10.1002/anie.201908308

Sokullu E, Gauthier MS, Coulombe B. 2021. Discovery of antivirals using phage display. *Viruses* 13: 1120. doi:10.3390/v13061120

Thie H, Voedisch B, Dübel S, Hust M, Schirrmann T. 2009. Affinity maturation by phage display. *Methods Mol Biol* 525: 309–322, xv. doi:10.1007/978-1-59745-554-1_16

Tiller T, Schuster I, Deppe D, Siegers K, Strohner R, Herrmann T, Berenguer M, Poujol D, Stehle J, Stark Y, et al. 2013. A fully synthetic human Fab antibody library based on fixed $V_H/V_L$ framework pairings with favorable biophysical properties. *MAbs* 5: 445–470. doi:10.4161/mabs.24218

Tiwari A, Abraham AK, Harrold JM, Zutshi A, Singh P. 2017. Optimal affinity of a monoclonal antibody: guiding principles using mechanistic modeling. *AAPS J* 19: 510–519. doi:10.1208/s12248-016-0004-1

Ubah O, Palliyil S. 2017. Monoclonal antibodies and antibody like fragments derived from immunised phage display libraries. *Adv Exp Med Biol* 1053: 99–117. doi:10.1007/978-3-319-72077-7_6

Valadon P, Pérez-Tapia SM, Nelson RS, Guzmán-Bringas OU, Arrieta-Oliva HI, Gómez-Castellano KM, Pohl MA, Almagro JC. 2019. ALTHEA GGold Libraries™: antibody libraries for therapeutic antibody discovery. *MAbs* 11: 516–531. doi:10.1080/19420862.2019.1571879

Wenzel EV, Roth KDR, Russo G, Führer V, Helmsing S, Frenzel A, Hust M. 2020. Antibody phage display: antibody selection in solution using biotinylated antigens. *Methods Mol Biol* 2070: 143–155. doi:10.1007/978-1-4939-9853-1_8

Wilkinson TC, Gardener MJ, Williams WA. 2015. Discovery of functional antibodies targeting ion channels. *J Biomol Screen* 20: 454–467. doi:10.1177/1087057114560698

Zhu Z, Dimitrov DS. 2009. Construction of a large naïve human phage-displayed Fab library through one-step cloning. *Methods Mol Biol* 525: 129–142, xv. doi:10.1007/978-1-59745-554-1_6

# Protocol 1

# Semisynthetic Phage Display Library Construction: Design and Synthesis of Diversified Single-Chain Variable Fragments and Generation of Primary Libraries

Juan C. Almagro[1,3] and Mary Ann Pohl[2]

[1]*GlobalBio, Inc., Cambridge, Massachusetts 02138, USA;* [2]*Ailux Biologics, Somerville, Massachusetts 02143, USA*

Display of antibody fragments on the surface of M13 filamentous bacteriophages is a well-established approach for the identification of antibodies binding to a target of interest. Here, we describe the first of a three-step method to construct Antibody Libraries for Therapeutic Antibody Discovery (ALTHEA) Libraries. The three-step method involves (1) primary library (PL) construction, (2) filtered library construction, and (3) secondary library construction. The first step, described here, entails design, synthesis, and cloning of four PLs. These PLs are designed with specific properties amenable to therapeutic antibody development using one universal variable heavy ($V_H$) scaffold and four distinct variable light ($V_L$) scaffolds. The scaffolds are diversified in positions that bind both protein and peptide targets identified in antibody–antigen complexes of known structure using the amino acid frequencies found in those positions in known human antibody sequences, avoiding residues that may lead to developability liabilities. The diversified scaffolds are combined with 90 synthetic neutral HCDR3 sequences designed with developable human diversity genes (IGHD) and joining heavy genes (IGHJ) in germline configuration, and assembled as single-chain variable fragments (scFvs) in a $V_L$–linker–$V_H$ orientation. The four designed PLs are synthesized using trinucleotide phosphoramidites (TRIMs) and cloned independently into a phagemid vector for M13 pIII display. Quality control of the cloning of the four PLs is also described, which involves sequencing scFvs in each library.

## MATERIALS

It is essential that you consult the appropriate Material Safety Data Sheets and your institution's Environmental Health and Safety Office for proper handling of equipment and hazardous materials used in this protocol.

RECIPES: Please see the end of this protocol for recipes indicated by <R>. Additional recipes can be found online at http://cshprotocols.cshlp.org/site/recipes.

### Reagents

Carbenicillin (Carb; 100 mg/mL) <R>
Deoxynucleotide (dNTP) solution (Thermo Scientific R0181)
DNA ladder, 100-bp (NEB N3231S) and 1-kb (NEB N3232S)
DreamTaq polymerase (Thermo EP0705)
*Escherichia coli* TG1 phage-competent cells (Antibody Design Laboratories PC001)
Gel Loading Dye, purple (6×; NEB B7024S)
GelRed Nucleic Acid Gel Stain, 10,000× (Biotium 41003)

---

[3]Correspondence: juan.c.almagro@globalbioinc.com

© 2026 Cold Spring Harbor Laboratory Press
Cite this protocol as *Cold Spring Harb Protoc*; doi:10.1101/pdb.prot108614

Glycerol (99%, v/v) Sigma-Aldrich G5516-4L (autoclaved)
Nuclease-free $H_2O$ (NEB B1500S)
pADL-23c phagemid vector (Antibody Design Laboratories PD0111)
QIAquick Gel Extraction Kit (QIAGEN 28706)
QIAquick PCR Purification Kit (QIAGEN 28106)
Sequencing oligonucleotides:
    For_scFv_seq (5′-GCGGATAACAATTTGAATTCAAGGAGACA-3′)
    Rev_scFv_seq (5′-TGTATGAGGTTTTGCTAAACAACTTTCA-3′)
SfiI restriction enzyme (20,000 units/mL; NEB R0123S) and 10× rCutSmart Buffer (included).
SOC recovery medium (NEB B9020)
T4 DNA ligase and 10× buffer (NEB M0202S)
Tris acetate-EDTA buffer 10× (Sigma-Aldrich T9650-1L)
UltraPure Agarose (Invitrogen 16500100)
YTCG liquid medium <R>
YTCG solid medium <R>

## Equipment

Agarose gel electrophoresis equipment, including tank, wires, gel molds, and combs
Cell scrapers (Nunc 179693PK)
Computer with DNA analysis software (e.g., Geneious Prime, DNASTAR, or equivalent)
Conical tubes (sterile, 15- and 50-mL, screw-top; Nunc 339650)
MicroPulser Electroporation cuvettes (0.1-cm gap; Bio-Rad 1652089)
Electroporator (Bio-Rad Gene Pulser Xcell Electroporation Systems)
Freezer at −20°C and −70°C
Glass spreader or beads, to plate bacteria
Incubator at 37°C (shaking and static)
PCR tubes (0.2-mL)
Power supply box
VWR, serological pipette, standard (sterile)
Thermocycler (96-well)
UV spectrophotometer

## METHOD

Based on the design of the diversified variable heavy ($V_H$) and variable light ($V_L$) scaffolds shown in Table 1 and neutral H3J fragments listed in Table 2, users will synthesize a pool of DNA fragments encoding for the sequences of the four distinct single-chain variable fragment (scFv) libraries described in Figure 1. The four synthetic scFv libraries will then be digested and ligated independently into the phagemid vector pADL23c as pIII fusion proteins. The ligation reactions will be transformed into E. coli, which allows for storage and propagation of the four primary libraries (PLs). Last, quality control (QC) is performed to assess the quality and diversity of the PLs.

Users will need access to DNA synthesis services such as for trinucleotide phosphoramidites (Suchsland et al. 2018), or Twist technology (https://www.twistbioscience.com/technology).

The design of the four PLs is based on one universal $V_H$ scaffold built with the sequences of the human germline genes IGHV3-23*01 and IGHJ4*01, combined with the sequences of four $V_L$ scaffolds in the germline gene configuration; namely, (1) IGKV4*01 + IGKJ*01, (2) IGKV3-20*01 + IGKJ*01, (3) IGKV3-11*01 + IGKJ*01, and (4) IGKV1-39*01 + IGKJ*01.

The nucleotide and amino acid sequences of the universal $V_H$ scaffold combined with the nucleotide and amino acid sequences of the four $V_L$ scaffolds in a $V_L$–linker–$V_H$ orientation are provided in Figure 1.

The positions to be diversified and the frequency of amino acids per position in the three CDRs of $V_L$ (LCDR1, LCDR2, and LCDR3) and two of the three CDRs of $V_H$ (HCDR1 and HCDR2) of each of the four PLs are outlined in Table 1.

## Chapter 9

**TABLE 1.** Diversification scheme for CDRs of $V_H$ and $V_L$ fragments

### Diversified positions in the CDRs of the $V_L$ 4-01 scaffold

| Amino acid | A | C | D | E | F | G | H | I | K | L | M | N | P | Q | R | S | T | V | W | Y | Total | |
|---|---|---|---|---|---|---|---|---|---|---|---|---|---|---|---|---|---|---|---|---|---|---|
| Y | | | 10 | | | 10 | 5 | | | | | | | | 10 | 15 | | | 50 | | 100 | LCDR1 |
| S | | | | | | | | | | | | | | | | | | | | | | |
| S | | | 10 | | 20 | | | | | | | 10 | | | | 40 | | | 20 | | 100 | |
| N | | | | | | | | | | | | | | | | | | | | | | |
| N | | | | | | | | | | | | | | | | | | | | | | |
| K | | | | 40 | | | | 50 | | | | | | | | | | | 10 | | 100 | |
| N | | | | | | | | | | | | | | | | | | | | | | |
| Y | | | 15 | | | | | | | | 15 | | | 20 | | | | | 50 | | 100 | |
| W | 20 | | 15 | | 15 | | | | | | | | | | | | 50 | | | | 100 | LCDR2 |
| Y | | | | | | 10 | 10 | | | | | | 10 | | 10 | | | | 10 | 50 | 100 | LCDR3 |
| Y | | | 20 | | | | | | | | | | 10 | | 10 | 10 | | | | 50 | 100 | |
| S | | | | 10 | | | | | | | | 25 | | | 10 | 45 | | | 10 | | 100 | |
| T | 10 | | | 10 | | | | | | | | 10 | | | 10 | 20 | 30 | | 10 | | 100 | |
| P | | | | | | | | | | | | | | | | | | | | | | |
| L | | | | | 20 | | | 20 | | 20 | | | | | | | | 20 | 20 | | 100 | |

### Diversified positions in the CDRs of the $V_L$ 3-20 scaffold

| Amino acid | A | C | D | E | F | G | H | I | K | L | M | N | P | Q | R | S | T | V | W | Y | Total | |
|---|---|---|---|---|---|---|---|---|---|---|---|---|---|---|---|---|---|---|---|---|---|---|
| S | | | 25 | | | | | | | | | | | 25 | | 50 | | | | | 100 | LCDR1 |
| S | | 25 | | | | | 25 | | | | | | | | | 50 | | | | | 100 | |
| S | | | | | | | | | | | | 30 | | | | 50 | 20 | | | | 100 | |
| Y | | 20 | | | | | | | | | | 10 | | | | 20 | | | | 50 | 100 | |
| G | | 20 | 15 | | | 50 | | | | | | | | | | | 15 | | | | 100 | LCDR2 |
| Y | | 10 | 10 | | 5 | 10 | 5 | | | | | | | | 10 | 10 | | | | 40 | 100 | LCDR3 |
| G | | | | 10 | | 40 | | | | | | | | | 10 | 30 | | | | 10 | 100 | |
| S | | | | 10 | | | | | | | | 10 | 10 | | 10 | 50 | | | | 10 | 100 | |
| S | 10 | | | 5 | | | | | | | | 10 | | 5 | | 50 | | 10 | 10 | | 100 | |
| P | | | | | | | | | | | | | | | | | | | | | | |
| L | | | | | 20 | | | 20 | | 20 | | | | | | | | 20 | 20 | | 100 | |

### Diversified positions in the CDRs of the $V_L$ 3-11 scaffold

| Amino acid | A | C | D | E | F | G | H | I | K | L | M | N | P | Q | R | S | T | V | W | Y | Total | |
|---|---|---|---|---|---|---|---|---|---|---|---|---|---|---|---|---|---|---|---|---|---|---|
| S | | | 15 | | | | | | | | | | | 15 | | 55 | 15 | | | | 100 | LCDR1 |
| S | | | | | | | | | | | | 25 | | | | 50 | 25 | | | | 100 | |
| Y | | | 13 | | | | | | | | | 13 | | | 13 | 13 | | | | 50 | 100 | |
| D | | | 50 | | 10 | | | | | | | 10 | | | | 10 | | 10 | 10 | | 100 | LCDR2 |
| R | 10 | | | 10 | 5 | 5 | | | | | | | | | 50 | 10 | | | 10 | | 100 | LCDR3 |
| S | | | 5 | 10 | 5 | 5 | | | | | | | | | 10 | 50 | | 5 | 10 | | 100 | |
| N | 5 | | | 10 | 5 | 5 | | | | | | 50 | | | 5 | 10 | | | 10 | | 100 | |
| W | 10 | | 10 | 10 | | 10 | 10 | | | | | 10 | | | 10 | 10 | | 10 | 10 | | 100 | |
| P | | | | | | | | | | | | | | | | | | | | | | |
| L | | | | | 20 | | | 20 | | 20 | | | | | | | | 20 | 20 | | 100 | |

### Diversified positions in the CDRs of the $V_L$ 1-39 scaffold

| Amino acid | A | C | D | E | F | G | H | I | K | L | M | N | P | Q | R | S | T | V | W | Y | Total | |
|---|---|---|---|---|---|---|---|---|---|---|---|---|---|---|---|---|---|---|---|---|---|---|
| S | | | 30 | | | | | | | | | | | | 15 | 55 | | | | | 100 | LCDR1 |
| S | | | | | | | | | | | | 25 | | | | 50 | 25 | | | | 100 | |
| Y | | | 15 | | | | | | | | | 15 | | 20 | 20 | | | | | 30 | 100 | |
| A | 50 | | | 5 | | 10 | | | | | | 5 | | | 10 | | | 10 | 10 | | 100 | LCDR2 |
| S | 5 | | 5 | 10 | | 10 | 5 | | | | | | | | 10 | 40 | | 5 | 10 | | 100 | LCDR3 |
| Y | 10 | | 10 | | | 10 | | | | | | | | | 10 | 20 | | | | 40 | 100 | |
| S | | | | 10 | | | | | | | | 10 | | 10 | 10 | 50 | | | 10 | | 100 | |
| T | | | 10 | 10 | | | 5 | | | | | 10 | | 10 | 10 | | 35 | | 10 | | 100 | |
| P | | | | | | | | | | | | | | | | | | | | | | |
| L | | | | | 20 | | | 20 | | 20 | | | | | | | | 20 | 20 | | 100 | |

(continued)

TABLE 1. Continued

| Amino acid | A | C | D | E | F | G | H | I | K | L | M | N | P | Q | R | S | T | V | W | Y | Total | |
|---|---|---|---|---|---|---|---|---|---|---|---|---|---|---|---|---|---|---|---|---|---|---|
| S | | | | | | | | | | | | | | | | 50 | 50 | | | | 100 | HCDR1 |
| S | | | 30 | | | | | | 20 | | | | | | | 50 | | | | | 100 | |
| Y | | | 20 | | | | | | | | | | | | | 10 | 20 | | | 50 | 100 | |
| A | 40 | | | | | 10 | | | | | | 10 | | | | | | 10 | | 30 | 100 | |
| A | 20 | 10 | 20 | 10 | | 5 | | | | | | | | | 20 | | 5 | | | 10 | 100 | HCDR2 |
| I | | | | | | | | | | | | | | | | | | | | | | |
| S | | | 25 | | | | | | | | | | | 15 | | 35 | | | | 25 | 100 | |
| G | | | | | | | | | | | | | | | | | | | | | | |
| S | | | 5 | 10 | | 20 | | | | | | | | | 10 | 30 | | | | 25 | 100 | |
| G | | | | | | | | | | | | | | | | | | | | | | |
| G | | | 15 | | | 60 | | | | | | | | | | 15 | | | | 10 | 100 | |
| S | | | 15 | | | 15 | | | | | | | | 10 | | 30 | 10 | | | 20 | 100 | |
| T | | | | | | | | | | | | | | | | | | | | | | |
| Y | | | 10 | | | | | 10 | | | 30 | | | 10 | 10 | | | | | 30 | 100 | |

Amino acids in the first column refer to the canonical residue in the germline gene sequence. Amino acids across the top row are variant residues to be introduced at the canonical residue's position in the CDR. Numbers represent the frequency (percentage) at which each variant residue should be represented when synthesizing the variable gene fragment.

TABLE 2. Neutral H3J (HCDR3/J$_H$) fragments to be incorporated into synthetic VH DNA fragments

| No. | Sequence | No. | Sequence | No. | Sequence |
|---|---|---|---|---|---|
| 1 | GYSGYDYAEYFQHWGQGTLVTVSS | 31 | GITGTDAFDVWGQGTMVTVSS | 61 | SIAARNWFDSWGQGTLVTVSS |
| 2 | GYSYGYAEYFQHWGQGTLVTVSS | 32 | GIVGATDAFDVWGQGTMVTVSS | 62 | VQLERNWFDSWGQGTLVTVSS |
| 3 | TTVTAEYFQHWGQGTLVTVSS | 33 | GTTGTDAFDVWGQGTMVTVSS | 63 | YYDILTGYYNNWFDSWGQGTLVTVSS |
| 4 | YSGSYYAEYFQHWGQGTLVTVSS | 34 | GYSSGYDAFDVWGQGTMVTVSS | 64 | YYYDSSGYYYNWFDSWGQGTLVTVSS |
| 5 | DYGDYAEYFQHWGQGTLVTVSS | 35 | LTGDAFDVWGQGTMVTVSS | 65 | YYYGSGSYYNNWFDSWGQGTLVTVSS |
| 6 | DYSNYAEYFQHWGQGTLVTVSS | 36 | VDIVATIDAFDVWGQGTMVTVSS | 66 | EYSSSSNWFDSWGQGTLVTVSS |
| 7 | SIAARAEYFQHWGQGTLVTVSS | 37 | GYSGYDYYFDYWGQGTLVTVSS | 67 | GITGTNWFDSWGQGTLVTVSS |
| 8 | VQLERAEYFQHWGQGTLVTVSS | 38 | GYSYGYYFDYWGQGTLVTVSS | 68 | GIVGATNWFDSWGQGTLVTVSS |
| 9 | YYDILTGYYNAEYFQHWGQGTLVTVSS | 39 | TTVTYFDYWGQGTLVTVSS | 69 | GTTGTNWFDSWGQGTLVTVSS |
| 10 | YYYDSSGYYYAEYFQHWGQGTLVTVSS | 40 | YSGSYYYFDYWGQGTLVTVSS | 70 | GYSSGYNWFDSWGQGTLVTVSS |
| 11 | YYYGSGSYYNAEYFQHWGQGTLVTVSS | 41 | DYGDYYFDYWGQGTLVTVSS | 71 | LTGNWFDSWGQGTLVTVSS |
| 12 | EYSSSSAEYFQHWGQGTLVTVSS | 42 | DYSNYYFDYWGQGTLVTVSS | 72 | VDIVATINWFDSWGQGTLVTVSS |
| 13 | GITGTAEYFQHWGQGTLVTVSS | 43 | SIAARYFDYWGQGTLVTVSS | 73 | GYSGYDYYGMDVWGQGTTVTVSS |
| 14 | GIVGATAEYFQHWGQGTLVTVSS | 44 | VQLERYFDYWGQGTLVTVSS | 74 | GYSYGYYGMDVWGQGTTVTVSS |
| 15 | GTTGTAEYFQHWGQGTLVTVSS | 45 | YYDILTGYYNYFDYWGQGTLVTVSS | 75 | TTVTYGMDVWGQGTTVTVSS |
| 16 | GYSSGYAEYFQHWGQGTLVTVSS | 46 | YYYDSSGYYYYFDYWGQGTLVTVSS | 76 | YSGSYYYGMDVWGQGTTVTVSS |
| 17 | LTGAEYFQHWGQGTLVTVSS | 47 | YYYGSGSYYNYFDYWGQGTLVTVSS | 77 | DYGDYYGMDVWGQGTTVTVSS |
| 18 | VDIVATIAEYFQHWGQGTLVTVSS | 48 | EYSSSSYFDYWGQGTLVTVSS | 78 | DYSNYYGMDVWGQGTTVTVSS |
| 19 | GYSGYDYDAFDVWGQGTMVTVSS | 49 | GITGTYFDYWGQGTLVTVSS | 79 | SIAARYGMDVWGQGTTVTVSS |
| 20 | GYSYGYDAFDVWGQGTMVTVSS | 50 | GIVGATYFDYWGQGTLVTVSS | 80 | VQLERYGMDVWGQGTTVTVSS |
| 21 | TTVTDAFDVWGQGTMVTVSS | 51 | GTTGTYFDYWGQGTLVTVSS | 81 | YYDILTGYYNYGMDVWGQGTTVTVSS |
| 22 | YSGSYYDAFDVWGQGTMVTVSS | 52 | GYSSGYYFDYWGQGTLVTVSS | 82 | YYYDSSGYYYYGMDVWGQGTTVTVSS |
| 23 | DYGDYDAFDVWGQGTMVTVSS | 53 | LTGYFDYWGQGTLVTVSS | 83 | YYYGSGSYYNYGMDVWGQGTTVTVSS |
| 24 | DYSNYDAFDVWGQGTMVTVSS | 54 | VDIVATIYFDYWGQGTLVTVSS | 84 | EYSSSSYGMDVWGQGTTVTVSS |
| 25 | SIAARDAFDVWGQGTMVTVSS | 55 | GYSGYDYNWFDSWGQGTLVTVSS | 85 | GITGTYGMDVWGQGTTVTVSS |
| 26 | VQLERDAFDVWGQGTMVTVSS | 56 | GYSYGYNWFDSWGQGTLVTVSS | 86 | GIVGATYGMDVWGQGTTVTVSS |
| 27 | YYDILTGYYNDAFDVWGQGTMVTVSS | 57 | TTVTNWFDSWGQGTLVTVSS | 87 | GTTGTYGMDVWGQGTTVTVSS |
| 28 | YYYDSSGYYYDAFDVWGQGTMVTVSS | 58 | YSGSYYNWFDSWGQGTLVTVSS | 88 | GYSSGYYGMDVWGQGTTVTVSS |
| 29 | YYYGSGSYYNDAFDVWGQGTMVTVSS | 59 | DYGDYNWFDSWGQGTLVTVSS | 89 | LTGYGMDVWGQGTTVTVSS |
| 30 | EYSSSSDAFDVWGQGTMVTVSS | 60 | DYSNYNWFDSWGQGTLVTVSS | 90 | VDIVATIYGMDVWGQGTTVTVSS |

Chapter 9

*The third CDR of V$_H$ (HCDR3) is built with a set of 90 "placeholder" HCDR3 sequences called neutral H3J fragments (Table 2). These neutral H3J fragments may confer flexibility during the library construction process to avoid any structural constraints that one HCDR3 imposes to residues in contact with the HCDR3. Such flexibility should accommodate the diversity of the natural H3J fragments (see Protocol 3: Semisynthetic Phage Display Library Construction: Generation of Single-Chain Variable Fragment Secondary Libraries [Almagro and Pohl 2024a]) in the final step of the ALTHEA Gold Plus Libraries construction process.*

## Synthesis of the Diversified scFv Libraries

1. Synthesize 1 µg of each of the four diversified scFv libraries using trinucleotide phosphoramidites (Suchsland et al. 2018) or Twist technology (https://www.twistbioscience.com/technology).

2. To increase the quality of the libraries, take into account the following considerations:

   a. Avoid the following restriction sites, except in the leader peptide and tags: SfiI, GGCCNNNNNGGCC; BglI, GCCNNNNNGGC; SpeI, ACTACG; MluI, ACGCGT; NcoI, CCATGG; KpnI, GGTACC; and BsaI, GGTCTC (both strands).

   b. Use acceptable codon usage for good expression in *E. coli* and mammalian cells.

   c. Avoid the following post-translational modifications or potential liabilities: *N*-glycosylation (site), *O*-glycosylation (site), and no alternate ribosome-binding site (RBS).

   d. Avoid splicing sites.

   *It should be noticed that the nucleotide sequences of the scFvs listed in Table 1 have been designed following the above considerations. The codons of the diversified positions and neutral H3J fragments should also be designed following the above considerations. Commonly, companies offering synthesis services such as GENEWIZ (https://www.genewiz.com/), GeneArt (https://www.thermofisher.com/mx/es/home/life-science/cloning/gene-synthesis/directed-evolution/geneart-combinatorial-libraries.html), or Twist (https://www.twistbioscience.com/) provide expert advice on the best way to implement a synthesis project and set expectations about the quality of the synthetic products. In some cases, QC of the synthetic fragments includes Sanger sequencing of 96 clones chosen at random from the pool of synthetic fragments. In other cases, the QC is performed by high-throughput sequencing (HTS). The turnaround time of the synthesis and HTS depends on the complexity of the project. For ALTHEA Gold Plus Libraries, it takes ~8 wk.*

   *The synthetic scFv libraries are delivered in TE buffer as nonamplified libraries (~300 ng) and amplified libraries (~10 ug). The nonamplified libraries are used to regenerate the libraries as needed. The amplified libraries are used for digestion and cloning of the synthetic scFv libraries.*

## Construction of PLs

### *Digestion of the Four Synthetic scFv Libraries for Cloning into the pADL-23c Vector*

*The amplified synthetic scFv libraries are digested and cloned as four separate PLs to maximize the final diversity of ALTHEA Gold Plus Libraries. By keeping the PLs separately as sublibraries (each unique V$_L$–V$_H$ pairing), one can use either individual sublibraries or the entire library pooled for selections, allowing for additional flexibility in the panning approaches for given projects and targets. Because each PL will require several electroporation reactions, it may be desirable to split this part of the protocol over several days, to make the workload manageable and to avoid any cross-contamination of the libraries.*

3. Prepare a reaction mix to digest each synthetic scFv library with SfiI as follows:

   | Component | Volume/amount |
   |---|---|
   | CutSmart Buffer (10×; NEB) | 10.0 µL |
   | Synthetic scFv library | 3.0 µg |
   | SfiI (20,000 units/mL) | 3.6 µL (72.0 U) |
   | Nuclease-free H$_2$O | To 150.0 µL |

4. Incubate for 2.5 h at 50°C.

5. Remove enzymes, salts, and subproducts of the digestion using the QIAquick PCR Purification Kit. Elute with 30 µL of nuclease-free H$_2$O to obtain a concentrated sample.

6. Determine DNA concentration by UV spectrophotometry and store it at −20°C until use.

## A  $V_L$ 4-01/VH 3-23

```
gct gga ttg tta tta ctc gcg gcc cag ccg gcc atg gca gac atc gtg atg acc cag tct
 A   G   L   L   L   A   A   Q   P   A   M   A   D   I   V   M   T   Q   S
cca gac tcc ctg tct gtg tct ctg ggc gaa cgt gcc acc atc aac tgc aag tcc agc cag
 P   D   S   L   S   V   S   L   G   E   R   A   T   I   N   C   K   S   S   Q
agt gtt tta tac agc tcc aac aat aag aac tac tta gct tgg tac cag cag aaa cca gga
 S   V   L   Y   S   S   N   N   K   N   Y   L   A   W   Y   Q   Q   K   P   G
cag cct cct aag ctg ctc att tac tgg gct tcc acc cgg gaa tcc ggg gtc cct gac cga
 Q   P   P   K   L   L   I   Y   W   A   S   T   R   E   S   G   V   P   D   R
ttc agt ggc agc ggg tct ggg aca gat ttc act ctc acc atc agc agc ctg cag gct gaa
 F   S   G   S   G   S   G   T   D   F   T   L   T   I   S   S   L   Q   A   E
gat gtg gca gtt tat tac tgt cag caa tat tat agt act cct acg ttc ggc caa ggt
 D   V   A   V   Y   Y   C   Q   Q   Y   Y   S   T   P   L   T   F   G   Q   G
acc aag gtg gaa atc aaa ggt ggt ggt ggt tca ggt ggt ggt tct ggc ggc ggc tcc
 T   K   V   E   I   K   G   G   G   G   S   G   G   G   G   S   G   G   G   S
ggt ggt ggt gga tcc gag gtg cag ctg ttg gag tct ggg gga ggc ttg gta cag cct ggg
 G   G   G   G   S   E   V   Q   L   L   E   S   G   G   G   L   V   Q   P   G
ggg tcc ctg cga ctc tcc tgt gca gcc tct gga ttc acc ttt agc agc tat gcc atg agc
 G   S   L   R   L   S   C   A   A   S   G   F   T   F   S   S   Y   A   M   S
tgg gtc cgc cag gct cca ggg aag ggg ctg gag tgg gtg tca gct att agt ggt agt ggt
 W   V   R   Q   A   P   G   K   G   L   E   W   V   S   A   I   S   G   S   G
ggt agc aca tac tac gca gac tcc gtg aag ggc cgg ttc acc atc tcc cgt gac aat tcc
 G   S   T   Y   Y   A   D   S   V   K   G   R   F   T   I   S   R   D   N   S
aag aac acg ctg tat ctg caa atg aac agc ctg cgt gcc gag gac acg gcc gtg tat tac
 K   N   T   L   Y   L   Q   M   N   S   L   R   A   E   D   T   A   V   Y   Y
tgt gcg aaa xxx ggc ccg gga ggc caa cac cat cac cac
 C   A   K   X   G   P   G   G   Q   H   H   H   H
```

Color code:
Leader peptide (Pel B); Positions to be diversified; GS Linker; Neutral H3J fragments; Tags

## B  $V_L$ 3-20/VH 3-23

```
gct gga ttg tta tta ctc gcg gcc cag ccg gcc atg gca gaa att gtg ttg acg cag tct
 A   G   L   L   L   A   A   Q   P   A   M   A   E   I   V   L   T   Q   S
cca ggc acc ctg tct ttg tct cca gga gaa cgt gcc acc ctc tcc tgc cgt gcc agt cag
 P   G   T   L   S   L   S   P   G   E   R   A   T   L   S   C   R   A   S   Q
agt gtt agc agc agc tac tta gcc tgg tac cag cag aaa cct ggc cag gct ccc cga ctc
 S   V   S   S   S   Y   L   A   W   Y   Q   Q   K   P   G   Q   A   P   R   L
ctc atc tat ggt gca tcc agc cgt gcc act ggc atc cca gac cgt ttc agt ggc agt ggg
 L   I   Y   G   A   S   S   R   A   T   G   I   P   D   R   F   S   G   S   G
tct ggg aca gac ttc act ctc acc atc agc aga ctg gag cct gaa gat ttt gca gtg tat
 S   G   T   D   F   T   L   T   I   S   R   L   E   P   E   D   F   A   V   Y
tac tgt cag cag tat ggt agc tca cct ctg acg ttc ggc caa ggg acc aag gtg gaa atc
 Y   C   Q   Q   Y   G   S   S   P   L   T   F   G   Q   G   T   K   V   E   I
aaa ggt ggt ggt ggt tca ggt ggt ggt tct ggc ggc ggc tcc ggt ggt ggt tcc
 K   G   G   G   G   S   G   G   G   G   S   G   G   G   S   G   G   G   S
gag gtg cag ctg ttg gag tct ggg gga ggc ttg gta cag cct ggg ggg tcc ctg cga ctc
 E   V   Q   L   L   E   S   G   G   G   L   V   Q   P   G   G   S   L   R   L
tcc tgt gca gcc tct gga ttc acc ttt agc agc tat gcc atg agc tgg gtc cgc cag gct
 S   C   A   A   S   G   F   T   F   S   S   Y   A   M   S   W   V   R   Q   A
cca ggg aag ggg ctg gag tgg gtg tca gct att agt ggt agt ggt ggt agc aca tac tac
 P   G   K   G   L   E   W   V   S   A   I   S   G   S   G   G   S   T   Y   Y
gca gac tcc gtg aag ggc cgg ttc acc atc tcc cgt gac aat tcc aag aac acg ctg tat
 A   D   S   V   K   G   R   F   T   I   S   R   D   N   S   K   N   T   L   Y
ctg caa atg aac agc ctg cgt gcc gag gac acg gcc gtg tat tac tgt gcg aaa xxx ggc
 L   Q   M   N   S   L   R   A   E   D   T   A   V   Y   Y   C   A   K   X   G
ccg gga ggc caa cac cat cac cac
 P   G   G   Q   H   H   H   H
```

Color code:
Leader peptide (Pel B); Positions to be diversified; GS Linker; Neutral H3J fragments; Tags

FIGURE 1. Sequences of the four different single-chain variable fragment (scFv) scaffolds used to construct the primary libraries. (A) Light chain scaffold IGKV4*01 + IGKJ*01 and heavy chain scaffold IGHV3-23*01 + IGHJ4*01. (B) Light chain scaffold IGKV3-20*01 + IGKJ*01, and heavy chain scaffold IGHV3-23*01 + IGHJ4*01. (C) Light chain scaffold IGKV3-11*01 + IGKJ*01 and heavy chain scaffold IGHV3-23*01 + IGHJ4*01. (D) Light chain scaffold IGKV1-39*01 + IGKJ*01 and heavy chain scaffold IGHV3-23*01 + IGHJ4*01. The pelB leader sequence is highlighted in yellow. The glycine–serine linker is underlined. Residues to be diversified according to the schemes in Table 1 are highlighted in gray. The site where neutral H3J fragments described in Table 2 are incorporated into the scaffolds is highlighted in purple. Tags are highlighted in teal. (*Figure 1 continues on following page.*)

**C** $V_L$ 3-11/VH 3-23

```
gct gga ttg tta tta ctc gcg gcc cag ccg gcc atg gca gaa att gtg ttg aca cag tct
 A   G   L   L   L   L   A   A   Q   P   A   M   A   E   I   V   L   T   Q   S
cca gcc acc ctg tct ttg tct cca ggg gaa aga gcc acc ctc tcc tgc agg gcc agt cag
 P   A   T   L   S   L   S   P   G   E   R   A   T   L   S   C   R   A   S   Q
agt gtt agc agc tac tta gcc tgg tac caa cag aaa cct ggc cag gct ccc agg ctc ctc
 S   V   S   S   Y   L   A   W   Y   Q   Q   K   P   G   Q   A   P   R   L   L
atc tat gat gca tcc aac agg gcc act ggc atc cca gcc agg ttc agt ggc agt ggg tct
 I   Y   D   A   S   N   R   A   T   G   I   P   A   R   F   S   G   S   G   S
ggg aca gac ttc act ctc acc atc agc agc cta gag cct gaa gat ttt gca gtt tat tac
 G   T   D   F   T   L   T   I   S   S   L   E   P   E   D   F   A   V   Y   Y
tgt cag cag cgt agc aac tgg cct ctg acg ttc ggc caa ggg acc aag gtg gaa atc aaa
 C   Q   Q   R   S   N   W   P   L   T   F   G   Q   G   T   K   V   E   I   K
ggt ggt ggt ggt tca ggt ggt ggt ggt tct ggc ggc ggc tcc ggt ggt ggt gga tcc gag
 G   G   G   G   S   G   G   G   G   S   G   G   G   S   G   G   G   G   S   E
gtg cag ctg ttg gag tct ggg gga ggc ttg gta cag cct ggg ggg tcc ctg cga ctc tcc
 V   Q   L   L   E   S   G   G   G   L   V   Q   P   G   G   S   L   R   L   S
tgt gca gcc tct gga ttc acc ttt agc agc tat gcc atg agc tgg gtc cgc cag gct cca
 C   A   A   S   G   F   T   F   S   S   Y   A   M   S   W   V   R   Q   A   P
ggg aag ggg ctg gag tgg gtg tca gct att agt ggt agt ggt ggt agc aca tac tac gca
 G   K   G   L   E   W   V   S   A   I   S   G   S   G   G   S   T   Y   Y   A
gac tcc gtg aag ggc cgg ttc acc atc tcc cgt gac aat tcc aag aac acg ctg tat ctg
 D   S   V   K   G   R   F   T   I   S   R   D   N   S   K   N   T   L   Y   L
caa atg aac agc ctg cgt gcc gag gac acg gcc gtg tat tac tgt gcg aaa xxx ggc ccg
 Q   M   N   S   L   R   A   E   D   T   A   V   Y   Y   C   A   K   X   G   P
gga ggc caa cac cat cac cac
 G   G   Q   H   H   H   H
```

**Color code:**
Leader peptide (Pel B); Positions to be diversified; GS Linker;
Neutrial H3J fragments; Tags

**D** $V_L$ 1-39/VH 3-23

```
gct gga ttg tta tta ctc gcg gcc cag ccg gcc atg gca gac atc cag atg acc cag tct
 A   G   L   L   L   L   A   A   Q   P   A   M   A   D   I   Q   M   T   Q   S
cca tcc tcc ctg tct gca tct gta gga gac aga gtc acc atc act tgc cgg gca agt cag
 P   S   S   L   S   A   S   V   G   D   R   V   T   I   T   C   R   A   S   Q
agc att agc agc tat tta aat tgg tat cag cag aaa cca ggg aaa gcc cct aag ctc ctg
 S   I   S   S   Y   L   N   W   Y   Q   Q   K   P   G   K   A   P   K   L   L
atc tat gct gca tcc agt ttg caa agt ggg gtc cca tca agg ttc agt ggc agt gga tct
 I   Y   A   A   S   S   L   Q   S   G   V   P   S   R   F   S   G   S   G   S
ggg aca gat ttc act ctc acc atc agc agt ctg caa cct gaa gat ttt gca act tac tac
 G   T   D   F   T   L   T   I   S   S   L   Q   P   E   D   F   A   T   Y   Y
tgt caa cag agt tac agt acc cct ctg acg ttc ggc caa ggg acc aag gtg gaa atc aaa
 C   Q   Q   S   Y   S   T   P   L   T   F   G   Q   G   T   K   V   E   I   K
ggt ggt ggt ggt tca ggt ggt ggt ggt tct ggc ggc ggc tcc ggt ggt ggt gga tcc gag
 G   G   G   G   S   G   G   G   G   S   G   G   G   S   G   G   G   G   S   E
gtg cag ctg ttg gag tct ggg gga ggc ttg gta cag cct ggg ggg tcc ctg cga ctc tcc
 V   Q   L   L   E   S   G   G   G   L   V   Q   P   G   G   S   L   R   L   S
tgt gca gcc tct gga ttc acc ttt agc agc tat gcc atg agc tgg gtc cgc cag gct cca
 C   A   A   S   G   F   T   F   S   S   Y   A   M   S   W   V   R   Q   A   P
ggg aag ggg ctg gag tgg gtg tca gct att agt ggt agt ggt ggt agc aca tac tac gca
 G   K   G   L   E   W   V   S   A   I   S   G   S   G   G   S   T   Y   Y   A
gac tcc gtg aag ggc cgg ttc acc atc tcc cgt gac aat tcc aag aac acg ctg tat ctg
 D   S   V   K   G   R   F   T   I   S   R   D   N   S   K   N   T   L   Y   L
caa atg aac agc ctg cgt gcc gag gac acg gcc gtg tat tac tgt gcg aaa xxx ggc ccg
 Q   M   N   S   L   R   A   E   D   T   A   V   Y   Y   C   A   K   X   G   P
gga ggc caa cac cat cac cac
 G   G   Q   H   H   H   H
```

**Color code:**
Leader peptide (Pel B); Positions to be diversified; GS Linker;
Neutrial H3J fragments; Tags

FIGURE 1. *Continued*


Cite this protocol as *Cold Spring Harb Protoc*; doi:10.1101/pdb.prot108614

## Digestion of the pADL-23c Vector for Cloning the Four Synthetic scFvs

7. Prepare a 500-µL reaction mix to digest the pADL-23c vector with SfiI as follows:

| Component | Volume/amount |
| --- | --- |
| CutSmart Buffer (10×; NEB) | 50.0 µL |
| pADL-23c phagemid vector | 10.0 µg |
| SfiI (20,000 units/mL) | 12.0 µL (240 units) |
| Nuclease-free $H_2O$ | To 500.0 µL |

8. Incubate for 4 h at 50°C.

9. Remove enzymes, salts, and subproducts of the digestion with the QIAquick PCR Purification Kit. Elute with 50 µL of nuclease-free $H_2O$ to obtain a concentrated sample.

10. Prepare a 1% (w/v) agarose gel and run the sample. To do this, do the following:
    i. Melt agarose in Tris acetate-EDTA buffer. Add GelRed to 1× and cast the gel, adding a comb.
    ii. After the gel has solidified, add 10 µL of gel loading dye to the sample to a final concentration of 1× and run it alongside a 1-kb DNA ladder at 90 V for 90 min using a power supply.

11. Purify the band corresponding to the digested pADL-23c vector (∼3900 bp) with the QIAquick Gel Extraction Kit. Elute with 50 µL of nuclease-free $H_2O$.

12. Determine DNA concentration by UV spectrophotometry and store it at −20°C until use.

## Ligation of Digested scFv Libraries with the Digested Phagemid Vector

13. Prepare the scFv–phagemid ligation mixture (2:1 molar ratio [insert:vector]) as follows:

| Component | Volume/amount |
| --- | --- |
| 10× Buffer for T4 DNA ligase | 37.0 µL |
| Digested pADL-23c vector | 3000 µg |
| Digested synthetic scFv library | 1225 µg |
| Nuclease-free $H_2O$ | To 370 µL |

14. Take an 11-µL (∼120 ng) aliquot of the mixture and place it in another tube. This will serve as the negative ligation control when running the gel in Step 18. Then, add 36.8 µL of T4 ligase to the tube from Step 13 to complete the ligation mixture.

15. Incubate the ligation mix for 2.5 h at 23°C.

16. Remove enzymes, salts, and subproducts of the ligation with the QIAquick PCR Purification Kit. Elute with 35 µL of nuclease-free $H_2O$ to obtain a concentrated sample.

17. Determine DNA concentration by UV spectrophotometry and store it at −20°C until use.

18. To assess the quality of the ligation, run a 1% (w/v) agarose gel (see Step 10) with the ligation control and ligation mix.

    *A single band of vector should be seen in the negative ligation control, and a ladder should be observed in the ligation mix.*

## Electroporation of the Ligation Mix

*For an estimated library size of >$10^9$ colony-forming units (cfu) of each PL, at least 15 separate electroporation reactions are required.*

19. Prechill electroporation cuvettes on ice.

20. Add 2 µL (~130 ng) of the ligation mix into 25 µL of TG1 electrocompetent cells and electroporate in 0.1-cm cuvettes at 1.8 kV, 200 Ω (resistance), and 25 µF (capacitance).

21. Immediately add 975 µL of SOC recovery medium into each cuvette and transfer the cells to a 15-mL screw-top conical tube. Repeat for at least 15 separate mixes.

22. Pool all the resultant electroporated cells into a 50-mL screw-top conical tube (15 mL in total for 15 reactions).

23. Incubate with shaking at 225 rpm for 45 min at 37°C.

24. To determine the estimated size of the library, take a 50-µL aliquot of the electroporated cells and dilute it in 450 µL of YTCG liquid medium. Generate 1:10 serial dilutions to $10^{-10}$. Plate 50 µL of the $10^{-5}$–$10^{-10}$ dilutions onto separate 100-mm YTCG solid medium plates and carefully spread the cells using a glass spreader or beads. Incubate overnight at 37°C.

25. In parallel, plate 1-mL aliquots of the electroporated cells onto large agar (150-mm × 15-mm) YTCG solid medium plates. Incubate overnight at 37°C.

26. The next day, count the colonies on the serial dilution plates. Consider plates "countable" when they have between 40 and 400 well-separated colonies. Use the following formula to estimate the size of the library after electroporation: cfu = [number of colonies × 10 (dilution in 500 µL of medium) × dilution factor)]/50 µL.

27. If the library size is >$10^9$ cfu, continue to the following steps. If the library size is <$10^9$ cfu, see Troubleshooting.

28. Scrape bacteria from the large culture plates (from Step 25) by washing each plate with 3 mL of YTCG liquid medium and loosening the bacterial lawn from the plate using disposable cell scrapers. Collect cell material with a serological pipette and transfer it to a 50-mL screw-top conical tube.

29. Make 1-mL glycerol stocks by mixing 28 mL of bacteria and 14 mL of 99% glycerol. Store the glycerol stocks at −70 °C.

    *These vials are now PL live stocks of one sublibrary.*

### QC of the PLs

*The quality of each PL sublibrary can be assessed by either sending a plate with colonies chosen at random from the titration plates for Sanger sequencing, or by sending PCR products of the colonies for sequencing. The following procedure is for sequencing PCR products.*

30. For each sublibrary, prepare a 25-µL colony PCR mix for 30 colonies as follows:

| Component | Volume/amount (one reaction) | Volume/amount (36 reactions)[a] |
|---|---|---|
| DreamTaq DNA Buffer (10×) | 5.00 µL | 180.00 µL |
| dNTPs (2 mM) | 2.50 µL | 90.00 µL |
| Primer For_scFv_seq (10 µM) | 2.00 µL | 72.00 µL |
| Primer Rev_scFv_seq (10 µM) | 2.00 µL | 72.00 µL |
| DNA template (random colonies) | 15.00 ng | 15.00 ng each |
| DreamTaq DNA polymerase | 0.25 µL | 9.00 µL |
| Nuclease-free H$_2$O | To 25.0 µL | To 900.00 µL |

[a]The reaction mix has an ~10% excess, to account for pipetting error.

31. Dispense 25 µL of the PCR mix into 32 separate PCR tubes.
32. Pick 30 colonies from the library titer (small) plates from Step 26 with sterile toothpicks or pipette tips, selecting different colony sizes to avoid biasing the selection, and add the cell material directly to each PCR tube. Keep one reaction mix without sample (negative control), and use the other for a positive control (a known DNA scFv template)
33. Run the PCR in a thermocycler with the following parameters:

| Step | Time | Temperature |
| --- | --- | --- |
| Initial denaturation | 300.0 sec | 95°C |
| 25 cycles | 30.0 sec | 95°C |
|  | 30.0 sec | 67°C |
|  | 60.0 sec | 72°C |
| Final extension | 10.0 min | 72°C |

34. Run a 1% (w/v) agarose gel (Step 10) with 5 µL of each PCR mix to confirm successful PCR amplification of each colony. If positive, a band of the approximate size of the insert should be observed.
35. Purify the remaining PCR product using the QIAquick PCR Purification Kit and send it for Sanger sequencing with the For_scFv_seq and Rev_scFv_seq primers.
36. Analyze DNA sequence data with the appropriate software, such as Geneious Prime, DNASTAR, or equivalent.

    *The expected result is that 50%–60% of the sequences would be in-frame and match the library design. No identical scFv sequences should be observed. Otherwise, it is likely that the diversity of the library is compromised. If these criteria are not met, see Troubleshooting.*

    *If the PLs meet the above criteria, proceed to Protocol 2: Semisynthetic Phage Display Library Construction: Generation of Filtered Libraries (Almagro and Pohl 2024b).*

## TROUBLESHOOTING

*Problem (Step 27):* The estimated size of the library is $<10^9$.
*Solution:* Repeat the cloning reaction and electroporation, or add more electroporation reactions.

*Problem (Step 36):* Some sequences are identical.
*Solution:* Check the PCR products and repeat the cloning steps.

## COMPETING INTEREST STATEMENT

M.A.P. is an employee of Xtalpi, Inc. J.C.A. is founder and CEO of GlobalBio, Inc., and has commercial interest in ALTHEA Gold Plus Libraries.

## ACKNOWLEDGMENTS

J.C.A. thanks Keyla Gomez Castellanos for helping in the preparation and review of the protocol described here, and Dr. Sonia Mayra Perez Tapia for her support and fruitful discussions.

# Chapter 9

## RECIPES

### Carbenicillin (Carb; 100 mg/mL)

Prepare 100 mg/mL carbenicillin disodium salt (Millipore-Sigma C1389-5G) in milli-Q-purified $H_2O$, filter sterilize.

Store indefinitely at −20°C.

### YTCG Liquid Medium

Mix 16 g of tryptone, 10 g of yeast extract, and 5 g of NaCl. Add $H_2O$ and mix to dissolve all reagents. Adjust pH to 7.0 with NaOH. Adjust volume to 950 mL with deionized $H_2O$. Sterilize by autoclaving for 20 min at 15 psi (1.05 kg/cm$^2$) on liquid cycle. Allow medium to cool down. Right before use, add 1 mL of a sterile carbenicillin stock solution <R>, for a final carbenicillin concentration of 100 µg/mL, and 50 mL of filter-sterilized 20% (w/v) glucose, for a final glucose concentration of 1% (w/v). YTCG liquid medium can be stored for 2 wk at 4°C.

### YTCG Solid Medium

Mix 16 g of tryptone, 10 g of yeast extract, 5 g of NaCl, and 15 g of agar. Add $H_2O$ and mix to dissolve all reagents. Adjust pH to 7.0 with NaOH. Adjust volume to 950 mL with deionized $H_2O$. Sterilize by autoclaving for 20 min at 15 psi (1.05 kg/cm$^2$) on liquid cycle. Allow medium to cool down to 60°C. Then, add 1 mL of a sterile carbenicillin stock solution <R>, for a final carbenicillin concentration of 100 µg/mL, and 50 mL of filter-sterilized 20% (w/v) glucose, for a final glucose concentration of 1% (w/v). Pour into both 150-mm × 15-mm (large) and 100-mm × 15-mm (small) Petri dishes. YTCG solid medium can be stored for 4 wk at 4°C.

## REFERENCES

Almagro JC, Pohl MA. 2024a. Semisynthetic phage display library construction: generation of single-chain variable fragment secondary libraries. *Cold Spring Harb Protoc* doi:10.1101/pdb.prot108616

Almagro JC, Pohl MA. 2024b. Semisynthetic phage display library construction: generation of filtered libraries. *Cold Spring Harb Protoc* doi:10.1101/pdb.prot108615

Suchsland R, Appel B, Müller S. 2018. Preparation of trinucleotide phosphoramidites as synthons for the synthesis of gene libraries. *Beilstein J Org Chem* 14: 397–406. doi:10.3762/bjoc.14.28

Protocol 2

# Semisynthetic Phage Display Library Construction: Generation of Filtered Libraries

Juan C. Almagro[1,3] and Mary Ann Pohl[2]

[1]GlobalBio, Inc., Cambridge, Massachusetts 02138, USA; [2]Ailux Biologics, Somerville, Massachusetts 02143, USA

Display of antibody fragments on the surface of M13 filamentous bacteriophages is a well-established approach for the identification of antibodies binding to a target of interest. Here, we describe the second of a three-step method to construct Antibody Libraries for Therapeutic Antibody Discovery (ALTHEA) Gold Plus Libraries. The three-step method involves (1) primary library (PL) construction, (2) filtered library (FL) construction, and (3) secondary library construction. The second step, described here, involves display of the PLs as single-chain variable fragment (scFv) fusions to protein pIII of the M13 phage, as well as heat shock treatment and subsequent selection of well-folded and thermostable scFvs via protein L binding, whereas unstable and defective scFvs are removed by washing steps and centrifugation. The quality of the filtration process is assessed by sequencing clones chosen at random from the FLs. These libraries, enriched with thermostable antibodies, are then ready to be used for the third and final step of the process: generation of secondary libraries.

## MATERIALS

It is essential that you consult the appropriate Material Safety Data Sheets and your institution's Environmental Health and Safety Office for proper handling of equipment and hazardous materials used in this protocol.

RECIPES: Please see the end of this protocol for recipes indicated by <R>. Additional recipes can be found online at http://cshprotocols.cshlp.org/site/recipes.

### Reagents

1000X Kanamycin stock <R>
Carbenicillin (Carb; 100 mg/mL) <R>
Deoxynucleotide (dNTP) solution (Thermo Scientific R0181)
DreamTaq polymerase (Thermo EP0705)
*Escherichia coli* TG1 Phage-Competent cells (Antibody Design Laboratories PC001)
Gel Loading Dye, purple (6×; NEB B7024S)
Glucose, 20% (w/v) (Sigma-Aldrich D9434-2.5KG) (filter-sterilized)
Glycerol, 99% (v/v) (Sigma-Aldrich G5516-4L) (autoclaved)
Helper phage CM13K (trypsin-sensitive; Antibody Design Laboratories PH010P)
Media containing agar or agarose <R>
MPBS (3%, w/v, nonfat dry milk in PBS, pH 7.4; filter-sterilized)

---

[3]Correspondence: juan.c.almagro@globalbioinc.com

© 2026 Cold Spring Harbor Laboratory Press
Cite this protocol as *Cold Spring Harb Protoc*; doi:10.1101/pdb.prot108615

NaCl (sodium chloride) <R>
Nuclease-free H$_2$O (NEB B1500S)
PEG/NaCl, 5× <R>
Phage elution buffer (200 mM glycine-HCL, pH 2.2)
Phage neutralization buffer (1.5 M Tris, pH 8.8)
Phosphate-buffered saline (PBS; pH 7.4; Gibco 70011-044)
Primary library live stocks from Protocol 1: Semisynthetic Phage Display Library Construction: Design and Synthesis of Diversified Single-Chain Variable Fragments and Generation of Primary Libraries (Almagro and Pohl 2024a).
Primers:
    For_scFv_seq (5′-GCGGATAACAATTTGAATTCAAGGAGACA-3′)
    Rev_scFv_seq (5′-TGTATGAGGTTTTGCTAAACAACTTTCA-3′)
Protein L Magnetic Beads (Pierce 88849)
QIAquick PCR Purification Kit (QIAGEN 28106)
TPBS (0.1%, v/v, Tween 20 in PBS, pH 7.4, filter-sterilized)
Tris, 100 mM, pH 8.0.
Trypsin, TPCK-treated (10 mg/mL; Sigma-Aldrich T1426-100MG)
YT <R>
YT agar medium (2×) <R>
YTCG liquid medium <R>
YTCG solid medium <R>
YTCK liquid medium <R>

## Equipment

Agarose gel electrophoresis equipment, including tank, wires, gel molds, and combs.
Cell scrapers (Nunc 179693PK)
Cell spreader or glass beads, to plate cells
Centrifuge (refrigerated, tabletop, with capacity for 50-mL conical tubes)
Computer with DNA analysis software (e.g., Geneious Prime, DNASTAR, or equivalent)
Conical tubes (sterile, 15- and 50-mL, screw-top; Nunc 339650)
Flasks (250-mL and 1-L, sterile, preferably disposable and baffle-bottom)
Freezer at −70°C
Ice bath
Incubators at 37°C (static and shaking) and 30°C (shaking)
Magnetic bead separator for microcentrifuge tubes (e.g., DynaMag-2; Thermo Fisher 12321D)
Microcentrifuge tubes (RNase-free, 1.5-mL)
PCR tubes (0.2-mL)
Power supply box
Refrigerated microcentrifuge
Rocker
Rotating tube mixer
Serological pipettes, 10- and 50-mL (sterile)
Spectrophotometer for measuring UV light
Syringe filter (0.45-μm)
Thermocycler
Vortex
Water bath at 60°C

## METHOD

*To generate the filtered libraries (FLs), the TG1 stock transformed with each primary library (PL) is cultured and rescued with helper phage to produce the antibody library displayed on phage virions. These PLs are then subjected to heat shock for 10 min at 60°C followed by protein L filtration. Because protein L binds the κ light chain of PL single-chain variable fragments (scFvs), phage displaying thermostable and properly folded scFvs are captured during the filtration process. Improperly folded and/or defective scFvs do not bind protein L and are therefore removed by centrifugation. After elution and amplification of the protein L binding scFv fused to the phage, quality control (QC) of the FLs is performed. If the quality of the FLs is deemed sufficient, then they can be used to create the secondary libraries (see Protocol 3: Semisynthetic Phage Display Library Construction: Generation of Single-Chain Variable Fragment Secondary Libraries [Almagro and Pohl 2024b]).*

### Preparation of the Phage-Displayed PLs for Filtration

1. Thaw a vial of the PL stock stored at −70°C.

    *See Protocol 1: Semisynthetic Phage Display Library Construction: Design and Synthesis of Diversified Single-Chain Variable Fragments and Generation of Primary Libraries (Almagro and Pohl 2024a).*

2. Inoculate 100 mL of YTCG medium in a 250-mL flask with 500–800 µL of the PL live stock to generate a culture with a starting $OD_{600}$ of ∼0.1.

3. Incubate with shaking at 225 rpm at 37°C until log phase is reached ($OD_{600}$ at ∼0.4).

4. Add 400 µL ($2 \times 10^{12}$ virions/mL) of helper phage CM13K to the cell culture.

5. Incubate without shaking for 30 min at 37°C, and then with shaking at 225 rpm for 30 min.

6. Transfer the culture to two 50-mL screw-top conical tubes and centrifuge at 3500g for 15 min at 4°C.

7. Discard the supernatant by decanting and resuspend the pellet in each tube with 50 mL of YTCK medium. Resuspended by pipetting or vortex.

8. Transfer the resuspended cells (100 mL) to a single 1-L culture flask.

9. Incubate with shaking at 225 rpm overnight (12–16 h) at 30°C.

### Phage Precipitation with PEG/NaCl

10. Transfer the overnight culture from Step 9 to two separate 50-mL conical tubes.

11. Centrifuge at 7000g for 30 min at 4°C.

12. Transfer 40 mL of the supernatant from each tube to a single 250-mL culture flask (total volume 80 mL).

13. Add 20 mL of 5× PEG/NaCl. Mix well by shaking the flask and incubate for 1 h in an ice bath.

    *The solution should start to appear cloudy, indicating precipitation of phage.*

14. Transfer the solution to 50-mL conical tubes and centrifuge at 7000g for 30 min at 4°C.

15. Remove the supernatant with a serological pipette and centrifuge at 7000g for 2 min at 4°C.

16. Remove the residual supernatant with a 200-µL pipette and add 40 mL of 1× PBS.

17. Resuspend the phage pellet by pipetting, and centrifuge at 7000g for 2 min at 4°C.

18. Transfer the suspension to a 250-mL culture flask, add 10 mL of 5× PEG/NaCl, and incubate for 1 h in an ice bath.

19. Repeat Steps 14–15.

20. Resuspend the pellet in 1 mL of 1× PBS, transfer to a microcentrifuge tube, and centrifuge at 6000g for 5 min at 4°C to remove the remaining bacterial debris.

21. Filter the supernatant through a 0.45-µm syringe filter.

Chapter 9

22. Quantify phage by measuring absorption at 269 and 320 nm and calculating the virion concentration (virions per milliliter) with the following formula:

$$\text{virions/mL} = \frac{(\text{OD}_{269} - \text{OD}_{320}) \times 6 \times 10^{16}}{\text{number of bases per virion}}$$

*$OD_{269}$ and $OD_{320}$ represent optical density at 269 and 320 nm, and the number of bases per virion represents the size of the phagemid vector with the insert (~4712 bp)*

*This sample constitutes the virion stock of the PLs. Store for up to 2 d at 4°C (do not add glycerol).*

### Magnetic Bead Equilibration and Blocking of Nonspecific Interactions

23. Transfer 525 µL of protein L magnetic beads to a 1.5-mL microcentrifuge tube and place the tube on the magnetic bead separator for 3 min at room temperature.

24. Carefully pipette away and discard the storage solution, taking care not to disturb the magnetic beads.

25. Take the tube and resuspend the magnetic beads in 525 µL of 1× PBS. Then, place tube back on the magnetic bead separator for 3 min at room temperature, and discard the supernatant by aspiration.

26. Take the tube and resuspended the magnetic beads in 525 µL of TPBS and incubate for 1 h at room temperature in a tube rotator.

27. Place the tube with the mix (magnetics beads and TPBS) on the magnetic bead separator for 3 min at room temperature and discard the supernatant by aspiration.

28. Block the protein L magnetic beads by adding 525 µL of MPBS and incubating for 1 h at room temperature with gentle end-over-end mixing at 40 rpm. Keep these beads at room temperature until use in Step 35.

### Selection for Thermostable scFvs

29. To prevent nonspecific binding of phage particles to the 15-mL conical tubes, prepare one blocked 15-mL tube per PL by adding 5 mL of MPBS to each tube and incubating with gentle end-over-end mixing at 40 rpm for 1 h at room temperature. Discard the MPBS solution and keep at room temperature.

30. To prevent nonspecific binding of phage particles to microcentrifuge tubes, prepare four blocked tubes per PL by adding 1.5 mL of MPBS to each tube and incubating with gentle end-over-end mixing at 40 rpm for 1 h at room temperature. Discard the MPBS solution and keep at room temperature.

31. In 15-mL screw-top conical tubes, prepare 5 mL of each PL (virion stocks from Step 22) at a concentration of $2 \times 10^{12}$ virions/mL in MPBS ($1 \times 10^{13}$ total virions).

32. Incubate this phage solution for 10 min at 60°C in a water bath, followed by incubation for 30 min at room temperature.
    *This step serves to select antibodies displayed on the phage surface that are more thermostable and can refold properly after heat shock.*

33. Centrifuge at 12,000g for 1 min at room temperature to remove aggregated phage.

34. Recover 4 mL of supernatant phage solution in previously blocked 15-mL conical tubes (from Step 29).

35. Add 500 µL of previously blocked protein L magnetic beads (from Step 28).

36. Incubate for 1.5 h at room temperature, rocking the tube.

37. Transfer 1.125 mL of the mixture of supernatant phage solution and magnetic beads to a microcentrifuge tube preblocked with MPBS (from Step 30). Set the tube on the magnetic bead separator for 3 min at room temperature and discard the supernatant by aspiration, taking care not to disturb the magnetic beads. Repeat this step three more times until all the beads (3.375 mL) are placed in a single microcentrifuge tube.

38. Wash magnetics beads five times with 1 mL of TPBS and then five times with 1 mL of PBS at room temperature. To wash, resuspend the beads in 1 mL of the corresponding buffer, set the tube on the magnetic bead separator for 3 min, and then discard the supernatant by aspiration to remove unbound phage. After the last wash with PBS, transfer the beads to a new microcentrifuge tube preblocked with MPBS (from Step 30).

## Elution

39. Prepare 1 mL of trypsin elution buffer: Dilute trypsin stock with 100 mM Tris to a final concentration of 1 mg/mL trypsin.

40. To elute phage from beads, add 500 µL of trypsin elution buffer to the microcentrifuge tube from Step 38.

41. Incubate the tube on a rocker for 10 min at room temperature. Set the tube on the magnetic bead separator for 3 min at room temperature and then recover the eluate by pipetting.

    *Do not discard the magnetics beads; keep on the bench until use in Step 43.*

42. Transfer the trypsin eluate to a 50-mL screw-top conical tube. Keep on ice until use in Step 46.

43. To elute additional phage from the bead mix by acidic pH, add 500 µL of phage elution buffer to the magnetic beads from Step 41.

44. Incubate the tube from Step 43 on a rocker for 10 min at room temperature and then set the tube on the magnetic bead separator for 3 min at room temperature.

45. Recover the eluate by pipetting and transferring it to a microcentrifuge tube. Neutralize the reaction by adding 500 µL of phage neutralization buffer.

46. Transfer the 1-mL acid elution mix to the 50-mL Falcon tube containing the trypsin elution (from Step 42).

    *Proceed with the total combined volume (~1.5 mL; 500 µL of trypsin elution + 1000 µL of acid elution) to phage amplification.*

## Amplification of Eluted Phage

47. Set up a TG1 preculture by inoculating 1 mL of YT medium with 10 µL of *E. coli* TG1 stock. Incubate the culture with shaking at 225 rpm overnight at 37°C.

48. The next day, inoculate 25 mL of YT medium with 250 µL of the TG1 preculture from Step 47. Incubate with shaking at 225 rpm at 37°C until an $OD_{600}$ of 0.4–0.5 is reached (2.0–2.5 h).

49. Add 10 mL of the culture to the ~1.5-mL phage eluate from Step 46.

50. Incubate for 45 min at 37°C without shaking. During the incubation, proceed with the following:
    i. Make serial 1:10 dilutions ($10^{-3}$, $10^{-4}$, $10^{-5}$, $10^{-6}$, $10^{-7}$, and $10^{-8}$) of the eluate by adding 20 µL of phage into a tube with 180 µL of YT medium. Add 1000 µL of log phase TG1 (from Step 48) into each tube and incubate without shaking for 30 min at 37°C. Finally, plate 100 µL of the phage-infected cells using a cell spreader or glass beads on YTCG solid medium plates to titrate the output. Incubate overnight at 37°C. Create a duplicate plate with the same dilution.
    ii. Titrate the PL input by plating serial dilutions ($10^{-4}$, $10^{-6}$, and $10^{-8}$) on YTCG solid medium plates to compare titers before and after heat shock and filtration. Incubate overnight at 37°C.

Chapter 9

51. After the 45-min incubation from Step 50, continue with amplification of phage by pelleting the cells at 6000g for 15 min at 4°C.

52. Remove the supernatant with a serological pipette and resuspend the pellet by adding 1 mL of YTCG liquid medium.

53. Plate the resuspended cells using a cell spreader or glass beads onto two YTCG solid medium plates (150 mm in diameter). Incubate overnight at 37°C.

54. The next day, scrape bacteria from the large culture plates from Step 53 by washing each plate with 4 mL of YTCG liquid medium and loosening the bacterial lawn from the plate using disposable cell scrapers. Collect cell material with a serological pipette and transfer it to a 50-mL screw-top conical tube.

55. Add 3 mL of 99% glycerol and store at −70°C in 1-mL aliquots.
    *This is now the FL live stock.*

56. Titer calculations: Count the number of colonies on each plate from Step 50. Take the average of the number of the same dilution plated on both plates.

$$CFU = \frac{[\# \text{colonies} \times \text{dilution factor} \times (\text{dilution of } 20\,\mu\text{L of phage solution and } 1000\,\mu\text{L of TG1 cells})]}{100\,\mu\text{L (plated volume)}}$$

### Quality Control of the FLs

*The quality of the FLs can be assessed by either sending a plate with 30 colonies chosen at random from the titration plates for Sanger sequencing, or by sending PCR products of the colonies for sequencing. The following procedure is for sequencing PCR products.*

57. For each library, prepare a 36-reaction PCR mix as follows:

| Component | Volume/amount(one reaction) | Volume/amount[a](36 reactions) |
|---|---|---|
| DreamTaq DNA Buffer (10×) | 5.00 μL | 180.00 μL |
| dNTPs (2 mM) | 2.50 μL | 90.00 μL |
| Primer For_scFv_seq (10 μM) | 2.00 μL | 72.00 μL |
| Primer Rev_scFv_seq (10 μM) | 2.00 μL | 72.00 μL |
| DNA template (random colonies) | 15.00 ng | 15.00 ng each |
| DreamTaq DNA Polymerase | 0.25 μL | 9.00 μL |
| Nuclease-free H$_2$O | To 25.0 μL | To 900.00 μL |

[a]The reaction mix considers two additional reactions, to account for pipetting error.

58. Dispense 25 μL of the PCR mix into 34 separate PCR tubes.

59. Pick 30 colonies from the library titer plates from Step 50 with sterile toothpicks or pipette tips, selecting different colony sizes to avoid biasing the selection, and add the cell material directly to each PCR tube. Add two reaction controls in duplicate: one plasmid containing a known scFv (positive control), and one reaction mix without plasmid (negative control).

60. Run the reactions in a thermocycler with the following parameters:

| Step | Time | Temperature |
| --- | --- | --- |
| Initial denaturation | 300.0 sec | 95°C |
| 25 cycles | 30.0 sec | 95°C |
| | 30.0 sec | 67°C |
| | 60.0 sec | 72°C |
| Final extension | 10.0 min | 72°C |

61. Run a 2% (w/v) agarose gel with 5 µL of each PCR mix to confirm successful PCR amplification of each colony. If positive, i.e., a band of the approximate size of that of the positive control is observed, purify the remaining PCR product using the QIAquick PCR Purification Kit and send it for Sanger sequencing with the For_scFv_seq and Rev_scFv_seq primers.

62. Analyze DNA sequence data with the appropriate software, such as Geneious Prime, DNASTAR, or equivalent.

    *Similar to the PLs (see Protocol 1: Semisynthetic Phage Display Library Construction: Design and Synthesis of Diversified Single-Chain Variable Fragments and Generation of Primary Libraries [Almagro and Pohl 2024a]), quality control of the FLs can be performed by sequencing individual clones or via high-throughput sequencing analysis to determine the level of diversity among individual antibody sequences.*

    *An additional quality control step can be performed with a protein L or protein A ELISA to assess the number of functional clones in the library after the filtration process, as described by Valadon et al. (2019).*

    *Provided that the FL demonstrates sufficient sequence diversity, library construction can proceed to the next step: construction of the secondary libraries (SLs) (see Protocol 3: Semisynthetic Phage Display Library Construction: Generation of Single-Chain Variable Fragment Secondary Libraries [Almagro and Pohl 2024b]).*

## COMPETING INTEREST STATEMENT

M.A.P. is an employee of Xtalpi, Inc. J.C.A. is founder and CEO of GlobalBio, Inc., and has commercial interest in ALTHEA Gold Plus Libraries.

## ACKNOWLEDGMENTS

J.C.A. thanks Keyla Gomez Castellanos for helping in the preparation and review of the protocol described here, and Dr. Sonia Mayra Perez Tapia for her support and fruitful discussions.

## RECIPES

### 1000× Kanamycin Stock

Kanamycin
Make kanamycin to a final concentration of 50 mg/ml in $H_2O$.

Filter-sterilize and store in aliquots at −20°C.

## Carbenicillin (Carb; 100 mg/mL)

Prepare 100 mg/mL carbenicillin disodium salt (Millipore-Sigma C1389-5G) in milli-Q-purified $H_2O$, filter sterilize.

Store indefinitely at −20°C.

## Media Containing Agar or Agarose

Prepare liquid media according to the recipe given. Just before autoclaving, add one of the following:

| | |
|---|---|
| Bacto agar (for plates) | 15 g/L |
| Bacto agar (for top agar) | 7 g/L |
| Agarose (for plates) | 15 g/L |
| Agarose (for top agarose) | 7 g/L |

Sterilize by autoclaving for 20 min at 15 psi (1.05 kg/cm$^2$) on liquid cycle. When the medium is removed from the autoclave, swirl it gently to distribute the melted agar or agarose evenly throughout the solution. *Be careful!* The fluid may be superheated and may boil over when swirled. Before adding thermolabile substances (e.g., antibiotics), allow the medium to cool to 50°C–60°C, and mix the medium by swirling to avoid producing air bubbles.

Before pouring the plates, set up a color code (e.g., two red stripes for LB-ampicillin plates; one black stripe for LB plates, etc.), and mark the edges of the plates with the appropriate colored markers. Pour plates directly from the flask; allow ∼30–35 mL of medium per 90-mm plate. To remove bubbles from the medium in the plate, flame the surface of the medium with a Bunsen burner before the agar or agarose hardens. When the medium has hardened completely, invert the plates and store them at 4°C until needed.

The plates should be removed from storage 1–2 h before they are used. If the plates are fresh, they will "sweat" when incubated at 37°C. When this condensation drops on the agar/agarose surface, it allows bacterial colonies or bacteriophage plaques to spread and increases the chances of cross-contamination. This problem can be avoided by wiping off the condensation from the lids of the plates and then incubating the plates for several hours at 37°C in an inverted position before they are used. Alternatively, remove the liquid by shaking the lid with a single, quick motion. To minimize the possibility of contamination, hold the open plate in an inverted position while removing the liquid from the lid.

## NaCl (Sodium chloride)

To prepare a 5 M solution: Dissolve 292 g of NaCl in 800 mL of $H_2O$. Adjust the volume to 1 L with $H_2O$. Dispense into aliquots and sterilize by autoclaving. Store the NaCl solution at room temperature.

## PEG/NaCl (5X)

200 g PEG-8000 (polyethylene glycol; Sigma)
150 g NaCl
Bring to 1 L total volume with $H_2O$ and stir until dissolved. Sterilize by autoclaving at 121°C for 20 min at 15 psi on liquid cycle.

### YT

Tryptone, 16 g
Yeast extract, 10 g
NaCl, 5 g
Deionized H$_2$O, to 900 mL

To prepare 2× YT medium, shake until the solutes have dissolved. Adjust the pH to 7.0 with 5 N NaOH. Adjust the volume of the solution to 1 liter with deionized H$_2$O. Sterilize by autoclaving for 20 min at 15 psi (1.05 kg/cm$^2$) on liquid cycle.

### YT Agar Medium (2×)

| Reagent | Quantity (for 1 L) |
| --- | --- |
| Agar | 15 g |
| Bacto Tryptone | 16 g |
| Yeast extract | 10 g |
| NaCl | 5 g |

Dissolve in 1 L of distilled H$_2$O; sterilize by autoclaving.

### YTCG Liquid Medium

Mix 16 g of tryptone, 10 g of yeast extract, and 5 g of NaCl. Add H$_2$O and mix to dissolve all reagents. Adjust pH to 7.0 with NaOH. Adjust volume to 950 mL with deionized H$_2$O. Sterilize by autoclaving for 20 min at 15 psi (1.05 kg/cm$^2$) on liquid cycle. Allow medium to cool down. Right before use, add 1 mL of a sterile carbenicillin stock solution <R>, for a final carbenicillin concentration of 100 µg/mL, and 50 mL of filter-sterilized 20% (w/v) glucose, for a final glucose concentration of 1% (w/v). YTCG liquid medium can be stored for 2 wk at 4°C.

### YTCG Solid Medium

Mix 16 g of tryptone, 10 g of yeast extract, 5 g of NaCl, and 15 g of agar. Add H$_2$O and mix to dissolve all reagents. Adjust pH to 7.0 with NaOH. Adjust volume to 950 mL with deionized H$_2$O. Sterilize by autoclaving for 20 min at 15 psi (1.05 kg/cm$^2$) on liquid cycle. Allow medium to cool down to 60°C. Then, add 1 mL of a sterile carbenicillin stock solution <R>, for a final carbenicillin concentration of 100 µg/mL, and 50 mL of filter-sterilized 20% (w/v) glucose, for a final glucose concentration of 1% (w/v). Pour into both 150-mm × 15-mm (large) and 100-mm × 15-mm (small) Petri dishes. YTCG solid medium can be stored for 4 wk at 4°C.

### YTCK Liquid Medium

Mix 16 g of tryptone, 10 g of yeast extract, and 5 g of NaCl. Add H$_2$O and mix to dissolve all reagents. Adjust pH to 7.0 with NaOH. Adjust volume to 1 L with deionized H$_2$O. Sterilize by autoclaving for 20 min at 15 psi (1.05 kg/cm$^2$) on liquid cycle. Allow the medium to cool down to 60°C. Immediately before use, add 1 mL of a sterile carbenicillin stock solution <R>, for a final carbenicillin concentration of 100 µg/mL, and 1 mL of kanamycin stock solution <R>, for a final kanamycin concentration of 50 µg/mL. YTCK liquid medium should always be prepared fresh.

## REFERENCES

Almagro JC, Pohl MA. 2024a. Semisynthetic phage display library construction: design and synthesis of diversified single-chain variable fragments and generation of primary libraries. *Cold Spring Harb Protoc* doi:10.1101/pdb.prot108614

Almagro JC, Pohl MA. 2024b. Semisynthetic phage display library construction: generation of single-chain variable fragment secondary libraries. *Cold Spring Harb Protoc* doi:10.1101/pdb.prot108616

Valadon P, Pérez-Tapia SM, Nelson RS, Guzmán-Bringas OU, Arrieta-Oliva HI, Gómez-Castellano KM, Pohl MA, Almagro JC. 2019. ALTHEA gold libraries™: antibody libraries for therapeutic antibody discovery. *MAbs* 11: 516–531. doi:10.1080/19420862.2019.1571879

Protocol 3

# Semisynthetic Phage Display Library Construction: Generation of Single-Chain Variable Fragment Secondary Libraries

Juan C. Almagro[1,3] and Mary Ann Pohl[2]

[1]GlobalBio, Inc., Cambridge, Massachusetts 02138, USA; [2]Ailux Biologics, Somerville, Massachusetts 02143, USA

Display of antibody fragments on the surface of M13 filamentous bacteriophages is a well-established approach for the identification of antibodies binding to a target of interest. Here, we describe the third and final step of a three-step method to construct Antibody Libraries for Therapeutic Antibody Discovery (ALTHEA) Libraries. The three-step method involves (1) primary library construction, (2) filtered library (FL) construction, and (3) secondary library (SL) construction. In the third step, described here, the nucleotide sequences encoding the single-chain variable fragments (scFvs) of FLs are amplified by PCR and combined with the heavy- chain CDR3 region (HCDR3) and joining fragments (H3J) obtained from a pool of donors to maximize diversity ("natural H3J fragments"). These natural H3J fragments are amplified with a set of primers designed to capture >95% of the natural H3J repertoire. The resultant fragments replace the neutral H3J fragments of the FLs, resulting in the final semisynthetic secondary libraries. The quality of these libraries is assessed by sequencing clones chosen at random from the libraries, typically 96 clones. These libraries are then ready to be used for phage selections on targets of interest, providing a robust antibody discovery platform.

## MATERIALS

It is essential that you consult the appropriate Material Safety Data Sheets and your institution's Environmental Health and Safety Office for proper handling of equipment and hazardous materials used in this protocol.

RECIPES: Please see the end of this protocol for recipes indicated by <R>. Additional recipes can be found online at http://cshprotocols.cshlp.org/site/recipes.

### Reagents

Carbenicillin (Carb; 100 mg/mL) <R>
Chloroform (Merck 25666)
Deoxynucleotide (dNTP) solution (Thermo Scientific R0181)
DNA ladder, 100-bp (NEB N3231S) and 1-kb (NEB N3232S)
DreamTaq polymerase with 10× buffer (Thermo EP0705)
Ethanol, 75% (v/v) in $H_2O$
*Escherichia coli* TG1 Phage-Competent cells (Antibody Design Laboratories PC001)
Gel Loading Dye, purple (6×; NEB B7024S)

---

[3]Correspondence: juan.c.almagro@globalbioinc.com

© 2026 Cold Spring Harbor Laboratory Press
Cite this protocol as *Cold Spring Harb Protoc*; doi:10.1101/pdb.prot108616

Chapter 9

GelRed Nucleic Acid Gel Stain, 10,000× (Biotium 41003)
Glycerol, 99% (v/v) (Sigma-Aldrich G5516-4L) (autoclaved)
Glycogen (optional; see Step 15; Thermo R0561)
Isopropanol
Lymphoprep (STEMCELL Technologies 07851)
Oligonucleotides:
    Natural H3J fragments:
        UFR3FOR (5′-GACACGGCCGTGTATTACTGTGC-3′)
        JHSfiIrev1 (5′-GTGGTGATGGTGTTGGCCTCCCGGGCCTGAGGAGACRGTGACCAGGG-3′)
        JHSfiIrev2 (5′-GTGGTGATGGTGTTGGCCTCCCGGGCCTGAAGAGACGGTGACCATTG-3′)
        JHSfiIrev3 (5′-GTGGTGATGGTGTTGGCCTCCCGGGCCTGAGGAGACGGTGACCGTGG-3′)
    Scaffold assembly:
        SfiIFOR (5′-CGCTGGATTGTTATTACTCGCG-3′)
        UFR3REV (5′-GCACAGTAATACACGGCCGTGTC-3′)
        SfiREV (5′-GTGGTGATGGTGTTGGCCTC-3′)
    Sequencing:
        For_scFv_seq (5′-GCGGATAACAATTTGAATTCAAGGAGACA-3′)
        Rev_scFv_seq (5′-TGTATGAGGTTTTGCTAAACAACTTTCA-3′)
PBMCs

*The natural H3J fragments are obtained from peripheral blood mononuclear cells (PBMCs) of healthy donors. The larger the donor pool, the less biased the set of H3J fragments, due to V region polymorphisms and the immunological history of the individuals. For ALTHEA Gold Libraries, 200 donors were used (Valadon et al. 2019). We used our own source of PBMCs (leukoplatelet concentrate) for building the libraries, and we describe the protocol using those (see Step 1), but a number of companies sell PBMCs; see, for instance, https://www.atcc.org/products/pcs-800-011 and https://www.stemcell.com/human-peripheral-blood-b-cells-frozen.html*

Purified plasmid DNA from filtered libraries (see Protocol 2: Semisynthetic Phage Display Library Construction: Generation of Filtered Libraries [Almagro and Pohl 2024a])
Nuclease-free $H_2O$ (NEB B1500S)
pADL-23c phagemid vector (Antibody Design Laboratories PD0111)
Phosphate-buffered saline (PBS; pH 7.4; Gibco 70011-044)
Phusion Hot Start Flex DNA Polymerase and Phusion HF Reaction Buffer (5×) (NEBM0535S)
PolyA Spin mRNA Isolation Kit (NEB S1560)
ProtoScript II First-Strand cDNA Synthesis Kit (NEB E6560S)
SfiI restriction enzyme (20,000 units/mL; NEB R0123S)
SOC recovery medium (NEB B9020)
QIAquick Gel Extraction Kit (QIAGEN 28706)
QIAquick PCR Purification Kit (QIAGEN 28106)
QIAGEN Plasmid Midi Kit (QIAGEN 12945)
T4 DNA ligase and 10× buffer (NEB M0202S)
Tris acetate-EDTA buffer (Sigma-Aldrich T9650-1L)
TRIzol Reagent (Invitrogen 15596026)
UltraPure Agarose (Invitrogen 16500100)
YTCG liquid medium <R>
YTCG solid medium <R>

## Equipment

Agarose gel electrophoresis equipment (tank, wires, gel molds, and combs)
Cell scrapers (Nunc 179693PK)

Centrifuge (tabletop, with capacity for 50-mL conical tubes and swinging rotor buckets)
Computer with DNA analysis software (e.g., Geneious Prime, DNASTAR, or equivalent)
Conical tubes (sterile, 15- and 50-mL, screw-top; Nunc 339650)
Electroporator (Bio-Rad Gene Pulser Xcell Electroporation Systems)
Freezers at −70°C and −20°C
Fume hood
Glass spreader or beads, to plate cells
Heat block
Ice
Incubator at 37°C (shaking and static)
Microcentrifuge (refrigerated, standard, benchtop)
Microcentrifuge tubes (RNase-free, 1.5-mL)
MicroPulser Electroporation cuvettes (0.1-cm gap; Bio-Rad 1652089)
Pasteur pipette
PCR tubes (0.2-mL)
Power supply box
Serological pipettes (5 mL, sterile)
Spectrophotometer for measuring UV light
Syringe (5-mL)
Thermocycler 96- wells
Vortex

## METHOD

*In this protocol, peripheral blood mononuclear cells (PBMCs) are isolated and used to extract total RNA, followed by complementary DNA (cDNA) synthesis. The resulting cDNA is used for PCR amplification of the heavy chain CDR3-J region (HCDR3-J) fragments. Subsequently, these fragments are assembled into the single-chain variable fragments (scFvs) from the filtered libraries (FLs) (see Protocol 2: Semisynthetic Phage Display Library Construction: Generation of Filtered Libraries [Almagro and Pohl 2024a]) via PCR, followed by cloning into the phagemid vector.*

### Isolation of Lymphocytes (Peripheral Blood Mononuclear Cells [PBMCs]) from Leukoplatelet Concentrate

*Steps 1–8 describe the isolation of PBMCs from a single leukoplatelet concentrate. Users should repeat these steps for the necessary number of concentrates. We commonly isolate PBMCs from 200 leukoplatelet concentrates.*

*As an alternative to the protocol described in this section, PBMCs can be purchased (one vial with $1 \times 10^7$ to $1 \times 10^8$ cells).*

1. Recover the leukoplatelet concentrate through a cut in one of the hoses of the leukoplatelet concentrate bag by taking ∼2 mL of sample with a 5-mL syringe.

2. Dilute the concentrate 1:2 with 1× PBS (total of 4 mL).

3. Place 2 mL of Lymphoprep in a conical tube and add the 4 mL of blood diluted with 1× PBS (from Step 2) through the walls of the tube, being careful to minimize mixing of blood with Lymphoprep.

4. Centrifuge at 800g for 30 min at 20°C in a swinging bucket rotor with brake off.

5. Using a Pasteur pipette, collect the middle fraction of the gradient and transfer it into a fresh 15-mL centrifuge tube.

    *Be careful not to take serum or erythrocytes, to maintain the purity of the cells.*

6. Using a serological pipette, add 5 mL of 1× PBS and centrifuge at 200g for 10 min at room temperature, to pellet the cells.

7. Decant the supernatant and resuspend the cell pellet by pipetting with 5 mL of cold 1× PBS. Centrifuge at 200g for 5 min at room temperature.

8. Decant the supernatant and resuspend the cell pellet by pipetting with 0.25 mL of 1× PBS. Transfer the cell suspension to a microcentrifuge tube.

## Total RNA Isolation from Lymphocytes

*At this point, users will have one cell suspension per leukoplatelet concentrate processed in Steps 1–8. Repeat the steps below for each one. In our standard workflow, users would have 200 suspensions to process.*

9. In a fume hood, add 0.75 mL of TRIzol Reagent per 0.25 mL of cell suspension (or per $1 \times 10^6$ cells).
10. Pipette the lysate up and down several times to homogenize, and incubate for 5 min at room temperature.
11. Add 200 µL of chloroform per 1 mL of TRIzol Reagent used for lysis, securely cap the tube, and thoroughly mix by shaking.
12. Incubate for 2–3 min at room temperature.
13. Centrifuge the sample at 12000$g$ for 15 min at 4°C.

    *The mixture will separate into a lower red phenol–chloroform phase, an interphase, and a colorless upper aqueous phase.*

14. Transfer the aqueous phase containing the RNA to a new microcentrifuge tube by angling the tube at 45° and pipetting the solution out.
15. (Optional) If the starting sample is small (<$10^6$ cells), add 5–10 µg of RNase-free glycogen as a carrier to the aqueous phase.
16. Add 500 µL of isopropanol to the aqueous phase for every 1 mL of TRIzol Reagent used for lysis.
17. Incubate for 10 min at 4°C.
18. Centrifuge at 12,000$g$ for 10 min at 4°C.

    *The total RNA precipitate will form a white gel-like pellet at the bottom of the tube.*

19. Discard the supernatant with a micropipette.
20. Add 1 mL of 75% (v/v) ethanol to the pellet for every 1 mL of TRIzol Reagent used for lysis.
21. Vortex the sample briefly and then centrifuge at 7500$g$ for 5 min at 4°C.
22. Discard the supernatant with a micropipette and air-dry the RNA pellet for 5–10 min at room temperature.
23. Resuspend the pellet in 30 µL of nuclease-free $H_2O$.
24. Incubate for 15 min in a heat block set at 55°C.
25. Using a spectrophotometer, measure absorbance at 260 and 280 nm to quantify the total RNA concentration of each sample. Verify the integrity of total RNA (see note). Keep total RNA at −70°C until use.

    *The most common method used to verify the integrity of total RNA is to run an aliquot of the RNA sample on a denaturing agarose gel. The 28S rRNA band should be approximately twice as intense as the 18S rRNA band. This 2:1 ratio (28S rRNA:18S rRNA) is a good indication that the RNA is intact. For a more rigorous sample assessment, instruments such as a Bioanalyzer can provide detailed information about the condition of RNA preparations.*

    *At this point, users will have 200 separate total RNA samples.*

## mRNA Isolation

*Although the following steps can be performed with one donor, we recommend the use of several donors, to avoid biases due to the immunological history of one individual.*

*We commonly start from 200 leukoplatelet concentrates (see Step 1), which means that users will have 200 total RNA samples at this point.*

26. From the 200 total RNA samples from Step 25, create 20 pools in separate 1.5-mL tubes, each containing total RNA from 10 different samples in equal amounts.
27. Extract mRNA separately from each of the 20 total RNA pools by using the PolyA Spin mRNA Isolation Kit, following the manufacturer's instructions. Resuspend each pellet in 20 µL of nuclease-free H$_2$O and incubate for 10 min in a heat block set at 55°C.
28. Using a spectrophotometer, measure absorbance at 260 and 280 nm to quantify the mRNA concentration of each sample.

## First-Strand cDNA Synthesis

*Each of the 20 pools of mRNA should be separately processed to synthesize cDNA using the ProtoScript II First-Strand cDNA Synthesis Kit, following the manufacturer's instructions.*

29. Thaw the components of the cDNA synthesis kit on ice and mix by inverting the tubes several times.
30. Mix each mRNA sample and primer d(T)23 VN (provided in the cDNA synthesis kit) in a separate sterile RNase-free microcentrifuge tube as described below:

| Component | Volume |
|---|---|
| mRNA | 1–6 µL (up to 1 µg) |
| d(T)23 VN (50 µM) | 2.0 µL |
| Nuclease-free H$_2$O | To 8.0 µL |

31. Denature each mix for 5 min at 65°C in a thermocycler. Centrifuge briefly at 8000g for 20 sec at room temperature and promptly place on ice.
32. Add the following components (provided in the kit) to each sample:

| Component | Volume |
|---|---|
| ProtoScript II Reaction Mix (2×) | 10.0 µL |
| ProtoScript II Enzyme Mix (10×) | 2.0 µL |

33. Incubate each 20-µL cDNA synthesis reaction for 1 h at 42°C in a thermocycler.
34. Inactivate the enzyme for 5 min at 80°C in a thermocycler.

   *At the end of this section, users will have 20 separate cDNA samples.*

   *Store the cDNA product at −20°C until use. The cDNA obtained will be used as a template to amplify the repertoire of HCDR3 sequences, which is a key element in defining the specificity and affinity of antibodies and is by far the most diverse region of the variable genes.*

## Amplification of the Natural HCDR3/J$_H$ (H3J) Fragments by PCR

*Double-stranded DNA containing the repertoire of natural H3J fragments is obtained by PCR using a universal forward primer (UFR3FOR) that was designed based on the finding that up to 95% of the circulating antibodies have a conserved motif (CM) at the nucleotide level in the FR3, which encodes the amino acid sequence "DTAVYYCA," just before the HCDR3. The reverse primers (JHSfilrev1, JHSfilrev2, and JHSfilrev3) are used to amplify all six human IGHJ germline genes (J$_H$).*

35. Set up a PCR master mix as follows (72× master mix for 63 PCR reactions, accounting for pipetting errors):

Chapter 9

| Component | Volume (1×) | Volume (72×) (master mix) |
|---|---|---|
| 10× Buffer for DreamTaq | 2.50 µL | 180 µL |
| dNTPs (2 mM) | 2.50 µL | 180 µL |
| Primer UFR3FOR (10 µM) | 2.50 µL | 180 µL |
| Primer JHSfiIrev 1, 2 or 3 (10 µM) | | — |
| DNA template (cDNA from Step 34) | | — |
| DreamTaq DNA Polymerase | 0.25 µL | 18 µL |
| Nuclease-free H$_2$O | 12.75 µL | 918 µL |

*Note that the master mix does NOT have a reverse primer; it will be added in Step 36*

36. Divide the master mix into three separate tubes, each with 492 µL, and then add to each of the three reactions the corresponding reverse primers as follows:
    - Mix 1: 60 µL of JHSfiIrev1 primer (10 µM)
    - Mix 2: 60 µL of JHSfiIrev2 primer (10 µM)
    - Mix 3: 60 µL of JHSfiIrev3 primer (10 µM)

37. Divide each mix into 21 separate 0.2-mL tubes, each with 23 µL. This will be used for each of the 20 cDNA samples (Step 34) and a negative control per mix.

38. Add 2 µL of each of the 20 cDNA samples (from Step 34) to the corresponding tubes, as template. Add H$_2$O to the negative control.

39. Perform the PCR amplification using the following program in a thermocycler:

| Step | PCR conditions | |
|---|---|---|
| | Time | Temperature |
| Initial denaturation | 300.0 sec | 95°C |
| 25 cycles | 30.0 sec | 95°C |
| | 60.0 sec | 67°C |
| | 60.0 sec | 72°C |
| Final extension | 10 min | 72°C |

40. Separate the PCR products by 2% (w/v) agarose gel electrophoresis:
    i. Melt agarose in Tris acetate-EDTA buffer to a final concentration of 2% (w/v). Add GelRed to 1× and cast the gel, adding a comb.
    ii. After the gel has solidified, add gel loading dye to the samples (5 µL per PCR product) to a final concentration of 1× and run it alongside a 100-bp DNA ladder at 90 V for 90 min using a power supply.

41. Pool all PCR reactions (total volume 1200 µL = 60 PCR products × 20 µL) and purify using 12 separate columns of the QIAquick PCR Purification Kit, following to the manufacturer's protocol (use 100 µL per column). Elute the sample in each column with 30 µL of nuclease-free H$_2$O and then pool the 12 eluates (total volume 360 µL)

    *This is called "natural H3J fragments."*

42. Determine DNA concentration by UV spectrophotometry and store at −20 °C until use.

## PCR Assembly of the Filtered Libraries with the Natural H3J Fragments

*The secondary libraries (SLs) are assembled from two PCR fragments: (1) natural H3J fragments (from Step 42) and (2) a fragment of the scFv synthetic scaffold obtained from the FLs encompassing the $V_L$, the GS19 linker, and the $V_H$ just before the conserved motif (fragment $V_L$–$V_H$_CM).*

*Amplification of the $V_L$–$V_H$_CM Fragment*

43. Do a plasmid preparation of the filtered libraries (FLs) obtained in Protocol 2: Semisynthetic Phage Display Library Construction: Generation of Filtered Libraries (Almagro and Polh 2024a). Using QIAGEN Plasmid Midi Kit according to the manufacturer's protocol.

44. Prepare a PCR mix as follows (15 reactions):

| Component | Volume/amount (one reaction) | Volume/amount (15 reactions) |
| --- | --- | --- |
| Phusion HF Reaction Buffer (5×) | 5.00 µL | 75.00 µL |
| dNTPs (2 mM) | 2.50 µL | 37.50 µL |
| Primer SfiIFOR (10 µM) | 1.25 µL | 18.75 µL |
| Primer UFR3REV (10 µM) | 1.25 µL | 18.75 µL |
| FL plasmid DNA (DNA template, from Step 43) | 15.00 ng | 225.00 ng |
| Phusion Hot Start Flex DNA Polymerase | 0.25 µL | 3.75 µL |
| Nuclease-free H$_2$O | To 25 µL | To 375 µL |

*A total of 225 ng of FL plasmid DNA is used as template for amplification.*

45. Aliquot the mix into 15 separate 0.2-mL tubes. Then, run the reactions in the thermocycler using the following conditions:

| Step | Time | Temperature |
| --- | --- | --- |
| Initial denaturation | 30.0 sec | 98°C |
| 10 cycles | 15.0 sec | 98°C |
|  | 30.0 sec | 67°C |
|  | 30.0 sec | 72°C |
| Final extension | 10.0 min | 72°C |

46. Pool PCR products (total volume 375 µL) and purify using two columns of QIAquick PCR Purification Kit according to the manufacturer's protocol (i.e., run 187.5 µL in each column). Elute the sample in each column with 30 µL of nuclease-free H$_2$O and then pool the eluates (total volume 60 µL).

47. Separate the total volume (60 µL) by 1% (w/v) agarose gel electrophoresis.
    i. Melt agarose in Tris acetate-EDTA buffer for a final concentration of 1% (w/v). Add GelRed to 1× and cast the gel, adding a comb.
    ii. After the gel has solidified, add gel loading dye to the sample to a final concentration of 1× and run it alongside a 1-kb DNA ladder at 90 V for 90 min using a power supply.
       *Users can run the sample across multiple wells or use combs appropriate for the sample volume listed.*

48. Purify the band corresponding to the $V_L$–$V_H$_CM fragment (∼726 bp) with the QIAquick Gel Extraction Kit according to the manufacturer's protocol. Elute with 40 µL of nuclease-free H$_2$O.
    *This is called the "$V_L$–$V_H$_CM fragments."*

49. Determine DNA concentration by UV spectrophotometry and store at −20°C until use.

*PCR Assembly of the $V_L$–$V_H$_CM Fragment with the Natural H3J Fragment to Generate the Secondary Libraries*

50. Prepare a PCR master mix for twenty 25-µL reactions containing the following reagents:

Chapter 9

| Component | Volume/amount (one reaction) | Volume/amount (20 reactions; master mix) |
|---|---|---|
| Phusion HF Reaction Buffer (5×) | 5.00 μL | 100 μL |
| dNTPs (2 mM) | 2.50 μL | 50 μL |
| Primer SfiIFOR (10 μM) | 2.00 μL | 40 μL |
| Primer SfiREV (10 μM) | 2.00 μL | 40 μL |
| Fragment $V_L$–$V_H$_CM | 12.75 ng | 255 ng |
| Natural H3J fragments | 1.00 μL (2.26 ng/μL) | 20 μL |
| Phusion Hot Start Flex DNA Polymerase | 0.25 μL | 5 μL |
| Nuclease-free $H_2O$ | To 25 μL | To 500 μL |

*The $V_L$–$V_H$_CM fragment (from FLs) and natural H3J fragments will be mixed in equimolar amounts.*

51. Split the master mix into 20 separate PCR tubes. Then, run the following program:

| Step | Time | Temperature |
|---|---|---|
| Initial denaturation | 30.0 sec | 98°C |
| 15 cycles | 15.0 sec | 98°C |
|  | 30.0 sec | 67°C |
|  | 30.0 sec | 72°C |
| Final extension | 10.0 min | 72°C |

52. Pool the 20 reactions and concentrate the PCR product with the QIAquick PCR Purification Kit according to the manufacturer's protocol. Elute with 50 μL of elution buffer.

    *These are the final semisynthetic scFvs to clone the SLs.*

53. Run a 1% (w/v) agarose gel (see Step 47) and purify the scFv assembled DNA with the QIAquick Gel Extraction Kit according to the manufacturer's protocol. Elute with 50 μL of elution buffer.

54. Determine DNA concentration by UV spectrophotometry and store at −20°C until use.

    *This is called the "semisynthetic scFvs."*

## Construction of Secondary Libraries

### Digestion of the Semisynthetic scFvs for Cloning into the pADL-23c Vector

*Semisynthetic scFvs are digested and cloned as four separate SLs to maximize the final diversity of ALTHEA Gold Plus Libraries. Because each SL will require several electroporation reactions, it may be desirable to split this part of the protocol over several days, to make the workload manageable and to avoid any cross-contamination of the libraries.*

*The process for one sublibrary is described below. Consider that there are four sublibraries in ALTHEA Gold Plus (see Protocol 1: Semisynthetic Phage Display Library Construction: Design and Synthesis of Diversified Single-Chain Variable Fragments and Generation of Primary Libraries [Almagro and Pohl 2024b]).*

55. Prepare a reaction mix to digest the semisynthetic scFvs with SfiI as follows:

| Component | Volume/amount |
|---|---|
| CutSmart Buffer (10×; NEB) | 10.0 μL |
| Semisynthetic scFvs | 3.0 μg |
| SfiI (20,000 units/mL) | 3.6 μL (72.0 units) |
| Nuclease-free $H_2O$ | To 150.0 μL |

56. Incubate the digestion reaction for 2.5 h at 50°C.

57. Remove enzymes, salts, and subproducts of the digestion using the QIAquick PCR Purification Kit. Elute with 30 µL of nuclease-free H$_2$O to obtain a concentrated sample.
58. Determine DNA concentration by UV spectrophotometry and store it at −20°C until use.

*Digestion of the pADL-23c Vector for Cloning the Semisynthetic scFvs*

59. Prepare a 500-µL reaction mix to digest the pADL-23c vector with SfiI as follows:

| Component | Volume/amount |
| --- | --- |
| CutSmart Buffer (10×; NEB) | 50.0 µL |
| pADL-23c phagemid vector | 10.0 µg |
| SfiI (20,000 units/mL) | 12.0 µL (240 units) |
| Nuclease-free H$_2$O | To 500.0 µL |

60. Incubate the digestion reaction for 4 h at 50°C.
61. Remove enzymes, salts, and subproducts of the digestion with the QIAquick PCR Purification Kit. Elute with 50 µL of nuclease-free H$_2$O to obtain a concentrated sample.
62. Prepare a 1% (w/v) agarose gel and run the sample.

    i. Melt agarose in Tris acetate-EDTA buffer. Add GelRed to 1× and cast the gel, adding a comb.
    ii. After the gel has solidified, add 10 µL of gel loading dye to the sample to a final concentration of 1× and run it alongside a 1-kb DNA ladder at 90 V for 90 min using a power supply.

63. Purify the band corresponding to the digested pADL-23c vector (∼3900 bp) with the QIAquick Gel Extraction Kit. Elute with 50 µL of nuclease-free H$_2$O.
64. Determine DNA concentration by UV spectrophotometry and store it at −20°C until use.

*Ligation of Digested Semisynthetic scFvs with the Digested Phagemid Vector*

65. Prepare the scFv–phagemid ligation mixture (2:1 molar ratio [insert:vector]) as follows:

| Component | Volume/amount |
| --- | --- |
| 10× Buffer for T4 DNA ligase | 49.3 µL |
| Digested pADL-23c vector | 4000 µg |
| Digested semisynthetic scFv | 1634 µg |
| Nuclease-free H$_2$O | To 493.2 µL |

66. Take a 10.5-µL (∼120 ng) aliquot of the mixture and place it in another tube. This will serve as the negative ligation control when running the gel in Step 70. Then, add 49.2 µL of T4 DNA ligase (400,000 units/mL) to the ligation reaction from Step 65.
67. Incubate the ligation mix for 2.5 h at 23°C.
68. Remove enzymes, salts, and subproducts of the ligation with the QIAquick PCR Purification Kit. Elute with 40 µL of nuclease-free H$_2$O to obtain a concentrated sample.
69. Determine DNA concentration by UV spectrophotometry and store it at −20°C until use.
70. To assess the quality of the ligation, run a 1% (w/v) agarose gel as described in Step 62 with the ligation control and ligation mix.

    *A single band of vector should be seen in the negative ligation control, and a ladder should be observed in the ligation mix.*

Chapter 9

*Electroporation of the Ligation Mix*

*For an estimated library size of >$10^{10}$ colony-forming units (cfu) of each PL, at least 17 separate electroporation reactions are required.*

71. Prechill electroporation cuvettes on ice before starting electroporations.
72. Add 2 µL (~160 ng) of the ligation mix into 25 µL of TG1 electrocompetent cells and electroporate in 0.1-cm cuvettes at 1.8 kV, 200 Ω (resistance), and 25 µF (capacitance). Repeat for a total of 17 separate electroporation reactions.
73. Immediately add 975 µL of SOC recovery medium into each cuvette and transfer the cells to a single 50-mL screw-top tube.
74. Incubate with shaking at 225 rpm for 45 min at 37°C.
75. To determine the estimated size of the library, take a 50-µL aliquot of the electroporated cells and dilute it in 450 µL of YTCG liquid medium. Generate 1:10 serial dilutions to $10^{-10}$. Plate 50 µL of the $10^{-5}$–$10^{-10}$ dilutions onto separate 100-mm YTCG solid medium plates and carefully spread the cells using a glass spreader or beads. Incubate overnight at 37°C.
76. In parallel, and for each mix, plate a 1-mL aliquot of the electroporated cells onto separate large agar (150-mm × 15-mm) YTCG solid medium plates. Incubate overnight at 37 °C.
77. The next day, count the colonies on the serial dilution plates. Consider plates "countable" when they have between 40 and 400 well-separated colonies. Use the following formula to estimate the size of the library after electroporation:
    cfu = number of colonies × 10 (dilution in 500 µL of medium) × dilution factor/50 µL.
78. If the library size is >$10^{10}$ cfu, continue to the following steps. If not, see Troubleshooting.
79. Scrape bacteria from the large culture plates (from Step 76) by washing each plate with 3 mL of YTCG liquid medium and loosening the bacterial lawn from the plate using disposable cell scrapers. Collect cell material with a serological pipette and transfer it to a 50-mL screw-top conical tube.
80. Make 1-mL glycerol stocks by mixing 28 mL of bacteria and 14 mL of 99% glycerol. Store the glycerol stocks at −70°C.

    *These vials are now SL live stocks of one sublibrary.*

## Quality Control of the SLs

*The quality of the SLs can be assessed by sequencing individual clones or performing high-throughput sequencing analysis. For sequencing of individual clones, this can be done by sending a plate with 50 colonies chosen at random from the titration plates for Sanger sequencing or by sending PCR products of the colonies for sequencing. The following procedure is for sequencing PCR products.*

81. Pick 50 colonies randomly from the secondary library titer plates from Step 76 with sterile toothpicks or pipette tips and resuspend in 30 µL of H$_2$O. Select different colony sizes to avoid biasing the selection.
82. For each sublibrary, prepare a 25-µL colony PCR mix as follows:

| Component | Volume/amount (one reaction) | Volume/amount (60 reactions)[a] |
|---|---|---|
| DreamTaq DNA Buffer (10×) | 5.00 µL | 300 µL |
| dNTPs (2 mM) | 2.50 µL | 150.00 µL |
| Primer For_scFv_seq (10 µM) | 2.00 µL | 120.00 µL |

(*continued*)

## Using Phage Display for Antibody Drug Discovery

*Continued*

| Component | Volume/amount (one reaction) | Volume/amount (60 reactions)[a] |
|---|---|---|
| Primer Rev_scFv_seq (10 μM) | 2.00 μL | 120.00 μL |
| DNA template (random colonies resuspended in $H_2O$) | 2.00 μL | 2.00 μL each |
| DreamTaq DNA Polymerase | 0.25 μL | 15.00 μL |
| Nuclease-free $H_2O$ | To 25.0 μL | To 1500.00 μL |

[a]The reaction mix has an excess of six reactions, to account for pipetting error.

83. Dispense 23 μL of the PCR mix into 54 PCR tubes. Then, add 2 μL of each colony resuspended in $H_2O$ from Step 81 to each reaction. In addition, in the remaining four tubes, consider two controls, each in duplicate: one with a plasmid containing a known scFv (positive control) and one without plasmid (negative control).

84. Run the PCR in a thermocycler with the following parameters:

| Step | Time | Temperature |
|---|---|---|
| Initial denaturation | 300.0 sec | 95°C |
| 25 cycles | 30.0 sec | 95°C |
|  | 30.0 sec | 67°C |
|  | 60.0 sec | 72°C |
| Final extension | 10.0 min | 72°C |

85. Run a 1% agarose gel with 5 μL of each reaction to confirm successful PCR amplification of each colony. If positive, i.e., a band of the approximate size of the positive control is observed, purify the remaining PCR product using the QIAquick PCR Purification Kit and send it for Sanger sequencing with the For_scFv_seq and Rev_scFv_seq primers.

86. Analyze the DNA sequence data with the appropriate software, such as Geneious Prime, DNASTAR, or equivalent.

> *The expected result is that 95% of the sequences are in-frame and match the library design. No identical scFv sequences should be observed. Otherwise, it is likely that the diversity of the library is compromised. See Troubleshooting.*
>
> *High-throughput sequencing analysis of the HCDR should yield a Gaussian distribution resembling the human HCDR3 repertoire. Test phage selections can also be performed with a common antigen, such as lysozyme or BSA, to validate the libraries, as described by Valadon et al. (2019).*

## TROUBLESHOOTING

*Problem (Step 78):* The estimated size of the library is $<10^{10}$ cfu.
*Solution:* Repeat the cloning reaction and electroporation, or add more electroporation reactions.

*Problem (Step 86):* Less than 95% of the sequences are in-frame and match the library design, or identical scFv sequences are observed.
*Solution:* Because the diversity of the library is compromised, the libraries should be generated again.

Chapter 9

## COMPETING INTEREST STATEMENT

M.A.P. is an employee of Xtalpi, Inc. J.C.A. is founder and CEO of GlobalBio, Inc., and has commercial interest in ALTHEA Gold Plus Libraries.

## ACKNOWLEDGMENTS

J.C.A. thanks Keyla Gomez Castellanos for helping in the preparation and review of the protocol described here, and Dr. Sonia Mayra Perez Tapia for her support and fruitful discussions.

## RECIPES

### Carbenicillin (Carb; 100 mg/mL)

Prepare 100 mg/mL carbenicillin disodium salt (Millipore-Sigma C1389-5G) in milli-Q-purified $H_2O$, filter sterilize.

Store indefinitely at −20°C.

### YTCG Liquid Medium

Mix 16 g of tryptone, 10 g of yeast extract, and 5 g of NaCl. Add $H_2O$ and mix to dissolve all reagents. Adjust pH to 7.0 with NaOH. Adjust volume to 950 mL with deionized $H_2O$. Sterilize by autoclaving for 20 min at 15 psi (1.05 kg/cm$^2$) on liquid cycle. Allow medium to cool down. Right before use, add 1 mL of a sterile carbenicillin stock solution <R>, for a final carbenicillin concentration of 100 μg/mL, and 50 mL of filter-sterilized 20% (w/v) glucose, for a final glucose concentration of 1% (w/v). YTCG liquid medium can be stored for 2 wk at 4°C.

### YTCG Solid Medium

Mix 16 g of tryptone, 10 g of yeast extract, 5 g of NaCl, and 15 g of agar. Add $H_2O$ and mix to dissolve all reagents. Adjust pH to 7.0 with NaOH. Adjust volume to 950 mL with deionized $H_2O$. Sterilize by autoclaving for 20 min at 15 psi (1.05 kg/cm$^2$) on liquid cycle. Allow medium to cool down to 60°C. Then, add 1 mL of a sterile carbenicillin stock solution <R>, for a final carbenicillin concentration of 100 μg/mL, and 50 mL of filter-sterilized 20% (w/v) glucose, for a final glucose concentration of 1% (w/v). Pour into both 150-mm × 15-mm (large) and 100-mm × 15-mm (small) Petri dishes. YTCG solid medium can be stored for 4 wk at 4°C.

## REFERENCES

Almagro JC, Pohl MA. 2024a. Semisynthetic phage display library construction: generation of filtered libraries. *Cold Spring Harb Protoc* doi:10.1101/pdb.prot108615

Almagro JC, Pohl MA. 2024b. Semisynthetic phage display library construction: design and synthesis of diversified single-chain variable fragments and generation of primary libraries. *Cold Spring Harb Protoc* doi:10.1101/pdb.prot108614

Valadon P, Pérez-Tapia SM, Nelson RS, Guzmán-Bringas OU, Arrieta-Oliva HI, Gómez-Castellano KM, Pohl MA, Almagro JC. 2019. ALTHEA Gold Libraries™: antibody libraries for therapeutic antibody discovery. *MAbs* 11: 516–531. doi:10.1080/19420862.2019.1571879

CHAPTER 10

# Beyond Single Clones: High-Throughput Sequencing in Antibody Discovery

Ahmed S. Fahad,[1,3] Matías F. Gutiérrez-Gonzalez,[1,3] Bharat Madan,[1] and Brandon J. DeKosky[1,2,4]

[1]The Ragon Institute of Massachusetts General Hospital, Massachusetts Institute of Technology and Harvard University, Cambridge, Massachusetts 02139, USA; [2]Department of Chemical Engineering, Massachusetts Institute of Technology, Cambridge, Massachusetts 02139, USA

Antibody repertoire sequencing and display library screening are powerful approaches for antibody discovery and engineering that can connect DNA sequence with antibody function. Antibody display and screening studies have made a tremendous impact on immunology and biotechnology over the last decade, accelerated by technological advances in high-throughput DNA sequencing techniques. Indeed, bioinformatic analysis of antibody DNA library data has now taken a central role in modern antibody drug discovery, and is also critical for many ongoing studies of human immune development. Here, we describe current trends in antibody DNA library screening and analysis, and introduce a selection of protocols describing fundamental bioinformatic techniques to enable scientists to efficiently study antibody DNA libraries.

## INTRODUCTION

Improved antibody high-throughput sequencing (HTS) technologies have made a tremendous impact on immunology and biotechnology over the last decade, transforming the technical landscape for antibody and T-cell receptor (TCR) repertoire analysis (Wang et al. 2018; Lagerman et al. 2019; Banach et al. 2021; Fahad et al. 2021, 2022; Madan et al. 2021). For instance, in basic immunology, large-scale studies of antibody responses have improved our understanding of the determinants of vaccine efficacy, the origin and development of broadly neutralizing antiviral antibodies, and the general elicitation and formation of antibody immune responses (Linnemann et al. 2013; Howie et al. 2015; Zhang et al. 2018; Ludwig et al. 2019; Pai and Satpathy 2021; Fahad et al. 2022, 2023). In translational studies, antibody HTS has enabled rapid and efficient antibody discovery campaigns using a variety of experimental discovery platforms (Hu et al. 2015; Wang et al. 2018; Fahad et al. 2021; Porebski et al. 2024). HTS has thus become a fundamental, flexible tool to study immune repertoire development and discover new antibody molecules (Fig. 1).

In particular, analysis of protein display library data is a key application of antibody-focused bioinformatic pipelines (Kwong et al. 2017). A protein display platform links the DNA sequence of a gene (in this case, an antibody gene) directly to the functional performance of the protein that it encodes, and enables the sequencing of genes encoding for proteins with a desired function (e.g., specific binding characteristics to an antigen of interest). Importantly, HTS expands our ability to analyze display

---

[3]These authors contributed equally to this work.
[4]Correspondence: dekosky@mit.edu

© 2026 Cold Spring Harbor Laboratory Press
Cite this introduction as *Cold Spring Harb Protoc*; doi:10.1101/pdb.top107772

# Chapter 10

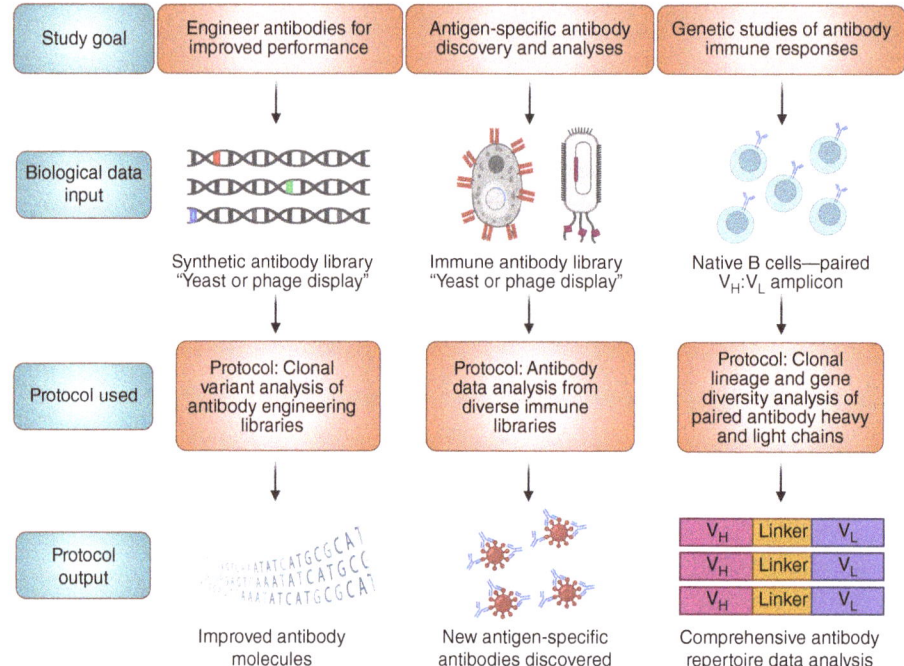

FIGURE 1. Applications of antibody high-throughput sequencing (HTS) analysis for antibody engineering and discovery, with associated bioinformatic protocols that support each analysis. Antibody HTS analysis is an important tool that supports various research and discovery goals. Each study goal requires a unique set of HTS sequence data inputs that are then analyzed using a distinct bioinformatic protocol, which results in a unique biological output that supports the study goal. All three approaches listed here (described in detail in the accompanying protocols) share some common elements with the others, such as performing sequence data comparisons between presorted and postsorted populations to detect enriched antibodies that represent antigen-specific candidates. The analysis of paired antibody sequencing data also requires specialized bioinformatic tools to analyze associated heavy and light chains.

library studies by orders of magnitude relative to prior low-throughput sequencing technologies (Frei and Lai 2016; Smith 2023). Protein display is now widely used for antibody discovery and engineering, and unlocks efficient selection, characterization, and modification of antibody genes based on desired functional properties (Wang et al. 2018; Oh et al. 2020; Fahad et al. 2021; Madan et al. 2021; Ehling et al. 2022; Yuan et al. 2022). Selection is often performed using fluorescence-activated cell sorting (FACS) (Wang et al. 2018) or, alternatively, by immunoprecipitation of phage particles against an antigen bound to a tube surface (Yuan et al. 2022). In some cases, phage can also be captured for their ability to bind to an antigen expressed directly on cells (Glaser et al. 2023).

HTS studies are often carried out by sampling B-cell-encoded genes from a donor or immunized animal of interest. It is also common to immortalize repertoires in phage, yeast, or mammalian display for high-throughput screening (Valldorf et al. 2022; Rader 2023; Smith 2023; Veggiani and Sidhu 2024). Antibody HTS uses targeted amplification of the V(D)J segments that encode the antibody, followed by preparation of DNA for HTS. As of 2024, the Illumina MiSeq and the Pacific Biosciences Sequel II sequencing platforms are often used for antibody HTS because they offer appropriate sequencing depth, read length, and cost for most antibody applications. Once antibody repertoires are sequenced on appropriate platforms to generate raw sequencing reads, the bioinformatic analysis of antibody data can begin.

## ANALYSIS OF ANTIBODY DATA

Most HTS analysis pipelines include a quality control step to preprocess the raw data and remove dominant sources of error. HTS data often have substitution and insertion/deletion errors in ~0.5% of bases, with the error profile varying across a read according to the sequencing platform used and the performance of a particular sequencing run. Therefore, preprocessing is usually tailored to a specific antibody HTS platform, and different sequencing platforms should have different raw data preprocessing based on the dominant error types (Loman et al. 2012; Amarasinghe et al. 2020). For 2 × 300-bp paired-end Illumina reads (commonly used for antibody gene sequencing due its appropriate read length, sufficient to characterize the ∼425- to 450-bp antibody variable regions), raw data preprocessing often first merges Reads 1 and 2 using an efficient and purpose-built sequence alignment and merging program like FLASH (Magoč and Salzberg 2011). Next, merged reads with low quality are efficiently removed in a subsequent quality-filtering step using tools such as fastq_quality_filter (http://hannonlab.cshl.edu/fastx_toolkit/). Successful preprocessing generates mostly high-quality sequences, with an enhanced probability that any observed mismatches to known antibody germline sequences comprise a true mutation rather than errors introduced by sequencing.

Following initial quality control, a purpose-built antibody gene alignment software is used next for sequence data annotation. IgBLAST is one popular and reliable alignment software (Ye et al. 2013), which was developed as a dedicated version of the BLAST algorithm (Altschul et al. 1990) for immunoglobulin gene annotation. IgBLAST's algorithm performs a sequence alignment to antibody germline genes, and reports the percentage of identity of the query sequence to the nearest germline V, (D), and J genes. IgBLAST and most other antibody gene alignment programs identify common antibody gene regions, such as the framework regions (FR1, FR2, FR3, and FR4) and the complementarity-determining regions (CDR1, CDR2, and CDR3). These gene annotations enable scientists to perform clonal antibody lineage tracking, particularly to track members of expanded B-cell lineages. Alignment data also supports somatic hypermutation detection, B-cell lineage clustering analyses, repertoire diversity statistics, and phylogenetic analyses to compare antibody immunity across individuals, time points, or immune compartments. Purpose-built antibody gene alignment programs are effectively used for large antibody repertoire analyses with diverse germline genes; however, we also note, from our experience, that analysis of lower-diversity synthetic libraries (e.g., template antibodies with engineered mutations, or transgenic mice expressing single antibody germline genes) can also be efficiently analyzed with direct alignments to the known template antibody (Madan et al. 2021; Banach et al. 2022, 2023).

## APPLICATIONS OF ANTIBODY SEQUENCE ANALYSIS

One of the most straightforward applications of antibody HTS is to evaluate the functional impact of mutations on a template antibody (Fig. 1). We previously reported a method to track the affinity impact of mutations in synthetic antibody gene libraries (Wrenbeck et al. 2016; Madan et al. 2021; Banach et al. 2022, 2023). Methods for high-throughput data analysis of engineered antibody mutation libraries are described in an accompanying protocol (see Protocol 1: Clonal Variant Analysis of Antibody Engineering Libraries [Fahad et al. 2024a]).

Natural immunity generates tremendous diversity for antibody discovery, and can be effectively screened to profile immune responses in disease models, study basic biology features, and discover new antibodies (Fig. 1). Because antigen-specific antibodies are usually very rare in starting libraries (i.e., less than one antigen-specific antibody per $10^5$–$10^{10}$ genes), it is often necessary to perform multiple rounds of enrichment to purify antibody clones with desired functional properties from a library. An accompanying protocol describes HTS analysis of antibody clonal prevalence to identify antigen-specific antibodies from antibody display screening experiments (see Protocol 2: Antibody Data Analysis from Diverse Immune Libraries [Fahad et al. 2024b]). Although we currently work mainly with Illumina platforms, the analysis described in this protocol can be used on data from any HTS platform to analyze DNA libraries throughout each round of the sorting process (Fahad et al. 2021; Madan et al. 2021).

As both heavy and light chains vary in natural antibody immunity, obtaining the correct heavy and light chain pairing is also critical to recapitulate functional properties using display experiments. Natively paired variable heavy ($V_H$):variable light ($V_L$) technologies thus allow natural antibody discovery based on the exact heavy and light chains elicited in vivo (Fig. 1). $V_H$:$V_L$ antibody tracking, lineage tracing, and other paired $V_H$:$V_L$ data analyses are presented in an accompanying protocol (see Protocol 3: Clonal Lineage and Gene Diversity Analysis of Paired Antibody Heavy and Light Chains [Fahad et al. 2024c]). Using this protocol, each antibody heavy chain remains associated with its native light chain for robust analysis of complete paired heavy and light chain antibody function.

## CLOSING REMARKS

HTS has revolutionized the study of adaptive immunity by offering a powerful method both to survey large pools of antibodies and to track gene performance in display studies for antibody discovery and engineering. Bioinformatic techniques enable clonal tracking across library enrichment rounds to discover antibodies with high accuracy. These HTS-based tools have provided remarkable opportunities to explore immune sequence–function landscapes in unprecedented detail, and to quickly and efficiently discover precision biologics for broad therapeutic applications.

## ACKNOWLEDGMENTS

This work was supported by National Institutes of Health grants DP5OD023118, R21AI166396, R21AI143407, U01AI169587, and R01AI181684; The Bill and Melinda Gates Foundation; and the Mark and Lisa Schwartz AI/ML/Immunology Initiative.

## REFERENCES

Altschul SF, Gish W, Miller W, Myers EW, Lipman DJ. 1990. Basic local alignment search tool. *J Mol Biol* 215: 403–410. doi:10.1016/S0022-2836(05)80360-2

Amarasinghe SL, Su S, Dong X, Zappia L, Ritchie ME, Gouil Q. 2020. Opportunities and challenges in long-read sequencing data analysis. *Genome Biol* 21: 30. doi:10.1186/s13059-020-1935-5

Banach BB, Cerutti G, Fahad AS, Shen C-H, Oliveira De Souza M, Katsamba PS, Tsybovsky Y, Wang P, Nair MS, Huang Y, et al. 2021. Paired heavy- and light-chain signatures contribute to potent SARS–CoV-2 neutralization in public antibody responses. *Cell Rep* 37: 109771. doi:10.1016/j.celrep.2021.109771

Banach BB, Tripathi P, Pereira LDS, Gorman J, Nguyen TD, Dillon M, Fahad AS, Kiyuka PK, Madan B, Wolfe JR, et al. 2022. Highly protective antimalarial antibodies via precision library generation and yeast display screening. *J Exp Med* 219: e20220323. doi:10.1084/jem.20220323

Banach BB, Pletnev S, Olia AS, Xu K, Zhang B, Rawi R, Bylund T, Doria-Rose NA, Nguyen TD, Fahad AS, et al. 2023. Antibody-directed evolution reveals a mechanism for enhanced neutralization at the HIV-1 fusion peptide site. *Nat Commun* 14: 7593. doi:10.1038/s41467-023-42098-5

Ehling RA, Weber CR, Mason DM, Friedensohn S, Wagner B, Bieberich F, Kapetanovic E, Vazquez-Lombardi R, Roberto RBD, Hong K-L, et al. 2022. SARS–CoV-2 reactive and neutralizing antibodies discovered by single-cell sequencing of plasma cells and mammalian display. *Cell Rep* 38: 110242. doi:10.1016/j.celrep.2021.110242

Fahad AS, Timm MR, Madan B, Burgomaster KE, Dowd KA, Normandin E, Gutiérrez-González MF, Pennington JM, De Souza MO, Henry AR, et al. 2021. Functional profiling of antibody immune repertoires in convalescent Zika virus disease patients. *Front Immunol* 12: 615102. doi:10.3389/fimmu.2021.615102

Fahad AS, Chung CY, Lopez Acevedo SN, Boyle N, Madan B, Gutiérrez-González MF, Matus-Nicodemos R, Laflin AD, Ladi RR, Zhou J, et al. 2022. Immortalization and functional screening of natively paired human T cell receptor repertoires. *Protein Eng Des Sel* 35: gzab034. doi:10.1093/protein/gzab034

Fahad AS, Chung CY, López Acevedo SN, Boyle N, Madan B, Gutiérrez-González MF, Matus-Nicodemos R, Laflin AD, Ladi RR, Zhou J, et al. 2023. Cell activation-based screening of natively paired human T cell receptor repertoires. *Sci Rep* 13: 8011. doi:10.1038/s41598-023-31858-4

Fahad AS, Gutiérrez-González MF, Madan B, DeKosky BJ. 2024a. Clonal variant analysis of antibody engineering libraries. *Cold Spring Harb Protoc* doi:10.1101/pdb.prot108626

Fahad AS, Gutiérrez-González MF, Madan B, DeKosky BJ. 2024b. Antibody data analysis from diverse immune libraries. *Cold Spring Harb Protoc* doi:10.1101/pdb.prot108627

Fahad AS, Gutiérrez-González MF, Madan B, DeKosky BJ. 2024c. Clonal lineage and gene diversity analysis of paired antibody heavy and light chains. *Cold Spring Harb Protoc* doi:10.1101/pdb.prot108628

Frei JC, Lai JR. 2016. Protein and antibody engineering by phage display. In *Methods in enzymology* (ed. Pecoraro VL), Vol. 580 of *Peptide, protein and enzyme design*, pp. 45–87, Academic Press, Cambridge, MA.

Glaser V, Karsli-Ünal Ü, Hagedorn M, Pieper T. 2023. Antibody selection on cells targeting membrane proteins. In *Phage display: methods and protocols* (ed. Hust M, Lim TS), pp. 315–325, Springer, New York.

Howie B, Sherwood AM, Berkebile AD, Berka J, Emerson RO, Williamson DW, Kirsch I, Vignali M, Rieder MJ, Carlson CS, et al. 2015. High-throughput pairing of T cell receptor α and β sequences. *Sci Transl Med* 7: 301ra131. doi:10.1126/scitranslmed.aac5624

Hu D, Hu S, Wan W, Xu M, Du R, Zhao W, Gao X, Liu J, Liu H, Hong J. 2015. Effective optimization of antibody affinity by phage display integrated with high-throughput DNA synthesis and sequencing technologies. *PLoS ONE* 10: e0129125. doi:10.1371/journal.pone.0129125

Kwong PD, Chuang G-Y, DeKosky BJ, Gindin T, Georgiev IS, Lemmin T, Schramm CA, Sheng Z, Soto C, Yang A-S, et al. 2017. Antibodyomics: bioinformatics technologies for understanding B-cell immunity to HIV-1. *Immunol Rev* 275: 108–128. doi:10.1111/imr.12480

Lagerman CE, López Acevedo SN, Fahad AS, Hailemariam AT, Madan B, DeKosky BJ. 2019. Ultrasonically-guided flow focusing generates precise emulsion droplets for high-throughput single cell analyses. *J Biosci Bioeng* 128: 226–233. doi:10.1016/j.jbiosc.2019.01.020

Linnemann C, Heemskerk B, Kvistborg P, Kluin RJC, Bolotin DA, Chen X, Bresser K, Nieuwland M, Schotte R, Michels S, et al. 2013. High-throughput identification of antigen-specific TCRs by TCR gene capture. *Nat Med* **19**: 1534–1541. doi:10.1038/nm.3359

Loman NJ, Misra RV, Dallman TJ, Constantinidou C, Gharbia SE, Wain J, Pallen MJ. 2012. Performance comparison of benchtop high-throughput sequencing platforms. *Nat Biotechnol* **30**: 434–439. doi:10.1038/nbt.2198

Ludwig J, Huber A, Bartsch I, Busse CE, Wardemann H. 2019. High-throughput single-cell sequencing of paired TCRα and TCRβ genes for the direct expression-cloning and functional analysis of murine T-cell receptors. *Eur J Immunol* **49**: 1269–1277. doi:10.1002/eji.201848030

Madan B, Zhang B, Xu K, Chao CW, O'Dell S, Wolfe JR, Chuang G-Y, Fahad AS, Geng H, Kong R, et al. 2021. Mutational fitness landscapes reveal genetic and structural improvement pathways for a vaccine-elicited HIV-1 broadly neutralizing antibody. *Proc Natl Acad Sci* **118**: e2011653118. doi:10.1073/pnas.2011653118

Magoč T, Salzberg SL. 2011. FLASH: fast length adjustment of short reads to improve genome assemblies. *Bioinformatics* **27**: 2957–2963. doi:10.1093/bioinformatics/btr507

Oh EJ, Liu R, Liang L, Freed EF, Eckert CA, Gill RT. 2020. Multiplex evolution of antibody fragments utilizing a yeast surface display platform. *ACS Synth Biol* **9**: 2197–2202. doi:10.1021/acssynbio.0c00159

Pai JA, Satpathy AT. 2021. High-throughput and single-cell T cell receptor sequencing technologies. *Nat Methods* **18**: 881–892. doi:10.1038/s41592-021-01201-8

Porebski BT, Balmforth M, Browne G, Riley A, Jamali K, Fürst MJLJ, Velic M, Buchanan A, Minter R, Vaughan T, et al. 2024. Rapid discovery of high-affinity antibodies via massively parallel sequencing, ribosome display and affinity screening. *Nat Biomed Eng* **8**: 214–232. doi:10.1038/s41551-023-01093-3

Rader C. 2023. The pComb3 phagemid family of phage display vectors. *Cold Spring Harb Protoc* doi:10.1101/pdb.over107756

Smith GP. 2023. Principles of affinity selection. *Cold Spring Harb Protoc* doi:10.1101/pdb.over107894

Valldorf B, Hinz SC, Russo G, Pekar L, Mohr L, Klemm J, Doerner A, Krah S, Hust M, Zielonka S. 2022. Antibody display technologies: selecting the cream of the crop. *Biol Chem* **403**: 455–477. doi:10.1515/hsz-2020-0377

Veggiani G, Sidhu SS. 2024. Beyond natural immune repertoires: synthetic antibodies. *Cold Spring Harb Protoc* doi:10.1101/pdb.top107768

Wang B, DeKosky BJ, Timm MR, Lee J, Normandin E, Misasi J, Kong R, McDaniel JR, Delidakis G, Leigh KE, et al. 2018. Functional interrogation and mining of natively paired human VH:VL antibody repertoires. *Nat Biotechnol* **36**: 152–155. doi:10.1038/nbt.4052

Wrenbeck EE, Klesmith JR, Stapleton JA, Adeniran A, Tyo KEJ, Whitehead TA. 2016. Plasmid-based one-pot saturation mutagenesis. *Nat Methods* **13**: 928–930. doi:10.1038/nmeth.4029

Ye J, Ma N, Madden TL, Ostell JM. 2013. IgBLAST: an immunoglobulin variable domain sequence analysis tool. *Nucleic Acids Res* **41**: W34–W40. doi:10.1093/nar/gkt382

Yuan TZ, Garg P, Wang L, Willis JR, Kwan E, Hernandez AGL, Tuscano E, Sever EN, Keane C, Soto C, et al. 2022. Rapid discovery of diverse neutralizing SARS–CoV-2 antibodies from large-scale synthetic phage libraries. *mAbs* **14**: 2002236. doi:10.1080/19420862.2021.2002236

Zhang S-Q, Ma K-Y, Schonnesen AA, Zhang M, He C, Sun E, Williams CM, Jia W, Jiang N. 2018. High-throughput determination of the antigen specificities of T cell receptors in single cells. *Nat Biotechnol* **36**: 1156–1159. doi:10.1038/nbt.4282

## Protocol 1

# Clonal Variant Analysis of Antibody Engineering Libraries

Ahmed S. Fahad,[1,2,3] Matías F. Gutiérrez-Gonzalez,[1,2,3] Bharat Madan,[1] and Brandon J. DeKosky[1,2,4]

[1]*The Ragon Institute of Massachusetts General Hospital, Massachusetts Institute of Technology, and Harvard University, Cambridge, Massachusetts 02139, USA;* [2]*Department of Chemical Engineering, Massachusetts Institute of Technology, Cambridge, Massachusetts 02139, USA*

In vitro antibody evolution is a powerful technique for improving monoclonal antibodies. This can be achieved by generating artificial diversity on an antibody template, which can be done using different in vitro diversification techniques. The resulting libraries consist of single- or multimutant variants of a defined antibody template that are screened for improved function using antibody display. Here, we describe a bioinformatic protocol for tracking synthetic antibody variants using high-throughput sequencing across screening rounds, enabling efficient high-throughput interpretation of the function of individual mutations in sorted antibody display libraries. The protocol enables a user to achieve precision analysis and interpretation of clonal antibody variant data for discovery purposes, especially for high-throughput antibody engineering or optimization against target antigens.

## MATERIALS

### Reagents

Bioinformatic tools:
- FastQC, available at https://www.bioinformatics.babraham.ac.uk/projects/fastqc/ (Babraham Institute)
- FASTX-Toolkit v0.0.14 (http://hannonlab.cshl.edu/fastx_toolkit/download.html)
- FLASH v1.2.11 (https://sourceforge.net/projects/flashpage/files/latest/download) (optional; see Step 2)
- IgBLAST v1.18, available at https://ftp.ncbi.nih.gov/blast/executables/igblast/release/ (Ye et al. 2013)
- Perl v5.30, can be downloaded at https://www.perl.org/get.html
- Python 3 or above (https://www.python.org/)
- usearch 5 (a program for sequence alignment and clustering; free 32-bit version can be downloaded from https://www.drive5.com/usearch/download.html)

Custom scripts and files, available from the GitHub repository associated with this protocol (visit https://github.com/dekoskylab/CSHL_protocols/, and then go to Scripts/Scripts for Clonal Variant Analysis of Antibody Engineering Libraries):

[3]These authors contributed equally to this work
[4]Correspondence: dekosky@mit.edu

aa_comparison_v3.py; parses the output of sorting.py and outputs a file reporting the number of reads found for each variant, the amino acid sequence of the variant, the position of the mutation, and the template and mutated residue

annotation_script_v1.0.sh; this script gathers the analysis performed in all samples and aggregates into a single file that includes the enrichment ratio for each variant detected in each round of sorting

annotation_job_names.txt; this is an accessory tab-separated text file that contains a list of files to be read by annotation_script_v.1.0.sh; the first column contains a short description of the file (e.g., round number and binding affinity sort condition), and the second column has the name of the file for that particular round

parallel_jobsubmit_v2-mod_sbatch.sh; used to submit multiple parallel jobs to SLURM

blank_jobsubmit; blank template for job submissions using parallel_jobsubmit_v2-mod_sbatch.sh

jobsumit.py; maps the translated sequences to the template sequence

sorting.py; used to accurately map antibody variants to the template sequence using usearch

query_aa.fasta; fasta file containing the amino acid sequence of the template antibody

tab_to_fasta.sh; converts nucleotide sequences from IgBLAST to fasta format

translate.pl; translates nucleotide sequences into amino acid sequences

MiSeq_analysis_v2.py; this script formats and prepares sequences extracted from the IgBLAST annotation results for variant counting

Database of antibody V, D, and J germline genes from mouse; the latest version can be downloaded from the IgBLAST website and is also available in the associated GitHub repository (visit https://github.com/dekoskylab/CSHL_protocols/ and then go to Scripts/Scripts for Clonal Variant Analysis of Antibody Engineering Libraries/ncbi-igblast-1.18.0/database)

High-throughput sequencing (HTS) data

*Data can be generated using commonly used HTS technologies for antibody analysis, such as Illumina MiSeq or Pacific Biosciences Sequel II. To analyze data with diversity across the antibody variable region, a technology that sequences the full variable domain must be used. MiSeq 2 × 300 bp is often used because it is long enough to cover the ~440 bp of the heavy chain at high quality and a comparatively lower cost. In the example data*

TABLE 1. Sample data used in this protocol

| Accession code | Library name | Sorting round | Antigen probe | Estimated affinity by FACS | Prefix |
|---|---|---|---|---|---|
| SRR13067939 | BM_vFP_VH_NNK-VL-pos | N/A | N/A | N/A | VL_pos |
| SRR13067917 | BM_vFP_R1_1_S1-VH | 1 | BG505 FP8v1 | Low | R1_v1_Low |
| SRR13067916 | BM_vFP_R1_2_S2-VH | 1 | BG505 FP8v1 | Medium | R1_v1_Med |
| SRR13067915 | BM_vFP_R1_3_S3-VH | 1 | BG505 FP8v1 | High | R1_v1_High |
| SRR13067914 | BM_vFP_R1_4_S4-VH | 1 | BG505 FP8v2 | Low | R1_v2_Low |
| SRR13067913 | BM_vFP_R1_5_S5-VH | 1 | BG505 FP8v2 | Medium | R1_v2_Med |
| SRR13067912 | BM_vFP_R1_6_S6-VH | 1 | BG505 FP8v2 | High | R1_v2_High |
| SRR13067943 | BM_vFP_R2_1_S11-VH | 1 | BG505 FP8v1 | Low | R2_v1_Low |
| SRR13067942 | BM_vFP_R2_2_S12-VH | 2 | BG505 FP8v1 | Medium | R2_v1_Med |
| SRR13067941 | BM_vFP_R2_3_S13-VH | 2 | BG505 FP8v1 | High | R2_v1_High |
| SRR13067940 | BM_vFP_R2_4_S14-VH | 2 | BG505 FP8v2 | Low | R2_v2_Low |
| SRR13067938 | BM_vFP_R2_5_S15-VH | 2 | BG505 FP8v2 | Medium | R2_v2_Med |
| SRR13067937 | BM_vFP_R2_6_S16-VH | 2 | BG505 FP8v2 | High | R2_v2_High |
| SRR13067930 | BM_vFP_R3_1_S1-VH | 3 | BG505 FP8v1 | Low | R3_v1_Low |
| SRR13067929 | BM_vFP_R3_2_S2-VH | 3 | BG505 FP8v1 | Medium | R3_v1_Med |
| SRR13067927 | BM_vFP_R3_3_S3-VH | 3 | BG505 FP8v1 | High | R3_v1_High |
| SRR13067926 | BM_vFP_R3_4_S4-VH | 3 | BG505 FP8v2 | Low | R3_v2_Low |
| SRR13067925 | BM_vFP_R3_5_S5-VH | 3 | BG505 FP8v2 | Medium | R3_v2_Med |
| SRR13067924 | BM_vFP_R3_6_S6-VH | 3 | BG505 FP8v2 | High | R3_v2_High |

Sample data can be downloaded from the Sequence Read Archive (SRA) database from NCBI. Data can be downloaded manually by searching for the accession codes or using the NCBI SRA Toolkit. See https://www.ncbi.nlm.nih.gov/sra for further information.

Chapter 10

*provided here, we use 2 × 300 bp MiSeq data obtained from Madan et al. (2021) that consists of a synthetic library generated from the anti-HIV antibody vFP16.02 that was diversified using site saturation mutagenesis and displayed in yeast. Functional improvement was done by selecting variants with improved binding against BG505 DS-SOSIP HIV-1 Env trimer harboring two variants of the fusion peptide target epitope (BG505 FP8v1 and BG505 FP8v2). These data consist of antibody heavy chain variable regions sequenced from yeast libraries before functional screening, and after three rounds of selection against the BG505 DS-SOSIP HIV-1 Env trimer probes. Sample data from Madan et al. (2021) are available on the NCBI Sequence Read Archive (see Table 1).*

Microsoft Excel or similar spreadsheet software

Sequence (amino acid) of the template antibody in FASTA format; in the example described here, the amino acid sequence of vFP16.02 is used in a fasta format in the file query_aa.fasta (available in the associated GitHub repository)

### Equipment

High-performance computer or cluster with a UNIX-like environment.

*Analysis of HTS data uses more resources than those available in a typical laptop or desktop machine. Typical data sets range from a few dozen megabytes of data to gigabytes per sample. As such, a high-performance computing (HPC) platform is often necessary to analyze the data in a reasonable amount of time.*

Slurm Workload Manager (https://slurm.schedmd.com/documentation.html); other workload managers can be used, but the code should be modified accordingly

## METHOD

*This protocol describes the steps to efficiently identify antibody gene variants, and the software used in this protocol consists of specialized tools for quality control and analysis of HTS runs. An overview of the steps needed for the analysis is shown in Figure 1.*

*The protocol uses a relatively stringent read quality threshold to ensure that called mutations correspond to a true biologically relevant mutation rather than a sequencing error. First, IgBLAST is used to accurately map the beginning*

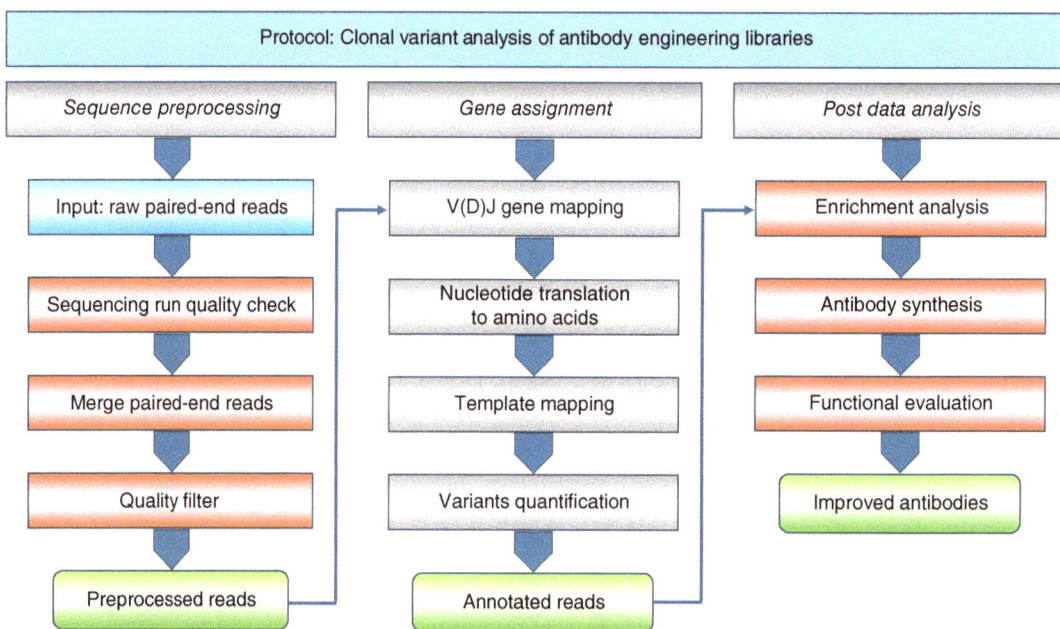

**FIGURE 1.** Identification of improved antibody genes from synthetic antibody library screening data as described in this protocol. (*Left* column) Sequence preprocessing: Illumina MiSeq raw data are preprocessed to remove erroneous reads. (*Middle* column) Gene assignment: Gene assignment includes V(D)J identification, nucleotide translation to amino acids, and variant quantification for all quality reads. (*Right* column) Post data analysis: Improved antibodies are selected from processed high-throughput sequencing data for gene synthesis and recombinant antibody production.

and the end of the template antibody on the reads. The reported nucleotide sequence is then translated into an amino acid sequence, mapped to the template sequence, and stored in a fasta file. The final step counts how many times each mutation observed appears in the data and reports that data to the user. The counts can be used to assess the quality of a mutagenesis library, and to compare samples across enrichment rounds and calculate an enrichment ratio (ER). The ER value allows for simplified tracking of mutation variants across sort conditions for the rapid identification of improved antibody variants according to a sorted library sample. For example, a clone with a high ER in a high-affinity sorted sample would correspond to a high-affinity clone. In our experience, for synthetic libraries, an ER of >5 often represents a clear signal of functional enrichment for any given mutation in a sorted sample.

## Software Preparation and Data Preprocessing

1. Install the required software:
    i. Download and install FastQC, FLASH, and FASTX-toolkit v0.014.
    ii. Add the directories containing the executable files to your PATH variable.
        ```
        export PATH=$PATH:/path/to/FLASH/
        export PATH=$PATH:/path/to/FASTX-toolkit v0.014/
        export PATH=$PATH:/path/to/igblast/bin
        ```
    iii. Next, load the new .profile to your environment by
        ```
        source .profile
        ```
        *The user needs to load the .profile file once. You can check whether you have successfully assigned the path to your executables by calling "flash -help", "fastq_quality_filter -help", or "igblastn -h" from your $HOME directory. The command-line shell interface will return the software's help if you have successfully loaded the new .profile.*

2. Perform quality control of sequence reads:
    i. Use FastQC to check the quality of the sequencing data. Run a postsequencing quality assessment as follows:
        ```
        fastqc exp-name_Read1.fastq.gz exp-name_Read2.fastq.gz
        ```
        *The FastQC program can be used as an optional preanalysis step that provides a visual representation of read quality for the raw HTS data along with run data quality statistics. The software generates a highly useful quality control report that can help users quickly identify any issues in sequencing data quality. FastQC results are output as an html link with a summary of per-base and overall quality score statistics. It also provides a color-coded visual evaluation of the quality score distribution for each cycle across a given HTS run. The html links need to be copied to the user's local computer and opened in a web browser.*

        *Usually, Read1 is of higher quality than Read2. Lower than average read quality may indicate problematic sequencing data, which can sometimes be caused by generally low-quality base-call scores, low library diversity, or certain overrepresented adapter sequences.*

    ii. (Optional, see note) If using data from a paired-end sequencing run, merge Read1 and Read2 using FLASH (Magoč and Salzberg 2011).
        ```
        flash -r 300 -f 440 -s 100 -c runcode_1.fastq exp-runcode_2.fastq
        > exp-name-linked_file.fastq
        ```
        *The flags -r, -f, and -s define the read length, fragment length, and the standard deviation of fragment length, respectively. The flag -c indicates to write the result in the standard output.*

        *FLASH combines the separate Read1 and Read2 from Illumina paired-end sequencing to generate a new, single merged read. When there is a mismatched base between Read1 and Read2, FLASH will use the nucleotide base call with the highest quality across the two reads.*

        *To process multiple data sets in parallel, a file prefix<exp-name_>is used to identify the output data derived from each of the 19 files of this example run. Use the prefixes provided in Table 1 to match the experiment names with the data from annotation_job_names.txt (see below). These file prefixes will ensure the scripts can read all the files from a given experiment. If different prefixes are used, the file annotation_job_names.txt should be changed accordingly.*

        *Paired-end sequencing is necessary to sequence complete ~440-bp heavy chain antibody variable regions when using the Illumina platform. Long-read sequencing methods such as Pacific Biosciences are also sometimes used, in which case a paired-end read-merging step is not required.*

        *Omit the FLASH pairing step when single-read sequence data are used.*

Chapter 10

iii. Remove low-quality reads using fastq_quality_filter:

```
fastq_quality_filter -Q33 -q 30 -p 90 -i exp-name_linked_file
.fastq -o exp-name_linked_q30p90.fastq
```

*Parameters: [-i] FAST[A/Q] input file; [-o] FAST[A/Q] output file; [-q N] minimum quality score to keep; [-p N] minimum percent of bases that must have the requested quality score; [-Q 33] option specifies the format of quality scores.*

*The user may adjust their quality-filtering thresholds according to the quality of the input raw reads. We recommend filtering at least the 5% lowest-quality reads from a data set. If >25% of sequences in a data set are being removed in quality filtering when using the above parameters, see Troubleshooting.*

## Gene Assignment and Translation to Amino Acid Sequences

3. Convert files to fasta format using fastq_to_fasta:

   ```
   fastq_to_fasta -Q33 -i exp-name_linked_q30p90.fastq -o exp-name_linked_q30p90.fasta
   ```

4. Run IgBLAST.

   i. For the sample data provided, it is advisable to split the exp-name_linked_q30p90.fastq file into smaller pieces. In our hands, a file size of 200,000 reads per file is an appropriate balance between speed and number of cluster jobs required. To split a file, run the following command:

      ```
      split -l 400000 exp-name_linked_q30p90.fasta exp-name_linked_q30p90.fasta
      ```

      *This command will create a number of smaller files with an alphabetical suffix, each with 200,000 or fewer sequences.*

   ii. Next, assemble the commands to run IgBLAST on all file chunks by using the following command:

       ```
       ls | grep exp-name_linked_q30p90.fastq_ | awk '{print
       "igblastn -germline_db_V path/to/your/Mouse_V.fasta -germ
       line_db_D path/to/your/Mouse_D.fasta -germline_db_J path/
       to/your/Mouse_J.fasta -outfmt 19 -auxiliary_data / path/to/
       your/optional_file/mouse_gl.aux -query " $1 " > "$1"_igblast"}'>
       igblast_commands
       ```

       *Note that the database for V, D, and J germline sequences from mouse is provided, along with the necessary auxiliary files, but it is necessary to point out their exact location within your file system. The example data are derived from a mouse antibody; germline databases for other organisms can also be used and downloaded from the IgBLAST repository.*

   iii. Next, use the file igblast_commands to send the list of commands to the HPC cluster for processing. The file contains one IgBLAST processing command on each line:

        ```
        bash parallel_jobsubmit_v2-mod_sbatch.sh igblast_commands
        ```

        *The script parallel_jobsubmit_v2-mod_sbatch.sh will read each line of the igblast_commands and send it to the cluster every 2 sec using the blank_jobsubmit template.*

   iv. Once the run is completed, concatenate back all the IgBLAST output files generated:

       ```
       cat exp-name*igblast > exp-name_igblast_all.txt
       ```

## Post Data Analysis: Mapping Antibody Sequences Calculating Enrichment Ratios

5. (Optional) If analyzing a different data set than the data provided in this protocol, create a new FASTA file that contains the template amino acid sequence used to originally construct the synthetic antibody library. Name it query_aa.fasta:

   ```
   cat your_template_aa_sequence > query_aa.fasta
   ```

   *The query_aa.fasta file provided in the GitHub repository contains the amino acid sequence for vFP16.02.*

6. Determine the raw amino acid mutation counts for each sequence. To do this, execute the script Miseq_analysis_v2.py as:

    ```
    python Miseq_analysis_V2.py exp-name_igblast_all.txt 118
    query_aa.fasta
    ```

    *The arguments of this script are the name of the IgBLAST output file, the number of amino acid residues in the template (in this case, 118 amino acids), and the template amino acid sequence.*

    *This script extracts amino acid sequences from the expt-name_igblast.txt file to a separate file name, VDJaa.txt_ff, which is necessary to divide the data into smaller files for processing using the jobsubmit .py script in a parallelized computational cluster. Miseq_analysis_v2.py extracts sequences constituting the entire length of the reported template sequence. Compilation and division of the data into dozens or hundreds of smaller files permits parallel analysis in hours (as opposed to weeks) using a high-throughput compute cluster.*

    *The Miseq_analysis_v2.py script automatically generates another shell script named \*sort_blast as an intermediate file to prepare for parallel analysis on the cluster. These scripts will be run in the next step.*

7. Submit the exp-name_VDJaa.txt_ff-sort_blast script as a sbatch job to the computer cluster. This script will generate a file (*user_out-AA) containing the raw count of single amino acid mutations in the sorted library:

    ```
    sbatch exp-name_VDJaa.txt_ff-sort_blast
    ```

8. Execute script aa_comparison_v3.py as shown below, to generate a file named exp-name_result_aa-comp.txt

    ```
    python aa_comparison_V3.py exp-name_igblast_all.txt 118
    ```

    *The arguments for this script are the name of the IgBLAST output file and the length of the antibody amino acid sequence (for the example data provided, 118 amino acids).*

9. Execute the script annotation_script_v1.0.sh as the following:

    ```
    bash annotation_script_v1.0.sh annotation_job_names.txt
    ```

    *The text file "annotation_job_names.txt" includes two tab-separated columns. The first tab-separated field contains the sequence file prefix, and the second one contains the file generated in the previous step. For the example files provided for this protocol, the file names included are shown below. Each prefix indicates the round of sorting (e.g., R1, R2, R3, or the VL_pos [presort]), and the affinity gate used for sorting (low, medium, or high), which is used to support clonal ER tracking across library sort conditions. This list of files and descriptions is used by annotation_script_v1.0.sh to load the files:*

    VL_pos VL_pos_result_aa-comp.txt

    R1_v1_Low R1_v1_Low_result_aa-comp.txt

    R1_v1_Med R1_v1_Med_result_aa-comp.txt

    R1_v1_high R1_v1_High_result_aa-comp.txt

    R1_v2_Low R1_v2_Low_result_aa-comp.txt

    R1_v2_Med R1_v2_Med_result_aa-comp.txt

    R1_v2_high R1_v2_High_result_aa-comp.txt

    R2_v1_Low R2_v1_Low_result_aa-comp.txt

    R2_v1_Med R2_v1_Med_result_aa-comp.txt

    R2_v1_high R2_v1_High_result_aa-comp.txt

    R2_v2_Low R2_v2_Low_result_aa-comp.txt

    R2_v2_Med R2_v2_Med_result_aa-comp.txt

    R2_v2_high R2_v2_High_result_aa-comp.txt

    R3_v1_Low R3_v1_Low_result_aa-comp.txt

    R3_v1_Med R3_v1_Med_result_aa-comp.txt

    R3_v1_high R3_v1_High_result_aa-comp.txt

    R3_v2_Low R3_v2_Low_result_aa-comp.txt

    R3_v2_Med R3_v2_Med_result_aa-comp.txt

    R3_v2_high R3_v2_High_result_aa-comp.txt

    *The script will generate Exp_name_report.txt, which includes amino acid sequences for each variant detected, along with frequencies and sequence enrichment ratio (ER) data across samples.*

Chapter 10

## Mining Clones from Enrichment Ratio Data

*Improved antibody variants can be selected and discovered by tracking their ER. We often import the database into Excel to examine the ER for each individual antibody sequence, which allows functional interpretation of each clone based on enrichment across different sort conditions. In the sequence data provided, ER is a proxy of binding to the HIV Env trimer. A high ER for a certain clone after sorting with the HIV Env trimer would indicate that the clone was enriched in the sorting operation and is therefore specific to HIV-1 Env. The thresholds for considering a variant as functionally improved vary according to the display system and the antigen probe(s) used.*

10. Open a new Excel sheet and import the Exp_name_report.txt file generated in Step 9. Select "Text file," "Delimited," "Delimiters (Tab)," and "General column data format." Then proceed as follows:

    i. Select the third row, which is labeled "Data_type," and choose sort and filter. Then, apply filters to all columns. Select the drop-down menu of the filters located at the third row and, under filter, select "Greater Than" and enter "1."

    *We often use data filters to exclude variant sequences with only single read count in the reference library, due to high probability of sequencing errors for single-read HTS sequences.*

    ii. Select the filter of the last antigen-specific HTS round data. Then, from the drop-down menu of the filters, select "Greater Than" and enter "5" to exclude all variants with an ER less than 5 and eliminate low-prevalence clones from late-stage screening libraries. Also, to display the data based on enrichment, select "Descending."

    iii. Synthesize genes for the selected variants to confirm functional improvement using wet-lab binding assays.

    *We normally use flow cytometry or ELISA to quickly measure improved binding performance.*

## TROUBLESHOOTING

**Problem (Step 2):** More than 25% of sequences in a data set are being removed in quality-filtering step.
**Solution:** Check the specific HTS run data to ensure that quality metrics are consistent with expectations for instrument performance (i.e., if the run had low quality, rerun the HTS analysis). If the run quality metrics were consistent with expectations, a researcher may alter their specified quality threshold to avoid removing excessive amounts of data and ensure good coverage of their antibody libraries. If the run quality is poor, the HTS run should be repeated, often by optimizing the amount of input DNA (and/or adjusting the percentage of PhiX sequencing control for Illumina runs) to improve sequence quality.

## ACKNOWLEDGMENTS

We thank Pedro Seber e Silva for helpful review and feedback on the protocol and associated code. This work was supported by National Institutes of Health grants DP5OD023118, R21AI166396, R21AI143407, U01AI169587, and R01AI181684; The Bill and Melinda Gates Foundation; and the Mark and Lisa Schwartz AI/ML/Immunology Initiative.

## REFERENCES

Madan B, Zhang B, Xu K, Chao CW, O'Dell S, Wolfe JR, Chuang GY, Fahad AS, Geng H, Kong R, et al. 2021. Mutational fitness landscapes reveal genetic and structural improvement pathways for a vaccine-elicited HIV-1 broadly neutralizing antibody. *Proc Natl Acad Sci* **118**: e2011653118. doi:10.1073/pnas.2011653118

Magoč T, Salzberg SL. 2011. FLASH: fast length adjustment of short reads to improve genome assemblies. *Bioinformatics* **27**: 2957–2963. doi:10.1093/bioinformatics/btr507

Ye J, Ma N, Madden TL, Ostell JM. 2013. IgBLAST: an immunoglobulin variable domain sequence analysis tool. *Nucleic Acids Res* **41**: W34–W40. doi:10.1093/nar/gkt382

# Protocol 2

# Antibody Data Analysis from Diverse Immune Libraries

Ahmed S. Fahad,[1,2,3] Matías F. Gutiérrez-Gonzalez,[1,2,3] Bharat Madan,[1] and Brandon J. DeKosky[1,2,4]

[1]The Ragon Institute of Massachusetts General Hospital, Massachusetts Institute of Technology, and Harvard University, Cambridge, Massachusetts 02139, USA; [2]Department of Chemical Engineering, Massachusetts Institute of Technology, Cambridge, Massachusetts 02139, USA

Antibody functional screening studies and next-generation sequencing require careful processing and interpretation of sequence data for optimal results. Here, we provide a detailed protocol for the functional analysis of antibody gene data, including antibody repertoire quantification and functional mapping of high-throughput screening data based on enrichment ratio values, which are a simple way to determine the enrichment of each sequenced antibody after sorting a display library against desired antigens. This protocol enables a user to apply a set of simple yet powerful bioinformatic tools for high-throughput analysis and interpretation of antibody data that is especially well suited for display library screening and for antibody discovery applications.

## MATERIALS

### Reagents

Bioinformatic tools:
- FastQC, available at https://www.bioinformatics.babraham.ac.uk/projects/fastqc/
- FASTX-Toolkit v0.0.14, available at http://hannonlab.cshl.edu/fastx_toolkit/download.html
- FLASH v1.2.11, available at https://sourceforge.net/projects/flashpage/files/latest/download (optional; see Step 2)
- IgBLAST v1.18, available at https://ftp.ncbi.nih.gov/blast/executables/igblast/release/ (Ye et al. 2013)
- Perl v5.30, available at https://www.perl.org/get.html
- Python 3 or above (https://www.python.org/)

Custom scripts and files, available at the GitHub repository associated with this protocol (visit https://github.com/dekoskylab/CSHL_protocols and then go to Scripts/Scripts for Antibody Data Analysis from Diverse Immune Libraries/):
- annotation_script_v1.0.sh; this script pairs the data analyzed from different rounds of antibody sorting and calculates the enrichment ratio for each clone detected in both sorting runs
- annotation_job_names.txt; contains the file names of the sorting runs analyzed, and is used by annotation_script_v1.0.sh to read the data
- igblast_processing.sh; converts the IgBLAST output into fasta format

---

[3]These authors contributed equally to this work.
[4]Correspondence: dekosky@mit.edu

© 2026 Cold Spring Harbor Laboratory Press
Cite this protocol as *Cold Spring Harb Protoc*; doi:10.1101/pdb.prot108627

TABLE 1. Sample data used in this protocol

| Accession code | Library name | Sorting round | Antigen probe | Estimated affinity by FACS | Prefix |
|---|---|---|---|---|---|
| SRR13729764 | 591_K_VL+_sorted | N/A | N/A | N/A | VL_pos |
| SRR13729700 | 591_K_5455_rd3_High_sorted | 1 | ZIKV VLP | High | R3_High |

Sample data can be downloaded from the Sequence Read Archive (SRA) database from NCBI. Data can be either downloaded manually by searching for the accession codes or using the NCBI SRA Toolkit. See https://www.ncbi.nlm.nih.gov/sra for further information.

parallel_jobsubmit_v2-mod_sbatch.sh; reads individual data related to each job and sends them to the cluster every 2 sec using the blank_jobsubmit template

Database of human V, D, and J germline genes; this database is included in the associated Github repository, in the IgBLAST database folder (visit https://github.com/dekoskylab/CSHL_protocols/ and then go to Scripts/Scripts for Antibody Data Analysis from Diverse Immune Libraries/ncbi-igblast-1.18.0/database/)

High-throughput sequencing (HTS) data

*Data can be generated using commonly used HTS technologies for antibody analysis, such as Illumina MiSeq or Pacific Biosciences Sequel II. To analyze data with diversity across the antibody variable region, a technology that sequences the full variable domain must be used. MiSeq 2 × 300 bp is often used because it is long enough to cover the ~440 bp of the heavy chain at high quality and a comparatively lower cost. In the example data provided here, we use 2 × 300 bp MiSeq data published previously (Fahad et al. 2021). Data consist of human immune libraries from convalescent human donors after symptomatic Zika virus infection. Immune libraries were expressed in a yeast Fab display format and sorted for light chain surface display (VL$^+$ or VL_pos), thereby selecting the population of yeast cells with complete Fab molecules expressed on the yeast surface. Libraries were next screened against Zika virus-like-particles, and enriched in three rounds of sorting. The sample data provided consist of heavy chain variable gene sequences from immune libraries displayed on yeast and selected against Zika virus-like particles. See Table 1 for accession code information.*

## Equipment

High-performance computer or cluster with a UNIX-like environment

*Analysis of high-throughput sequencing (HTS) data uses more resources than those available in a typical laptop or desktop machine. Typical data sets range from a few dozen megabytes of data to gigabytes per sample. As such, a high-performance computing (HPC) platform is often necessary to analyze the data in a reasonable amount of time.*

Slurm Workload Manager (https://slurm.schedmd.com/documentation.html); other workload managers can be used, but the code provided will have to be modified accordingly

## METHOD

*An overview of the method is shown in Figure 1. Briefly, the user will perform a quality control step on the obtained reads using FastQC. The aim of this step is to ensure that the sequence quality is enough to proceed with the analysis. Next, paired-end reads will be merged, and low-quality sequences will be removed. The obtained reads are mapped to germline antibody sequences using IgBLAST. Next, the clonal prevalence of each antibody CDR-H3 is compared across sorting rounds to detect which clones are enriched, and thus potentially represent antigen-binding signals in sort gates. The clonal prevalence is tracked by an enrichment ratio (ER) value, which is a simple way to determine the enrichment of a particular antibody in a given sorted sample. These high-throughput analysis techniques thus allow facile tracking of antibody specificity in a library by calculating the ER for each clone in the library after sorting a display library against desired antigens.*

1. Install the required software:

    i. Download and install FastQC, FLASH, and FASTX-toolkit v0.014

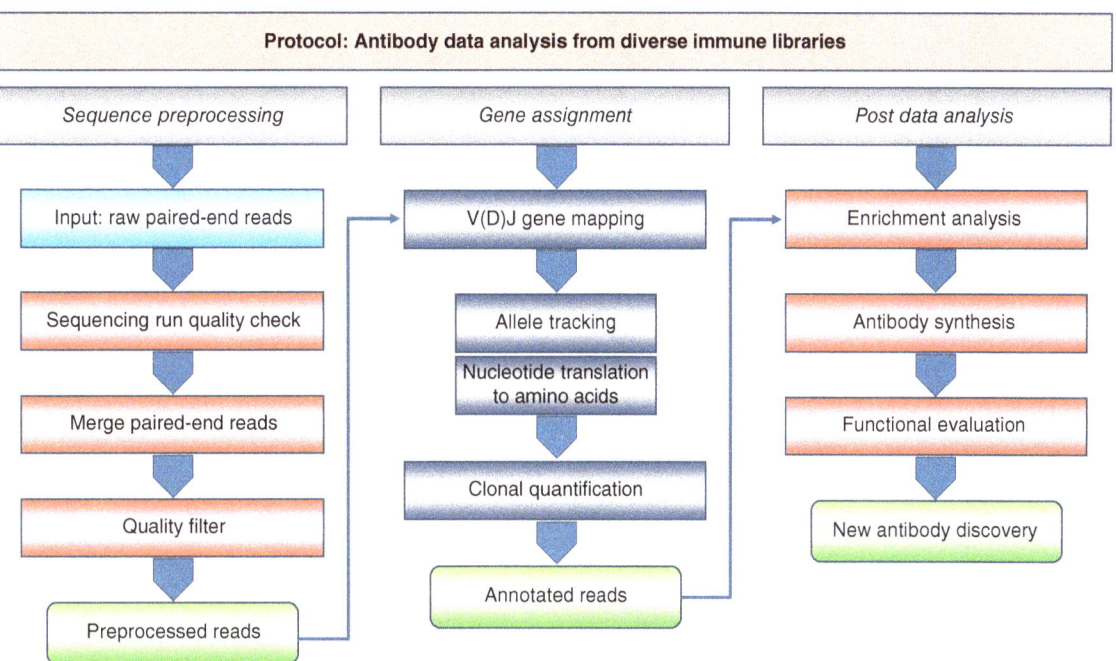

FIGURE 1. Antigen-specific antibody identification from diverse immune library data as described in this protocol. (*Left* column) Preprocessing: Illumina MiSeq raw data are preprocessed to remove erroneous reads. (*Middle* column) Gene assignment: Gene assignment includes V(D)J identification, nucleotide translation to amino acids, and clonal quantification. (*Right* column) Post data analysis: New antibody candidates are selected for gene synthesis and recombinant antibody production to enable functional evaluation.

    ii. Add the directories containing the executable files to your PATH variable. First, add the following lines to your .profile file:

```
export PATH=$PATH:/path/to/FLASH/
export PATH=$PATH:/path/to/FASTX-toolkit v0.014/
export PATH=$PATH:/path/to/igblast/bin
```

    iii. Next, load the new .profile to your environment by:

```
source .profile
```

*The user needs to load the .profile file once. You can check whether you have successfully assigned the path to your executables by calling "flash -help", "fastq_quality_filter -help", or "igblastn -h" from your $HOME directory. The command-line shell interface will return the software's help if you have successfully loaded the new .profile.*

2. Perform quality control of sequence reads:

    i. Use FastQC to check the quality of the sequencing run. Run the postsequencing quality assessment as follows:

```
fastqc exp-name_Read1.fastq.gz exp-name_Read2.fastq.gz
```

*The FastQC program can be used as an optional preanalysis step that provides a visual representation of read quality for the raw HTS along with run data quality statistics. The software generates a highly useful quality control report that can help users quickly identify any issues in sequencing data quality. FastQC results are output as an html file with a summary of per-base and overall quality score statistics. It also provides a color-coded visual evaluation of the quality score distribution for each cycle across a given HTS run. The html files need to be copied to the user's local computer and opened in a web browser.*

*Usually, Read1 tends to have higher quality than Read2. Lower than average read quality may indicate problematic sequencing data, such as a sequencing run with low-quality base-call scores, less diverse libraries, or overrepresented adapter sequences.*

    ii. (Optional) Merge Read1 and Read2 from a paired-end sequencing run using FLASH (Magoč and Salzberg 2011).

```
flash -r 300 -f 440 -s 100 -c exp-name_read1.fastq exp-name_
read2.fastq>exp-name-linked_file.fastq
```

The flags -r, -f, and -s define the read length, fragment length, and the standard deviation of fragment length, respectively. The flag -c indicates to write the result in the standard output.

*FLASH combines the separate Read1 and Read2 from Illumina paired-end sequencing to generate a new, single merged read. When there is a mismatched base between Read1 and Read2, FLASH will use the nucleotide base call with the highest quality across the two reads.*

*To process multiple data sets in parallel, a file prefix <exp-name_> is used to identify the output data derived from each file. Use the prefixes VL_pos_ and R3_High to match the file names to the annotation_job_names.txt file (see below). These file prefixes will ensure the scripts can read all the files from a given experiment. If different prefixes are used, the file annotation_job_names.txt should be changed accordingly.*

*Paired-end sequencing is needed to sequence the ∼440-bp heavy chain antibody variable regions when using the Illumina platform. Long-read sequencing methods such as Pacific Biosciences are also sometimes used, in which case a paired-end read-merging step is not required.*

*Omit the FLASH pairing step when single-read sequence data are used.*

iii. Remove low-quality reads using fastq_quality_filter:
```
fastq_quality_filter -Q33 -q 30 -p 50 -i exp-name_linked_file
.fastq -o exp-name_linked_q30p50.fastq
```

The flags -q 30 and -p 50 indicate the minimum quality score and the minimum percent of bases to keep, respectively.

*The user may adjust their quality filtering thresholds according to the quality of the input raw reads. We recommend filtering at least the lowest-quality 5% of reads from a data set. If >25% of sequences in a data set are being removed in quality filtering, see Troubleshooting.*

## Gene Annotation and Clonal Quantification

3. Convert files to fasta format using fastq_to_fasta:
   ```
   fastq_to_fasta -Q33 -i exp-name_linked_q30p50.fastq -o exp-
   name_linked_q30p50.fasta
   ```

4. Run IgBLAST.

   i. Given the number of reads obtained in the sample data provided, it is advisable to split the exp-name_linked_q30p50.fastq file. In our hands, a file size of 200,000 reads per file is an appropriate balance between speed and number of cluster jobs required. To do this, run the following command:
   ```
   split -l 400000 exp-name_linked_q30p50.fasta exp-name_lin-
   ked_q30p50.fasta
   ```
   *This command will create a number of files with an alphabetical suffix, each with 200,000 sequences.*

   ii. Next, assemble the commands to run IgBLAST on all file chunks by using the following command:
   ```
   ls | grep exp-name_linked_q30p50.fastq_ | awk '{print "igblastn
   -germline_db_V path/to/your/Human_V.fasta -germline_db_D path/
   to/your/Human_D.fasta  -germline_db_J  path/to/your/Human_
   J.fasta -outfmt 19 -auxiliary_data / path/to/your/optional_
   file/human_gl.aux -query " $1 "> "$1"_igblast"}'>igblast_
   commands
   ```
   *Note that the database for V, D, and J germline sequences from human is provided, along with the associated auxiliary files, but it is necessary to point out their exact location within your file system.*

iii. Next, use the file igblast_commands to send the list of commands to the HPC cluster for processing. The file contains one IgBLAST processing command on each line:
```
bash parallel_jobsubmit_v2-mod_sbatch.sh igblast_commands
```
*If IgBLAST is taking too long or jobs are terminated after a time limit, see Troubleshooting.*

iv. Once the run is completed, concatenate back all the IgBLAST output files generated:
```
cat exp-name*igblast>exp-name_igblast_all.txt
```

v. Aggregate the data at a clonal level. Although there are different ways to define what a clone is, in most cases a clone is defined by tracking the nucleotide sequence of the CDR-H3. As such, aggregate and count each line with the same CDR-H3. To do this, run the igblast_processing.sh script:
```
bash igblast_processing.sh exp-name_igblast_all.txt
```
*This command generates a file named exp-name_igblast_all.txt_processed, which contains the clonal counts and nucleotide and amino acid sequences of CDR-H3s.*

## Enrichment Ratio Calculation

5. Measure antibody clonal performance across sorting rounds using the script annotation_script_v1.0.sh, which reads the IgBLAST outputs and calculates the prevalence of each antibody clone. The text file annotation_job_names.txt contains the file names and descriptions in a tab-separated format, as follows:

   | | |
   |---|---|
   | VL+_Library | VL_pos_igblast_all.txt_processed |
   | R3_High | R3_High_igblast_all.txt_processed |

   *The script annotation_script_v1.0.sh calculates the prevalence of each clone by dividing the number of reads for that clone by the sum of all quality reads in a given sample. Next, the script compiles clonal data based on exact CDR-H3 nucleotide sequence and calculates the enrichment ratio using the following formula:*

   $$ER = \frac{\text{prevalence of a clone in each round of sorting}}{\text{prevalence of the clone in input library}}.$$

   *In the provided example, ER is calculated from the prevalence of antibodies clones as defined by their CDR-H3 genes. Run the annotation script as follows:*
   ```
   bash annotation_script_v1.0.sh annotation_job_names.txt
   ```

   *The output file now contains the CDR-H3 sequence of each clone, followed by the number of reads and frequencies for each sorting round. As the final step, use the above formula to calculate the enrichment ratio of each clone in the prevalence data output files for quantitative ER analysis. If you receive a "There is no such a file or directory" error after running the command, see Troubleshooting.*

## TROUBLESHOOTING

***Problem (Step 2):*** More than 25% of sequences in a data set are being removed in quality filtering.
***Solution:*** Check the sequencing run parameters to ensure good run quality. Low read quality can have several possible causes, including barcode overlap, high or low input DNA concentrations, machine or kit/reagent errors, or low-quality input DNA. If helpful, consult with the HTS instrument supplier or another expert in HTS instrument operation to improve run performance.

***Problem (Step 4):*** IgBLAST is taking too long or jobs are terminated after a time limit.
***Solution:*** Split submission into smaller files with the following command:
```
split -l 100000 exp-name_linked_q30p50.fasta exp-name_linked_q30p50.fasta
```

*Problem (Step 5):* After running the command you receive the following error: "There is no such a file or directory."
*Solution:* Ensure that the annotation_job_names.txt file is tab-separated, and that there are no white spaces between filenames and file descriptions.

## DISCUSSION

In our experience with yeast display (Fahad et al. 2021; Madan et al. 2021; Banach et al. 2023), we find that an enrichment ratio (ER) of 5 or more after two rounds of yeast display has a high probability (~95%) of identifying antigen-specific clones. We have found that the magnitude of ER is weakly correlated with the degree of performance improvement (e.g., a clone with an ER of 250 might only be marginally higher performance than a clone with an ER of 7). These quantitative efforts enable mapping and recovery of antibodies that are highly enriched across rounds but remain at a relatively low prevalence in the sorted libraries, and therefore are often missed by standard techniques such as colony picking. Experienced members of our laboratory often manipulate the compiled ER data for antigen-specific and affinity-specific clonal mapping in command-line formats. A visual, Excel-based analysis is also recommended for exploring the compiled data from yeast display screening studies, particularly for new users. Visual spreadsheet-based analyses are easily achieved by exporting processed data into a tab-separated format to explore antibody data with commonly available productivity software.

## ACKNOWLEDGMENTS

We thank Pedro Seber e Silva and Sunwoo Lee for helpful review and feedback on the protocol and associated code. This work was supported by National Institutes of Health grants DP5OD023118, R21AI166396, R21AI143407, U01AI169587, and R01AI181684; The Bill and Melinda Gates Foundation; and the Mark and Lisa Schwartz AI/ML/Immunology Initiative.

## REFERENCES

Banach BB, Pletnev S, Olia AS, Xu K, Zhang B, Rawi R, Bylund T, Doria-Rose NA, Nguyen TD, Fahad AS, et al. 2023. Antibody-directed evolution reveals a mechanism for enhanced neutralization at the HIV-1 fusion peptide site. *Nat Commun* 14: 7593. doi:10.1038/s41467-023-42098-5

Fahad AS, Timm MR, Madan B, Burgomaster KE, Dowd KA, Normandin E, Gutiérrez-González MF, Pennington JM, De Souza MO, Henry AR, et al. 2021. Functional profiling of antibody immune repertoires in convalescent Zika virus disease patients. *Front Immunol* 12: 615102. doi:10.3389/fimmu.2021.615102

Madan B, Zhang B, Xu K, Chao CW, O'Dell S, Wolfe JR, Chuang G-Y, Fahad AS, Geng H, Kong R, et al. 2021. Mutational fitness landscapes reveal genetic and structural improvement pathways for a vaccine-elicited HIV-1 broadly neutralizing antibody. *Proc Natl Acad Sci* 118: e2011653118. doi:10.1073/pnas.2011653118

Magoč T, Salzberg SL. 2011. FLASH: fast length adjustment of short reads to improve genome assemblies. *Bioinformatics* 27: 2957–2963. doi:10.1093/bioinformatics/btr507

Ye J, Ma N, Madden TL, Ostell JM. 2013. IgBLAST: an immunoglobulin variable domain sequence analysis tool. *Nucleic Acids Res* 41: W34–W40. doi:10.1093/nar/gkt382

Protocol 3

# Clonal Lineage and Gene Diversity Analysis of Paired Antibody Heavy and Light Chains

Ahmed S. Fahad,[1,2,3] Matías F. Gutiérrez-Gonzalez,[1,2,3] Bharat Madan,[1] and Brandon J. DeKosky[1,2,4]

[1]The Ragon Institute of Massachusetts General Hospital, Massachusetts Institute of Technology, and Harvard University, Cambridge, Massachusetts 02139, USA; [2]Department of Chemical Engineering, Massachusetts Institute of Technology, Cambridge, Massachusetts 02139, USA

Antibodies consist of unique variable heavy ($V_H$) and variable light ($V_L$) chains, and both are required to fully characterize an antibody. Methods to detect paired heavy and light chain variable regions ($V_H$:$V_L$) using high-throughput sequencing (HTS) have recently enabled large-scale analysis of complete functional antibody responses. Here, we describe an HTS computational pipeline to analyze paired $V_H$:$V_L$ antibody sequences and obtain a comprehensive profile of immune diversity landscapes, including gene usage, antibody isotypes, and clonal lineage analysis. This protocol uses Illumina MiSeq $2 \times 300$-bp sequencing data and integrates with several different computational tools for flexible analyses of paired $V_H$:$V_L$ gene repertoire data to enable efficient antibody discovery.

## MATERIALS

### Reagents

Custom scripts and files, available at https://github.com/dekoskylab/CSHL_protocols, under Scripts/ Scripts for Clonal Lineage and Gene Diversity Analysis of Paired Antibody Heavy and Light Chains:
blank_jobsubmit and large_jobsubmit (for the submission of parallel jobs on a compute cluster managed by Slurm); parameters in these files can be modified to fit the user's compute cluster environment.
CDR3motif_search_part1_v2.1.sh
CDR3motif_search_part2_v2.1.sh
CDR3motif_search_analysis_v3.2.sh
check_hold-mod_sbatch.sh
compile_igblast_isotype.sh
human_barcodes.txt; contains human constant region primer sequences for isotype identification
parallel_jobsubmit_v2-mod_sbatch.sh
part2_VHVL_commands-mod_sbatch.sh
qualityfilter_script_human_v1.0.sh
translate.pl
vpairhumanalysis-igblast-isotype_BDmod.pl

---

[3]These authors contributed equally to this work
[4]Correspondence: dekosky@mit.edu

© 2026 Cold Spring Harbor Laboratory Press
Cite this protocol as Cold Spring Harb Protoc; doi:10.1101/pdb.prot108628

Chapter 10

TABLE 1. Sample data used in this protocol

| Accession code | Library name |
| --- | --- |
| SRX10117706 | Donor1_paired_VHVL_GAMKL_PB_sequences |
| SRX10117705 | Donor1_paired_VHVL_GAKL_PB_sequences |
| SRX10117748 | Donor2_paired_VHVL_Standard_GAMKL_PB_sequences |
| SRX10117747 | Donor2_paired_VHVL_Standard_GAKL_PB_sequences |

Sample data can be downloaded from the Sequence Read Archive (SRA) database from NCBI. Data can be downloaded manually by searching for the accession codes or using the NCBI SRA Toolkit. See https://www.ncbi.nlm.nih.gov/sra for further information.

High-throughput sequencing (HTS) data

*HTS data can be generated using common technologies for antibody analysis, such as Illumina MiSeq or Pacific Biosciences Sequel II. In the example described here, we use 2 × 300 bp MiSeq data obtained from Fahad et al. (2021) that consists of a natural paired heavy and light chain ($V_H$:$V_L$) immune library. The immune libraries were obtained from convalescent human donors after symptomatic Zika virus infection. Sample data from Fahad et al. (2021) are available in the NCBI Sequence Read Archive (Table 1).*

Software:

FastQC, available at https://www.bioinformatics.babraham.ac.uk/projects/fastqc/
FASTX-Toolkit v0.0.14, available at http://hannonlab.cshl.edu/fastx_toolkit/download.html
IgBLAST v1.18, available at https://ftp.ncbi.nih.gov/blast/executables/igblast/release/
  (Ye et al. 2013)
Perl v5.30, available at https://www.perl.org/get.html
USEARCH v5, available at https://www.drive5.com/usearch/ (Edgar 2010)

## Equipment

High-performance computer or cluster with a UNIX-like environment.

*We strongly recommend working under a UNIX-based operating system, such as Ubuntu or Mac OS X. If the user uses a Windows operating system, then we recommend installing Ubuntu or Cygwin; both are free and open-source software. Ubuntu is available at https://ubuntu.com/desktop/wsl and Cygwin is available at http://www.cygwin.com/.*

## METHOD

This protocol provides a flexible pipeline to process, annotate, and analyze paired antibody $V_H$:$V_L$ genes in a high-throughput manner (Fig. 1). The pipeline takes MiSeq Illumina 2 × 300-bp results, i.e., Read1 and Read2 fastq files, as input data. Computational analysis is usually performed at the command line, and the core of the $V_H$:$V_L$ bioinformatic process includes four major operations:

(A) Quality filtering to remove low-quality reads, performed by FASTX-Toolkit command written inside qualityfilter_script_human_v1.0.sh script.

(B) Read alignment to a reference germline gene database, sequence annotation, and CDR3 identification. This step is performed by the IgBLAST command inside vpairhumanalysis-igblast-isotype_BDmod.pl script.

(C) Grouping and clustering similar CDR-H3 sequences to estimate library diversity. This is performed by the USEARCH command within CDR3motif_search_analysis_v3.2.sh script.

(D) Translation of nucleotide sequences to amino acid sequences. Translation is performed by the translate.pl script, which is within CDR3motif_search_part2_v2.1.sh script. The final unique CDR-H3:L3 results are written into a text file named "<exp-name>_unique_final.txt."

The quality filtering operation (A) is described in the "Data Preprocessing" steps below, and the other three major operations (B–D) are performed subsequently in an automated manner. Begin with an initial quality assessment of sequence data before initiating the analysis.

# High-Throughput Sequencing in Antibody Discovery

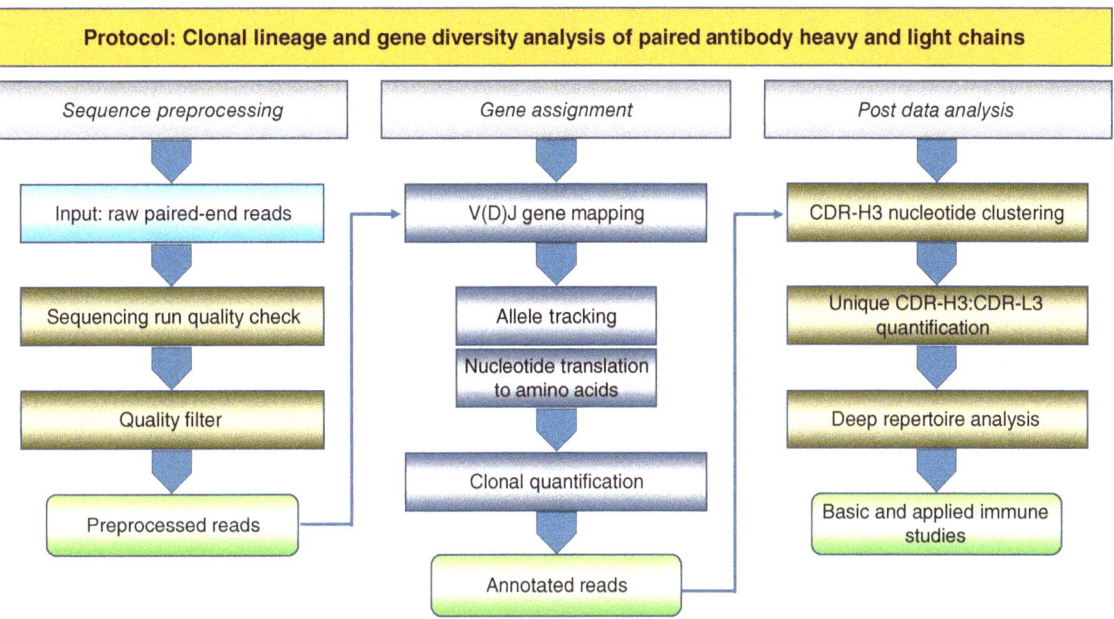

FIGURE 1. Paired $V_H$:$V_L$ repertoire lineage analysis from natively paired antibody immune libraries. (*Left* column) Sequence preprocessing: Illumina MiSeq raw data are preprocessed to remove erroneous reads on either heavy or light chains. (*Middle* column) Gene assignment: Gene assignment includes V(D)J identification, nucleotide translation to amino acids, and clonal variant quantification for all quality reads. (*Right* column) Post data analysis: CDR-H3 clustering and quantification are performed to enable deep CDR-H3:CDR-L3 repertoire analysis, including statistical and biophysical metrics related to antibody repertoire composition.

1. Perform postsequencing quality assessment:

    i. Create a new directory at a desired location for the project:
       ```
       mkdir project1
       ```
       *After logging into a Unix-based system, the user may create a new directory and copy MiSeq raw fastq files, all the above scripts, and any accessory files into it.*

    ii. Run the FastQC command to check quality score of a few of the sequence files in your HTS run:
       ```
       fastqc exp-name_Read1.fastq.gz exp-name_Read2.fastq.gz
       ```
       *The FastQC program can be used as an optional preanalysis step that provides a visual representation of read quality for the raw HTS data along with run data quality statistics. The software generates a quality control report that can help users quickly identify any issues in sequencing data quality. FastQC results are output as an html link with a summary of per-base and overall quality score statistics. FastQC also provides a color-coded visual evaluation of the quality score distribution for each cycle across a given MiSeq run. The html files can be copied to the user's local computer and opened in a web browser.*

## Data Preprocessing

*Users can either unzip the compressed sequencing files individually (Step 2) or create a bulk set of unzip job commands for submission to a cluster (Step 3).*

2. Unzip the raw fastq MiSeq compressed files individually using the following command:
   ```
   gunzip <exp-name>_read1.fastq.gz <exp-name>_read2.fastq.gz
   ```
   *Depending on the size of the files, the command prompt will be on hold for a few minutes before it returns when running this command. Fastq files will then be unzipped and ready for downstream processing.*

3. Alternatively, create a bulk set of unzip job commands for submission to a cluster:

    i. Generate an unzip command for each compressed fastq file in the current folder, and store it in a text file.
       ```
       ls | grep fastq.gz | awk '{print "gunzip " $1}' > for_gunzip.txt
       ```

Chapter 10

   ii. Run the bulk unzip job as follows:

   `bash parallel_jobsubmit_v2-mod_sbatch.sh for_gunzip.txt`

   *The parallel_jobsubmit_v2-mod_sbatch.sh is a custom job submitter bash script for parallel submission of jobs to the cluster. The script will submit individual unzipping commands to the user's clustering for faster processing. This step will take few minutes to complete.*

4. Create a text file and name it as "pairlist.txt".

   i. Run this command to create the pairlist.txt file:

   ```
   ls | grep fastq | grep R2 | sed 's/_R2_001.fastq//' | awk '{print $1 "_R1_001.fastq " $1 "_R2_001.fastq " $1 " human_barcodes.txt"}' > pairlist.txt
   ```

   *This file provides the MiSeq data file names, Read1 and Read 2 file names, and the human constant region barcode sequences as an input for the main downstream scripts.*

   ii. Check the human_barcodes.txt file (located in the GitHub repository) by running the following command:

   `cat human_barcodes.txt`

   *The output that should be observed is as follows:*

   *IgG AAAAATGGGCCCTGCGATGGGCCCTTGGTGGAGG*
   *IgM AAAAATGGGCCCTGGGTTGGGGCGGATGCACTCC*
   *IgA AAAAATGGGCCCTGCTTGGGGCTGGTCGGGGATG*
   *IgK AAAAGTGCGGCCGCAGATGGTGCAGCCACAGTTC*
   *IgL AAAAGTGCGGCCGCGAGGGCGGGAACAGAGTGAC*

   *The human_barcodes.txt file is designed to work with the primers used for paired heavy and light chain sequencing during data generation. The examples above are designed for compatibility with paired $V_H$:$V_L$ primers reported previously (DeKosky et al. 2013, 2014; McDaniel et al. 2016). The sequences in the human_barcodes.txt file can be adjusted if different primers are used.*

## Perform Automated $V_H$:$V_L$ Data Analysis

5. Submit the complete $V_H$:$V_L$ pipeline analysis as follows:

   `bash check_hold-mod_sbatch.sh`

   *The check_hold-mod_sbatch.sh coordinates the analysis by performing quality filtering, read alignment, and clustering. The script identifies the input files from the pairlist.txt file.*

## Interpret the Final Results

6. Perform line counts of the original fastq files and quality-filtered results from the outputs of the job in Step 5 ("<exp-name>_q20p50.fasta" files) to calculate the percentage of reads that were removed due to low read quality. Around 5% of reads should be excluded with the provided "-q 20 -p 50" filtering parameters if the MiSeq run quality is sufficient. If lower- or higher-quality scores are desired, change the quality filtering stringency performed in Step 5 by modifying the qualityfilter_script_human_v1.0.sh script:

   ```
   fastq_quality_filter -Q33 -q 20 -p 50 -i $file1 -o $file1"_q20p50.fastq"
   fastq_quality_filter -Q33 -q 20 -p 50 -i $file2 -o $file2"_q20p50.fastq"
   ```

   *Parameters: [-i] FAST[A/Q] input file; [-o] FAST[A/Q] output file; [-q N] minimum quality score to keep; [-p N] minimum percent of bases that must have the requested quality score; [-Q 33] option specifies the format of quality scores.*

7. If desired, change the reference germline gene database that IgBLAST uses to perform read alignments by modifying the [vpairhumanalysis-igblast-isotype_BDmod.pl] script (which is called by the check_hold-mod_sbatch.sh command). The read alignment results are written into the output files "<exp-name>__paired_igblast_isotype_all.txt." Following the IgBLAST instructions, you can include any standard or user-custom germline gene databases, such as

those derived from antibody germline genes deposited at the International Immunogenetics Information System (IMGT) (Lefranc 2014). As one example, download current NCBI germline gene databases (i.e., human_gl_V, mouse_gl_V, etc.) from the NCBI ftp site (https://ftp.ncbi.nih.gov/blast/executables/igblast/release/database/).

```
$igblastpre=dirname(abs_path($0)).'/./igblastn      -germline_db_V
'.dirname(abs_path($0)).'/database/human_gl_V       -germline_db_D
'.dirname(abs_path($0)).'/database/human_gl_D       -germline_db_J
'.dirname(abs_path($0)).'/database/human_gl_J       -auxiliary_data
'.dirname(abs_path($0)).'/optional_file/human_gl.aux    -domain_
system imgt -num_alignments_V 4 -num_alignments_D 4 -num_align-
ments_J 4 -outfmt 3 -query ';
```

*Parameters: [igblastn] IgBLAST command for nucleotide sequences; [database/human] provides human gene database; [optional_file/human_gl.aux] auxiliary database for CDR3/FWR4 reading frame identification; [germline_db_V] variable V region database; [germline_db_J] variable J region database; [germline_db_D] variable D region database; [-outfmt 3] output sequences in a tab-separated format.*

*A user may adjust the IgBLAST command options inside [vpairhumanalysis-igblast-isotype_BDmod.pl script: line 74] according to their desired analysis parameters.*

*Learn more about IgBLAST search database customization at https://www.ncbi.nlm.nih.gov/igblast/.*

8. Change the clustering parameters inside CDR3motif_search_analysis_v3.2.sh [lines 19 and 38] as desired for your analyses. The final operation of the check_hold-mod_sbatch.sh command calls CDR3motif_search_analysis_v3.2.sh to cluster CDR-H3 junction nucleotide sequences using the USEARCH program and write the clustering output to "<exp-name>_unique_final.txt." It runs two clustering commands with different optimized parameters based on sequence length—one command clusters longer CDR-H3 junction sequences, and the second command performs optimized clustering for shorter CDR-H3 junction sequences (<20 nt):

```
usearch -cluster "$EXPTNAME"heavynt_junctions_over1read.fasta -w
4 --maxrejects 0 -usersort --iddef 2 --nofastalign -id 0.96 -minlen
11 -uc "$EXPTNAME" results.uc -seedsout "$EXPTNAME"seeds.uc
```

*Parameters: [-cluster] Usearch clustering command; [-w] the word length used to seed new alignments (shorter word lengths provide a more thorough search, but require more computational resources than longer word lengths); [--maxrejects] search for a match terminates after specified number of failed attempts to match a target sequence (default value is 32; 0 indicates that this function is turned off); [-usersort] this option is used for clustering only and allows the user to define a sort order for the input sequences; [--iddef] this option defines the sequence identity metric to be used; [--nofastalign] the heuristics algorithm is disabled for maximum search sensitivity but requires more computational resources; [-id] identity threshold (this parameter is specified as a fractional identity that ranges from 0.0 to 1.0); [-minlen] minimum sequence length; [-uc] produce a tab-separated file format with clustering information; [-seedsout] output cluster seed sequences into a fasta file; [--queryalnfract] fraction of the query sequence that is aligned to the target sequence, ranges from 0.0 to 1.0 (0% to 100%); [--targetalnfract] fraction of the target sequence that is aligned to the query sequence, ranges from 0.0 to 1.0 (0% to 100%).*

*The automated steps for $V_H:V_L$ analysis are the most resource-intensive, and most processors will require between 4 and 6 h to complete the job. After analysis, the scripts output a text file named "<exp-name>_unique_final.txt" that contains a list of unique CDR-H3:CDR-L3 pairs with two or more reads for each observed $V_H:V_L$ cluster in the output data set. The script also provides all (i.e., preclustered) CDR-H3: CDR-L3 paired reads with two or more reads in the "<exp-name>_CDR3_nt_over1read.txt" files. We have observed some single-read $V_H:V_L$ sequences to be functional in antibody display and screening studies; however, we find that many of the single-read $V_H:V_L$ sequences are likely sequence errors and therefore we do not include single-read clusters in most immune repertoire statistical analyses.*

## DISCUSSION

This protocol enables detailed analyses of native antibody $V_H:V_L$ repertoire sequences, including paired CDR-H3 and CDR-L3 nucleotide and amino acid sequences, gene usage and isotype information, and CDR-H3 clustering operations to estimate natural antibody repertoire diversity. These data can be used to study the biological features of human antibody repertoires, including for peripheral and naive B-cell subsets, and to provide insight regarding antibody selection and maturation characteristics for paired antibody heavy and light chains in unprecedented depth (DeKosky et al. 2016). Natively paired $V_H$:$V_L$ information is increasingly important to examine levels of B-cell somatic hypermutation (SHM) in response to natural infection (Fahad et al. 2021; Pan et al. 2023), which has major implications for vaccine design against difficult targets like HIV, where a high level of SHM is required for antibody-mediated protection (Kwong et al. 2017; Duan et al. 2018). Short-read Illumina analyses can often be supplemented with complete $V_H$:$V_L$ sequencing in parallel using longer-read technologies (e.g., Pacific Biosciences sequencing platforms), which accelerates the study of expanded clonal lineages for in-depth antibody discovery and engineering.

We note that the computational protocol provided here is a modular pipeline, with adjustable parameters at key steps that can be modified to support paired heavy and light chain antibody analyses that are customized to a user's preferences. These protocols thus serve as a starting point for paired $V_H$:$V_L$ studies both to improve biological understanding and to advance translational antibody discovery.

## ACKNOWLEDGMENTS

We thank Pedro Seber e Silva and Sunwoo Lee for helpful review and feedback on the protocol and associated code. This work was supported by National Institutes of Health grants DP5OD023118, R21AI166396, R21AI143407, U01AI169587, and R01AI181684; The Bill and Melinda Gates Foundation; and the Mark and Lisa Schwartz AI/ML/Immunology Initiative.

## REFERENCES

DeKosky BJ, Ippolito GC, Deschner RP, Lavinder JJ, Wine Y, Rawlings BM, Varadarajan N, Giesecke C, Dörner T, Andrews SF, et al. 2013. High-throughput sequencing of the paired human immunoglobulin heavy and light chain repertoire. *Nat Biotechnol* 31: 166–169. doi:10.1038/nbt.2492

DeKosky BJ, Kojima T, Rodin A, Charab W, Ippolito GC, Ellington AD, Georgiou G. 2014. In-depth determination and analysis of the human paired heavy- and light-chain antibody repertoire. *Nat Med* 21: 86–91. doi:10.1038/nm.3743

DeKosky BJ, Lungu OI, Park D, Johnson EL, Charab W, Chrysostomou C, Kuroda D, Ellington AD, Ippolito GC, Gray JJ, et al. 2016. Large-scale sequence and structural comparisons of human naive and antigen-experienced antibody repertoires. *Proc Natl Acad Sci* 113: E2636–E2645. doi:10.1073/pnas.1525510113

Duan H, Chen X, Boyington JC, Cheng C, Zhang Y, Jafari AJ, Stephens T, Tsybovsky Y, Kalyuzhniy O, Zhao P, et al. 2018. Glycan masking focuses immune responses to the HIV-1 CD4-binding site and enhances elicitation of VRC01-class precursor antibodies. *Immunity* 49: 301–311.e5. doi:10.1016/j.immuni.2018.07.005

Edgar RC. 2010. Search and clustering orders of magnitude faster than BLAST. *Bioinformatics* 26: 2460–2461. doi:10.1093/bioinformatics/btq461

Fahad AS, Timm MR, Madan B, Burgomaster KE, Dowd KA, Normandin E, Gutiérrez-González MF, Pennington JM, De Souza MO, Henry AR, et al. 2021. Functional profiling of antibody immune repertoires in convalescent Zika virus disease patients. *Front Immunol* 12: 615102. doi:10.3389/fimmu.2021.615102

Kwong PD, Chuang G-Y, DeKosky BJ, Gindin T, Georgiev IS, Lemmin T, Schramm CA, Sheng Z, Soto C, Yang A-S, et al. 2017. Antibodyomics: bioinformatics technologies for understanding B-cell immunity to HIV-1. *Immunol Rev* 275: 108–128. doi:10.1111/imr.12480

Lefranc M-P. 2014. Immunoglobulins: 25 years of immunoinformatics and IMGT-ONTOLOGY. *Biomolecules* 4: 1102–1139. doi:10.3390/biom4041102

McDaniel JR, DeKosky BJ, Tanno H, Ellington AD, Georgiou G. 2016. Ultra-high-throughput sequencing of the immune receptor repertoire from millions of lymphocytes. *Nat Protoc* 11: 429–442. doi:10.1038/nprot.2016.024

Pan X, López Acevedo SN, Cuziol C, De Tavernier E, Fahad AS, Longjam PS, Rao SP, Aguilera-Rodríguez D, Rezé M, Bricault CA, et al. 2023. Large-scale antibody immune response mapping of splenic B cells and bone marrow plasma cells in a transgenic mouse model. *Front Immunol* 14: 1137069. doi:10.3389/fimmu.2023.1137069

Ye J, Ma N, Madden TL, Ostell JM. 2013. IgBLAST: An immunoglobulin variable domain sequence analysis tool. *Nucleic Acids Res* 41: W34–W40. doi:10.1093/nar/gkt382

CHAPTER 11

# Insights from the Study of B-Cell Epitopes of a Microbial Pathogen by Phage Display

Gregg J. Silverman[1]

*Division of Rheumatology, Department of Medicine, New York University Grossman School of Medicine, New York, New York 10016, USA*

The human immune system evolved to defend against the panoply of microbial threats. By harnessing such ability, vaccines have cumulatively saved hundreds of millions of lives. Despite such tremendous success, there have also been remarkable failures, such as the lack of a clinically proven vaccine against *Staphylococcus aureus* (SA), which continues to pose an urgent public health threat. In practice, it has proven challenging to identify the molecular basis for relevant epitopes for this pathogen. Here, we summarize our experience implementing an integrated approach using phage display technology for the identification of B-cell epitopes of microbial virulence factors, which we developed with a focus on SA. This approach was used to define minimal B-cell epitopes of the staphylococcal leucocidin family of pore-forming toxins (PFTs) that have been implicated in staphylococcal clinical infection. Our methodology provides proof of principle for an approach well suited for the rapid and efficient generation of modular protein-based vaccines for protection from clinical infection, which can be used to target pathogens for which no vaccine is currently available.

## INTRODUCTION

Phage display expression cloning is a powerful tool for the study of protein–protein interactions (Smith 1985). Indeed, this technology was recognized with the 2018 Nobel Prize in Chemistry (see Smith 2019).

Most reports on applications of phage display have focused on the use of this technology for antibody domain isolation and optimization for diagnostic and therapeutic use, and our laboratory has previously also had this focus (Sasano et al. 1993; Barbas et al. 1995; Silverman et al. 1995a,b; Roben et al. 1996; Cary et al. 2000; Shaw et al. 2001; Berry et al. 2003; Meyer et al. 2004; Gronwall et al. 2014). In other efforts, phage display systems have shown utility for peptide display (Zwick et al. 1998; Noren and Noren 2001), cDNA expression cloning, and epitope characterization (see Williamson and Silverman 2001; Rhyner et al. 2004; Cariccio et al. 2016; Ryvkin et al. 2018; Fuhner et al. 2019; He et al. 2019, among others). Fewer studies, however, have focused on the identification of the structural and functional properties of the antigenic sites targeted by antibodies (for review, see Williamson and Silverman 2001).

In the last few years, our studies have focused on the identification and testing of antigenic determinants, because these are essential for developing multimeric vaccines that protect from invasive infection. To accelerate the development of such clinical vaccines, it was therefore a natural choice

---

[1]Correspondence: gregg.silverman@nyulangone.org

© 2026 Cold Spring Harbor Laboratory Press
Cite this introduction as *Cold Spring Harb Protoc*; doi:10.1101/pdb.top107777

for us to turn to phage display technology to develop an integrated approach for the selection of immunodominant B-cell epitopes from gene fragment libraries, given our vast experience with this technique. Due to its great clinical importance, we have focused our studies on the opportunistic microbial pathogen *Staphylococcus aureus* (SA). Here, we outline our experience using this approach to identify relevant SA B-cell epitopes, which we hope will also help others implement similar projects for other pathogens.

In the following sections, we highlight both the scientific challenges posed by the multitude of countermeasures of this preeminent microbial pathogen and our methodical approach to identifying minimal epitopes (i.e., the smallest protein fragments containing the site recognized by a monoclonal antibody [mAb]), intended to facilitate the rapid development of modular vaccine systems for the prevention of serious infection. To highlight the challenges inherent to the biology of this pathogen, we first describe the clinical syndromes caused by these infections and provide an overview of the genomic diversity within clinical SA isolates that harbor diverse gene sets encoding potential virulence factors contributing to invasive infection. We also explain that within our studies, our early steps were intended to deconvolute and rank the hierarchy within the diversity of SA antigens that are recognized within convalescent immune responses in patients who survived serious infections.

We then introduce the pComb-Opti8 phage display system, which is based on the classical pComb3 phagemid. Like the pComb3 phagemid, the pComb-Opti8 phagemid is also used to display proteins in phage display experiments. We discuss the essential features of the pComb-Opti8 phage display system that make it ideal for discovering the minimal antibody binding sites (i.e., epitopes) on functionally important microbial antigens (Fig. 1). An overview is provided of the sequential steps in the generation and application of gene fragment libraries in the pComb-Opti8 vector, as well as the determination of the minimal structural features for antibody binding and the testing of the immunogenicity of the epitopes relevant to the development of protective vaccines (Fig. 2). Cumulatively, we provide an overview of our interim progress in developing an integrated phage display approach, outlined in a validated protocol (see Protocol 1: Cloning and Selection from Antigen Fragment Libraries for Epitope Identification [Silverman 2025]), designed to advance our overarching goal of developing better clinical vaccines that target the protein antigens of microbial pathogens.

## STAPHYLOCOCCAL INFECTIONS POSE A SUBSTANTIAL AND INCREASING CLINICAL BURDEN

Methicillin-resistant *Staphylococcus aureus* (MRSA) kills an estimated 19,000 people in the United States per year—more than HIV, and more than *Streptococcus pneumoniae*, *Neisseria meningitidis*, *Hemophilus influenzae*, and *Streptococcus pyogenes* combined (Klevens et al. 2007). In contrast to the success seen with vaccines against some of these common pathogenic bacteria (Peltola et al. 1977; Eskola et al. 1987; Shinefield and Black 2000; Ramsay et al. 2001), and despite extensive efforts, a successful preventative vaccine against *S. aureus* has remained elusive (for reviews, see Pier 2013; Fowler and Proctor 2014).

While in healthy individuals, *S. aureus* is typically an innocuous member of the human cutaneous and mucosal microbiome, it is also a common opportunistic pathogen responsible for a wide range of diseases, in both community and hospital settings (Lowy 1998). During the past two decades, there has been an alarming expansion in the burden of invasive staphylococcal disease caused by the emergence of MRSA strains, which, due to their antibiotic resistance, have severely and progressively restricted antimicrobial treatment options. Community-associated MRSA (CA-MRSA) infections, which are the invasive SA infections that begin outside of a healthcare facility in an otherwise healthy individual, account for much of this burden (Moran et al. 2006; Klevens et al. 2007; Limbago et al. 2009; Talan et al. 2011; Ammerlaan et al. 2013) and pose a major public health threat. Indeed, MRSA is on the World Health Organization's list of bacteria posing the greatest threats to human health (Willyard 2017). Besides cases of skin and soft tissue infection (SSTI), SA is also a frequent cause of

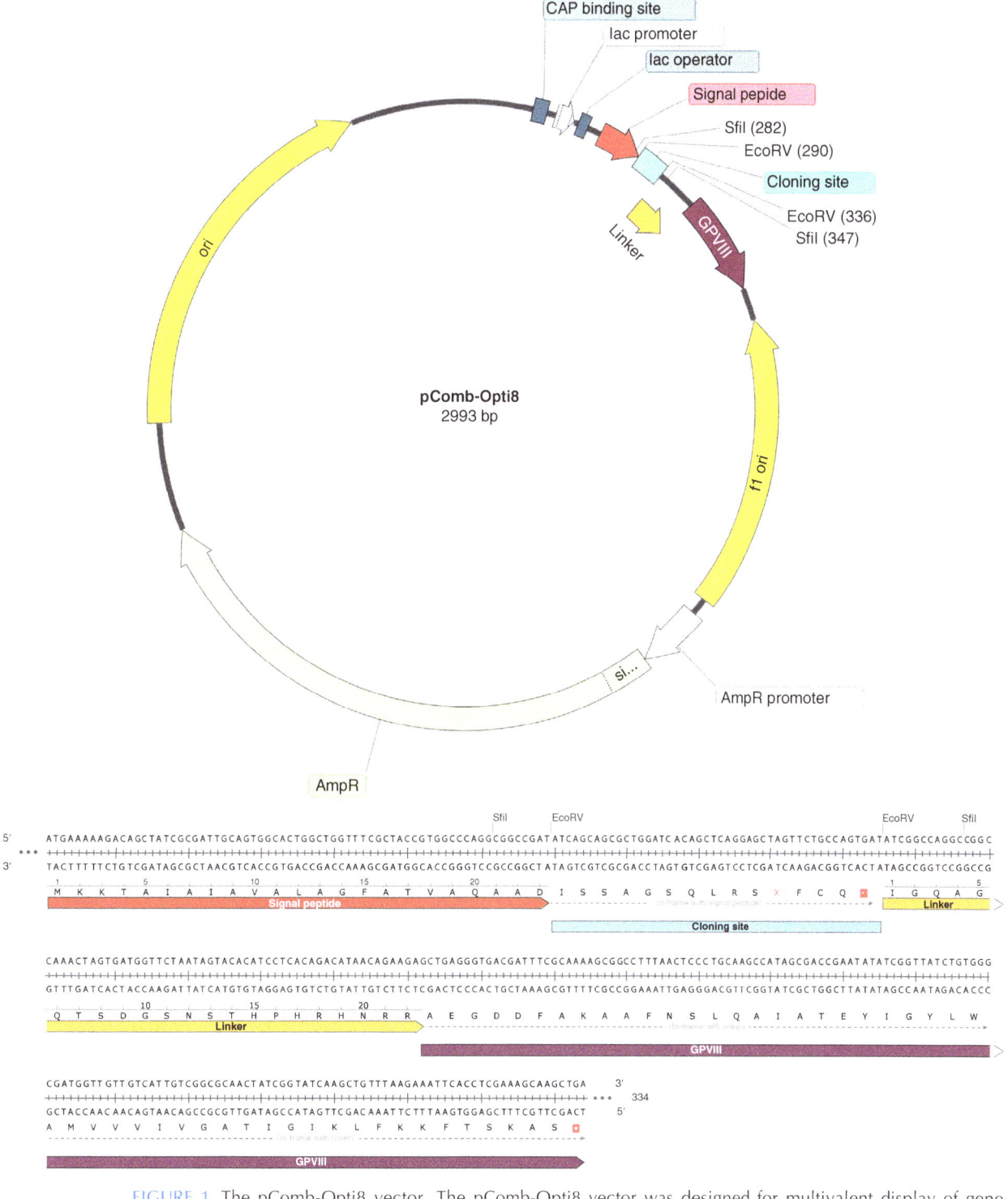

FIGURE 1. The pComb-Opti8 vector. The pComb-Opti8 vector was designed for multivalent display of gene fragment products on filamentous bacteriophage (Hernandez et al. 2020a). The cloning site, linker region, and modified coat protein VIII (pVIII) were specially designed for this gene fragment phage display vector system (Hernandez et al. 2020a). This vector was generated by modification of the pComb3X phagemid system designed for monomeric gene fusion display (see Chapter 4, Overview: The pComb3 Phagemid Family of Phage Display Vectors [Rader 2024]). For dedicated use for the selection of the products of gene fragments, the pComb-Opti8 vector was generated with custom oligonucleotides that introduced a flexible linker, followed by the introduction of a truncated gVIII optimized for expression of a fusion protein with the major coat protein VIII (Weiss et al. 2000). Figure adapted from Hernandez et al. (2020a), © 2020 American Society for Microbiology, reproduced with permission.

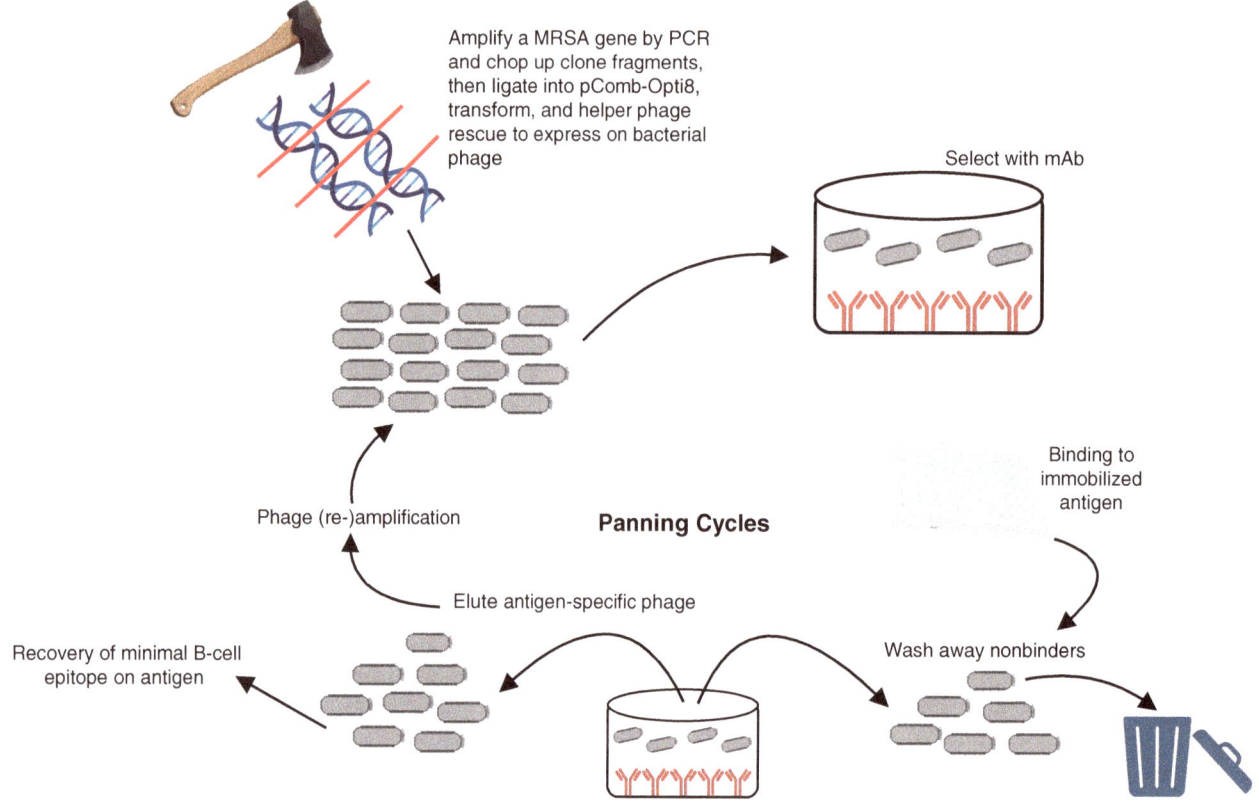

FIGURE 2. Use of the pComb-Opti8 system for cloning of gene fragment-encoded minimal epitopes. In the example shown here, the gene of interest in *Staphylococcus aureus*, such as that for a leucocidin toxin, is amplified via PCR and then enzymatically digested into gene fragments. Following digestion of the pComb-Opti8 phagemid vector, gene fragments are ligated into the vector's multiple cloning site (MCS). Following electroporation into the TG-1 strain of *Escherichia coli*, expansion and rescue with VCSM13 helper phage are performed. Then, bio-panning is conducted using an individual monoclonal antibody (mAb) of interest immobilized onto the precoats of microtiter wells. After blocking and incubation with the phage library, nonbinders are washed away and discarded, and then bound phage is eluted with an acid solution, which is then neutralized and used to infect a fresh sample of bacteria. Generally, four sequential rounds of library selection are performed for each experiment (Hernandez et al. 2020a,b). PCR is used for amplification and barcoding, which enables high-throughput sequencing of the original library and sequentially selected libraries (not shown).

life-threatening invasive infections, including bacteremia-associated pneumonia. SA bacteremia has a mortality rate estimated at 20% (Corey 2009; van Hal et al. 2012), and can also result in metastatic infections that frequently relapse, despite optimal treatment (Lautenschlager et al. 1993; Holland et al. 2014).

A key challenge in *S. aureus* research has been the genetic diversity within the genomes of different clinical SA strains and the associated significant differences in gene content and the toxins the genes encode (Fitzgerald et al. 2001; Lindsay and Holden 2004; El Garch et al. 2009). While some strains express a capsular polysaccharide, other clinical strains, including the well-known clinical isolates of USA300, lack a common polysaccharide capsule that could be used as an antigenic target in a vaccine. In contrast, many other common clinical bacterial pathogenic species have capsular polysaccharides that have been exploited in now well-validated vaccines. However, the cumulative experience argues that a capsular-alone vaccine approach is unlikely to be sufficient to confer protection against all *S. aureus* clinical infection syndromes.

Skin and soft tissue infection by SA are linked to strains expressing specific toxins, such as the Panton–Valentine leucocidin (PVL), a member of the pore-forming toxin (PFT) family (Lina et al. 1999; Piewngam et al. 2018). The targeting of these virulence factors has therefore been considered an attractive strategy to combat these infections through the development of vaccines or protein-based treatments for passive immunotherapy, which impart greatly enhanced antibody-mediated immune responses.

In our laboratory, we set out to accelerate progress in the development of a clinically efficacious SA vaccine by first exploring the immune recognition of SA protein antigens by the human immune system in the circulation of those recovering from serious infections, as we anticipated such analyses would effectively highlight the set of staphylococcal proteins that might be useful in a vaccine that can elicit protective immunity. We summarize key aspects of our experience below.

## SA INFECTIONS ARE ASSOCIATED WITH A RANGE OF IMMUNOGENIC ANTIGENS

Clinical infection by SA is associated with the release of dozens of exotoxins and other virulence factors, which in many cases have documented mechanisms for diminishing host defenses (Spaan et al. 2017). To date, the selection of candidate antigens—and associated epitopes—for experimental vaccine development has relied heavily on the successful in vitro expression of proteins and animal models of infection, but the prominence of an antigen in these settings may differ significantly in humans for both relative protein expression during clinical infection and roles of specific factors in virulence.

To address this issue, we first performed studies that comprehensively characterized the human immune response to staphylococcal protein antigens. Our studies used sera from patients recovering from invasive disease, and we were especially interested in understanding the immunogenicity of protein virulence factors of SA infection syndromes associated with great clinical morbidity and mortality. We characterized the relative immunodominance of SA antigenic targets represented in printed arrays of all approximately 2600 open reading frame (ORF)-encoded proteins from an MRSA USA300 clinical isolate. In patients recovering from skin and soft tissue infections, IgG antibodies to leucocidins generally peaked at ∼6 wk but declined thereafter, often with parallel decreases in the frequency of antigen-specific circulating memory B cells (Pelzek et al. 2018a). We therefore have also considered boosting the in vivo representation of these antigen-specific memory B cells, which include the cellular sources of these immune responses that should provide great protection from infection.

## HUMAN IMMUNOPROTEOME OF *S. AUREUS* GENE PRODUCTS

To help guide the development of effective adaptive immune protective and therapeutic agents, we used an approach that enabled an unbiased survey of human immune responsiveness to all potential SA protein antigens (Radke et al. 2018), which used longitudinal samples obtained from patients with the three clinical syndromes: invasive cutaneous infection, systemic pediatric infection, and prosthetic joint infection. Using serum IgG from patients who recovered from these types of serious and potentially fatal infections, and based on the relative binding signals and representation of sera in the collection of sera tested, our investigations identified a hierarchy of serum SA protein antigens, most with known toxin activity. Members of the leucocidin family of SA virulence factors were among the most broadly immunodominant in the circulation of those who recovered from these SA infections; hence, developing antibody responses to these proteins in those previously unexposed or immunodeficient could be considered in future vaccine formulations.

Among the top 25 SA proteins most commonly recognized by circulating antibody responses in sera from those who recovered from invasive infection, five are members of the PFT family that includes the staphylococcal leucocidins. Our surveys also identified two Ig-binding proteins, staphylococcal protein A (SpA) (Graille et al. 2000) and a protein called the second immunoglobulin-binding protein (Sbi), of S. aureus. These proteins can bind IgG molecules through nonimmune mechanisms. These results confirmed that both secreted and surface SA proteins are dominant targets of human immune responses (Radke et al. 2018). Moreover, the majority of these antigen "top hits" have reported enzymatic or cell-specific activities that have been postulated to contribute to SA virulence, in part by impairing the integrity of host barriers and immune defenses (Foster 2005).

In the arms race between the human immune system and SA, PFT proteins are thought to facilitate a preemptive attack by the bacteria on the host. Indeed, these toxins dismember layers of cellular defenses. Certain toxins also cause the lysis of red cells, making available a nutritional bounty to these microbial invaders, especially from hemoglobin, as there is only limited availability of iron to a blood-borne pathogen (Bennett et al. 2019). Hence, PFT functional neutralization and clearance may be central to the defense against SA infection.

Over the past decade, the cellular receptors for all of the bicomponent leucocidins (Luks) have been identified, which, to a large degree, has helped rationalize the cellular tropism and species specificity of these toxins (for review, see Spaan et al. 2017). During infection, the leucocidins are secreted as inactive monomeric subunits, but upon binding of a subunit to the membrane receptors of a targeted host cell, these subunits oligomerize to form pores that act as cell-killing machines that break down epithelial barriers, disable immune cells, and aid the scavenging of nutrients (Reyes-Robles and Torres 2016; Badarau et al. 2017). Certain PFTs are associated with specific clinical infection syndromes. For example, the Panton–Valentine leucocidin (PVL) is associated with primary skin and soft tissue infection and pneumonia (Lina et al. 1999), and these PFTs also represent important antigens recognized by the host immunity (Pelzek et al. 2018; Radke et al. 2018).

Membership in the PFT family is defined by structural homology that is shared by both hemolysins and leucocidins, which are among the immunodominant targets for host immune defenses (Radke et al. 2018). Due to their conserved overall structures, there are epitopes on different toxins that are bound by cross-reactive antibodies, which could represent the Achilles' heel of the SA pathogen (discussed further below). In part, we saw these conserved structural features as potential opportunities that prompted our investigations of the antigenic molecular hot spots in the PFTs.

## STAPHYLOCOCCAL PFTS ARE ATTRACTIVE ANTIGENS FOR PHAGE DISPLAY–BASED INVESTIGATIONS

Members of the PFT family and the structurally related hemolysin A (HlgA) toxin are secreted as water-soluble monomeric proteins (each of ∼30 kDa) and share a common β-barrel structure (for review, see Spaan et al. 2017). This shared overall structure provides an effective structural strategy for generating oligomeric pores that assemble on host cell membranes to mediate targeted cell intoxication and execution (Spaan et al. 2017). Except for α-hemolysin, which forms a homoheptamer (Song et al. 1996), leukotoxins form bipartite structures with noncovalent association of two distinct proteins, a class S and a class F component, each subunit type of ∼31 and 34 kDa, respectively. Together, these assemble into what are likely hetero-octameric structures (for review, see Badarau et al. 2017).

Over the past two decades, the crystal structures of these water-soluble monomers and the membrane-embedded oligomeric pores formed by the PFTs have been solved. The host molecular targets of the SA leucocidin family have also been characterized by a variety of techniques (Nguyen and Kamio 2004; Yamashita et al. 2011, 2014; Sugawara et al. 2015; for review, see Spaan et al. 2017). In the past, a number of computational approaches have been developed to identify B-cell epitopes, yet in practice, these in silico methods have fallen short (Blythe and Flower 2005). However, more recently developed computational methods may be powerful (Jumper et al. 2021) but will still require experimental verification. Hence, to identify minimal B-cell epitopes of members of this family of structurally related leucocidin toxins, we developed a means using phage display to sequentially interrogate individual leukotoxins for minimal epitopes that could potentially be used in experimental vaccine formulations. We discuss this approach below.

## A PHAGEMID-BASED DISPLAY SYSTEM FOR TOXIN EPITOPE RECOVERY

To characterize the structural features of B-cell epitopes of the PFTs that are recognized by antibody responses, we developed an integrated experimental system. We used a phage display approach because it affords a proven means to efficiently define the structural basis of protein–protein interactions, such as antigen binding by a specific antibody. The filamentous phage III and VIII coat proteins (termed pIII and pVIII, respectively) can each be used for the construction of fusion proteins via their amino termini (see Chapter 4, Overview: The pComb3 Phagemid Family of Phage Display Vectors [Rader 2024]).

For our studies, we developed the pComb-Opti8 phagemid vector (Fig. 1; Hernandez et al. 2020a), which was generated from an updated version of the pComb vector (Weiss et al. 2000). This vector was further engineered for cloning and display of the fragments of the genes that encode for fusion proteins with the filamentous phage major protein, pVIII (Fig. 1). This design allows for the multivalent presentation of the product of a cloned gene fragment, and this vector incorporates a specialized linker that facilitates fusion of the product of the cloned gene with the coat protein (Weiss et al. 2000). During assembly of the viral particle within the *Escherichia coli* host cell, the fusion protein is mixed with a larger quantity of native pVIII molecules, encoded by the native pVIII gene (*gVIII*) provided by infection with a suitable helper phage, which is crucial for maintaining the overall phage structure.

Within our experimental studies, we refined a comprehensive technical approach for the phage display of toxin fragments, which we describe in detail in an accompanying protocol (see Protocol 1: Cloning and Selection from Antigen Fragment Libraries for Epitope Identification [Silverman 2025]) (Fig. 2). To generate an antigen-specific library, the approach involves the design of sequence-specific flanking oligonucleotides to amplify the toxin genes of interest. With a standard PCR, such full-length genes of interest are individually amplified without signal leader sequences or stop codons. To confirm the predicted product size and homogeneity, these amplimer products are then evaluated by agarose electrophoresis. The purified amplimer products are then individually fragmented into smaller DNA pieces with a dsDNA-specific fragmentase, which we have used to produce gene fragments of an average size of ∼50 bp, encoding random amino acid subregions of the toxin. Fragmentation is followed by polishing of the DNA fragments with a Quick Blunting Kit (NEB). These fragments are then ligated into the cloning site of a pComb-Opti8 vector that has been previously restriction enzyme–prepared to generate blunt ends for cloning into the vector. By electroporation, a library is then generated that can be rescued in phage form with helper phage that is suitable for antibody-mediated rounds of selection.

After panning of the gene fragment phage library with a specific antitoxin mAb, clonal analyses using high-throughput sequencing (HTS) are performed, which enable the identification of gene fragments encoding selected protein fragments. In our earlier phage display studies, we used methods to assess diversity in an antibody library, which involved interrogating a set of picked clones from titration plates, but these efforts were overly laborious and generally inadequate to evaluate the true diversity within the original library. Such methods could not accurately assess the shifts within the composition of sublibraries obtained with each sequential round of antibody panning. HTS addresses this issue. Indeed, to apply a better approach to library compositional analysis, each of our libraries that contained the hemolysin A (*HlgA*) gene fragments (i.e., each pComb-Opti8-HlgA fragment library) was separately amplified with a pair of specially designed oligonucleotides that also introduce a library-specific barcode. The original library and each sequentially selected library were then subjected to HTS. Hence, the overall scheme involves library generation, which is followed by phage library clonal selection and then sublibrary expansion. Thereafter, additional rounds of selection of epitope-related gene fragments are performed with the same mAb (Fig. 2). Selection is later followed by HTS-based library analyses. This approach has proven to be efficient, practical, and informative (Hernandez et al. 2020a,b).

Our HTS approach revealed that the original HlgA library comprises diverse gene fragment clones that vary in size, as per our design. These fragments are distributed throughout the entire parental gene, in all three reading frames, and with fragments cloned in both orientations within the pComb-Opti8 vector (Hernandez et al. 2020a,b). Although most inserts might be predicted to be functionally defective and do not produce a polypeptide, the library is sufficiently large and also has a very high level of redundancy for fragment coverage over the entire toxin gene. Taken together, the representation of relevant clones in the correct orientation and reading frame proved sufficient to ensure the success of our efforts to select functional small epitopes with cognate antibody-mediated clonal selection (Hernandez et al. 2020a,b).

## IDENTIFICATION OF LEUCOCIDIN-ENCODED B-CELL EPITOPES BY PHAGE DISPLAY

In these proof-of-concept studies, we followed an experimental plan designed to identify primary sequences within a PFT protein that embodied a small epitope responsible for antibody cross-reactivity between different members of the PFT family, which we found to share homology at an amino acid level, presumably due to having arisen from a common evolutionary ancestor. These small epitopes have also been evaluated for their functional utility, relevant to future incorporation in a protective vaccine.

In our first experiments, to determine whether we could recover epitopes responsible for the cross-reactivity of antibodies between different PFT proteins, we adapted our standard phage display panning protocol and used a murine mAb termed anti-LukE3 mAb. This antibody was generated by immunization with native recombinant leucocidin E (LukE), a member of the PFT family, which shares considerable primary sequence identity with the HlgA toxin protein (Hernandez et al. 2020a). In these panning studies with the HlgA gene fragment library, after the third and fourth rounds of selection with the anti-LukE3 mAb, the ratios of output phage titers were significantly higher with the anti-LukE3 mAb than the output documented from panning of the same library with an irrelevant IgG2a antibody (i.e., negative control). This finding was the first evidence that clonal focusing can be attained with a leucocidin toxin-specific antibody.

To investigate the diversity of antigen fragment clones that were selected, and as described above, each of the sequentially selected sublibraries was amplified with a set of flanking library-specific barcoding oligonucleotides. By this amplification, library-specific barcodes were integrated into every read with each library after amplification. The libraries were then subjected to HTS. The sequences in the original library and the selected sublibraries were aligned to the sequence of the gene used to make the fragment library (Fig. 3). Notably, after the first round of panning, the sublibrary displayed a massive decrease in clonal diversity, while there was concordant relative expansion in certain clones that had overlapping gene sequences, all derived from a limited sequence subregion of the toxin gene. Within these clones, an overall preferential representation of clones included codons 123–133 of the *HlgA* gene. Notably, this clonal predominance was also documented in the sublibraries obtained with each subsequent selection round, and this pattern of clonal focusing was most pronounced following the fourth round of selection (i.e., pan 4) (Fig. 3). These panning experiments successfully recovered gene fragment products that had specific binding interactions with the selecting mAb. Indeed, analysis of the polyclonal mixture of phage of the fourth selected sublibrary by ELISA showed strong concentration-dependent binding interactions with the selecting anti-LukE3 mAb (Hernandez et al. 2020a).

After the last round of selection, individual colonies displayed a high level of clone duplication and limited diversity. There were only 15 unique clones, each of them containing only 15–42 in-frame codons per selected clone. Importantly, these sequences encoded overlapping portions of the same HlgA toxin subregion, which varied only in their amino- and/or carboxy-terminal ends. These 15 clones all included the conserved 11-amino-acid sequence YLPKNKIDSAD from the parental HlgA protein (Fig. 3). In the solved HlgA crystallographic structure, this candidate epitope sequence is localized to a solvent-exposed loop subregion that connects the two adjacent β-strands (Hernandez et al. 2020a).

## MOLECULAR CHARACTERIZATION OF A CROSS-REACTIVE LEUCOCIDIN EPITOPE

As described above, the anti-LukE3 mAb was used in our phage display panning studies because this antibody binds to different Luk toxins that display sequence similarity at an amino acid level

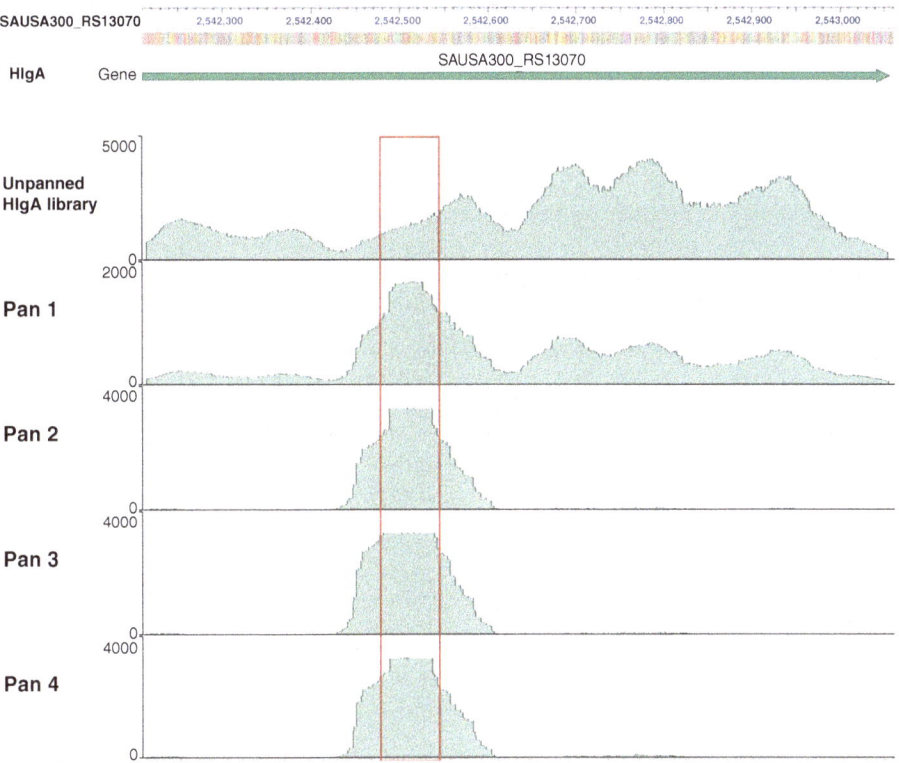

FIGURE 3. High-throughput sequencing of the original (unpanned) and sequentially selected sublibraries. In the example shown, selection was performed by panning of the *HlgA* gene fragment library with the anti-LukE3 monoclonal antibody. The *top* sequences represent the parental *HlgA* gene sequence. Selected phage fragment clones were distributed within one *HlgA* gene subregion, which is shown to have undergone preferential sequential enrichment with each round of anti-LukE3 monoclonal antibody (mAb) panning, indicated by Pan 1, Pan 2, Pan 3, and Pan 4. The red box identifies the minimal B-cell epitope selected by this mAB. Figure adapted from Hernandez et al. (2020a), © 2020 American Society for Microbiology, reproduced with permission.

(i.e., cross-reactive mAb). To further investigate the functional properties of minimal B-cell epitope protein structures recovered from the above-described phage display sublibrary, we first evaluated the minimal requirements for binding, which were independently confirmed in studies of synthetic peptides representing the LukE108–118 and HlgA123–133 sequences. Modeling studies indicated that in the LukS and HlgA toxins, there are conserved residues, Tyr108 and Lys111, that are solvent-exposed and therefore potentially accessible for protein–protein interactions. In alanine scanning studies, the reactivity of the anti-LukE3 mAb was significantly diminished with the LukE108–118 Y108A and HlgA123–133 Y123A mutant peptides. Furthermore, the anti-LukE3 mAb had no detectable reactivity with either LukE108–118 K111A or HlgA123–133 K126A variant peptides, which further established which key residues were required for recognition of this immunodominant epitope (Hernandez et al. 2020a). We sought a deeper understanding of structure–function relationships associated with this minimal epitope.

## VARIATIONS OF A CROSS-REACTIVE EPITOPE FOLD IN A CONSERVED β-LOOP IN THE LEUCOCIDIN PROTEIN

Whereas we had shown that there was primary sequence relatedness in the toxin subdomains recognized by the exemplary anti-LukE3 mAb, we next sought to learn more about where there were shared three-dimensional structure conformations. To assess the conformations that these subregions fold into, we performed computational ab initio folding studies. Indeed, these studies predicted that this minimal epitope folded into a β-loop turn within the overall β-barrel structure of the full-length toxins, and this structure would not be significantly changed by the introduction of any of the natural sequence variations in the different PFT members (Hernandez et al. 2020a). Hence, these studies support the notion that antibody binding interaction within this minimal B-cell epitope requires direct interactions with the side chains of the two specific amino acid positions, as discussed above. Thus, our phage display gene fragment experimental approach illuminated that the core binding of antibody–epitope interaction could be deconvoluted to a very small conserved surface on the leucocidin protein (Hernandez et al. 2020a,b). While this particular set of experiments characterized a cross-reactive epitope that might facilitate antibody-mediated clearance of the toxin from the body during an infection, we wondered about the nature of an epitope that could be responsible for antibody-mediated inhibition of toxin function.

## IDENTIFICATION OF A LUK NEUTRALIZATION EPITOPE IN A β-HAIRPIN LOOP IN THE STEM SUBDOMAIN OF THE LEUCOCIDIN HOLOPROTEIN

We next turned our attention to defining the minimal epitope associated with the PFT functional toxin activity, as the binding of certain anti-Luk mAbs neutralized the functional activity of the toxin. For these studies, we used a panel of mAbs that had been generated by immunization with the HlgC leucocidin subunit (Hernandez et al. 2020b). These anti-HlgC mAbs were of special interest, as each could inhibit the cell-killing activity of this toxin in a well-validated ex vivo cytotoxicity assay, in which 0.85 μg/mL (i.e., 24 nM) of a HlgCB holotoxin killed 90% of neutrophils (Hernandez et al. 2020b).

Due to the high shared primary sequence identity between HlgC and LukS, we repurposed the pComb-Opti8-LukS gene fragment library by panning with the above-described anti-HlgC mAbs that had been generated by immunization with the HlgC leucocidin subunit (Hernandez et al. 2020b). We thereby identified the LukS240–259 subregion, which we showed had specific binding interactions with the anti-HlgC mAbs. Synthetic peptide studies revealed that the binding interactions of these mAbs all require the LukS His252 and Tyr253 residues, and binding was also dependent on the homologous residues in the context of the homologous HlgC subregion. As anticipated, these small subregion peptides also efficiently inhibited the binding of the mAbs with the LukS and HlgC full-length proteins. As documented in crystallographic-based modeling studies, this subregion represents a hairpin loop in the resting conformation of the toxin (Fig. 4).

The immunogenicity of this minimal epitope was demonstrated in studies of immune responses in mice that were immunized with synthetic peptides derived from this epitope generated as keyhole limpet hemocyanin (KLH) conjugates. High-titer IgG responses were induced that were cross-reactive with the recombinant proteins LukS and HlgC. The binding interactions with these proteins were also dependent on the critical amino acid residues His252 and Tyr253, and conservation of the same His and Tyr residues in LukS and HlgC were required for immune recognition by all of the anti-HlgC (Hernandez et al. 2020b). Indeed, when these critical residues were presented, this same peptide was also recognized by IgG antibodies in the sera of patients recovering from *S. aureus* infections (Hernandez et al. 2020b). With peptide variants, the His and Tyr residues at the tip of the inherently stable hairpin loop were shown to be required for the binding interaction with each of the anti-HlgC mAbs (Hernandez et al. 2020b).

The capacity of these mAbs to inhibit the cytotoxicity of HlgC was diminished by preincubation with the LukS248–258 peptide in solution, while incubation with the mutant peptide with substitutions for the critical His and Tyr residues (LukS248–258mutHY-GP) (Hernandez et al. 2020b) did not interfere with this neutralization capacity (Hernandez et al. 2020b). These findings therefore confirmed that all of these anti-HlgC mAbs bind an epitope with the above-described conserved His and Tyr residues within the same hairpin conformation.

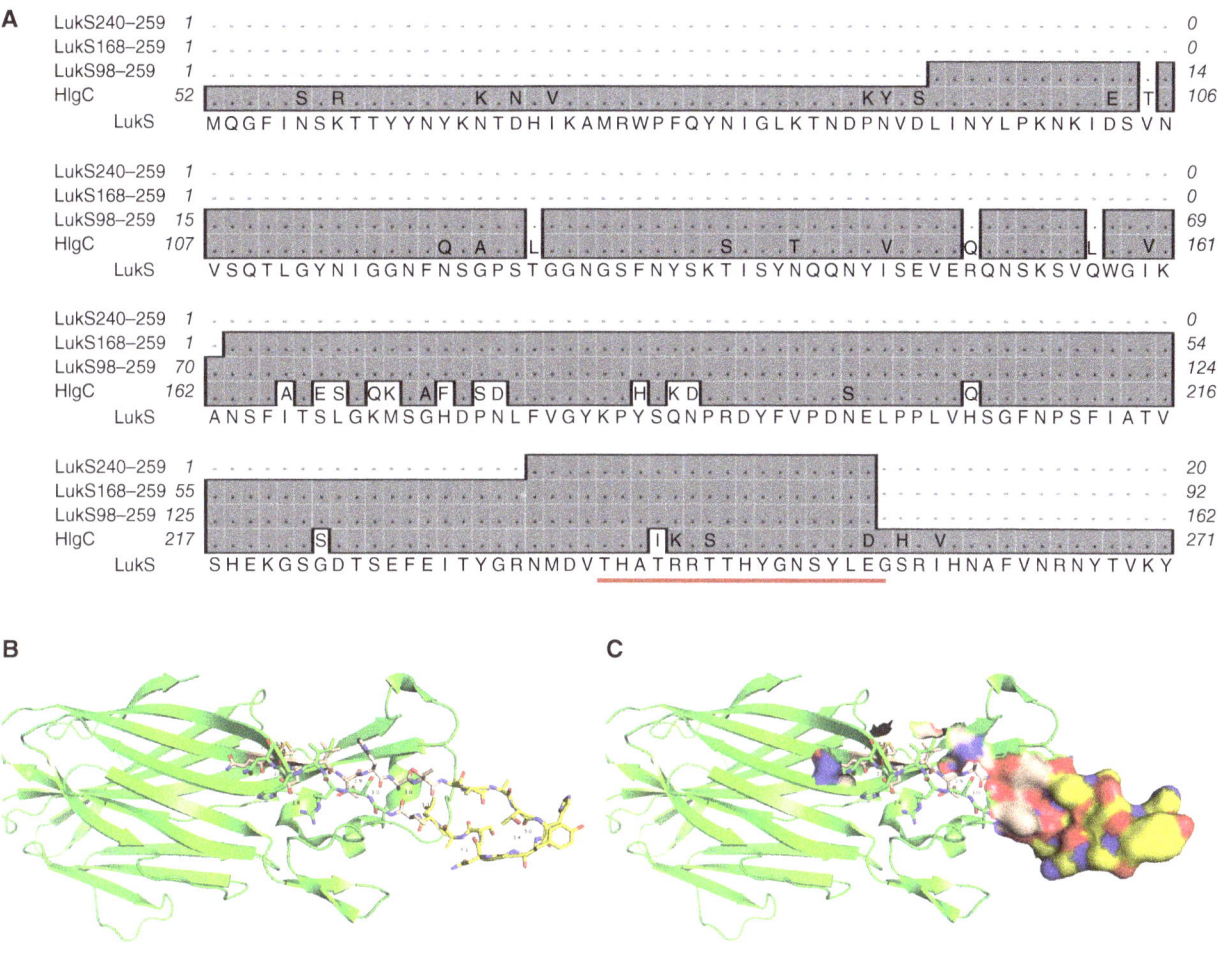

FIGURE 4. Sequences of phage fragment clones selected by the anti-HlgC2 monoclonal antibody (mAb) from the LukS phage display fragment library. (A) Deduced amino acid sequences of recurrent *LukS* gene fragment phage clones selected by anti-HlgC mAbs (depicted in gray box) aligned to a conserved region of the PVL LukS homologs, LukS and HlgC. These findings provide a structural basis for mAb cross-reactivity. The LukS244–260 subregion is underlined. (B) Structural visualization of the LukS crystal structure (PDB: 4IYA), with the solvent-exposed structure conserved among the clones highlighted in yellow, with charged surfaces in red and blue. In this solved structure, residues LukS240–242 are not solvent-exposed. (C) The solvent-exposed surface of LukS244–260 on the LukS crystal structure is depicted. Figure adapted from Hernandez et al. (2020b) under a Creative Commons Attribution 4.0 International License (CC BY 4.0); https://creativecommons.org/licenses/by/4.0/

## CONVALESCENT SERUM ANTIBODIES COMMONLY RECOGNIZE A PRIMARY SEQUENCE–DEPENDENT B-CELL EPITOPE

To investigate the clinical relevance of the neutralization-associated LukS246–260 and HlgC235–257 minimal epitopes of the PFT family identified using a phage display system, we performed binding assays with sera from patients recovering from invasive SA infection (Pelzek et al. 2018; Hernandez et al. 2020b). From longitudinal surveys of these patients, we found that, in most cases, the highest titers of SA-specific antibodies were detected in convalescent sera collected ∼6 wk after an individual patient's initial clinical presentation (Pelzek et al. 2018). Using a validated multiplex bead-based assay, we performed binding studies that included full-length Luk subunit proteins and the isolated minimal LukS246–260 and HlgC235–257 subregion peptides, along with the control LukS248–258mutHY-GP peptide (Hernandez et al. 2020b). Most convalescent sera were enriched for IgG antibodies to the full-length proteins and also to these minimal subregion epitopes but not to the mutant peptides. Strikingly, we observed a statistically significant strong correlation between the level of IgG reactivity with the LukS246–260 and the HlgC239–257 peptides in paired *t*-tests for each individual tested, which could indicate that the same antibodies in these sera were cross-reactive between these two minimal epitopes. In addition, antibody reactivity with the full-length HlgC had a significant correlation with reactivity with the LukS246–260-related epitope, which required Tyr and His residues that are conserved within the hairpin structures in the HlgC and LukS holoproteins (Hernandez et al. 2020b).

Independent studies have implicated the HlgC235–255 subregion in the binding interaction of the LukS homologous toxin subunit with the host cell receptor (Laventie et al. 2014). We hypothesized that the interaction between an antibody and this subregion (Fig. 4), which represents the "rim" portion of the native conformation of the leucocidin protein, as depicted in the solved crystal structure, stabilizes the subunit in a quiescent state that does not convey the cell-killing function of the holotoxin. In parallel studies, we have identified several other Luk-neutralizing antibodies that interact with epitopes localized to the same subregion of the stem region (Fig. 5). These findings therefore suggest a molecular basis by which an antibody can interfere with toxin activity by binding to a small primary amino acid sequence (identified via phage display) that is central to mounting effective host defenses. In several instances, these minimal loop-based epitopes were mimicked by ~10–20 amino acid peptides, with a stable conformation of the same epitope exhibited in the tertiary structural context of the much larger native protein. Binding interactions in a solution of these minimal epitopes can block the recognition by a mAb of the native protein. Such epitopes can therefore be described as "validated conformational epitopes."

The PFT fold represents a conserved β-barrel conformational structure. From our applications of antibody-mediated interrogation of pComb-Opti8 fragment libraries, we have identified many epitopes on the solvent-exposed surfaces of these toxins (Hernandez et al. 2020a,b; Protocol 1: Cloning and Selection from Antigen Fragment Libraries for Epitope Identification [Silverman 2025]). However, we found that these recognized epitopes were not randomly distributed. In the β-barrel structures, based on localization in the solved crystallographic structure, these minimal epitopes—identified by phage display—were often assigned to β-folds (Figs. 4, 5; Table 1).

Our studies confirmed that the structure–function relationships of subregion(s) in the Luk molecule responsible for toxin activity are complex. The LukE3 epitope localizes to a portion of the stem domain in isolated Luk monomers, which is folded into a hairpin with features akin to a clasped safety

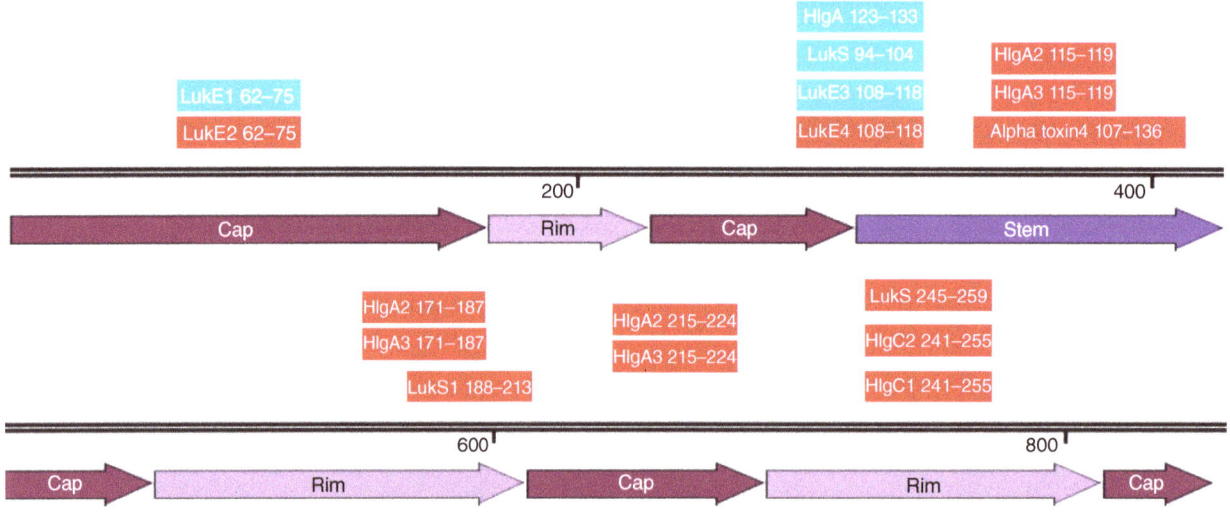

FIGURE 5. Localization of minimal B-cell epitopes within the common leucocidin structure. The primary sequence of the consensus for pore-forming toxins (PFTs), including leucocidins, is shown, with an overlay of the minimal epitopes in specific PFT proteins that have been identified using the pComb-Opti8 system, as described in the text. Cap, rim, and stem refer to domains identified in the folded three-dimensional structure of the toxin. These epitopes are generally found at β-folds in the subunit, as documented in reported crystallographic studies.

TABLE 1. Minimal B-cell epitopes identified with phage display gene fragment libraries described here

| | IEDB submission ID no. | Sa protein | Antibody name | Clone designation | Synthetic peptide/toxin codon/protein subregion |
|---|---|---|---|---|---|
| 1 | 1000736 | LukS | Anti-S1 | 3-2.2.12 | LukScyc188–213 |
| 2 | 1000740 | LukE | Anti-E1 | 1-18.2.1 | LukE28–85 |
| 3 | 14746233 | LukE | Anti-E3 | 1-26.1.11.6 | LukE108–118 |
| 4 | 1000741 | HlgA | Anti-E3 | 1-26.1.11.6 | HlgA94–1004 |
| 5 | 14746232 | LukS | Anti-E3 | 1-26.1.11.6 | LukS94–104 |
| 6 | 14746231 | HlgC | Anti-E4 | 1-26.1.11.6 | HlgC96–106 |
| 7 | 1000773 | HlgA | Anti-gA1 | 1-5.1.8 | HlgA200–216,244–255 |
| 8 | 1000774 | HlgA | Anti-gA2 | 1-11.2.17 | HlgA200–216,244–255 |
| 9 | 1000775 | HlgA | Anti-gA3 | 1-22.1.6 | HlgA200–216,244–255 |
| 10 | 1000776 | HlgA | Anti-gA4 | 1-37.3.3 | HlgA200–216,244–255 |
| 11 | 14746230 | HlgC | Anti-gC1 | 1-5.1.4.4 | HlgC241–255 |
| 12 | 1000777 | LukS | Anti-gC1 | 1-5.1.4.4 | LukS245–259 |
| 13 | 1000739 | LukS | Anti-gC2 | 1-30.1.9 | LukS245–259 |
| 14 | 1000780 | HlgC | Anti-gC2 | 1-30.1.9 | HlgC241–255 |
| 15 | 14746230 | HlgC | Anti-gC3 | 1-40.1.2 | HlgC241–255 |
| 16 | 1000778 | LukS | Anti-gC3 | 1-40.1.2 | LukS245–259 |
| 17 | 1000106 | HlgC | Anti-gC4 | 1-42.25 | HlgC241–255 |
| 18 | 1000079 | LukS | Anti-gC4 | 1-42.25 | LukS245–259 |
| 19 | 1000104 | LukE | Anti-E4 | 1-31.3.3.3 | LukE108–118 |
| 20 | 1000105 | LukE | Anti-E2 | 1-20.1.3 | LukE62–75 |
| 21 | 1000112 | HLA | Anti-HLA4 | 1-3.2.22 | HLA107–136 |

(IEDB) Immune Epitope Database, (Sa) *Staphylococcus aureus*. All of these epitopes were identified experimentally. Data are from Hernandez et al. (2020a,b).

pin. After assembly of the full oligomeric complex on the targeted host cell membrane, the stem domains in the subunits undergo a conformational change to an extended conformation, which is stabilized by intramolecular interactions with the cap domain, the functional equivalent of triggering a mouse trap. However, the anti-LukE antibody freezes the toxin in the inactive conformation, thereby blocking the cytolytic functional properties of Luk in which the fully assembled oligomeric superstructure punches a pore/hole in the host cell membrane, causing intoxication and cell death (for reviews, see Alonzo and Torres 2014; Reyes-Robles and Torres 2016). Notably, this inhibitory anti-LukE3 antibody was shown to bind to a primary sequence comprising two antiparallel β-strands connected by a hairpin loop (Fig. 4) that fixes the holoprotein in the nonfunctional conformation (Hernandez et al. 2020b).

From selection of gene fragment libraries with a diverse panel of postimmunization monoclonal antibodies, we found that the polyclonal serum antibodies induced by immunization with different PFT proteins in some cases cross-reacted with the same toxin. By phage display, we documented examples in which cross-reactivity was due to antibody binding to highly homologous, near-identical, subregions in different toxins that are immunogenic after immunization.

We speculate that in the fully assembled pore structure, the antigen receptor of a B cell may not be able to gain access to all regions on the surface of molecules in the extended stem conformation, and this may, in part, explain the greater immunogenicity of some subregions. β-strands themselves compose the greatest proportion of the surface of a PFT molecule, but we did not identify any antibodies that bound solely to a primary sequence assignable to an individual β-strand. Importantly, other antibodies may instead be able to access and bind sites only expressed in molecules in the inactive conformation. Such a binding interaction may stabilize and fix the overall structure to prevent the executioner phase responsible for cytotoxic activity. Alternatively, an antibody may neutralize leucocidin activity by interfering with the sites essential to dimerization or by blocking subsequent intermolecular interactions essential for cell killing (Hernandez et al. 2020b). Although not all epitopes characterized were associated with toxin function, we speculate that many of the antibodies that we characterized, including the antibodies cross-reactive between different PFT proteins, may aid clearance from the host and thereby limit their toxicity.

Chapter 11

## CONCLUDING REMARKS

Here we have described our rationale for developing a phage display vector system based on the pComb family, which has provided a practical and time-efficient means to identify a multitude of minimal B-cell epitopes expressed on a protein antigen (Hernandez et al. 2020a,b; Protocol 1: Cloning and Selection from Antigen Fragment Libraries for Epitope Identification [Silverman 2025]). Furthermore, we discussed our first steps in developing this approach, which aimed to identify epitopes on clinically important toxins produced by an SA isolate recovered from a patient with clinical infection (Radke et al. 2018).

With our pComb-Opti8 system, we used a panel of mAbs to interrogate PFT gene fragment libraries, an approach that was successful in isolating fragment clones that were bound by a mAb in about half of our experiments (Hernandez et al. 2020b). From the epitope data, as described above, we generated synthetic peptides in murine immunization studies, which directly assessed the capacity of such epitopes for immunogenicity in an immunologically naive host. This peptide conjugate elicited IgG responses that were highly reactive with both the autologous synthetic peptide and the full-length Luk toxin homologs.

Our experience to date with the pComb-Opti8 system provides a different perspective on the classical teaching for the structural nature of B-cell epitopes. It has previously been taught that 80% of antibody clones recognize conformational surfaces, made up of composite surfaces composed of amino acid contributions from noncontiguous portions of a protein antigen. The rest were thought to represent B-cell epitopes that are primary amino acid sequence–dependent linear epitopes without real three-dimensional structure. This model for "primary sequence–dependent epitopes" dates to the work of Mario Geysen in the 1980s and should be reconsidered (Geysen et al. 1984). Our experimental design is very different from that of Geysen, in which synthetic peptides were immobilized directly onto a solid phase for binding studies (Geysen et al. 1984), and this format likely reduced any capacity for conformational folding. In our studies, peptides were synthesized and attached to a small (SerGlySerGly) linker to connect to a carboxy-terminal biotin. With such constructs, avidin is placed directly on the precoat and captures the biotinylated tag, while we assume the peptide itself, in the solution phase, adopts thermodynamically favored conformations that overlap with those adopted in the native protein.

Notably, from each successful panning experiment with toxin fragment libraries, we found that binding was based on a primary amino acid sequence. We also found evidence that antibody binding required the recognized peptide to fold into a conformation, as antibody binding to the native toxin protein was blocked by the selected phagemid clone or a synthetic peptide with the same primary amino acid sequence when present in solution phase. Hence, based on our studies, we conclude that gene fragment phage display libraries can identify primary sequence–dependent epitopes recapitulated by small oligopeptides. Even when such an epitope was composed of only 11 sequential amino acids, this minimal epitope adopted a conformation recapitulating that expressed in the native fold of the entire protein, as the capacity to inhibit binding of the antibody to the native protein in solution suggested functional equivalence with the native epitope within the conformational full-length protein structure. Moreover, in our pilot surface plasmon resonance studies, we determined a $K_d$ of 27–96 n$M$ for interactions with a 15-mer peptide, an affinity within the range of typical antibody–antigen interactions (Liang et al. 2007; GJ Silverman unpubl.).

With a gene fragment– and phage display–based pipeline, we have identified and validated immunogenic B-cell epitopes that are cross-reactive between members of the pore-forming leucocidin family. This approach could be harnessed to identify novel epitopes for a much-needed *S. aureus*–protective subunit vaccine. Taken together, we envision that this type of immunogenic, minimal B-cell epitope, linked to toxin functional neutralization, could have utility as a component in a multimodular vaccine, although it may also require other components, such as an adjuvant that conveys Th17-biased responses, which may further enhance host protection (Proctor 2012).

The pComb-Opti8 system discussed here has provided new perspectives on B-cell epitopes. These findings further confirm that such epitopes can be described as examples of validated "primary sequence–defined conformational epitopes" that retain key features shared with the much larger toxin protein. Notably, these validated peptide epitopes can also be used as metrics to guide preclinical reverse vaccinology efforts that seek to mold inducible antibody responses with enhanced protective features against SA infections.

We have recently documented that such primary sequence–dependent B-cell epitopes are also prominent contributors to antibody-mediated neutralization of HlgCB cytotoxicity for human polymorphonuclear leukocytes (Hernandez et al. 2020b). Our studies provide further validation of newly developed tools that should be widely applicable for the evaluation of the "fitness" of potential vaccine components of an antistaphylococcal antibody response for recognition of neutralization-associated epitopes.

This and other independently developed practical phage display tools (Wang and Yu 2004) can be harnessed to accelerate the generation of vaccine modules targeting subdomains from a range of pathogen proteins and virulence factors. These generated vaccines can induce immunity to functionally inactivate more toxins and virulence factors, enhance the clearance of bacteria, and provide immunity to the broadest range of patients. All efforts to develop a protective SA vaccine have failed to date, and it can also be argued that single-component vaccines may be inadequate to overcome the multitiered assault associated with SA invasive infection. By considering epitopes that interact with diverse genetic immune response elements (e.g., MHC II alleles), this approach aims to induce immunity beyond that provided by the antibody repertoire produced by B lymphocytes and to counter different forms of inherited genetic factors that increase susceptibility to current and emerging pathogens.

## ACKNOWLEDGMENTS

I appreciate the contributions to the development of the ideas discussed here and the technical contributions of three New York University (NYU) students: Adam Pelzek, Emily Radke, and David Hernandez. I especially thank Bo Shopin, Victor Torres, Tim Cardozo, Beatrix Uberheide, and all of our collaborators and colleagues. PFT neutralization studies were performed in the laboratory of Victor Torres (formerly at NYU). The earlier development of phage display technology in our research group was refined by Paul Roben, Stephen Cary, Carl Goodyear, and Caroline Grönwall, each of whom subsequently went on to establish their research laboratories. The inspiration for many of these studies came from conversations and encounters at our annual course on "Antibody Engineering and Display Technologies" at Cold Spring Harbor Laboratory. I appreciate the contributions over the years of Dennis Burton, Don Siegel, our coinstructors, our technical assistants, and all of our students at the course. I am indebted to the contributions and vision of my close friend, Carlos F. Barbas III, born Maurice Bernstein. The work discussed here was supported in part by National Institutes of Health (NIH)-National Institute of Allergy and Infectious Diseases HHSN272201400019C, "B-Cell Epitope Discovery and Mechanisms of Antibody Protection" (G.J.S.), and NIH-National Institute of Arthritis and Musculoskeletal and Skin Diseases P50 AR070591-01A1 and T32GM66704. Flow cytometry and genomics support was provided by NYU Langone's Cytometry and Cell Sorting Laboratory and the NYU Langone Health Genome Technology Center, supported in part by NIH-National Cancer Institute P30CA016087.

## REFERENCES

Alonzo F III, Torres VJ. 2014. The bicomponent pore-forming leucocidins of *Staphylococcus aureus*. *Microbiol Mol Biol Rev* **78:** 199–230. doi:10.1128/MMBR.00055-13

Ammerlaan HS, Harbarth S, Buiting AG, Crook DW, Fitzpatrick F, Hanberger H, Herwaldt LA, van Keulen PH, Kluytmans JA, Kola A, et al. 2013. Secular trends in nosocomial bloodstream infections: antibiotic-resistant bacteria increase the total burden of infection. *Clin Infect Dis* **56:** 798–805. doi:10.1093/cid/cis1006

Badarau A, Trstenjak N, Nagy E. 2017. Structure and function of the two-component cytotoxins of *Staphylococcus aureus* - learnings for designing novel therapeutics. *Adv Exp Med Biol* **966:** 15–35. doi:10.1007/5584_2016_200

Barbas SM, Ditzel HJ, Salonen EM, Yang WP, Silverman GJ, Burton DR. 1995. Human autoantibody recognition of DNA. *Proc Natl Acad Sci* **92:** 2529–2533. doi:10.1073/pnas.92.7.2529

Bennett MR, Bombardi RG, Kose N, Parrish EH, Nagel MB, Petit RA, Read TD, Schey KL, Thomsen IP, Skaar EP, et al. 2019. Human mAbs to *Staphylococcus aureus* IsdA provide protection through both heme-blocking and Fc-mediated mechanisms. *J Infect Dis* **219:** 1264–1273. doi:10.1093/infdis/jiy635

Berry JD, Rutherford J, Silverman GJ, Kaul R, Elia M, Gobuty S, Fuller R, Plummer FA, Barbas CF III. 2003. Development of functional human monoclonal single-chain variable fragment antibody against HIV-1 from human cervical B cells. *Hybrid Hybridomics* **22:** 97–108. doi:10.1089/153685903321948021

Blythe MJ, Flower DR. 2005. Benchmarking B cell epitope prediction: underperformance of existing methods. *Protein Sci* **14:** 246–248. doi:10.1110/ps.041059505

Cariccio VL, Domina M, Benfatto S, Venza M, Venza I, Faleri A, Bruttini M, Bartolini E, Giuliani MM, Santini L, et al. 2016. Phage display revisited: epitope mapping of a monoclonal antibody directed against *Neisseria meningitidis* adhesin A using the PROFILER technology. *MAbs* **8:** 741–750. doi:10.1080/19420862.2016.1158371

Cary SP, Lee J, Wagenknecht R, Silverman GJ. 2000. Characterization of superantigen-induced clonal deletion with a novel clan III-restricted avian monoclonal antibody: exploiting evolutionary distance to create antibodies specific for a conserved VH region surface. *J Immunol* **164:** 4730–4741. doi:10.4049/jimmunol.164.9.4730

Corey GR. 2009. *Staphylococcus aureus* bloodstream infections: definitions and treatment. *Clin Infect Dis* **48:** S254–S259. doi:10.1086/598186

El Garch F, Hallin M, De Mendonca R, Denis O, Lefort A, Struelens MJ. 2009. StaphVar-DNA microarray analysis of accessory genome elements of community-acquired methicillin-resistant *Staphylococcus aureus*. *J Antimicrob Chemother* **63:** 877–885. doi:10.1093/jac/dkp089

Eskola J, Peltola H, Takala AK, Kayhty H, Hakulinen M, Karanko V, Kela E, Rekola P, Ronnberg PR, Samuelson JS, et al. 1987. Efficacy of *Haemophilus influenzae* type b polysaccharide-diphtheria toxoid conjugate vaccine in infancy. *N Engl J Med* **317:** 717–722. doi:10.1056/NEJM198709173171201

Fitzgerald JR, Sturdevant DE, Mackie SM, Gill SR, Musser JM. 2001. Evolutionary genomics of *Staphylococcus aureus*: insights into the origin of methicillin-resistant strains and the toxic shock syndrome epidemic. *Proc Natl Acad Sci* **98:** 8821–8826. doi:10.1073/pnas.161098098

Foster TJ. 2005. Immune evasion by staphylococci. *Nat Rev Microbiol* **3:** 948–958. doi:10.1038/nrmicro1289

Fowler VG Jr, Proctor RA. 2014. Where does a *Staphylococcus aureus* vaccine stand? *Clin Microbiol Infect* **20:** 66–75. doi:10.1111/1469-0691.12570

Fuhner V, Heine PA, Zilkens KJC, Meier D, Roth KDR, Moreira G, Hust M, Russo G. 2019. Epitope mapping via phage display from single-gene libraries. *Methods Mol Biol* **1904:** 353–375. doi:10.1007/978-1-4939-8958-4_17

Geysen HM, Meloen RH, Barteling SJ. 1984. Use of peptide synthesis to probe viral antigens for epitopes to a resolution of a single amino acid. *Proc Natl Acad Sci* **81:** 3998–4002. doi:10.1073/pnas.81.13.3998

Graille M, Stura EA, Corper AL, Sutton BJ, Taussig MJ, Charbonnier JB, Silverman GJ. 2000. Crystal structure of a *Staphylococcus aureus* protein A domain complexed with the Fab fragment of a human IgM antibody: structural basis for recognition of B-cell receptors and superantigen activity. *Proc Natl Acad Sci* **97:** 5399–5404. doi:10.1073/pnas.97.10.5399

Gronwall C, Charles ED, Dustin LB, Rader C, Silverman GJ. 2014. Selection of apoptotic cell specific human antibodies from adult bone marrow. *PLoS One* **9:** e95999. doi:10.1371/journal.pone.0095999

He B, Dzisoo AM, Derda R, Huang J. 2019. Development and application of computational methods in phage display technology. *Curr Med Chem* **26:** 7672–7693. doi:10.2174/0929867325666180629123117

Hernandez DN, Tam K, Shopsin B, Radke EE, Kolahi P, Copin R, Stubbe FX, Cardozo T, Torres VJ, Silverman GJ. 2020a. Unbiased identification of immunogenic *Staphylococcus aureus* leukotoxin B-cell epitopes. *Infect Immun* **88:** e00785-19. doi:10.1128/IAI.00785-19

Hernandez DN, Tam K, Shopsin B, Radke EE, Law K, Cardozo T, Torres VJ, Silverman GJ. 2020b. Convergent evolution of neutralizing antibodies to *Staphylococcus aureus* γ-hemolysin C that recognize an immunodominant primary sequence-dependent B-cell epitope. *mBio* **11:** e00460-20. doi:10.1128/mBio.00460-20

Holland TL, Arnold C, Fowler VG Jr. 2014. Clinical management of *Staphylococcus aureus* bacteremia: a review. *JAMA* **312:** 1330–1341. doi:10.1001/jama.2014.9743

Jumper J, Evans R, Pritzel A, Green T, Figurnov M, Ronneberger O, Tunyasuvunakool K, Bates R, Žídek A, Potapenko A, et al. 2021. Highly accurate protein structure prediction with AlphaFold. *Nature* **596:** 583–589. doi:10.1038/s41586-021-03819-2

Klevens RM, Morrison MA, Nadle J, Petit S, Gershman K, Ray S, Harrison LH, Lynfield R, Dumyati G, Townes JM, et al. 2007. Invasive methicillin-resistant *Staphylococcus aureus* infections in the United States. *JAMA* **298:** 1763–1771. doi:10.1001/jama.298.15.1763

Lautenschlager S, Herzog C, Zimmerli W. 1993. Course and outcome of bacteremia due to *Staphylococcus aureus*: evaluation of different clinical case definitions. *Clin Infect Dis* **16:** 567–573. doi:10.1093/clind/16.4.567

Laventie BJ, Guerin F, Mourey L, Tawk MY, Jover E, Maveyraud L, Prevost G. 2014. Residues essential for Panton-Valentine leukocidin S component binding to its cell receptor suggest both plasticity and adaptability in its interaction surface. *PLoS One* **9:** e92094. doi:10.1371/journal.pone.0092094

Liang M, Klakamp SL, Funelas C, Lu H, Lam B, Herl C, Umble A, Drake AW, Pak M, Ageyeva N, et al. 2007. Detection of high- and low-affinity antibodies against a human monoclonal antibody using various technology platforms. *Assay Drug Dev Technol* **5:** 655–662. doi:10.1089/adt.2007.089

Limbago B, Fosheim GE, Schoonover V, Crane CE, Nadle J, Petit S, Heltzel D, Ray SM, Harrison LH, Lynfield R, et al. 2009. Characterization of methicillin-resistant *Staphylococcus aureus* isolates collected in 2005 and 2006 from patients with invasive disease: a population-based analysis. *J Clin Microbiol* **47:** 1344–1351. doi:10.1128/JCM.02264-08

Lina G, Piemont Y, Godail-Gamot F, Bes M, Peter MO, Gauduchon V, Vandenesch F, Etienne J. 1999. Involvement of Panton–Valentine leukocidin-producing *Staphylococcus aureus* in primary skin infections and pneumonia. *Clin Infect Dis* **29:** 1128–1132. doi:10.1086/313461

Lindsay JA, Holden MT. 2004. *Staphylococcus aureus*: superbug, super genome? *Trends Microbiol* **12:** 378–385. doi:10.1016/j.tim.2004.06.004

Lowy FD. 1998. *Staphylococcus aureus* infections. *N Engl J Med* **339:** 520–532. doi:10.1056/NEJM199808203390806

Meyer M, Belke DD, Trost SU, Swanson E, Dieterle T, Scott B, Cary SP, Ho P, Bluhm WF, McDonough PM, et al. 2004. A recombinant antibody increases cardiac contractility by mimicking phospholamban phosphorylation. *FASEB J* **18:** 1312–1314. doi:10.1096/fj.03-1231fje

Moran GJ, Krishnadasan A, Gorwitz RJ, Fosheim GE, McDougal LK, Carey RB, Talan DA. 2006. Methicillin-resistant *S. aureus* infections among patients in the emergency department. *N Engl J Med* **355:** 666–674. doi:10.1056/NEJMoa055356

Nguyen VT, Kamio Y. 2004. Cooperative assembly of β-barrel pore-forming toxins. *J Biochem* **136:** 563–567. doi:10.1093/jb/mvh160

Noren KA, Noren CJ. 2001. Construction of high-complexity combinatorial phage display peptide libraries. *Methods* **23:** 169–178. doi:10.1006/meth.2000.1118

Peltola H, Makela H, Kayhty H, Jousimies H, Herva E, Hallstrom K, Sivonen A, Renkonen OV, Pettay O, Karanko V, et al. 1977. Clinical efficacy of meningococcus group A capsular polysaccharide vaccine in children three months to five years of age. *N Engl J Med* **297:** 686–691. doi:10.1056/NEJM197709292971302

Pelzek AJ, Shopsin B, Radke EE, Tam K, Ueberheide BM, Fenyo D, Brown SM, Li Q, Rubin A, Fulmer Y, et al. 2018. Human memory B cells targeting *Staphylococcus aureus* exotoxins are prevalent with skin and soft tissue infection. *mBio* **9:** e02125-17. doi:10.1128/mBio.02125-17

Pier GB. 2013. Will there ever be a universal *Staphylococcus aureus* vaccine? *Hum Vaccin Immunother* **9:** 1865–1876. doi:10.4161/hv.25182

Piewngam P, Zheng Y, Nguyen TH, Dickey SW, Joo HS, Villaruz AE, Glose KA, Fisher EL, Hunt RL, Li B, et al. 2018. Pathogen elimination by probiotic *Bacillus* via signalling interference. *Nature* **562:** 532–537. doi:10.1038/s41586-018-0616-y

Proctor RA. 2012. Is there a future for a *Staphylococcus aureus* vaccine? *Vaccine* **30:** 2921–2927. doi:10.1016/j.vaccine.2011.11.006

Rader C. 2024. The pComb3 phagemid family of phage display vectors. *Cold Spring Harb Protoc* doi:10.1101/pdb.over107756

Radke EE, Brown SM, Pelzek AJ, Fulmer Y, Hernandez DN, Torres VJ, Thomsen IP, Chiang WK, Miller AO, Shopsin B, et al. 2018. Hierarchy of human IgG recognition within the *Staphylococcus aureus* immunome. *Sci Rep* **8**: 13296. doi:10.1038/s41598-018-31424-3

Ramsay ME, Andrews N, Kaczmarski EB, Miller E. 2001. Efficacy of meningococcal serogroup C conjugate vaccine in teenagers and toddlers in England. *Lancet* **357**: 195–196. doi:10.1016/S0140-6736(00)03594-7

Reyes-Robles T, Torres VJ. 2016. *Staphylococcus aureus* pore-forming toxins. *Curr Top Microbiol Immunol* **409**: 121–144. doi:10.1007/82_2016_16

Rhyner C, Weichel M, Fluckiger S, Hemmann S, Kleber-Janke T, Crameri R. 2004. Cloning allergens via phage display. *Methods* **32**: 212–218. doi:10.1016/j.ymeth.2003.08.003

Roben P, Barbas SM, Sandoval L, Lecerf JM, Stollar BD, Solomon A, Silverman GJ. 1996. Repertoire cloning of lupus anti-DNA autoantibodies. *J Clin Invest* **98**: 2827–2837. doi:10.1172/JCI119111

Ryvkin A, Ashkenazy H, Weiss-Ottolenghi Y, Piller C, Pupko T, Gershoni JM. 2018. Phage display peptide libraries: deviations from randomness and correctives. *Nucleic Acids Res* **46**: e52. doi:10.1093/nar/gky077

Sasano M, Burton DR, Silverman GJ. 1993. Molecular selection of human antibodies with an unconventional bacterial B cell antigen. *J Immunol* **151**: 5822–5839. doi:10.4049/jimmunol.151.10.5822

Shaw PX, Horkko S, Tsimikas S, Chang MK, Palinski W, Silverman GJ, Chen PP, Witztum JL. 2001. Human-derived anti-oxidized LDL autoantibody blocks uptake of oxidized LDL by macrophages and localizes to atherosclerotic lesions in vivo. *Arterioscler Thromb Vasc Biol* **21**: 1333–1339. doi:10.1161/hq0801.093587

Shinefield HR, Black S. 2000. Efficacy of pneumococcal conjugate vaccines in large scale field trials. *Pediatr Infect Dis J* **19**: 394–397. doi:10.1097/00006454-200004000-00036

Silverman GJ. 2025. Cloning and selection from antigen fragment libraries for epitope identification. *Cold Spring Harb Protoc* doi:10.1101/pdb.prot108660

Silverman GJ, Barbas S, Roben P, Burton DR. 1995a. Repertoire cloning of human lupus autoantibodies. *Ann N Y Acad Sci* **764**: 565–566. doi:10.1111/j.1749-6632.1995.tb55882.x

Silverman GJ, Roben P, Bouvet JP, Sasano M. 1995b. Superantigen properties of a human sialoprotein involved in gut-associated immunity. *J Clin Invest* **96**: 417–426. doi:10.1172/JCI118051

Smith GP. 1985. Filamentous fusion phage: novel expression vectors that display cloned antigens on the virion surface. *Science* **228**: 1315–1317. doi:10.1126/science.4001944

Smith GP. 2019. Phage display: simple evolution in a Petri dish (Nobel lecture). *Angew Chem Int Ed Engl* **58**: 14428–14437. doi:10.1002/anie.201908308

Song L, Hobaugh MR, Shustak C, Cheley S, Bayley H, Gouaux JE. 1996. Structure of staphylococcal α-hemolysin, a heptameric transmembrane pore. *Science* **274**: 1859–1866. doi:10.1126/science.274.5294.1859

Spaan AN, van Strijp JAG, Torres VJ. 2017. Leukocidins: staphylococcal bicomponent pore-forming toxins find their receptors. *Nat Rev Microbiol* **15**: 435–447. doi:10.1038/nrmicro.2017.27

Sugawara T, Yamashita D, Kato K, Peng Z, Ueda J, Kaneko J, Kamio Y, Tanaka Y, Yao M. 2015. Structural basis for pore-forming mechanism of staphylococcal α-hemolysin. *Toxicon* **108**: 226–231. doi:10.1016/j.toxicon.2015.09.033

Talan DA, Krishnadasan A, Gorwitz RJ, Fosheim GE, Limbago B, Albrecht V, Moran GJ. 2011. Comparison of *Staphylococcus aureus* from skin and soft-tissue infections in US emergency department patients, 2004 and 2008. *Clin Infect Dis* **53**: 144–149. doi:10.1093/cid/cir308

van Hal SJ, Jensen SO, Vaska VL, Espedido BA, Paterson DL, Gosbell IB. 2012. Predictors of mortality in *Staphylococcus aureus* bacteremia. *Clin Microbiol Rev* **25**: 362–386. doi:10.1128/CMR.05022-11

Wang LF, Yu M. 2004. Epitope identification and discovery using phage display libraries: applications in vaccine development and diagnostics. *Curr Drug Targets* **5**: 1–15. doi:10.2174/1389450043490668

Weiss GA, Wells JA, Sidhu SS. 2000. Mutational analysis of the major coat protein of M13 identifies residues that control protein display. *Protein Sci* **9**: 647–654. doi:10.1110/ps.9.4.647

Williamson RA, Silverman GJ. 2001. Gene fragment libraries and genomic and cDNA expression cloning. In *Phage display: a laboratory manual* (ed. Barbas CF III, et al.), pp. 6.1–6.11. Cold Spring Harbor Laboratory Press, Cold Spring Harbor, NY.

Willyard C. 2017. The drug-resistant bacteria that pose the greatest health threats. *Nature* **543**: 15. doi:10.1038/nature.2017.21550

Yamashita K, Kawai Y, Tanaka Y, Hirano N, Kaneko J, Tomita N, Ohta M, Kamio Y, Yao M, Tanaka I. 2011. Crystal structure of the octameric pore of staphylococcal γ-hemolysin reveals the β-barrel pore formation mechanism by two components. *Proc Natl Acad Sci* **108**: 17314–17319. doi:10.1073/pnas.1110402108

Yamashita D, Sugawara T, Takeshita M, Kaneko J, Kamio Y, Tanaka I, Tanaka Y, Yao M. 2014. Molecular basis of transmembrane β-barrel formation of staphylococcal pore-forming toxins. *Nat Commun* **5**: 4897. doi:10.1038/ncomms5897

Zwick MB, Bonnycastle LL, Noren KA, Venturini S, Leong E, Barbas CF 3rd, Noren CJ, Scott JK. 1998. The maltose-binding protein as a scaffold for monovalent display of peptides derived from phage libraries. *Anal Biochem* **264**: 87–97. doi:10.1006/abio.1998.2793

Protocol 1

# Cloning and Selection from Antigen Fragment Libraries for Epitope Identification

Gregg J. Silverman[1]

*Department of Medicine, New York University Grossman School of Medicine, New York, New York 10016, USA*

To understand what drives an immune response, it is important to characterize, at a molecular level, the site(s) on an immunogenic antigen that is directly contacted by a soluble antibody or B-cell antigen receptor (BCR) on the surface of a B lymphocyte. Moreover, antibody binding interactions with a microbial protein can interfere with the functional activity of a toxin (i.e., neutralization) and/or can aid in the clearance of the microbial protein from the body, further underscoring the importance of such characterization. Phage display technology is a potent tool that can be used to study any type of protein–protein interaction. In recent years, we have refined methods for the identification of the minimal binding contact sites of an antibody with an antigen. Here, we describe a workflow for optimizing antibody-mediated selection and for the identification and characterization of antigen-specific epitopes. This workflow includes (1) the generation of large libraries of random fragments of a gene of interest cloned into the validated pComb-Opti8 phagemid expression cloning vector system; (2) electroporation of these libraries into electrocompetent bacterial cells and subsequent recovery of viral particles, each of which displays the cloned gene fragment product as a fusion protein with the filamentous phage major coat protein VIII (pVIII); (3) recovery of individual phagemid clones that express the smallest functional epitopes recognized by an experimental antibody; (4) an efficient means of using high-throughput DNA sequencing to interrogate sequentially selected libraries to rapidly identify the gene subregions encoding epitopes of interest; and (5) means for the further characterization of potential antibody–epitope binding interactions.

## MATERIALS

You must consult the appropriate Material Safety Data Sheets and your institution's Environmental Health and Safety Office for proper handling of equipment and hazardous materials used in this protocol.

RECIPES: Please see the end of this protocol for recipes indicated by <R>. Additional recipes can be found online at http://cshprotocols.cshlp.org/site/recipes.

### Reagents

Agarose, molecular biology–grade (e.g., Bio-Rad 1613100EDU)
Blocker Casein (Thermo Fisher 37583)
*This item is a ready-to-use solution of purified casein for blocking steps.*

Bovine serum albumin (BSA; Sigma-Aldrich A3294)

---

[1]Correspondence: gregg.silverman@nyulangone.org

© 2026 Cold Spring Harbor Laboratory Press
Cite this protocol as *Cold Spring Harb Protoc*; doi:10.1101/pdb.prot108660

BSA (2% w/v; in 1× PBS)
> *Filter-sterilize and store for up to 3 d at 4°C.*

Carbenicillin stock solution <R>
Custom oligonucleotides
> *These oligonucleotides will be designed and ordered from a vendor for use in PCR-based amplification of the gene of interest (see Section A). Other oligonucleotides will be needed for the amplification of sequential libraries for DNA sequence determination (see Section E). Many vendors are available, and we have used Integrated DNA Technologies (IDT).*

DEPC-treated water (Thermo Scientific AM9916) or similar nuclease-free water
EcoRV-HF restriction enzyme and rCutSmart buffer (NEB R3195S)
*Escherichia coli* TG1 strain, electrocompetent (Lucigen LGC 60502-1)
*Escherichia coli* XL-1 Blue (Agilent 200249)
Ethanol (100%, 80%, and 10% v/v; in double-distilled water [ddH$_2$O])
Gel loading dye, blue (6×; NEB B7021S)
Glucose (1% w/v; in water) (optional; see the note above Step 45)
Glycine HCl buffer (1 M, pH 2.5) (Bioworld 40125039)
Glycogen, molecular biology–grade (Thermo Scientific R0561)
Helper phage, VCSM13 (Agilent Technologies 200251)
HRP-tagged anti-M13 antibody (Thermo Fisher MA5-36126)
Immunoglobulin isotype control
> *The isotype must match the antibody used for library selection.*

Kanamycin stock solution <R>
LB-carbenicillin agar plates <R>
Monarch DNA Gel Extraction Kit (NEB T1020L)
Monarch PCR & DNA Cleanup Kit (NEB T1030)
Monoclonal antibody to a protein immunogen of interest, used for library selection
> *For an antigen-specific project, consider using a panel of independent, antigen-specific monoclonal antibodies to your target antigen, which may be used singly or in combination for simultaneous selection.*

NaHCO$_3$ (0.1 M, pH 8.6)
NEBNext dsDNA Fragmentase (NEB M0348S)
NEBNext High-Fidelity 2× PCR Master Mix (NEB M0541)
Phosphate-buffered saline (PBS; 10×), sterile (Fisher Scientific AM9624)
Plasmid or genomic DNA containing the gene of interest to be used in phage display (see note under Method heading for details)
Plasmid prep kit (QIAprep Spin Miniprep Kit 27106, or similar)
Polyethylene glycol (PEG)-8000
Prestained molecular weight markers for electrophoresis
Protein G–coated paramagnetic beads (Pierce 88847)
pComb-Opti8 vector (Creative Biogene, special order)
> *See Hernandez et al. (2020a) for details on the construction of the pComb-Opti8 vector.*

Quick Blunting Kit (NEB E1201S)
SB medium <R>
Shrimp alkaline phosphatase, recombinant, and rCutSmart buffer (NEB M0371S)
SOC medium <R>
Sodium acetate (3 M, pH 5.2) <R>
Sodium azide solution (10% w/v; in water)
Sodium chloride (NaCl)
SuperSignal ELISA Pico Chemiluminescent Substrate (Thermo Scientific 37069)
T4 DNA ligase and buffer (NEB M0202SX)

Chapter 11

TE buffer <R>
Tetracycline stock solution <R>
Tris–acetate–EDTA buffer (TAE; 50×) (Thermo Fisher B49)
Tris-buffered saline, with 1% (w/v) BSA (pH 8.0) (1% BSA–TBS) (Millipore Sigma T6789)
Tris HCl (50 mM, pH 9.1)
Tween 20 (Promega PRH5152)
Tween 20 (0.05% and 0.1% v/v; in PBS)
Tween 20 (0.05% v/v; in PBS) with 1% (w/v) BSA

## Equipment

Aspirator pipette on flexible tubing attached to a waste collection bottle on a vacuum line
Cabinet incubator (static) for bacterial plates, at 37°C
Centrifuge floor unit with refrigeration (Beckman with JA-10 rotor)
Conical tubes, 50-mL, sterile
DNA electrophoresis system; e.g., Mini-Sub Cell GT Horizontal System (Bio-Rad 1704405)
Electroporation cuvettes (2-mm gap)
Electroporator unit (e.g., Bio-Rad Micropulse 1652100)
ELISA 96-well plates with lid (Costar, high binding Corning 3361, Fisher Scientific 07-200-721)
Erlenmeyer flasks, 500-mL, sterile (covered with aluminum foil)
Freezers at −20°C and −80°C
GelDoc EZ System (Bio-Rad system 1708270), or equivalent, for imaging of DNA gel
Heating blocks at 37°C, 42°C, and 65°C
Ice bucket
Litmus paper
Microcentrifuge, tabletop, with refrigeration
Microcentrifuge tubes, 1.5- and 2-mL
Micropipettes, 10-, 20-, and 200-μL
Micropipette sterile tips, 10- and 100-μL (Fisher Scientific)
Multichannel 12-tip pipette that uses 100-μL pipette tips
NanoDrop instrument or equivalent spectrophotometric device
NEBNEXT magnetic separation stand (NEB S1515S)
Paper towels
Pipet-Aid, battery-powered, for 1-, 2-, and 10-mL serologic pipettes
Plate sealer sheets (Millipore BR701365)
Plate spreaders, disposable (Fisher 14-665-230)
Platform rocker
PowerPac Basic Power Supply (Bio-Rad 1645050), or equivalent
Refrigerator at 4°C
Rotating mixer
Serological plastic pipettes, 2- and 10-mL (Fisher 170372N and 170356N)
Shaking bacterial incubator at 37°C
Single-edge razor blade (single-use)
Spectrophotometer (with compatible cuvettes to read $OD_{600}$ from ELISA)
Syringe push filter, for aggregate and viable bacteria removal, 0.2-mm (Thermo Scientific 723-2520)
TapeStation 4200 system, for DNA fragment analysis (Agilent G2991BA)
Thermocycler
Vortex mixer

## METHOD

*For the identification of antigens expressed in proteins, we have developed an integrated approach for the generation and selection from gene fragment libraries using a variant of the pComb system that enables fusions with the pVIII gene (gVIII) from the M13 filamentous phage. This approach has been used to identify minimal epitopes on antigen proteins. The genes that encode these protein antigens are first fragmented, and the fragments are then cloned into the custom-built pComb-Opti8 vector (Fig. 1). In our associated article (Introduction: Insights from the Study of B-Cell Epitopes of a Microbial Pathogen by Phage Display [Silverman 2025]), we summarize findings from earlier reports using this system (Radke et al. 2018; Hernandez et al. 2020a,b).*

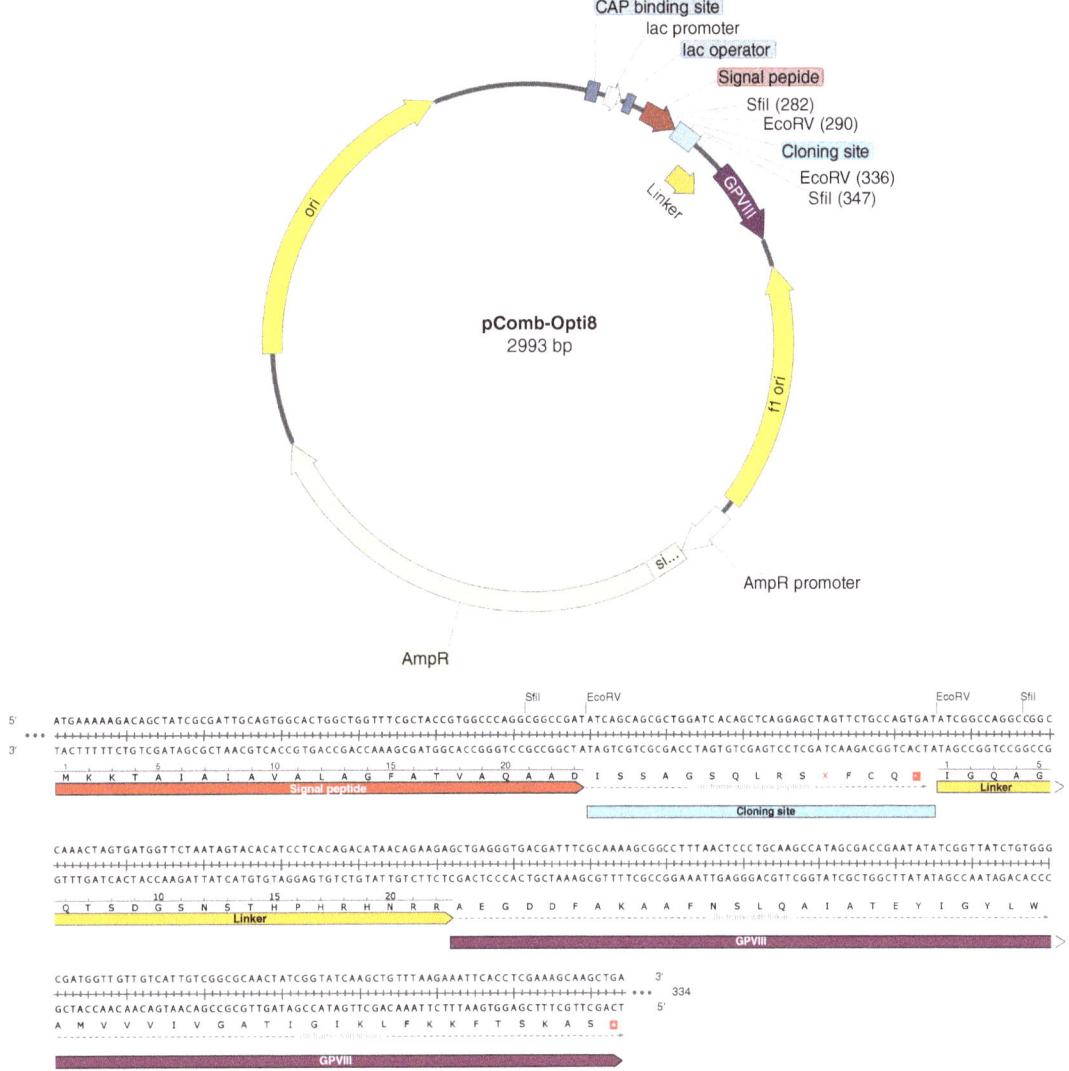

**FIGURE 1.** pComb-Opti8 vector. The pComb-Opti8 vector is designed for multivalent display of gene fragment products on filamentous bacteriophage (Hernandez et al. 2020a). The cloning site, linker region, and modified coat protein VIII (pVIII) were specially designed for this gene fragment phage display vector system (Hernandez et al. 2020a). This vector was generated by modification of the pComb3X phagemid system, designed for monomeric gene fusion display (Andris-Widhopf et al. 2000; see Chapter 4, Overview: The pComb3 Phagemid Family of Phage Display Vectors [Rader 2024]). For the dedicated use of cloning *Staphylococcal* epitopes, the pComb-Opti8 vector was generated using custom oligonucleotides to introduce a flexible linker, followed by introduction of a truncated *gVIII* optimized for the expression of a fusion protein with the major coat protein VIII (Weiss et al. 2000). In pilot studies, the functionality of this system was confirmed in an experiment by cloning a previously described high-affinity antitetanus toxoid (TT) human Fab (Andris-Widhopf et al. 2000). In later applications, the gene for this Fab was excised, and libraries of gene fragments were cloned in dedicated applications focused on the recovery and characterization of B-cell epitopes. Figure adapted from Hernandez et al. (2020a), © 2020 American Society for Microbiology, reproduced with permission.

# Chapter 11

*In this protocol, we describe a workflow to use antigen-specific monoclonal antibodies for the selection of gene fragment-encoded protein clones from antigen gene fragment libraries. In theory, polyclonal antibodies can be used instead.*

*In a pComb-Opti8 phagemid library, each clone has a gene fragment inserted into the cloning site of the vector that is packaged into a viral particle, which results in expression of a fusion protein of the major coat protein in physical linkage with the protein encoded by the gene fragment. Hence, selection is based on antibody recognition of its cognate epitope. The entire workflow has been organized here into five sections (A–E), which we outline below. For an overview of the procedure for the generation of the library and the selection of clones, see Figure 2.*

*In Section A, we explain how the gene of interest is first amplified using the polymerase chain reaction (PCR), how random gene fragments of small size are made from the PCR products by enzymatic digestion, and how these fragments are then prepared for cloning. The gene of interest can be in the form of a plasmid containing the cloned gene. Alternatively, one can start with purified genomic DNA from the microbial organism of interest. Custom oligonucleotides are designed and ordered for specific PCR-mediated amplification of the gene of interest. For the design of the oligonucleotides, apply well-validated rules of PCR and omit stop codons and leader sequences. A sense orientation oligonucleotide should be designed from the 5' end of the gene of interest, with a reverse antisense complementary oligonucleotide from the 3' end. After PCR and fragmentation of the PCR products, the gene fragments are end-polished to be compatible with the prepared vector.*

*Section B describes how the pComb-Opti8 vector is first prepared by EcoRV enzyme restriction digestion to generate a cloning site and then phosphatase-treated to prevent self-ligation.*

*In Section C, we describe how the prepared vector is mixed in different ratios with the gene fragments prepared in Section A. Parallel ligation reactions are performed with different molar ratios of the prepared inserts with a fixed amount of the prepared vector. Test ligations are then performed with ~10% of each of the ligation reactions to assess transformation efficiency. Electroporation of these ligated plasmids into electrocompetent bacteria is performed, and the resulting titer of the transformed bacterial cells will enable identification of the molar ratio of gene fragment DNA to prepared vector that yields the highest efficiency of transformation per microgram of DNA equivalent. This ratio is then used in scaled-up reactions, and larger libraries (containing more transformants) are generated.*

*This approach does not use restriction sites with sticky ends that would fix the DNA reading frame and orientation of the clones, and, therefore, half of the inserts will be ligated into the vector in an incorrect orientation in relation to the upstream lac promoter in the vector. The genes in these particular clones will not be expressed as a protein (discussed in Silverman 2001). If performed optimally, each microgram of ligated gene fragment vector can yield a library that comprises $10^7$ or more members, and the system will nonetheless be permissive of the clones in the library that are defective, due to orientation or reading frame, because the content of potentially functional clones will be sufficient to enable successful recovery of in-frame clones expressing the minimal epitope.*

*The vector includes an ampicillin resistance gene, and after these vector insert ligations are used to transform electro-competent E. coli (such as the TG1 strain of E. coli), the bacteria are grown in media containing ampicillin (or carbenicillin, which is more temperature-stable). At this step, helper phage, which provides all other phage proteins, enables recovery of the fragment library, with each gene fragment member packaged into individual filamentous phage particles.*

*On the surface of such a phage particle, multiple copies of that gene fragment protein product are displayed as a fusion with the solvent-exposed pVIII phage major coat protein. In practice, the library will have great clonal diversity, and if a large enough library is made, many copies of each member will be represented in the library. As mentioned above, by this approach, the library is more than sufficient for enabling the goal of recovering clones with in-frame functional antigenic epitopes.*

*In Section D, we describe an approach for gene fragment clonal selection that uses antibody-based binding of the products of the gene fragment, which selects specific phage clones based on protein–protein interactions.*

*Here, the user coats the wells of a 96-well polystyrene immunoassay plate with a selecting antibody for recovering phage clones. After blocking any residual nonspecific binding sites in the wells, the phage form of the library from Section C, in which phage particles display gene fragment protein products, of which a subset will be antibody-reactive, is added to the wells. Acid elution is used to subsequently release the bound phage particles from these binding interactions with the selecting antibody. The selected phage particles can be used to infect freshly grown E. coli cells, and the resultant phage collection is considered to be a sublibrary. These sublibraries now include highly represented clone members that express a small, structurally minimal B-cell epitope. With the addition of helper phage, these clones are "rescued" in the phage form as a selected sublibrary. To enable the isolation of the highest-affinity phagemid-containing clones of interest, iterative rounds of selection (also termed panning) with an antigen-specific antibody are then performed. From each round of panning, bacterial pellets, which contain plasmids that are later subjected to DNA sequence analysis, are isolated. We briefly discuss using an ELISA assay to evaluate the binding of phage clones isolated from the phage library.*

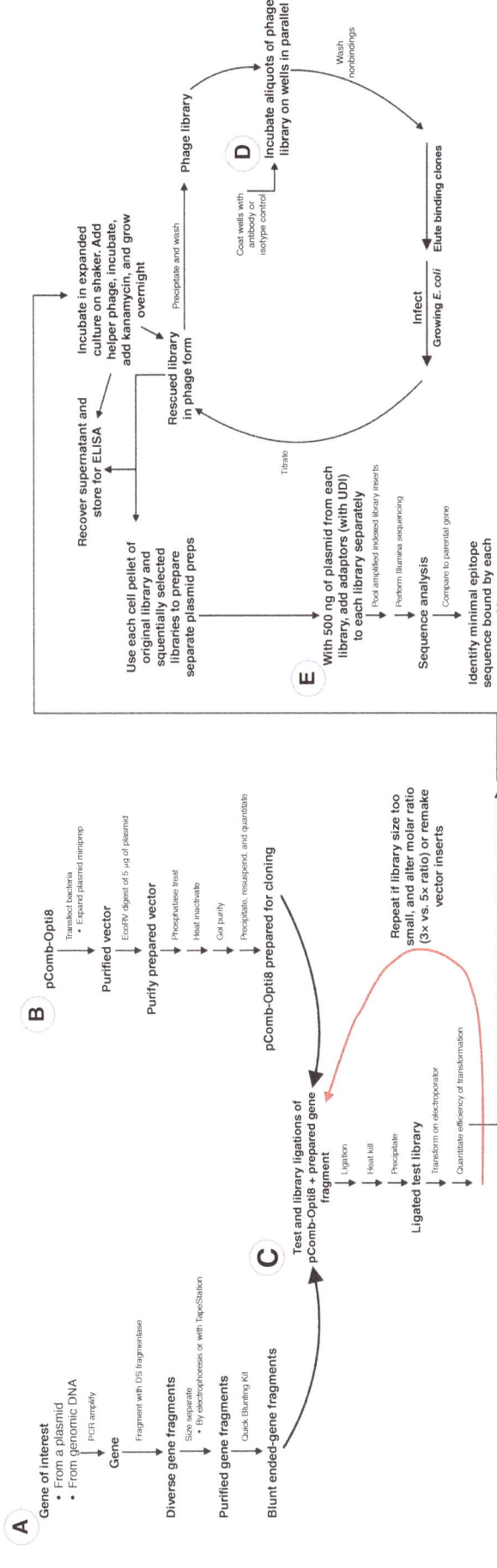

FIGURE 2. Flowchart for sequential steps in generating and selecting from a random gene fragment library using the pComb-Opti8 vector as described in this protocol. Sections A–E are as indicated in the text.

Chapter 11

*In Section E, we describe a method to tag every potential DNA sequence in a sublibrary using barcoded oligonucleotides that flank the cloning site in each individual pComb-Opti8 (sub)library (Fig. 3). Here, the user integrates a unique 10-residue oligonucleotide library-specific barcode into PCR products generated from the inserts in each sublibrary. With this tagging, each gene fragment sequence is linked to the sublibrary from which it came. These oligonucleotides reflect unique dual index (UDI) DNA sequences, which are depicted in Table 1.*

*Following this tagging, the PCR products of many libraries are pooled and then subjected to high-throughput DNA sequencing together. Based on the integrated barcode/UDI that identifies each source library, fragment genes can then later be assigned to the sublibrary from which they were isolated.*

*Comparisons with the sequentially selected libraries enable the identification of the minimal sequences that encode a functional epitope, bound by the selecting monoclonal antibody (mAb), that is localized within the protein encoded by the gene of interest. Thereby, the relative representation of subregion sequence variations across each of the selected phage sublibraries is determined.*

## Section A. Preparation of Gene Fragments for Cloning into a Phagemid Library

*Here, toward the goal of characterizing minimal B-cell epitopes, we first describe the initial step of the generation of a DNA library composed of fragments of a gene of interest. These fragments will be cloned into the pComb-Opti8 vector, which enables the expression of the fragments as fusion proteins with a portion of the filamentous phage major coat protein VIII on the phage surface.*

1. Design oligonucleotides that omit signal leader sequences and stop codons for the gene of interest. Using 20 ng of the full-length gene of interest in a 100-μL final reaction volume, conduct PCR using NEBNext High-Fidelity 2× PCR Master Mix according to the manufacturer's instructions. Prepare five or more individual PCR reactions, as at least 10 μg of PCR product will be needed to optimize the fragmentation method.

    *As an example, use oligonucleotide primers complementary to the 5′ end and antisense reverse for the 3′ end of the coding region of the full-length gene sequence of LukS of* Staphylococcus aureus, *as previously reported (for strain LACUSA300) (Hernandez et al. 2020a).*

    *For each PCR reaction for a gene of interest, determine the best reaction conditions and gene-specific oligonucleotides empirically.*

2. Pool the products of the separate amplification reactions into a single sterile 1.0-mL microcentrifuge tube.

    *As yield from a PCR can vary greatly, the yield of each reaction will need to be determined empirically using a NanoDrop device.*

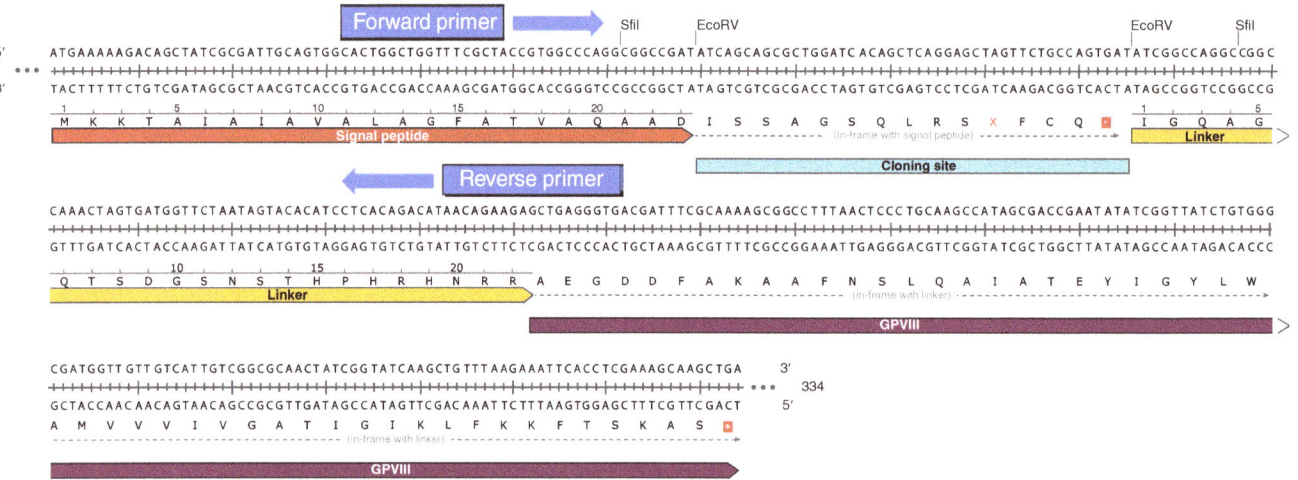

FIGURE 3. Design of PCR primers for embedding unique paired unique dual indexes (UDIs) into each pComb-Opti8 library. For individual library amplification, paired oligonucleotides with UDI sequences from NEB are shown in Table 1. The sequence shown here corresponds to a portion within the pComb-Opti8 vector, into which the library is cloned.

TABLE 1. Unique dual indexes (UDIs) for oligonucleotide-based library labeling

| Index name | Reverse 3′ UDIs for sample | Forward 5′ UDIs for sample |
|---|---|---|
| UDP0001 | GAACTGAGCG | TCGTGGAGCG |
| UDP0002 | AGGTCAGATA | CTACAAGATA |
| UDP0003V3 | CGACATCCGA | TACGTTCATT |
| UDP0004 | ATTCCATAAG | TGCCTGGTGG |
| UDP0005V3 | CACAATAGGA | TCCATCCGAG |
| UDP0006 | AACATCGCGC | GTCCACTTGT |
| UDP0007 | CTAGTGCTCT | TGGAACAGTA |
| UDP0008 | GATCAAGGCA | CCTTGTTAAT |
| UDP0009 | GACTGAGTAG | GTTGATAGTG |
| UDP0010 | AGTCAGACGA | ACCAGCGACA |
| UDP0011 | CCGTATGTTC | CATACACTGT |
| UDP0012 | GAGTCATAGG | GTGTGGCGCT |
| UDP0013 | CTTGCCATTA | ATCACGAAGG |
| UDP0014 | GAAGCGGCAC | CGGCTCTACT |
| UDP0015 | TCCATTGCCG | GAATGCACGA |
| UDP0016 | CGGTTACGGC | AAGACTATAG |
| UDP0017 | GAGAATGGTT | TCGGCAGCAA |
| UDP0018 | AGAGGCAACC | CTAATGATGG |
| UDP0019 | CCATCATTAG | GGTTGCCTCT |
| UDP0020 | GATAGGCCGA | CGCACATGGC |
| UDP0021 | ATGGTTGACT | GGCCTGTCCT |
| UDP0022 | TATTGCGCTC | CTGTGTTAGG |
| UDP0023 | ACGCCTTGTT | TAAGGAACGT |
| UDP0024 | TTCTACATAC | CTAACTGTAA |
| UDP0025 | AACCATAGAA | GGCGAGATGG |

These unique dual indexes (UDIs) are incorporated into primers used to amplify library inserts. Clones from each sublibrary should be amplified with primers containing different UDIs. UDIs allow the inserts from multiple sublibraries to be mixed together for DNA sequencing while still retaining information about their origin.

3. Run a fraction of the pooled products on a TapeStation (or evaluate them by agarose electrophoresis with prestained molecular weight [MW] markers selected based on the predicted size of the gene of interest).

   *This step is performed to evaluate yield and homogeneity, and to document the absence of DNA products of other MW sizes that might be contaminants.*

4. Purify the pooled PCR fragments by running them on an appropriate percentage agarose gel in 1× TAE. With a clean, single-edge razor blade, cut out the band containing the PCR product of the desired size and use the Monarch DNA Gel Extraction Kit (or similar) to isolate the PCR product, following the manufacturer's instructions.

5. To generate fragments of ~50- to 100-bp average size required to generate the library, use 2 µg of the purified PCR amplicons for fragmentation using the dsDNA Fragmentase (NEB). Follow the manufacturer's instructions.

   *To generate random gene fragments of ~50- to 100-bp average size, you will need to empirically determine the duration of incubation of your PCR fragments with the dsDNA fragmentase, and use the TapeStation system to visualize the size and yield of the resulting fragments. Alternatively, electrophoretic separation and an appropriate MW marker can be used to visualize the products.*

6. To prepare the inserts for subsequent blunt-end cloning (Section B), polish them with the Quick Blunting Kit (NEB). Follow the manufacturer's instructions.

   *After the blunt ending, the phosphate groups on the 3′ ends of the gene fragments are retained, which is essential for subsequent ligation into the vector prepared in Section B.*

Chapter 11

7. Concentrate the fragmented products via ethanol precipitation:

    i. Transfer the fragmented DNA into a sterile 1.5-mL microcentrifuge tube. Add 0.1 vol of 3 M sodium acetate (pH 5.2) and 3 vol of ice-cold 100% ethanol. To mix thoroughly, securely close the cap of the tube, and vortex gently.

    ii. Incubate the mix for 1 h or overnight at −20°C, or for 1 h at −80°C.

    *Overnight incubation will yield more precipitation and may be required if the DNA concentration is <100 ng/mL.*

    iii. Centrifuge the tube at maximum speed (13,000 rpm) in a microcentrifuge for 30 min at 4°C. Carefully aspirate off the supernatant.

    iv. Wash the DNA pellet with 0.5 mL of ice-cold 80% ethanol and then centrifuge at maximum speed for 10 min at 4°C. Aspirate the supernatant, and then repeat the wash and centrifugation step an additional time. Remove the supernatant again, and then centrifuge the pellet briefly (at top speed for 10 sec). Aspirate any remaining trace amount of liquid and discard.

    v. Air-dry the pellet, and resuspend in 100 µL of nuclease-free water with gentle vortexing.

8. Quantify the DNA using a NanoDrop spectrophotometer. Store the fragments at −20°C until the prepared vector is ready to be included in a DNA ligation reaction.

    *Expect a 30% yield, or 600 ng. If the yield is <400 ng, repeat the steps in Section A.*

    *These fragments will be used for blunt-end ligation into the restriction enzyme–prepared pComb-Opti8 vector.*

## Section B. Preparation of the Phage Display Vector for Cloning

*Perform all bacterial manipulations under sterile conditions. Avoid cross-contamination of clones between the different libraries.*

9. Thaw a 50-µL aliquot of TG1 electrocompetent bacterial cells for 10 min on ice, and examine carefully to assess whether thawing is complete (i.e., when all ice crystals have disappeared). Mix gently and carefully pipette these chemically competent cells into a sterile 2-mL microcentrifuge tube. Place on ice.

10. Add 1–5 µL containing 100 pg to 1 µg of purified pComb-Opti8 plasmid DNA into the cell mixture. Carefully flick the tube four to five times to mix the DNA with the cells. Do not vortex. Place the mixture for 30 min on ice. Do not mix further.

11. Carefully place the transformation tube in a heat block or use a water bath, set at 42°C, for exactly 30 sec. Do not mix, as this may shear the DNA. Then, place the tube for 5 min on ice.

12. With a sterile 1-mL pipette, add 950 µL of SOC medium at room temperature into the 2-mL tube.

13. Place the tube at 37°C and shake vigorously at 250 rpm, or rotate for 60 min.

14. Mix the cells thoroughly by flicking the tube and inverting. With a sterile spreader, distribute 10% of this volume onto a prewarmed LB-carbenicillin agar plate, for antibiotic-based selection of the transformed bacteria, and incubate the plate overnight at 37°C.

    *If this attempt results in no colonies on the plate, repeat Steps 9–13. To yield more colonies per plate, centrifuge the mixture in the tube at 4000 rpm for 10 min in a microcentrifuge at room temperature after Step 13. Then, discard 900 µL of the supernatant and resuspend the bacterial pellet in the remaining 100 µL of supernatant. Spread 95 µL onto one LB-carbenicillin agar plate, and mix the remaining 5 µL of the supernatant with 90 µL of SOC medium. Spread this dilution onto another LB-carbenicillin agar plate. Incubate both plates overnight at 37°C.*

15. The next day, carefully examine the plate and identify a region on the surface where individual colonies can be distinguished. Pick a single colony and inoculate it into 2 mL of SB medium in a 50-mL disposable conical tube. Incubate with shaking at 300 rpm for 1 h at 37°C. Then, add 10 mL of SB medium prewarmed to 37°C and 12 µL of 100 mg/mL carbenicillin stock solution. Grow with shaking at 300 rpm overnight at 37°C.

16. Take the now turbid bacterial culture and with a plasmid prep column kit (QIAprep kit), extract the DNA, following the manufacturer's directions. To remove any remaining contaminants, ethanol-precipitate the eluted plasmid (as described in Step 7) and resuspend the pellet in 100 µL of TE buffer. Determine the plasmid concentration using a NanoDrop instrument (or similar spectrophotometer).

17. To prepare the pComb-Opti8 vector (see Fig. 1) for cloning, digest with the EcoRV restriction enzyme, which recognizes the nucleotide sequence GAT^ATC within the cloning site in the vector (Fig. 1). Prepare the following mix, adding the components in the listed order:

| Component | Amount |
| --- | --- |
| Plasmid DNA | 5 µg |
| 10× rCutSmart buffer | 25 µL |
| Nuclease-free water | 222.5 µL |
| EcoRV-HF (20 units/µL) | 2.5 µL (Add last. Keep the enzyme in the freezer until just before use. Mix by pipetting up and down several times.) |

18. Place the reaction in a heat block or a water bath for 1 h at 37°C.

19. Halt the reaction by incubation for 20 min at 65°C.

20. Isolate the digested plasmid using a spin column DNA cleanup kit, following the manufacturer's directions.

    *At this point, users will obtain a pComb-Opti8 vector without overhanging nucleotides, which is suitable for the cloning of the blunt-ended gene fragments generated in Section A.*

    *The methylation of plasmid DNA that can occur in certain strains of* E. coli *does not affect EcoRV digestion.*

21. To remove possible contaminants, precipitate the digested plasmid as follows:
    i. Add 0.1 vol of 3 M sodium acetate (pH 5.2) and 3 vol of ice-cold 100% ethanol. Vortex briefly to mix thoroughly. Precipitate overnight at −20°C, or for 1 h at −80°C.
    ii. Centrifuge at maximum speed for 10 min in a microcentrifuge at 4°C, and then carefully aspirate the supernatant.
    iii. Add 50 µL of ice-cold 80% ethanol and mix by vortexing briefly.
    iv. Centrifuge at maximum speed for 10 min in a microcentrifuge at 4°C, and then carefully aspirate the supernatant. Centrifuge again at maximum speed for 10 min in a microcentrifuge at 4°C, carefully aspirate residual supernatants, and then air-dry for 1 h.
    v. Resuspend the pellet in 20 µL of TE buffer by gentle flicking.

22. Subject the purified vector to recombinant shrimp alkaline phosphatase (SAP) treatment by preparing the following reaction:

| Component | Amount |
| --- | --- |
| Prepared vector (from Step 21) | 5 µg in 20 µL |
| rCutSmart Buffer (10×) | 10 µL |
| rSAP (1000 U/mL) | 5 µL |
| Nuclease-free water | To 100 µL final volume |

*The vector is treated with a phosphatase, as described here, to greatly reduce the occurrence of self-ligation of the empty vector. Thereby, the relative efficiency of the ligation of the blunt-ended fragment inserts into the prepared vector is enhanced.*

23. Incubate the reaction mixture in a heat block or water bath for 30 min at 37°C.

24. Stop the reaction by heat inactivation for 5 min at 65°C.

Chapter 11

25. Mix the digested plasmid with gel loading buffer and run the reaction on a 0.6% (w/v) agarose gel in 1× TAE buffer. For size reference, include a 1-kb molecular weight ladder in one lane. With a clean single-edge razor blade, cut out the ~3.3-kb band while retaining as little as possible of the agarose gel. To reduce unwanted impurities and contaminants, extract the plasmid DNA with the Monarch DNA Gel Extraction Kit, per the manufacturer's instructions.

26. After gel extraction, perform ethanol precipitation of the extracted vector as follows:
    i. Add 1 µL of glycogen, 20 µL (0.1 vol) of 3 M sodium acetate (pH 5.2), and 440 µL (2.2 vol) of ice-cold 80% ethanol. Incubate the tube for 1 h at −80°C.
    ii. Centrifuge the tube at maximum speed for 10 min in a microcentrifuge at 4°C. Aspirate the supernatant carefully.
    iii. Add 50 µL of 80% ice-cold ethanol and centrifuge at maximum speed for 10 min in a microcentrifuge at 4°C.
    iv. Aspirate the supernatant carefully, and air-dry for 1 h.

27. Resuspend the precipitated DNA in 10 µL of sterile TE buffer. Measure the concentration of DNA in nanograms per microliter using a NanoDrop spectrophotometer. Store the vector at −20°C until use.

### Section C. Test and Library Ligations of Gene Fragment Inserts into the Prepared Vector

*The calculated number of moles of prepared gene fragment inserts (from Step 8) and that of the prepared pComb-Opti8 vector (from Step 27) are used to estimate the insert:vector ratio for the ligation reaction. Here, the inserts should have an average size of 50 bp, and the pComb-Opti8 vector is 3300 bp. While a molar ratio of inserts to vector of 5:1 may be optimal, empirical studies with different molar ratios may be required to attain the greatest transformation efficiency (i.e., the one that yields the greatest number of colonies per microgram of total DNA equivalent), and such tests of ligation reaction efficiency are described below. The goal is to generate a gene fragment library that is the largest and most diverse possible.*

#### Ligations

28. Perform small-scale test ligations by proceeding up to Step 43. Here, small amounts of prepared vector and gene fragments are used, in various molar ratios, to confirm the suitability of the prepared vector and inserts for high-efficiency ligation and transformation. Use only a fraction (e.g., <20%) of the prepared vector in a test ligation reaction. Always compare to a negative control of ligation of the prepared vector alone, which should have a transformant yield of <10% of that of the prepared vector into which gene fragments have been ligated. For test ligations, mix the following in a 1.5-mL sterile microcentrifuge tube:

| Compound | Amount |
| --- | --- |
| Digested vector (from Step 27) | 800 ng |
| Gene fragment inserts (from Step 8) | 25 ng (if a product with a mean size of 50 bp is used); this number should be varied for different insert to vector ratios |
| 10× ligase buffer | 20 µL |
| T4 DNA ligase | 10 µL |
| Nuclease-free water | To a final volume of 200 µL |

*The above is one example of a test ligation. Different ratios of insert to vector should be tested. The goal is to generate a library with a total of over 10 million members.*

*As mentioned above, include a vector-only control without the addition of gene fragments. Transformation of the religated, prepared vector alone provides the background of the reaction. If the plasmid was cut and purified correctly, ligation of the vector alone should yield few, if any, transformants, because the cut vector was treated with phosphatase.*

*A more in-depth discussion of test ligations is presented in a protocol described by Andris-Widhopf et al. (2011).*

29. Incubate the reactions at the temperature and duration suggested by the manufacturer of the ligase.

30. Heat-kill the residual ligase enzyme by incubating the reactions in a 65°C water bath or heat block for 10 min.

## Cleanup

31. Precipitate the DNA in each reaction by adding 1 µL of glycogen, 20 µL (0.1 vol) of 3 M sodium acetate (pH 5.2), and 440 µL (2.2 vol) of ice-cold 80% ethanol. Mix each reaction gently and then store upright overnight at −80°C.

    *Precipitated library DNA ligations are stable for months when stored at −80°C.*

32. Centrifuge each tube from Step 31 in a microcentrifuge at maximum speed for 15 min at 4°C. Without disturbing the DNA pellet, remove the supernatant by pipetting and discard it. Rinse the pellet gently with 1 mL of 80% ethanol. Then, centrifuge each tube at maximum speed for 15 min in a microcentrifuge at 4°C. Again, remove this supernatant. Repeat the ethanol rinse and centrifugation step, and remove and carefully discard the supernatant. To air-dry, carefully place the inverted tubes on a clean paper towel on the benchtop for 1 h at room temperature.

    *Do not overdry the pellet, or it will no longer be easily dissolvable in ddH$_2$O or TE.*

33. Dissolve the pellet in 200 µL of TE buffer with brief warming in a 37°C water bath, followed by a brief gentle burst of vortexing to put the pellet into solution. Do not overvortex, as the DNA will be damaged.

## Electroporation of the Test Ligations

*Before starting, prewarm 1 mL of SOC medium to 37°C for every planned electroporation reaction. To avoid contamination, wipe the outside of the sealed tubes of medium with 10% ethanol before and after placing them in the incubator.*

*Here, the purified outputs of ligation reactions are gently mixed with an aliquot of the electrocompetent bacteria. During electroporation, if you notice a visible (and audible) spark in the electroporator, this means that washing of the ligation product was inadequate, and there was too much salt still present. The high conductivity in the solution will result in a spark that will kill the bacteria in the cuvette and may also damage the electronics of the electroporator. In such cases, further washes in ice-cold 80% ethanol are needed. For this reason, to minimize the potential for the loss of the ligated library, it is prudent to divide up your ligation reaction product and then perform multiple separate electroporation reactions. We recommend dividing the library into four aliquots for this purpose. This careful approach, with test ligations, lessens the risk of loss of these precious materials.*

34. Thaw on an ice slush 50 µL of electrocompetent *E. coli* strain TG1 per electroporation reaction to be performed. Do not keep the TG1 *E. coli* on ice for >20 min before use.

35. To prepare for electroporation, use a sterile 1-mL pipette tip, with a snipped-off end, to divide the chilled ligation reaction into four chilled microcentrifuge tubes. Each tube should contain 100–600 ng of ligated DNA from Step 33. To each tube, add 50 µL of electrocompetent *E. coli* (TG1 strain). Mix by carefully pipetting up and down once, and then transfer the mix into a prechilled 0.2-mm electroporation cuvette. Incubate the cuvette containing the bacteria and the ligation mixture for 1 min on ice. Then, carefully and quickly dry off the outside of the cuvette and place it into the instrument. Electroporate at 2.5 kV, 25 mF, and 200 Ω.

    *Expect the run time (termed t) in the electroporator to be ~4.0 msec.*

36. After the electroporation, flush each cuvette immediately with 950 μL of SOC medium prewarmed to 37°C. After electroporation, place the contents of each cuvette into separate sterile 50-mL disposable polypropylene conical tubes.

37. To best recover all of the bacteria, rinse out each cuvette twice with an additional 1 mL of SOC, for a total of ~2 mL of medium, and combine all together in the same tube. With the cap firmly screwed on, shake at 300 rpm for 1 h at 37°C.

    *As an alternative, to flush the library out of the electroporation cuvette, one can instead use the buffer provided by Lucigen.*

### Titration of the Library

38. Remove each tube from the shaker. To determine the transformation efficiency and size of each resulting library, remove 2 μL of each 2-mL culture with a sterile pipette and mix with 100 μL of sterile SB medium. If a library has been split into four, this plating need only be performed for one of the four tubes in the library. Save the remainder of each 2-mL culture at 4°C.

39. For plating, prepare two plates, which are here referred to as plate A and plate B:
    i. For plate A, remove 10 μL of the diluted culture to use for plate B. Plate the remaining 92 μL and spread (with a disposable spreader) onto a prewarmed LB-carbenicillin agar plate.
    ii. For plate B, take the 10 μL saved from plate A and add it to 90 μL of sterile prewarmed (37°C) SB medium. Spread the dilution onto a prewarmed LB-carbenicillin agar plate.

40. With the lids on, place the plates upright overnight in the 37°C incubator cabinet.

41. Calculate the number of transformants as follows:
    i. To determine the total number of colonies resulting from each ligation, multiply the number of colonies by 1000 for plate A (this is because you sampled 2 μL of a total of 2000 μL).
    ii. As plate B came from only 10% of the diluted sample, multiply the number of colonies on plate B by a factor of 10,000.

    *If needed, further dilutions can be performed; the goal is to have plates with approximately 100–200 colonies that can be accurately counted on a plate. Always consider the volumes of the culture medium and dilutions used for each plate to estimate the library size.*

42. Determine and report the size of each library generated as follows: transformation efficiency = number of colonies/micrograms of DNA in the transformation reaction.

43. To confirm that the library contains inserted DNA, choose a colony from the plate in Step 40 (test one colony from each library tested) and grow in 2 mL of SB medium with 2 μL of the 100 mg/mL carbenicillin stock, for a final concentration of 100 μL/mL. After 4 h to overnight growth at 37°C, use a miniprep kit to purify the plasmid (using the manufacturer's directions). Digest the purified plasmid with the EcoRV enzyme as in Steps 17 and 18. Precipitate the digested plasmid as in Step 21. Run ~200 ng of the digested material with gel loading buffer in one well of a 0.6% (w/v) agarose gel in 1× TAE, along with DNA standards of defined 1 kb and smaller sizes to enable evaluation of the insert.

    *After electrophoretic separation, a comparison of the DNA in the different lanes will document that a successful ligation of an insert of the predicted size had been attained.*

### Generation of the Library Based on Optimized Ratios

44. When you succeed in attaining more than 10 million transformants per microgram of prepared gene fragments ligated into the prepared vector from a test ligation, follow Steps 28–43 to prepare the library based on that optimal ratio of vector to insert. Perform at least four ligations with this molar ratio of inserts to prepared vector.

    *In general, the success of this project is largely dependent on whether the generated library is of sufficient size (i.e., the number of transformant members in a library as a measure of diversity). A total library with more than 1 million members is generally acceptable, while 10 million members is preferred. The larger a* library, the more likely there will be complete coverage of the entire DNA sequence, in-frame and in the correct orientation, that encodes the protein of interest.

To make a larger library, the plasmid forms generated during different attempts at the transformation of ligation reactions can be pooled and used together in a bacterial transformation.

Alternatively, the phage form of these smaller libraries can be pooled to increase overall library diversity before performing a clonal selection of the phage form of the library with an antigen-reactive antibody.

## Expanding the Library in Bacteria

There is always a concern that there can be clone dropout during library propagation, potentially because certain clone protein products are toxic to the transformed bacteria.

Here, to suppress bacterial expression in transformed bacteria, bacterial libraries can be grown in SB media (along with antibiotics) that also contains a final concentration of filter-sterilized 1% (w/v) glucose. In this approach, the glucose reduces the leakiness of the lac promoter and reduces or completely blocks the expression of the clone protein.

45. Warm the rest of the 2-mL culture from Step 38 to 37°C. Add it to 10 mL of SB medium prewarmed to 37°C and add 12 µL of 100 mg/mL carbenicillin stock. On a shaking 37°C incubator, securely place each culture and shake at 250 rpm. Incubate for 1.5 h.

46. Transfer each culture into an autoclaved, sterile 500-mL Erlenmeyer flask. Add 183 mL of 37°C prewarmed SB medium and 200 µL of 100 mg/mL carbenicillin stock. Cover with sterile aluminum foil and crimp at the neck to avoid contamination. Shake this culture at 300 rpm for 1.5–2 h at 37°C.

47. To recover the bacterial library in a phage particle form, add 2 mL of VCSM13 helper phage that has $10^{12}$–$10^{13}$ plaque-forming units (pfu)/mL to each culture (for further information, see Protocol 10.2 in Rader and Barbas 2001). Incubate on a shaker for an additional 3 h at 37°C.

    While we suggest using a helper phage compatible with the pComb system from a vendor, the user can instead expand the helper phage and aliquot for library generation in their laboratory using a validated protocol (see Protocol 10.2 in Rader and Barbas 2001).

48. To lyse any nontransformed bacterial cells, add 200 µL of 50 mg/mL kanamycin stock to each culture and continue shaking at 300 rpm overnight at 37°C.

49. Transfer each overnight bacterial culture into a clean 500 mL polypropylene centrifuge bottle. To recover the transformed bacteria as a pellet, centrifuge the bottles at 3000$g$ (e.g., 4000 rpm in a Beckman JA-10 rotor) for 15 min at 4°C. For future library sequence analysis, recover the bacterial pellet on the bottom of each of the bottles. These pellets contain dead bacteria with their pComb-Opti8 constructs in plasmid form. Pellets can be stored for months to years at −20°C, or the plasmids can be recovered immediately (see Step 87). Additionally, store the phage-containing supernatants with sodium azide at 4°C until used.

    To preserve a spent culture, which contains the phage form of the library, add sodium azide to the supernatant. A common concentration range is 0.02%–0.1% (w/v). This range is generally effective for preventing microbial contamination in various biological samples and solutions. It is important to remember that sodium azide is toxic, so always handle it with the appropriate safety precautions.

## Library Recovery in Phage Format

50. For this section, make sure your centrifuge is prechilled to 4°C, with the rotor in place overnight. Precipitate the phage as follows: Transfer the ∼200 mL of each phage-containing supernatant from Step 49 into a separate clean 500-mL centrifuge bottle. Add 8 g of PEG-8000 (4% w/v final) and 6 g of sodium chloride (3% w/v final). Dissolve the solids by shaking at 300 rpm for 5 min at 37°C. Then, incubate the bottles for 30 min on ice.

51. Centrifuge the bottles at 15,000g (e.g., 9000 rpm in the Beckman JA-10 rotor) for 15 min at 4°C. Discard the phage-depleted supernatant, drain each bottle by inverting it on a paper towel for at least 1 min, and then do an additional centrifugation at 15,000g for 2 min. Carefully pipette out the residual media and discard. Wipe off the remaining liquid from the upper part of each centrifuge bottle with a clean paper towel.

52. Resuspend the phage pellet in 2 mL of 1% BSA-TBS by pipetting up and down along the side of the centrifuge bottle where the line of phage precipitate is visible. Transfer each phage-containing suspension into a sterile 2-mL microcentrifuge tube. Resuspend further by pipetting up and down with a 1-mL pipette tip. Centrifuge at maximum speed in a microcentrifuge for 5 min at 4°C. To remove bacterial debris, pass the phage-containing solution (supernatant) through a 0.2-µm filter into a new sterile 2-mL microcentrifuge tube. Store the phage-containing supernatant at 4°C until use.

*For antibody-mediated selection from the phage form of the library, proceed immediately to Section D.*

*Note that the phage must be freshly prepared within 1–2 d of when it is used for panning. If the phage preparation has been stored for more than a couple of days, it may not be suitable, as proteins on the surface of the phage degrade over time, which impairs attempts at antibody-mediated selection of phage. Therefore, the library will need to be freshly reamplified before panning (i.e., selection). To do this, take 10 mL of fresh TG1 or XL-1 Blue culture grown overnight at 37°C in SB medium, add at least 20 µL of a previously stored phage library, and after shaking at 250 rpm for 1.5 h, add 10 µL of the 100 mg/mL carbenicillin stock. On a shaking 37°C incubator, incubate for 1.5 h. Proceed to Step 46.*

### Section D. Panning on 96-Well High-Binding-Capacity Microtiter Plates

*Panning is a term that refers to selecting specific gene fragment clones from a phage library that displays fusion proteins based on a specific protein–protein interaction. Biopanning was first coined to refer to the use of biotinylated selecting ligands, but here is used more generally.*

*The primary approach is to use 96-well high-binding-capacity microtiter plates for recovering antibody-reactive gene fragment phage particles in the library. An alternative approach, using paramagnetic beads coated with Protein G, provides a means to use a much greater amount of the monoclonal antibody for selection from a library, and a brief protocol for Protein G–based selection is presented at the end of this section. If conducting the alternate protocol, please proceed to Step 61 instead of this section.*

53. On the afternoon prior to panning, start an overnight culture of fresh XL-1 Blue or TG1 bacteria in 20 mL of SB medium. Shake this culture at 300 rpm overnight at 37°C.

    *For each day of a round of biopanning, fresh XL-1 Blue or TG1 bacteria must be newly grown. Fresh growth reduces the chance that these bacterial samples have been contaminated by an experimental or environmental bacteriophage, which would prevent infection with the eluted phage, because each E. coli cell can be infected only once by a filamentous phage.*

54. Also on the day before panning, prepare 500 µL of a suspension of the antigen-specific purified monoclonal antibody (mAb) at 2 µg/mL in 1× PBS. Similarly, prepare 500 µL of a suspension of the isotype control antibody at 2 µg/mL in 1× PBS. For greater effective surface area for interactions, place 100 µL of mAb into each of four replicate wells on a microtiter plate, and place 100 µL of isotype control antibody in each of four different replicate wells on the same plate. Use an indelible marker to indicate which wells you coated. Seal the plate and incubate overnight at 4°C on a rocking platform or similar.

    *As there should be no specific interactions with the isotype control antibody, these wells should not recover phage that express a functional epitope, and hence this provides a valuable negative control.*

55. The next morning, discard the liquid from the plate. Invert the plate and tamp twice onto a paper towel to remove the remaining liquid. To each antibody-coated well, add 100 µL of 2% BSA-PBS and then seal the plate. Incubate with gentle agitation on a rocking platform for 60 min at room temperature.

56. Tamp out the blocking solution from the inverted plate, and then add 100 µL of the recovered phage form of the fragment library (from Step 52) in 1% BSA-TBS to each of the four mAb and four control wells. Incubate on a rocking platform for 2 h at room temperature.

57. Tamp out any liquid in the wells and use a micropipette to wash all experimental wells with 100 µL of 0.05% Tween 20-PBS. Tamp out the residual liquid.

    *These instructions are suitable for the first round of panning. For later rounds, use two or more rounds of washing with 0.05% Tween 20-PBS.*

58. To recover the bound phage from each antibody-coated well, add 100 µL of 1 M glycine HCl (pH 2.5) to each well and gently pipette up and down to disrupt any antibody–antigen binding interactions. To remove the selected phage, carefully pipette the eluted unbound phage from each well. Pool the phage selected by the mAb in a sterile 50-mL conical tube. Similarly, pool the phage selected by the isotype control antibody. In each tube, neutralize the acid with 150 µL of 50 mM Tris HCl (pH 9.1). Add 2 mL of a fresh aliquot of XL-1 Blue bacteria taken directly from the shaking 37°C incubator and incubate with shaking at 250 rpm for 1 h at 37°C.

    *Depending on the number of tubes pooled, more Tris HCl may need to be added. Confirm that a neutral pH has been attained using litmus paper.*

59. To each of these 50-mL tubes, add 10 mL of SB medium prewarmed to 37°C and add 12 µL of 100 mg/mL carbenicillin stock. Securely place each culture on a shaking 37°C incubator and shake at 250 rpm for 1.5 h.

60. Transfer each culture into a sterile 500-mL Erlenmeyer flask. Add 183 mL of SB medium prewarmed to 37°C and 200 µL of the 100 mg/mL carbenicillin stock. Place the sterile aluminum foil over the top of the flask and crimp at the neck to avoid contamination. Shake this culture at 300 rpm for 1.5–2 h at 37°C.

    *This step is a repeat from Step 45. Following this step, proceed through Steps 46–60 to perform repeated rounds of selection of the phage library. In most cases, four rounds of selection are sufficient and further library focusing is not needed. However, this will need to be determined empirically by library sequencing. See Discussion.*

## Optional: Protein G Bead–Based Library Selection by a Monoclonal Antibody

*As an alternative to using microtiter wells, which have a limited binding capacity, to perform library selection, a bead-based approach can be used. Protein G binds a site on all subclasses of human IgG with a subnanomolar $K_D$ binding affinity. With 1 mL of a slurry of paramagnetic beads coated with Protein G, >280 µg of purified IgG can be bound. Hence, this system allows efficient loading of any purified monoclonal IgG of interest.*

61. On the afternoon prior to selection, start an overnight culture of fresh XL-1 Blue or TG1 bacteria in 20 mL of SB medium. Shake this culture at 300 rpm overnight at 37°C.

62. Prepare two aliquots of magnetic protein G–coated beads in microcentrifuge tubes (20 µL each of the slurry provided by the vendor): one for mAb binding and one for the isotype antibody control. Place each tube onto the magnetic separation stand, and remove the liquid with a pipette. The beads will be immobilized on the side of the tube due to magnetic force. Wash each aliquot twice with 100 µL of 0.1% Tween 20-PBS, removing the liquid on the magnetic separation stand between each wash as described above.

63. Remove the second 0.1% Tween 20-PBS wash from each tube of beads. Add 0.5 mL of the purified mAb of interest at 2 µg/mL in 1× PBS to one tube. To the control tube, add 0.5 mL of the isotype control antibody at 2 µg/mL in 1× PBS. Incubate both tubes with agitation for 20 min at room temperature.

64. Place each tube onto the magnetic separation stand, and then carefully pipette off the supernatant and discard it.

65. Add 1 mL of Blocker Casein to each tube of isolated beads. Incubate with gentle agitation on a rotating wheel for 30 min at room temperature. Place each tube onto the magnetic separation

stand, and then remove the supernatant carefully by pipetting. Add 1 mL of 0.05% Tween 20-PBS to each tube, vortex for 10 sec to wash, and then place each tube onto the magnetic separation stand. Remove the supernatant by pipetting. Repeat this wash once on each tube. Remove the supernatant after the last wash.

*These antibody-coated beads can then be used to recover fragment protein-expressing phage particles in the phage form of the library.*

66. To each aliquot of washed beads, add 100 µL of phage library from Step 52 in 1 mL of Blocker Casein. Place this tube to mix on a rotating wheel gently for 2 h at room temperature.

67. Place each tube onto the magnetic separation stand, and then carefully pipette off the supernatant and discard it. Wash the beads with 200 µL of Blocker Casein. Place each tube onto the magnetic separation stand, and then carefully pipette off the supernatant and discard it. Repeat the wash, remove the supernatant, and add 200 µL of Blocker Casein to each tube.

68. Mix slowly on a rotating wheel for 30 min, and then wash beads three times with 200 µL of 0.05% Tween 20-PBS with 1% BSA, isolating the beads on the magnetic stand between washes.

69. Drain the tubes by inversion against the magnetic separation stand to retain the beads in the tube. When the tube is drained but beads remain damp, place the tube upright in a tube holder.

70. To elute the bound phage, add 100 µL of 1 M glycine HCl (pH 2.5) to each tube of beads. Close the top of the tube, and mix by flicking with your finger five times. Incubate for 5 min. Place the tubes onto the magnetic separation stand, and gently pipette out the supernatants. Place the supernatants into 50-mL conical tubes, and then neutralize each with 150 µL of 50 mM Tris HCl (pH 9.1). To each tube, add 2 mL of a fresh aliquot of XL-1 Blue bacteria taken from the shaking incubator and incubate with shaking at 250 rpm for 1 h at 37°C.

71. To each of these 50-mL tubes, add 10 mL of SB medium prewarmed to 37°C and add 12 µL of 100 mg/mL carbenicillin stock. Securely place each culture on a shaking 37°C incubator and shake at 250 rpm for 1.5 h.

*Following this step, proceed through Steps 46–52, followed by the desired selection method, to perform repeated rounds of selection of the phage library.*

## Optional: Phage Binding ELISA Protocol

*A phage binding ELISA (enzyme-linked immunosorbent assay) is a common method for evaluating the binding of phage clones derived from phage display libraries to a specific target molecule, and is used here to test the success of panning. In this assay, microtiter wells are prepared by coating with the monoclonal antibody used to select the gene fragment/epitope clones. After blocking of the wells, serial dilutions of the original library in phage form are incubated on these wells. On parallel wells, serial dilutions of each sequentially selected library (sublibraries) are also incubated. Binding of the phage is detected using an HRP-tagged anti-M13 antibody. By this approach, the functional properties of gene fragment clones selected for binding to an antibody in this system are demonstrated. Adjust coating concentration, incubation times, and washing steps as needed for specific selecting monoclonal antibodies, control antibodies, and phage libraries.*

72. Plan your plate with appropriate dilutions of the original phage library and each sublibrary after panning. We recommend testing four wells each with the original library and each sublibrary at 1:10, 1:50, and 1:250 dilutions. Each of these dilutions should be tested on wells coated with the monoclonal antibody and the isotype control antibody, as well as the blank control wells.

73. Dilute the monoclonal antibody used for library selection to 5 µg/mL in 0.1 M NaHCO$_3$ (pH 8.6). For a control, similarly dilute an irrelevant isotype control antibody. Prepare 30 µL of antibody solution for each well to be coated.

74. Place the diluted antibodies (25 µL per well) into an ELISA microtiter plate based on the number of wells planned in Step 72. Include control wells that will be blocked but do not contain antibody, to assess background binding to the plastic of the well. Seal the plate and incubate overnight at 4°C.

75. Tamp out the coating solution. Add 150 µL of 2% BSA-PBS (ELISA blocking buffer) to each well. Seal the plate and incubate for 1 h at 37°C.

76. During blocking, dilute phage stocks from the original library and each sublibrary 1:5 in 0.05% Tween 20-PBS. Further dilute phage stocks as described in Step 72. Prepare 55 µL per well.

77. Tamp out the blocking buffer from the plate. Wash the coated plate multiple times with 150 µL of 0.05% Tween 20-PBS using a multichannel pipet.

78. Add 50 µL of appropriately diluted phage to the monoclonal antibody wells, the control antibody wells, and the blocked-only wells. Seal the plate and incubate for 1–2 h at room temperature with agitation.

79. During the incubation, dilute HRP-conjugated anti-M13 1:2000 in 0.05% Tween 20-PBS. Prepare 55 µL per well.

80. Tamp out the phage from the plate. Wash the plate 10 times with 150 µL in 0.05% Tween 20-PBS using a multichannel pipet.

81. Add 50 µL of diluted HRP-conjugated anti-M13 antibody to each well. Seal the plate and incubate for 1 h at 37°C.

82. Prepare the SuperSignal ELISA Pico Chemiluminescent Substrate per the manufacturer's instructions. Prepare 55 µL per well.

83. Tamp out the anti-M13 antibody from the plate. Wash the plate 10 times with 150 µL in 0.05% Tween 20-PBS using a multichannel pipet.

84. Add 50 µL of SuperSignal ELISA Pico Chemiluminescent Substrate to each well.

85. Incubate the plate at room temperature and measure the absorbance at 450 nm with an ELISA plate reader at two to three time points, usually starting at 30 min. Adjust as needed based on preliminary results.

    *If the signal does not drop with the original dilutions, additional 1:5 dilutions can be tested.*

    *See Troubleshooting.*

86. Based on the ELISA assay, identify which round of library selection resulted in the greatest ELISA signal. Retrieve the corresponding bacterial pellet from Step 49 and proceed to Step 87.

## Section E. Preparation of Libraries for High-Throughput Sequencing (HTS)

*Insights into the sequence diversity in the original library and how sequential rounds of selection affect that diversity is always a fundamental question in phage display studies. In earlier approaches, a "representative" number of clones were chosen from the original and the selected libraries, and each of these was assessed for DNA sequence diversity and binding of the expressed protein. However, if a library is estimated, based on the efficiency of electroporation transformation, to be immense (e.g., more than $10^7$ members), one can never accurately assess diversity by assessing a selection of individual colonies.*

*Alternatively, high-throughput sequencing data can be useful to identify the minimal epitope bound to an antibody and is described here. This approach also allows us to evaluate, by DNA sequencing, whether selection has resulted in clonal focusing. Briefly, plasmid DNA will be isolated from bacteria infected with a phage library that has gone through rounds of antibody selection. Tailored oligonucleotides for each library will be used to barcode each round of selected sublibrary. The amplimers will be isolated and prepared for DNA sequencing.*

87. After the desired number of rounds of library selection using antibodies bound to ELISA plates or protein G beads, and based on ELISA results if desired, retrieve bacterial pellets from infection of bacteria with the selected library that were frozen in Step 49. Use a miniprep kit to prepare purified plasmid. Determine the concentration using a NanoDrop spectrophotometer.

    *Separately extract the plasmid DNA from each bacterial pellet of a library or sublibrary.*

## PCR-Based Addition of Oligonucleotides that Document the Library Source of a Selected Clone

As described above, PCR products are generated and sent for sequencing of the library clone inserts. In this section, we describe how a PCR-based step is used to enable identification of every potential DNA sequence in a sublibrary using barcoded oligonucleotides (Fig. 3) that flank the cloning site in each individual pComb-Opti8 (sub)library. This identification is made possible by incorporating unique dual index (UDI) sequence (see Table 1) into the primers used to amplify the gene fragments in the library prior to DNA sequencing.

88. Order the following PCR primers, which will be used both to amplify the individual phage library inserts for sequencing and to add a UDI sequence, such that the DNA sequence will indicate from which sublibrary the insert came. Order two additional primers comprising only the underlined 5′ end of each of the library primers, which will later be used as DNA sequencing primers. Order enough primers to allow identification of each sublibrary.

    Primers:

    ALL-LIBRARY forward primer: 5′-<u>AATGATACGGCGACCACCGAGATCTACACT</u>AXXXXXXXXXXCACTGGC TGGTTTCGCTAC-3′

    ALL-LIBRARY reverse primer: 5′-<u>CAAGCAGAAGACGGCATACGAGAT</u>XXXXXXXXXXXGTCACCCTCAGC TCTTCTGT-3′

    *These pairs of indexing nucleotide sequences were taken from a NEB protocol intended for Illumina-based DNA sequencing. An overview of HTS for Illumina is provided in Illumina 2025a.*

    *The embedded UDI sequences (here designated "XXXXXXXXXX") are unique for each library, and enable electronic assignment by the UDI of each read to its source library. An overview of UDIs for Illumina for library barcoding is provided in Illumina 2025b.*

    *For these primers, only the 3′ ends, after the XXXXXXXXXX, anneal on flanking sides of the multicloning site (MCS) and amplify the cloned gene fragment sequences and the immediate regions in pComb-Opti8 (Fig. 3). Each has a 55%–58% GC content, and an estimated annealing temperature of 59°C.*

### Library Amplification

89. Prepare the following PCR mix for each library to be tested, adding the master mix first to a 0.2-mL microcentrifuge tube:

| Component | Volume needed for 50-μL reaction | Final concentration |
|---|---|---|
| 10 μM Forward UDI-containing primer | 1 μL | 0.2 μM |
| 10 μM Reverse UDI-containing primer | 1 μL | 0.2 μM |
| Plasmid DNA from Step 87 | Variable | 500 ng |
| NEBNext High-Fidelity 2× PCR Master Mix | 25 μL | |
| Nuclease-free water | To 50-μL final volume | |

*We recommend assembling all reaction components on ice directly in the tube that will later be placed into the thermal cycler and, when ready, mix by flicking and quickly transfer each of the tubes to a thermocycler. Also include a control reaction with no plasmid DNA.*

90. Run the samples using the following program in a thermocycler:
    2 min at 94°C;
    followed by five cycles of
      45 sec at 90°C,
      1 min at 56°C, and
      1.5 min at 72°C;
    10 min at 72°C; and
    hold at 4°C until ready for next steps.

    *Afterward, store samples at 4°C until analysis.*

*Note that we have recommended five PCR cycles as a starting point, but the correct number may need to be determined empirically and may range from three to seven cycles. Samples prepared with a different method before library preparation may require reoptimization of the number of PCR cycles. The use of a higher number of cycles will yield more product amplimer but may skew and cause artifacts in the library composition.*

91. Take 10% of each reaction including the "no DNA" control, mix with gel loading buffer, and run on a 2% (w/v) agarose gel in 1× TAE. Run with an appropriate MW marker in an adjacent lane. Run at 80 V for 1 h. The "no DNA" control PCR reaction should not have a product.

### Library Preparation for DNA Sequencing

92. Purify the PCR products by ethanol precipitation as in Step 7. Resuspend each pellet in 50 µL of TE.

93. Run 1 µL of the purified library on a TapeStation chip following the manufacturer's instructions. Check that the library size shows a narrow distribution. If ~50 bp was the minimum size selected after fragmentation, there might be an expected maximum size of ~120 bp.

94. Send pooled library amplimers to your preferred provider for sequencing on a MiSeq HTS (Illumina) instrument.

*With the resulting data set, analyses of the composition of the original library and of antibody-selected libraries indicate the specific DNA sequences (i.e., genotype) in each library to evaluate for clonal selection. For a comment on the DNA sequence analysis of the resulting data, see Discussion.*

## TROUBLESHOOTING

*Problem (Step 85):* A lack of ELISA signal is observed.
*Solution:* Such a result could indicate that weak binding clones have been isolated or that the assay is of insufficient sensitivity. Incorrect reagent preparation or order, incorrect dilution, or washing problems can also cause difficulties.

## DISCUSSION

We have used the workflow outlined above to identify minimal B-cell epitopes for toxins from *Staphylococcus aureus* (see also Introduction: Insights from the Study of B-Cell Epitopes of a Microbial Pathogen by Phage Display [Silverman 2025]). Our experience has shown that minimal B-cell epitopes associated with β-folds in the overall structure of a bacterial leukocidin toxin are common within members of the PFT toxin family of *S. aureus*, and it is worth noting that we did not recover epitopes associated with the β-strands themselves. Phage display systems may also be better suited for recapitulating the β-folds of antigenic proteins.

In applying this technology, we found that multiple rounds of selection reduced the background of nonbinders, and by this approach, we may only have needed to sequence a relatively limited number of clones. However, when millions of clones are sequenced by high-throughput sequencing (HTS), it is readily apparent that we can detect clonal focusing after the first round of selection. The high correlation between the relative abundance of clones after the first round, which became more focused after subsequent rounds, may suggest that further rounds of enrichment are not always necessary. Artifacts can also arise due to differences in phage clone competence and propagation rates. Indeed, there can be a concern that multiple rounds of selection may even result in the exclusion—or down-weighting—of interesting phages from the pool. In some cases, it therefore may be preferred to perform the first round of selection in duplicate or triplicate and compare the outcome of varying binding conditions and stringency of washing iterations.

The success of epitope selection with the pComb-Opti8 vector has been rigorously validated by DNA sequence analysis of the newly generated fragment libraries and the selected epitopes using HTS (with the MiSeq Illumina instrument) of a library, which yields many millions of reads per run. From a practical perspective, several barcoded gene fragment libraries can be run together in a single lane. From the resulting data set, the reads for each library can be separately identified, owing to the multiplexing identifiers that flank the cloning site in each sublibrary. Each of the complete reads will represent a sequence from the parental gene of interest that is represented in a range of sizes, from cloning of these gene fragment sequences into the pComb-Opti8 vector.

The reads from these sequencing runs are assembled with appropriate software, for instance, ZEGA global sequence alignment software (Abagyan and Batalov 1997), with great redundancy, which will ensure the accuracy of each of the read sequences.

DNA sequence analysis software will enable alignment based on the similarity of the sequences of each clone isolated from a phage display library. For example, comparisons of these data and visualization can also be performed with SeqMan NGen Genome Assembly Software (DNAStar), which enables the generation of assemblies by comparing sequentially selected libraries. An example is shown in Figure 4 of the associated article (Introduction: Insights from the Study of B-Cell Epitopes of a Microbial Pathogen by Phage Display [Silverman 2025]). Here, each of the sequential rounds of library selection is associated with a reduced diversity in the represented clones. When selection is successful, the clones in the library should display a progressively narrower range of sequences from the parental gene in each successive round of selection, and these focused clones will identify the subregion of the antigenic protein that is bound by the selecting antibody.

A focused library should be composed of highly related sequences, each with some additional upstream and downstream adjoining residues from the parental gene, which are selected together from a library. The presence of some differences between clones provides evidence that each of the sequence-unique sets of clones had been individually selected based on protein–protein interactions of the displayed protein fragment with the selecting antibody. In contrast, if there is only one monotonous clone sequence, this may be evidence of a parental library of limited diversity or that a defective clone may have acquired a growth advantage for a reason unrelated to the protein sequence it expresses. Such defective clones could arise, for instance, from spontaneous mutations in the origin of replication of the plasmid, among other reasons.

Comparisons of selected clones will readily enable the identification of the smallest recovered sequence (i.e., minimal epitope). If the crystallographic structure of the parental protein has been solved, a potential epitope "hit" should be a continuous sequence within the clone that is identical to a subregion in the protein of interest. This epitope-associated subregion can be identified and visualized by PyMOL software (Schrödinger and DeLano 2025), which depicts the 3D structural location of the primary amino acid subsequence hit within the crystallographic structure if available in the Protein Data Bank (PDB).

Users can apply these methods to identify and characterize short continuous polypeptides that are functionally equivalent to the epitope bound in the native protein. A synthetic peptide can then be synthesized with this short sequence with the addition of carboxy-terminal biotin, which can be used in direct binding and inhibition studies, as described in the associated article (see Introduction: Insights from the Study of B-Cell Epitopes of a Microbial Pathogen by Phage Display [Silverman 2025]). In addition, NMR, cryoEM, and other types of studies can be performed to assess 3D structure and conformation. Taken together, this approach enables the identification and characterization of small peptide sequences bound by an antibody of interest. We have speculated that these sequences would be well suited for incorporation into a modular multiepitope protective vaccine.

## RECIPES

### Carbenicillin Stock Solution

Dissolve 2 g of carbenicillin (Thermo Scientific 10177012) into 20 mL of double-distilled (dd) sterile $H_2O$ (final 100 mg/mL stock). Sterilize the solution using a 0.2-mm filter (Thermo Scientific 11208710) and store in 1.5-mL aliquots in microcentrifuge tubes for 1 year at −20°C.

### Kanamycin Stock Solution

Weigh 0.5 g of kanamycin (GoldBio K-120 [CAS25389-94-0, MW = 582.6]). Dissolve into 10 mL of double-distilled sterile $H_2O$ (sterile dd$H_2O$). Prewet a 0.22-µm syringe filter by drawing through 5–10 mL of sterile dd$H_2O$, and then discard the water. Sterilize the kanamycin stock by passage through the prepared 0.22-µm syringe filter. Make 1-mL aliquots in sterile microcentrifuge tubes. Stock may be kept for 1 yr at −20°C.

### LB-Carbenicillin Agar Plates

Prepare 1 L of LB agar (Lennox L agar) (Thermo Fisher 22700025) in 1 L of water per the manufacturer's instructions. After autoclaving, cool to ∼50°C. Add 1 mL of carbenicillin stock solution <R> to a final concentration of 100 µg/mL. Pour media into Petri dishes and allow to solidify. Store plates at 4°C.

### SB Medium

In a large, sterile 4- to 5-L Erlenmeyer flask, add the following ingredients: 30 g of MOPS [3(N-Morpholino) propane sulfonic acid] and 90 g of tryptone. In a ventilated hood, add 60 g of yeast extract. Add 2 L of double-distilled (dd) $H_2O$ and mix on a magnetic stir plate while keeping covered.

Once the mixture goes into solution, add the remaining 1 L of dd$H_2O$ and let stir for another 3 min. Place the pH probe into the flask and titrate the media with either HCl or NaOH to attain a pH of 7. Together, this makes 3 L. Divide the contents among four individual 1-L flasks. Sterilize by autoclaving at 15 psi on a liquid cycle for 20 min at 121°C.

### SOC Medium

| Reagent | Final concentration |
| --- | --- |
| Tryptone | 2% (w/v) |
| Yeast extract | 0.5% (w/v) |
| NaCl | 10 mM |
| KCl | 2.5 mM |

Add these ingredients to water in a flask and autoclave at 121°C. Allow the solution to cool. Add sterile $MgCl_2$ to a final concentration of 10 mM and sterile glucose to a final concentration of 20 mM. Store at room temperature.

### Sodium Acetate (3 M, pH 5.2)

Dissolve 246.1 g of sodium acetate in 500 mL of deionized $H_2O$. Adjust the pH to 5.2 with glacial acetic acid. Allow the solution to cool overnight. Adjust the pH once more to 5.2 with glacial acetic acid. Adjust the final volume to 1 L with deionized $H_2O$, and filter-sterilize with a 0.5-µm filter into an autoclaved bottle. Store for up to 2 mo at room temperature.

*TE Buffer*

| Reagent | Quantity (for 100 mL) | Final concentration |
|---|---|---|
| EDTA (0.5 M, pH 8.0) | 0.2 mL | 1 mM |
| Tris-Cl (1 M, pH 8.0) | 1 mL | 10 mM |
| H$_2$O | to 100 mL | |

*Tetracycline (10 mg/mL)*

Prepare 10 mg/mL tetracycline hydrochloride (Millipore-Sigma T7660-25G) in ethanol (Millipore-Sigma 493511-1L). Protect the solution from light and store it indefinitely at −20°C.

## ACKNOWLEDGMENTS

The earlier development of phage display technology in our research group was refined by Paul Roben, Stephen Cary, Carl Goodyear, and Caroline Grönwall, each of whom subsequently went on to establish their research laboratories. The inspiration for many of these studies came from conversations and encounters at our annual fall course on "Antibody Engineering and Display Technologies" at Cold Spring Harbor Laboratory. The work discussed here was supported in part by National Institutes of Health (NIH)-National Institute of Allergy and Infectious Diseases HHSN272201400019C, "B-Cell Epitope Discovery and Mechanisms of Antibody Protection," and NIH-National Institute of Arthritis and Musculoskeletal and Skin Diseases P50 AR070591-01A1 and T32GM66704. I also acknowledge the efforts of David Hernandez in refining this protocol as part of his thesis work.

## REFERENCES

Abagyan RA, Batalov S. 1997. Do aligned sequences share the same fold? *J Mol Biol* **273**: 355–368. doi:10.1006/jmbi.1997.1287

Andris-Widhopf J, Rader C, Steinberger P, Fuller R, Barbas CF III. 2000. Methods for the generation of chicken monoclonal antibody fragments by phage display. *J Immunol Methods* **242**: 159–181. doi:10.1016/S0022-1759(00)00221-0

Andris-Widhopf J, Steinberger P, Fuller R, Rader C, Barbas CF III. 2011. Generation of human Fab antibody libraries: PCR amplification and assembly of light- and heavy-chain coding sequences. *Cold Spring Harb Protoc* **2011**: pdb.prot065565. doi:10.1101/pdb.prot065565

Hernandez DN, Tam K, Shopsin B, Radke EE, Kolahi P, Copin R, Stubbe FX, Cardozo T, Torres VJ, Silverman GJ. 2020a. Unbiased identification of immunogenic *Staphylococcus aureus* leukotoxin B-cell epitopes. *Infect Immun* **88**: e00785-19. doi:10.1128/IAI.00785-19

Hernandez DN, Tam K, Shopsin B, Radke EE, Law K, Cardozo T, Torres VJ, Silverman GJ. 2020b. Convergent evolution of neutralizing antibodies to *Staphylococcus aureus* γ-hemolysin C that recognize an immunodominant primary sequence-dependent B-cell epitope. *mBio* **11**(3): e00460-20. doi:10.1128/mBio.00460-20

Illumina. 2025a. NGS workflow steps. https://www.illumina.com/science/technology/next-generation-sequencing/beginners/ngs-workflow.html [accessed July 3, 2025].

Illumina. 2025b. Illumina unique dual sequences. Document #1000000002694 v21. https://support-docs.illumina.com/SHARE/AdapterSequences/Content/SHARE/AdapterSeq/Illumina_DNA/IlluminaUDIndexes.htm [accessed July 3, 2025].

Rader C. 2024. The pComb3 phagemid family of phage display vectors. *Cold Spring Harb Protoc* **2024**: pdb.over107756. doi:10.1101/pdb.over107756

Rader C, Barbas CF III. 2001. Selection from antibody libraries. In *Phage display: a laboratory manual* (ed. Barbas CF III, et al.), pp. 13.1–13.15. Cold Spring Harbor Laboratory Press, Cold Spring Harbor, NY.

Radke EE, Brown SM, Pelzek AJ, Fulmer Y, Hernandez DN, Torres VJ, Thomsen IP, Chiang WK, Miller AO, Shopsin B, et al. 2018. Hierarchy of human IgG recognition within the *Staphylococcus aureus* immunome. *Sci Rep* **8**: 13296. doi:10.1038/s41598-018-31424-3

Schrödinger L, DeLano W. 2025. PyMOL. http://www.pymol.org [accessed July 1, 2025].

Silverman GJ. 2001. Functional domains and scaffolds. In *Phage display: a laboratory manual* (ed. Barbas CF III, et al.), pp. 5.1–5.25. Cold Spring Harbor Laboratory Press, Cold Spring Harbor, NY.

Silverman GJ. 2025. Insights from the study of B-cell epitopes of a microbial pathogen by phage display. *Cold Spring Harb Protoc* doi:10.1101/pdb.top107777

Weiss GA, Wells JA, Sidhu SS. 2000. Mutational analysis of the major coat protein of M13 identifies residues that control protein display. *Protein Sci* **9**: 647–654. doi:10.1110/ps.9.4.647

CHAPTER 12

# Use of Phage Display and Other Molecular Display Methods for the Development of Monobodies

Akiko Koide[1,2] and Shohei Koide[1,3,4]

[1]Laura and Isaac Perlmutter Cancer Center, New York University Langone Health, New York, New York 10016, USA; [2]Department of Medicine, [3]Department of Biochemistry and Molecular Pharmacology, New York University School of Medicine, New York, New York 10016, USA

Synthetic binding proteins are human-made binding proteins that use non-antibody proteins as the starting scaffold. Molecular display technologies, such as phage display, enable the construction of large combinatorial libraries and their efficient sorting and, thus, are crucial for the development of synthetic binding proteins. Monobodies are the founding system of a set of synthetic binding proteins based on the fibronectin type III (FN3) domain. Since the original report in 1998, the monobody and related FN3-based systems have steadily been refined, and current methods are capable of rapidly generating potent and selective binding molecules to even challenging targets. The FN3 domain is small (∼90 amino acids) and autonomous and is structurally similar to the conventional immunoglobulin (Ig) domain. Unlike the Ig domain, however, the FN3 lacks a disulfide bond but is highly stable. These attributes of FN3 present unique opportunities and challenges in the design of phage and other display systems, combinatorial libraries, and library sorting strategies. This article reviews key technological innovations in the establishment of our monobody development pipeline, with an emphasis on phage display methodology. These give insights into the molecular mechanisms underlying molecular display technologies and protein–protein interactions, which should be broadly applicable to diverse systems intended for generating high-performance binding proteins.

## INTRODUCTION

Developing custom reagents that selectively and potently bind a target molecule of interest is a core biotechnological methodology with broad utility. The ability of the adaptive immune system to generate antibodies that can bind diverse antigens after animal immunization has been exploited for decades for the development of research reagents and therapeutics, and analysis of the immune system and antibody–antigen interactions has provided key insights into the underlying molecular mechanisms. The key aspects of this process involve the generation of molecular diversity (Tonegawa 1983) and the selection of antibody clones of interest (Burnet 1976; Hodgkin et al. 2007). Each B cell expresses a single antibody on its surface, which establishes the linkage between the sequence of an antibody (i.e., genotype) and its function (i.e., phenotype) (Nossal and Lederberg 1958; Melchers and Andersson 1974). The invention of phage display made it possible to effectively apply this principle of genotype–phenotype linkage at the level of protein function outside the context of the immune system (Smith 1985).

[4]Correspondence: shohei.koide@nyulangone.org

© 2026 Cold Spring Harbor Laboratory Press
Cite this overview as *Cold Spring Harb Protoc*; doi:10.1101/pdb.over107982

# Chapter 12

The ability of the natural immune system and early successes of phage display in generating functional antibody fragments and short peptides have inspired the genesis of the field of synthetic binding proteins (Skerra 2000; Binz et al. 2005; Koide 2010). Similar to how natural antibodies with different binding specificity are generated primarily with different complementarity-determining regions that collectively form a contiguous surface of the otherwise mostly invariant immunoglobulin molecule (Mariuzza et al. 1987), synthetic binding proteins are generated by altering portions of a functionally inert protein, referred to as a protein scaffold. One might envision that, by starting with a well-chosen scaffold, one can achieve high binding function with additional desirable properties, such as small size, high stability, ease of production, and ease of use as a building block for constructing multidomain, multifunction proteins. A common goal of developing a synthetic binding protein system is to have the ability to generate proteins that bind to diverse target molecules, similar to that of the natural immune system.

Over the years, our community has collectively solved the challenge of generating highly functional synthetic binding proteins. There are now a number of well-established systems, including anticalin (Richter et al. 2014), affibody (Feldwisch and Tolmachev 2012), DARPin (Plückthun 2015), and monobody (Hantschel et al. 2020). In this article, we provide a historical account of the development of the monobody system, with particular emphasis on illustrating technological challenges and breakthroughs, which help us understand the molecular mechanisms underlying phage display and protein–protein interaction and should be applicable to diverse synthetic binding protein systems.

## The Monobody System

Monobody is a synthetic binding protein that is built from the molecular scaffold of the fibronectin type III (FN3) domain. The use of FN3 as a scaffold was first reported in 1998 by our group (Koide et al. 1998). Our vision was that, whereas immunoglobulins are effective scaffolds for creating binding proteins, that is, antibodies, it may be possible to establish a simpler system for generating binding proteins against a protein of interest. At the time of the inception of the monobody system, FN3 appeared to be a particularly attractive scaffold, as described below.

FN3 is a small, autonomously folded domain with a β-sandwich architecture (Fig. 1). It is among the most commonly occurring domains in mammalian proteomes, present in both extracellular and intracellular proteins. Indeed, according to the SMART database, the human proteome has 4104 FN3 domains in 673 proteins (smart.embl-heidelberg.de/smart/do_annotation.pl?DOMAIN=FN3). FN3 is a member of the immunoglobulin (Ig) superfold, with seven β-strands, and its β-strand topology is similar to that of the Ig variable domain, although FN3 has only seven β-strands (Fig. 1A). A key difference between FN3 and the conventional Ig domain is that FN3 lacks an intradomain disulfide bond that characterizes conventional Ig domains (Fig. 1A). Nevertheless, many FN3 domains have high conformational stability (Plaxco et al. 1996; Koide et al. 1998; Cota et al. 2000; Hamill et al. 2000; Jacobs et al. 2012). Therefore, we originally anticipated that it would be easier to produce FN3-based binding proteins, for example, by cytoplasmic overexpression in *Escherichia coli*. This is because the reducing environment of the cytoplasm makes it challenging to produce conventional Ig molecules, as their folding usually depends on the formation of disulfide bonds. Indeed, we and others have shown that monobodies can be expressed at a high level in their functional form in the cytoplasm of *E. coli* as well as in eukaryotic cells (Koide et al. 1998, 2002; Wojcik et al. 2010; Grebien et al. 2011).

Despite the many attractive attributes of FN3, it is not used by the adaptive immune system. Therefore, there were no immediate sources of sequence diversity equivalent to the natural repertoires of B-cell receptors (including antibodies) and T-cell receptors. Consequently, key innovations in the history of monobody development included the establishment of the phenotype–genotype linkage using molecular display systems and the design of effective combinatorial libraries (Koide et al. 1998, 2012a; Wojcik et al. 2010).

One could, in principle, use any FN3 domain from the thousands available in databases to build a new FN3-based system. In reality, however, only a small number of natural and designed FN3 molecules have been adopted as molecular scaffolds for developing synthetic binding proteins. In

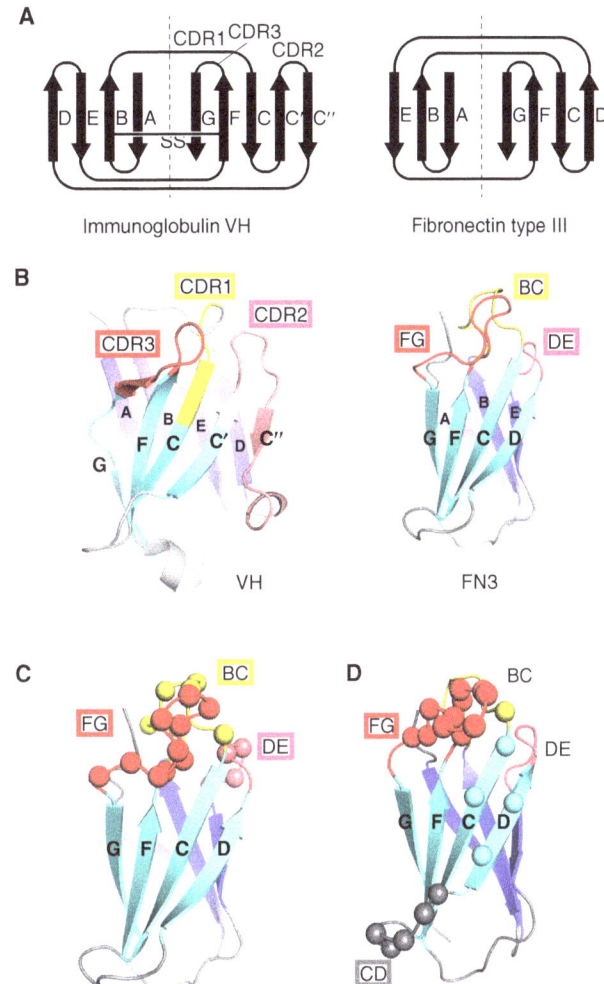

FIGURE 1. Comparison of immunoglobulin heavy-chain variable (VH) and fibronectin type III (FN3) domains and monobody library designs. (A) β-Strand and loop topology. The two β-sheets of the corresponding domains are shown in an open book manner with the dashed lines indicating the boundary between the two β-sheets. SS denotes an intradomain disulfide bond. (B) Cartoon representation of VH (PDB ID: 1VFB) and FN3 (3CSB). The complementarity-determining regions (CDRs) are according to the Kabat definition (Wu et al. 1993). (A,B, Drawn based on Fig. 1 in Koide et al. 1998.) (C,D) Two distinct types of monobody library designs. An antibody-like library design is depicted in C, and a non-antibody-like, "side" library is shown in D. (C,D, Modified from Fig. 1 in Koide et al. 2012a, with permission.)

addition to the tenth FN3 of human fibronectin used for monobodies and adnectins (Xu et al. 2002), a consensus FN3 domain and one from human tenascin have been used for constructing the centyrin and TN3 systems, respectively (Jacobs et al. 2012; Oganesyan et al. 2013). This limited variety is not surprising, because it takes substantial effort to establish an effective system for generating binding proteins even when using a single scaffold, as we discuss below for the monobody system. Furthermore, now that there already are well-established systems, including monobody, DARPin, anticalin, and affibody (Sha et al. 2017), it is our opinion that the field has reached the point of diminishing returns in terms of developing a synthetic binding protein system using a novel scaffold.

## Phage Display of Monobodies

At the time of the inception of the monobody concept, that is, the use of FN3 as a scaffold for the development of synthetic binding proteins, phage display was an obvious choice as a molecular display platform for constructing libraries and performing library sorting. We started with the then-standard system that fused the FN3 scaffold to a carboxy-terminal fragment of p3 and p8 of the M13 phage (Koide et al. 1998; Sidhu et al. 2000; Richards et al. 2003). This fusion gene was placed under the control of the *lac* promoter (Koide et al. 1998). This system was considered sufficiently effective for generating the first set of monobodies, which showed what was considered at the time to be moderate affinity. In retrospect, however, the affinity and selectivity of the monobodies from early studies are nowhere near those that we can achieve using current technologies (Sha et al. 2017; Hantschel et al. 2020; Akkapeddi et al. 2021). Subsequent studies identified challenges in phage display associated with high stability and rapid folding of FN3, and developed strategies to overcome them, as discussed below.

For effective recovery of functional clones, that is, clones that bind to a target of interest, it is crucial that all clones encoded by a phage display library are displayed on the phage surface, with minimal bias. We noticed low levels of surface display of monobodies using our original vector, which likely limited our ability to identify functional clones. Phage display systems based on M13 or fd phage require that the protein to be displayed is secreted, along with the extracellular portion of a phage coat protein, that is, p3 and p8, into the periplasm of host *E. coli* (Petrenko and Smith 2005). Such secretion is enabled by the attachment of a signal sequence to the amino terminus of the protein of interest. The early generations of phage display vectors used signal sequences that mediate Sec-dependent, post-translational secretion, which requires the fully translated protein to be unfolded as it is translocated across the membrane. These vectors were developed for linear peptides, disulfide-constrained peptides, and antibody fragments, which are either natively disordered or largely disordered until disulfide bonds are formed in the oxidizing environment of the extracellular milieu and the lumen of the secretion pathway, that is, the ER and Golgi. In other words, these molecules are disordered until they have been transferred across the plasma membrane, and are unlikely to obstruct the translocation process. In contrast, FN3 is highly stable and rapidly folds into its native globular conformation (Plaxco et al. 1996), and it is likely to present a substantial energetic barrier for the translocation process.

An important innovation came from Steiner et al. (2006), who showed that replacing a Sec-dependent signal sequence with a signal recognition particle (SRP)-dependent sequence dramatically increases the surface display level of another highly stable, rapidly folding protein, DARPin. We found that replacing a Sec-dependent signal sequence with an SRP-dependent signal sequence from *E. coli* DsbA similarly increases the display level of the monobody-p3 fusion on M13 phage by 100-fold (Wojcik et al. 2010). We also found that one can further increase the display level by the use of a mutant helper phage called hyperphage (Rondot et al. 2001) in conjunction with fusing the monobody to the full-length p3 instead of only to the carboxy-terminal half of p3, which is required for making the resulting phage particles infectious.

Using this "SRP phage display" system, we systematically examined conditions that potentially affect the display level. As expected, the number of phage particles in the culture supernatant depended nonlinearly on the length of culture and, consequently, it is important to experimentally determine an appropriate duration of phage propagation. Interestingly, the aeration of the *E. coli* culture for propagating phages strongly affected the display level. Phage particles produced with a non-baffled flask and slow shaking, that is, shaking just sufficient to maintain cell suspension, showed up to 50 times higher levels of monobody display than those produced using a baffled flask and vigorous shaking (Wojcik et al. 2010). In contrast, we did not observe such dependence on aeration of the display level of a Fab phage vector. Although we have not elucidated the molecular mechanism underlying this dependence of the display level on culture conditions, these results further underscore that phage display of a highly stable protein imposes stress on the host *E. coli* cells and that culture conditions suitable for rapid *E. coli* growth do not necessarily produce phage particles suitable for effective library sorting.

## Monobody Library Designs

The most common approach to generating synthetic binding proteins is to design a combinatorial library in which amino acid diversity is introduced at positions of a scaffold in such a way that the diversified positions are expected to form a contiguous surface for interacting with a target of interest. The absence of a natural, immune-like repertoire for FN3 presents both challenges and opportunities in combinatorial library design. The amino acid diversity needs to provide new interactions and also minimize negative impacts on biophysical properties such as stability. The size of a molecular display library that can be experimentally interrogated in a meaningful manner (up to approximately $10^{13}$ depending on display method) is much smaller than the total number of possible sequences that can be encoded even in a small library design, for example, $3 \times 10^{19}$ for a total of 15 fully randomized positions using one codon for each amino acid. Consequently, it is unlikely that the positions and amino acid diversity at each position can be optimized by an exhaustive search of fully randomized libraries.

There are two important factors for designing libraries: positions to be diversified and chemical diversity (which amino acid residues to be used and their ratio). Given the structural homology of FN3 with immunoglobulin G (IgG), our initial approach was to introduce diversity in the loops that are structurally equivalent to the complementarity-determining regions (CDRs) of immunoglobulins (Koide et al. 1998). However, it was unclear which loops of FN3 can be extensively modified without a drastic reduction in stability. For example, one poorly chosen mutation could denature the protein and make it nonfunctional and, thus, such a mutation should be avoided in library designs. We systematically tested the effects of altering loop regions of the monobody scaffold by insertion mutagenesis, which revealed that loops except for the EF loop can tolerate mutations (Batori et al. 2002). The EF loop has a feature called tyrosine corner that is important for the structural integrity of FN3 (Hamill et al. 2000). Therefore, we have kept the EF loop as wild-type in our libraries. This restriction makes it difficult to construct a library with diversification in multiple loops using the "bottom" end of the molecule (Koide et al. 2002). In parallel, we identified a stabilizing mutation at a site distal to the loops (Fig. 1B) (Koide et al. 2001). These biophysical studies provided foundational knowledge for designing libraries.

Three loops located at the "top" part of the FN3 protein, BC, DE, and FG (Fig. 1B), tolerate extensive mutations (Batori et al. 2002) and have thus been used as sites for constructing combinatorial libraries (Fig. 1C). Monobodies are stable even after mutation at more than 20% of the amino acid residues (Parker et al. 2005; Koide et al. 2012a).

Early libraries using diversification schemes with the NNK codons that encode all 20 amino acids, produced only low-affinity monobodies (Koide et al. 1998, 2002), which prompted us to examine the importance of chemical diversity. An important breakthrough was the finding that a monobody library using a binary code consisting of only Tyr and Ser but with varied loop lengths, produced high-affinity monobodies with good specificity (Koide et al. 2007). This work was inspired by pioneering studies by the Sidhu group that established the effectiveness of "reduced codes," including the Tyr/Ser binary code, in the Fab scaffold (Fellouse et al. 2004, 2005). The effectiveness of the Tyr/Ser binary code in the much smaller monobody scaffold was still surprising. Structural studies of these molecules defined the dominant roles of Tyr in making contacts in the interface (Koide and Sidhu 2009). Because the size of a combinatorial library using a binary code (e.g., $2^{20} = \sim 1$ million) is small enough to be fully sampled in a phage display library, studies using such libraries determine the effectiveness of library designs, as they are not affected by limited, stochastic sampling of encoded sequences that plague libraries that use more expanded codes.

The simplicity of the binary library enabled us to examine the effects of expanding the code. As one might expect, including additional amino acids at lower frequencies than Tyr and Ser produced more functional monobodies (Gilbreth et al. 2008). Structural studies revealed that Tyr still plays the dominant role in the interface of a monobody with an expanded code, and that the additional amino acid types mainly contribute to improving the shape complementarity of the interface (Gilbreth et al. 2008; Gilbreth and Koide 2012). An independent study confirmed the utility of additional amino acid types (Hackel and Wittrup 2010). Subsequent iterations including parallel studies of Fab libraries (Fellouse et al. 2007; Wojcik et al. 2010; Miller et al. 2012) have resulted in a design that has proven to be highly effective in producing monobodies for diverse targets. Such a design uses the following composition: 30% Tyr, 15% Ser, 10% Gly, 5% Phe, 5% Trp, and 2.5% each of all the other amino acids except for Cys (Koide et al. 2012a). Subsequently, others have developed similar "antibody-like" libraries (Xu et al. 2002; Olson and Roberts 2007; Hackel et al. 2008; Wensel et al. 2017; Lipovšek et al. 2018; Kondo et al. 2020).

Despite these advances, we found a unique challenge in the production of phage display libraries of monobodies. Large combinatorial libraries in the phage display format are often constructed using Kunkel mutagenesis (Kunkel 1985; Kunkel et al. 1987) and electroporation of the SS320 E. coli strain as the host (Sidhu et al. 2000). Because the efficiency of Kunkel mutagenesis is typically 50%–80%, a library constructed with this method contains a substantial fraction of the template construct used for mutagenesis. A construct that contains a termination codon, which is eliminated upon mutagenesis, for example, in a CDR, is often used as the template. In this manner, the "stop template," which does not display the protein of interest on the phage, is depleted in a library sorting process. In the case of phage display of monobodies, phage particles containing a stop template propagated more rapidly than monobody-displaying phage particles during the amplification step, dominating the amplified library and overriding the enrichment of target-binding clones. This growth bias is probably due to the stress of producing monobody-p3 fusion proteins on *E. coli*. As a workaround, we used a template construct that does display a monobody clone but it can be selectively removed by restriction enzyme digestion of purified DNA.

The use of synthetic libraries rather than natural immune libraries offers complete freedom in library design. Whereas antibody-like libraries have produced highly functional monobodies, structural studies of these monobodies in complex with their respective targets have revealed mismatches between the library design and the actual surfaces of monobodies used for contacting their targets. A subset of monobodies use the FG loop and β-strand regions of the scaffold for making contacts with their targets, rather than the three diversified loops (Gilbreth and Koide 2012; Koide et al. 2012a). These observations inspired us to design a distinct library in which we diversify residues in the CD and FG loops that are located on the opposite ends of the monobody molecule (Fig. 1D) and in the strands that connect these loops. This "side" library shows distinct interface topography from that of the antibody-like, "loop" libraries and, thus, is capable of generating monobodies that bind to flatter surfaces in target molecules (Koide et al. 2012a). In retrospect, it is not surprising that we have been able to develop monobodies that use this surface for target recognition, because it corresponds approximately to the heterodimerization interfaces in immunoglobulin G (IgG), one between the heavy-chain variable (VH) and light-chain variable (VL) domains and one between the heavy-chain constant 1 (CH1) and light-chain constant (CL) domains, as well as the homodimerization interface of the heavy-chain constant 3 (CH3) domain, which are all high-affinity interactions. In the conventional IgG molecules, this surface is already taken, and not available for antigen recognition. Other groups have followed our design and developed similar libraries (Diem et al. 2014; Wensel et al. 2017; Chan et al. 2019). Having two distinct types of libraries greatly improves the likelihood of developing highly functional monobodies to diverse targets.

## Other Display Systems for Monobody Discovery

The robustness of FN3 proteins has made monobody and other FN3 scaffolds compatible with virtually all molecular display systems. In addition to phage display, mRNA display (Xu et al. 2002; Olson and Roberts 2007; Kondo et al. 2020), yeast two hybrid (Koide et al. 2002), DNA display (Diem et al. 2014), and yeast display (Hackel et al. 2008; Koide et al. 2012a) have all been used.

Yeast display, for instance, has been used for iterative improvement of affinity, which have yielded molecules with low pM $K_D$ values (Hackel et al. 2008). Yeast display has also been used as the secondary screening method for clones enriched from larger primary libraries constructed in the phage and mRNA display formats (Koide et al. 2012a,b). The combination of phage/mRNA and yeast display complements their strengths and minimizes their weaknesses, such as the difficulty in quantitative sorting with phage/mRNA display and the difficulty in constructing large libraries with yeast display. We find the combination of phage display and yeast display to be particularly effective in comprehensively characterizing an enriched pool from sorting of a phage display library and in

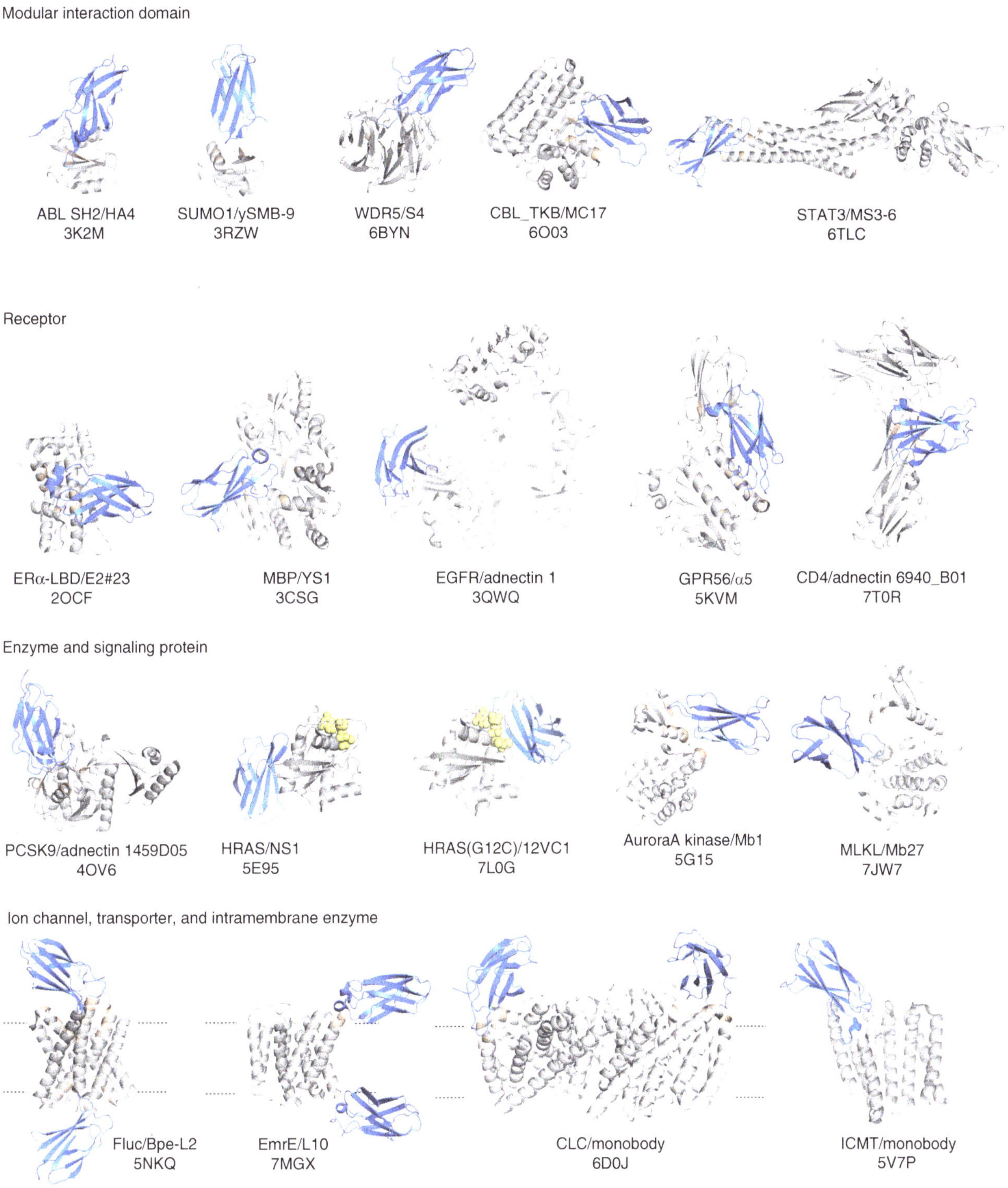

Display Methods for the Development of Monobodies

Modular interaction domain

ABL SH2/HA4
3K2M

SUMO1/ySMB-9
3RZW

WDR5/S4
6BYN

CBL_TKB/MC17
6O03

STAT3/MS3-6
6TLC

Receptor

ERα-LBD/E2#23
2OCF

MBP/YS1
3CSG

EGFR/adnectin 1
3QWQ

GPR56/α5
5KVM

CD4/adnectin 6940_B01
7T0R

Enzyme and signaling protein

PCSK9/adnectin 1459D05
4OV6

HRAS/NS1
5E95

HRAS(G12C)/12VC1
7L0G

AuroraA kinase/Mb1
5G15

MLKL/Mb27
7JW7

Ion channel, transporter, and intramembrane enzyme

Fluc/Bpe-L2
5NKQ

EmrE/L10
7MGX

CLC/monobody
6D0J

ICMT/monobody
5V7P

FIGURE 2. Examples of crystal structures of monobodies and adnectins in complex with their respective target proteins. The examples are divided based on the type of target proteins. Monobodies and adnectins are shown in blue and the targets in gray, with the epitope in orange. For HRAS, the bound nucleotide is show in yellow. The target name, binder name, and Protein Data Bank (PDB) ID are shown for each structure. (ABL) Abelson tyrosine-protein kinase, (SH2) src homology 2, (SUMO) small ubiquitin-related modifier, (WDR5) WD repeat-containing protein 5, (CBL) E3 ubiquitin-protein ligase CBL, (TKB) tyrosine kinase binding, (STAT3) signal transducer and activator of transcription 3, (ERα) estrogen receptor α, (LBD) ligand-binding domain, (MBP) maltose-binding protein, (EGFR) epidermal growth factor receptor, (GRP56) G-protein coupled receptor 56, (PCSK9) proprotein convertase subtilisin/kexin type 9, (MLKL) mixed-lineage kinase domain-like protein, (Fluc) fluoride channel, (CLC) chloride channel, (ICMT) protein-S-isoprenylcysteine O-methyltransferase.

discovering clones with exquisite selectivity. In phage display, enriched clones, typically dozens to hundreds, are individually produced and screened using phage ELISA. This process may miss rare clones with desirable properties. Yeast display coupled with a fluorescence-activated cell sorter can screen millions of cells and, thus, can interrogate the entire pool of clones in an enriched phage-display library. Furthermore, it is straightforward to implement both positive and negative sorting in yeast display. We have successfully developed monobodies with exquisite specificity and high potency using this approach (Akkapeddi et al. 2021; Teng et al. 2022).

## CONCLUSIONS

Phage display and other molecular display technologies continue to play central roles in the discovery of monobodies and other FN3-based synthetic binding proteins. Numerous molecules have been reported for diverse biomedical applications, including therapeutics that have successfully completed phase II clinical trials (NCT02515669, NCT03549260; see also Stein et al. 2019) and over 75 crystal structures of monobody-target complexes available in the Protein Data Bank (Fig. 2). Describing specific details of these individual molecules is beyond the scope of this article. The reader, however, is referred to recent reviews for a discussion on monobodies for therapeutic and diagnostic applications (Chandler and Buckle 2020), intracellular target discovery and validation (Akkapeddi et al. 2021), and structural and mechanistic investigation (Hantschel et al. 2020). The expanding efforts and innovations made by the community will define applications that exploit unique strengths of the monobody molecules, including small size, high stability and rapid folding, the absence of disulfide bonds, and the ease of constructing fusion proteins (Fulcher et al. 2017; Donnelly et al. 2018; Kulemzin et al. 2018; Ludwicki et al. 2019; Röth et al. 2020; Lim et al. 2021; Robu et al. 2021). These successes in developing novel synthetic binding proteins may now appear routine, but they are the culmination of multidisciplinary research. The ambition of establishing a robust system that consistently generates potent and selective binding proteins to diverse targets has challenged the protein engineering community both intellectually and technologically. Indeed, to establish such a system, one needs to become adept at display technologies, protein–protein interactions, protein stability and folding, and microbiology. Practitioners of display technologies and synthetic binding protein development would be well served by acquiring a deep understanding of the mechanistic underpinnings of these powerful technologies.

## REFERENCES

Akkapeddi P, Teng KW, Koide S. 2021. Monobodies as tool biologics for accelerating target validation and druggable site discovery. *RSC Med Chem* 12: 1839–1853. doi:10.1039/D1MD00188D

Batori V, Koide A, Koide S. 2002. Exploring the potential of the monobody scaffold: effects of loop elongation on the stability of a fibronectin type III domain. *Protein Eng* 15: 1015–1020. doi:10.1093/protein/15.12.1015

Binz HK, Amstutz P, Plückthun A. 2005. Engineering novel binding proteins from nonimmunoglobulin domains. *Nat Biotechnol* 23: 1257–1268. doi:10.1038/nbt1127

Burnet FM. 1976. A modification of Jerne's theory of antibody production using the concept of clonal selection. *CA Cancer J Clin* 26: 119–121. doi:10.3322/canjclin.26.2.119

Chan R, Buckley PT, O'Malley A, Sause WE, Alonzo F, Lubkin A, Boguslawski KM, Payne A, Fernandez J, Strohl WR, et al. 2019. Identification of biologic agents to neutralize the bicomponent leukocidins of *Staphylococcus aureus*. *Sci Transl Med* 11: eaat0882. doi:10.1126/scitranslmed.aat0882

Chandler PG, Buckle AM. 2020. Development and differentiation in monobodies based on the fibronectin type 3 domain. *Cells* 9: 610. doi:10.3390/cells9030610

Cota E, Hamill SJ, Fowler SB, Clarke J. 2000. Two proteins with the same structure respond very differently to mutation: the role of plasticity in protein stability. *J Mol Biol* 302: 713–725. doi:10.1006/jmbi.2000.4053

Diem MD, Hyun L, Yi F, Hippensteel R, Kuhar E, Lowenstein C, Swift EJ, O'Neil KT, Jacobs SA. 2014. Selection of high-affinity Centyrin FN3 domains from a simple library diversified at a combination of strand and loop positions. *Protein Eng Des Sel* 27: 419–429. doi:10.1093/protein/gzu016

Donnelly DJ, Smith RA, Morin P, Lipovsek D, Gokemeijer J, Cohen D, Lafont V, Tran T, Cole EL, Wright M, et al. 2018. Synthesis and biologic evaluation of a novel (18)F-labeled adnectin as a PET radioligand for imaging PD-L1 expression. *J Nucl Med* 59: 529–535. doi:10.2967/jnumed.117.199596

Feldwisch J, Tolmachev V. 2012. Engineering of affibody molecules for therapy and diagnostics. *Methods Mol Biol* 899: 103–126. doi:10.1007/978-1-61779-921-1_7

Fellouse FA, Wiesmann C, Sidhu SS. 2004. Synthetic antibodies from a four-amino-acid code: a dominant role for tyrosine in antigen recognition. *Proc Natl Acad Sci* 101: 12467–12472. doi:10.1073/pnas.0401786101

Fellouse FA, Li B, Compaan DM, Peden AA, Hymowitz SG, Sidhu SS. 2005. Molecular recognition by a binary code. *J Mol Biol* 348: 1153–1162. doi:10.1016/j.jmb.2005.03.041

Fellouse FA, Esaki K, Birtalan S, Raptis D, Cancasci VJ, Koide A, Jhurani P, Vasser M, Wiesmann C, Kossiakoff AA, et al. 2007. High-throughput generation of synthetic antibodies from highly functional minimalist phage-displayed libraries. *J Mol Biol* **373:** 924–940. doi:10.1016/j.jmb.2007.08.005

Fulcher LJ, Hutchinson LD, Macartney TJ, Turnbull C, Sapkota GP. 2017. Targeting endogenous proteins for degradation through the affinity-directed protein missile system. *Open Biol* **7:** 170066. doi:10.1098/rsob.170066

Gilbreth RN, Koide S. 2012. Structural insights for engineering binding proteins based on non-antibody scaffolds. *Curr Opin Struct Biol* **22:** 413–420. doi:10.1016/j.sbi.2012.06.001

Gilbreth RN, Esaki K, Koide A, Sidhu SS, Koide S. 2008. A dominant conformational role for amino acid diversity in minimalist protein–protein interfaces. *J Mol Biol* **381:** 407–418. doi:10.1016/j.jmb.2008.06.014

Grebien F, Hantschel O, Wojcik J, Kaupe I, Kovacic B, Wyrzucki AM, Gish GD, Cerny-Reiterer S, Koide A, Beug H, et al. 2011. Targeting the SH2-kinase interface in Bcr-Abl inhibits leukemogenesis. *Cell* **147:** 306–319. doi:10.1016/j.cell.2011.08.046

Hackel BJ, Wittrup KD. 2010. The full amino acid repertoire is superior to serine/tyrosine for selection of high affinity immunoglobulin G binders from the fibronectin scaffold. *Protein Eng Des Sel* **23:** 211–219. doi:10.1093/protein/gzp083

Hackel BJ, Kapila A, Wittrup KD. 2008. Picomolar affinity fibronectin domains engineered utilizing loop length diversity, recursive mutagenesis, and loop shuffling. *J Mol Biol* **381:** 1238–1252. doi:10.1016/j.jmb.2008.06.051

Hamill SJ, Cota E, Chothia C, Clarke J. 2000. Conservation of folding and stability within a protein family: the tyrosine corner as an evolutionary cul-de-sac. *J Mol Biol* **295:** 641–649. doi:10.1006/jmbi.1999.3360

Hantschel O, Biancalana M, Koide S. 2020. Monobodies as enabling tools for structural and mechanistic biology. *Curr Opin Struct Biol* **60:** 167–174. doi:10.1016/j.sbi.2020.01.015

Hodgkin PD, Heath WR, Baxter AG. 2007. The clonal selection theory: 50 years since the revolution. *Nat Immunol* **8:** 1019–1026. doi:10.1038/ni1007-1019

Jacobs SA, Diem MD, Luo J, Teplyakov A, Obmolova G, Malia T, Gilliland GL, O'Neil KT. 2012. Design of novel FN3 domains with high stability by a consensus sequence approach. *Protein Eng Des Sel* **25:** 107–117. doi:10.1093/protein/gzr064

Koide S. 2010. Design and engineering of synthetic binding proteins using nonantibody scaffolds. In *Protein engineering and design* (ed. Park SJ, Cochran JR), pp. 109–130. CRC, Boca Raton, FL.

Koide S, Sidhu SS. 2009. The importance of being tyrosine: lessons in molecular recognition from minimalist synthetic binding proteins. *ACS Chem Biol* **4:** 325–334. doi:10.1021/cb800314v

Koide A, Bailey CW, Huang X, Koide S. 1998. The fibronectin type III domain as a scaffold for novel binding proteins. *J Mol Biol* **284:** 1141–1151. doi:10.1006/jmbi.1998.2238

Koide A, Jordan MR, Horner SR, Batori V, Koide S. 2001. Stabilization of a fibronectin type III domain by the removal of unfavorable electrostatic interactions on the protein surface. *Biochemistry* **40:** 10326–10333. doi:10.1021/bi010916y

Koide A, Abbatiello S, Rothgery L, Koide S. 2002. Probing protein conformational changes by using designer binding proteins: application to the estrogen receptor. *Proc Natl Acad Sci* **99:** 1253–1258. doi:10.1073/pnas.032665299

Koide A, Gilbreth RN, Esaki K, Tereshko V, Koide S. 2007. High-affinity single-domain binding proteins with a binary-code interface. *Proc Natl Acad Sci* **104:** 6632–6637. doi:10.1073/pnas.0700149104

Koide A, Wojcik J, Gilbreth RN, Hoey RJ, Koide S. 2012a. Teaching an old scaffold new tricks: monobodies constructed using alternative surfaces of the FN3 scaffold. *J Mol Biol* **415:** 393–405. doi:10.1016/j.jmb.2011.12.019

Koide S, Koide A, Lipovšek D. 2012b. Target-binding proteins based on the 10th human fibronectin type III domain ($^{10}$Fn3). *Methods Enzymol* **503:** 135–156. doi:10.1016/B978-0-12-396962-0.00006-9

Kondo T, Iwatani Y, Matsuoka K, Fujino T, Umemoto S, Yokomaku Y, Ishizaki K, Kito S, Sezaki T, Hayashi G, et al. 2020. Antibody-like proteins that capture and neutralize SARS-CoV-2. *Sci Adv* **6:** eabd3916. doi:10.1126/sciadv.abd3916

Kulemzin SV, Gorchakov AA, Chikaev AN, Kuznetsova VV, Volkova OY, Matvienko DA, Petukhov AV, Zaritskey AY, Taranin AV. 2018. VEGFR2-specific FnCAR effectively redirects the cytotoxic activity of T cells and YT NK cells. *Oncotarget* **9:** 9021–9029. doi:10.18632/oncotarget.24078

Kunkel TA. 1985. Rapid and efficient site-specific mutagenesis without phenotypic selection. *Proc Natl Acad Sci* **82:** 488–492. doi:10.1073/pnas.82.2.488

Kunkel TA, Roberts JD, Zakour RA. 1987. Rapid and efficient site-directed mutagenesis without phenotypic selection. *Methods Enzymol* **154:** 367–382. doi:10.1016/0076-6879(87)54085-X

Lim S, Khoo R, Juang YC, Gopal P, Zhang H, Yeo C, Peh KM, Teo J, Ng S, Henry B, et al. 2021. Exquisitely specific anti-KRAS biodegraders inform on the cellular prevalence of nucleotide-loaded states. *ACS Cent Sci* **7:** 274–291. doi:10.1021/acscentsci.0c01337

Lipovšek D, Carvajal I, Allentoff AJ, Barros A, Brailsford J, Cong Q, Cotter P, Gangwar S, Hollander C, Lafont V, et al. 2018. Adnectin-drug conjugates for Glypican-3-specific delivery of a cytotoxic payload to tumors. *Protein Eng Des Sel* **31:** 159–171. doi:10.1093/protein/gzy013

Ludwicki MB, Li J, Stephens EA, Roberts RW, Koide S, Hammond PT, DeLisa MP. 2019. Broad-spectrum proteome editing with an engineered bacterial ubiquitin ligase mimic. *ACS Cent Sci* **5:** 852–866. doi:10.1021/acscentsci.9b00127

Mariuzza RA, Phillips SE, Poljak RJ. 1987. The structural basis of antigen-antibody recognition. *Annu Rev Biophys Biophys Chem* **16:** 139–159. doi:10.1146/annurev.bb.16.060187.001035

Melchers F, Andersson J. 1974. Immunoglobulin production in B-lymphocytes: synthesis of the membrane-bound receptor and the secreted serum glycoprotein immunoglobulin M. *Biochem Soc Symp* 73–85.

Miller KR, Koide A, Leung B, Fitzsimmons J, Yoder B, Yuan H, Jay M, Sidhu SS, Koide S, Collins EJ. 2012. T cell receptor-like recognition of tumor in vivo by synthetic antibody fragment. *PLoS ONE* **7:** e43746. doi:10.1371/journal.pone.0043746

Nossal GJ, Lederberg J. 1958. Antibody production by single cells. *Nature* **181:** 1419–1420. doi:10.1038/1811419a0

Oganesyan V, Ferguson A, Grinberg L, Wang L, Phipps S, Chacko B, Drabic S, Thisted T, Baca M. 2013. Fibronectin type III domains engineered to bind CD40L: cloning, expression, purification, crystallization and preliminary X-ray diffraction analysis of two complexes. *Acta Crystallogr Sect F Struct Biol Cryst Commun* **69:** 1045–1048. doi:10.1107/S1744309113022847

Olson CA, Roberts RW. 2007. Design, expression, and stability of a diverse protein library based on the human fibronectin type III domain. *Protein Sci* **16:** 476–484. doi:10.1110/ps.062498407

Parker MH, Chen Y, Danehy F, Dufu K, Ekstrom J, Getmanova E, Gokemeijer J, Xu L, Lipovšek D. 2005. Antibody mimics based on human fibronectin type three domain engineered for thermostability and high-affinity binding to vascular endothelial growth factor receptor two. *Protein Eng Des Sel* **18:** 435–444. doi:10.1093/protein/gzi050

Petrenko VA, Smith GP. 2005. Vectors and modes of display. In *Phage display in biotechnology and drug discovery* (ed. Sidhu SS), pp. 63–110. CRC, San Diego.

Plaxco KW, Spitzfaden C, Campbell ID, Dobson CM. 1996. Rapid refolding of a proline-rich all-β-sheet fibronectin type III module. *Proc Natl Acad Sci* **93:** 10703–10706. doi:10.1073/pnas.93.20.10703

Plückthun A. 2015. Designed ankyrin repeat proteins (DARPins): binding proteins for research, diagnostics, and therapy. *Annu Rev Pharmacol Toxicol* **55:** 489–511. doi:10.1146/annurev-pharmtox-010611-134654

Richards J, Miller M, Abend J, Koide A, Koide S, Dewhurst S. 2003. Engineered fibronectin type III domain with a RGDWXE sequence binds with enhanced affinity and specificity to human αvβ3 integrin. *J Mol Biol* **326:** 1475–1488. doi:10.1016/S0022-2836(03)00082-2

Richter A, Eggenstein E, Skerra A. 2014. Anticalins: exploiting a non-Ig scaffold with hypervariable loops for the engineering of binding proteins. *FEBS Lett* **588:** 213–218. doi:10.1016/j.febslet.2013.11.006

Robu S, Richter A, Gosmann D, Seidl C, Leung D, Hayes W, Cohen D, Morin P, Donnelly DJ, Lipovšek D, et al. 2021. Synthesis and preclinical evaluation of a $^{68}$Ga-labeled adnectin, $^{68}$Ga-BMS-986192, as a PET agent for imaging PD-L1 expression. *J Nucl Med* **62:** 1228–1234. doi:10.2967/jnumed.120.258384

Rondot S, Koch J, Breitling F, Dubel S. 2001. A helper phage to improve single-chain antibody presentation in phage display. *Nat Biotechnol* **19:** 75–78. doi:10.1038/83567

Röth S, Macartney TJ, Konopacka A, Chan KH, Zhou H, Queisser MA, Sapkota GP. 2020. Targeting endogenous K-RAS for degradation through the affinity-directed protein missile system. *Cell Chem Biol* **27:** 1151–1163.e1156. doi:10.1016/j.chembiol.2020.06.012

Sha F, Salzman G, Gupta A, Koide S. 2017. Monobodies and other synthetic binding proteins for expanding protein science. *Protein Sci* **26:** 910–924. doi:10.1002/pro.3148

Sidhu SS, Lowman HB, Cunningham BC, Wells JA. 2000. Phage display for selection of novel binding peptides. *Methods Enzymol* **328:** 333–363. doi:10.1016/S0076-6879(00)28406-1

Skerra A. 2000. Engineered protein scaffolds for molecular recognition. *J Mol Recognit* **13:** 167–187. doi:10.1002/1099-1352(200007/08)13:4<167::AID-JMR502>3.0.CO;2-9

Smith GP. 1985. Filamentous fusion phage: novel expression vectors that display cloned antigens on the virion surface. *Science* **228:** 1315–1317. doi:10.1126/science.4001944

Stein E, Toth P, Butcher MB, Kereiakes D, Magnu P, Bays H, Zhou R, Turner TA. 2019. Safety, tolerability and Ldl-C reduction with a novel anti-Pcsk9 recombinant fusion protein (Lib003): results of a randomized, double-blind, placebo-controlled, phase 2 study. *Atherosclerosis* **287:** E7. doi:10.1016/j.atherosclerosis.2019.06.019

Steiner D, Forrer P, Stumpp MT, Plückthun A. 2006. Signal sequences directing cotranslational translocation expand the range of proteins amenable to phage display. *Nat Biotechnol* **24:** 823–831. doi:10.1038/nbt1218

Teng KW, Koide A, Koide S. 2022. Engineering binders with exceptional selectivity. *Methods Mol Biol* **2491:** 143–154. doi:10.1007/978-1-0716-2285-8_8

Tonegawa S. 1983. Somatic generation of antibody diversity. *Nature* **302:** 575–581. doi:10.1038/302575a0

Wensel D, Sun Y, Li Z, Zhang S, Picarillo C, McDonagh T, Fabrizio D, Cockett M, Krystal M, Davis J. 2017. Discovery and characterization of a novel CD4-binding adnectin with potent anti-HIV activity. *Antimicrob Agents Chemother* **61:** e00508-17. doi:10.1128/AAC.00508-17

Wojcik J, Hantschel O, Grebien F, Kaupe I, Bennett KL, Barkinge J, Jones RB, Koide A, Superti-Furga G, Koide S. 2010. A potent and highly specific FN3 monobody inhibitor of the Abl SH2 domain. *Nat Struct Mol Biol* **17:** 519–527. doi:10.1038/nsmb.1793

Wu TT, Johnson G, Kabat EA. 1993. Length distribution of CDRH3 in antibodies. *Proteins* **16:** 1–7. doi:10.1002/prot.340160102

Xu L, Aha P, Gu K, Kuimelis RG, Kurz M, Lam T, Lim AC, Liu H, Lohse PA, Sun L, et al. 2002. Directed evolution of high-affinity antibody mimics using mRNA display. *Chem Biol* **9:** 933–942. doi:10.1016/S1074-5521(02)00187-4

CHAPTER 13

# Phage-Displayed SH2 Domain Libraries: From Ultrasensitive Tyrosine Phosphoproteome Probes to Translational Research

Gregory D. Martyn[1,2] and Gianluca Veggiani[1,3,4]

[1]*Donnelly Centre for Cellular and Biomolecular Research, University of Toronto, Toronto, Ontario M5S 3E1, Canada;* [2]*Department of Molecular Genetics, University of Toronto, Toronto, Ontario M5S 1A8, Canada;* [3]*Department of Pathobiological Sciences, School of Veterinary Medicine, Louisiana State University, Baton Rouge, Louisiana 70803, USA*

Tyrosine phosphorylation is a critical regulator of cell signaling. A large fraction of the tyrosine phosphoproteome, however, remains uncharacterized, largely due to a lack of robust and scalable methods. The Src homology 2 (SH2) domain, a structurally conserved protein domain present in many intracellular signal-transducing proteins, naturally binds phosphorylated tyrosine (pTyr) residues, providing an ideal scaffold for the development of sensitive pTyr probes. Its modest affinity, however, has greatly limited its application. Phage display is an in vitro technique used for identifying ligands for proteins and other macromolecules. Using this technique, researchers have been able to engineer SH2 domains to increase their affinity and customize their specificity. Indeed, highly diverse phage display libraries have enabled the engineering of SH2 domains as affinity-purification (AP) tools for proteomic analysis as well as probes for aberrant tyrosine signaling detection and rewiring, and represent a promising class of novel diagnostics and therapeutics. This review describes the unique structure–function characteristics of SH2 domains, highlights the fundamental contribution of phage display in the development of technologies for the dissection of the tyrosine phosphoproteome, and highlights prospective uses of SH2 domains in basic and translational research.

## INTRODUCTION

Tyrosine phosphorylation is one of the most important posttranslational modifications (PTMs), playing a central role in regulating signal transduction pathways that are essential for cell growth, migration, replication, and apoptosis (Hunter 2014). Phosphorylated tyrosine (pTyr) residues are crucial players in cell signaling and exert their function by modulating the activity, localization, and interaction of cellular proteins that contain modular domains that specifically recognize them (Yaffe 2002). Among the hundreds of pTyr-recognition domains in humans (Seet et al. 2006), the best-characterized ones are Src homology 2 (SH2) domains. First identified in 1986, SH2 domains were initially described as a small (∼100 amino acids; 12-kDa) noncatalytic regulatory domain of the v-Fps/Fes protein kinase that was homologous in sequence to Src-related kinases (Sadowski et al. 1986). Subsequent studies showed that SH2 domains can bind activated receptor tyrosine kinases (RTKs) (Moran et al. 1990) and, therefore, play a pivotal role in signal transduction pathways that control cell physiology (Koch et al. 1992). Given the prominent role of tyrosine

[4]Correspondence: gveggiani@lsu.edu

© 2026 Cold Spring Harbor Laboratory Press
Cite this overview as *Cold Spring Harb Protoc*; doi:10.1101/pdb.over107981

phosphorylation in signal transduction, it is unsurprising that dysregulation of SH2 domain–mediated signaling has been observed in numerous human pathologies, including neurodegeneration and cancer (Li et al. 2012; Tenreiro et al. 2014; Aschner and Downey 2018; Du and Lovly 2018). Therefore, mapping protein phosphorylation events and understanding their functional implications are relevant for uncovering pathogenic signaling pathways and identifying novel therapeutic strategies.

Despite significant technological advances over the past two decades, the dynamic range and detection sensitivity of pTyr-containing proteins and peptides is limited both by the rapid pTyr turnover and its low cellular abundance relative to phosphorylated serine/threonine (pSer/pThr). In fact, tyrosine phosphorylation constitutes <1% of the total phosphoproteome in a mammalian cell and, consequently, pTyr sites are often underrepresented in samples (Sharma et al. 2014), as pTyr-containing proteins and peptides are typically isolated from complex mixtures using immobilized metal-affinity chromatography (IMAC), which enriches indiscriminately for pSer/pThr/pTyr peptides. The use of pTyr-specific antibodies enables precise analysis of the tyrosine phosphoproteome, but this approach is poorly scalable, suffers from batch-to-batch variations, has limited efficiency, and cannot be used for in vivo studies (Nita-Lazar et al. 2008; Bian et al. 2016). Conversely, the use of SH2 domains represent an attractive alternative, as their intrinsic specificity for pTyr residues and ability to fold intracellularly make them amenable probes for studying the pTyr proteome, both in vitro and in vivo. However, the intrinsic weak affinity of SH2 domains requires their engineering before their use.

In recent years, several groups have developed highly functional SH2 domain libraries to enhance their affinity or alter their binding specificity. Engineered SH2 variants have enabled the development of several novel methodologies and biotechnologies that have potential applications in translational research. The scope of this article is to provide an overview of the structure–function of the SH2 domain, describe how this information has been used to engineer the domain via phage display, and, finally, highlight diverse applications of SH2 domains in the dissection of cell signaling, as well as in the development of diagnostics and therapeutics.

## THE STRUCTURE–FUNCTION OF SH2 DOMAINS

### SH2 Domains in Signal Transduction Networks

The tyrosine phosphorylation signaling system relies on the tightly regulated activity of (1) kinases (writers) that catalyze the transfer of phosphoryl groups to target proteins, (2) phosphatases (erasers) that remove pTyr moieties, and (3) pTyr-binding domains (readers) that decode and transduce pTyr signals (Fig. 1A; Liu et al. 2012). The pTyr signaling networks typically start with growth factors originating in the extracellular matrix (ECM), which binds to specialized cell surface receptors such as receptor tyrosine kinases (RTKs) (Lemmon et al. 2014). Upon ligand binding, RTKs typically undergo dimerization, leading to the autophosphorylation of tyrosine residues in the carboxy-terminal regions of the same RTK (Schlessinger et al. 2014). Such pTyr marks act as signals for recruiting proteins containing pTyr-binding domains, which act either as adaptors for the recruitment of effector proteins or directly exert catalytic activities. The human genome encodes hundreds of phosphotyrosine-binding domains (PTBs), but the SH2 domains, of which 122 have been annotated, are the best characterized and represent the prototypical pTyr-binding module (Liu et al. 2006).

The SH2 domain is typically associated with other modular-binding domains (i.e., Src homology 3 [SH3] phosphotyrosine-binding [PTBs], and pleckstrin homology [PH] domains) or with the catalytic domain in a protein (Haslam and Shields 2012). In this way, by localizing proteins to the correct subcellular location in response to precise pTyr marks at the appropriate time, SH2 domain–containing proteins can phosphorylate, dephosphorylate, or act as docking sites for the recruitment of additional proteins, propagating signaling cascades that govern cellular behavior.

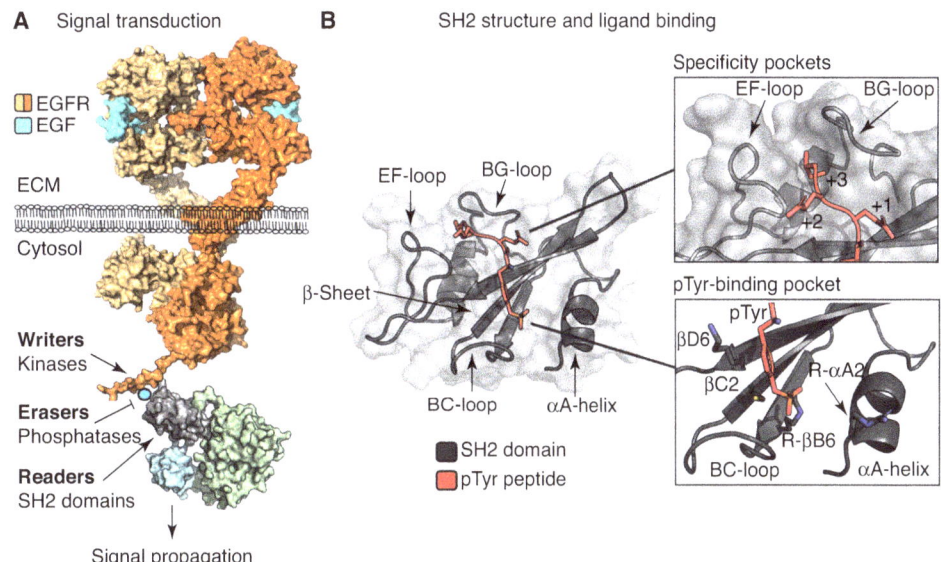

FIGURE 1. Human Src homology 2 (SH2) domain structure–function in signal transduction pathways. (*A*) Phosphotyrosine-dependent signal transduction pathways are composed of kinases (writers), phosphatases (erasers), and readers (SH2 domains). Together, these components selectively phosphorylate or dephosphorylate specific Tyr residues. SH2 domains can selectively bind phosphorylated tyrosine (pTyr) residues, thereby localizing proteins to the correct cellular location and mediating important protein–protein interactions. When correctly operating, this signal-transduction system relays important growth, stress state, and apoptotic signals to the cell nucleus that ultimately result in a change in transcription and biological activity. In this example, we show the membrane-embedded epidermal growth factor receptor (EGFR, orange), which upon interaction with its ligand EGF (cyan) dimerizes, activating its tyrosine kinase domain, resulting in *trans*-autophosphorylation of multiple tyrosine residues (blue circle). Phosphorylation of Tyr residues provides docking sites for SH2 domains that transduce this signal via mitogen-activated protein kinases (MAPKs). Conversely, phosphatases catalyze EGFR dephosphorylation, thereby suppressing its signaling. (*B*) The human SH2 fold consists of a central β-sheet flanked by two α-helices. The BC-loop, αA-helix, and one side of the β-sheet comprise the pTyr-binding pocket, while the EF- and BG-loops form specificity-determining pockets. Depicted here is the structure of the SH2 domain of the tyrosine kinase Src bound to a pTyr peptide (PDB ID: 1HCT). The specificity-determining pockets are formed by the EF- and BG-loops and interact with residues on the carboxyl-terminus side of the pTyr peptide. The pTyr-binding pocket coordinates the pTyr moiety, using a highly conserved Arg (R-βB6). A less conserved R-αA2 residue has also been shown to be important in pTyr coordination in some SH2 domains and has proved to be crucial for engineering high-affinity SH2 domain variants. The backside of the pTyr-binding pocket is formed, in part, by the βC2 and βD6 residues, which interact with the phenyl ring of pTyr. Residues in the BC-loop are highly variable in both sequence and length in different human SH2 domains. Together, these components of the pTyr-binding pocket cooperate to specifically interact with pTyr targets.

## SH2 Domain Structure and Its Role in Molecular Recognition

Structurally, the SH2 domain fold is highly conserved and is comprised of a three-stranded antiparallel β-sheet flanked by two α-helices (Waksman et al. 1992), which ensures specific and tightly controlled interactions with pTyr targets (Fig. 1B). As a result of such structural conservation, the vast majority of SH2 domains bind pTyr peptides in a two-pronged mode, in which one side of the domain directly coordinates the pTyr moiety controlling the strength of the interaction, while a second binding site dictates binding specificity (Marasco and Carlomagno 2020).

## Structure of the pTyr-Binding Pocket

The pTyr-binding pocket is situated on one side of the central β-sheet ("backside" of the pTyr-binding pocket), toward the αA-helix (Fig. 1B). The βC2 and βD6 strands comprise part of the backside of the pTyr-binding pocket and are crucial for contacting the phenyl ring of the pTyr side chain. These residues can vary greatly between different human SH2 domains and have been shown to be essential for contributing to SH2-binding affinity (Kaneko et al. 2012; Martyn et al. 2022). Using a highly conserved arginine (Arg; R) found within a common "FLVR" motif located in the βB-strand, SH2 domains establish an interaction with the phosphoryl moiety of pTyr, and such interaction contributes the majority of the total free energy of the SH2 ligand binding (Bradshaw and Waksman 1999). Although most human SH2 domains have a conserved Arg in the βB-strand that contacts the pTyr, the SH2 domains of Rin2, SH2D5, and Tyk2 lack this residue. Other human SH2 domains, such as SH21A (SAP/SH2D1A), are instead capable of binding non-phosphorylated peptides (Chan et al. 2003), and form a three-pronged binding mode in some specialized circumstances (Hwang et al. 2002).

Recent studies have identified 93 new SH2 domains in the bacteria *Legionella pneumophila* (Kaneko et al. 2018). Unlike their eukaryotic orthologs, these domains are unique in that they lack specificity-determining regions but still retain the core fold of the pTyr-binding pocket. This includes a conserved Arg residue that contacts the phosphoryl moiety of the pTyr residue. Remarkably, these SH2 domains, despite not engaging the residues flanking the pTyr residue, can still bind pTyr-containing targets with micromolar affinity.

## The Structure of Specificity-Determining Pockets

Adjacent to the pTyr-binding pocket, there are a series of specificity-determining pockets formed by the EF- and BG-loops that interact with side chains of residues carboxy terminal to the pTyr residue (Fig. 1B; Kaneko et al. 2010). Residues found in these loops dictate the charge, hydrophobicity, and overall conformation of these pockets, and establish favorable or unfavorable contacts with ligand side chains carboxy terminal to the pTyr residue, thereby dictating specificity. To understand the binding preferences of SH2 domains, many studies have been conducted using a variety of approaches, including molecular dynamic simulations (Gan and Roux 2009), oriented peptide array libraries (OPALs) (Li et al. 2008), protein microarrays (Jones et al. 2006), high-throughput fluorescence polarization assays (Hause et al. 2012; Leung et al. 2014), far western blot analysis (Machida et al. 2007), and affinity purification-mass spectrometry (AP-MS) (Höfener et al. 2014). Results from such studies have revealed key insights. For example, the Grb2-SH2 domain contains a bulky tryptophan (Trp; W) residue in the EF-loop that constrains the contacts that the domain can make with specific pTyr peptides (Rondeau et al. 1999). To accommodate this steric blockage, an asparagine (Asn; N) at the +2 position in the target pTyr peptide allows Grb2 binding. Additional studies have also highlighted the importance of EF- and BG-loops and the effect of nonpermissive residues (responsible for decreased binding) in binding, and allowed further classification of SH2 domain specificities (Liu et al. 2010).

To organize the plethora of SH2 ligand–binding data available, a probabilistic interaction network database (PepSpotDB) has been developed to classify these complex SH2–ligand interactions (Tinti et al. 2013). This work probed SH2 interactions for a large subset of the human pTyr proteome using peptide microarrays and AP-MS and combined these data with other publicly available data sets, developing a broad classification system comprising 17 different specificity classes for human SH2 domains (Tinti et al. 2013). There are, however, some exceptions to these canonical binding modes, as some SH2 domains use noncanonical binding modes for pTyr peptide recognition (Wagner et al. 2013). A notable example is the SH2 domain of the E3 ubiquitin ligase Cbl. Cbl-SH2 is characterized by a tyrosine kinase binding (TKB) domain adjacent to the SH2 domain that makes significant contacts with pTyr targets, thereby dictating its binding specificity (Hu and Hubbard 2005). Additionally, the Cbl-SH2 domain, unlike other SH2 domains, can also interact with pTyr ligands in the reverse orientation, with amino-terminal residues contacting the specificity-determining pockets (Ng et al. 2008).

## Summary

Although a useful descriptor, the two-pronged binding mode of SH2 domains is likely inadequate for describing binding events in vivo. This is highlighted by recent studies showing SH2 domain–phospholipid interactions (Park et al. 2016). Other studies and reviews based on structure–function analysis highlight the importance of the interplay between SH2 domains and other domains present in the full-length protein for efficient interaction with pTyr targets (Hubbard 1999; Filippakopoulos et al. 2008). Nonetheless, the continuously expanding wealth of available structure–function and protein–protein interaction data provide a valuable resource for discerning general principles controlling pTyr binding and SH2 domain engineering.

## ENGINEERING SH2 DOMAINS USING PHAGE DISPLAY

### Engineering SH2 Domain Affinity

Systematic and large-scale investigation of the tyrosine phosphoproteome using pTyr-specific antibodies is limited due to their cost of production and their inability to be efficiently folded intracellularly. Conversely, the natural binding specificity and intracellular localization of SH2 domains make them an ideal scaffold for pTyr recognition. However, the binding affinity of SH2 domains for pTyr peptides is moderate, ranging from $10^{-5}$ M to $10^{-8}$ M (Gan and Roux 2009), and thus unsuitable for the development of highly efficient probes. To enhance the interaction strength for pTy-containing targets, several groups have used in vitro protein evolution techniques to fine-tune the binding affinity and specificity of different SH2 domains (Table 1). In this review, we will focus only on SH2 domains that have been engineered via phage display.

Taking advantage of the wealth of structural information available, Kaneko et al. constructed a large phage display library of the human Fyn-SH2 domain, in which 15 residues were randomized and subjected to affinity selections against 33 pTyr peptides that contained different amino acids flanking the pTyr residue (Fig. 2). Such an approach enabled the isolation of numerous Fyn-SH2 domain variants with high binding affinity and unaltered specificity relative to the wild-type domain (Kaneko et al. 2012). This study showed that many amino acid substitutions could be introduced into the pTyr-binding pocket without affecting its structure and function (Kaneko et al. 2012). Additionally, the authors identified a variant containing only three amino acid substitutions relative to the wild-type

TABLE 1. List of mutational studies of the Src homology 2 (SH2) domain

| Engineered SH2 domain | Mutated positions | Method | Binding modulation | Reference |
|---|---|---|---|---|
| PLCγ | βD5, βE4, EF1, BG2, and BG3 | Random mutagenesis | Dual specificity | Malabarba et al. 2001 |
| Src | βD3 | Incorporation of lysyl derivatives via site-directed mutagenesis | Specificity | Virdee et al. 2010 |
| Src | EF1 | Site-directed mutagenesis | Specificity | Marengere et al. 1994 |
| Fyn | EF1 | Site-directed mutagenesis | Specificity | Kaneko et al. 2010 |
| BRGD1 | EF2 | Site-directed mutagenesis | Specificity | Kaneko et al. 2010 |
| Src | βB5, βB6, βB7, BC1, and BC2 | Phage display | Specificity for sTyr | Ju et al. 2016 |
| Src | βB5, βB7, BC1, and BC2 | Phage display | Specificity for sTyr | Lawrie et al. 2021 |
| Fyn | EF-loop (EF1, EF2, and EF3), BG-loop (BG1, BG2, and BG3) | Phage display | Specificity | Veggiani et al. 2019 |
| Fyn | αA2, αA3, αA5, βB5, βB7, BC-loop (BC1, BC2, BC3, BC4, BC5 and BC6), βC1, βC3, βC4, βC5, βD2, βD3, and βD5 | Phage display | Affinity | Kaneko et al. 2012 |
| Src | BC4, βC2, and βD6 | Site-directed mutagenesis | Affinity | Kaneko et al. 2012 |
| Grb2 | BC4, βC2, and βD6 | Site-directed mutagenesis | Affinity | Kaneko et al. 2012 |
| Fes | αA2, αA5, βB3, BC-loop (BC1, BC2, BC3, BC4, BC5, and BC6), βC2, βC4, βD4, and βD6 | Phage display | Affinity | Martyn et al. 2022 |
| 17 different SH2 domains | BC-loop, βC2, and βD6 | Loop grafting via site-directed mutagenesis | Affinity | Martyn et al. 2022 |

The residues and the method to introduce genetic diversity are indicated, as well as their effect on SH2 domain–binding affinity or specificity (binding modulation).

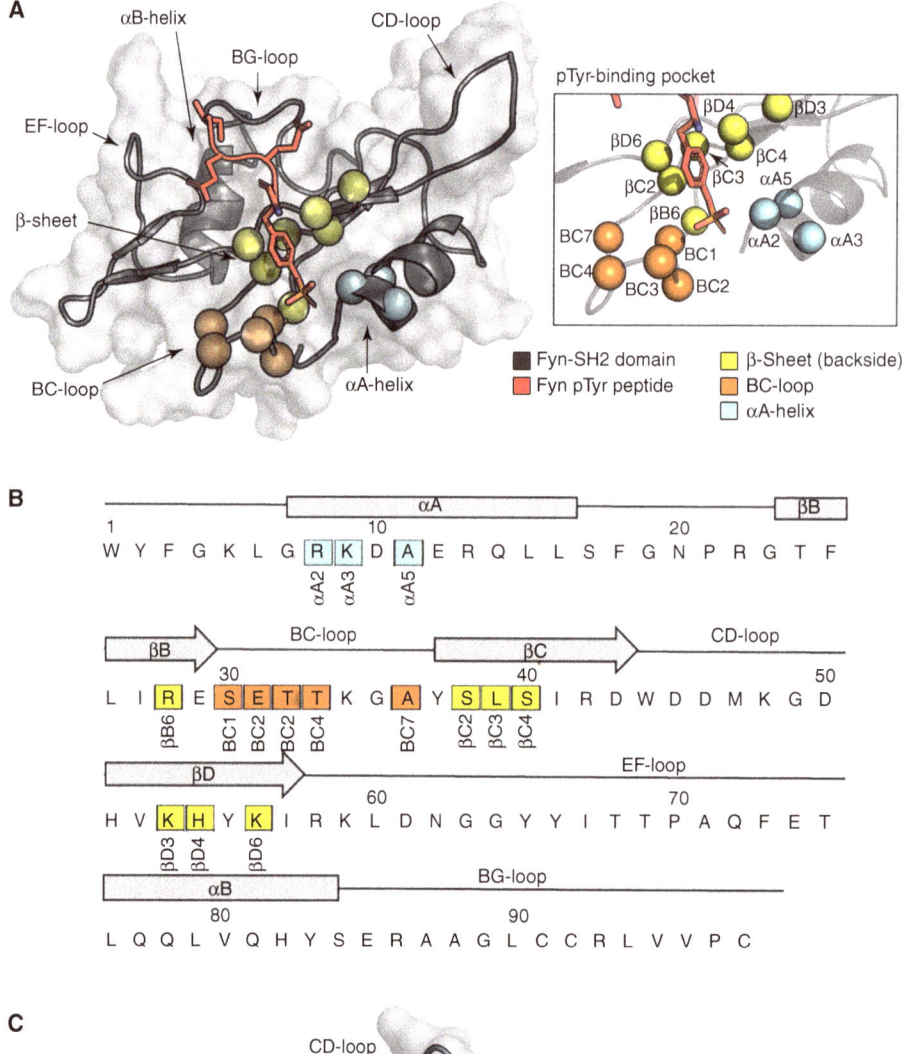

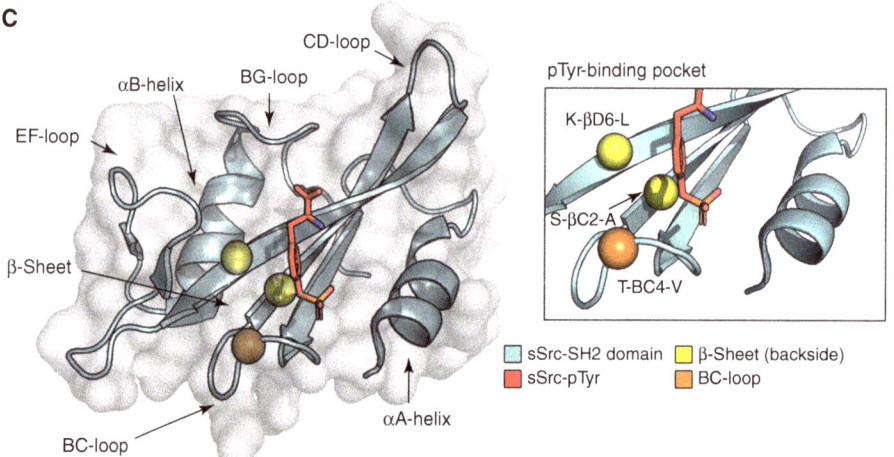

**FIGURE 2.** Design of a Fyn-Src homology 2 (SH2) phage display library and selection of superFyn and superSrc. (*A*) The structure of wild-type Fyn (PDB ID: AOT) bound to a pTyr peptide was used to design the Fyn-SH2 phage display library. Residues with side chains oriented toward and proximal to (within 10 Å) the pTyr residue were selected for mutagenesis. Randomized residues are shown as spheres and color-coded based on their position in the domain. (*B*) Primary amino acid sequence of the Fyn-SH2 domain, with residues selected for mutagenesis colored as in *A*. (*C*) Structure of the superSrc SH2 domain bound to a pTyr peptide (PDB ID: 4F5B), with positions of mutagenized residues shown as spheres and colored as in *A*. The resulting superbinder substitutions discovered from the Fyn-SH2 library were grafted into the Src-SH2 domain to create superSrc, which displayed a remarkable increase in affinity relative to the wild-type domain (Kaneko et al. 2012).

domain that enabled an affinity enhancement of up to two orders of magnitude. More remarkably, when such affinity-enhancing substitutions were introduced in the SH2 domain of the tyrosine kinase Src (Src-SH2), which is characterized by an identical BC-loop to that of Fyn-SH2, the affinity of Src-SH2 domain was significantly increased, thereby generating an SH2 domain superbinder, superSrc.

Inspired by this work, we recently developed a novel phage display library based on the SH2 domain of the tyrosine protein kinase Fes, which shares only limited sequence homology with Src- and Fyn-SH2 domains (Fig. 3; Martyn et al. 2022). This library randomized 13 residues in the pTyr-binding pocket, including all six residues comprising the BC-loop, and yielded superFes, a Fes-SH2 domain variant with 490-fold greater affinity than the parental domain (Martyn et al. 2022). Phage display was essential in discovering novel amino acid substitutions in the pTyr-binding pockets of the Fes-, Fyn-, and Src-SH2 domains. Using the sets of substitutions discovered in these SH2 domains, we showed that residues positioned in the backside of the pTyr-binding pocket ($\beta$C2 and $\beta$D6 residues) contribute to the enhancement of binding affinity in a BC-loop sequence-specific manner (Martyn et al. 2022). Additionally, we also noted the importance of an Arg residue in the $\alpha$A-helix ($\alpha$A2) for engagement of the pTyr. Remarkably, when the superSrc or superFes binding-enhancing mutations, along with an Arg at the $\alpha$A2 position, were grafted into other human SH2 domains, we observed a substantial increase in their affinity. Therefore, by using phage display to discover essential sets of binding-enhancing mutations, we could develop a modular strategy to enable the rapid and facile enhancement of SH2 affinity using rational engineering.

A recent study added additional complexity to the SH2–pTyr-binding interaction by demonstrating that a non-phosphorylated peptide, discovered using a phage display peptide library, could bind to an exosite that is distinct from the pTyr-binding pocket of the SOCS2-SH2 domain (Linossi et al. 2021). This peptide could enhance the affinity of the SOCS2-SH2 domain to its ligand ~20-fold. This study further showed that the binding affinity of SH2 domains can also be allosterically modulated, indicating the presence of additional molecular determinants involved in establishing SH2 domain–ligand interactions.

## Engineering SH2 Domain Specificity

The specificity-determining pockets of the human Fyn-SH2 domain (Fyn-SH2) have also been engineered to alter its binding specificity (Table 1; Marengere et al. 1994; Malabarba et al. 2001; Virdee et al. 2010; Liu et al. 2019; Veggiani et al. 2019). We developed phage display libraries of Fyn-SH2 and randomized residues in each of the EF- and BG-loops to alter the binding selectivity of the domain (Fig. 4). Affinity selection experiments were performed using pTyr peptides with different residues flanking the pTyr moiety, thereby enabling us to sample the sequence space of permissive and nonpermissive residues in the SH2 domain. Such a strategy yielded Fyn-SH2 variants with altered and diverse binding specificities in domains that shared the same binding preference profile as other human SH2 domains (Veggiani et al. 2019), in agreement with previous studies (Kaneko et al. 2010). More importantly, when used in phosphoproteomic assays, Fyn-SH2 variants with diverse binding specificities enabled a more comprehensive analysis of the human phosphoproteome, suggesting that the use of SH2 domains with complementary specificity profiles might expand the repertoire of identified pTyr targets in proteomes.

## Engineering SH2 Domains for Binding to Noncanonical PTMs

Tyrosine O-sulfation (sTyr) is a common yet understudied PTM (Baeuerle and Huttner 1985). Exclusively present in numerous integral membrane and secreted proteins, sTyr plays a crucial role in controlling their interaction with physiological partners as well as pathogens (Stone et al. 2009). Tyrosine sulfation is predicted to occur in about 1% of all tyrosine residues (Huttner 1988), yet our understanding of such a modification is still limited. This is primarily due to the lack of probes for sTyr enrichment and characterization, as the efficacy of currently available sTyr-specific antibodies is limited (Hoffhines et al. 2006; Kehoe et al. 2006).

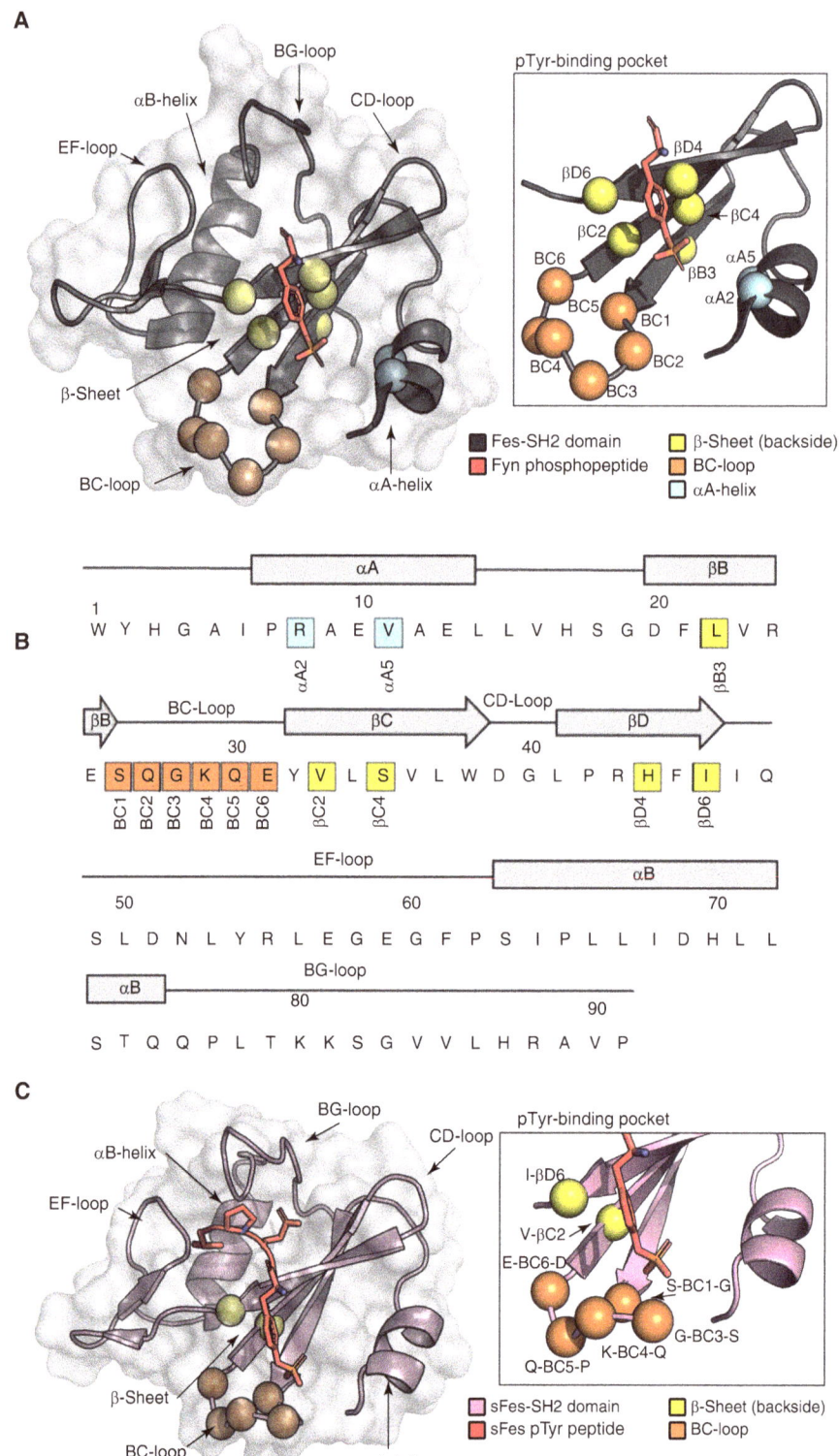

FIGURE 3. Design of the Fes-Src homology 2 (SH2) library and selection of superFes. (A) The structure of wild-type Fes (PDB ID: 1WQU) was aligned to that of wild-type Fyn (PDB ID: 1AOT) bound to a phosphorylated tyrosine (pTyr) peptide to guide Fes-SH2 library design. Here we show, in gray, the structure of the Fes-SH2 domain. Residues with side chains pointing toward the pTyr and within 10 Å distance were selected for mutagenesis. Amino acids in the BC-loop (orange), in the αA-helix (cyan), and in the βB, βC, and βD strands (yellow) that were subjected to diversification are shown as spheres. (B) Primary amino acid sequence of the Fes-SH2 domain with residues selected for mutagenesis colored as in A. (C) Structure of the superFes (sFes)-SH2 domain bound to a pTyr peptide (PDB ID: 7T1K), with positions of randomized residues shown as spheres and colored as in A.

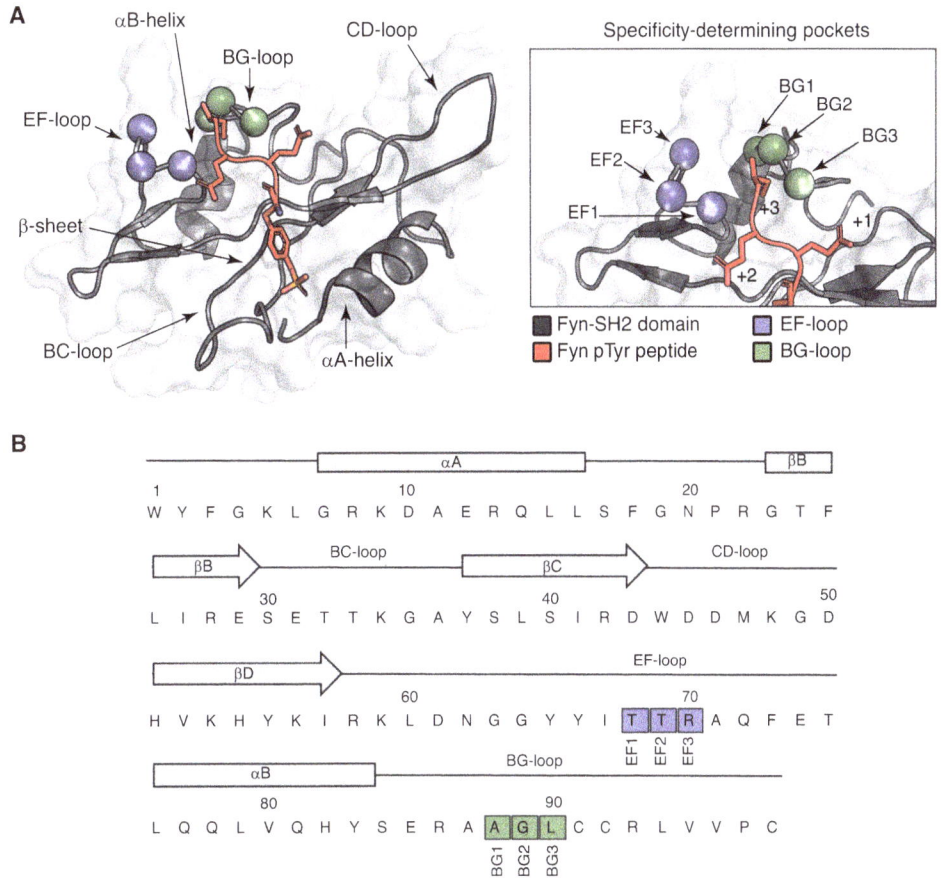

FIGURE 4. Design of a Fyn-Src homology 2 (SH2) phage display library for selection of variants with altered binding specificity. (A) Shown is the structure of wild-type Fyn-SH2 domain (PDB ID: AOT, gray) in complex with a phosphorylated tyrosine (pTyr) peptide, which was used to design the phage display library. Residues in the EF- and BG-loops that were selected for mutagenesis are shown as blue (BG-loop) and green (EF-loop) spheres, respectively. (B) Primary amino acid sequence of the Fyn-SH2 domain with residues selected for mutagenesis colored as in A.

To overcome this limitation, Ju et al. (2016) developed a phage display library of the Src-SH2 domain and altered its binding specificity for precise recognition of sTyr-modified peptides. Further engineering of SH2 affinity and specificity toward sTyr sequences was also done using phage display (Lawrie et al. 2021). Using affinity selection, Ju and coworkers were able to isolate SH2 domain variants with a minimal set of substitutions in the pTyr-binding pocket that, despite their highly similar physiochemical properties, displayed preferred recognition of sTyr- over pTyr-containing targets. Furthermore, the sTyr-specific Src-SH2 domain variants enabled the detection of sTyr on the surface of cells (Ju et al. 2016) and allowed the enrichment of sTyr-modified proteins from cell lysates (Lawrie et al. 2021). This work further highlights the structural malleability of SH2 domains as well as the power of phage display to engineer SH2 domains capable of discriminating among distinct targets with nearly identical physiochemical properties.

## BIOTECHNOLOGICAL APPLICATIONS OF SH2 SUPERBINDERS

### Affinity Purification Tools for Mass Spectrometry (MS)

Despite significant technological advances and the discovery of numerous phosphopeptides in human cells and tissues (Von Stechow et al. 2015), the detection of pTyr-containing peptides remains

Chapter 13

challenging due to the low stoichiometry of phosphorylation events and low cellular pTyr abundance (Sharma et al. 2014). As mentioned earlier, pTyr-containing proteins (pTyr proteins) and pTyr peptides are typically enriched from tissues and cells after proteolytic digestion using IMAC, which results in the indiscriminate enrichment of pSer, pThr, and pTyr peptides (Humphrey et al. 2018). As an alternative, pTyr-specific antibodies can be used, but they are expensive and are often characterized by low binding efficiency, making them unsuitable for systematic and large-scale analyses (Bian et al. 2016). Therefore, there is a need for robust, scalable, and cost-effective technologies that can enable highly efficient analysis of the pTyr proteome. Indeed, as discussed throughout the article, SH2 domains represent attractive probes for the deep analysis of tyrosine phosphorylation events, as they can specifically recognize pTyr-containing proteins in complex mixtures, such as cell lysates and biological fluids, albeit with low efficiency (Höfener et al. 2014). Moreover, SH2 domains can be produced at scale and inexpensively, making them accessible to any laboratory able to express recombinant proteins.

High-affinity SH2 domains have been shown to be a superior tool to enrich pTyr peptides from cells and analyze whole pTyr proteomes using MS (Bian et al. 2016; Tong et al. 2017; Veggiani et al. 2019). In particular, superSrc enabled ultra-deep coverage of the Tyr phosphoproteome and the identification of novel pTyr sites that were undetectable using anti-pTyr antibodies (Bian et al. 2016). Furthermore, we recently showed that the use of pools of high-affinity SH2 domains with different binding specificities allows the isolation of a more diverse set of pTyr proteins, thereby expanding the depth and coverage of the human pTyr proteome (Veggiani et al. 2019; Martyn et al. 2022).

## Synthetic Biology

The central role of SH2 domains in mediating signal transduction has allowed scientists to rewire signaling circuitry to induce apoptosis (Howard et al. 2003), monitor receptor activation (Barnea et al. 2008), and even fine-tune stem-cell-fate determination (Findlay et al. 2013). By using engineered Grb2-SH2 domain variants characterized by altered binding specificities (Findlay et al. 2013) or architecture (Yasui et al. 2014), it has been possible to induce the differentiation of mouse embryonic stem cells (mESCs) to the primitive endoderm lineage. In fact, SH2 domains with tailored binding specificities can precisely recruit proteins containing them to defined pTyr sites, initiating mESCs fate decision and driving lineage specification. Another use of SH2 domains in synthetic biology is described by Sun and coworkers, who developed SH2 domains–integrated sensing and activating proteins (iSNAPs) to visualize the dynamics of signal transduction in live cells and to rewire the CD47-SIRPα signaling axis, enhancing phagocytosis of tumor cells by macrophages (Sun et al. 2017). Therefore, engineered SH2 domains can be used to recruit the pTyr-signaling machinery, thereby modulating and/or altering a palette of diverse cellular activities.

## Biosensors and Cancer Diagnostics

SH2 domains have also been used to develop biosensors for probing signaling pathways of interest in cells. The Grb2- and Fyn-SH2 domains have been used to develop highly sensitive probes of signal transduction events in response to growth factors (Freeman et al. 2012; Fujioka et al. 2017). A FRET biosensor employing the Crkl-SH2 domain has been used to monitor the response of the oncogenic tyrosine kinase Bcr-Abl to drug treatment in patient-derived chronic myeloid leukemia cells (Horiguchi et al. 2017). More recently, Farrell and coworkers used multiple SH2 domains conjugated to fluorophores to investigate T-cell signaling in super-resolution microscopy experiments (Farrell et al. 2022), enabling the visualization of signaling nodes of the immunological synapse.

Given their ability to discern minimal changes in pTyr proteomes, SH2 profiling (the use of SH2 domains to survey the global tyrosine phosphorylation state in proteomes) is being explored as a sensitive and novel means for cancer diagnosis (Machida et al. 2007). For example, SH2 profiling not only enabled accurate classification of lung cancer cell lines with different genetic backgrounds, but also allowed analysis of cancer cell sensitivity to the tyrosine kinase inhibitor (TKI) erlotinib (Machida et al. 2010), suggesting that SH2 domains could be used as predictive biomarkers for TKI treatment of cancer patients. Additionally, SH2 domain–mediated analysis of pTyr signaling has been shown to improve the classification of tumors and the stratification of leukemia patients (Dierck et al. 2006), demonstrating that SH2 domain–dependent detection of aberrant pTyr signaling pathways might contribute to the development of personalized cancer diagnostics.

### SH2 Domains as Potential Cancer Therapeutics

Small molecule drugs targeting tyrosine kinase catalytic activity have been successful for the treatment of a number of different malignancies. These TKIs, however, can be promiscuous, binding to many different tyrosine kinases, resulting in limited efficacy and off-target effects (Knapp and Sundström 2014). The SH2 domain pTyr-binding site (Nioche et al. 2002; Page et al. 2012), as well as the interface between the SH2 and the catalytic domain of the oncogenic Brc-Abl fusion and of the tyrosine phosphatase SHP2 (Grebien et al. 2011; Garcia Fortanet et al. 2016), have also been the target of small molecules attempting to disrupt aberrant pTyr signaling in cancer (Kraskouskaya et al. 2013). These efforts showed that the SH2 domain can represent an alternative target for therapeutic intervention and have the potential to expand the druggable space for TKI-resistant cancers.

In addition to providing alternative druggable targets, SH2 domains have also been explored as a novel class of biologics. The Src-SH2 superbinder showed a dominant-negative effect on EGFR signaling in cancer cells, highlighting the ability of high-affinity SH2 domains to suppress signaling pathways essential for cancer cell proliferation (Kaneko et al. 2012). Based on this study, Liu et al. fused a cell-penetrating peptide to superSrc for intracellular delivery of the domain, and observed anticancer activity in melanoma xenograft mouse models (Liu et al. 2018). Therefore, not only do SH2 domains represent suitable probes for the investigation of the pTyr proteome and visualization of signaling events but might also represent a novel class of cancer therapeutics.

## CONCLUDING REMARKS

The discovery, characterization, engineering, and application of SH2 domain variants in basic and translational research is a scientific success story that spans several decades. The SH2 domain has already taught us a great deal about Tyr phosphorylation, and engineering efforts directed at improving and/or altering its properties continue to expand our understanding of Tyr phosphorylation-based signaling networks in both health and disease. Engineered SH2 domains are now finding their role in diagnostic applications and might contribute to the development of a novel class of cancer therapeutics.

Undoubtedly, phage display has proven its worth in dissecting the structure–function relationship of SH2 domains, identifying targets of tyrosine kinase activity, and modulating pTyr target recognition. Recent studies have highlighted the relevance of allosteric modulation of the domain for establishing specific and high-affinity interactions. Therefore, the development of large phage display libraries exploring the sequence space of residues distal to the pTyr-binding and specificity-determining pockets will likely extend our understanding of SH2 domain dynamics and pTyr target recognition.

Regarding the analysis of tyrosine phosphorylation events in cells, the continuous development of novel engineered SH2 domain variants will greatly expand the repertoire of characterized pTyr sites, allowing the generation of innovative diagnostics and the identification of novel drug targets. Additionally, engineered high-affinity SH2 domains will enable more sensitive biosensors that could address whether RTKs exert their activity directly, by translocating into the cell nucleus, or indirectly, via the activation of intracellular signaling molecules that would act as effectors.

Overall phage display–mediated analysis of the SH2 domain represents a useful tool for the study of the tyrosine phosphoproteome, furthering our understanding of tyrosine phosphorylation in eukaryotic biology and its role in human pathology, thereby providing a platform for the discovery of novel therapies.

Chapter 13

## ACKNOWLEDGMENTS

We acknowledge and thank Dr. Sachdev Sidhu for his advice and comments on this work.

## REFERENCES

Aschner Y, Downey GP. 2018. The importance of tyrosine phosphorylation control of cellular signaling pathways in respiratory disease: pY and pY not. *Am J Respir Cell Mol Biol* 59: 535–547. doi:10.1165/rcmb.2018-0049TR

Baeuerle PA, Huttner WB. 1985. Tyrosine sulfation of yolk proteins 1, 2, and 3 in *Drosophila melanogaster*. *J Biol Chem* 260: 6434–6439. doi:10.1016/S0021-9258(18)88991-8

Barnea G, Strapps W, Herrada G, Berman Y, Ong J, Kloss B, Axel R, Lee KJ. 2008. The genetic design of signaling cascades to record receptor activation. *Proc Natl Acad Sci* 105: 64–69. doi:10.1073/pnas.0710487105

Bian Y, Li L, Dong M, Liu X, Kaneko T, Cheng K, Liu H, Voss C, Cao X, Wang Y, et al. 2016. Ultra-deep tyrosine phosphoproteomics enabled by a phosphotyrosine superbinder. *Nat Chem Biol* 12: 959–966. doi:10.1038/nchembio.2178

Bradshaw JM, Waksman G. 1999. Calorimetric examination of high-affinity Src SH2 domain-tyrosyl phosphopeptide binding: dissection of the phosphopeptide sequence specificity and coupling energetics. *Biochemistry* 38: 5147–5154. doi:10.1021/bi982974y

Chan B, Lanyi A, Song HK, Griesbach J, Simarro-Grande M, Poy F, Howie D, Sumegi J, Terhorst C, Eck MJ. 2003. SAP couples Fyn to SLAM immune receptors. *Nat Cell Biol* 5: 155–160. doi:10.1038/ncb920

Dierck K, Machida K, Voigt A, Thimm J, Horstmann M, Fiedler W, Mayer BJ, Nollau P. 2006. Quantitative multiplexed profiling of cellular signaling networks using phosphotyrosine-specific DNA-tagged SH2 domains. *Nat Methods* 3: 737–744. doi:10.1038/nmeth917

Du Z, Lovly CM. 2018. Mechanisms of receptor tyrosine kinase activation in cancer. *Mol Cancer* 17: 58. doi:10.1186/s12943-018-0782-4

Farrell MV, Nunez AC, Yang Z, Pérez-Ferreros P, Gaus K, Goyette J. 2022. Protein-PAINT: superresolution microscopy with signaling proteins. *Sci Signal* 15: 1–10. doi:10.1126/scisignal.abg9782

Filippakopoulos P, Kofler M, Hantschel O, Gish GD, Grebien F, Salah E, Neudecker P, Kay LE, Turk BE, Superti-Furga G, et al. 2008. Structural coupling of SH2-kinase domains links Fes and Abl substrate recognition and kinase activation. *Cell* 134: 793–803. doi:10.1016/j.cell.2008.07.047

Findlay GM, Smith MJ, Lanner F, Hsiung MS, Gish GD, Petsalaki E, Cockburn K, Kaneko T, Huang H, Bagshaw RD, et al. 2013. Interaction domains of Sos1/Grb2 are finely tuned for cooperative control of embryonic stem cell fate. *Cell* 152: 1008–1020. doi:10.1016/j.cell.2013.01.056

Freeman J, Kriston-Vizi J, Seed B, Ketteler R. 2012. A high-content imaging workflow to study Grb2 signaling complexes by expression cloning. *J Vis Exp* 68: 1–6. doi:10.3791/4382

Fujioka M, Asano Y, Nakada S, Ohba Y. 2017. SH2 domain-based FRET biosensor for measuring BCR-ABL activity in living CML cells. *Methods Mol Biol* 1555: 513–534. doi:10.1007/978-1-4939-6762-9_30

Gan W, Roux B. 2009. Binding specificity of SH2 domains: insight from free energy simulations. *Proteins* 74: 996–1007. doi:10.1002/prot.22209

Garcia Fortanet J, Chen CHT, Chen YNP, Chen Z, Deng Z, Firestone B, Fekkes P, Fodor M, Fortin PD, Fridrich C, et al. 2016. Allosteric inhibition of SHP2: identification of a potent, selective, and orally efficacious phosphatase inhibitor. *J Med Chem* 59: 7773–7782. doi:10.1021/acs.jmedchem.6b00680

Grebien F, Hantschel O, Wojcik J, Kaupe I, Kovacic B, Wyrzucki AM, Gish GD, Cerny-Reiterer S, Koide A, Beug H, et al. 2011. Targeting the SH2-kinase interface in Bcr-Abl inhibits leukemogenesis. *Cell* 147: 306–319. doi:10.1016/j.cell.2011.08.046

Haslam NJ, Shields DC. 2012. Peptide-binding domains: are limp handshakes safest? *Sci Signal* 5: pe40. doi:10.1126/scisignal.2003372

Hause R, Jr, Leung KK, Barkinge JL, Ciaccio MF, Chuu CP, Jones RB. 2012. Comprehensive binary interaction mapping of SH2 domains via fluorescence polarization reveals novel functional diversification of ErbB receptors. *PLoS ONE* 7: e44471. doi:10.1371/journal.pone.0044471

Höfener M, Heinzlmeir S, Kuster B, Sewald N. 2014. Probing SH2-domains using inhibitor affinity purification (IAP). *Proteome Sci* 12: 41. doi:10.1186/1477-5956-12-41

Hoffhines AJ, Damoc E, Bridges KG, Leary JA, Moore KL. 2006. Detection and purification of tyrosine-sulfated proteins using a novel anti-sulfotyrosine monoclonal antibody. *J Biol Chem* 281: 37877–37887. doi:10.1074/jbc.M609398200

Horiguchi M, Fujioka M, Kondo T, Fujioka Y, Li X, Horiuchi K, Satoh A O, Nepal P, Nishide S, Nanbo A, et al. 2017. Improved FRET biosensor for the measurement of BCR-ABL activity in chronic myeloid leukemia cells. *Cell Struct Funct* 42: 15–26. doi:10.1247/csf.16019

Howard PL, Chia MC, Del Rizzo S, Liu F, Pawson T. 2003. Redirecting tyrosine kinase signaling to an apoptotic caspase pathway through chimeric adaptor proteins. *Proc Natl Acad Sci* 100: 11267–11272. doi:10.1073/pnas.1934711100

Hu J, Hubbard SR. 2005. Structural characterization of a novel Cbl phosphotyrosine recognition motif in the APS family of adapter proteins. *J Biol Chem* 280: 18943–18949. doi:10.1074/jbc.M414157200

Hubbard SR. 1999. Src autoinhibition: let us count the ways. *Nat Struct Biol* 6: 711–714. doi:10.1038/11468

Humphrey SJ, Karayel O, James DE, Mann M. 2018. High-throughput and high-sensitivity phosphoproteomics with the EasyPhos platform. *Nat Protoc* 13: 1897–1916. doi:10.1038/s41596-018-0014-9

Hunter T. 2014. The genesis of tyrosine phosphorylation. *Cold Spring Harb Perspect Biol* 6: a020644. doi:10.1101/cshperspect.a020644

Huttner WB. 1988. Tyrosine sulfation and the secretory pathway. *Annu Rev Physiol* 50: 363–376. doi:10.1146/annurev.ph.50.030188.002051

Hwang PM, Li C, Morra M, Lillywhite J, Muhandiram DR, Gertler F, Terhorst C, Kay LE, Pawson T, Forman-Kay JD, et al. 2002. A "three-pronged" binding mechanism for the SAP/SH2D1A SH2 domain: structural basis and relevance to the XLP syndrome. *EMBO J* 21: 314–323. doi:10.1093/emboj/21.3.314

Jones RB, Gordus A, Krall JA, MacBeath G. 2006. A quantitative protein interaction network for the ErbB receptors using protein microarrays. *Nature* 439: 168–174. doi:10.1038/nature04177

Ju T, Niu W, Guo J. 2016. Evolution of Src homology 2 (SH2) domain to recognize sulfotyrosine. *ACS Chem Biol* 11: 2551–2557. doi:10.1021/acschembio.6b00555

Kaneko T, Huang H, Zhao B, Li L, Liu H, Voss CK, Wu C, Schiller MR, Li SS-C. 2010. Loops govern SH2 domain specificity by controlling access to binding pockets. *Sci Signal* 3: ra34. doi:10.1126/scisignal.2000796

Kaneko T, Huang H, Cao X, Li X, Li C, Voss C, Sidhu SS, Li SSC. 2012. Superbinder SH2 domains act as antagonists of cell signaling. *Sci Signal* 5: ra68. doi:10.1126/scisignal.2003021

Kaneko T, Stogios PJ, Ruan X, Voss C, Evdokimova E, Skarina T, Chung A, Liu X, Li L, Savchenko A, et al. 2018. Identification and characterization of a large family of superbinding bacterial SH2 domains. *Nat Commun* 9: 4549. doi:10.1038/s41467-018-06943-2

Kehoe JW, Velappan N, Walbolt M, Rasmussen J, King D, Lou J, Knopp K, Pavlik P, Marks JD, Bertozzi CR, et al. 2006. Using phage display to select antibodies recognizing post-translational modifications independently of sequence context. *Mol Cell Proteomics* 5: 2350–2363. doi:10.1074/mcp.M600314-MCP200

Knapp S, Sundström M. 2014. Recently targeted kinases and their inhibitors—the path to clinical trials. *Curr Opin Pharmacol* 17: 58–63. doi:10.1016/j.coph.2014.07.015

Koch CA, Moran MF, Anderson D, Liu X, Mbamalu G, Pawson T. 1992. Multiple SH2-mediated interactions in v-src-transformed cells. *Mol Cell Biol* 12: 1366–1374. doi:10.1128/mcb.12.3.1366-1374.1992

Kraskouskaya D, Duodu E, Arpin CC, Gunning PT. 2013. Progress towards the development of SH2 domain inhibitors. *Chem Soc Rev* 42: 3337–3370. doi:10.1039/c3cs35449k

Lawrie J, Waldrop S, Morozov A, Niu W, Guo J. 2021. Engineering of a small protein scaffold to recognize sulfotyrosine with high specificity. *ACS Chem Biol* 16: 1508–1517. doi:10.1021/acschembio.1c00382

Lemmon MA, Schlessinger J, Ferguson KM. 2014. The EGFR family: not so prototypical receptor tyrosine kinases. *Cold Spring Harb Perspect Biol* 6: 1–18. doi:10.1101/cshperspect.a020768

Leung KK, Hause RJ, Barkinge JL, Ciaccio MF, Chuu C-P, Jones RB. 2014. Enhanced prediction of Src homology 2 (SH2) domain binding potentials using a fluorescence polarization-derived c-Met, c-Kit, ErbB, and androgen receptor interactome. *Mol Cell Proteomics* 13: 1705–1723. doi:10.1074/mcp.M113.034876

Li L, Wu C, Huang H, Zhang K, Gan J, Li SSC. 2008. Prediction of phosphotyrosine signaling networks using a scoring matrix-assisted ligand identification approach. *Nucl Acids Res* 36: 3263–3273. doi:10.1093/nar/gkn161

Li L, Tibiche C, Fu C, Li L, Tibiche C, Fu C, Kaneko T, Moran MF, Schiller MR, Li SS, et al. 2012. The human phosphotyrosine signaling network: evolution and hotspots of hijacking in cancer. *Genome Res* 22: 1222–1230. doi:10.1101/gr.128819.111

Linossi EM, Li K, Veggiani G, Tan C, Dehkhoda F, Hockings C, Calleja DJ, Keating N, Feltham R, Brooks AJ, et al. 2021. Discovery of an exosite on the SOCS2-SH2 domain that enhances SH2 binding to phosphorylated ligands. *Nat Commun* 12: 7032. doi:10.1038/s41467-021-26983-5

Liu BA, Jablonowski K, Raina M, Pawson T, Nash PD, Arce M. 2006. The human and mouse complement resource of SH2 domain proteins: establishing the boundaries of phosphotyrosine signaling. *Mol Cell* 22: 851–868. doi:10.1016/j.molcel.2006.06.001

Liu BA, Jablonowski K, Shah EE, Engelmann B, Jones RB, Nash PD. 2010. SH2 domains recognize contextual peptide sequence information to determine selectivity. *Mol Cell Proteomics* 9: 2391–2404. doi:10.1074/mcp.M110.001586

Liu BA, Engelmann BW, Nash PD. 2012. The language of SH2 domain interactions defines phosphotyrosine-mediated signal transduction. *FEBS Lett* 586: 2597–2605. doi:10.1016/j.febslet.2012.04.054

Liu AD, Xu H, Gao YN, Luo DN, Li ZF, Voss C, Li SSC, Cao X. 2018. (Arg)$_9$-SH2 superbinder: a novel promising anticancer therapy to melanoma by blocking phosphotyrosine signaling. *J Exp Clin Cancer Res* 37: 1–13. doi:10.1186/s13046-017-0664-4

Liu H, Huang H, Voss C, Kaneko T, Qin WT, Sidhu S, Li SSC. 2019. Surface loops in a single SH2 domain are capable of encoding the spectrum of specificity of the SH2 family. *Mol Cell Proteomics* 18: 372–382. doi:10.1074/mcp.RA118.001123

Machida K, Thompson CM, Dierck K, Jablonowski K, Kärkkäinen S, Liu B, Zhang H, Nash PD, Newman DK, Nollau P, et al. 2007. High-throughput phosphotyrosine profiling using SH2 domains. *Mol Cell* 26: 899–915. doi:10.1016/j.molcel.2007.05.031

Machida K, Eschrich S, Li J, Bai Y, Koomen J, Mayer BJ, Haura EB. 2010. Characterizing tyrosine phosphorylation signaling in lung cancer using SH2 profiling. *PLoS ONE* 5: e13470. doi:10.1371/journal.pone.0013470

Malabarba MG, Milia E, Faretta M, Zamponi R, Pelicci PG, Di Fiore PP. 2001. A repertoire library that allows the selection of synthetic SH2s with altered binding specificities. *Oncogene* 20: 5186–5194. doi:10.1038/sj.onc.1204654

Marasco M, Carlomagno T. 2020. Specificity and regulation of phosphotyrosine signaling through SH2 domains. *J Struct Biol X* 4: 100026. doi:10.1016/j.yjsbx.2020.100026

Marengere LEM, Songyang Z, Gish GD, Schaller MD, Parsons JT, Stern MJ, Cantley LC, Pawson T. 1994. SH2 domain specificity and activity modified by a single residue. *Nature* 369: 502–505. doi:10.1038/369502a0

Martyn GD, Veggiani G, Kusebauch U, Morrone SR, Yates BP, Singer AU, Tong J, Manczyk N, Gish G, Sun Z, et al. 2022. Engineered SH2 domains for targeted phosphoproteomics. *ACS Chem Biol* 17: 1472–1484. doi:10.1021/acschembio.2c00051

Moran MF, Koch CA, Anderson D, Ellis C, England L, Martin GS, Pawson T. 1990. Src homology region 2 domains direct protein–protein interactions in signal transduction. *Proc Natl Acad Sci* 87: 8622–8626. doi:10.1073/pnas.87.21.8622

Ng C, Jackson RA, Buschdorf JP, Sun Q, Guy GR, Sivaraman J. 2008. Structural basis for a novel intrapeptidyl H-bond and reverse binding of c-Cbl-TKB domain substrates. *EMBO J* 27: 804–816. doi:10.1038/emboj.2008.18

Nioche P, Liu W, Broutin I, Latreille Á, Vidal M, Charbonnier F, Roques B, Garbay C, Ducruix A. 2002. Crystal structures of the SH2 domain of Grb2: highlight on the binding of a new high-affinity inhibitor. *J Mol Biol* 315: 1167–1177. doi:10.1006/jmbi.2001.5299

Nita-Lazar A, Saito-Benz H, White FM. 2008. Quantitative phosphoproteomics by mass spectrometry: past, present, and future. *Proteomics* 8: 4433–4443. doi:10.1002/pmic.200800231

Page BDG, Khoury H, Laister RC, Fletcher S, Vellozo M, Manzoli A, Yue P, Turkson J, Minden MD, Gunning PT. 2012. Small molecule STAT5-SH2 domain inhibitors exhibit potent antileukemia activity. *J Med Chem* 55: 1047–1055. doi:10.1021/jm200720n

Park MJ, Sheng R, Silkov A, Jung DJ, Wang ZG, Xin Y, Kim H, Thiagarajan-Rosenkranz P, Song S, Yoon Y, et al. 2016. SH2 domains serve as lipid-binding modules for pTyr-signaling proteins. *Mol Cell* 62: 7–20. doi:10.1016/j.molcel.2016.01.027

Rondeau JM, Sabio M, Topiol S, Weidmann B, Zurini M, Bair KW, Ettmayer P, France D, Gounarides J, Jarosinski M, et al. 1999. Structural and conformational requirements for high-affinity binding to the SH2 domain of Grb2. *J Med Chem* 42: 971–980. doi:10.1021/jm9811007

Sadowski I, Stone JC, Pawson T. 1986. A noncatalytic domain conserved among cytoplasmic protein-tyrosine kinases modifies the kinase function and transforming activity of Fujinami sarcoma virus P130$^{gag\text{-}fps}$. *Mol Cell Biol* 6: 4396–4408. doi:10.1128/mcb.6.12.4396-4408.1986

Schlessinger J, Schlessinger J, Adrain C, Freeman M, Wagner MJ, Stacey MM, Bernard A, Lisabeth EM, Falivelli G, Elena B, et al. 2014. Receptor tyrosine kinases: legacy of the first two decades. *Cold Spring Harb Perspect Biol* 6: 1–13. doi:10.1101/cshperspect.a008912

Seet BT, Dikic I, Zhou M-M, Pawson T. 2006. Reading protein modifications with interaction domains. *Nat Rev Mol Cell Biol* 7: 473–483. doi:10.1038/nrm1960

Sharma K, D'Souza RCJ, Tyanova S, Schaab C, Wiśniewski J, Cox J, Mann M. 2014. Ultradeep human phosphoproteome reveals a distinct regulatory nature of Tyr and Ser/Thr-based signaling. *Cell Rep* 8: 1583–1594. doi:10.1016/j.celrep.2014.07.036

Stone MJ, Chuang S, Hou X, Shoham M, Zhu JZ. 2009. Tyrosine sulfation: an increasingly recognised post-translational modification of secreted proteins. *New Biotechnol* 25: 299–317. doi:10.1016/j.nbt.2009.03.011

Sun J, Lei L, Tsai CM, Wang Y, Shi Y, Ouyang M, Lu S, Seong J, Kim TJ, Wang P, et al. 2017. Engineered proteins with sensing and activating modules for automated reprogramming of cellular functions. *Nat Commun* 8: 477. doi:10.1038/s41467-017-00569-6

Tenreiro S, Eckermann K, Outeiro TF. 2014. Protein phosphorylation in neurodegeneration: friend or foe? *Front Mol Neurosci* 7: 42. doi:10.3389/fnmol.2014.00042

Tinti M, Kiemer L, Costa S, Miller ML, Sacco F, Olsen JV, Carducci M, Paoluzi S, Langone F, Workman CT, et al. 2013. The SH2 domain interaction landscape. *Cell Rep* 3: 1293–1305. doi:10.1016/j.celrep.2013.03.001

Tong J, Cao B, Martyn GD, Krieger JR, Taylor P, Yates B, Sidhu SS, Li SSC, Mao X, Moran MF. 2017. Protein-phosphotyrosine proteome profiling by superbinder-SH2 domain affinity purification mass spectrometry, sSH2-AP-MS. *Proteomics* 17: 1–5. doi:10.1002/pmic.201600360

Veggiani G, Huang H, Yates BP, Tong J, Kaneko T, Joshi R, Li SSC, Moran MF, Gish G, Sidhu SS. 2019. Engineered SH2 domains with tailored specificities and enhanced affinities for phosphoproteome analysis. *Protein Sci* 28: 403–413. doi:10.1002/pro.3551

Virdee S, Macmillan D, Waksman G. 2010. Semisynthetic Src SH2 domains demonstrate altered phosphopeptide specificity induced by incorporation of unnatural lysine derivatives. *Chem Biol* 17: 274–284. doi:10.1016/j.chembiol.2010.01.015

von Stechow L, Francavilla C, Olsen JV. 2015. Recent findings and technological advances in phosphoproteomics for cells and tissues. *Expert Rev Proteomics* 12: 469–487. doi:10.1586/14789450.2015.1078730

Wagner MJ, Stacey MM, Liu BA, Pawson T. 2013. Molecular mechanisms of SH2- and PTB-domain-containing proteins in receptor tyrosine kinase signaling. *Cold Spring Harb Perspect Biol* 5: a008987. doi:10.1101/cshperspect.a008987

Waksman G, Kominos D, Robertson SC, Pant N, Baltimore D, Birge RB, Cowburn D, Hanafusa H, Mayer BJ, Overduin M, et al. 1992. Crystal structure of the phosphotyrosine recognition domain of SH2 of v-*src* complexed with tyrosine-phosphorylated peptides. *Nature* **358:** 646–653. doi:10.1038/358646a0

Yaffe MB. 2002. Phosphotyrosine-binding domains in signal transduction. *Nat Rev Mol Cell Biol* **3:** 177–186. doi:10.1038/nrm759

Yasui N, Findlay GM, Gish GD, Hsiung MS, Huang J, Tucholska M, Taylor L, Smith L, Boldridge WC, Koide A, et al. 2014. Directed network wiring identifies a key protein interaction in embryonic stem cell differentiation. *Mol Cell* **54:** 1034–1041. doi:10.1016/j.molcel.2014.05.002

CHAPTER 14

# Generation and Characterization of Engineered Ubiquitin Variants to Modulate the Ubiquitin Signaling Cascade

Chen T. Liang,[1,3] Olivia Roscow,[1,3] and Wei Zhang[1,2,4]

[1]Department of Molecular and Cellular Biology, College of Biological Science, University of Guelph, Guelph, Ontario N1G2W1, Canada; [2]CIFAR Azrieli Global Scholars Program, Canadian Institute for Advanced Research, MaRS Centre, Toronto, Ontario M5G1M1, Canada

The ubiquitin signaling cascade plays a crucial role in human cells. Consistent with this, malfunction of ubiquitination and deubiquitination is implicated in the initiation and progression of numerous human diseases, including cancer. Therefore, the development of potent and specific modulators of ubiquitin signal transduction has been at the forefront of drug development. In the past decade, a structure-based combinatorial protein-engineering approach has been used to generate ubiquitin variants (UbVs) as protein-based modulators of multiple components in the ubiquitin–proteasome system. Here, we review the design and generation of phage-displayed UbV libraries, including the processes of binder selection and library improvement. We also provide a comprehensive overview of the general in vitro and cellular methodologies involved in characterizing UbV binders. Finally, we describe two recent applications of UbVs for developing molecules with therapeutic potential.

## INTRODUCTION

Protein–protein interactions (PPIs) govern all biological functions and cellular processes. Many of these interactions are transient and are mediated by posttranslational modifications (PTMs). The predominant PTMs, such as phosphorylation, ubiquitination, and acetylation, typically mediate weak and transitory interactions with several substrates, allowing a modified protein to interact with many binding partners and to have widespread biological effects (Perkins et al. 2010; Maculins et al. 2020). Dysregulation of PTMs and, therefore, PPIs, often results in dysregulation of signaling pathways, thereby leading to disease (de Las Rivas and Fontanillo 2010), with Alzheimer's disease, Parkinson's disease, amyotrophic lateral sclerosis, and numerous types of cancer as notable examples (Atkin and Paulson 2014). Consequently, generating potent and specific modulators of PPIs is of great clinical interest because they may allow both the properties of important PPIs to be probed and novel therapeutic drugs with fewer off-target interactions to be developed (Liang et al. 2021). While small molecule inhibitors have been developed and proven to successfully disrupt certain PPIs, they pose notable disadvantages, including nonspecific binding to multiple unintended targets, low bioavailability, and high risk of toxicity (Arkin and Wells 2004; Curigliano and Criscitiello 2014).

A recent approach for rapidly and efficiently generating novel PPI modulators uses protein-engineered combinatorial phage display libraries (Liang et al. 2021). In this process, protein domains and

---
[3]These authors contributed equally to this work.
[4]Correspondence: weizhang@uoguelph.ca

© 2026 Cold Spring Harbor Laboratory Press
Cite this overview as *Cold Spring Harb Protoc*; doi:10.1101/pdb.over107784

# Chapter 14

motifs that mediate a wide array of PPIs are used as scaffolds or templates for the basis of a phage display library, whereby residues that are suspected of contributing significantly to binding interactions are diversified and panned against a target protein such that improved binders can be isolated and enriched (Arkin and Wells 2004). A phage display library typically contains billions of diversified polypeptide variants and phage display selection filters out the strongest binders from these variants, which can then be isolated for further characterization (Liang et al. 2021). In this approach, a diversified variant of a scaffold protein is expressed externally on a phage particle via fusion to a phage coat protein, establishing a direct relationship between the variant's phenotype and its genotype contained within the phage particle. Through a series of panning steps, low-affinity phage binders are removed, while high-affinity phage binders are retained, eluted, and enriched in subsequent rounds of selection (Tonikian et al. 2007). This allows for one of the proteins in an interaction pair to be engineered to obtain enhanced binding to one or more binding partners, thereby modulating PPIs (Lavanya et al. 2014).

Phage display is a powerful tool for engineering binding proteins that can be used to modulate and elucidate signaling pathways (Lavanya et al. 2014). In the past decade, the laboratory of Dr. Sachdev Sidhu pioneered the development of combinatorial ubiquitin variant (UbV) libraries to create synthetic proteins that can modulate the ubiquitin (Ub)-signaling pathway (Ernst et al. 2013). Ub is expressed in all eukaryotic cells and participates in a wide variety of biological pathways, ranging from those mediating proteasomal degradation to those regulating cell-signaling cascades (Komander and Rape 2012). Due to its multiple binding partners, Ub is, therefore, an excellent candidate for diversification and confirming PPIs.

In this article, we first discuss the Ub-signaling pathway and then introduce the design and generation of phage-displayed UbV libraries. We then provide a comprehensive overview of methods for the general characterization of selected UbVs. These are divided into (1) binding assays, for determining binding affinity and enzyme kinetics, (2) functional assays, for determining and validating the effects of binding in mammalian cell lines in vitro, and (3) structural determination assays, for understanding the modulating mechanisms of UbVs and for further improving UbVs (Ernst et al. 2013; Zhang et al. 2016; Gabrielsen et al. 2017). Finally, we introduce two newly developed methods that can explore the potential of UbVs for therapeutic drug development.

## THE UBIQUITIN SIGNALING CASCADE

Ub, named for its nearly ubiquitous presence in eukaryotic cells, is a highly conserved and stable 76-amino-acid protein that can be enzymatically tagged on to target proteins, which can trigger various effects and responses. This PTM participates in a myriad of crucial biological functions, including the DNA damage response, immune and inflammatory responses, stress responses, and cell-cycle progression (Pickart and Eddins 2004). Dysregulation of ubiquitination is implicated in many human diseases (Komander and Rape 2012; Popovic et al. 2014; Beck et al. 2022), making it a focal point of disease research. The structure of Ub consists of a tight β-fold with a flexible carboxy-terminal tail that facilitates conjugation to both the enzymes in the ubiquitination cascade and substrate proteins (Pickart and Eddins 2004). This structure is highly conserved among eukaryotes, with only four residues differing among yeast, plants, and animals (Catic and Ploegh 2005).

The ubiquitination process is often referred to as an enzymatic cascade because of the series of stepwise enzyme reactions that take place (Fig. 1). First, Ub is transferred to a Ub-activating enzyme, known as E1, via a two-step process: Ub is activated by adenylation of its carboxy-terminal tail by E1, the hydrolysis of which attaches Ub to E1 via a thioester bond. Next, Ub is transferred to a cysteinyl residue on a Ub-conjugating enzyme, known as E2, via transthiolation, whereby E2 replaces E1. Lastly, Ub is transferred to a substrate protein via a Ub-protein ligase, known as E3 or E3 ligase (Ciechanover 1998), and this transfer can occur through two different mechanisms, depending on the architecture of the E3 in question. While there are relatively few E1 and E2 isozymes, there are up to 1000 E3

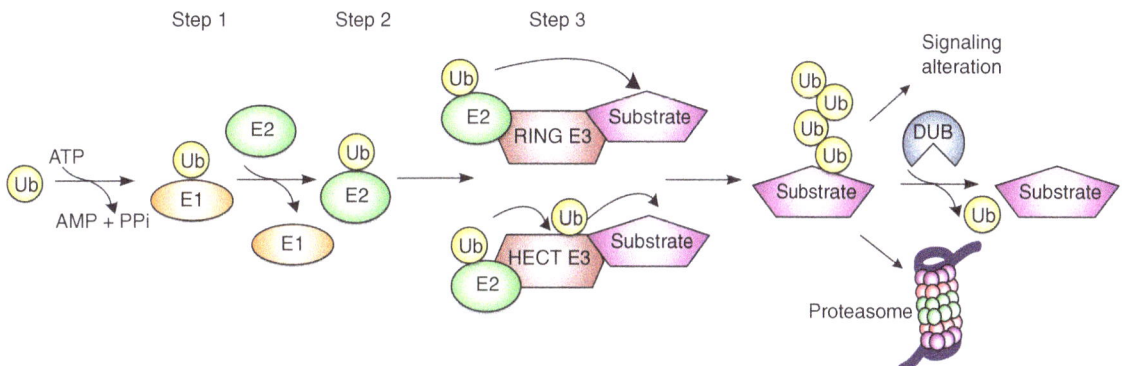

FIGURE 1. The ubiquitin (Ub) signaling cascade. Ubiquitination comprises a three-step enzymatic reaction cascade. Step 1: First, E1-activating enzyme adenylates Ub to form a thioester bond between the carboxy-terminal carboxyl group of Ub and the thiol group of the E1 active site cysteine residue. Step 2: Next, the activated Ub is transferred to the cysteine residue of an E2-conjugating enzyme by a thioester linkage. Step 3: Lastly, an E3 ligase recruits a charged E2 and transfers Ub or poly-Ub chains to specific protein substrates. Different E3 ligases (e.g., HECT or RING family) have distinct catalytic mechanisms. Substrate ubiquitination can lead to various cellular responses, including specific signaling events or degradation via the proteasome. Deubiquitinases (DUBs) catalyze cleavage of Ub from ubiquitinated proteins to reverse the ubiquitination process.

ligases, which have differential specificity for target proteins (Nakayama and Nakayama 2006) and can be categorized based on three major catalytic domains: homologous to E3-AP carboxyl terminus (HECT), really interesting new gene (RING), and ring-between-ring (RBR) (George et al. 2018). RING E3 ligases are predominant and directly transfer Ub from E2 to a target substrate by spatially coordinating the E2 and substrate. The less common HECT E3 ligases act as intermediates and transfer Ub from E2 to itself via another thioester bond before transferring Ub to a target substrate (Fig. 1; Metzger et al. 2012). RBR E3 ligases are a hybrid of both HECT and RING domains in that they bind E2-Ub complexes, similarly to RING E3s, but also form E3-Ub intermediates before passing Ub on to substrates, similarly to HECT E3s (Dove and Klevit 2017). Ub is then transferred to substrates via an isopeptide bond with a lysine residue of the target substrate. E3 ligases additionally possess the ability to undergo autoubiquitination, whereby they catalyze the transfer of Ub onto themselves (Chen et al. 2002). Lastly, ubiquitination can be reversed in a process termed deubiquitination, which is catalyzed by deubiquitinases (DUBs). This process facilitates the removal of Ub chains from ubiquitinated substrates (Komander et al. 2009).

## DESIGN AND GENERATION OF THE FIRST UbV LIBRARIES

Ubiquitination interactions use common Ub surfaces that are recognized by many proteins with low affinity yet very high specificity. These critical Ub regions make ideal candidates for diversification, which has the aim of creating UbVs with increased affinity over wild-type Ub for specific binding partners (Ernst et al. 2013; Liang et al. 2021). This approach depends on mutations within the binding surfaces of Ub that can improve molecular contacts with proteins, such that both affinity and specificity for Ub-interacting proteins are increased. In theory, tighter binding of variants can outcompete binding of wild-type Ub and result in inhibition or activation of a target protein, whether by competing for a protein's active site or binding allosterically at distal sites. This results in targeted modulation of PPIs for that target protein. Phage display has taken advantage of these features of the Ub scaffold to become one of the most successful methods for identifying protein-based modulators of Ub-interacting complexes.

Computational and bioinformatics methods were initially used to identify key residues for diversification (Tonikian et al. 2007). Zhang et al. (2013) analyzed the structure of Ub complexed with Ub-specific protease 7 (USP7), a DUB whose dysregulation is implicated in multiple cancer types

(Valles et al. 2020; Lu et al. 2021). They determined the best residues to mutate and generated a library targeting seven residues in the β1–β2 loop buried in the core of Ub to determine whether repacking the core and altering residues that contact surface residues of the target protein could improve binding. Through phage display, this library generated multiple UbV binders that showed enhanced binding and ultimately resulted in a functional inhibitor for USP7 (U7Ub25.2540) (Zhang et al. 2013). A drawback of this binder, however, was that it interacted nonspecifically with 10 other DUBs, limiting its utility as a specific inhibitor of USP7.

At the same time, Ernst et al. (2013) designed two libraries consisting of three diversified regions targeting solvent-exposed Ub-surface residues (Fig. 2A). They diversified ∼30 residues on the surface of Ub and panned against multiple Ub-interacting protein families, including DUBs, E2 enzymes, and E3 ligases, to produce an array of promising new binders. These two libraries also used a "soft randomization" strategy, whereby wild-type Ub residues are favored over mutations in the diversification process, helping to preserve protein conformation (Ernst et al. 2013). This approach was found to be very efficient at generating high-affinity modulators for Ub-interacting complexes. The success of this strategy at generating potent and specific binders is owed to reducing the likelihood of significant structural changes, which ultimately allowed the effects of mutating the binding pocket and PPI contact points to be investigated (Ernst et al. 2013). As such, this strategy became standardized for library design and was eventually used to generate more USP7 binders and produced an improved inhibitor for USP7 (UbV.7.2) that inhibited USP7 with substantially higher potency than the previous inhibitor, and did so more specifically, as it interacted with only three off-target DUBs instead of 10 (Zhang et al. 2017a). The limited success of the Zhang et al. (2013) library design, which generated a UbV (U7Ub25.2540) with enhanced affinity but poor specificity, may be attributed to diversification of core residues having deleterious effects on the structural integrity of the UbV, potentially compromising its conformation and longevity. In contrast, the Ernst et al. (2013) library targeted solvent-exposed surface residues that are more resilient to mutation.

Consequently, the most critical step in phage display is library construction (Fig. 2B). Library construction relies on an oligonucleotide-directed mutagenesis method, in which mutations are introduced into a single-stranded DNA (ssDNA) phagemid template (Kunkel 1985; Huang et al. 2012). Detailed protocols for UbV library generation have been described previously (Zhang and Sidhu 2018). Briefly, the scaffold for diversification is fused to the M13 bacteriophage pIII coat protein in a phagemid, which allows fused proteins to be displayed externally on the phage surface upon infection of a host cell and superinfection with M13K07 helper phage. The phagemid also contains an origin of replication from f1 phage to allow phage packing and secretion when coupled with M13K07 helper phage. Phagemid ssDNA is purified from *Escherichia coli* CJ236, which has a *dut-/ung-* genotype. The *dut* gene encodes dUTPase, responsible for degrading deoxyuridine triphosphate (dUTP), and the *ung* gene encodes uracil N-glycosylase, which removes uracil from DNA. In the absence of these two genes, dUTP is incorporated into phagemid ssDNA instead of deoxythymine triphosphate (dTTP) and is not repaired, resulting in the generation of uracil-containing ssDNA (dU-ssDNA). Soft-randomized sets of mutagenic oligonucleotides corresponding to the regions-of-interest for diversification are phosphorylated and annealed to the dU-ssDNA template strand, after which addition of T7 DNA polymerase, T4 DNA ligase, deoxyadenine triphosphate (dATP), deoxyguanine triphosphate (dGTP), deoxycytosine triphosphate (dCTP), and dTTP drive generation of covalently closed circular double-stranded DNA (CCC-dsDNA). This phagemid CCC-dsDNA is a heteroduplex of the dU-ssDNA template and a newly synthesized strand that contains the mutagenic oligonucleotide sequences and thymine instead of uracil. Transformation of *dut$^+$/ung$^+$ E. coli* SS320 with CCC-dsDNA preferentially degrades the dU strand and replicates the mutated strand, resulting in production of a highly diversified dsDNA phagemid library. At this point, library diversity can be determined. The library is then amplified to increase total phagemid concentration and quantified (Fig. 2B; Kunkel 1985; Fellouse and Pal 2005; Tonikian et al. 2007; Huang et al. 2012; Zhang and Sidhu 2018). The library is then ready for the phage-display selection process.

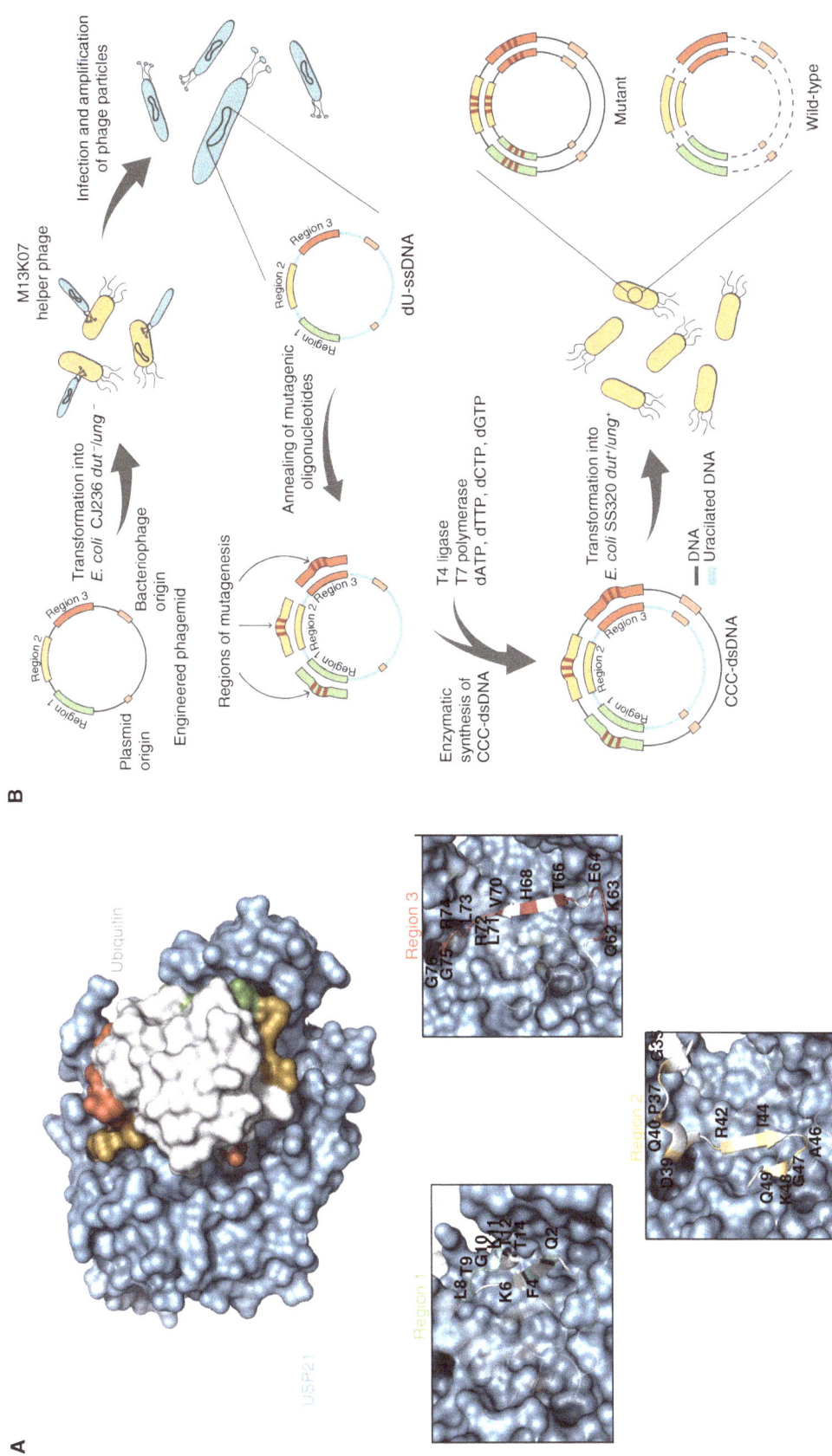

FIGURE 2. Overview of ubiquitin variant (UbV) library residue selection and library generation. (A) Pymol diagram of USP21 (blue) in complex with Ub (white), and Ub surface regions selected for randomization: region 1 (green), region 2 (yellow), and region 3 (red) (PDB entry 3I3T). (B) A pIII phagemid containing full-length wild-type Ub is transformed into *Escherichia coli* CJ236. M13K07 helper phage is used to superinfect the transformed cells and is amplified to generate uracilated single-stranded DNA (dU-ssDNA). dU-ssDNA acts as a template to anneal the three mutagenic oligonucleotides. In addition to the mutagenic oligonucleotides, T4 ligase, T7 polymerase, and deoxynucleotides (dATP, dTTP, dCTP, and dGTP) are added to complete the synthesis of heteroduplex covalently closed circular double-stranded DNA (CCC-dsDNA). CCC-dsDNA is transformed into *E. coli* SS320 and results in the propagation of the mutated strand while wild-type dU-ssDNA is degraded.

Chapter 14

## SELECTION OF UbV BINDERS BY PHAGE DISPLAY

UbV phage display libraries typically contain more than $10^{10}$ unique variants, and these can be screened through a plate-based selection procedure to isolate the strongest binders that outcompete wild-type Ub for binding to a target protein (Ernst et al. 2013; Wu et al. 2016). In this process, the desired target protein is immobilized on a plate by adsorption (Fig. 3). The phage display library is then added to allow variants to bind. A washing step then removes unbound phages, to leave bound phages associated with the target protein. These bound phages are then eluted from the target protein, amplified through an *E. coli* host, and incubated on a new plate coated with the target protein in a second round of selection. This process is repeated multiple times to enrich selected phages and to ensure that the strongest binders remain. The entire phage display selection typically requires five rounds to yield sufficiently enriched binders. The enriched binders consist of phage particles containing the phagemid corresponding to the variant expressed on the phage surface. Binders can then be eluted, transduced into *E. coli*, and plated on a medium that selects for infected cells. Each colony on the plate carries a phagemid corresponding to a binder that was enriched through phage display and can be isolated for characterization to determine binder diversity, binding characteristics, and, ultimately, the success of the phage display. Typically, analysis of 96 colonies is sufficient to obtain a representative sample of the enriched phage pool. To sequence the enriched phages, PCR is performed using primers that flank the entire mutated insert in the phagemid, followed by Sanger sequencing. This reveals the binder diversity and also shows whether sequences are full-length or truncated. Truncated sequences typically represent an unsuccessful attempt at phage display, which may be caused by the candidate proteins being incompatible with this experimental method or because of human error. To determine whether the cause is human error, phage display for a target protein that has been previously documented to produce binders can be performed as a positive control. Obtaining binder sequences allows for specific mutations to be investigated and permits binders to be synthesized and reinvestigated if stocks get contaminated or otherwise compromised before full characterization is completed. Binding assays can then be performed through an enzyme-linked immunosorbent assay (ELISA) to provide initial crude binding data before more in-depth analyses (Fellouse and Pal 2005; Tonikian et al. 2007; Ernst et al. 2013; Zhang and Sidhu 2018). The combination of sequencing and binding data for a pool of 96 UbV clones is sufficient to determine the outcome of phage display and ideally yields one to five candidate binders, which can be further evaluated to probe their effects on the target protein's activity, structure, and related PPIs in mammalian cells via functional assays.

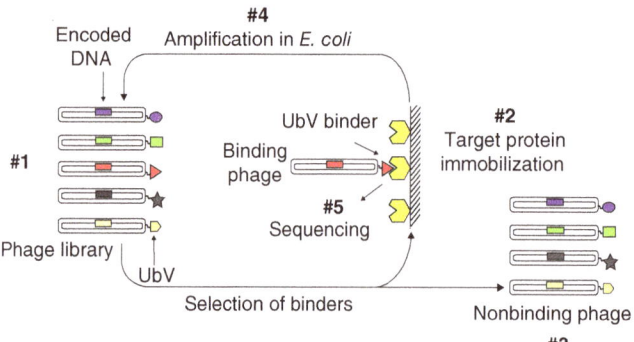

FIGURE 3. Phage display selections. Typically, five rounds of UbV phage display selection are performed, including the following five steps: (1) Each M13 phage particle of the phage-displayed UbV library encapsulates the UbV-encoding DNA in a phagemid and displays a unique UbV on its surface. (2) The phage-displayed UbV library is then applied to immobilized target proteins in a multiwell plate. (3) Target-binding phages are captured on the bottom of the plate while nonbinding phages are washed away. (4) Bound phages are eluted from the plate and further amplified through infection of an *Escherichia coli* bacterial host. (5) Phages from rounds 4 and 5 are isolated and undergo DNA sequencing, followed by binding assays, functional assays, and structural determination for full characterization.

## LIBRARY DESIGN IMPROVEMENTS

Phage display library design is crucial for generating improved binders because selecting critical binding residues for diversification will yield the best results. During residue selection, rationalized approaches must be established to ensure that diversification of the selected residues does not compromise the stability of the protein (Liang et al. 2021). While protein evolution tends to favor increased binding and specificity of PPIs, many interactions are low affinity and/or low specificity ones, to permit quick dissociation of the binder and diverse interactions, respectively. This leaves room for creating binders with both increased affinity and specificity to modulate activities of target proteins. A critical stage in developing UbV modulators is the evaluation of currently identified binders and the use of this data to inform library design. Previous characterization of UbV binders provides a strong foundation for deciding which residues to target in phage display libraries.

An example of gradual and informed library modification is the generation of a UbV for the Skp1–Cul1–F-box (SCF) ligase complex, an E3 ligase composed of three core subunits, S-phase kinase-associated protein 1 (Skp1), Cullin 1 (Cul1), and Ring-box protein 1 (Rbx1), and one of the 69 F-box proteins, which plays a role in substrate specificity (Gorelik and Sidhu 2017). Using the library generated by Ernst et al. (2013), four UbV binders (UbV.Fw7.1, 7.2, 7.3, and 7.4) were generated for two F-box target proteins, Fbw7 and Fbw11. Structural characterization of UbV.Fw7.1 in complex with Skp1–F-box$^{Fbw7}$ hinted that binding could be further improved by avoiding unwanted interaction with a negatively charged loop near the amino terminus of Skp1 (Gorelik et al. 2016). This prompted the design of a modified library (library 2) based on UbV.Fw7.1, whereby G8, R10, and T11 were left intact, and the remaining residues were randomized. This led to binder UbV.Fw7.5. Although UbV.Fw7.5 showed strong binding affinity for Fbw7, it also showed binding to other related F-box proteins (Gorelik et al. 2016). To alleviate cross-reactivity, another library (library 3) was generated based on UbV.Fw7.5 in which residues G8, A9, R10, and T11 were replaced with 11–13 hard randomized residues intended to increase possible contact area with the complex. Library 3 was panned against the target Fbw11 to determine whether binding preference could be switched without altering the library design. As a result, a high-affinity binder, UbV.Fw11.1, was generated and showed no binding to Fbw7 or any other tested Fbw domains except Fbw1, which bound weakly. A final library (library 4) was generated based on UbV.Fw11.1 to further improve the affinity of Fbw11 by soft randomization of region 1. This produced a significantly stronger and more specific binder for Skp1–F-box$^{Fw11}$, UbV.Fw11.2 (Gorelik et al. 2016, 2018).

These studies show how incremental improvements in library design have been used to target multiple Skp1–F-box complexes to increase affinity and specificity with each iteration. Further library improvements were made by using the previously generated UbV.Fw.7.5 binder as a scaffold for diversification because it showed high affinity for Fwb7 but also bound closely related off-target proteins (Gorelik et al. 2018). In contrast to the previous libraries, known key contact residues were left unaltered and, instead, six to eight residues were inserted into the β1–β2 loop and 10–12 loop residues were hard randomized. This new library was panned against 32 F-box family member proteins and produced strong binders for 11 F-boxes, each of which had little to no cross-reactivity with other closely related F-box proteins. This approach shows that contact residues are not the only candidates to consider for diversification and that flexible loop structures can be significantly changed without deleterious effects on binder stability. Ultimately, these studies show how improving past library designs using an incremental approach can aid in generating potent and highly specific binders (Gorelik et al. 2016, 2018; Gorelik and Sidhu 2017).

## CHARACTERIZATION OF BINDERS

### Overview

The application of UbV technology for designing novel binders of E3 ligases and DUBs to modulate the Ub signaling cascade has been extensively discussed in the literature (Miersch and Sidhu 2016;

Veggiani et al. 2019; Laplante and Zhang 2021; Liang et al. 2021; Liu et al. 2021). However, a comprehensive review of the methodology involved in the characterization of binders is lacking. Below, we provide an overview of key experimental procedures used to investigate the properties and functions of UbV binders (Fig. 4). First, binding kinetics for the binders can be determined through a variety of binding assays, including ELISA, isothermal titration calorimetry (ITC), biolayer interferometry (BLI), and surface plasmon resonance (SPR). These assays not only determine the degree of binding and kinetic properties, but also the specificity of binders for their target protein and off-target proteins, which is a crucial step in paring down large pools of binders. Binding assays, however, are not sufficient to determine whether binders have any effect on the activity of their target protein. Therefore, it is vital that binders with promising results in binding assays, such as those showing high affinity and/or specificity for their target protein, be investigated further using functional assays. The most widely used and well-established techniques are ubiquitination assays, auto-ubiquitination assays, and deubiquitination assays. Next, structural analysis of binder–protein complexes using X-ray crystallography or mass spectrometry (MS) are conducted to determine precise binding sites, spatial orientation of proteins, and the positions of key residues. This provides a molecular basis for describing putative mechanisms of protein activity modulation. Moreover, structural data can facilitate generation of improved libraries for new target proteins by identifying pivotal residues and their binding orientations.

Binders that have shown promise in both binding and functional assays can be further analyzed with cell-based assays to determine their function in a physiologically relevant context. Here, mammalian cell lines are transfected with binders to allow transient or stable expression. Common cell-based assays include coimmunoprecipitation (Co-IP) and mammalian cell ubiquitination assays. Last, characterization in organoids is critical for assessing binder effects in complex three-dimensional tissues, which may reveal unforeseen side effects that are not evident in cellular assays.

In summary, the best way to narrow down the results of phage display is to use a funnel approach for characterization, whereby more general and inexpensive assays, such as binding assays, are performed first to narrow down the pool of binders. Remaining binders are then analyzed in more specialized and costly assays, such as functional, structural, and cell-based assays and, ultimately, in complex tissue models. We describe these assays below.

## Binding Assays

### Enzyme-Linked Immunosorbent Assay (ELISA)

ELISA is a colorimetric immunoassay used to assess the degree of interaction between two proteins (Valadon and Scharff 1996). A target protein is immobilized or coated on a solid phase, typically in wells of a microplate. Next, the plate is blocked with nonreactive protein solutions to avoid nonspecific antibody binding. Blocking solutions of either bovine serum albumin (BSA) or nonfat dried milk are commonly used for this (Xiao 2013). Samples containing binding proteins are then added to wells, followed by addition of an HRP-linked antibody specific to an antigen associated with the binding protein of interest. For assessing phage-displayed UbVs, an HRP-linked M13 bacteriophage monoclonal antibody is used to detect bound phage and the associated UbV binders. A chromogenic substrate, such as 3,3′,5,5′-tetramethylbenzidine (TMB), is added and reacts with the HRP linked to the bound antibody to yield a color change that is linearly proportional to the degree of binding (Engvall 2010; Xiao 2013; Aydin 2015). In the case of TMB, a blue color is produced. Phosphoric acid is added to stop the reaction from going to completion, transforming the blue color to a yellow color that can be quantified with an absorbance plate reader at 450 nm. Without an available standard curve, absorbances obtained from the target protein are usually normalized to those from negative control proteins (e.g., BSA and epitope tags) to determine the relative increase in binding. Negative controls consisting of phage binders in wells containing no target protein can also help remove false-positive binders early in the characterization process by identifying phage binders that bind nonspecifically to plate wells and not the target protein.

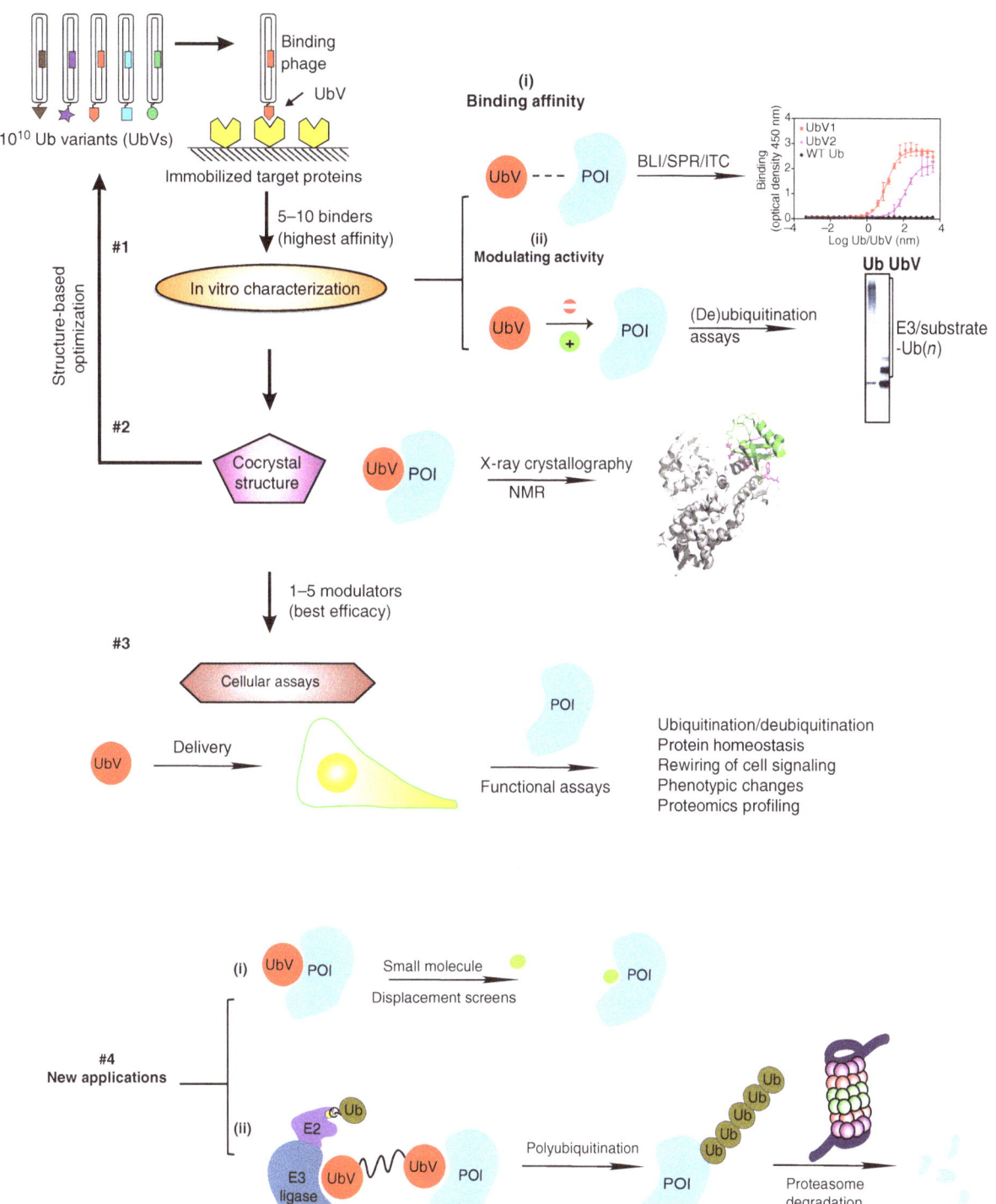

FIGURE 4. Characterization of ubiquitin variant (UbV) binders. Schematic representation of the UbV development and characterization pipeline. (1) Five to ten binders from phage display are analyzed through in vitro characterization. (i) UbVs should be characterized using binding assays, such as enzyme-linked immunosorbent assay (ELISA), isothermal titration calorimetry (ITC), biolayer interferometry (BLI), and surface plasmon resonance (SPR) to determine affinity, kinetics, and specificity. (ii) Any UbVs with promising binding characteristics should be further assessed via functional assays, such as ubiquitination, deubiquitination, and autoubiquitination assays, to assess whether binding induces any change in activity of target and off-target proteins. (2) UbVs with promising binding and/or modulatory characteristics can be structurally analyzed via methods like X-ray crystallography and mass spectrometry (MS) to investigate protein conformation and possible binding modes. (3) One to five promising UbVs can be delivered into cells for characterization using cellular assays, such as coimmunoprecipitation (Co-IP) and ubiquitination assays. UbVs that possess significant binding and/or functional properties should be analyzed in complex tissue models, such as organoids. (4) New applications for UbVs are constantly emerging. These include (i) small-molecule displacement screens to interrogate UbV potency, and (ii) incorporation into PROTAC molecules, termed UbVIPs (ubiquitin variant–induced proximity). (POI) Protein of interest.

ELISA is an easy method for identifying candidate binders for further study because it requires small sample volumes, can be performed in a high-throughput manner, and is relatively quick to execute. As such, it is the primary step for narrowing down candidate UbV binders for further characterization. ELISA, however, does have some notable constraints, which include a high possibility of false positives, particularly caused by bubbles that can interfere with absorbance readings if not removed, inconsistency in washing between steps, and cross-contamination between wells leading to inaccurate signals. Also, ELISA cannot measure binding interactions in real time because it requires an enzyme reaction to produce a signal; therefore, binding kinetics cannot be determined. Consequently, this method is recommended as an initial step for removing definitively weak binders from phage display output pools but not for decisive validation of binders (Valadon and Scharff 1996; Xiao 2013; Zhang and Sidhu 2018).

The detection range and sensitivity of ELISA is highly dependent on the specific antibody–antigen interaction involved but can be enhanced by using substrates that produce a fluorescent or chemiluminescent signal. Care must be taken, however, to ensure that measurements fall within the linear range of the instrument and reagents, otherwise the absorbance will level off and no longer increase linearly with the degree of binding, making very strong binders appear as if they bind less strongly.

One of the pioneering phage display studies on UbVs, which designed the most successful UbV library to date, screened over 140 unique UbVs via ELISA (Ernst et al. 2013). This facilitated identification of UbVs that showed markedly stronger binding and increased specificity over wild-type Ub as well as screening out poor binders that may have nonspecifically bound to plates throughout phage display. In a follow-up study that developed UbV inhibitors and activators for HECT E3 ligases (Zhang et al. 2016), ELISA was used to evaluate binding affinity of 69 UbVs to their cognate HECT domains, as well as their specificity against a panel of 20 different HECT E3s, aiding in narrowing down the large pool of UbVs to only those with the most promising characteristics.

### *Isothermal Titration Calorimetry (ITC)*

ITC determines binding affinity based on the principles of thermodynamics, in which interaction between two molecules results in the release or absorption of heat (Velazquez-Campoy et al. 2015). Two identical calorimetric cells are used, one as a reference cell and the other as a sample cell (Srivastava and Yadav 2019). The reference cell contains the target protein buffer without any target protein and serves as a reference for any changes that occur in the sample cell. The target protein is loaded into the sample cell and the sample solution containing the binder of interest is placed in a syringe, which is injected into the sample cell in small volumes at specified time intervals. Both cells are subjected to the same temperature, pressure, and volume changes so that as the binder solution is injected into sample and reference cells, heat of dilution can be accounted for by subtracting the heat change in the reference cell from that of the sample cell, and, thus, the remaining temperature change in the sample cell corresponds to the degree of binding occurring.

The calorimeter aims to keep the conditions of the sample cell the same as the reference cell, and the work done to restore the heat to that of the reference cell is measured (Canny et al. 2018; Srivastava and Yadav 2019). Thus, for an exothermic reaction, the calorimeter heats the sample cell less, and for an endothermic reaction, the calorimeter heats the sample cell more. The power required to equalize the heat between the two cells over time is recorded as a thermogram, typically showing sharp peaks for each sample injection as power usage is increased or decreased temporarily before returning to the baseline. Peaks from the sample cell become smaller over time as the target protein becomes saturated by the binders and less heat exchange occurs, whereas peaks from the reference cell remain equal in magnitude throughout the experiment. Injections continue until heat change is no longer detected, indicating that the target protein has been saturated. Any heat changes after saturation are due to nonbinding-related phenomena, such as the heat of dilution. The power usage peaks are integrated to give the total heat exchanged, which is plotted against the ratio of binder concentration to the concentration of target protein, typically resulting in a sigmoidal isotherm. If ITC is performed correctly and produces a nonsigmoidal isotherm, this indicates the possibility of either a multisubstrate binding event, no significant binding occurring, or binding enthalpies being too small to measure, as is sometimes the case with noncovalent complexes.

ITC is more accurate than ELISA but also considerably more involved. ITC does not require protein immobilization, which reduces the likelihood of destabilizing the target protein. ITC is also highly sensitive, being able to detect heat changes smaller than 1 µJ, and a large range of binding affinities can be explored (down to picomolar values) through experimental optimization. An advantage that this method has over many others is that it is compatible with optically dense or otherwise irregular solutions. In addition, samples are not consumed or destroyed during measurement. It is a label-free method because the signal that is measured is simply power input required to equalize the heat of the sample cell with the reference cell. A drawback of ITC, however, is that interactions unrelated to binding between the binder and its target protein may also give off heat and represent a potentially significant source of error and inaccuracy. This can be avoided by making sure all buffer compositions in cells and syringes match as closely as possible by degassing, desalting, and matching pH, all of which can be sources of ionization enthalpy that distort measurements. Additionally, larger amounts of samples are required for ITC, and it is a low throughput method relative to ELISA and BLI (Freire et al. 1990).

ITC has been used to probe the interaction between a UbV (UbV-G08) and the Tudor domain of 53BP1. This analysis showed that the pair bound with affinity two orders of magnitude higher than that of the 53BP1 Tudor domain with a wild-type interaction partner, dimethylated histone H4 Lys20 (Canny et al. 2018). To determine whether the mutated residues were responsible for the increased affinity, key residues were reverted back to wild-type Ub residues, and binding with the 53BP1 Tudor domain was again assessed via ITC. This showed that UbV binding was eliminated altogether, indicating the strong contribution of these residues toward mediating binding with 53BP1. Similarly, UbVs generated against the F-box proteins Fbl11 (UbV.Fl11.1) and Fbl10 (UbV.Fl10.1) were investigated via ITC and determined to have $K_D$'s against their target proteins of 25 and 91 nM, respectively (Gorelik et al. 2018), demonstrating the utility of this method for interrogating binding kinetics of UbVs and their target proteins. Interestingly, an ITC experiment where UbV.Fl10.1 was titrated into Fbl11, a protein that is highly similar to Fbl10, resulted in a much lower affinity, with a $K_D$ of 421 nM, demonstrating marked specificity for its target protein even when presented with a highly similar off-target protein. In this way, ITC can also be used to determine binding specificity in addition to affinity.

### Biolayer Interferometry (BLI)

BLI is an optical method for measuring binding interactions that works similarly to ITC (Müller-Esparza et al. 2020). In BLI, the target protein is immobilized onto a solid phase, in this case a fiber-optic biosensor that contains an internal uncoated reference layer, and samples containing binding proteins are placed in solution (Kumaraswamy and Tobias 2015). The coated biosensor is dipped into a sample solution to allow binding interactions to occur between the binder and the target protein. As the binder interacts with the protein coated on the biosensor, a layer of bound protein forms on the biosensor surface and increases in thickness in proportion to the degree of binding. The thickness of this protein layer causes thin-film interference to occur, which shifts the interference pattern of reflected light. Thus, when white light is shone on both the protein layer and the reference layer, the protein layer and the reference layer will produce two different reflection patterns. This difference is measured as the wavelength shift over time and recorded as a sensorgram that shows the degree of binding that has occurred between the coated protein and proteins present in the sample (Zhang et al. 2016; Petersen 2017).

Because there are many samples to analyze post-phage display, typically the target protein is immobilized on the biosensor tip while the solution contains the binders. However, if the signal is weak or significant nonspecific binding occurs, the positions of the proteins can be reversed. Immobilized proteins ideally have higher valency than the sample/analyte, which reduces avidity effects and increases stability (Shah and Duncan 2014; Kumaraswamy and Tobias 2015). It is critical that the immobilized protein does not change in any way that would influence its strength or affinity toward the binder of interest; therefore, careful attention needs to be paid to immobilization conditions and methods. While increasing the density of the coated protein on the biosensor leads to increased signals in the binding step, it may also introduce noise and nonspecific binding, and optimization is thus required to find a balance between maximizing sensitivity and minimizing noise. To account for noise caused by buffer components that influence the biosensor, a negative control of buffer alone is measured and subtracted from all sample data.

BLI is more accurate than ELISA but also more intricate, much like ITC. It is a high-throughput method that can evaluate binding affinity precisely and avoids the difficulties of microfluidic systems, on which methods like SPR are based. It is also a label-free method and binding can be monitored in real-time; therefore, binding kinetics can also be determined through this method. However, like ITC, BLI cannot distinguish between interactions with the binder of interest and other molecules present in the sample; as a result, samples must be relatively pure to produce reliable binding data. It also has poor reproducibility when compared with other binding assays, such as SPR or ITC (Yang et al. 2016; Petersen 2017). Another downside of BLI is the necessity of an immobilization step. This introduces the possibility of altering the conformation of the target protein, which might interfere with binding (Shah and Duncan 2014).

Despite these limitations, BLI has proven useful for probing UbV affinity and specificity. Phillips et al. (2013) used BLI to show that UbVs generated against USP14 not only bound it with 30- to 250-fold higher affinity, but also had 2- to 200-fold decreased affinity for off-target proteins UCHL1 and UCHL3 compared with wild-type Ub, demonstrating that both affinity and specificity can be refined significantly in a single phage display experiment. In another study, dissociation constants for Ub and UbV binding from BLI titrations were determined, which confirmed that many UbVs bound to their cognate HECT domains 500- to 1000-fold more strongly than Ub (Zhang et al. 2016).

### Surface Plasmon Resonance (SPR)

SPR is similar to BLI in that it is an optical, label-free method for measuring protein interactions (Tang et al. 2010). SPR is dependent on the binding of molecules to a surface, which alters the refractive index of the surface and the reflection of incident light (Bakhtiar 2013). Instead of a probe-like structure, the biosensor takes the form of a chip consisting of a thin layer of inert gold and glass that, when excited by photons, allows for total internal reflection to occur, which propagates as an evanescent wave on the gold surface. When a polarized light beam strikes the gold layer at a specific angle, the wave can excite delocalized electrons, known as plasmons, and cause a sharp drop in the intensity of the reflected light. The angle at which this occurs is known as the SPR angle and is dependent on the refractive index of the glass, which itself is dependent on the density of molecules bound to the gold surface. Therefore, the SPR angle is directly related to the mass and density of molecules bound at the gold surface.

Target proteins can be immobilized and coupled to the gold surface using an increasing variety of surface chemistries, either directly on the sensor or indirectly via a protein tag, the choice of which will vary depending on the architecture of the proteins in question. In the association phase, a sample solution containing the binder of interest is injected into a microfluidic channel that is run over the coated sensor and allowed to bind to target proteins. As with BLI, binding events change the refractive index of the sensor such that when polarized light is shone onto the chip, the change in SPR angle can be measured by a detector that records the change over time to produce a sensorgram. Once the immobilized protein is saturated and a steady state is reached, the dissociation phase begins and continues until all binders have dissociated from the immobilized protein, such that the sensorgram returns to the pre-injection baseline or until a regeneration step is performed. A regeneration step is used to remove all binders interacting with the immobilized proteins so that binding sites are available for interaction with the same or a different binder in subsequent injections, which is crucial for consistent results. The final sensorgram shows the change in SPR angles as response units (RUs) detected through all phases of binding and allows detailed interrogation of binding kinetics.

SPR resembles BLI in many ways, to the point that advantages and disadvantages of SPR are generally similar to those of BLI. SPR can measure affinities in the low nanomolar to high micromolar range; however, a trade-off between throughput and consistency needs to be considered because SPR has lower throughput than BLI but tends to produce more accurate results (Yang et al. 2016). Another disadvantage of SPR is that it can require larger sample volumes, similar to ITC.

After initial validation via ELISA, SPR has been used to evaluate the binding affinity of UbVs generated against the U-box domain of the UBE4B E3 ligase (UbV.E4B), and the RING domains of Y371-phosphorylated c-CBL and XIAP E3 ligases (UbV.pCBL and UbV.XR, respectively) (Gabrielsen et al. 2017). Performing SPR with various truncations and substitutions in UBE4B indicated where binding may have occurred. SPR was also used to identify the pCBL and XIAP residues critical for engaging in interactions with UbVs, as well as to assess UbV-binding specificity (for UbV.pCBL) and UbV dimerization mutants (for UbV.XR). Collectively, this study by Gabrielsen et al. (2017) highlighted the use of SPR for interrogating specific binding regions, which can complement available structural data and help delineate functional modulation mechanisms.

## Functional Assays

### Ubiquitination Assay

Functional assays vary greatly depending on the kind of binder being engineered through phage display and the kind of PPIs they engage in. For UbVs, ubiquitination assays can be used to determine the effect of UbV binding on E3 ligase activity and Ub processing, and to determine the specificity preferences of a UbV for different proteins (Choo and Zhang 2009). These assays mimic ATP-dependent covalent attachment of Ub substrates in vitro and require all the typical ubiquitination components, primarily an E1 enzyme, an E2 enzyme, an E3 ligase, wild-type Ub, ATP, and $Mg^{2+}$ as a cofactor. A target protein is also required, which may be one of the E1/E2/E3 enzymes, in addition to the UbV(s) of interest. The E1/E2/E3 enzymes used will vary depending on the specific assay being performed and will have to be compatible with one another for the assay to work.

Generally, there are two kinds of assays that can be performed. One assesses E3 autoubiquitination activity, whereby certain E3 ligases act upon themselves and form poly-Ub chains. The other assesses the activity of an E3 ligase on specific substrates. For both assays, reaction components are combined with a range of UbV concentrations to determine its optimal concentration for future assays. Reaction samples are typically taken at various time points and analyzed by sodium dodecyl-sulfate polyacrylamide gel electrophoresis (SDS-PAGE) and anti-Ub western blotting to probe E3/substrate ubiquitination levels (Ernst et al. 2013; Watson et al. 2019). Ubiquitination typically presents as smears or high-molecular weight ladders, the intensities of which will vary based on potential modulation by UbVs. This assay is primarily qualitative but can be made semiquantitative by using densitometry to normalize bands to loading controls and comparing the relative intensities between bands and lanes. Advantages of this assay are that it is quick, high-throughput, can be automated, requires small volumes of reagents, and does not require proprietary equipment. A limitation of this assay, however, is that it does not give insight into binder toxicity to host cells. In addition, observed effects may not carry over to cellular assays, where the reaction environment is more complex.

The utility of ubiquitination assays was shown when a UbV ($UbV^w$) was generated for anaphase-promoting complex/cyclosome (APC/C), a RING E3 ligase complex primarily involved in cell division, to investigate its potential to inhibit autoubiquitination at the APC/C catalytic site (Yang et al. 2021). A ubiquitination assay using APC/C, the E1 enzyme UBE1, the E2 enzyme UBE2C, and $UbV^w$ showed inhibition of APC/C autoubiquitination as a direct result of the $UbV^w$ competing with UBE2C for a shared binding surface (Watson et al. 2019). In another example, a UbV was generated for neural precursor cell expressed developmentally down-regulated protein 4 (NEDD4), an HECT E3 ligase primarily involved in the regulation of membrane proteins and the endosomal–lysosomal pathway. A ubiquitination assay comprising NEDD4, the E1 enzyme UBE1, the E2 enzyme UbcH7, ATP, Ub, and the UbV was analyzed by immunoblotting using an anti-NEDD4 antibody and showed decreased ubiquitination of the NEDD4 target substrate (Ernst et al. 2013).

Chapter 14

*Deubiquitination Assay*

DUBs function in opposition to E3 ligases by removing Ub chains from ubiquitinated proteins through cleavage of isopeptide bonds both within Ub chains and between chains and substrate proteins. DUBs also possess varying levels of specificity for different linkages of poly-Ub chains (Petroski 2008; Canadeo and Huibregtse 2016). Deubiquitination assays take two forms: probe-based or non-probe-based (Cho et al. 2020). Non-probe-based assays proceed similarly to ubiquitination assays, with notable differences being the replacement of E1/E2/E3 enzymes with a DUB, and Ub being replaced with a linked Ub, such as di-Ub, tetra-Ub, or poly-Ub. The hydrolysis of poly-Ub chains to smaller chains and mono-Ub can be monitored over time and visualized via SDS-PAGE and anti-Ub western blotting. Probe-based assays use a fluorescent or chemiluminescent probe consisting of a fluorophore conjugated to the carboxy-terminal tail of Ub. Probe choice varies considerably depending on experimental design and availability, with examples of common probes being ubiquitin-7-amino-4-methylcoumarin (Ub-AMC), ubiquitin-aminoluciferin (Ub-AML), and ubiquitin-rhodamine110 (Ub-Rho110). DUBs catalyze the release of the carboxy-terminal probe from the Ub conjugate, which, depending on the probe, either dequenches the fluorescence or releases a chemiluminescent substrate, and the degree of fluorescence/chemiluminescence correlates with the degree of deubiquitination. Generally, for both assays, a DUB, a UbV, and a Ub substrate (either a poly-Ub chain or a probe) are combined. A range of UbV concentrations is recommended to determine the optimal concentration for future assays. The reaction is assessed over a time course by SDS-PAGE and western blotting using an anti-Ub antibody to assess deubiquitination activity. For non-probe-based assays, deubiquitination is represented by the appearance of bands at lower molecular weight than that of the poly-Ub substrate used, such as di- and mono-Ub bands. For probe-based assays, the increase in fluorescence or chemiluminescence is indicative of DUB activity. Note that the general methodology described here will vary for each assay, particularly for probe-based assays.

These assays generally have the same advantages and limitations as ubiquitination assays; however, the use of fluorescent probes necessitates careful experimental design and planning. For example, Ub-AMC has a narrow excitation wavelength range in the UV region, which can excite other molecules and lead to false-positives (Cho et al. 2020). As such, probe choice needs to be considered carefully. Another limitation to consider is that many probes are not attached via covalent lysine isopeptide bonds but are, instead, covalently attached to the carboxy-terminal backbone (Li et al. 2013); therefore, the reaction does not directly represent the activity of the DUB for Ub isopeptide bonds, although the activities are often similar.

In a recent example of a non-probe-based assay, Manczyk et al. (2019) generated a UbV targeting the USP37 DUB and assessed the ability of USP37 to process tetra- and di-Ub chains in the presence of this UbV. SDS-PAGE and Coomassie blue staining showed that USP37 processivity was markedly decreased by the UbV (Manczyk et al. 2019). Probe-based deubiquitination assays have also proved useful for validating the inhibitory effects of UbVs. In one example, a Ub-AMC assay showed considerable inhibitory effects of a UbV (UbV.2.6) on the DUB, USP2 (Pascoe et al. 2019).

## Structural Studies

Investigating the structure and spatial orientation of proteins and their binding partners is central to characterizing PPIs because it provides critical information about binding modes and mechanisms. Elucidating the structure of protein complexes helps delineate specific domains and motifs involved in binding, which can be used to inform the generation of better modulators. For UbVs, observing the binder while complexed with its target protein allows a better understanding of which modulatory mechanisms are most effective for the target protein in question. Structural analyses complement binding assays to help elucidate the binding mechanisms that confer improved affinity and specificity, which cannot be determined through binding assays alone. These structural details of UbV-binding sites in target proteins may also prove useful for improved library design and further selections against the same target protein to yield even better binders. We discuss some of the assays for structural analyses below.

## X-Ray Crystallography

X-ray crystallography is one of the preeminent methods for atomic-resolution structural determination, as it provides a clear three-dimensional molecular structure (Drenth 2007; Messerschmidt 2007; Dessau and Modis 2010; McPherson and Gavira 2014). This method hinges on protein crystallization, which remains a trial-and-error process that is still poorly understood. Generally, crystallization requires highly purified proteins to be supersaturated so that, upon gradual addition of precipitating reagents such as salts, polyethylene glycol, or organic solvents, they can precipitate out of solution and form crystal nucleation sites. Parameters such as concentration of the precipitating reagent, pH, and temperature can be altered to optimize crystallization on a per-protein basis. Crystal structures consist of highly ordered arrangements of molecules and require slow growth to achieve this arrangement; therefore, extreme care must be taken with respect to supersaturation and speed of precipitation. If precipitation occurs too quickly, too many nucleation sites arise, resulting in smaller and less ordered crystals. Crystals are kept in solution, saturated vapors, or at low temperatures to prevent destabilization. The crystal structure is interrogated by mounting it in front of an X-ray source that directs a beam at the crystal, which is diffracted by electrons of the crystallized proteins. Destructive and constructive interference of the diffracted waves results in a repeating diffraction pattern. Through mathematical operations, such as Bragg's law and Fourier transformation, this diffraction pattern can be decoded to reveal the structure of the fundamental repeating unit of the protein crystal (Drenth 2007; Messerschmidt 2007; Dessau and Modis 2010; McPherson and Gavira 2014).

Crystallization is the limiting step of X-ray crystallography, and this had led to the development of various crystallization techniques. Batch crystallization, for instance, is the oldest and simplest method for crystallization. This method consists of adding a precipitating reagent and leaving the supersaturated protein solution undisturbed while crystals form. Microbatch crystallization is another common method and proceeds similarly, with a droplet containing both protein and precipitating reagent immersed in oil to prevent evaporation. Crystallization can be aided by seeding, a process in which a crystal is transferred from a solution that favors formation of nucleation sites to one that favors crystal growth. In the liquid–liquid diffusion method of crystallization, the protein solution is layered on top of the precipitating reagent in a small capillary tube, which permits slow diffusion of the precipitating reagent into the protein solution. In the vapor diffusion method, an unsaturated protein-precipitant solution is suspended as a small droplet over a reservoir, such that vapor equilibration eventually supersaturates the protein-precipitant solution. Last, dialysis can be used to crystallize proteins by dialyzing the precipitating reagent through a semipermeable membrane into the protein solution. None of the numerous crystallization techniques currently available, however, dominates over the others, and the most suitable one must be determined and adopted on a per-protein basis. While X-ray crystallography is an incredibly useful technique for elucidating true biological structures (as opposed to predictive computational methods), its difficulty has limited the accessibility and potential utility of the technique.

Over 20 protein-UbV cocrystal structures have been solved to date (Ernst et al. 2013; Brown et al. 2016; Gorelik et al. 2016, 2018; Yamaguchi et al. 2016; Zhang et al. 2016, 2017a; Gabrielsen et al. 2017, 2019; Leung et al. 2017; Manczyk et al. 2017, 2019; Wiechmann et al. 2017; Canny et al. 2018; Pascoe et al. 2019; Teyra et al. 2019; Watson et al. 2019; Garg et al. 2020; Maculins et al. 2020; Middleton et al. 2021; Tang et al. 2021) and the target proteins span all major protein families in the Ub-signaling cascade (e.g., USP and OTU DUB families, HECT and RING E3 families, and E2- and Ub-binding domains). As expected, many UbV mutations lead to improved hydrophobic contacts and optimization of hydrogen bond networks, such that they bind to their target protein much like wild-type Ub (Ernst et al. 2013; Zhang et al. 2016) but with higher affinity and specificity. Intriguingly, there are also many cases in which UbVs have adopted new binding mechanisms and surfaces with their target proteins. For example, UbV.8.2, a UbV generated against USP8, was crystallized and shown to bind to an allosteric site and inhibit USP8 function by stabilizing an autoinhibited state in which the active site is held in a closed conformation (Ernst et al. 2013). In another example, analysis of a crystal containing a Ubv.B1.1 bound to OTUB1 showed that the UbV bound to a distal Ub-binding site and allosterically inhibited OTUB1 activity, thereby revealing the true mode of inhibition performed by this UbV (Ernst et al. 2013). Last, Ubv.Fw7.1, a UbV generated against the F-box protein Fbw7, was found to unexpectedly bind at the interface of the Skp1 and F-box domains so that the binding surface for the UbV overlapped extensively with the binding surface for Cul1. Consequently, the UbV inhibited E3 activity by disrupting Cul1 binding. This revealed a previously uncharacterized inhibitory site in SCF E3 ligases (Gorelik et al. 2016). Such examples of structural clarification highlight the importance and necessity of investigating binding modes of phage display–generated binders because they can target varied surfaces on target proteins.

### Mass Spectrometry (MS)

MS is an analytical chemistry technique that can measure the mass-to-charge ratios ($m/z$) of fragmented proteins to determine their mass, and ultimately, their structure, primary sequence, interactions, and modifications. A purified protein sample can be prepared for MS analysis by digesting the protein into smaller peptides that are separated, typically via high-performance liquid chromatography, and then sent to a mass spectrometer. Within the mass spectrometer, samples are ionized so that they become fragmented and positively charged. These ions are accelerated through an electric or magnetic field that deflects the ions based on their $m/z$ ratio and causes them to hit a detector at different points (Aebersold and Mann 2003). The detector then produces a spectrum depicting the abundances of ionized fragments corresponding to different $m/z$ ratios. Spectra are then compared against databases of validated mass spectra.

There are many factors to consider when designing an MS experiment for analyzing proteins, such as preparation, digestion, and separation of proteins, how to ionize the protein sample, and the kind of mass analyzer to use. The most frequently used methods to interrogate proteins via MS are proteolysis, chemical cross-linking, hydrogen–deuterium exchange, and hydroxyl radical protein footprinting (Konermann and Simmons 2003; Vandermarliere et al. 2016; Kiselar and Chance 2018; Richards et al. 2021).

MS is very sensitive, fast, and versatile because it can be combined with many other techniques for protein purification and separation. Because of its sensitivity, it also requires small sample sizes. MS can be used to identify PTMs, including Ub chains attached to a target protein. However, identification of Ub chains on a target protein can be improved if a more specific type of MS is used, tandem MS (MS/MS or $MS^2$), which requires two or more mass spectrometers used back-to-back. Here, ionized peptides are separated in the first mass spectrometer based on their $m/z$ ratio, after which peptide ions with a specific $m/z$ ratio are further dissected by sending them to a second mass spectrometer that dissociates the peptide ion into fragments that differ in mass by a single amino acid (Coon et al. 2005). This allows for large proteins to be broken apart piece-by-piece so that the entire sequence can be determined. Immunoprecipitation MS (IP-MS) is an adaptation of MS that can be used to evaluate UbV-binding partners following Co-IP and has been used to determine specificity for multiple UbVs (Zhang et al. 2017b; Canny et al. 2018). Another adaptation of MS, known as ubiquitin absolute quantification (UB-AQUA), helps identify Ub chains and linkage types on modified target proteins (Ordureau et al. 2016; Rose and Mayor 2018). In UB-AQUA, tryptic digestion of Ub linkages creates a unique di-Gly Ub peptide that indicates which lysine residue(s) were previously conjugated. Synthetically labeled internal peptide standards that are otherwise identical to wild-type Ub, known as AQUA peptides, are spiked into the digested sample and then analyzed via MS. The spiked peptide standards coelute with digested Ub peptides, thereby allowing quantification of each Ub linkage in the sample and detailed interrogation of Ub branching (Phu et al. 2011). UB-AQUA has been used to investigate the specific effects of UbVs on the abundance of different kinds of Ub chains. For example, this method confirmed the effects of a NEDD4L UbV activator (NL.1), which increased the total abundance of auto-ubiquitinated K63-linked poly-Ub chains in cells by ∼20% (Zhang et al. 2016).

Despite its many advantages, costs associated with MS are relatively high and the technique requires a skilled operator, which makes it somewhat impractical and difficult to incorporate into general research.

## Cellular Assays

UbVs that possess promising properties following in vitro assays should then be validated in mammalian cell lines to determine their effects on target proteins in a more complex environment. This is crucial to determine whether the in vitro effects of UbVs can be observed in mammalian cell lines and also to determine whether off-target effects arise. In a cellular environment, binders have to contend with convoluted PPIs and PTMs that are not considered in in vitro assays; as such, assessing their effects in mammalian cell lines is critical, particularly for drug development.

### Coimmunoprecipitation (Co-IP)

Co-IP is a technique used to probe cell-wide PPIs that uses antibodies to capture proteins and analyze their binding interactions. The difference between Co-IP and regular IP experiments is that Co-IP aims to pull down a protein's binding partner(s) along with it. In a typical UbV Co-IP experiment, a UbV is expressed in a target cell line, after which cells are lysed under nondenaturing conditions to preserve protein interactions and the extract is incubated with target-specific antibodies. Beads coupled to protein A and/or protein G are then allowed to interact with and bind to the target antibody. The choice of which protein to couple to the beads depends on the isotype of the antibody used to bind the target protein. Either a magnet or centrifugation is then used to collect the beads, depending on the type of beads used. Protein complexes are then eluted from the beads and analyzed by SDS-PAGE and western blotting with two different antibodies, one against a tag on the UbV and another against the target protein, which will help evaluate the interaction between the two proteins (Zhang et al. 2016; Lin and Lai 2017). Probing with antibodies for off-target proteins is also recommended, to confirm the specificity of the binder. If the UbV and target protein interact, they should both show bands when probed with their respective antibodies, whereas lack of interaction is indicated by reduced strength of one of these bands relative to a positive control known to pull down that protein, or the band being absent altogether.

A significant advantage of this method is that protein interactions occur intracellularly in their natural state. Additionally, the entire interactome can be probed and PTMs are accounted for in all interactions. Co-IP alone, however, cannot determine whether the interaction is direct or indirect; this must be determined via in vitro binding assays and/or by solving the cocrystal structure of the protein complex.

This technique has been extensively used to probe and confirm UbV binding to target proteins in physiologically relevant contexts. For example, suspected binding of a UbV generated against UBE4B, termed UbV.E4B, was shown via Co-IP to interact with its target protein, consistent with previous noncellular assays (Gabrielsen et al. 2017). Co-IP has also been used to confirm cellular interaction between UbVIP-i53#5, a UbV modified to become a proteolysis-targeting chimera (PROTAC), and its target, 53BP1 (Aminu et al. 2022).

### Cellular Ubiquitination Assays

Ubiquitination assays can also be performed in mammalian cell lines (Choo and Zhang 2009) to validate the intracellular effects of UbVs. Once a UbV is expressed in a target cell line and interactions between the UbV and cellular proteins are allowed to occur, cells are lysed and ubiquitinated target proteins detected by SDS-PAGE and western blotting with Ub-specific antibodies (Ernst et al. 2013; Zhang et al. 2016). As with in vitro ubiquitination assays, ubiquitinated proteins appear as smears and high molecular weight bands, the strength of which indicate the degree of ubiquitination. For example, RhoB ubiquitination was assessed upon NL.1 expression, confirming that RhoB is a substrate of NEDD4L and that NL.1 acts as an activator of NEDD4L. Reduced RhoB abundance was also shown to cause reduced cell migration in NL.1-expressing cells (Zhang et al. 2016).

An advantage of performing ubiquitination assays in mammalian cell lines, rather than in vitro, is that the cellular environment permits the correct folding of proteins in their native state, thereby making the results more clinically significant with respect to drug development. Cell lines, however, require frequent maintenance, and transfecting cells and assessing results is much more demanding than in vitro approaches (Potter and Heller 2011; Makrides 2020). Nonetheless, cellular assays are highly important for validating binding assays and modulatory activities of UbVs.

### Cellular Deubiquitination Assays

Deubiquitination assays in mammalian cells are also highly recommended to validate intracellular effects of UbVs (Cho et al. 2020). In a typical cellular deubiquitination assay, a target DUB and His-tagged Ub are overexpressed in cells. This results in the substrate being ubiquitinated with the His-tagged Ub, which can then be purified from the cell lysate by immobilized metal ion affinity chromatography or by pull-down methods with nickel-charged resins that extract His-tagged proteins. These proteins are then analyzed by SDS-PAGE and western blotting with anti-Ub antibodies to assess the degree of ubiquitination. After introduction of a UbV into cells, this assay can reveal potential modulatory effects by comparing the change in intensity of smears and high molecular weight bands, which represent ubiquitinated proteins. Increased intensity of these smears/bands indicates inhibition of the DUB, whereas lower intensity indicates activation of the DUB.

Another kind of cellular deubiquitination assay involves activity-based probes (ABPs). ABPs consist of a mono-Ub moiety conjugated to a reporter tag (such as FLAG or TAMRA) on one end, and a cysteine-reactive electrophilic group that will form a covalent bond with DUB active sites (such as propargylamide or vinylmethyl ester) on the other. ABPs are incubated with cell lysate to form covalent bonds with DUB active sites and analyzed via SDS-PAGE and western blotting with antibodies against the ABP tag. If there is ABP binding, the band for the DUB appears at a higher molecular weight. The degree of binding between DUBs and ABPs indicates the activity of the DUB; therefore, the intensity of the higher molecular weight DUB-ABP band correlates with the activity of the DUB. In the presence of a UbV, this change in intensity can indicate a potential modulatory role of the UbV on the DUB.

The advantages and disadvantages of the nickel pull-down method are similar to those of the cellular ubiquitination assays: it is more laborious than in vitro assays but provides a better basis for assessing UbV functions. ABP assays, on the other hand, are an attractive alternative that do not require chromatography or pull-down methods, but they are restricted to use in vitro with cell lysates because no ABP is cell permeable, likely because of their large sizes. Chemical and physical methods for delivery of ABPs into cells have been successful, but it is possible that these methods affect ubiquitination and cell homeostasis and induce apoptosis and necrosis in cells. Last, ABPs can label non-DUB proteins and not all ABPs react equally with individual DUBs, meaning ABPs need to be selected and optimized on a per-DUB basis (Borodovsky et al. 2002).

### UbV Characterization in Organoids

Organoids are simplified organ models generated by growing cell clusters in a defined three-dimensional arrangement that closely mimics human anatomy (Drost and Clevers 2018). Organoids can originate from either pluripotent stem cells, organ-specific adult stem cells, or cancer cells, and their growth can be unguided, with intrinsic factors determining differentiation, or guided, with differentiation induced in a targeted manner. Cells are seeded in a three-dimensional extracellular matrix with growth factors that help induce growth and differentiation. As organoids grow, the medium is refreshed for growth to continue (De Souza 2018; Kim et al. 2020). Organoids have been successfully developed for many organ types, and they have aided characterization of diseases and the development of therapeutics (Li and Belmonte 2019).

Organoids offer many benefits over cell lines because they more accurately replicate the natural physiological environment in which PPIs take place (Liu et al. 2022). Organoids replicate tissue/tumor structure, cell-type composition, cancer metastatic properties, and tumor progression much more accurately than cell lines, making them highly valuable for modeling drug interactions (Ren et al. 2022). There is, however, a high barrier to using this technology in terms of cost and time required;

organoid growth rate will vary based on the tissue type being grown and it can take a considerable amount of time before a desired stage of development is reached. Another drawback is poor control in replicating conditions.

A mouse distal colon organoid, known to express NEDD4L, was used to explore the effects of two UbVs, NL.1 and NL.3. Both have appreciable binding affinity for NEDD4L, and NL.1 activated and NL.3 inhibited NEDD4L in the organoid (Zhang et al. 2016). Interestingly, after the UbVs were introduced into the organoids via lentiviral transfection, NL.1 caused considerable luminal swelling while NL.3 did not. These effects would not have been observable in other cell models, demonstrating the utility and importance of organoids for drug development. This study also provides a good example of the widespread effects that can arise from modulating PPIs, even with a very targeted approach, and highlights the utility of organoids for performing more pharmacologically relevant drug assays.

## FUTURE OUTLOOK

Combinatorial library designs coupled with phage display have generated UbVs capable of modulating many important PPIs, providing an avenue for developing novel therapeutics that can rescue dysregulated PPIs. The inherently weak binding affinity between many native proteins and their binding partners means it is difficult to identify binding regions necessary for designing strong chemical binders. Phage display is a tool that circumvents this difficulty, and continued advances in its methodologies and applications have expanded the utility of this technique beyond just modulating PPIs. Recently, Maculins et al. (2020) integrated UbVs into a robust high-throughput screening assay to identify chemical binders. In this approach, a UbV and its target protein were conjugated to separate parts of a split nanoluciferase that becomes functional when brought together through interaction between the UbV and its target protein. A panel of 14,784 chemicals was screened to assess inhibition of this interaction and 41 compounds of interest were identified, all of which corresponded to just two clusters of chemical structures. Such an assay demonstrates the utility of phage display for developing diverse small-molecule inhibitors.

Additionally, UbVs with significant binding but lacking notable inhibitory activity have great potential for use in targeted protein degradation via PROTACs and UbV-induced proximity (UbVIP). Aminu et al. (2022) showed this potential by developing UbVIP molecules comprised of two UbVs, a noninhibitory one that binds to 53BP1 and another noninhibitory one that binds to an E3 ligase, connected by a polypeptide linker. This construct was able to bring 53BP1 into close proximity with an E3 ligase such that ubiquitination and eventual degradation of 53BP1 occurred, allowing highly targeted protein degradation to be achieved. Another potential use of UbVs, and phage display binders in general, is as novel ABPs to probe protein biochemical mechanisms. Small-molecule inhibitors have been previously used in such a way to interrogate the activity of various DUBs (Kooij et al. 2020; Panyain et al. 2020). These examples show the untapped potential and utility of phage display and resultant binders for novel applications, whether in drug development or fundamental biological discovery.

## COMPETING INTEREST STATEMENT

We declare no conflicts of interest.

## ACKNOWLEDGMENTS

The ubiquitin variant technology was developed by the Sidhu laboratory. We thank the members of the Zhang laboratory for helpful discussions. Work in our laboratory is supported by a Canadian Institutes for Health Research (CIHR) Project Grant (PJT-162249) awarded to W.Z.

# Chapter 14

## REFERENCES

Aebersold R, Mann M. 2003. Mass spectrometry-based proteomics. *Nature* 422: 198–207. doi:10.1038/nature01511

Aminu B, Fux J, Mallette E, Petersen N, Zhang W. 2022. Targeted degradation of 53BP1 using ubiquitin variant induced proximity. *Biomolecules* 12: 1–16. doi:10.3390/biom12040479

Arkin MR, Wells JA. 2004. Small-molecule inhibitors of protein–protein interactions: progressing towards the dream. *Nat Rev Drug Discov* 3: 301–317. doi:10.1038/nrd1343

Atkin G, Paulson H. 2014. Ubiquitin pathways in neurodegenerative disease. *Front Mol Neurosci* 7: 63. doi:10.3389/fnmol.2014.00063

Aydin S. 2015. A short history, principles, and types of ELISA, and our laboratory experience with peptide/protein analyses using ELISA. *Peptides* 72: 4–15. doi:10.1016/j.peptides.2015.04.012

Bakhtiar R. 2013. Surface plasmon resonance spectroscopy: a versatile technique in a biochemist's toolbox. *J Chem Educ* 90: 203–209. doi:10.1021/ed200549g

Beck DB, Werner A, Kastner DL, Aksentijevich I. 2022. Disorders of ubiquitylation: unchained inflammation. *Nat Rev Rheumatol* 6: 1–13. doi:10.1038/s41584-022-00778-4

Borodovsky A, Ovaa H, Kolli N, Gan-Erdene T, Wilkinson KD, Ploegh HL, Kessler BM. 2002. Chemistry-based functional proteomics reveals novel members of the deubiquitinating enzyme family. *Chem Biol* 9: 1149–1159. doi:10.1016/S1074-5521(02)00248-X

Brown NG, VanderLinden R, Watson ER, Weissmann F, Ordureau A, Wu KP, Zhang W, Yu S, Mercredi PY, Harrison JS, et al. 2016. Dual RING E3 architectures regulate multiubiquitination and ubiquitin chain elongation by APC/C. *Cell* 165: 1440–1453. doi:10.1016/j.cell.2016.05.037

Canadeo LA, Huibregtse JM. 2016. A billion ubiquitin variants to probe and modulate the UPS. *Mol Cell* 62: 2–4. doi:10.1016/j.molcel.2016.03.023

Canny MD, Moatti N, Wan LCK, Fradet-Turcotte A, Krasner D, Mateos-Gomez PA, Zimmermann M, Orthwein A, Juang YC, Zhang W, et al. 2018. Inhibition of 53BP1 favors homology-dependent DNA repair and increases CRISPR-Cas9 genome-editing efficiency. *Nat Biotechnol* 36: 95–102. doi:10.1038/nbt.4021

Catic A, Ploegh HL. 2005. Ubiquitin: conserved protein or selfish gene? *Trends Biochem Sci* 30: 600–604. doi:10.1016/j.tibs.2005.09.002

Chen A, Kleiman FE, Manley JL, Ouchi T. 2002. Autoubiquitination of the BRCA1·BARD1 RING ubiquitin ligase. *J Biol Chem* 277: 22085–22092. doi:10.1074/jbc.M201252200

Cho J, Park J, Kim EE. 2020. Assay systems for profiling deubiquitinating activity. *Int J Mol Sci* 21: 1–16. doi:10.3390/ijms21165638

Choo YS, Zhang Z. 2009. Detection of protein ubiquitination. *J Vis Exp* 30: 10–11. doi:10.3791/1293-v

Ciechanover A. 1998. The ubiquitin: proteasome pathway: on protein death and cell life. *EMBO J* 17: 7151–7160. doi:10.1093/emboj/17.24.7151

Coon JJ, Syka JEP, Shabanowitz J, Hunt DF, Electron T, Jose S. 2005. Tandem mass spectrometry for peptide and protein sequence analysis. *Biotechniques* 38: 2–4. doi:10.2144/05384TE01

Curigliano G, Criscitiello C. 2014. Successes and limitations of targeted cancer therapy in breast cancer. *Prog Tumor Res* 41: 15–35. doi:10.1159/000355896

de Las Rivas J, Fontanillo C. 2010. Protein–protein interactions essentials: key concepts to building and analyzing interactome networks. *PLoS Comput Biol* 6: 1–8. doi:10.1371/journal.pcbi.1000807

De Souza N. 2018. Organoids. *Nat Methods* 15: 23. doi:10.1038/nmeth.4576

Dessau MA, Modis Y. 2010. Protein crystallization for X-ray crystallography. *J Vis Exp* 9: 1–6. doi:10.1167/9.8.1

Dove KK, Klevit RE. 2017. RING-between-RING E3 ligases: emerging themes amid the variations. *J Mol Biol* 429: 3363–3375. doi:10.1016/j.jmb.2017.08.008

Drenth J. 2007. Crystallizing a protein. In *Principles of protein X-ray crystallography*, pp. 1–20. Springer, New York.

Drost H, Clevers H. 2018. Organoids in cancer research. *Nat Rev Cancer* 18: 408–418. doi:10.1038/s41568-018-0007-6

Engvall E. 2010. The ELISA, enzyme-linked immunosorbent assay. *Clin Chem* 56: 319–320. doi:10.1373/clinchem.2009.127803

Ernst A, Avvakumov G, Tong J, Fan Y, Zhao Y, Alberts P, Persaud A, Walker JR, Neculai AM, Neculai D, et al. 2013. A strategy for modulation of enzymes in the ubiquitin system. *Science* 339: 590–595. doi:10.1126/science.1230161

Fellouse FA, Pal G. 2005. Methods for the construction of phage-displayed libraries. In *Phage display in biotechnology and drug discovery* (ed. Sidhu SS, Geyer CR), Vol. 1, pp. 111–142. CRC, Boca Raton, Florida.

Freire E, Obdulio L, Straume M. 1990. Isothermal titration calorimetry. *J Am Chem Soc* 62: 950–959. doi:10.1021/ac00217a002

Gabrielsen M, Buetow L, Nakasone MA, Ahmed SF, Sibbet GJ, Smith BO, Zhang W, Sidhu SS, Huang DT. 2017. A general strategy for discovery of inhibitors and activators of RING and U-box E3 ligases with ubiquitin variants. *Mol Cell* 68: 456–470. doi:10.1016/j.molcel.2017.09.027

Gabrielsen M, Buetow L, Kowalczyk D, Zhang W, Sidhu SS, Huang DT. 2019. Identification and characterization of mutations in ubiquitin required for non-covalent dimer formation. *Structure* 27: 1452–1459. doi:10.1016/j.str.2019.06.008

Garg P, Ceccarelli DF, Keszei AFA, Kurinov I, Sicheri F, Sidhu SS. 2020. Structural and functional analysis of ubiquitin-based inhibitors that target the backsides of E2 enzymes. *J Mol Biol* 432: 952–966. doi:10.1016/j.jmb.2019.09.024

George AJ, Hoffiz YC, Charles AJ, Zhu Y, Mabb AM. 2018. A comprehensive atlas of E3 ubiquitin ligase mutations in neurological disorders. *Front Genet* 9: 1–17. doi:10.3389/fgene.2018.00029

Gorelik M, Sidhu SS. 2017. Specific targeting of the deubiquitinase and E3 ligase families with engineered ubiquitin variants. *Bioeng Transl Med* 2: 31–42. doi:10.1002/btm2.10044

Gorelik M, Orlicky S, Sartori MA, Tang X, Marcon E, Kurinov I, Greenblatt JF, Tyers M, Moffat J, Sicheri F, et al. 2016. Inhibition of SCF ubiquitin ligases by engineered ubiquitin variants that target the Cul1 binding site on the Skp1-F-box. *Proc Natl Acad Sci* 113: 3527–3532. doi:10.1073/pnas.1519389113

Gorelik M, Manczyk N, Pavlenco A, Kurinov I, Sidhu SS, Sicheri F. 2018. A structure-based strategy for engineering selective ubiquitin variant inhibitors of Skp1-Cul1-F-box ubiquitin ligases. *Structure* 26: 1226–1236. doi:10.1016/j.str.2018.06.004

Huang R, Fang P, Kay BK. 2012. Improvements to the Kunkel mutagenesis protocol for constructing primary and secondary phage-display libraries. *Methods* 58: 10–17. doi:10.1016/j.ymeth.2012.08.008

Kim J, Koo BK, Knoblich JA. 2020. Human organoids: model systems for human biology and medicine. *Nat Rev Mol Cell Biol* 21: 571–584. doi:10.1038/s41580-020-0259-3

Kiselar J, Chance MR. 2018. High-resolution hydroxyl radical protein footprinting: biophysics tool for drug discovery. *Annu Rev Biophys* 47: 315–333. doi:10.1146/annurev-biophys-070317-033123

Komander D, Rape M. 2012. The ubiquitin code. *Annu Rev Biochem* 81: 203–229. doi:10.1146/annurev-biochem-060310-170328

Komander D, Clague MJ, Urbé S. 2009. Breaking the chains: structure and function of the deubiquitinases. *Nat Rev Mol Cell Biol* 10: 550–563. doi:10.1038/nrm2731

Konermann L, Simmons DA. 2003. Protein-folding kinetics and mechanisms studied by pulse-labeling and mass spectrometry. *Mass Spectrom Rev* 22: 1–26. doi:10.1002/mas.10044

Kooij R, Liu S, Sapmaz A, Xin BT, Janssen GMC, van Veelen PA, Ovaa H, ten Dijke P, Geurink PP. 2020. Small-molecule activity-based probe for monitoring ubiquitin C-terminal hydrolase L1 (UchL1) activity in live cells and zebrafish embryos. *J Am Chem Soc* 142: 16825–16841. doi:10.1021/jacs.0c07726

Kumaraswamy S, Tobias R. 2015. Label-free kinetic analysis of an antibody–antigen interaction using biolayer interferometry. *Methods Mol Biol* 1278: 165–182. doi:10.1007/978-1-4939-2425-7_10

Kunkel TA. 1985. Rapid and efficient site-specific mutagenesis without phenotypic selection. *Proc Natl Acad Sci* 82: 488–492. doi:10.1073/pnas.82.2.488

Laplante G, Zhang W. 2021. Targeting the ubiquitin–proteasome system for cancer therapeutics by small-molecule inhibitors. *Cancers* 13: 1–43. doi:10.3390/cancers13123079

Lavanya V, Adil M, Ahmed N, Rishi AK, Jamal S. 2014. Small molecule inhibitors as emerging cancer therapeutics. *Integr Cancer Sci Ther* 1: 39–46.

Leung I, Jarvik N, Sidhu SS. 2017. A highly diverse and functional naïve ubiquitin variant library for generation of intracellular affinity reagents. *J Mol Biol* **429:** 115–127. doi:10.1016/j.jmb.2016.11.016

Li M, Belmonte JCI. 2019. Organoids: preclinical models of human disease. *N Engl J Med* **380:** 569–579. doi:10.1056/NEJMra1806175

Li YT, Liang J, Li JB, Fang GM, Huang Y, Liu L. 2013. New semi-synthesis of ubiquitin C-terminal conjugate with 7-amino-4-methylcoumarin. *Peptide Sci* **20:** 102–107.

Liang CT, Roscow OMA, Zhang W. 2021. Recent developments in engineering protein–protein interactions using phage display. *Protein Eng Des Sel* **34:** 1–13. doi:10.1093/protein/gzab014

Lin JS, Lai EM. 2017. Protein–protein interactions: co-immunoprecipitation. *Methods Mol Biol* **1615:** 211–219. doi:10.1007/978-1-4939-7033-9_17

Liu Q, Aminu B, Roscow O, Zhang W. 2021. Targeting the ubiquitin signaling cascade in tumor microenvironment for cancer therapy. *Int J Mol Sci* **22:** 1–26. doi:10.3390/ijms22020791

Liu J, Huang X, Huang L, Huang J, Liang D, Liao L, Deng Y, Zhang L, Zhang B, Tang W. 2022. Organoid: next-generation modeling of cancer research and drug development. *Front Oncol* **11:** 1–9. doi:10.3389/fonc.2021.826613

Lu J, Zhao H, Yu C, Kang Y, Yang X. 2021. Targeting ubiquitin-specific protease 7 (USP7) in cancer: a new insight to overcome drug resistance. *Front Pharmacol* **12:** 1–9. doi:10.3389/fphar.2021.648491

Maculins T, Garcia-Pardo J, Skenderovic A, Gebel J, Putyrski M, Vorobyov A, Busse P, Varga G, Kuzikov M, Zaliani A, et al. 2020. Discovery of protein–protein interaction inhibitors by integrating protein engineering and chemical screening platforms. *Cell Chem Biol* **27:** 1441–1451. e1447. doi:10.1016/j.chembiol.2020.07.010

Makrides SC. 2020. Vectors for gene expression in mammalian cells. *New Comp Biochem* **38:** 9–26. doi:10.1016/S0167-7306(03)38002-0

Manczyk N, Yates BP, Veggiani G, Ernst A, Sicheri F, Sidhu SS. 2017. Structural and functional characterization of a ubiquitin variant engineered for tight and specific binding to an α-helical ubiquitin interacting motif. *Protein Sci* **26:** 1060–1069. doi:10.1002/pro.3155

Manczyk N, Veggiani G, Teyra J, Strilchuk AW, Sidhu SS, Sicheri F. 2019. The ubiquitin interacting motifs of USP37 act on the proximal Ub of a di-Ub chain to enhance catalytic efficiency. *Sci Rep* **9:** 4119. doi:10.1038/s41598-019-40815-z

McPherson A, Gavira JA. 2014. Introduction to protein crystallization. *Acta Crystallogr Sect F Structural Biol Commun* **70:** 2–20. doi:10.1107/S2053230X13033141

Messerschmidt A. 2007. *In X-ray crystallography of biomacromolecules*. Wiley-VCH, Germany.

Metzger MB, Hristova VA, Weissman AM. 2012. HECT and RING finger families of E3 ubiquitin ligases at a glance. *J Cell Sci* **125:** 531–537. doi:10.1242/jcs.091777

Middleton AJ, Teyra J, Zhu J, Sidhu SS, Day CL. 2021. Identification of ubiquitin variants that inhibit the E2 ubiquitin conjugating enzyme, Ube2k. *ACS Chem Biol* **16:** 1745–1756. doi:10.1021/acschembio.1c00445

Miersch S, Sidhu SS. 2016. Intracellular targeting with engineered proteins. *F1000Res* **5:** 1–12. doi:10.12688/f1000research.8915.1

Müller-Esparza H, Osorio-Valeriano M, Steube N, Thanbichler M, Randau L. 2020. Bio-layer interferometry analysis of the target binding activity of CRISPR-Cas effector complexes. *Front Mol Biosci* **7:** 98. doi:10.3389/fmolb.2020.00098

Nakayama KI, Nakayama K. 2006. Ubiquitin ligases: cell-cycle control and cancer. *Nat Rev Cancer* **6:** 369–381. doi:10.1038/nrc1881

Ordureau A, Münch C, Harper JW. 2016. Quantifying ubiquitin signaling. *Mol Cell* **176:** 100–106. doi:10.1016/j.molcel.2015.02.020

Panyain N, Godinat A, Lanyon-Hogg T, Lachiondo-Ortega S, Will EJ, Soudy C, Mondal M, Mason K, Elkhalifa S, Smith LM, et al. 2020. Discovery of a potent and selective covalent inhibitor and activity-based probe for the deubiquitylating enzyme UCHL1, with antifibrotic activity. *J Am Chem Soc* **142:** 12020–12026. doi:10.1021/jacs.0c04527

Pascoe N, Seetharaman A, Teyra J, Manczyk N, Satori MA, Tjandra D, Makhnevych T, Schwerdtfeger C, Brasher BB, Moffat J, et al. 2019. Yeast two-hybrid analysis for ubiquitin variant inhibitors of human deubiquitinases. *J Mol Biol* **431:** 1160–1171. doi:10.1016/j.jmb.2019.02.007

Perkins JR, Diboun I, Dessailly BH, Lees JG, Orengo C. 2010. Transient protein–protein interactions: structural, functional, and network properties. *Structure* **18:** 1233–1243. doi:10.1016/j.str.2010.08.007

Petersen RL. 2017. Strategies using bio-layer interferometry biosensor technology for vaccine research and development. *Biosensors* **7:** 49. doi:10.3390/bios7040049

Petroski MD. 2008. The ubiquitin system, disease, and drug discovery. *BMC Biochem* **9:** S7. doi:10.1186/1471-2091-9-S1-S7

Phillips AH, Zhang Y, Cunningham CN, Zhou L, Forrest WF, Liu PS, Steffek M, Lee J, Tam C, Helgason E, et al. 2013. Conformational dynamics control ubiquitin–deubiquitinase interactions and influence in vivo signaling. *Proc Natl Acad Sci* **110:** 11379–11384. doi:10.1073/pnas.1302407110

Phu L, Izrael-Tomasevic A, Matsumoto ML, Bustos D, Dynek JN, Fedorova AV, Bakalarski CE, Arnott D, Deshayes K, Dixit VM, et al. 2011. Improved quantitative mass spectrometry methods for characterizing complex ubiquitin signals. *Mol Cell Proteomics* **10:** M110 003756. doi:10.1074/mcp.M110.003756

Pickart CM, Eddins MJ. 2004. Ubiquitin: structures, functions, mechanisms. *Biochim Biophys Acta Mol Cell Res* **1695:** 55–72. doi:10.1016/j.bbamcr.2004.09.019

Popovic D, Vucic D, Dikic I. 2014. Ubiquitination in disease pathogenesis and treatment. *Nat Med* **20:** 1242–1253. doi:10.1038/nm.3739

Potter H, Heller R. 2011. Transfection by electroporation. *Curr Protoc Cell Biol* **52:** 20.5.1–20.5.10. doi:10.1002/0471143030.cb2005s52

Ren X, Chen W, Yang Q, Li X, Lixia X. 2022. Patient-derived cancer organoids for drug screening: basic technology and clinical application. *J Gastroenterol Hepatol* **37:** 1446–1454. doi:10.1111/jgh.15930

Richards AL, Eckhardt M, Krogan NJ. 2021. Mass spectrometry-based protein–protein interaction networks for the study of human diseases. *Mol Syst Biol* **17:** e8792. doi:10.15252/msb.20188792

Rose A, Mayor T. 2018. Exploring the rampant expansion of ubiquitin proteomics. *Methods Mol Biol* **1844:** 345–362. doi:10.1007/978-1-4939-8706-1_22

Shah NB, Duncan TM. 2014. Bio-layer interferometry for measuring kinetics of protein–protein interactions and allosteric ligand effects. *J Vis Exp* **84:** e51383. doi:10.3791/51383

Srivastava VK, Yadav R. 2019. *Isothermal titration calorimetry*. Elsevier, Amsterdam.

Tang Y, Zeng X, Liang J. 2010. Surface plasmon resonance: an introduction to a surface spectroscopy technique. *J Chem Educ* **87:** 742–746. doi:10.1021/ed100186y

Tang JQ, Veggiani G, Singer A, Teyra J, Chung J, Sidhu SS. 2021. A panel of engineered ubiquitin variants targeting the family of domains found in ubiquitin specific proteases (DUSPs). *J Mol Biol* **433:** 167300. doi:10.1016/j.jmb.2021.167300

Teyra J, Singer AU, Schmitges FW, Jaynes P, Kit Leng Lui S, Polyak MJ, Fodil N, Krieger JR, Tong J, Schwerdtfeger C, et al. 2019. Structural and functional characterization of ubiquitin variant inhibitors of USP15. *Structure* **27:** 590–605. doi:10.1016/j.str.2019.01.002

Tonikian R, Zhang Y, Boone C, Sidhu SS. 2007. Identifying specificity profiles for peptide recognition modules from phage-displayed peptide libraries. *Nat Protoc* **2:** 1368–1386. doi:10.1038/nprot.2007.151

Valadon P, Scharff MD. 1996. Enhancement of ELISAs for screening peptides in epitope phage display libraries. *J Immunol Methods* **197:** 171–179. doi:10.1016/0022-1759(96)00133-0

Valles GJ, Bezsonova I, Woodgate R, Ashton NW. 2020. USP7 is a master regulator of genome stability. *Front Cell Dev Biol* **8:** 717. doi:10.3389/fcell.2020.00717

Vandermarliere E, Stes E, Gevaert K, Martens L. 2016. Resolution of protein structure by mass spectrometry. *Mass Spectrom Rev* **35:** 653–665. doi:10.1002/mas.21450

Veggiani G, Gerpe MCR, Sidhu SS, Zhang W. 2019. Emerging drug development technologies targeting ubiquitination for cancer therapeutics. *Pharmacol Ther* **199:** 139–154. doi:10.1016/j.pharmthera.2019.03.003

Velazquez-Campoy A, Leavitt SA, Freire E. 2015. Characterization of protein–protein interactions by isothermal titration calorimetry. *Methods Mol Biol* **1278:** 183–204. doi:10.1007/978-1-4939-2425-7_11

Watson ER, Grace CRR, Zhang W, Miller DJ, Davidson IF, Rajan Prabu J, Yu S, Bolhuis DL, Kulko ET, Vollrath R, et al. 2019. Protein engineering of a ubiquitin-variant inhibitor of APC/C identifies a cryptic K48 ubiquitin chain binding site. *Proc Natl Acad Sci* **116:** 17280–17289. doi:10.1073/pnas.1902889116

Wiechmann S, Gärtner A, Kniss A, Stengl A, Behrends C, Rogov VV, Rodriguez MS, Dötsch V, Müller S, Ernst A. 2017. Site-specific inhibition of the small ubiquitin-like modifier (SUMO)-conjugating enzyme Ubc9 selectively impairs SUMO chain formation. *J Biol Chem* **292:** 15340–15351. doi:10.1074/jbc.M117.794255

Wu CH, Liu IJ, Lu RM, Wu HC. 2016. Advancement and applications of peptide phage display technology in biomedical science. *J Biomed Sci* **23:** 8. doi:10.1186/s12929-016-0223-x

Xiao Y. 2013. Enzyme-linked. *J Immunol Methods* **384:** 148–151. doi:10.1016/j.jim.2012.06.009

Yamaguchi M, VanderLinden R, Weissmann F, Qiao R, Dube P, Brown NG, Haselbach D, Zhang W, Sidhu SS, Peters JM, et al. 2016. Cryo-EM of mitotic checkpoint complex-bound APC/C reveals reciprocal and conformational regulation of ubiquitin ligation. *Mol Cell* **63:** 593–607. doi:10.1016/j.molcel.2016.07.003

Yang D, Singh A, Wu H, Kroe-Barrett R. 2016. Comparison of biosensor platforms in the evaluation of high affinity antibody-antigen binding kinetics. *Anal Biochem* **508:** 78–96. doi:10.1016/j.ab.2016.06.024

Yang Q, Zhao J, Chen D, Wang Y. 2021. E3 ubiquitin ligases: styles, structures and functions. *Mol Biomed* **2:** 23. doi:10.1186/s43556-021-00043-2

Zhang W, Sidhu SS. 2018. Generating intracellular modulators of E3 ligases and deubiquitinases from phage-displayed ubiquitin variant libraries. *Methods Mol Biol* **1844:** 101–119. doi:10.1007/978-1-4939-8706-1_8

Zhang Y, Zhou L, Rouge L, Phillips AH, Lam C, Liu P, Sandoval W, Helgason E, Murray JM, Wertz IE, et al. 2013. Conformational stabilization of ubiquitin yields potent and selective inhibitors of USP7. *Nat Chem Biol* **9:** 51–58. doi:10.1038/nchembio.1134

Zhang W, Wu KP, Sartori MA, Kamadurai HB, Ordureau A, Jiang C, Mercredi PY, Murchie R, Hu J, Persaud A, et al. 2016. System-wide modulation of HECT E3 ligases with selective ubiquitin variant probes. *Mol Cell* **62:** 121–136. doi:10.1016/j.molcel.2016.02.005

Zhang W, Bailey-Elkin BA, Knaap RCM, Khare B, Dalebout TJ, Johnson GG, van Kasteren PB, McLeish NJ, Gu J, He W, et al. 2017a. Potent and selective inhibition of pathogenic viruses by engineered ubiquitin variants. *PLoS Pathog* **13:** e1006372. doi:10.1371/journal.ppat.1006372

Zhang W, Sartori MA, Makhnevych T, Federowicz KE, Dong X, Liu L, Nim S, Dong A, Yang J, Li Y, et al. 2017b. Generation and validation of intracellular ubiquitin variant inhibitors for USP7 and USP10. *J Mol Biol* **429:** 3546–3560. doi:10.1016/j.jmb.2017.05.025

CHAPTER 15

# Engineering of Affibody Molecules

Stefan Ståhl,[1] Hanna Lindberg, Linnea Charlotta Hjelm, John Löfblom, and Charles Dahlsson Leitao

*Department of Protein Science, KTH Royal Institute of Technology, 106 91 Stockholm, Sweden*

Affibody molecules are small, robust, and versatile affinity proteins currently being explored for therapeutic, diagnostic, and biotechnological applications. Surface-exposed residues on the affibody scaffold are randomized to create large affibody libraries from which novel binding specificities to virtually any protein target can be generated using combinatorial protein engineering. Affibody molecules have the potential to complement—or even surpass—current antibody-based technologies, exhibiting multiple desirable properties, such as high stability, affinity, and specificity, efficient tissue penetration, and straightforward modular extension of functional domains. It has been shown in both preclinical and clinical studies that affibody molecules are safe, efficacious, and valuable alternatives to antibodies for specific targeting in the context of in vivo diagnostics and therapy. Here, we provide a general background of affibody molecules, give examples of reported applications, and briefly summarize the methodology for affibody generation.

## INTRODUCTION

Affibody molecules are small (6- to 7-kDa, 58-amino acid) affinity proteins of a tightly packed three-helical bundle architecture with a hydrophobic core, engineered to bind specific molecular targets (Fig. 1; Ståhl et al. 2017). The scaffold is based on the Z-domain, which originates from the introduction of two point mutations in the immunoglobulin (Ig) G-binding B-domain of staphylococcal protein A. Target specificity is achieved by combinatorial protein engineering from randomization of typically 13 surface-exposed residues on helices 1 and 2, followed by selection or high-throughput screening using technologies such as phage, bacterial, or yeast display. Affibody molecules can be recombinantly produced in bacterial hosts such as *Escherichia coli*, both as single domains and as more complex fusion constructs, greatly reducing the cost and complexity of manufacturing. They are easily engineered and are amenable for modular extension of functional or binding moieties through simple genetic fusion. The small size of the affibody scaffold provides alternative avenues and distinct advantages that have the potential to circumvent many of the challenges typically associated with full-length antibodies.

In this review, we provide a brief overview of affibody molecules, discussing various applications, the current state of the affibody technology, and methodologies for affibody generation. First, we discuss examples of biomedical applications that have been investigated using affibody molecules and translational successes thus far in clinical evaluations. We also discuss affibody molecules in the context of radionuclide molecular imaging. Last, we discuss methods by which affibody molecules are generated, looking specifically at approaches for library design and common selection strategies.

[1]Correspondence: ssta@kth.se

© 2026 Cold Spring Harbor Laboratory Press
Cite this introduction as *Cold Spring Harb Protoc*; doi:10.1101/pdb.top107760

# Chapter 15

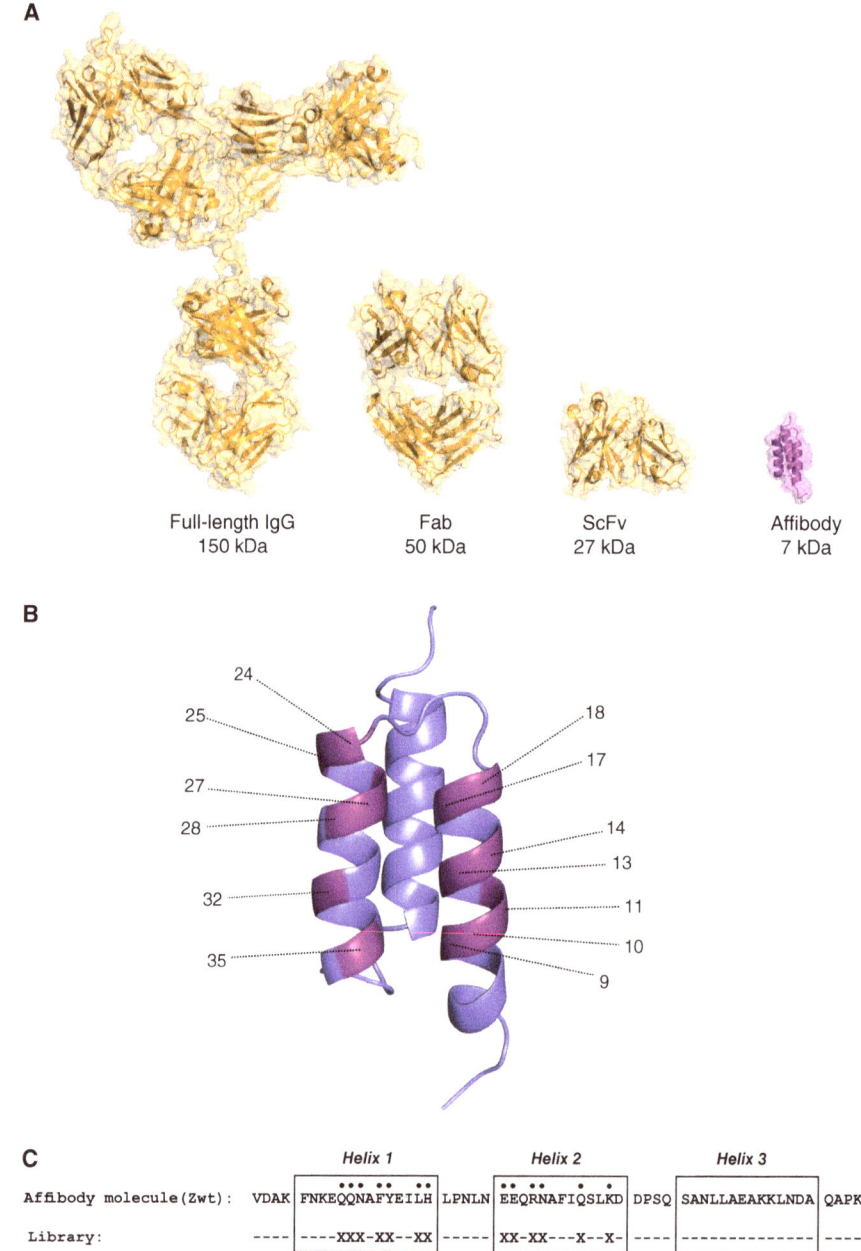

FIGURE 1. Structure of the affibody molecule. (A) Three-dimensional structures, showing a full-length IgG antibody (PDB ID 1HZH), a Fab antibody fragment (PDB ID 6MH2), an scFv antibody fragment (PDB ID 1X9Q), and an affibody molecule (PDB ID 2B88). The proteins are shown in scale, and the molecular weights are indicated below the corresponding protein name. (B) Three-dimensional structure of an affibody molecule, with the 13 randomized positions on helices 1 and 2 indicated. (C) Amino acid sequence in single-letter code of the protein A–derived $Z_{wt}$ domain, with the three α-helices boxed and the randomized positions indicated.

## BIOMEDICAL APPLICATIONS OF AFFIBODY MOLECULES

The small size of affibody molecules offers great advantages for biomedical applications. For instance, it allows for rapid blood clearance by renal excretion, which offers a favorable pharmacokinetic profile for diagnostic imaging (Tolmachev and Orlova 2020). Additionally, affibodies readily extravasate from blood and efficiently penetrate tissues, offering favorable tumor-targeting properties (Schmidt and Wittrup 2009). Furthermore, affibody molecules may alleviate logistical challenges and improve patient compliancy due to the potential for alternative administration routes such as subcutaneous, pulmonary, oral, and ocular delivery, which is being explored with promising results (Ståhl et al. 2017). For therapeutic applications, which often necessitate longer blood circulation for prolonged drug exposure and reduced dosing frequency, the serum half-life can be readily extended by various technologies. The most well-investigated half-life extension strategy for affibody molecules is genetic fusion to an engineered albumin-binding domain, conferring a biodistribution profile and half-life similar to that of serum albumin (Jonsson et al. 2008).

Research in the field has primarily been focused on medical applications, with more than 650 publications to date describing affibody molecules versus more than 50 molecular targets relevant for oncology, neurodegenerative diseases, and inflammatory disorders (Ståhl et al. 2017). The studies span from engineering efforts and in vitro cell assays to preclinical and clinical evaluations. Several affibody molecules have been engineered to bind medically relevant targets with subnanomolar—and even subpicomolar—affinity, such as platelet-derived growth factor receptor beta (PDGFRβ) (400 pM) (Lindborg et al. 2011), vascular endothelial growth factor receptor 2 (VEGFR2; 200 pM) (Fleetwood et al. 2016), epidermal growth factor receptor (EGFR; 160 pM) (Andersson et al. 2016), human epidermal growth factor receptor 2 (HER2) (22 pM) (Orlova et al. 2006), HER3 (21 pM) (Malm et al. 2013), amyloid β peptide (20 pM) (Lindberg et al. 2015; Boutajangout et al. 2019), and interleukin (IL)-17A (0.3 pM) (Ståhl et al. 2017). These high-affinity binders have been used to explore various treatment strategies, including diagnostic molecular imaging of tumors, receptor signaling inhibition, ligand trapping, and cytotoxic payload delivery. Currently, three affibody molecules have proceeded to different stages of clinical development and testing. A HER2-targeting affibody molecule, ABY-025, labeled with the radioactive compound gallium-68, has been used, together with positron-emission tomography (PET) imaging, to stratify patients with HER2-positive primary and metastatic breast cancer in a phase I clinical study (Velikyan et al. 2019), and patients are currently being recruited for phase II/III. In addition, a fluorescently labeled EGFR-targeting affibody molecule, ABY-029, is being evaluated for guided resection of tumors in patients with recurrent glioma (Sexton et al. 2013). A treatment for psoriatic arthritis is being investigated using a dimeric affibody, ABY-035, to neutralize IL-17A, which has shown a favorable safety profile and efficacy over the standard of care in phase II clinical trials (https://acelyrin.com/press/acelyrin-affibody-ab-inmagene-announce-data-from-global-phase-2-trial-of-izokibep). Indeed, results from clinical studies have thus far shown that affibody molecules are safe and tolerable in humans. ABY-035 has shown rapid and sustained efficacy in patients, as well as favorable safety and tolerability data, and the majority of adverse events have been mild and resolved during treatment.

The small size and robustness of affibody molecules are particularly useful in the context of radionuclide molecular imaging. More than 200 publications concerning preclinical affibody-based tumor imaging are currently available. The most extensively studied cancer biomarkers using affibody-based tracers are the HER family of receptors (Rinne et al. 2021), including EGFR, HER2, and HER3. The fast pharmacokinetics associated with these molecules offers improved contrast and more sensitive scans at earlier time points, usually just a few hours after injection. Affibody molecules generally show high thermal and chemical stability, which facilitates downstream processes such as purification, conjugation, and radionuclide labeling. Their small size also has implications for the design of affibody-based imaging tracers. Minor alterations may considerably affect their overall charge and hydrophilicity, meaning that biodistribution is tunable by optimization of various parameters, such as chelators, nuclides, tags, and amino acid composition, which has the potential to minimize off-target interactions without noticeably affecting affinity for the cancer-associated target antigen. High affinity and, thus, longer tumor retentions, which are usually attainable for affibody molecules, are necessary to achieve high contrast for imaging molecular targets with slow internalization, in particular for the detection of targets with low expression levels (Tolmachev et al. 2012).

## AFFIBODY GENERATION AND LIBRARY DESIGN

Affibody libraries are typically constructed by randomizing 13–15 positions on helices 1 and 2. By using a predefined set of codons, a large naive repertoire of tens of billions of affibody variants is created, from which unique binding specificities can be selected (Fig. 1). Degenerate NNK codons (where N = A/C/G/T and K = G/T) have traditionally been used for the construction of affibody libraries. Most libraries today, however, are made using a mixture of trinucleotide codons during DNA library synthesis, providing more controlled diversification, decreased bias, and limited introduction of stop codons (Arunachalam et al. 2012). Cysteine residues are absent in the affibody scaffold and are usually omitted in the library design, to enable post-selection introduction of a solitary cysteine for site-specific thiol-coupling of functionalizing groups, such as metal chelators, fluorophores, biotin, and cytotoxic compounds. Proline is also often excluded due to its helix-breaking propensity. Moreover, the small size of the affibody gene (174 bp) makes synthesis and molecular cloning of DNA-encoded affibody libraries straightforward. The randomized positions are located within a sequence of merely 78 bp on helices 1 and 2 (Fig. 1). The DNA library can thus be synthesized as a single randomized oligonucleotide, that includes, for example, restriction enzyme sites or overlapping sequences for restriction-free gene assembly to the display plasmid of choice, as described, for instance, in Protocol 1: Cloning of Affibody Libraries for Display Methods (Ståhl et al. 2023).

Affibody molecules generated from naive selections typically have affinities in the nanomolar range. If higher affinities are desired, a maturation library is usually constructed to improve affinity. When designing the maturation library, amino acid frequencies and alanine scanning can be used to elucidate the importance of certain amino acids and positions for the binding interaction. For example, a maturation library was created for a HER3-binding affibody, in which the binding contributions of the 13 residues were weighed based on the results from alanine scanning (Malm et al. 2013). Each of the 13 positions were randomized using an equal mixture of 17 trinucleotide codons, but the original residues were preserved to a degree proportionate to their binding contribution (a method often denoted as "soft mutagenesis"), resulting in binders with affinities in the low picomolar range.

Selections of affibody molecules are performed by panning against the target antigen in a setting appropriate for the choice of display system. Phage display is still the most commonly used method for generation of new affibody molecules, and detailed instructions for construction of phage-displayed affibody libraries and phage-display biopanning using such libraries are described in Protocol 1: Cloning of Affibody Libraries for Display Methods (Ståhl et al. 2023) and Protocol 2: Selection of Affibody Molecules Using Phage Display (Hjelm et al. 2023), respectively (Fig. 2A). In addition to phage display, selection of affibody molecules has also been shown using other display methods, such as yeast display (Stern et al. 2019), ribosome display (Grimm et al. 2011), *E. coli* display (Andersson et al. 2019), and staphylococcal display (Malm et al. 2013). Methods for generating libraries in *E. coli* and *Staphylococcus carnosus*, together with details on selection and screening from such libraries are described in Protocol 3: Selection of Affibody Molecules Using *Escherichia coli* Display (Dahlsson Leitao et al. 2023) and Protocol 4: Selection of Affibody Molecules Using Staphylococcal Display (Löfblom et al. 2023). The main advantage of using cell-based display systems is that the selection can be performed using fluorescence-activated cell sorting (FACS) (Löfblom 2011). For FACS, the target antigen is fluorescently labeled and incubated with cells that each display $10^4$–$10^5$ copies of a unique affibody variant. During selection, binding populations can be separated from nonbinding populations in real time based on fluorescent signals, allowing enrichment to be observed and followed throughout each selection round. Surface expression of the affibody variants is generally assessed by the inclusion of an albumin-binding domain in the displayed protein construct, followed by incubation with fluorescently labeled serum albumin (Löfblom et al. 2005). For cell-based display systems, the complexity of naive libraries is typically reduced by using magnetic-activated cell sorting in a preenrichment step before library sorting using FACS (Fig. 2B), which is described as part of Protocol 3: Selection of Affibody Molecules Using *Escherichia coli* Display (Dahlsson Leitao et al. 2023).

## CONCLUSIONS

Affibody molecules are an interesting and versatile class of affinity proteins that are being explored for therapeutic and diagnostic applications for several diseases, as well as for other biotechnological

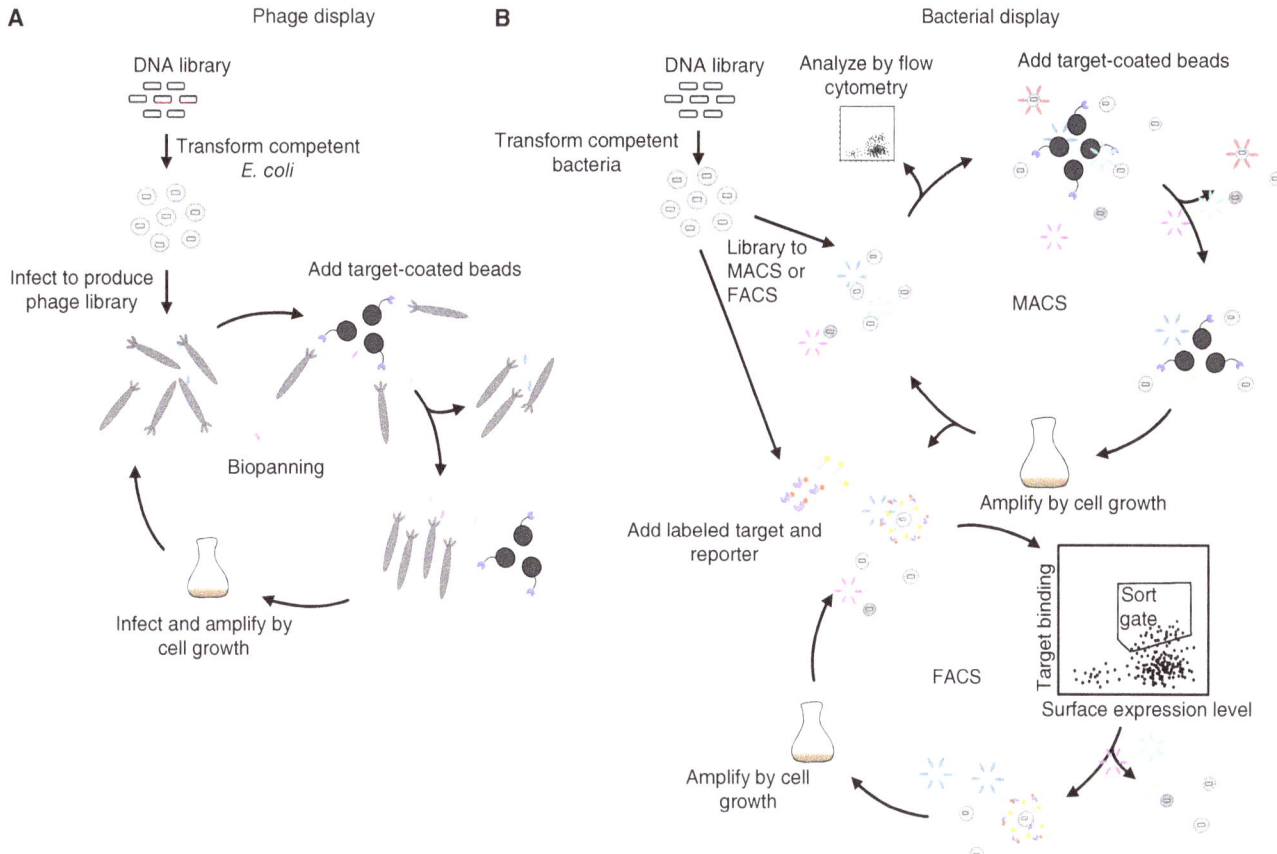

FIGURE 2. Common selection strategies for generating affibody molecules. Schematic overview of the selection processes in (A) phage display and (B) bacterial display. In phage display, the affibody library is displayed on bacteriophages and panned against the target coated on a solid support, such as paramagnetic beads. The binding population associated with the beads is isolated by washing away the nonbinding population. The binding population is amplified and the procedure is repeated two to four times to enrich for affibody variants that bind to the target. In bacterial cell display, the affibody library is displayed on the surface of bacterial cells and incubated with fluorescently labeled target and reporter molecules, followed by the isolation of binding populations using fluorescence-activated cell sorting (FACS). The binding population is amplified, and the procedure is repeated two to four times to enrich for affibody variants that bind to the target. Note that, for larger libraries, magnetic-activated cell sorting (MACS) is advisable to reduce the library sizes to allow efficient FACS.

applications. The unique characteristics of the affibody molecule have the potential to complement antibody-based therapies and improve on current cancer diagnostic imaging technologies. Generation of novel binding specificities from large affibody libraries against virtually any target is possible and straightforward.

# ACKNOWLEDGMENTS

The research on affibodies and their selection systems was funded by the Swedish Agency for Innovation VINNOVA (2019/00104 and the CellNova Center 2017/02105), the Knut and Alice Wallenberg Foundation through the Wallenberg Center for Protein Technology (KAW 2019.0341 and KAW 2021.0197), the Swedish Cancer Society (grants 201090 PjF; 222023 Pj01H), the Swedish Research Council (2019-05115), the Swedish Brain Foundation (grants FO2018-0094, FO2021-0407, FO2022-0253), the Tussilago foundation (FL-0002.025.551-7), the Strategic Research Area Neuroscience (StratNeuro), as well as the Schörling Family Foundation via the Swedish FTD Initiative.

# Chapter 15

## REFERENCES

Andersson KG, Oroujeni M, Garousi J, Mitran B, Ståhl S, Orlova A, Löfblom J, Tolmachev V. 2016. Feasibility of imaging of epidermal growth factor receptor expression with ZEGFR:2377 affibody molecule labeled with 99mTc using a peptide-based cysteine-containing chelator. *Int J Oncol* 49: 2285–2293. doi:10.3892/ijo.2016.3721

Andersson KG, Persson J, Ståhl S, Löfblom J. 2019. Autotransporter-mediated display of a naïve affibody library on the outer membrane of *Escherichia coli*. *Biotechnol J* 14: e1800359. doi:10.1002/biot.201800359

Arunachalam TS, Wichert C, Appel B, Müller S. 2012. Mixed oligonucleotides for random mutagenesis: best way of making them. *Org Biomol Chem* 10: 4641–4650. doi:10.1039/c2ob25328c

Boutajangout A, Lindberg H, Awwad A, Paul A, Baitalmal R, Almokyad I, Höidén-Guthenberg I, Gunneriusson E, Frejd FY, Härd T, et al. 2019. Affibody-mediated sequestration of amyloid β demonstrates preventive efficacy in a transgenic Alzheimer's disease mouse model. *Front Aging Neurosci* 11: 64. doi:10.3389/fnagi.2019.00064

Dahlsson Leitao C, Hjelm LC, Ståhl S, Löfblom J, Lindberg H. 2023. Selection of affibody molecules using *Escherichia coli* display. *Cold Spring Harb Protoc* doi:10.1101/pdb.prot108400

Fleetwood F, Güler R, Gordon E, Ståhl S, Claesson-Welsh L, Löfblom J. 2016. Novel affinity binders for neutralization of vascular endothelial growth factor (VEGF) signaling. *Cell Mol Life Sci* 73: 1671–1683. doi:10.1007/s00018-015-2088-7

Grimm S, Yu F, Nygren PÅ. 2011. Ribosome display selection of a murine IgG1 fab binding affibody molecule allowing species selective recovery of monoclonal antibodies. *Mol Biotechnol* 48: 263–276. doi:10.1007/s12033-010-9367-1

Hjelm LC, Dahlsson Leitao C, Ståhl S, Löfblom J, Lindberg H. 2023. Selection of affibody molecules using phage display. *Cold Spring Harb Protoc* doi:10.1101/pdb.prot108399

Jonsson A, Dogan J, Herne N, Abrahmsén L, Nygren PÅ. 2008. Engineering of a femtomolar affinity binding protein to human serum albumin. *Protein Eng Des Sel* 21: 515–527. doi:10.1093/protein/gzn028

Lindberg H, Härd T, Löfblom J, Ståhl S. 2015. A truncated and dimeric format of an Affibody library on bacteria enables FACS-mediated isolation of amyloid-beta aggregation inhibitors with subnanomolar affinity. *Biotechnol J* 10: 1707–1718. doi:10.1002/biot.201500131

Lindborg M, Cortez E, Höidén-Guthenberg I, Gunneriusson E, Von Hage E, Syud F, Morrison M, Abrahmsén L, Herne N, Pietras K, et al. 2011. Engineered high-affinity affibody molecules targeting platelet-derived growth factor receptor β in vivo. *J Mol Biol* 407: 298–315. doi:10.1016/j.jmb.2011.01.033

Löfblom J. 2011. Bacterial display in combinatorial protein engineering. *Biotechnol J* 6: 1115–1129. doi:10.1002/biot.201100129

Löfblom J, Wernérus H, Ståhl S. 2005. Fine affinity discrimination by normalized fluorescence activated cell sorting in staphylococcal surface display. *FEMS Microbiol Lett* 248: 189–198. doi:10.1016/j.femsle.2005.05.040

Löfblom J, Hjelm LC, Dahlsson Leitao C, Ståhl S, Lindberg H. 2023. Selection of affibody molecules using staphylococcal display. *Cold Spring Harb Protoc* doi:10.1101/pdb.prot108401

Malm M, Kronqvist N, Lindberg H, Gudmundsdotter L, Bass T, Frejd FY, Höidén-Guthenberg I, Varasteh Z, Orlova A, Tolmachev V, et al. 2013. Inhibiting HER3-mediated tumor cell growth with affibody molecules engineered to low picomolar affinity by position-directed error-prone PCR-like diversification. *PLoS ONE* 8: e62791. doi:10.1371/journal.pone.0062791

Orlova A, Magnusson M, Eriksson TLJ, Nilsson M, Larsson B, Höidén-Guthenberg I, Widström C, Carlsson J, Tolmachev V, Ståhl S, et al. 2006. Tumor imaging using a picomolar affinity HER2 binding Affibody molecule. *Cancer Res* 66: 4339–4348. doi:10.1158/0008-5472.CAN-05-3521

Rinne SS, Orlova A, Tolmachev V. 2021. PET and SPECT imaging of the EGFR family (RTK class I) in oncology. *Int J Mol Sci* 22: 3663. doi:10.3390/ijms22073663

Schmidt MM, Wittrup KD. 2009. A modeling analysis of the effects of molecular size and binding affinity on tumor targeting. *Mol Cancer Ther* 8: 2861–2871. doi:10.1158/1535-7163.MCT-09-0195

Sexton K, Tichauer K, Samkoe KS, Gunn J, Hoopes PJ, Pogue BW. 2013. Fluorescent affibody peptide penetration in glioma margin is superior to full antibody. *PLoS ONE* 8: e60390. doi:10.1371/journal.pone.0060390

Ståhl S, Gräslund T, Eriksson Karlström A, Frejd FY, Nygren P-Å, Löfblom J. 2017. Affibody molecules in biotechnological and medical applications. *Trends Biotechnol* 35: 691–712. doi:10.1016/j.tibtech.2017.04.007

Ståhl S, Hjelm LC, Dahlsson Leitao C, Löfblom J, Lindberg H. 2023. Cloning of affibody libraries for display methods. *Cold Spring Harb Protoc* doi:10.1101/pdb.prot108398

Stern LA, Lown PS, Kobe AC, Abou-Elkacem L, Willmann JK, Hackel BJ. 2019. Cellular-based selections aid yeast-display discovery of genuine cell-binding ligands: targeting oncology vascular biomarker CD276. *ACS Comb Sci* 21: 207–222. doi:10.1021/acscombsci.8b00156

Tolmachev V, Orlova A. 2020. Affibody molecules as targeting vectors for PET imaging. *Cancers (Basel)* 12: 651. doi:10.3390/cancers12030651

Tolmachev V, Tran TA, Rosik D, Sjöberg A, Abrahmsén L, Orlova A. 2012. Tumor targeting using affibody molecules: interplay of affinity, target expression level, and binding site composition. *J Nucl Med* 53: 953–960. doi:10.2967/jnumed.111.101527

Velikyan I, Schweighöfer P, Feldwisch J, Seemann J, Frejd FY, Lindman H, Sörensen J. 2019. Diagnostic HER2-binding radiopharmaceutical, [$^{68}$Ga]Ga-ABY-025, for routine clinical use in breast cancer patients. *Am J Nucl Med Mol Imaging* 9: 12–23.

Protocol 1

# Cloning of Affibody Libraries for Display Methods

Stefan Ståhl, Linnea Charlotta Hjelm, Charles Dahlsson Leitao, John Löfblom, and Hanna Lindberg[1]

*Department of Protein Science, KTH Royal Institute of Technology, 106 91 Stockholm, Sweden*

Affibody molecules are small (6-kDa) affinity proteins folded in a three-helical bundle and generated by directed evolution for specific binding to various target molecules. The most advanced affibody molecules are currently tested in the clinic, and data from more than 300 subjects show excellent activity and safety profiles. The generation of affibody molecules against a particular target starts with the generation of an affibody library, which can then be used for panning using multiple methods and selection systems. This protocol describes the molecular cloning of DNA-encoded affibody libraries to a display vector of choice, for either phage, *Escherichia coli*, or *Staphylococcus carnosus* display. The DNA library can come from different sources, such as error-prone polymerase chain reaction (PCR), molecular shuffling of mutations from previous selections, or, more commonly, from DNA synthesis using various methods. Restriction enzyme-based subcloning is the most common strategy for affibody libraries of higher diversity (e.g., $>10^7$ variants) and is described here.

## MATERIALS

It is essential that you consult the appropriate Material Safety Data Sheets and your institution's Environmental Health and Safety Office for proper handling of equipment and hazardous materials used in this protocol.

RECIPES: Please see the end of this protocol for recipes indicated by <R>. Additional recipes can be found online at http://cshprotocols.cshlp.org/site/recipes.

### Reagents

Affibody library DNA encoding randomized helices 1 and 2 (Andersson et al. 2019)
Biotinylated target protein (prebiotinylated or biotinylated using, e.g., the Biotin-XX Microscale Protein Labeling Kit, Molecular Probes B30010)
Carbenicillin (Thermo Scientific 10177012, 100 mg/mL stock in sterile $H_2O$, filter sterilized, 0.2-μm)

*Stock can be stored in the freezer until first use (up to several years). Store opened aliquots for up to 6 mo at 4°C. Use at 100 μg/mL final concentration for all experiments.*

CutSmart buffer (10×) (NEB, B6004S)
D-glucose (Sigma-Aldrich G8270, 20%, w/v, in water, filter sterilized 0.2-μm)
Display vectors pAffi1 (for phage display) (Grönwall et al. 2007), pBadv2.2 (for *Escherichia coli* display, modified from paraBAD) (Andersson et al. 2019), and pSCZ2 (for *Staphylococcus carnosus* display, modified from pSCXm) (Wernérus and Ståhl 2002)

---

[1]Correspondence: hanli@kth.se

© 2026 Cold Spring Harbor Laboratory Press
Cite this protocol as *Cold Spring Harb Protoc*; doi:10.1101/pdb.prot108398

DNA loading dye (Thermo Scientific 11541575, and included in QIAGEN Plasmid kits and QIAquick PCR purification kits, see below)
DNA mass ladder (e.g., GeneRuler DNA Ladder Mix; Thermo Scientific SM0334)
dNTPs (200-mM, Thermo Scientific R0182)
DpnI restriction endonuclease (NEB R0176S)
Dynabeads M-280 Streptavidin (Invitrogen 11205D)
Electrocompetent *E. coli* cells for phage display (e.g., XL1-Blue, Agilent 200228), *E. coli* display (e.g., BL21 [DE] E. cloni EXPRESS Lucigen 60300-1), and *S. carnosus* subcloning (e.g., TOP10, Thermo Scientific C404050)

*These cells are sold with recovery medium. Prewarm media before electroporation according to supplier's instructions.*

Ethanol (ice-cold, 70%, v/v, and 96%, v/v, in water)
GelCode Blue Safe Protein Stain (Thermo Scientific 24596)
GelRed Nucleic Acid Gel stain (Biotium 41003)
Glycerol (85%, v/v, in water, sterilize by autoclaving)
High-fidelity DNA polymerase and HF buffer (5×) (e.g., Phusion polymerase, NEB M0530)
High and low DNA mass ladder

*For concentration determination, use, for example, MassRuler Express forward DNA ladder mix; Thermo Scientific SM1283 (High DNA mass ladder: SM0311. Low DNA mass ladder SM0333).*

LB liquid medium (for example, LB liquid medium II <R>)

*This is for* E. coli *display.*

Low-fidelity polymerase and DreamTaq polymerase buffer (10×) (e.g., DreamTaq polymerase, Thermo Scientific EP0702)
Molten agarose <R>

*Prepare a 1% (w/v) agarose gel.*

PCR product-purification kit (e.g., QIAquick PCR Purification Kit; QIAGEN 28104)
Phosphate-buffered saline (PBS), pH 7.4 <R>
Phosphate-buffered saline with Tween (0.1%) (PBST) <R>
Plasmid DNA Maxiprep kit (e.g., QIAGEN 12162, for *S. carnosus* display)
Protein gel (e.g., NuPAGE 4 to 12%, Bis-Tris 12%, Invitrogen NP0321BOX)
Protein low-molecular-weight ladder (VWR 17-0446-01)
Reducing loading dye for sodium dodecyl sulfate-polyacrylamide gel electrophoresis (SDS-PAGE) (NEB, B7703S)
SDS-PAGE running buffer (e.g., 20× MES buffer for NuPAGE Bis-Tris gels; Thermo Fisher NP0002)

*Use 1× MES buffer for experiments.*

Sodium acetate (3 M, pH 5.5) <R>
Sterile H$_2$O (autoclaved MilliQ)
T4 Ligase and DNA ligase buffer (10×) (NEB M0202L)
TAE <R>
Tryptic soy broth (TSB) <R>

*This is for phage display.*

Tryptose blood agar base (TBAB) plates <R>

*Prepare TBAB plates with 100 µg/mL carbenicillin. Store for up to 2 wk at 4°C.*

TSB + Y medium <R>

*This is for* S. carnosus *display.*

XhoI, NheI, and SacI restriction endonucleases (NEB, R0146L, R3131L, R3156L)

# Equipment

Aspirator
Centrifuge (refrigerated)
Cryovials (1.5-mL)
DNA electrophoresis equipment (Bio-Rad Sub-Cell GT Horizontal Electrophoresis System 1704483, Bio-Rad Sub-Cell GT Casting Gates 1704415, and Bio-Rad PowerPac Basic Power Supply 1645050, Bio-Rad GelDoc EZ System 1708270)
Electroporation cuvettes (2-mm gap, prechilled)
End-over-end rotamixer for 1.5-mL microcentrifuge tubes, set to 15 rpm
Erlenmeyer flasks
   *Use flasks with volume 10× larger than the culture.*

Freezers at −20°C and −80°C
Glass spreader or glass beads (sterile)
   *Choose according to user's preference.*

Heating block at 96°C and 16°C
Incubator set at 37°C (static and shaking)
Magnetic rack for 1.5-mL microcentrifuge tubes
Microcentrifuge
Microcentrifuge tubes (1.5-mL)
Micropulser electroporator (e.g., Bio-Rad MicroPulser 1652100)
PCR plates (96-well, VWR 732-5003)
SDS-PAGE equipment (Invitrogen STM4001)
Spectrophotometer
Thermocycler

# METHOD

*This protocol describes the creation of an affibody library displayed on either phage, E. coli, or S. carnousus. The final display format is determined by the user, and specific steps and material are indicated for each system. The approach is divided into three parts: primer design (Steps 1–6), cloning of affibody libraries (Steps 7–28), and determining the bead-binding capacity of biotinylated target proteins for selection (Steps 29–36). Steps 7–16 are repeated depending on the desired library size and yields for each step. Go through the steps once with a small reaction to verify yields and procedure before scaling up to a calculated number of reactions to create a large library.*

*Restriction-free subcloning is an alternative option for smaller libraries and is performed in multiple parallel reactions according to the guidelines provided by the corresponding supplier. Such workflow, however, is not described here.*

*The created library is used in selections in which it is panned against biotinylated target protein. Within the library selection, magnetic beads are used to capture and select out desired clones. The density and possible binding capacity of the beads to each protein differs and needs to be determined before continuing with selections. Therefore, the binding capacity is tested in Steps 29–36.*

## Primer Design

### Primers for Display Vector Amplification

For phage and E. coli *display, primers should include sites for XhoI and NheI. For* S. carnosus *display, primers should instead include sites for XhoI and SacI.*

1. Design a forward primer spanning the NheI restriction enzyme (RE) site, with six overhanging base pairs (bp) before the RE site, and ~20 bp downstream from the RE site.
   *This is just an example for design of primers for display vector amplification.*

Chapter 15

2. Design a reverse primer spanning the XhoI RE site, with six overhanging bp before the RE site, and ~20 bp in the primer direction.

3. Match melting temperatures by adjusting the length of the primers. Order them in 25 nmole scale.

### Sequencing Primers

4. Obtain sequencing primers recommended by the vendor, depending on the vector used.
   - For phage display, use the M13rev primer.
   - For *E. coli* and *Staphylococcus* display, design a primer 150 bp upstream of the insert, annealing to the plasmid constant region. Order them in 100 nmole scale.

### Library-Amplification Primers

5. Design primers with six overhanging bp before the RE site and include the sequence corresponding to 15–20 amino acids of the insert region or as many residues as possible until the first randomized position is reached in the library design. Match melting temperatures by adjusting the length of the primers. Order them in 25 nmole scale.

   *For phage and* E. coli *display, primers should include sites for XhoI and NheI. For* S. carnosus *display, primers should include sites for XhoI and SacI.*

### PCR Screening Primers

6. Design primers with a suitable $T_m$ about 100–150 bp from the restriction site, flanking both sides of the insert. Order them in 100 nmole scale.

## Cloning of Affibody Libraries

7. Set up five test polymerase chain reactions (PCRs) for both the library and vector DNA, each containing 1 µL of high-fidelity proofreading polymerase, 10 µL of polymerase buffer HF (5×), 200 µM dNTPs, 10 pmol of both forward and reverse primers, 1–5 ng of DNA, and Milli-Q $H_2O$ up to a total of 50 µL/well, in PCR plates.

8. Amplify the library and the vector in parallel using the following programs in two separate PCR machines:

   | Library | | | Vector | | |
   |---|---|---|---|---|---|
   | 1 cycle | 3 min | 98°C | 1 cycle | 3 min | 98°C |
   | 12 cycles | 10 sec | 98°C | 35 cycles | 10 sec | 98°C |
   | | 30 sec | Primer $T_m$ | | 30 sec | Primer $T_m$ |
   | | 30 sec | 72°C | | 3.5–4.5 min depending on vector | 72°C |
   | 1 cycle | 10 min | 72°C | 1 cycle | 10 min | 72°C |
   | 1 cycle | ∞ | 4°C | 1 cycle | ∞ | 4°C |

   *PCR products may be stored, preferably at −20°C, or short-term at 4°C.*

9. Verify amplification by gel electrophoresis.
   i. Supplement a molten 1% (w/v) agarose gel (1× TAE) with 1/100th of the agarose volume of GelRed.
   ii. Cast the 1% (w/v) agarose gel, add a comb, and allow it to solidify.
   iii. Mix 1 µL of PCR product with 4 µL of DNA loading dye.

iv. Load 5 µL of the mixture into the gel alongside 1 µL of an appropriate DNA (high for vector and low for library) mass ladder as a reference.

v. Run gel at 150 V for 20 min in 1× TAE buffer.

*Depending on the primers, the theoretical lengths of the affibody library, pAffi1, pBadv2.2, and pScZ2 should be ~140, ~4400, ~5350, and ~7550 bp, respectively.*

*We do not recommend performing additional cycles of PCR of the library, as it can introduce errors.*

*Approximately 100 µg of ligated DNA (Step 16) is needed to create libraries of $10^7$–$10^{11}$ complexity; thus, scale the number of reactions accordingly and repeat Steps 7–9 depending on the efficacy of the PCR amplification.*

10. Perform a DpnI digestion of the amplified (and methylated) vector DNA from Step 8. To do this, add 1 µL of DpnI to each 50 µL PCR of vector DNA and 5 µL of 10× CutSmart digestion buffer. Digest for 1–2 h at 37°C.

    *Alternatively, pool the vector application reactions into one tube and perform the digestion in a single tube, scaling reagents appropriately. This approach is more convenient during large amplification after the first test amplification has been made.*

    *Users can also pool the library reactions into a separate tube in this step as well.*

11. Purify the vector and library DNA separately using a PCR product-purification kit, according to the supplier's instructions.

    *The product is stable for years, preferably at −80°C but also at −20°C.*

12. Determine the DNA concentration by gel electrophoresis and with appropriate (high or low DNA, and concentration ladder) mass ladders as references by repeating Step 9, or by measuring absorbance at 260 nm.

13. Perform a small-scale test of the products in Step 11 to verify complete digestion before Step 14, using sufficient amount of DNA to see the result on a gel and depending on how much DNA can be spared.

    i. Set up a reaction containing 1 µg of cleaved and purified vector DNA, 5 µL of CutSmart buffer, and 10–20 U (minimum of 1 µL) of each restriction enzyme. Bring up to a final volume of 50 µL with sterile H₂O.

    ii. Incubate for at least 3 h at 37°C.

    iii. Verify the digestion by gel electrophoresis as described in Step 9.

    *If the vector appears to be incompletely linearized or contains impurities as low molecular bands, see Troubleshooting.*

14. Purify the cleaved vector and affibody DNA library by using a PCR purification kit, according to the supplier's recommendations. Separately pool the eluates of inserts and vectors. Determine the DNA concentration as described in Step 9.

    *Mix DNA with the provided binding buffer from the kit and apply it to the column before washing with ethanol and eluting.*

    *For the QIAquick column from QIAGEN, the DNA-binding capacity is 10 µg per column, and several columns will commonly be required for purification when a full library is prepared. Calculate the number of columns needed from the result in Step 12.*

15. Test the optimal ligation ratio for the insert and vector. Perform ligation at various molar ratios before the large-scale reaction in Step 17.

    i. Prepare separate 20 µL reactions, each containing 50 ng of vector, insert DNA at varying molar ratios (1:3, 1:5, 1:6; vector:insert), 4000 U of T4 ligase, 10× reaction buffer (to 1×), and sterile H₂O.

    *The different vector:insert ratios are used to determine the optimal one.*

    ii. As control, include a reaction without library DNA, to assess potential vector re-ligation.

    iii. Incubate reactions overnight at 16°C in a thermocycler.

    *The product can be frozen until use in Step 16.*

Chapter 15

16. Do a small-scale transformation to assess optimal ligation conditions before proceeding to large-scale reactions.

   i. Thaw electrocompetent bacterial cells on ice and prewarm transformation media according to the supplier's recommendations.

   ii. Add 25 µL of electrocompetent cells into prechilled cuvettes. Add 2 µL of the ligated DNA from Step 15 to the cells.

   iii. Transform the DNA into the electrocompetent cells by electroporation according to the supplier's instructions.

   iv. Immediately add prewarmed recovery medium (provided by the supplier) up to a final volume of 1 mL. Transfer to microcentrifuge tubes.

   v. Incubate for 1 h at 37°C with shaking at 150 rpm, or as recommended by the supplier.

   vi. Plate 100 µL of the cells on TBAB carbenicillin plates at different dilutions, from $10^1$ to $10^6$, to assess transformation efficiency. Incubate overnight at 37°C.

   vii. Inspect plates the next day and count plates containing approximately 100 colonies or the $10^1$ plate if few colonies are yielded. Choose the condition with the highest number of colonies. Ensure that the vector only reaction is not yielding any colonies. If colonies are obtained in the control, repeat the process from Step 10 and onward.

17. After verifying that insert and vector can be ligated and everything is correctly designed, proceed to scale up the library. Calculate the yield from each step and the final amount obtained in Step 14. Repeat Steps 7–14, and multiply reactions in all steps by the number of reactions needed to obtain ∼100 µg ligated vector with the chosen ligation molar ratio determined from Step 16vii. Double-digest the large-scale amplified DNA affibody library and vector in separate 500 µL reactions. Scale number of reactions depending on how many ng of DNA is available of both vector and library insert.

   i. Add 10 µg of DNA, 10 U of each RE (pAffi1 and pBadv2.2, XhoI and NheI; pScZ2, XhoI and SacI), 50 µL of 10× CutSmart digestion buffer, and sterile H$_2$O to a final volume of 500 µL.

   ii. Incubate for at least 3 h at 37°C.

   iii. Verify the digestion by gel electrophoresis as described in Step 12.

   iv. Purify DNA as in Step 14, and determine concentration as in Step 12.

      *If the vector background is >4%, see Troubleshooting.*

18. Ligate library and vector DNA with T4 ligase at an appropriate molar ratio, as determined in Step 16vii. Scale the number of reactions depending on the amount of purified and cleaved PCR product yielded in Step 17; prepare approximately 16–20 reactions. Start with one small test reaction to assess vector background (<4% of the total number of colonies of ligated vector with library insert) as in Steps 15–16, before continuing with the full library. If vector background is high, redo vector digestion.

   i. For 500 µL reactions, add 5 µg of vector, insert in the optimal molar ratio as decided in Step 16vii, 10,000 U of T4 ligase, 10× Cutsmart buffer (to 1×), and sterile H$_2$O.

      *Approximately 80 µg of vector DNA is recommended to be ligated for library size of $10^8$–$10^{11}$ clones, and depending on the selection system.*

   ii. As control, include a reaction without library DNA to assess potential vector re-ligation.

   iii. Incubate reactions overnight at 16°C in a heat block. Check vector background and ligation as described in Step 16 before proceeding.

19. Purify and concentrate the ligated DNA by ethanol precipitation.

   i. Pool together all ligation reactions. Add 0.1× of the reaction volume of sodium acetate and 2.5× the reaction volume of ice-cold 96% (v/v) ethanol. Mix and incubate for 1 h at −80°C.

   ii. Pellet the precipitated DNA by centrifugation at maximum speed (e.g., 15,000 rpm) for 30 min at 4°C.

# Engineering of Affibody Molecules

iii. Discard the supernatant by pipetting. Wash the pellet with ice-cold 70% (v/v) ethanol and repeat the centrifugation.

iv. Discard the supernatant by pipetting and let the DNA pellet air-dry at room temperature.

v. Dissolve the DNA in sterile $H_2O$ to a concentration of 200–500 ng/µL.

20. Do a small-scale transformation to assess transformation efficiency and optimal DNA amount for the number of reactions needed to reach the desired library size before performing large-scale transformations in Step 21.

    i. Thaw cells on ice according to the supplier's recommendations. Prewarm media per suppliers' recommendation.

    ii. Add 25–50 µL of electrocompetent cells into prechilled cuvettes. Add 100–600 ng of DNA from Step 19 to the cells.
    
    *Perform several electroporation reactions with varying amounts of DNA.*

    iii. Transform the DNA into the electrocompetent cells by electroporation according to the supplier's instructions.

    iv. Immediately add prewarmed recovery medium (provided by the supplier) up to a final volume of 1 mL. Transfer to microcentrifuge tubes.

    v. Incubate for 1 h at 37°C with shaking at 150 rpm, or as recommended by the supplier.

    vi. Plate 100 µL on TBAB carbenicillin plates at different dilutions, from $10^1$ to $10^6$, to assess transformation efficiency. Incubate overnight at 37°C.

    vii. Inspect plates the next day and count plates containing approximately 100 colonies or the $10^1$ plate if few colonies are yielded. Calculate the number of transformations needed to create the desired library and the transformation efficiency as cfu/µg of transformed ligated DNA. Choose the condition with the best transformation efficiency in relation to the number of reactions to create the library.

21. Transform the ligated and purified DNA from Step 19 into electrocompetent cells according to the supplier's instructions and using the optimal conditions determined in Step 20.

    *To ensure high transformation efficiency and to improve recovery of the transformed library, do not prepare more than 10–20 transformations at a time. Depending on the library size desired and efficiency, a $10^9$ sized library may need approximately 10–100 transformations.*

22. Immediately after transformation, pool the cell suspensions from 10–20 transformations into an Erlenmeyer shake flask of 10× the transformed volume and incubate for 1 h at 37°C with shaking at 150 rpm or as recommended by the supplier. Repeat Steps 20 and 21 until the desired library size is covered as calculated in Step 20.

23. After 1 h of incubation, pool the transformed cells. Take out 100 µL of the pool and titer the library by plating 100 µL of serial dilutions ($10^3$–$10^6$) in the recommended medium (LB medium for *E. coli*, TSB for phage, and TSB + Y for *S. carnosus* display) on TBAB carbenicillin plates using a glass spreader or beads. Incubate overnight at 37°C.

24. Use the remaining cell pool to inoculate 10× the culture volume of the recommended cultivation medium (LB medium for *E. coli*, TSB for phage, and TSB + Y for *S. carnosus* display) supplemented with carbenicillin, and, in case of phage vector, 2% (w/v) glucose, into new Erlenmeyer flasks. Incubate for 16 h at 37°C with shaking at 150 rpm.

25. The next day, measure the optical density at 600 nm ($OD_{600}$) of the cultures to estimate cell density. Calculate the library complexity from the titration plates from Step 23, and calculate the copy number of the library obtained as the ratio of the number of cells estimated by the optical density over the complexity of the library.

    *The relationship between $OD_{600}$ and the number of cells (cfu/mL) can be obtained from the cell supplier and varies by E. coli strain.*

26. Harvest the cells by centrifugation at 2500g for 15 min at 4°C.
    - For phage and *E. coli* display, pool cells and prepare glycerol stocks by mixing the cells with sterile glycerol to a final concentration of 15% (v/v) by pipetting. Distribute in 1-mL aliquots and store at −80°C in cryovials or microcentrifuge tubes.
    - For *S. carnosus* display, pool cells and use 0.25× of the culture volume to prepare glycerol stocks by mixing the cells with sterile glycerol to a final concentration of 15% (v/v) by pipetting. Distribute in 1-mL aliquots and store at −80°C in cryovials or microcentrifuge tubes. For the remaining *S. carnosus* cell culture, isolate plasmid using a plasmid DNA Maxiprep kit, according to the supplier's instructions. Dissolve the final DNA pellet in sterile $H_2O$ to a final concentration of >1 µg/µL (preferably >5 µg/µL). Determine the plasmid concentration as described in Step 12. Store until further use at −20°C.

    *Plasmid DNA can be stored for years at −20°C.*

27. Determine the viable cell concentration of the frozen glycerol stock by plating 100 µL of serial dilutions ($10^6$–$10^8$) in the appropriate medium (LB for *E. coli*, TSB for phage, and TSB + Y for *S. carnosus* display) on TBAB carbenicillin plates using a glass spreader or beads. Incubate overnight at 37°C.

    *Calculate the viable cell concentration from the titration numbers in the serial dilution, using plates with approximately 100 colonies. The viable concentration is the number of colonies on the plate multiplied by the dilution factor and divided by 100 µL of plated volume, and should be converted to cfu/mL. The viable cell concentration should cover the library complexity and is used to calculate the volume to inoculate during library amplification in the first step of the selection procedure.*

28. Analyze the library fragment length distribution and ligation efficiency by colony PCR screening and DNA sequencing. Use colonies from the titration plates in Step 27. Screen approximately 96–288 colonies in a 96-well plate format. Include the empty vector as control.

    i. Prepare PCR reactions of 25 µL/well, using 0.75 U of low-fidelity polymerase, 10× DreamTaq polymerase buffer (to 1×), 200 µM dNTPs, and 5 pmol of forward and reverse primers to amplify the affibody region of the vector in 35 cycles:

    | 1 cycle   | 3 min  | 95°C        |
    | --------- | ------ | ----------- |
    | 35 cycles | 30 sec | 95°C        |
    |           | 30 sec | Primer $T_m$ |
    |           | 1 min  | 72°C        |
    | 1 cycle   | 10 min | 72°C        |
    | 1 cycle   | ∞      | 4°C         |

    *Cycle parameters depend on the polymerase and primers. Those above are for DreamTaq.*

    ii. Verify amplification by gel electrophoresis as described in Step 9 using the GeneRuler DNA Ladder Mix. Calculate the percentage of affibody containing colonies in the library to background.

    iii. Identify the DNA library distribution by Sanger sequencing or high-throughput sequencing.

## Determination of Bead-Binding Capacity of Biotinylated Target Proteins for Selection

29. Homogenize streptavidin-coated magnetic beads for 5 min at room temperature and resuspend by gentle pipetting.

30. Add 30 µL of streptavidin-coated magnetic beads to each of two microcentrifuge tubes and 60 µL of beads to a third microcentrifuge tube. Wash each tube four times at room temperature by adding 500 µL of PBST, capturing the beads on the magnet rack for 1 min, and aspirating the supernatant.

31. Perform a bead-binding test of the biotinylated target. To one of the tubes with 30 µL of beads, add 1.5 µg of biotinylated protein and add PBST to a final volume of 300 µL. Incubate with end-over-end rotation for 1 h at room temperature.

32. Capture beads on the magnet for 1 min and transfer the supernatant to the other 30 µL bead aliquot. Repeat the incubation described in Step 31 and capture the beads for 1 min at room temperature on the magnet.

33. Transfer the supernatant to the tube containing the 60 µL bead aliquot. Repeat the incubation described in Step 31 and capture the beads on the magnet for 1 min at room temperature. Save the supernatant from the last bead incubation in a fourth microcentrifuge tube.

34. Take a new aliquot of the protein (1.5 µg) into a fresh tube. Dry the supernatant from the last tube and this new aliquot of the protein in a heat block at 96°C with open lids.

    *This usually takes 20–30 min. Dry until <10 µL of the volume is left. The new protein aliquot is used as a reference for protein amount captured by the beads.*

35. Dissolve the dried samples from Step 34 and the three tubes with beads in a final volume of 10 µL of PBS. Add protein loading dye to all tubes, and boil for 10 min at 96°C. Load the whole mix from each tube (including beads) onto an SDS-PAGE protein gel along with a protein ladder and run the samples at 210 V for 30 min in running buffer as recommended by the supplier.

36. Stain the gel with GelCode Blue Safe Protein Stain for 1 h at room temperature followed by destaining in deionized $H_2O$ overnight and analyze the results by eye.

    *Streptavidin is released from the beads upon boiling and is detectable as monomers at ~14 kDa. If no bands at 14 kDa are observed, the protein has not been released properly from the beads. Repeat the experiment and ensure beads are boiled properly in Step 35. Inspect if the 1.5 µg protein is captured by the first, first and second, or first through third bead sample tubes by comparing to the protein band size from the reference 1.5 µg aliquot. Ideally, no protein should be detected in the supernatant sample, as it indicates poor binding. Repeat the biotinylation step of the protein (if in-house labeled) if poor binding is achieved, or consider changing vendor. Calculate the average capture of the protein as µg/mL beads.*

    *The library is now created and the bead-binding capacity for the target protein in selection is established. Proceed with selections as described in Protocol 2: Selection of Affibody Molecules Using Phage Display (Hjelm et al. 2023), Protocol 3: Selection of Affibody Molecules Using* Escherichia coli *Display (Dahlsson Leitao et al. 2023), or Protocol 4: Selection of Affibody Molecules Using Staphylococcal Display (Löfblom et al. 2023), depending on the display vector used. The referenced protocols cover the expansion of the library to use in selection and how to perform a selection campaign toward the selected target protein.*

## TROUBLESHOOTING

*Problem (Step 13):* The vector is incompletely linearized or contains impurities as low molecular bands.
*Solution:* In case of incomplete linearization of the vector or impurities, isolate the digested DNA fragments by agarose gel extraction and purification with a DNA gel extraction kit (e.g., QIAGEN 28704), according to the supplier's recommendations.

*Problem (Step 17):* The vector background is >4%.
*Solution:* Ensure enzymes are stored properly. In case of supercoiled plasmid, evidenced by DNA electrophoresis as bands of lower expected mass, heat the plasmid for 20 min at 70°C before cleavage. Recleave vector if background is high, consider retreating the vector with DnpI enzyme as described in Step 10.

Chapter 15

# RECIPES

### EDTA

EDTA (ethylenediamenetetraacetic acid)
NaOH

To prepare EDTA at 0.5 M (pH 8.0): Add 186.1 g of disodium EDTA•2H$_2$O to 800 mL of H$_2$O. Stir vigorously on a magnetic stirrer. Adjust the pH to 8.0 with NaOH (∼20 g of NaOH pellets). Dispense into aliquots and sterilize by autoclaving. The disodium salt of EDTA will not go into solution until the pH of the solution is adjusted to ∼8.0 by the addition of NaOH.

### LB Liquid Medium II

| Reagent | Quantity (for 1 L) |
| --- | --- |
| NaCl | 5 g |
| Bacto Tryptone | 10 g |
| Yeast extract | 5 g |

Dissolve in 1 L of distilled H$_2$O; sterilize by autoclaving.

### Molten Agarose

1. Add either 0.8 or 1.5 g of agarose (Fisher Scientific BP160) to 100 mL of 1× TAE <R> electrophoresis running buffer in a 250-mL glass bottle to make a 0.8% or 1.5% solution, respectively.
2. Microwave until boiling starts, and then carefully swirl the bottle.
3. Continue microwaving at reduced power with occasional swirling until the agarose is fully dissolved.
4. Cool to 60°C before use.

    *The molten 0.8% agarose can be stored for up to 1 wk at 60°C, but the 1.5% agarose should be used fresh. Single-use aliquots of either can be stored for up to 6 mo at room temperature, and then microwaved to remelt.*

### Phosphate-Buffered Saline (PBS), pH 7.4

0.1 M sodium phosphate buffer (pH 7.4) <R>
0.15 M NaCl

Filter sterilize with a 0.45-µm filter. Store at room temperature.

### Phosphate-Buffered Saline with Tween (0.1%) (PBST)

| Reagent | Final concentration |
| --- | --- |
| NaCl | 150 mM |
| Na$_2$HPO$_4$ | 8 mM |
| NaH$_2$PO$_4$ | 2 mM |
| Tween 20 | 0.1% (w/v) |

Adjust to pH 7.4 with NaOH or HCl and filter-sterilize with a 0.45-µm filter. Store at room temperature.

## Sodium Acetate (3 M, pH 5.5)

1. Dissolve sodium acetate (Sigma-Aldrich S2889) in $H_2O$ to a concentration of 3 M.
2. Adjust pH to 5.5 with acetic acid.
3. Filter-sterilize (0.2 µm).

*Store for up to 1 yr at room temperature.*

## Sodium Phosphate

### 1 M sodium phosphate buffer (pH 6.0–7.2)

Mixing 1 M $NaH_2PO_4$ (monobasic) and 1 M $Na_2HPO_4$ (dibasic) stock solutions in the volumes designated in the table below results in 1 L of 1 M sodium phosphate buffer of the desired pH. To prepare the stock solutions, dissolve 138 g of $NaH_2PO_4 \cdot H_2O$ (monobasic; m.w. = 138) in sufficient $H_2O$ to make a final volume of 1 L and dissolve 142 g of $Na_2HPO_4$ (dibasic; m.w. = 142) in sufficient $H_2O$ to make a final volume of 1 L.

| Volume (mL) of 1 M $NaH_2PO_4$ | Volume (mL) of 1 M $Na_2HPO_4$ | Final pH |
|---|---|---|
| 877 | 123 | 6.0 |
| 850 | 150 | 6.1 |
| 815 | 185 | 6.2 |
| 775 | 225 | 6.3 |
| 735 | 265 | 6.4 |
| 685 | 315 | 6.5 |
| 625 | 375 | 6.6 |
| 565 | 435 | 6.7 |
| 510 | 490 | 6.8 |
| 450 | 550 | 6.9 |
| 390 | 610 | 7.0 |
| 330 | 670 | 7.1 |
| 280 | 720 | 7.2 |

### 0.1 M sodium phosphate buffer (pH 7.4)

Add 3.1 g of $NaH_2PO_4 \cdot H_2O$ and 10.9 g of $Na_2HPO_4$ (anhydrous) to distilled $H_2O$ to make a volume of 1 L. The pH of the final solution will be 7.4. This buffer can be stored for up to 1 mo at 4°C.

### 0.1 M sodium phosphate buffer (from 1 M stocks) at 25°C

To prepare 1 L of 0.1 M sodium phosphate buffer of the desired pH, the following mixtures should be diluted to 1 L (final volume) with $H_2O$.

| pH | Volume (mL) of 1 M $Na_2HPO_4$ | Volume (mL) of 1 M $NaH_2PO_4$ |
|---|---|---|
| 5.8 | 7.9 | 92.1 |
| 6.0 | 12.0 | 88.0 |
| 6.2 | 17.8 | 82.2 |
| 6.4 | 25.5 | 74.5 |
| 6.6 | 35.2 | 64.8 |
| 6.8 | 46.3 | 53.7 |
| 7.0 | 57.7 | 42.3 |
| 7.2 | 68.4 | 31.6 |
| 7.4 | 77.4 | 22.6 |
| 7.6 | 84.5 | 15.5 |
| 7.8 | 89.6 | 10.4 |
| 8.0 | 93.2 | 6.8 |

Chapter 15

*TAE*

Prepare a 50× stock solution in 1 L of $H_2O$:
  242 g of Tris base
  57.1 mL of acetic acid (glacial)
  100 mL of 0.5 M EDTA (pH 8.0)

The 1× working solution is 40 mM Tris-acetate/1 mM EDTA.

*Tryptic Soy Broth (TSB)*

1. Dissolve 30 g of TSB powder (BD 211825) in a final volume of 1 L of $H_2O$ and autoclave for 15 min at 121°C under standard conditions.

2. (Optional) Once cooled to room temperature, add (if required and immediately before use) the necessary antibiotics to the desired concentration from a 100×–1000× stock solution.

3. Store the medium without antibiotics for up to several months at room temperature.

*Tryptose Blood Agar Base (TBAB) Plates*

1. Mix 40 g of blood agar base (Merck 1.10886.0500) with 1 L of deionized $H_2O$.
2. Autoclave at 15 PSI for 15 min at 121°C.
3. (Optional) Cool to 50°C, and then add antibiotics as needed. Swirl the bottle to ensure even distribution of the antibiotics throughout the agar.
4. Aliquot into Petri dishes.

   *Store upside-down for up to weeks at 4°C, preferably in the dark because of light sensitivity of antibiotics.*

*TSB + Y Medium*

| Reagent | Amount |
| --- | --- |
| TSB powder (Merck 105458) | 30 g |
| Yeast extract (Merck Y1625) | 5 g |

1. Dissolve TSB powder and yeast extract in a final volume of 1 L of $H_2O$ and autoclave at 15 PSI for 20 min at 121°C.
2. (Optional) Once cooled to room temperature, add (if required and immediately before use) the necessary antibiotics to the desired concentration from a 1000× stock solution.

   *Store the medium without antibiotics for up to several months at room temperature.*

## ACKNOWLEDGMENTS

The research on affibodies and their selection systems was funded by the Swedish Agency for Innovation VINNOVA (2019/00104 and the CellNova Center 2017/02105), the Knut and Alice Wallenberg Foundation through the Wallenberg Center for Protein Technology (KAW 2019.0341 and KAW 2021.0197), the Swedish Cancer Society (grants 201090 PjF; 222023 Pj01H), the Swedish Research Council (2019-05115), the Swedish Brain Foundation (grants FO2018-0094, FO2021-0407, FO2022-0253), the Tussilago foundation (FL-0002.025.551-7), the Strategic Research Area Neuroscience (StratNeuro), as well as the Schörling Family Foundation via the Swedish FTD Initiative.

## REFERENCES

Andersson KG, Persson J, Ståhl S, Löfblom J. 2019. Autotransporter-mediated display of a naïve affibody library on the outer membrane of *E. coli*. *Biotechnol J* 14: e1800359. doi:10.1002/biot.201800359

Dahlsson Leitao C, Hjelm LC, Ståhl S, Löfblom J, Lindberg H. 2023. Selection of affibody molecules using *Escherichia coli* display. *Cold Spring Harb Protoc* doi:10.1101/pdb.prot108400

Grönwall C, Jonsson A, Lindström S, Gunneriusson E, Ståhl S, Herne N. 2007. Selection and characterization of affibody ligands binding to Alzheimer amyloid ß peptides. *J Biotechnol* 128: 162–183. doi:10.1016/j.jbiotec.2006.09.013

Hjelm LC, Dahlsson Leitao C, Ståhl S, Löfblom J, Lindberg H. 2023. Selection of affibody molecules using phage display. *Cold Spring Harb Protoc* doi:10.1101/pdb.prot108399

Löfblom J, Hjelm LC, Dahlsson Leitao C, Ståhl S, Lindberg H. 2023. Selection of affibody molecules using staphylococcal display. *Cold Spring Harb Protoc* doi:10.1101/pdb.prot108401

Wernérus H, Ståhl S. 2002. Vector engineering to improve a staphylococcal surface display expression system. *FEMS Microbiol Lett* 212: 47–54. doi:10.1016/S0378-1097(02)00689-4

# Protocol 2

# Selection of Affibody Molecules Using Phage Display

Linnea Charlotta Hjelm, Charles Dahlsson Leitao, Stefan Ståhl, John Löfblom, and Hanna Lindberg[1]

*Department of Protein Science, KTH Royal Institute of Technology, 106 91 Stockholm, Sweden*

Affibody molecules are small (6-kDa) affinity proteins generated by directed evolution for specific binding to various target molecules. The first step in this workflow involves the generation of an affibody library. This is then followed by amplification of the library, which can then be used for biopanning using multiple methods. This protocol describes amplification of affibody libraries, followed by biopanning using phage display and analysis of the selection output. The general procedure is mainly for selection of first-generation affibody molecules from large naive (unbiased) libraries, typically yielding affibody hits with affinities in the low nanomolar range. For selection from affinity maturation libraries with the aim of isolating variants of even higher affinities, the procedure is similar, but parameters such as target concentration and washing are adjusted to achieve the proper stringency.

## MATERIALS

It is essential that you consult the appropriate Material Safety Data Sheets and your institution's Environmental Health and Safety Office for proper handling of equipment and hazardous materials used in this protocol.

RECIPES: Please see the end of this protocol for recipes indicated by <R>. Additional recipes can be found online at http://cshprotocols.cshlp.org/site/recipes.

### Reagents

3,3′,5,5′-tetramethylbenzidine (TMB) substrate solution (Thermo Fisher Scientific N301)
Affibody library in the pAffi1 vector, as prepared in Protocol 1: Cloning of Affibody Libraries for Display Methods (Ståhl et al. 2023)
Anti-M13 bacteriophage antibody HRP (e.g., Sino Biological 11973-MM05T-H)
Carbenicillin (Thermo Scientific 10177012) (100 mg/mL stock in sterile $H_2O$, filter-sterilized, 0.2-μm)

*Stock can be stored in the freezer until first use (up to several years). Store opened aliquots for up to 6 mo at 4°C. Use at 100 μg/mL for all experiments.*

Dummy-expressing phagemid (i.e., pAffi1 without affibody library)
*Escherichia coli* (e.g., XL1-Blue, Agilent 200228)
*E. coli* cells harboring the pAffi1 affibody library, prepared as described in Protocol 1: Cloning of Affibody Libraries for Display Methods (Ståhl et al. 2023)
EDTA (2 M, pH 8.0) <R>
D-glucose (20%, w/v, in water, filter-sterilized, 0.2-μm)
Dynabeads M-280 Streptavidin (Invitrogen 11205D)
GelRed Nucleic Acid Gel stain (Millipore SCT123)

---

[1]Correspondence: hanli@kth.se

© 2026 Cold Spring Harbor Laboratory Press
Cite this protocol as *Cold Spring Harb Protoc*; doi:10.1101/pdb.prot108399

Glycerol (85%, v/v, in water, sterilized by autoclaving)
Glycine-HCl (0.1 M, pH 3.0) <R>
High and low DNA mass ladder (for concentration determination, e.g., MassRuler Express forward DNA ladder mix; Thermo Scientific [low ladder, SM1283; high ladder, SM0311])
Human serum albumin (HSA, Sigma-Aldrich SRP6182) in PBS (5 µg/mL, filter-sterilized, 0.2-µm)
IPTG (Thermo Scientific 34060) (0.5 M, filter-sterilized, 0.2-µm)

*Use at 0.1 M for all experiments.*

Kanamycin stock (for example, 1000× Kanamycin stock <R>)

*Use at 25 µg/mL for all cultivation experiments and 50 µg/mL for TBAB plates.*

Low-fidelity polymerase and DreamTaq polymerase buffer (10×) (e.g., DreamTaq polymerase, Thermo Scientific EP0702)
M13K07 helper phage (e.g., NEB N0315S)
Milli-Q $H_2O$, autoclaved 20 psi, 15 min
Molten agarose <R>

*Prepare a 1% (w/v) agarose gel.*

PEG–NaCl <R>, ice-cold
Phosphate-buffered saline (PBS), pH 7.4 <R>
Phosphate-buffered saline with 1% BSA (PBSB) <R>
Phosphate-buffered saline with Tween (0.1%) (PBST) <R>
Phosphate-buffered saline with Tween and BSA (PBSTB) <R>, freshly prepared

*Prepare solutions with 1%, 3%, and 5% (w/v) of BSA.*

PMSF (Thermo Scientific 36978) (1 M, filter-sterilized, 0.2-µm)
Sulfuric acid ($H_2SO_4$, 2 M, sterile)
Target protein; biotinylated, preferably stored at −80°C until use as by supplier's recommendation.

*Protein amount varies depending on the molecular weight; ∼100 µg is sufficient for a 75-kDa protein.*

Tetracycline (Sigma-Aldrich T3258) (10 mg/mL stock solution in absolute ethanol, filter-sterilized, 0.2-µm)

*Store stock for up to 6 mo at −20°C in the dark.*

*Use at 10 µg/mL for all experiments.*

Tris-HCl (1 M, pH 8.0)
Tris-HCl (25 mM, pH 7.4)
Tryptic soy broth (TSB) <R>
Tryptose blood agar base (TBAB) plates <R>

*Prepare TBAB plates with 100 µg/mL carbenicillin, 10 µg/mL tetracycline, and 1% (w/v) D-glucose using the relevant stocks in square BioAssay dishes. Store plates up to 1 wk at 4°C in the dark.*

*Prepare TBAB plates with 50 µg/mL kanamycin using the kanamycin stock solution in regular Petri dishes. Store for up to 1 mo at 4°C.*

*Prepare TBAB plates with 100 µg/mL carbenicillin using 100 mg/mL carbenicillin in regular Petri dishes. Store for up to 2 wk at 4°C.*

TSB + Y medium <R>
Virkon (Thermo Scientific 12328667)

## Equipment

Aspirator
BioAssay dishes (square, 245-mm × 245-mm; e.g., Corning 431111)
DNA electrophoresis equipment (Bio-Rad Sub-Cell GT Horizontal Electrophoresis System 1704483, Bio-Rad Sub-Cell GT Casting Gates 1704415, and Bio-Rad PowerPac Basic Power Supply 1645050, Bio-Rad GelDoc EZ System 1708270)

Centrifugation tubes (sterile; GSA, GS3)
: *Autoclave tubes before use.*

Centrifuge (refrigerated)
End-over-end rotamixer for 1.5-mL microcentrifuge tubes, set to 15 rpm
Filters (1.2-μm and 0.45-μm)
Flasks (50, 500-mL; 1, 5-L; unbaffled)
: *Use flasks with volume 10× larger than the culture.*

Freezer at −80°C
Glass spreader or glass beads (sterile)
: *Choose according to the user's preference.*

Ice
Incubator set at 37°C (shaking and static)
Magnetic rack for 1.5-mL microcentrifuge tubes
Microcentrifuge (refrigerated)
Microcentrifuge tubes (1.5-mL)
Multichannel pipette
Paper towels
Petri dishes (10-cm)
pH stick
Plate reader for measuring absorbance at 450 nm
Plates (deep-well, 96-well)
Plates (low protein binding, F-bottom, clear, 96- or 384-well; e.g., Greiner Bio-One 781901)
Spectrophotometer
Transfer pipette (2-mL)

## METHOD

*This method is divided into various sections. In Steps 1–13, we describe the amplification of a phagemid affibody library contained within a glycerol stock of bacteria. In Steps 14–32, the biopanning process toward a biotinylated protein is described, which is repeated for three to six cycles. Included is a negative-preselection step to remove potential binders to materials used in selection. Amplification of the new phage eluate is described in Steps 33–37, to be used in new selection cycles. Finally, In Step 38, the analysis of the selection output is described.*

*Phages spread easily to other cultures and between selection campaigns. Only work with phages in specific areas or rooms, to minimize the risk of infecting other cultures. Wipe pipettes with ethanol before every use. Wipe materials that have been present in phage work areas with ethanol if transferring to nonphage areas, as during cultivation. Bleach or virkonise bench before and after phage work.*

*Prepare media and agar plates in advance of starting the protocol.*

1. Use the phage library, containing the phagemid library vector and displaying the affibody (Fig. 1), from Protocol 1: Cloning of Affibody Libraries for Display Methods (Ståhl et al. 2023). Inoculate 100× the calculated library size by estimating the volume of glycerol stocks needed based on the determined viable cell concentration to cover the library. In a 5-L unbaffled flask (10× the cultivation volume) containing 500 mL of TSB with carbenicillin, tetracycline, and 2% (w/v) glucose, inoculate the stock to yield a starting optical density at 600 nm ($OD_{600}$) of 0.1. Adjust the number of flasks inoculated depending on the required number of glycerol stocks and library size.

2. Grow cultures at 37°C with shaking at 150 rpm. Monitor the cultivation by measuring the $OD_{600}$ until it reaches 0.5–0.8.
   : *The growth should take ∼2–3 h, but can be longer if cell viability is low in the glycerol stock.*

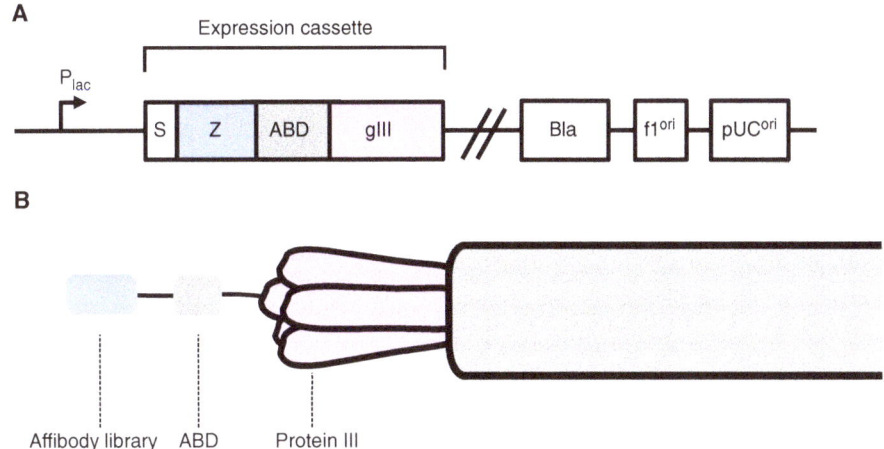

FIGURE 1. Construct displayed on bacteriophages containing the affibody library. (A) Schematic presentation of the coding regions of pAffi1 (Grönwall et al. 2007), showing the expression cassette encoding, under regulation of the *Escherichia coli lac* promoter ($p_{lac}$), a fusion protein comprising the OmpA leader sequence (S), the first six amino acids of native protein A in frame with the affibody library (Z), an albumin-binding domain (ABD), and the gene III protein (gIII). The gene fragment conferring resistance to ampicillin (Bla), the f1 origin of replication (f1$^{ori}$), and an *E. coli* origin of replication (pUC$^{ori}$) originate from pUC19 (Vieira and Messing 1987). (B) Schematic illustration of the expressed fusion protein attached to the phage. Note that the OmpA leader sequence is processed upon secretion.

3. Start an overnight culture of XL1-blue *E. coli* cells from a glycerol stock. Cultivate in TSB with tetracycline at 37°C, 150 rpm. The following morning, reinoculate in 50 mL TSB to an OD$_{600}$ of 0.1 AU. Grow until an OD$_{600}$ of 0.5–0.8 AU is reached.

   *The second inoculation should take 2–2.5 h to complete and should be timed so that the cells are ready to be used by Step 12.*

   *A culture at the right OD$_{600}$ can be stored on ice for 2 h without losing infectivity.*

4. Calculate the number of cells present at the measured OD$_{600}$ in each flask from Step 2, add 5× the number of cells of M13K07 helper phage, and swirl gently. Incubate for 25 min at 37°C statically, and then for 15 min at 37°C with shaking at 70 rpm.

   *The relationship between OD$_{600}$ and the number of cells (cfu/mL) can be obtained from the cell supplier and varies by E. coli strain.*

5. Harvest by centrifugation at 3500g for 10 min at 4°C. Discard the supernatant by decanting and resuspend the pellet from each flask in 5 mL of TSB + Y by gently pipetting.

6. Inoculate the resuspended pellet in the same volume of TSB + Y supplemented with carbenicillin as used in Step 1. Grow for 2 h at 37°C with shaking at 90 rpm. Add IPTG and kanamycin to final concentrations of 0.1 M and 25 µg/mL, respectively. Incubate for 16–18 h at 30°C with rotation at 90 rpm.

7. Pellet phage-infected cells by centrifugation at 4000g for 20 min at 4°C. Transfer the supernatant to a new centrifugation tube containing 0.25× the supernatant volume of ice-cold PEG–NaCl. Discard the pellet. Incubate for 1 h on ice.

   *The solution mixes efficiently by pouring the supernatant into the tube containing PEG–NaCl.*

8. Centrifuge at 15,000g for 45 min at 4°C. Discard the supernatant by decanting and recover the phage pellet by resuspending it in 10 mL per 100 mL of cultivation volume of a freshly prepared solution of ice-cold 25 mM Tris-HCl, 2 mM EDTA, and 1 mM PMSF.

9. Recentrifuge to remove the remaining bacteria as in Step 7. Precipitate phage by transferring the supernatant to a new tube containing 0.25× the supernatant volume of PEG–NaCl. If precipitation volume is <20 mL, incubate for 30 min on ice. Precipitate phage by centrifuging at 4500g for 40 min at 4°C.

   *If precipitation volume is >20 mL, follow precipitation centrifugation settings described in Step 8.*

Chapter 15

10. Discard the supernatant by decanting and resuspend the phage pellet in PBS (or PBSTB if continuing with selection directly) by pipetting. Resuspend phage to a stock concentration of $10^{11}$–$10^{13}$ plaque-forming units (pfu)/mL. Prewet 1.2-μm and 0.45-μm filters with PBS and then pass the suspension sequentially through the 1.2-μm and 0.45-μm filters.

    *A concentration of $10^{11}$ pfu/mL is yielded by overnight cultivation of phage. Calculate the resuspension volume from the used cultivation volume.*

    *Phage concentrations above $10^{13}$ pfu/mL will lead to formation of a gel that is difficult to dissolve.*

11. Prepare glycerol stocks of the phage that will not be used for selection the next day or for titration in Step 12 by mixing phage with glycerol to a final concentration of 40% (v/v) glycerol. Store at −80°C.

    *A selection should be started with a phage aliquot covering the library complexity 10×–100×. A 1 mL phage aliquot of $10^{11}$–$10^{13}$ pfu/mL is usually efficient for selection.*

12. Titer the phage in dilution series (library preparation or input, $10^1$–$10^{12}$; eluate, $10^1$–$10^6$) in Milli-Q $H_2O$ by infecting 90 μL of *E. coli* cells (Step 2) at an $OD_{600}$ of 0.5–0.8 with 10 μL of phage. Using a glass spreader or beads, spread the mixture on both TBAB carbenicillin and TBAB kanamycin plates overnight at 37°C.

    *Kanamycin titers allow estimation of superinfection rates and of the quality of the produced phage stock. Carbenicillin titers yields phagemid containing colonies. Calculate the phage concentration from titrations by calculating the number of colonies on a dilution plate containing ∼100 cfu. Multiply the number of colonies by the dilution factor and divide by 10, as 10 μL of phage were used in infection. This yields the number of pfu/mL in the phage stock.*

    *If the kanamycin titers are more than 10× the number of carbenicillin titer, see Troubleshooting.*

13. Analyze library amplification quality or selection output by colony polymerase chain reaction (PCR) and DNA sequencing, as described in Protocol 1: Cloning of Affibody Libraries for Display Methods (Ståhl et al. 2023).

14. One day before selections (described in Steps 14–32), start an overnight culture of XL1-blue *E. coli* cells from a glycerol stock, as described in Step 2. The following day, on the day of selection, measure $OD_{600}$ and inoculate an aliquot to asset up a starting culture with an $OD_{600}$ of 0.1 in a 500 mL flask containing 50 mL of TSB with tetracycline. This culture will be used for infection and titration. Grow at 37°C with shaking at 150 rpm until the $OD_{600}$ reaches 0.5–0.8, and keep on ice until use for a maximum of 2 h.

    *Growth to expected $OD_{600}$ should take 2–2.5 h to complete. Cells can be kept on ice for a maximum of 2 h when in log phase.*

    *If possible, time the start of inoculation so that the culture is ready to use in Step 33.*

15. Block the number of microcentrifuge tubes needed for preselection, selection, and washes by adding 0.5 mL of PBSTB with 1% (w/v) BSA and incubating them for >1 h at room temperature, with end-over-end rotation at 15 rpm.

16. Thaw one or more vials of phage on ice, prepared in Step 11. If proceeding with freshly prepared phage from Step 10 continue directly with Step 19. Precipitate phage from phage-glycerol stocks from Step 11, by adding 0.25× of the starting volume of ice-cold PEG–NaCl and mix by gentle pipetting. If volumes up to 20 mL are used, follow centrifugation instructions in Step 9 and proceed to Step 18. If volumes of 1 mL are used in microcentrifuge tubes, follow Step 17.

    *Precipitate an amount of phage that covers the affibody library complexity between 10× and 100×. For later selection rounds, the same titer in all panning experiments, for example, $10^{12}$ pfu in 1 mL, can be used.*

17. Incubate the microcentrifuge tube with phage for 30 min on ice, followed by centrifugation at 13,000g for 30 min at 4°C in a microcentrifuge.

18. Discard the supernatant by decanting and dissolve the phage pellet in PBSTB with 3% (w/v) BSA to a concentration of $10^{11}$–$10^{13}$ pfu/mL, in a maximal volume of 1 mL.

    *Use preblocked tubes (Step 15) for all further steps.*

*Calculate volume to dissolve in from phage titers as determined in Step 12.*

19. Prepare magnetic beads for a negative preselection.

    i. Homogenize the magnetic beads for 5 min at room temperature, on the bench. Gently resuspend by pipetting. Transfer 200 µL of beads to a preblocked microcentrifuge tube and wash twice by adding 1 mL of PBST, capturing beads on the magnet for 1 min, and aspirating the supernatant.

    ii. Block by adding 1 mL of PBSTB with 5% (w/v) BSA and incubate for 1 h at room temperature with end-over-end rotation at 15 rpm. These will be used in Step 24. Aspirate the blocking solution immediately before use.

    *The same number of beads used for selection is used during the negative selection in a minimum of 50 µL. If the solution continues to be brown from the beads, continue adsorption of beads on the magnetic rack until clear. Always keep beads in solution.*

20. Prepare beads for the positive selection, based on the bead-binding capacity of the biotinylated target as determined in Protocol 1: Cloning of Affibody Libraries for Display Methods (Ståhl et al. 2023).

    *The density of protein bound per bead can be decreased with each selection round to minimize avidity effects.*

21. Dilute the target protein in PBST.

    *Calculate a molar excess of target compared to the number of phage particles in the selection, with a recommended target concentration of 150–300 nM in the first round. Increase stringency in further rounds by lowering concentration or incorporating, for instance, off-rate selection strategies.*

22. Wash beads twice, as described in Step 19.i. Precapture target onto beads by incubating the beads in 500 µL of PBST containing the target protein (from Step 21) with end-over-end rotation at 15 rpm for 1 h at room temperature.

23. Recover all beads from Step 22 by capturing for ~1–5 min (until solution is clear) and discard the aspirated the supernatant. Post-block in PBSTB as described in Step 19ii. Keep the beads in blocking solution until the phage in Step 25 is ready to be added. Remove supernatant before use.

24. For negative selection, add phage ($1 \times 10^{13}$ pfu) to the blocked beads prepared in Step 19 in a final volume of 1 mL in PBSTB with 3% (w/v) BSA. Incubate for 1 h at room temperature with end-over-end rotation at 15 rpm.

25. Recover the supernatant from the beads of the negative selection by magnetic capture for 3–5 min (until solution is whitish). Add supernatant to beads prepared in Step 23. Incubate with end-over-end rotation at 15 rpm for 1 h at room temperature.

    *All brown beads should be captured. Complete capture of the beads to the magnet can take up to 5 min in highly concentrated phage solutions.*

    *Discard negative selection beads from Step 25 unless investigation of streptavidin binders is of interest.*

26. Perform magnetic bead capture 3–5 min until solution is whitish, and discard the supernatant by pipetting. Add 1 mL of PBST and transfer the beads and PBST mix to a new 1.5-mL preblocked tube for washing.

27. For the first panning round, incubate with end-over-end rotation at 15 rpm for 1 min at room temperature. Recover the beads by magnetic capture and discard aspired supernatant. Repeat the wash four times with PBST. Move samples to new tubes after the second and fourth wash. Perform the fifth wash with PBS.

    *Complete capture of the beads to the magnet can take up to 5 min in highly concentrated phage solutions. After three washes, a shorter capture time of 1 min is sufficient. Resuspend beads by gentle pipetting for each new wash.*

28. In the meantime, prepare a mix containing 450 µL of PBS and 50 µL of 1 M Tris-HCl pH 8.0.

29. After the fifth wash, transfer the sample to a new 1.5 mL tube and capture the beads for 1 min. Discard supernatant by aspiration. Elute phage bound to beads by adding 500 µL of 0.1 M glycine-

HCl. Gently mix by pipetting to resuspend beads and incubate statically for 10 min at room temperature.

*Alternative elution strategies such as trypsinization or use of desthiobiotinylated target with biotin can also be considered (Panagides et al. 2022).*

30. Neutralize the eluted phage by adding the PBS/Tris-HCl mixture prepared in Step 28. Capture beads on the magnet for 1 min. Transfer the supernatant to a new tube and discard the beads.

31. Check the pH neutralization of the reaction with a pH stick.

    *The pH should be 7.0–8.0.*

32. Save 50 µL of the eluate in a new tube, on ice, for the titration (to be done at a convenient time during the same day). Follow Step 12 for instructions on titration and use cells prepared in Step 14. Calculate eluate complexity from titer agar plates the next day and preform a PCR screening of selection output (Step 13) to ensure amplification of affibody containing phage. If not continuing directly with Step 33, prepare glycerol stocks by mixing the phage eluate with glycerol to a final concentration of 25% (v/v) glycerol and store for up to years at −80°C.

33. Add 20 mL of XL1-Blue cells from Step 14 to a 50-mL flask containing 750 µL of the phage eluate. Incubate for 25 min at 37°C statically, and then incubate for 15 min at 37°C with rotation at 70 rpm. Create glycerol stocks from remaining phage eluate, as described in Step 32.

    *It is desirable to have a 10×–100× excess of cells to phage.*

    *The cell and phage eluate volume can be decreased in subsequent rounds to 250–330 µL of phage eluate.*

34. Centrifuge cells at 3500g in a sterile centrifugation tube for 10 min at 4°C and remove the supernatant by aspiration. Gently resuspend the pellets in 1–2 mL of TSB and spread on square BioAssay Dish TBAB plates supplemented with carbenicillin, tetracycline, and 1% (w/v) glucose using a glass spreader or beads. Work under a flame to keep plate sterile.

35. The next day, recover the cells by adding ∼5 mL of TSB to the plate and scraping off the colonies with a sterile spreader. Measure $OD_{600}$ of the recovered cells. If not proceeding directly with Step 36, prepare glycerol stocks of cells by mixing them with glycerol at a final concentration of 15% (v/v) glycerol. Store at −80°C. If proceeding directly to Step 36, prepare glycerol stocks of the fraction of cells that is left after inoculation.

36. Calculate the number and volume of cells to be inoculated from the $OD_{600}$ measurement in Step 35: the number of cells should cover 100× eluate complexity (as calculated form Step 32). Inoculate the calculated volume of cells in a minimum of 50 mL of TSB with carbenicillin, tetracycline, and 2% (w/v) glucose to a starting $OD_{600}$ of 0.1 in a unbaffled flask of 10× the size of the cultivation volume. Grow at 37°C with shaking at 150 rpm until the $OD_{600}$ reaches 0.5–0.8. Continue with Steps 4–13.

    *The relationship between $OD_{600}$ and the number of cells (cfu/mL) can be obtained from the cell supplier and varies by E. coli strain.*

    *Growth should take 2–3 h. Longer cultivation times are common for the first rounds of selections.*

37. Repeat selection as described in Steps 14–36 for three to six cycles. In later selection cycles, target concentration and density of target on beads can be decreased to induce a selection pressure. Additionally, increase the time and number of washes used in Step 27 to introduce a selection pressure. Repeat selection until an enrichment of the phage eluate titers as determined in Step 32 in relation to the number of input phage is seen. An enrichment factor of above 3×–5× between two cycles should be observed. Additionally, the PCR selection should indicate an increase (up to 90%–100%) of affibody containing clones, as analyzed by gel electrophoresis (Step 13). The last cycles in the selection campaign promotes the enrichment of high affinity clones, and may be needed even if titers and PCR screening look good in an earlier cycle.

    *Other selection criteria, such as pH dependency, temperature, and off rate can be used (Zahnd et al. 2004; Pershad and Kay 2013; Seijsing et al. 2014).*

*If the PCR screening analysis reveals no affibody containing phages, or if phage eluate titers are low, see Troubleshooting.*

38. Analyze selection output from all cycles by colony PCR and analyze the last rounds of selection by thorough DNA sequencing, as described in Protocol 1: Cloning of Affibody Libraries for Display Methods [Ståhl et al. 2023]).

39. Perform a polyclonal ELISA on all selection outputs and starting library, and a monoclonal ELISA from the cycle with most positive clones from the polyclonal ELISA results.

    i. Analyze all polyclonal pools and monoclonal cultures in an ELISA for binding to BSA (1%, w/v, in PBS), HSA (5 µg/mL in PBS), and target antigen (1–10 µg/mL in PBS) all in the same 384-well microplate. Add 30 µL per well of antigen and incubate overnight at 4°C. Invert the plate, strike it on a paper towel, and wash once with PBST by pouring from the bottle. Block the plate with 60 µL of PBSTB with 5% (w/v) BSA per well for 1 h at room temperature with slight orbital rotation. Invert the plate and strike it on a paper towel to remove the blocking solution, and proceed to add phage.

    ii. In case of a polyclonal ELISA, use phage pools from each cycle at a final concentration of $10^6$–$10^8$ pfu/mL, diluted in PBST. For a monoclonal ELISA, cultivate single phage colonies and use supernatant from an overnight culture. Follow the procedure in Steps 1, 2, and 4–6, with cultivation in a 96-deep-well plate with 1 mL of culture volume. Spin down culture as Step 7 and use supernatant. Add 30 µL of phage stock or 10 µL of phage cultivation supernatant with an additional 20 µL of PBST with a multichannel pipette to a well containing each type of protein analyzed (prepared in Step 39i) in duplicate, and incubate for 1–2 h at room temperature with slow rotation.

        *Include controls in the ELISA for helper phage, buffer, cultivation medium, dummy-expressing phagemid, and, if available, phagemid clones against a different target.*

    iii. Wash the ELISA plate by pouring PBST over the plate five times. Invert and strike on a paper towel between washes.

    iv. Detect binding with 30 µL of anti-M13(HRP) monoclonal antibody. Incubate for 30 min at room temperature. Wash three times with PBST and once with PBS by pouring.

    v. Develop the signal by adding 30 µL of freshly mixed TMB substrate to each well according to the manufacturer's instructions. Let the color develop for 1–30 min at room temperature until a blue color is generated in the samples and background controls are only faintly blue. Add an equal volume of sulfuric acid to stop the TMB reaction and measure absorbance at 450 nm.

    vi. Perform a blank subtraction of absorbance values from those of cultivation medium wells. Normalize phage binding to target antigen to HSA expression when comparing the output values.

        *Clones binding BSA and SA wells to a higher degree than target well and HSA are sticky and should be discarded.*

## TROUBLESHOOTING

*Problem (Step 12):* The titer on kanamycin plates is 10× higher than phagemid titers (carbenicillin plates).
*Solution:* Consider producing a new phage batch by repeating Steps 1–12. Ensure a 5× molar ratio of infection M13K07 phage to cells in Step 4.

Chapter 15

*Problem (Step 37):* The PCR screening analysis reveals no affibody containing phages in eluate or very low phage eluate titers.

*Solution:* Consider going back one cycle in the selection campaign and reduce selection pressure by increasing target concentration in Step 21 and reducing time of washes in Step 27.

## DISCUSSION

A successful selection is indicated by enrichment of affibody clones as determined via colony PCR, enrichment of certain sequences as determined by DNA sequencing, and positive signals on phage ELISA. The polyclonal ELISA will indicate the phage binding pool for the selected target for each selection round. In a successful selection, the polyclonal ELISA signal will increase with the number of cycles performed, correlating with an enrichment of clones usually observed from phage titers. If low enrichment is seen in DNA sequencing, and a polyclonal ELISA shows a peak of binding that is not located in the last selection cycles, consider adding an extra panning round with mild conditions, to enrich for target protein binding clones. Perform the monoclonal ELISA on the last cycle(s) with high binding in the polyclonal ELISA. Select enriched clones from DNA sequencing with a positive binding profile in monoclonal ELISA for further characterization of the affibody. Continue with, for example, subcloning to a soluble format and analysis by circular dichroism, surface plasmon resonance, and other relevant characteristics for the target protein.

## RECIPES

### 1000× Kanamycin Stock

Kanamycin
Make kanamycin to a final concentration of 50 mg/ml in $H_2O$.

Filter-sterilize and store in aliquots at −20°C.

### EDTA (2 M, pH 8.0)

1. Dissolve EDTA in $H_2O$ to a concentration of 2 M.
2. Adjust pH to 8 with NaOH.
3. Sterilize by autoclaving at 15 PSI for 20 min at 121°C.

   *Store for up to 1 yr at 4°C.*

### Glycine-HCl (0.1 M, pH 3.0)

1. Dissolve glycine-HCl (Sigma-Aldrich G2879) in $H_2O$ to a concentration of 0.1 M.
2. Adjust the pH to 3.0 using HCl.
3. Filter with a 0.45-µm filter.

   *Store for up to 1 yr at room temperature.*

*Molten Agarose*

1. Add either 0.8 or 1.5 g of agarose (Fisher Scientific BP160) to 100 mL of 1× TAE <R> electrophoresis running buffer in a 250-mL glass bottle to make a 0.8% or 1.5% solution, respectively.
2. Microwave until boiling starts, and then carefully swirl the bottle.
3. Continue microwaving at reduced power with occasional swirling until the agarose is fully dissolved.
4. Cool to 60°C before use.

    *The molten 0.8% agarose can be stored for up to 1 wk at 60°C, but the 1.5% agarose should be used fresh. Single-use aliquots of either can be stored for up to 6 mo at room temperature, and then microwaved to remelt.*


*Phosphate-Buffered Saline (PBS), pH 7.4*

0.1 M sodium phosphate buffer (pH 7.4) <R>
0.15 M NaCl

Filter sterilize with a 0.45-µm filter. Store at room temperature.


*Phosphate-Buffered Saline with 1% BSA (PBSB)*

1. Combine the following salt components in distilled $H_2O$.

    | Reagent | Final concentration |
    | --- | --- |
    | NaCl | 150 mM |
    | $Na_2HPO_4$ | 8 mM |
    | $NaH_2PO_4$ | 2 mM |

2. Adjust salt components and distilled $H_2O$ to reach pH 7.4 by using NaOH or HCl for a final volume of 1 L.
3. Autoclave the solution at 15 PSI for 20 min at 121°C on a liquid cycle. Verify the pH again.
4. Dissolve BSA (e.g., SaveenWerner B2001-500) to final concentration of 1% (w/v). Adjust the volume to 1 L and sterilize with a 0.2-µm filter. Prepare fresh before use. Keep at 4°C.


*Phosphate-Buffered Saline with Tween (0.1%) (PBST)*

| Reagent | Final concentration |
| --- | --- |
| NaCl | 150 mM |
| $Na_2HPO_4$ | 8 mM |
| $NaH_2PO_4$ | 2 mM |
| Tween 20 | 0.1% (w/v) |

Adjust to pH 7.4 with NaOH or HCl and filter-sterilize with a 0.45-µm filter. Store at room temperature.

## Chapter 15

### Phosphate-Buffered Saline with Tween and BSA (PBSTB)

1. Combine the following salt components in distilled $H_2O$.

   | Reagent | Final concentration |
   |---|---|
   | NaCl | 150 mM |
   | $Na_2HPO_4$ | 8 mM |
   | $NaH_2PO_4$ | 2 mM |

2. Adjust salt components and distilled $H_2O$ to reach pH 7.4 by using NaOH or HCl for a final volume of 1 L.
3. Autoclave the solution at 15 PSI for 20 min at 121°C on a liquid cycle. Verify the pH again.
4. Add Tween 20 and dissolve BSA to final concentrations of 0.1% (w/v) and 5% (w/v), respectively. Adjust the volume to 1 L and sterilize with a 0.2-µm filter. Prepare fresh before use. Keep at 4°C.

   *Different concentrations of BSA (1%–5%, w/v) are used.*

5. When required, dilute PBSTB (5% BSA) with PBST to prepare solutions with lower concentrations of BSA.

### PEG–NaCl

| Reagent | Amount |
|---|---|
| PEG600 | 200 g |
| NaCl | 2.5 M (final concentration) |

1. Completely dissolve PEG600 and NaCl in a final volume of 1 L of $dH_2O$ with brief heating to 65°C.
2. Filter-sterilize with a 0.2-µm filter.
3. Autoclave at 15 PSI for 20 min at 121°C on a liquid cycle.

   *Store the solution for up to 1 yr at 4°C.*

## Sodium Phosphate

### 1 M sodium phosphate buffer (pH 6.0–7.2)

Mixing 1 M $NaH_2PO_4$ (monobasic) and 1 M $Na_2HPO_4$ (dibasic) stock solutions in the volumes designated in the table below results in 1 L of 1 M sodium phosphate buffer of the desired pH. To prepare the stock solutions, dissolve 138 g of $NaH_2PO_4 \cdot H_2O$ (monobasic; m.w. = 138) in sufficient $H_2O$ to make a final volume of 1 L and dissolve 142 g of $Na_2HPO_4$ (dibasic; m.w. = 142) in sufficient $H_2O$ to make a final volume of 1 L.

| Volume (mL) of 1 M $NaH_2PO_4$ | Volume (mL) of 1 M $Na_2HPO_4$ | Final pH |
|---|---|---|
| 877 | 123 | 6.0 |
| 850 | 150 | 6.1 |
| 815 | 185 | 6.2 |
| 775 | 225 | 6.3 |
| 735 | 265 | 6.4 |
| 685 | 315 | 6.5 |
| 625 | 375 | 6.6 |
| 565 | 435 | 6.7 |
| 510 | 490 | 6.8 |
| 450 | 550 | 6.9 |
| 390 | 610 | 7.0 |
| 330 | 670 | 7.1 |
| 280 | 720 | 7.2 |

### 0.1 M sodium phosphate buffer (pH 7.4)

Add 3.1 g of $NaH_2PO_4 \cdot H_2O$ and 10.9 g of $Na_2HPO_4$ (anhydrous) to distilled $H_2O$ to make a volume of 1 L. The pH of the final solution will be 7.4. This buffer can be stored for up to 1 mo at 4°C.

### 0.1 M sodium phosphate buffer (from 1 M stocks) at 25°C

To prepare 1 L of 0.1 M sodium phosphate buffer of the desired pH, the following mixtures should be diluted to 1 L (final volume) with $H_2O$.

| pH | Volume (mL) of 1 M $Na_2HPO_4$ | Volume (mL) of 1 M $NaH_2PO_4$ |
|---|---|---|
| 5.8 | 7.9 | 92.1 |
| 6.0 | 12.0 | 88.0 |
| 6.2 | 17.8 | 82.2 |
| 6.4 | 25.5 | 74.5 |
| 6.6 | 35.2 | 64.8 |
| 6.8 | 46.3 | 53.7 |
| 7.0 | 57.7 | 42.3 |
| 7.2 | 68.4 | 31.6 |
| 7.4 | 77.4 | 22.6 |
| 7.6 | 84.5 | 15.5 |
| 7.8 | 89.6 | 10.4 |
| 8.0 | 93.2 | 6.8 |

## Chapter 15

*TAE*

Prepare a 50× stock solution in 1 L of $H_2O$:
  242 g of Tris base
  57.1 mL of acetic acid (glacial)
  100 mL of 0.5 M EDTA (pH 8.0)

The 1× working solution is 40 mM Tris-acetate/1 mM EDTA.

*Tryptic Soy Broth (TSB)*

1. Dissolve 30 g of TSB powder (BD 211825) in a final volume of 1 L of $H_2O$ and autoclave for 15 min at 121°C under standard conditions.
2. (Optional) Once cooled to room temperature, add (if required and immediately before use) the necessary antibiotics to the desired concentration from a 100×–1000× stock solution.
3. Store the medium without antibiotics for up to several months at room temperature.

*Tryptose Blood Agar Base (TBAB) Plates*

1. Mix 40 g of blood agar base (Merck 1.10886.0500) with 1 L of deionized $H_2O$.
2. Autoclave at 15 PSI for 15 min at 121°C.
3. (Optional) Cool to 50°C, and then add antibiotics as needed. Swirl the bottle to ensure even distribution of the antibiotics throughout the agar.
4. Aliquot into Petri dishes.

   *Store upside-down for up to weeks at 4°C, preferably in the dark because of light sensitivity of antibiotics.*

*TSB + Y Medium*

| Reagent | Amount |
|---|---|
| TSB powder (Merck 105458) | 30 g |
| Yeast extract (Merck Y1625) | 5 g |

1. Dissolve TSB powder and yeast extract in a final volume of 1 L of $H_2O$ and autoclave at 15 PSI for 20 min at 121°C.
2. (Optional) Once cooled to room temperature, add (if required and immediately before use) the necessary antibiotics to the desired concentration from a 1000× stock solution.

   *Store the medium without antibiotics for up to several months at room temperature.*

## ACKNOWLEDGMENTS

The research on affibodies and their selection systems was funded by the Swedish Agency for Innovation VINNOVA (2019/00104 and the CellNova Center 2017/02105), the Knut and Alice Wallenberg Foundation through the Wallenberg Center for Protein Technology (KAW 2019.0341 and KAW 2021.0197), the Swedish Cancer Society (grants 201090 PjF; 222023 Pj01H), the Swedish Research Council (2019-05115), the Swedish Brain Foundation (grants FO2018-0094, FO2021-0407, FO2022-0253), the Tussilago foundation (FL-0002.025.551-7), the Strategic Research Area Neuroscience (StratNeuro), as well as the Schörling Family Foundation via the Swedish FTD Initiative.

## REFERENCES

Grönwall C, Jonsson A, Lindström S, Gunneriusson E, Ståhl S, Herne N. 2007. Selection and characterization of Affibody ligands binding to Alzheimer amyloid β peptides. *J Biotechnol* **128**: 162–183. doi:10.1016/j.jbiotec.2006.09.013

Panagides N, Zacchi LF, De Souza MJ, Morales RAV, Karnowski A, Liddament MT, Owczarek CM, Mahler SM, Panousis C, Jones ML, et al. 2022. Evaluation of phage display biopanning strategies for the selection of anti-cell surface receptor antibodies. *Int J Mol Sci* **23**: 8470. doi:10.3390/ijms23158470

Pershad K, Kay BK. 2013. Generating thermal stable variants of protein domains through phage display. *Methods (San Diego, Calif)* **60**: 38–45. doi:10.1016/j.ymeth.2012.12.009

Seijsing J, Lindborg M, Höidén-Guthenberg I, Bönisch H, Guneriusson E, Frejd FY, Abrahmsén L, Ekblad C, Löfblom J, Uhlén M, et al. 2014. Engineered affibody molecule with pH-dependent binding to FcRn mediates extended circulatory half-life of a fusion protein. *Proc Natl Acad Sci* **111**: 17110–17115. doi:10.1073/pnas.1417717111

Ståhl S, Hjelm LC, Dahlsson Leitao C, Löfblom J, Lindberg H. 2023. Cloning of affibody libraries for display methods. *Cold Spring Harb Protoc* doi:10.1101/pdb.prot108398

Vieira M, Messing J. 1987. Production of single-stranded plasmid DNA. *Methods Enzymol* **153**: 3–11. doi:10.1016/0076-6879(87)53044-0

Zahnd C, Spinelli S, Luginbühl B, Amstutz P, Cambillau C, Plückthun A. 2004. Directed *in vitro* evolution and crystallographic analysis of a peptide-binding single chain antibody fragment (scFv) with low picomolar affinity. *J Biol Chem* **279**: 18870–18877. doi:10.1074/jbc.M309169200

# Protocol 3

# Selection of Affibody Molecules Using *Escherichia coli* Display

Charles Dahlsson Leitao, Linnea Charlotta Hjelm, Stefan Ståhl, John Löfblom, and Hanna Lindberg[1]

*Department of Protein Science, KTH Royal Institute of Technology, 106 91 Stockholm, Sweden*

Affibody molecules are small (6-kDa) affinity proteins generated by directed evolution for specific binding to various target molecules. The first step in this workflow involves the generation of an affibody library, which can then be used for selection via multiple display methods. This protocol describes selection from affibody libraries by *Escherichia coli* cell surface display. With this method, high-diversity libraries of $10^{11}$ can be displayed on the cell surface. The method involves two steps for selection of binders from high-diversity libraries: magnetic-activated cell sorting (MACS) and fluorescence-activated cell sorting (FACS). MACS is used first to enrich the library in target-binding clones and to decrease diversity to a size that can be effectively screened and sorted in the flow cytometer in a reasonable time (typically $<10^7$ cells). The protocol is based on methodology using an AIDA-I autotransporter for display on the outer membrane, but the general procedures can also be adjusted and used for other types of autotransporters or alternative *E. coli* display methods.

## MATERIALS

It is essential that you consult the appropriate Material Safety Data Sheets and your institution's Environmental Health and Safety Office for proper handling of equipment and hazardous materials used in this protocol.

RECIPES: Please see the end of this protocol for recipes indicated by <R>. Additional recipes can be found online at http://cshprotocols.cshlp.org/site/recipes.

### Reagents

Affibody library in pBAD2.2 vector, as prepared in Protocol 1: Cloning of Affibody Libraries for Display Methods (Ståhl et al. 2023)

Alexa Fluor 647-conjugated human serum albumin (HSA-AF647)

*Label 20 mg of HSA (Sigma-Aldrich A9511) with AF647 carboxylic acid succinimidyl ester (Invitrogen A37573) according to supplier's instructions and determine the concentration and degree of labeling in a spectrophotometer.*

Biotinylated target protein, prepared as in Protocol 1: Cloning of Affibody Libraries for Display Methods (Ståhl et al. 2023)

Carbenicillin (Thermo Scientific 10177012) (100 mg/mL stock in sterile $H_2O$, filter-sterilized, 0.2-μm)

*Stock can be stored in the freezer until first use (up to several years). Store opened aliquots for up to 6 mo at 4°C. Use at 100 μg/mL for all experiments.*

---

[1]Correspondence: hanli@kth.se

© 2026 Cold Spring Harbor Laboratory Press
Cite this protocol as *Cold Spring Harb Protoc*; doi:10.1101/pdb.prot108400

*Escherichia coli* library, prepared as in Protocol 1: Cloning of Affibody Libraries for Display Methods (Ståhl et al. 2023)

Glycerol (85%, v/v, sterilized by autoclaving)

L-(+)-arabinose (20%, w/v, in water, filter-sterilized, 0.2-μm)
*Use at 0.6% (w/v) for all experiments.*

LB liquid medium (e.g., LB liquid medium II <R>)

Low-fidelity DNA polymerase and DreamTaq polymerase buffer (10×) (e.g., DreamTaq polymerase, NEB EP0702)

PBSP buffer <R> (room-temperature and ice-cold)

Streptavidin-coated paramagnetic beads (e.g., Dynabeads MyOne Streptavidin C1, Thermo Fisher 65001)

Streptavidin, R-Phycoerythrin Conjugate (SA-PE) (Invitrogen S866)
*Use at 2 μg/mL in all experiments.*

Tryptose blood agar base (TBAB) plates <R>
*Prepare TBAB plates with 100 μg/mL carbenicillin from 100 mg/mL carbenicillin.*

## Equipment

Aspirator

Centrifuge (refrigerated)

Centrifuge tubes (50-mL)

Culture flasks (5, 1-L; 50, 10-mL)
*Use flasks with volume 10× larger than the culture.*

End-over-end rotamixer for 1.5-mL microcentrifuge tubes, set to 15 rpm

Flow cytometer (e.g., MoFlo Astrios, Beckman Coulter with band pass filters 660 and 575 nm for detection of HSA-AF647 with 640 nm laser; and SA-PE with 488 nm laser)

Flow cytometry tubes

Freezer at −80°C

Glass spreader or glass beads (sterile)
*Choose according to the user's preference.*

Ice

Incubator set at 37°C (static and shaking)

Magnetic rack for 50-mL centrifuge tubes and 1.5-mL microcentrifuge tubes

Microcentrifuge (refrigerated)

Microcentrifuge tubes (1.5-mL)

Spectrophotometer for measuring optical density at 600 nm ($OD_{600}$) and cuvettes

Thermocycler

Vortex

## METHOD

*This method describes the display of affibody libraries on the surface of E. coli cells, followed by screening and selection of binders using magnetic-activated cell sorting (MACS) and fluorescence-activated cell sorting (FACS). First, the library (prepared as a glycerol stock, as described in Protocol 1: Cloning of Affibody Libraries for Display Methods) is amplified and induced for protein surface expression using the AIDA-I autotransporter for display on the outer membrane (Fig. 1) (Steps 1–3). Thereafter, the displayed library is panned against a biotinylated protein of choice (in this case, target bound to beads, prepared as described in Protocol 1: Cloning of Affibody Libraries for Display Methods [Ståhl et al. 2023]), using MACS (Steps 4–15) followed by FACS (Steps 21–28). MACS is commonly used in two to three cycles to enrich the binding population in the library before screening and sorting of high-affinity candidates by FACS. Flow cytometry is used to estimate the degree of enrichment (Steps 19 and 20).*

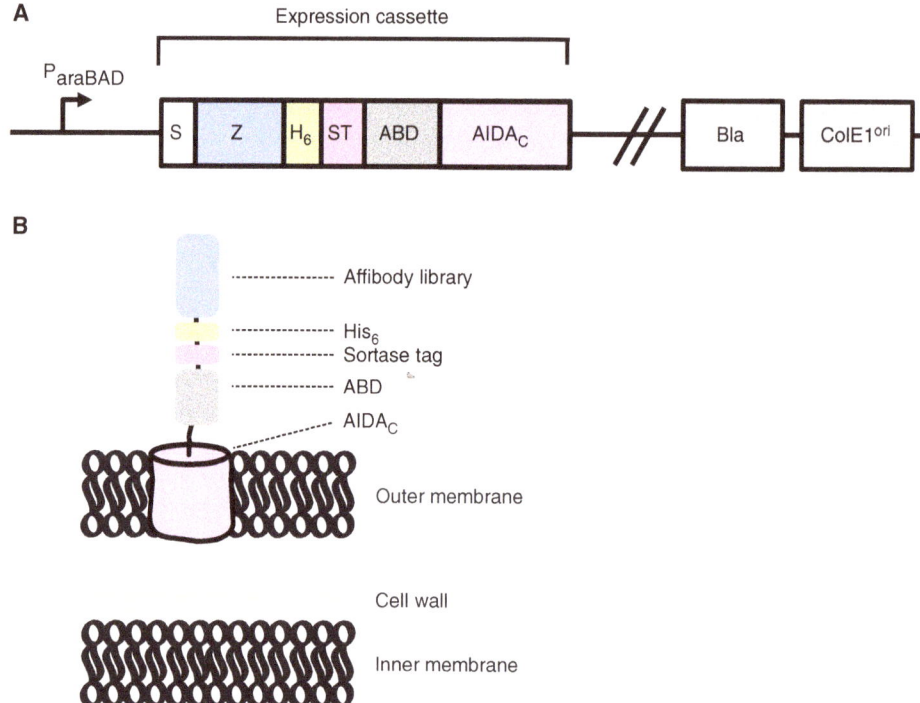

FIGURE 1. Construct displayed on *Escherichia coli* expressing the affibody library. (*A*) Schematic illustration of the expression cassette of the *E. coli* surface display vector (Andersson et al. 2019). The vector encodes, under control of the *araBAD* promoter ($P_{araBAD}$), a fusion protein consisting of the AIDA-I signal peptide (S), in frame with the affibody library (Z), a hexahistidyl peptide ($H_6$), a sortase A site (ST), an albumin-binding domain (ABD), and the β-barrel domain of AIDA-I ($AIDA_C$). (*B*) Schematic illustration of the expressed fusion protein attached to the outer membrane of the *E. coli* cell, with the signal peptide processed upon translocation to the outer cell surface.

### Culture Expansion and Induction of Protein Expression

1. With a pipette and sterile tips, inoculate 1×–100× the *E. coli* library titer from a glycerol stock into a 5-L flask containing 500 mL of LB medium supplemented with carbenicillin, to obtain a starting $OD_{600}$ of 0.1. Grow for 18–20 h at 37°C, with shaking at 150 rpm.

   *Several flasks might be needed to cover the library.*

   *The library complexity that was calculated in Step 27 from Protocol 1: Cloning of Affibody Libraries for Display Methods (Ståhl et al. 2023) is given in cfu/mL. By multiplying this number by, for instance, 100, it is possible to calculate the inoculation volume for 100-fold coverage of the library.*

2. From the overnight culture, reinoculate 1×–100× the library size into a 5-L flask containing 500 mL of LB medium supplemented with carbenicillin, to obtain a starting $OD_{600}$ of 0.1. Incubate the culture at 37°C, with shaking at 150 rpm.

3. Check the $OD_{600}$ using a spectrophotometer, and when it reaches 0.5, induce the cultures by adding arabinose to a final concentration of 0.6% (w/v). Incubate overnight at 25°C, with shaking at 150 rpm.

### MACS of *E. coli*-Displayed Affibody Libraries

4. The next day, prepare magnetic beads for both the negative and the positive selection.
   - For the negative selection, vigorously vortex streptavidin-coated beads and aliquot 100 µL in 1 mL of PBSP.
   - For the positive selection, prepare streptavidin-coated beads by coating them with biotinylated target in PBSP, based on the bead-binding capacity of the biotinylated target, as

described in Steps 29–36 of Protocol 1: Cloning of Affibody Libraries for Display Methods (Ståhl et al. 2023) or according to the supplier's recommendations.

5. Perform a magnetic capture of the beads for 1 min at room temperature using a large (commonly for the first round of MACS) or small (commonly for subsequent cycles) magnetic rack, and then aspirate the supernatant and wash in 1 mL of PBSP. Repeat the wash twice.
   - Transfer the beads for the negative selection to a 50-mL Falcon tube and add 20 mL of ice-cold PBPS and keep at 4°C until use.
     *Fifty-milliliter Falcon tubes are used for first selections.*
   - To the beads for the positive selection, precapture the biotinylated target on the beads as described in Steps 29–36 of Protocol 1: Cloning of Affibody Libraries for Display Methods (Ståhl et al. 2023), aiming for a coating density of ∼70%–100% (based on the bead-binding capacity). Prepare a number of beads for a bead-to-cell ratio of around 1:50 in the first rounds, and increase the proportion of beads in later rounds based on the flow-cytometric assessment of enrichment in Step 19. Incubate for 1 h at room temperature with rotation at 15 rpm. Immediately before positive selection, transfer beads to a 50-mL Falcon tube, wash beads with 20 mL of room-temperature PBSP, and resuspend in 20 mL of ice-cold PBSP in a 50-mL Falcon tube.

6. Measure the $OD_{600}$ of the library (from Step 3) and transfer cells covering the library 1×–100× to a 50-mL Falcon tube. Pellet cells by centrifugation at 4000$g$ for 10 min at 4°C and discard the supernatant by decanting. Wash the cells in 50 mL of ice-cold PBSP by gently vortexing until resuspended. Repeat the wash twice.
   *The $OD_{600}$ of the overnight culture should be >4.*

7. Perform a negative selection of the library by resuspending the pelleted cells in the 20 mL bead suspension from Step 5. Incubate with end-over-end rotation at 15 rpm for 30 min at room temperature.

8. Capture the beads in the negative selection on the large magnetic rack for 10 min at room temperature, rotating the magnet holder gently end-over-end twice. Collect the supernatant and transfer it to a new tube. Discard beads with captured cells.

9. Repeat the capture of beads from the supernatant by once again rotating the magnet holder gently end-over-end twice during the 10 min incubation at room temperature and collect the supernatant. Discard the captured beads. Save 10 µL of the supernatant on ice for the titration of the library input. Pellet the *E. coli* cells in the supernatant by centrifugation at 4000$g$ for 10 min at 4°C.

10. Resuspend the pelleted cells by pipetting with the 20 mL of target-coated magnetic beads to a final concentration of ∼$5 \times 10^{10}$ cells/mL. Incubate with end-over-end rotation at 15 rpm for 2 h at room temperature.

11. Capture beads with bound affibody–cell complexes on the magnet for 10 min at room temperature. Aspirate the supernatant and incubate on ice until titration.

12. Wash beads four times with 10 mL of ice-cold PBSP by inverting the tubes and incubating for 3 min on ice before magnetic capture for 10 min at room temperature. Centrifuge supernatants from the washes at 4000$g$ for 10 min at 4°C, collect the supernatants by pipetting, and save them on ice for the titration.

13. Resuspend the bead–affibody–cell output (from Step 12) in 50 mL of LB medium with carbenicillin and grow the cells in a 500-mL flask overnight at 37°C with shaking at 150 rpm.

14. Inoculate 10 mL of LB medium with carbenicillin with an aliquot of the unsorted library to use as a control for flow-cytometric analysis of the enrichment. Grow overnight with shaking at 150 rpm.

# Chapter 15

15. Titrate aliquots from all steps (Step 9, input; Step 10, unbound fraction; Step 12, washes; Step 13, output). To do this, prepare dilution series using LB medium (input, $10^4$–$10^6$; other, $10^6$–$10^9$), spread 100 µL of each on TBAB carbenicillin plates using a glass spreader or beads, and incubate overnight at 37°C.

16. The next day, inoculate cells from the overnight culture (Step 14) into a new small flask with 5 mL of LB medium with carbenicillin, yielding a starting $OD_{600}$ of 0.1, and repeat Step 2.
    *Cultures will be used for both new rounds of MACS and for flow-cytometric analysis of the enrichment.*

17. As controls, inoculate unsorted cells (Step 1) and cells from the output (Step 13) to small flasks with 5 mL LB medium with carbenicillin, without induction.

18. Prepare glycerol stocks of the library output (from Step 13) by mixing (pipetting or careful vortexing) cells with sterile glycerol to a final concentration of 15% (v/v) glycerol and store them at −80°C. Calculate the enrichment based on the colony number on the titration plates, as described in Step 27 of Protocol 1: Cloning of Affibody Libraries for Display Methods (Ståhl et al. 2023).

19. Analyze enrichment by flow cytometry:
    i. Transfer 10 µL of each culture (induced input and output from selection, non-induced input and output from selection) to 1 mL of PBSP in a 1.5-mL tube, mix, and centrifuge at 2000g for 4 min at 4°C.
    ii. Wash twice with 1 mL of PBSP by centrifuging and discarding the supernatant. After the washes, discard the supernatant using a pipette and resuspend cells by pipetting with 50 µL of PBSP containing 100 nM of biotinylated target.
    iii. Mix thoroughly and incubate for 1 h at room temperature with end-over-end rotation at 15 rpm.
    iv. Pellet cells by centrifugation (4000 rpm, 10 min, at 4°C), discard the supernatant with a pipette, and resuspend cells by careful pipetting with 1 mL of ice-cold PBSP. Centrifuge and resuspend by pipetting in 100 µL of PBSP containing 40 nM HSA-AF647 and 2 µg/mL SA-PE. Incubate for 30 min on ice in the dark.
    v. Wash twice by adding 1 mL of PBSP and centrifuging at 2000g for 4 min at 4°C. Resuspend the cells by pipetting in 200 µL of ice-cold PBSP, transfer to flow cytometry tubes, and keep them on ice in the dark until analysis.
    vi. Analyze simultaneous cell surface expression (monitored through HSA-AF647) and target binding (monitored through SA-PE) of $>5 \times 10^4$ events in a flow cytometer using appropriate excitation lasers and emission filters (HSA-AF647, 640-nm laser, 660-nm bandpass filter; SA-PE, 488-nm laser, 575-nm bandpass filter).

20. Based on the titrations in Step 18, perform additional rounds of MACS by repeating Steps 2–18. When the output titer corresponds to $10^6$–$10^7$ cfu, proceed to FACS for selection.
    *Typically, two or three rounds of MACS are used as preenrichment steps before FACS selection.*

## FACS of *E. coli*-Displayed Affibody Libraries

21. Inoculate 100 mL of LB carbenicillin in a 1-L flask with 10×–100× of the library output from the MACS selection (from Step 13 or 18), to obtain a starting $OD_{600}$ of 0.1. Repeat Steps 2–3.

22. Pipette 10×–100× the library output from the culture to a microcentrifuge tube, and wash as described in Steps 19i-iii. Label cells in 500 µL of PBSP containing 100 nM biotinylated target. Mix thoroughly and incubate with end-over-end rotation at 15 rpm for 1 h at room temperature.
    *Target concentration and volume, as well as time and temperature for the selection, will vary with library size and desired stringency.*

23. Proceed with washes and secondary incubations as described in Step 19iv, this time with volumes of 500 μL.

24. Wash the labeled cells (described in step 19v) but resuspend the pelleted cells in 600 μL of ice-cold PBSP by pipetting and transfer them to a flow cytometry tube. Keep on ice in the dark until flow-cytometric analysis.

25. Analyze simultaneous cell surface expression (monitored through HSA-AF647) and target binding (monitored through SA-PE) in a cell sorter using appropriate excitation lasers and emission filters (HSA-AF647, 640-nm laser, 660-nm bandpass filter; SA-PE, 488-nm laser, 575-nm bandpass filter). Gate cells showing surface expression and target binding, and screen cells corresponding to ∼10× the library size. Sort cells into 1 mL of LB medium.

26. Incubate the sorted cells with end-over-end rotation at 15 rpm for 1 h at 37°C. Inoculate them into 100 mL of LB medium supplemented with carbenicillin in a 1-L flask and incubate overnight at 37°C with shaking at 150 rpm.

    *Inoculation of sorted cells to smaller volumes of LB (down to 5 mL) could be beneficial in later rounds of FACS.*

27. Perform additional rounds of sorting by repeating Steps 21–26. Prepare glycerol stocks of the sorted cells by mixing them with sterile glycerol to a final concentration of 15% (v/v) by pipetting or careful vortexing. Store at −80°C.

28. After the last round of FACS, plate the sorted cells onto TBAB carbenicillin plates (aim for 100 cfu per plate) using a glass spreader or beads and incubate overnight at 37°C. Also sort cells directly into 1 mL LB media and repeat Steps 26–27.

29. Randomly pick colonies to a PCR plate to analyze library fragment length distribution and sequence enrichment by colony PCR screening and DNA sequencing, as described in Protocol 1: Cloning of Affibody Libraries for Display Methods (Ståhl et al. 2023).

## DISCUSSION

The possibility to combine MACS and FACS selections allows the user to identify high-affinity binders from large and complex libraries. FACS provides a great advantage to the method, as it can be used to monitor the selection procedure in real time and thus often indicates the success of the selection, which is further verified by colony screening and DNA sequencing of the sorted clones. Enriched clones with identical DNA profiles are often chosen for further on-cell characterization. In such characterization, individual clones are grown in 5 mL of LB media supplemented carbenicillin and induced, before labeling with target and secondary reagents and analysis in the flow cytometer. Ideally, candidates with highest target binding over expression level signals are chosen for more detailed characterization of the affibody molecule. Such characterization includes recloning to an expression vector, protein production, and analysis by circular dichroism, surface plasmon resonance, or other methods relevant for the purpose.

# RECIPES

### LB Liquid Medium II

| Reagent | Quantity (for 1 L) |
| --- | --- |
| NaCl | 5 g |
| Bacto Tryptone | 10 g |
| Yeast extract | 5 g |

Dissolve in 1 L of distilled $H_2O$; sterilize by autoclaving.

### PBSP Buffer

| Reagent | Amount or final concentration |
| --- | --- |
| Pluronic F108 NF surfactant (BASF) | 10 mL |
| NaCl | 150 mM |
| $Na_2HPO_4$ | 8 mM |
| $NaH_2PO_4$ | 2 mM |

1. Add reagents to sterile $H_2O$.
2. Adjust pH to 7.4 using HCl or NaOH.
3. Bring the volume to 1 L.
4. Filter-sterilize (0.2 μm).

   *Store for up to 2 mo at room temperature; discard if it becomes turbid.*

### Tryptose Blood Agar Base (TBAB) Plates

1. Mix 40 g of blood agar base (Merck 1.10886.0500) with 1 L of deionized $H_2O$.
2. Autoclave at 15 PSI for 15 min at 121°C.
3. (Optional) Cool to 50°C, and then add antibiotics as needed. Swirl the bottle to ensure even distribution of the antibiotics throughout the agar.
4. Aliquot into Petri dishes.

   *Store upside-down for up to weeks at 4°C, preferably in the dark because of light sensitivity of antibiotics.*

# ACKNOWLEDGMENTS

The research on affibodies and their selection systems was funded by the Swedish Agency for Innovation VINNOVA (2019/00104 and the CellNova Center 2017/02105), the Knut and Alice Wallenberg Foundation through the Wallenberg Center for Protein Technology (KAW 2019.0341 and KAW 2021.0197), the Swedish Cancer Society (grants 201090 PjF; 222023 Pj01H), the Swedish Research Council (2019-05115), the Swedish Brain Foundation (grants FO2018-0094, FO2021-0407, FO2022-0253), the Tussilago foundation (FL-0002.025.551-7), the Strategic Research Area Neuroscience (StratNeuro), as well as the Schörling Family Foundation via the Swedish FTD Initiative.

# REFERENCES

Andersson KG, Persson J, Ståhl S, Löfblom J. 2019. Autotransporter-mediated display of a naïve affibody library on the outer membrane of *Escherichia coli*. *Biotechnol J* **14:** e1800359. doi:10.1002/biot.201800359

Ståhl S, Hjelm LC, Dahlsson Leitao C, Löfblom J, Lindberg H. 2023. Cloning of affibody libraries for display methods. *Cold Spring Harb Protoc* doi:10.1101/pdb.prot108398

## Protocol 4

# Selection of Affibody Molecules Using Staphylococcal Display

John Löfblom, Linnea Charlotta Hjelm, Charles Dahlsson Leitao, Stefan Ståhl, and Hanna Lindberg[1]

Department of Protein Science, KTH Royal Institute of Technology, 106 91 Stockholm, Sweden

Affibody molecules are small (6-kDa) affinity proteins generated by directed evolution for specific binding to various target molecules. The first step in this workflow involves the generation of an affibody library, which can then be used for biopanning using multiple display methods. This protocol describes selection from affibody libraries using display on *Staphylococcus carnosus*. Display of affibodies on staphylococci is very efficient and straightforward because of the single cell membrane and the use of a construct with a constitutive promoter. The workflow involves display of affibody libraries on the surface of *S. carnosus* cells, followed by screening and selection of binders using fluorescence-activated cell sorting (FACS). The transformation of DNA libraries into *S. carnosus* is less efficient and more complicated than for *Escherichia coli*. Because of this, staphylococcal display is suitable for affinity maturation or other protein-engineering efforts that are not dependent on very high diversity, and thus magnetic-activated cell sorting (MACS) is often not required before FACS. However, MACS is an option, and MACS procedures used for *E. coli* can easily be adapted for use in *S. carnosus* if needed.

## MATERIALS

It is essential that you consult the appropriate Material Safety Data Sheets and your institution's Environmental Health and Safety Office for proper handling of equipment and hazardous materials used in this protocol.

RECIPES: Please see the end of this protocol for recipes indicated by <R>. Additional recipes can be found online at http://cshprotocols.cshlp.org/site/recipes.

### Reagents

Affibody library plasmid pScZ2[$Z_{Library}$] (>1 µg/µL), prepared as in Protocol 1: Cloning of Affibody Libraries for Display Methods (Ståhl et al. 2023)

B2 medium <R>, at room temperature or prewarmed to 37°C

Biotinylated target protein, prepared as in Protocol 1: Cloning of Affibody Libraries for Display Methods (Ståhl et al. 2023)

Chloramphenicol (Sigma-Aldrich C0378) (20 mg/mL in sterile $H_2O$, filter-sterilized, 0.2-µm)
  *Stock solution is stable for years at −20°C.*
  *Use at 10 µg/mL for all experiments.*

Glycerol (10%, v/v, and 85%, v/v, in water, sterilized by autoclaving, ice-cold)

Low-fidelity DNA polymerase and DreamTaq polymerase buffer (10×) (e.g., DreamTaq polymerase, NEB EP0702)

---

[1]Correspondence: hanli@kth.se

© 2026 Cold Spring Harbor Laboratory Press
Cite this protocol as *Cold Spring Harb Protoc*; doi:10.1101/pdb.prot108401

Chapter 15

PBSP buffer <R>
*Staphylococcus carnosus* TM300 cells (Götz 1990), streaked on a TBAB plate (Götz lab)
Sucrose (0.5 M)/glycerol <R> (10%, v/v, in water)
Tryptose blood agar base (TBAB) plates <R>
> Prepare TBAB plates as described in the recipe.
>
> Prepare TBAB plates with 10 µg/mL chloramphenicol using 20 mg/mL chloramphenicol. Can be stored for up to a few weeks at +4°C.

TSB + Y medium <R>

## Equipment

Autoclave
Centrifuge (refrigerated and room-temperature)
Cold room
Dry ice
Electroporation cuvettes (1-mm-gap, prechilled)
End-over-end rotamixer for 1.5-mL microcentrifuge tubes, set to 15 rpm
Flow cytometer (e.g., MoFlo Astrios, Beckman Coulter)
Flow cytometry tubes (Falcon 734-0000)
Freezer at −80°C
GSA tubes (Beckman 368454)
Heating block filled with $H_2O$ at 56°C
Ice
Incubator (static and shaking) set at 37°C
Microcentrifuge
Microcentrifuge tubes (1.5-mL)
Micropulser electroporator (e.g., Bio-Rad MicroPulser)
Shake flasks (500-mL and 5-L)
> Use flasks with volume 10× larger than the culture volume.

Spectrophotometer measuring optical density at 578 nm ($OD_{578}$) and cuvettes
Thermocycler
Vortex

## METHOD

*This method describes the display of affibody libraries on the surface of Gram-positive S. carnosus cells, followed by screening and selection of binders using fluorescence-activated cell sorting (FACS). First, electrocompetent S. carnosus TM300 cells are prepared (Steps 1–11), followed by transformation of the DNA library (Steps 12–21, prepared as described, for instance, in Protocol 1: Cloning of Affibody Libraries for Display Methods [Ståhl et al. 2023]). Display of affibodies on staphylococci is very efficient and straightforward because of the single cell membrane in this organism and constitutive promoter used (Fig. 1). The library is subsequently panned against a biotinylated protein of choice (target bound to beads, prepared, for instance, as described in Protocol 1: Cloning of Affibody Libraries for Display Methods [Ståhl et al. 2023]) using FACS [Steps 22–30]).*

### Preparation of Electrocompetent *S. carnosus* TM300 Cells

1. Autoclave two 500-mL shake flasks containing 60 mL of B2 medium, two 5-L shake flasks containing 500 mL of B2 medium, six GSA centrifugation tubes, 1 L of 10% (v/v) glycerol, and 3 L of distilled $H_2O$ all at 15 PSI for 20 min at 121°C.
   > *The high salt concentration of B2 medium gives a selective advantage to S. carnosus over many other bacteria such as E. coli and reduces the risk of contamination.*
   >
   > *One of the flasks will serve as a control for contamination, used only in Step 2.*

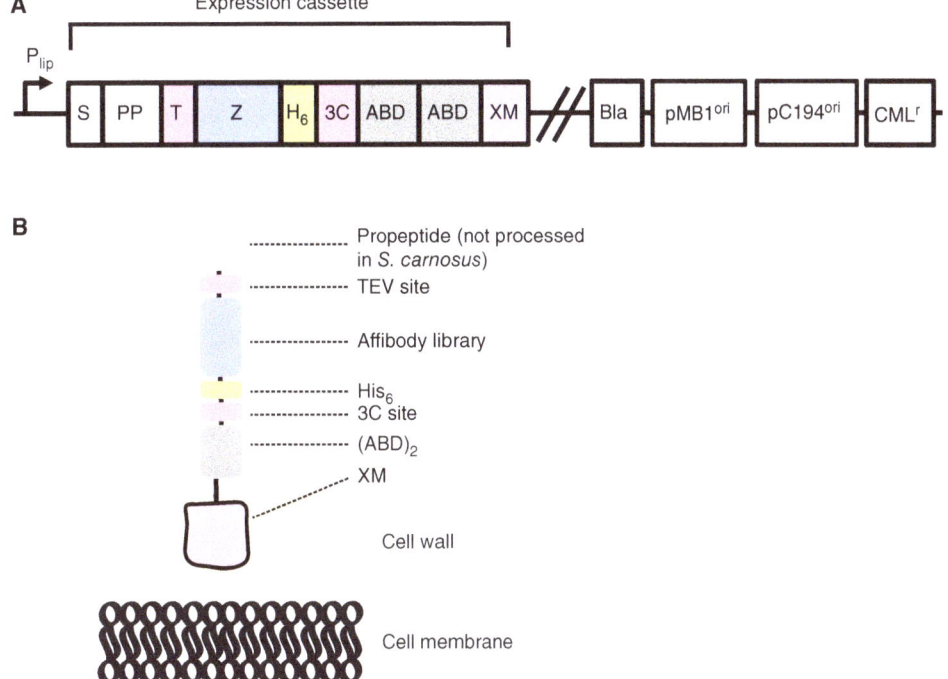

FIGURE 1. Construct displayed on the surface Staphylococcus carnosus containing the affibody library. (A) Schematic illustration of the expression cassette of the S. carnosus surface display vector (Kronqvist et al. 2008; Löfblom et al. 2017). The vector encodes, under control of the Staphylococcus hyicus lipase promoter ($P_{lip}$), a fusion protein consisting of the S. hyicus secretion signal (S) and propeptide (PP) in frame, a TEV protease cleavage site (T), the affibody library (Z), a hexahistidyl peptide ($H_6$), a 3C protease cleavage site (3C), two albumin-binding domains (ABD), and the cell wall–anchoring regions of staphylococcal protein A (XM). (B) Schematic illustration of the expressed fusion protein attached to the peptidoglycan layer of the S. carnosus cell wall, with the signal peptide processed upon secretion to the outer cell surface, while the propeptide is not processed.

2. Using a sterile pipette tip, inoculate a single colony of S. carnosus TM300 from a freshly streaked plate into a flask containing 60 mL of B2 medium. Prepare another flask with media but without cells, as a control. Incubate both flasks for at least 20 h at 37°C, with shaking at 150 rpm.

3. Precool distilled water, 10% (v/v) glycerol, GSA tubes, rotor, vortex, sterile pipette tips and 1.5-mL centrifugation tubes in the cold room overnight.

4. The next day, measure the $OD_{578}$ of the overnight culture from Step 2 using a spectrophotometer. Inoculate cells from the overnight culture into two flasks, each with 500 mL of B2 medium, to obtain a starting $OD_{578}$ of 0.5. Incubate at 37°C, with shaking at 150 rpm.

    *$OD_{578}$ of the overnight culture should be >8. The control should show no growth.*

5. When the $OD_{578}$ of the new cultures reaches ~4, immediately place cultures on ice for 15 min in the cold room to stop cell growth. Swirl the cultures once during this time.

    *It takes ~3–4 h for the culture to reach $OD_{578}$ ~ 4.*

6. In the cold room, distribute the cultures to six sterile GSA tubes and centrifuge at 3000g for 10 min at 4°C. Decant the supernatant and completely resuspend the pellets in the residual medium by careful vortexing. Dissolve each pellet in a small volume (50–100 mL) of $H_2O$. Make sure it is homogenous before adding more $H_2O$, until a final volume of 1 L has been distributed over the six tubes.

    *Carry out Steps 6–9 in the cold room.*

    *Perform all cell-washing steps with sterile ice-cold $H_2O$ or 10% (v/v) glycerol.*

7. Centrifuge at 4000g for 10 min at 4°C and repeat the wash with a total of 1 L of $H_2O$ distributed over the six GSA tubes. Centrifuge at 4000g for 10 min at 4°C. Carefully decant the supernatant and resuspend pellets by careful vortexing in a total of 540 mL of $H_2O$ across all tubes.

8. Centrifuge at 5000g for 10 min at 4°C. Carefully decant the supernatant and resuspend pellets by careful vortexing in a total of 540 mL of 10% (v/v) glycerol (across all six tubes), and pool into three GSA tubes.

9. Centrifuge at 5500g for 10 min at 4°C and carefully decant the supernatant. Resuspend the pellets by careful vortexing in a total of 180 mL of 10% (v/v) glycerol (across all three tubes), and pool into one tube.

10. Centrifuge at 5500g for 10 min at 4°C and carefully decant the supernatant. Add 5 mL of 10% (v/v) glycerol and resuspend carefully by vortexing. Prepare 240 μL aliquots, snap-freeze on dry ice, and quickly transfer to −80°C for storage until electroporation of the library (see the section Electroporation of Affibody Libraries into *S. carnosus*)

    *Each tube of cells can be used for four electroporation reaction, as each transformation requires 60 μL of cells.*

11. Use one of the frozen cell aliquots to prepare serial dilutions in B2 medium ($10^6$–$10^8$), and plate 100 μL of each dilution on TBAB plates using a glass spreader or beads. Incubate overnight at 37°C and count colonies the next day. Determine the viable cell concentration of the stock (cfu/mL) by multiplying the colony number by the dilution factor, and divide it by the 100 μL of plated volume. This number is used for inoculation and amplification of the library during selection.

    *The cell density should be ~$10^{11}$ cells/mL.*

## Electroporation of Affibody Libraries into *S. carnosus*

*Prior to large-scale library transformation, assess the transformation efficiency and number of reactions needed to reach the desired library size by carrying out small-scale pilot experiments, based on Steps 12–16 and 18. To generate libraries of $10^7$, approximately 100 electroporation reactions are required.*

12. Thaw the library plasmid (from Protocol 1: Cloning of Affibody Libraries for Display Methods [Ståhl et al. 2023]), and electrocompetent *S. carnosus* cells for 5 min on ice, followed by incubation of cells for 25 min at room temperature.

    *Preferably, do not prepare more than 20 transformations at a time, to retain the transformation efficiency of the untransformed cells and to boost the recovery of the transformed library.*

13. Heat cells for 90 sec in an $H_2O$-filled heating block at 56°C. Immediately add 1 mL of sucrose/glycerol solution and gently mix by pipetting.

14. Centrifuge at 4500g for 6 min at room temperature and discard the supernatant by pipetting. Gently resuspend the pelleted cells in 140 μL of sucrose/glycerol solution by pipetting. Ensure that the cells are completely resuspended before proceeding with electroporation.

15. Add at least 4 μL of library plasmid (~>4 μg) to each tube (the volume is ~240 μL, which includes 140 μL sucrose/glycerol solution and cells). Mix by pipetting and centrifuge for 1 sec at room temperature in a tabletop microcentrifuge. Incubate for 10 min at room temperature.

16. Transfer 60 μL of the cell suspension to a prechilled 1-mm-gap electroporation cuvette. Electroporate cells at 2.3 kV for 1.1 msec. Immediately add 940 μL of B2 medium prewarmed to 37°C and pipette to mix.

    *These settings may need to be tweaked for other equipment.*

17. Pool cells from each round of transformation (<20 transformations) in a 500-mL shake flask and incubate for 2 h at 37°C, with shaking at 150 rpm. Repeat Steps 12–17 until the desired library size is reached.

18. After 2 h of incubation, titer the library by preparing serial dilutions in B2 medium ($10^2$–$10^5$) and plating 100 µL of each dilution on TBAB chloramphenicol plates using a glass spreader or beads. Incubate for 48 h at 37°C before counting colonies to determine the titer, as described in Step 27 of Protocol 1: Cloning of Affibody Libraries for Display Methods (Ståhl et al. 2023).

19. Inoculate the remaining cell pool into 10× the volume of TSB + Y with chloramphenicol and incubate for 18 h at 37°C with shaking at 150 rpm.

20. The next day, pellet the library by centrifugation at 2000g for 10 min at 4°C. Prepare glycerol stocks by mixing cells with sterile 85% (v/v) glycerol to a final concentration of 15% glycerol, distribute the whole library into 1 mL aliquots, and store at −80°C until performing selections.

21. Take one of the frozen cell aliquots from Step 20 and prepare serial dilutions in B2 medium ($10^6$–$10^8$). Plate 100 µL of each dilution on TBAB plates using a glass spreader or beads. Incubate 48 h at 37°C before counting colonies. Determine the viable cell concentration of the stock (cfu/mL) by multiplying the colony number by the dilution factor and dividing it by the 100 µL plated volume. This number is used for inoculation and amplification of the library during selections.

    *The cell density should be ∼$10^{10}$–$10^{11}$ cells/mL.*

## FACS of *S. carnosus*-Displayed Affibody Libraries

22. With a pipette and sterile tips, inoculate 10×–100× the *S. carnosus* library size into a 5-L flask containing 500 mL of TSB + Y with chloramphenicol. Grow for 18 h at 37°C, with shaking at 150 rpm.

    *The library complexity that was calculated in Step 27 of Protocol 1: Cloning of Affibody Libraries for Display Methods (Ståhl et al. 2023) is given in cfu/mL. By multiplying this number by, for instance, 100, it is possible to calculate what volume to inoculate to cover the library 100-fold.*

    *To ensure proper expression of the library, preferably perform an initial expression test before FACS selection. Follow the procedure described in Step 19 of Protocol 3: Selection of Affibody Molecules Using* Escherichia coli *Display (Dahlsson Leitao et al. 2023); however, use only the secondary reagent HSA-AF647.*

23. The next day, measure the $OD_{578}$ of the library and transfer cells covering the library 10-fold to 100-fold to a microcentrifuge tube containing 1 mL of PBSP.

    *The $OD_{578}$ of the overnight culture should be >8. An $OD_{578}$ of 1 corresponds to ∼$10^9$ cfu/mL.*

24. Pellet cells by centrifugation at 2000g for 6 min at 4°C. Wash three times by resuspending in 1 mL of PBSP by pipetting, pelleting the cells by centrifugation, and discarding the supernatant by pipetting.

25. Prepare a 1-mL mix of PBSP containing 100 nM of biotinylated target and add it to the cells. Mix thoroughly and incubate with end-over-end rotation at 15 rpm for 1 h at room temperature.

    *Target concentration and volume, as well as time and temperature for the selection, will vary with library size and the desired stringency.*

26. Pellet cells by centrifugation, discard the supernatant with a pipette, and resuspend cells by careful pipetting with 1 mL of ice-cold PBSP. Centrifuge and resuspend by pipetting in 500 µL of PBSP containing 40 nM HSA-AF647 and 2 µg/mL SA-PE. Incubate for 30 min on ice in the dark.

27. Gently wash the labeled cells twice in 1 mL of PBSP, resuspend the cells by pipetting in 600 µL of ice-cold PBSP, transfer to flow cytometry tubes, and keep them on ice in the dark until flow-cytometric analysis.

    *Analyze simultaneous cell surface expression (monitored through HSA-AF647) and target binding (monitored through SA-PE) in a cell sorter using appropriate excitation lasers and emission filters (HSA-AF647, 640-nm laser, 660-nm bandpass filter; SA-PE, 488-nm laser, 575-nm bandpass filter). Gate cells showing surface expression and target binding, and screen cells corresponding to ∼10× the library size. Sort cells into 1 mL of TSB + Y medium.*

28. Incubate the sorted cells for 2 h at 37°C with end-over-end rotation at 15 rpm. Inoculate the cells into 100 mL of TSB + Y with chloramphenicol in a 1-L flask. Incubate overnight at 37°C, with shaking at 150 rpm.

# Chapter 15

*Inoculation of sorted cells to smaller volumes of LB (down to 5 mL) could be beneficial in later rounds of FACS.*

29. Perform additional rounds of sorting by repeating Steps 23–29. Prepare glycerol stocks of the sorted cells by mixing them with sterile glycerol to a final concentration of 15% (v/v) by pipetting or careful vortexing. Store at −80°C.

30. After the last round of FACS, plate the sorted cells onto TBAB chloramphenicol plates (aim for ∼100 cfu/plate) using a glass spreader or beads and incubate the plates for 2 d at 37°C. Also sort cells directly into 1 mL TSB + Y media and repeat Steps 29–30.

31. Randomly pick colonies to a PCR plate to analyze library fragment length distribution and sequence enrichment by colony PCR screening and DNA sequencing, as described in Protocol 1: Cloning of Affibody Libraries for Display Methods (Ståhl et al. 2023).

## DISCUSSION

Gram-positive *S. carnosus* cells are used in this protocol for surface display of affibody libraries followed by screening and selection of binders using FACS. Transformation of DNA libraries into *S. carnosus* is less efficient and more complicated than for *E. coli*. Thus, libraries are often more suitable for affinity maturation purposes or other protein-engineering efforts that are not dependent on very high diversity. Magnetic-activated cell sorting (MACS) is hence often not required, but can be an option and easily adopted to use in *S. carnosus* selection efforts. Such protocols can be adopted from Steps 4–18 of Protocol 3: Selection of Affibody Molecules Using *Escherichia coli* Display (Dahlsson Leitao et al. 2023). FACS screening is a great advantage for the selection procedure, as the possibility to monitor the selection in real time can indicate the success of the selection, which is further verified by colony screening and DNA sequencing of the sorted clones. Enriched clones with identical DNA profiles are often chosen for further on-cell characterization. In such characterization, individual clones are grown in 5 mL TSB + Y media supplemented with chloramphenicol before labeling with target and secondary reagents and analysis in the flow cytometer. Ideally, candidates with highest target-binding over expression level signals are chosen for more detailed characterization of the affibody molecule. Such characterization includes recloning to an expression vector, protein production, and analysis by circular dichroism, surface plasmon resonance, and other methods relevant for purpose.

## RECIPES

### B2 Medium

| Reagent | Amount |
| --- | --- |
| Casein hydrolysate | 20 g |
| Yeast extract | 50 g |
| NaCl | 50 g |
| $K_2HPO_4 \cdot 2H_2O$ | 2 g |

1. Dissolve the above reagents in 1.9 L of distilled $H_2O$.
2. Adjust pH to 7.5 using NaOH or HCl and add distilled $H_2O$ to 1.95 L.
3. Autoclave at 15 PSI for 20 min at 121°C.
4. Dissolve 10 g of glucose in 50 mL of distilled $H_2O$, sterilize with a 0.2-μm filter, and add it aseptically to the autoclaved medium.

*Store for up to months at room temperature or 4°C.*

### PBSP Buffer

| Reagent | Amount or final concentration |
| --- | --- |
| Pluronic F108 NF surfactant (BASF) | 10 mL |
| NaCl | 150 mM |
| $Na_2HPO_4$ | 8 mM |
| $NaH_2PO_4$ | 2 mM |

1. Add reagents to sterile $H_2O$.
2. Adjust pH to 7.4 using HCl or NaOH.
3. Bring the volume to 1 L.
4. Filter-sterilize (0.2 μm).

   *Store for up to 2 mo at room temperature; discard if it becomes turbid.*

### Sucrose (0.5 M)/Glycerol (10%)

1. Mix 57.5 mL of 85% (v/v) glycerol in $H_2O$ with 500 mL of distilled $H_2O$.
2. Autoclave at 15 PSI for 20 min at 121°C.
3. Add 85.6 g of sucrose and sterilize with a 0.2-μm filter.

   *Store for up to months at room temperature or 4°C.*

### Tryptose Blood Agar Base (TBAB) Plates

1. Mix 40 g of blood agar base (Merck 1.10886.0500) with 1 L of deionized $H_2O$.
2. Autoclave at 15 PSI for 15 min at 121°C.
3. (Optional) Cool to 50°C, and then add antibiotics as needed. Swirl the bottle to ensure even distribution of the antibiotics throughout the agar.
4. Aliquot into Petri dishes.

   *Store upside-down for up to weeks at 4°C, preferably in the dark because of light sensitivity of antibiotics.*

### TSB + Y Medium

| Reagent | Amount |
| --- | --- |
| TSB powder (Merck 105458) | 30 g |
| Yeast extract (Merck Y1625) | 5 g |

1. Dissolve TSB powder and yeast extract in a final volume of 1 L of $H_2O$ and autoclave at 15 PSI for 20 min at 121°C.
2. (Optional) Once cooled to room temperature, add (if required and immediately before use) the necessary antibiotics to the desired concentration from a 1000× stock solution.

   *Store the medium without antibiotics for up to several months at room temperature.*

## ACKNOWLEDGMENTS

The research on affibodies and their selection systems was funded by the Swedish Agency for Innovation VINNOVA (2019/00104 and the CellNova Center 2017/02105), the Knut and Alice Wallenberg Foundation through the Wallenberg Center for Protein Technology (KAW 2019.0341 and KAW

2021.0197), the Swedish Cancer Society (grants 201090 PjF; 222023 Pj01H), the Swedish Research Council (2019-05115), the Swedish Brain Foundation (grants FO2018-0094, FO2021-0407, FO2022-0253), the Tussilago foundation (FL-0002.025.551-7), the Strategic Research Area Neuroscience (StratNeuro), as well as the Schörling Family Foundation via the Swedish FTD Initiative.

## REFERENCES

Dahlsson Leitao C, Hjelm LC, Ståhl S, Löfblom J, Lindberg H. 2023. Selection of affibody molecules using *Escherichia coli* display. *Cold Spring Harb Protoc* doi:10.1101/pdb.prot108400

Götz F. 1990. *Staphylococcus carnosus*: a new host organism for gene cloning and protein production. *Soc Appl Bacteriol Symp Ser* **19:** 49S–53S.

Kronqvist N, Löfblom J, Severa D, Ståhl S, Wernérus H. 2008. Simplified characterization through site-specific protease-mediated release of affinity proteins selected by staphylococcal display. *FEMS Microbiol Lett* **278:** 128–136. doi:10.1111/j.1574-6968.2007.00990.x

Löfblom J, Rosenstein R, Nguyen MT, Ståhl S, Götz F. 2017. *Staphylococcus carnosus*, from starter culture to protein engineering platform. *Appl Microbiol Biotechnol* **101:** 8293–8307. doi:10.1007/s00253-017-8528-6

Ståhl S, Hjelm LC, Dahlsson Leitao C, Löfblom J, Lindberg H. 2023. Cloning of affibody libraries for display methods. *Cold Spring Harb Protoc* doi:10.1101/pdb.prot108398

## Appendix

# General Safety and Hazardous Material Information

Cold Spring Harbor Laboratory Manuals should be used by laboratory personnel with experience in laboratory and chemical safety or students under the supervision of trained personnel. The procedures, chemicals, and equipment referenced in these manuals may be hazardous and could cause serious injury unless performed, handled, and used with care and in a manner consistent with safe laboratory practices. Students and researchers using the procedures in these manuals do so at their own risk. It is essential for your safety that you consult the appropriate Material Safety Data Sheets, the manufacturers' manuals provided with the relevant equipment, and your institution's Environmental Health and Safety Office (hereafter referred to as safety office), as well as the General Safety and Disposal Cautions in this appendix for proper handling of hazardous materials. Cold Spring Harbor Laboratory makes no representations or warranties with respect to the material set forth in its manuals and has no liability in connection with the use of these materials.

All registered trademarks, trade names, and brand names mentioned in these manuals are the property of the respective owners. Readers should consult individual manufacturers and other resources for current and specific product information.

Appropriate sources for obtaining safety information and general guidelines for laboratory safety are provided in the General Safety and Hazardous Material Information Appendix below.

Users should always consult individual manufacturers, the manufacturers' safety guidelines, and other resources, including local safety offices, for current and specific product information and for guidance regarding the use and disposal of hazardous materials.

## THE PRIMARY SAFETY INFORMATION RESOURCES FOR LABORATORY PERSONNEL ARE THE FOLLOWING

*Institutional Safety Office.* The best source of toxicity, hazard, storage, and disposal information is your institutional safety office, which maintains and makes available the most current information. Always consult this office for proper use and disposal procedures.

Post the phone numbers for your local safety office, security office, poison control center, and laboratory emergency personnel in an obvious place in your laboratory.

*Material Safety Data Sheets (MSDSs).* The Occupational Safety and Health Administration (OSHA) requires that MSDSs accompany all hazardous products that are shipped. These data sheets contain detailed safety information. MSDSs should be filed in the laboratory in a central location as a reference guide.

# Appendix

## GENERAL SAFETY AND DISPOSAL CAUTIONS

The guidance offered here is intended to be generally applicable. However, proper waste disposal procedures vary among institutions; therefore, always consult your local safety office for specific instructions. All chemical waste must be disposed of in a suitable container clearly labeled with the type of material it contains and the date the waste was initiated.

It is essential for laboratory workers to be familiar with the potential hazards of materials used in laboratory experiments, and to follow recommended procedures for their use, handling, storage, and disposal.

The following general cautions should always be observed.

- **Before beginning the procedure**, become completely familiar with the properties of substances to be used.
- **The absence of a warning** does not necessarily mean that the material is safe, because information may not always be complete or available.
- **If exposed** to toxic substances, contact your local safety office immediately for instructions.
- **Use proper disposal procedures** for all chemical, biological, and radioactive waste.
- **For specific guidelines on appropriate gloves to use**, consult your local safety office.
- **Handle concentrated acids and bases** with great care. Wear goggles and appropriate gloves. A face shield should be worn when handling large quantities.

  *Do not mix strong acids with organic solvents because they may react. Sulfuric acid and nitric acid especially may react highly exothermically and cause fires and explosions.*

  *Do not mix strong bases with halogenated solvents because they may form reactive carbenes that can lead to explosions.*

- **Handle and store pressurized gas containers** with caution because they may contain flammable, toxic, or corrosive gases; asphyxiants; or oxidizers. For proper procedures, consult the Material Safety Data Sheet that is required to be provided by your vendor.
- **Never pipette** solutions using mouth suction. This method is not sterile and can be dangerous. Always use a pipette aid or bulb.
- **Keep halogenated and nonhalogenated** solvents separately (e.g., mixing chloroform and acetone can cause unexpected reactions in the presence of bases). Halogenated solvents are organic solvents such as chloroform, dichloromethane, trichlorotrifluoroethane, and dichloroethane. Nonhalogenated solvents include pentane, heptane, ethanol, methanol, benzene, toluene, *N,N*-dimethylformamide (DMF), dimethylsulfoxide (DMSO), and acetonitrile.
- **Laser radiation**, visible or invisible, can cause severe damage to the eyes and skin. Take proper precautions to prevent exposure to direct and reflected beams. Always follow the manufacturer's safety guidelines and consult your local safety office. See caution below for more detailed information.
- **Flash lamps**, because of their light intensity, can be harmful to the eyes. They also may explode on occasion. Wear appropriate eye protection and follow the manufacturer's guidelines.
- **Photographic fixatives, developers, and photoresists** contain chemicals that can be harmful. Handle them with care and follow the manufacturer's directions.
- **Power supplies and electrophoresis equipment** pose serious fire hazard and electrical shock hazards if not used properly.
- **Microwave ovens and autoclaves** in the laboratory require certain precautions. Accidents have occurred involving their use (e.g., when melting agar or Bacto Agar stored in bottles or when sterilizing). If the screw top is not completely removed and there is inadequate space for the steam

to vent, the bottles can explode and cause severe injury when the containers are removed from the microwave or autoclave. Always completely remove bottle caps before microwaving or autoclaving. An alternative method for routine agarose gels that do not require sterile agar is to weigh out the agar and place the solution in a flask.

- **Ultrasonicators** use high-frequency sound waves (16–100 kHz) for cell disruption and other purposes. This "ultrasound," conducted through air, does not pose a direct hazard to humans, but the associated high volumes of audible sound can cause a variety of effects, including headache, nausea, and tinnitus. Direct contact of the body with high-intensity ultrasound (not medical imaging equipment) should be avoided. Use appropriate ear protection and display signs on the door(s) of laboratories where the units are used.

- **Use extreme caution when handling cutting devices**, such as microtome blades, scalpels, razor blades, or needles. Microtome blades are extremely sharp. Use care when sectioning. If unfamiliar with their use, have an experienced user demonstrate proper procedures. For proper disposal, use the "sharps" disposal container in your laboratory. Discard used needles *unshielded*, with the syringe still attached. This prevents injuries and possible infections when manipulating used needles because many accidents occur while trying to replace the needle shield. Injuries may also be caused by broken Pasteur pipettes, coverslips, or slides.

- **Procedures for the humane treatment of animals** must be observed at all times. Consult your local animal facility for guidelines. Animals such as rats are known to induce allergies that can increase in intensity with repeated exposure. Always wear a lab coat and gloves when handling these animals. If allergies to dander or saliva are known, wear a mask.

## DISPOSAL OF LABORATORY WASTE

There are specific regulatory requirements for the disposal of all medical waste and biological samples. In the United States, these are mandated by the U.S. Environmental Protection Agency (see http://www.epa.gov/hw) and regulated by the individual states and territories (see http://www.epa.gov/hw/state-universal-waste-programs-united-states). Medical and biological samples that require special handling and disposal are generally termed Medical Pathological Waste (MPW), and medical, veterinary, and biological facilities will have programs for the collection of MPW and its disposal. Restrictions on how radioactive waste can be disposed of as regulated in the United States by the U.S. Nuclear Regulatory Commission can be found in 10 CFR 20.2001, General requirements for waste disposal (see http://www.nrc.gov/reading-rm/doc-collections/cfr/part020/part020-2001.html) or the individual **Agreement States**. The preferred method for the disposal of radioactively contaminated MPW is decay-in-storage (see http://www.nrc.gov/reading-rm/doc-collections/cfr/part035/part035-0092.html). Contact your institution's safety office and local regulatory agencies for instructions and requirements.

Waste and any materials contaminated with biohazardous materials must be decontaminated and disposed of as regulated medical waste. No harmful substances should be released into the environment in an uncontrolled manner. This includes all tissue samples, needles, syringes, scalpels, etc. Be sure to contact your institution's safety office concerning the proper practices associated with the handling and disposal of biohazardous waste.

Some basic rules are outlined below. For treatment of radioactive and biological waste, see sections on Radioactive Safety Procedures and Biological Safety Procedures.

- In practice, only **neutral aqueous solutions** without heavy metal ions and without organic solvents can be poured down the drain (e.g., most buffers). Acid and basic aqueous solutions need to be neutralized cautiously before their disposal by this method.

- For proper disposal of **strong acids and bases**, dilute them by placing the acid or base onto ice and neutralize them. Do not pour water into them. If the solution does not contain any other toxic compound, the salts can be flushed down the drain.

# Appendix

- For disposal of **other liquid waste**, similar chemicals can be collected and disposed of together, whereas chemically different wastes should be collected separately. This avoids chemical reactions between components of the mixture (see above). Collect at least inorganic aqueous waste, non-halogenated solvents, and halogenated solvents separately.
- Waste **from photo processing and automatic developers** should be collected separately to recycle the silver traces found in it.

## RADIOACTIVE SAFETY PROCEDURES

In the United States and other countries, the access to radioactive substances is strictly controlled. You may be required to become a registered user (e.g., by attending a mandatory seminar and receiving a personal dosimeter). A convenient calculator to perform routine radioactivity calculations can also be found at http://www.graphpad.com/quickcalcs/ChemMenu.cfm.

If you have never worked with radioactivity before, follow the steps below.

- **Avoid it when possible.** Many experiments that are traditionally performed with the help of radioactivity can now be done using alternatives based on fluorescence or chemiluminescence and colorimetric assays, including, for example, DNA sequencing, Southern and northern blots, and protein kinase assays. However, in other cases (e.g., metabolic labeling of cells), use of radioactivity cannot be avoided.
- **Be informed.** While planning an experiment that involves the use of radioactivity, consider the physicochemical properties of the isotope (half-life, emission type, and energy), the chemical form of the radioactivity, its radioactive concentration (specific activity), total amount, and its chemical concentration. Order and use only as much as is really needed.
- **Familiarize yourself** with the designated working area. Perform a mental and practical dry run (replacing radioactivity with a colored solution) to make sure that all equipment needed is available and to get used to working behind a shield. Handle your samples as if sterility would be required to avoid contamination.
- **Always wear appropriate gloves**, lab coat, and safety goggles when handling radioactive material.
- **Check the work area** for contamination before, during, and after your experiment (including your lab coat, hands, and shoes).
- **Localize your radioactivity.** Avoid formation of aerosols or contamination of large volumes of buffers.
- **Liquid scintillation cocktails** are often used to quantitate radioactivity. They contain organic solvents and small amounts of organic compounds. Try to avoid contact with the skin. After use, they should be regarded as radioactive waste; the filled vials are usually collected in designated containers, separate from other (aqueous) liquid radioactive waste.
- **Dispose of radioactive waste** only into designated, shielded containers (separated by isotope, physical form [dry/liquid], and chemical form [aqueous/organic solvent phase]). Always consult your safety office for further guidance in the appropriate disposal of radioactive materials.
- Among the experiments requiring **special precautions** are those that use [$^{35}$S]methionine and $^{125}$I, because of the dangers of airborne radioactivity. [$^{35}$S]methionine decomposes during storage into sulfoxide gases, which are released when the vial is opened. The isotope $^{125}$I accumulates in the thyroid and is a potential health hazard. $^{125}$I is used for the preparation of Bolton–Hunter reagent to radioiodinate proteins. Consult your local safety office for further guidance in the appropriate use and disposal of these radioactive materials before initiating any experiments. Wear appropriate gloves when handling potentially volatile radioactive substances, and work only in a radioiodine fume hood.

General Safety and Hazardous Material Information

## BIOLOGICAL SAFETY PROCEDURES

Biological safety fulfills three purposes: to avoid contamination of your biological sample with other species; to avoid exposure of the researcher to the sample; and to avoid release of living material into the environment. Biological safety begins with the receipt of the living sample; continues with its storage, handling, and propagation; and ends only with the proper disposal of all contaminated materials. A catalog of operations known as "sterile handling" is usually employed in manipulating living matter. However, the correct procedures largely depend on the samples, which can be quite diverse: *Escherichia coli* and other bacterial strains, yeasts, tissues of animal or plant origin, cultures of mammalian cells, or even derivatives from human blood are routinely handled in a biological laboratory. Two of these, bacteria and human blood products, are discussed in more detail below.

The U.S. Department of Health, Education, and Welfare (HEW) has classified various bacteria into different categories with regard to shipping requirements (see Sanderson and Zeigler, *Methods Enzymol* **204**: 248–264 [1991]). Nonpathogenic strains of *E. coli* (such as K12) and *Bacillus subtilis* are in Class 1 and are considered to present no or minimal hazard under normal shipping conditions. However, *Salmonella*, *Haemophilus*, and certain strains of *Streptomyces* and *Pseudomonas* are in Class 2. Class 2 bacteria are "Agents of ordinary potential hazard: agents which produce disease of varying degrees of severity… but which are contained by ordinary laboratory techniques." Contact your institution's safety office concerning shipping biological material.

Human blood, blood products, and tissues may contain occult infectious materials such as hepatitis B virus and human immunodeficiency virus (HIV) that may result in laboratory-acquired infections. Investigators working with lymphoblast cell lines transformed by Epstein–Barr virus (EBV) are also at risk of EBV infection. Any human blood, blood products, or tissues should be considered a biohazard and should be handled accordingly until proved otherwise. Wear appropriate disposable gloves, use mechanical pipetting devices, work in a biological safety cabinet, protect against the possibility of aerosol generation, and disinfect all waste materials before disposal. Autoclave contaminated plasticware before disposal; autoclave contaminated liquids or treat with bleach (10% [v/v] final concentration) for at least 30 minutes before disposal (this is valid also for used bacterial media).

Always consult your local institutional safety officer about specific handling and disposal procedures for samples. Further information can be found in the Frequently Asked Questions of the ATCC homepage (http://www.atcc.org) and is also available from the U.S. National Institute of Environmental Health and Human Services, Biological Safety (http://www.niehs.nih.gov/about/stewardship).

## GENERAL PROPERTIES OF COMMON HAZARDOUS CHEMICALS

The hazardous materials list can be summarized in the following categories:

- Inorganic acids, such as hydrochloric, sulfuric, nitric, or phosphoric, are colorless liquids with stinging vapors. Avoid spills on skin or clothing. Spills should be diluted with large amounts of water. The concentrated forms of these acids can destroy paper, textiles, and skin and cause serious injury to the eyes.
- Inorganic bases, such as sodium hydroxide, are white solids that dissolve in water and under heat development. Concentrated solutions will slowly dissolve skin and even fingernails.
- Salts of heavy metals are usually colored, powdered solids that dissolve in water. Many of them are potent enzyme inhibitors and, therefore, toxic to humans and the environment (e.g., fish and algae).
- Most organic solvents are flammable volatile liquids. Avoid breathing the vapors, which can cause nausea or dizziness. Also avoid skin contact.
- Other organic compounds, including organosulfur compounds such as mercaptoethanol or organic amines, can have very unpleasant odors. Others are highly reactive and should be handled with appropriate care.

Appendix

- If improperly handled, dyes and their solutions can stain not only your sample but also your skin and clothing. Some are also mutagenic (e.g., ethidium bromide), carcinogenic, and toxic.

- Nearly all names ending with "ase" (e.g., catalase, β-glucuronidase, or zymolyase) refer to enzymes. There are also other enzymes with nonsystematic names such as pepsin. Many of them are provided by manufacturers in preparations containing buffering substances, etc. Be aware of the individual properties of materials contained in these substances.

- Toxic compounds are often used to manipulate cells. They can be dangerous and should be handled appropriately.

- Be aware that several of the compounds listed have not been thoroughly studied with respect to their toxicological properties. Handle each chemical with appropriate care. Although the toxic effects of a compound can be quantified (e.g., $LD_{50}$ values), this is not possible for carcinogens or mutagens, where one single exposure can have an effect. Also realize that dangers related to a given compound may also depend on its physical state (fine powder vs. large crystals/diethyl ether vs. glycerol/dry ice vs. carbon dioxide under pressure in a gas bomb). Anticipate under which circumstances during an experiment exposure is most likely to occur and how best to protect yourself and your environment.

Cold Spring Harbor Laboratory Press (CSHLP) has used its best efforts in collecting and preparing the material contained herein but does not assume, and hereby disclaims, any liability for any loss or damage caused by errors and omissions in the publication, whether such errors and omissions result from negligence, accident, or any other cause. CSHLP does not assume responsibility for the user's failure to consult more complete information regarding the hazardous substances listed in this publication.

## REFERENCE

Sanderson KE, Zeigler DR. 1991. Storing, shipping, and maintaining records on bacterial strains. *Methods Enzymol* **204**: 248–264. doi:10.1016/0076-6879(91)04012-D

## WWW RESOURCES

ATCC Home page. http://www.atcc.org
ATCC, for Sample Handling (in Frequently Asked Questions). http://www.atcc.org/CulturesandProducts/TechnicalSupport/FrequentlyAskedQuestions/tabid/469/Default.aspx
GraphPad Software, Radioactivity Calculations. http://www.graphpad.com/quickcalcs/ChemMenu.cfm
National Institute of Environmental Health and Human Services, Biological Safety (NIEHS). http://www.niehs.nih.gov/about/stewardship
U.S. Environmental Protection Agency (EPA), Federal waste disposal regulations, Laboratory. http://www.epa.gov/epawaste/hazard/tsd/index.htm
U.S. Environmental Protection Agency (EPA), Individual States and Territories. http://www.epa.gov/epawaste/wyl/stateprograms.htm
U.S. Nuclear Regulatory Commission (NRC), Medical Pathological Radioactively Contaminated Waste (Decay-in-Storage). http://www.nrc.gov/reading-rm/doc-collections/cfr/part035/part035-0092.html
U.S. Nuclear Regulatory Commission (NRC), Radioactive Waste Disposal Regulations: General Requirements. http://www.nrc.gov/reading-rm/doc-collections/cfr/part020/part020-2001.html

# Index

Note: Page numbers followed by *f*, *t*, or *b* denote a figure, table, or box, respectively, on the corresponding page.

## A

Ab. *See* Antibody (Ab)
Abciximab, 79
ABPs. *See* Activity-based probes (ABPs)
Absolute quantification (AQUA), 396
Academic researchers, 245
Activation-induced cytidine deaminase (AID), 161
Activity-based probes (ABPs), 398
Adalimumab, 79
Adaptive immune system, 357
ADCC. *See* Antibody-dependent cellular cytotoxicity (ADCC)
Adenosine 5′-triphosphate (ATP), 226
Affibody
  generation, 405–406
  technology, 403
Affibody libraries, 405
  cloning of, 412–416
  electroporation of affibody libraries into *S. carnosus*, 446–447
  FACS
    *E. coli*-displayed affibody libraries, 440–441
    *S. carnosus*-displayed affibody libraries, 447–448
  MACS of *E. coli*-displayed affibody libraries, 438–440
  materials, 409
    equipment, 411
    reagents, 409–411
  method, 411
    cloning of affibody libraries, 412–416
    determination of bead-binding capacity of biotinylated target proteins for selection, 416–417
    primer design, 411–412
  troubleshooting, 417
Affibody molecules, 403
  biomedical applications of, 404–405
  materials, 422
    equipment, 423–424
    reagents, 422–423
  method, 424–429
    bacteriophages, 425*f*
  troubleshooting, 429
Affinity, 257
  maturation, 14–15
  purification tools for MS, 375–376
  selection, 15, 49, 79, 375
    display systems, 50*b*
    experiments, 373
    HTS as primary readout in, 16–17
    NGS as primary readout, 57–59
    process, 49–54
    regulate stringency, 59–62
    scourge of SUPs, 55–57
    single round of, 52*f*
    SUPs, 16
    two rounds of, 53*f*
    yield vs. stringency, 15–16, 54–55
Affinity purification-mass spectrometry (AP-MS), 370
Agarose gel electrophoresis equipment, 141
AID. *See* Activation-induced cytidine deaminase (AID)
Alexa Fluor 647-conjugated human serum albumin (HSA-AF647), 436
Alignment data, 295
Alkaline phosphatase (*PhoA*), 41, 238
All-purpose phage antibody libraries, 11–14
α-helix, 29
α-hemolysin, 322
αA-helix (αA2), 373
ALTHEA. *See* Antibody Libraries for Therapeutic Antibody Discovery (ALTHEA)
Amino acid, 79, 214, 217
  diversity, 360–361
  gene assignment and translation to, 302
Amino-terminal domains, 30
Amino termini, 30
Amphiphilic polymer, 254
Amplification, 54, 58
AMPure XP purification kit, 198
Analytical chemistry technique, 396
Anaphase promoting complex/cyclosome (APC/C), 393
Animal immunizations, 223
Antibody (Ab), 2, 212*f*, 218*f*, 245, 293
  analysis of Ab residues involved in antigen interaction, 207–214
  antibody–antigen interaction, 12, 357
  antibody-like libraries, 362
  applications of antibody sequence analysis, 295–296
  design, 248
  discovery, 245, 247–248, 251, 253
    process, 245–249
  display
    epitope discovery, 17–18
    nonantibody antigen-binding domains, 18–19
    peptides and protein domains, 17
    specialized protein domains, 19–21
  domain, 3, 8–9
  drug, 246
  functions, 224
  generation methods, 249
  HTS, 293–294
  humanization, 250
  immune responses, 293
  lead identification, 248*f*
  libraries, 63, 79–80, 323
  molecule, 77, 79
  sequences calculating enrichment ratios, 302–303
  small-scale preparation of antibody fragment-displaying phage, 194–195
  V regions, 18
Antibody data analysis, 294–295
  materials, 305
    equipment, 306
    reagents, 305–306
    sample data, 306*t*
  method, 306–308
    antigen-specific antibody identification, 307*f*
    enrichment ratio calculation, 309
    gene annotation and clonal quantification, 308–309
    troubleshooting, 309–310
Antibody-dependent cellular cytotoxicity (ADCC), 247
Antibody engineering, 162
  and display technologies, 2
  materials, 298
    equipment, 300
    reagents, 298–300
    sample data, 299*t*
  method, 300–301
    antibody genes from synthetic antibody library screening data, 300*f*
    gene assignment and translation to amino acid sequences, 302

Index

Antibody engineering (Continued)
  mapping antibody sequences
    calculating enrichment ratios, 302–303
    mining clones from enrichment ratio data, 304
    post data analysis, 302–303
    software preparation and data preprocessing, 301–302
  troubleshooting, 304
Antibody Libraries for Therapeutic Antibody Discovery (ALTHEA), 14
  gold library, 14
  gold plus libraries, 14, 251–253
Anti-CD19 CAR, 21
Antigen binders
  materials, 188
    chicken scFv HTS primers, 189
    equipment, 190–191
    forward and reverse index list, 190
    reagents, 188–190
  method, 191
    bio-panning, 192–194
    high-throughput sequencing analysis of chicken scFv library for in silico selection, 196–199
    immobilized antigens, 191
    phage ELISA, 195–196
    small-scale preparation of antibody fragment-displaying phage, 194–195
  troubleshooting, 199
Antigen-binding fragment (Fab), 2–3
  fragments, 9
Antigen(s), 82
  analysis of Ab residues involved in antigen interaction, 207–214
  antigen-binding clones, 163
  antigen-specific antibodies, 295
  antigen-specific library, 323
  fragment clones, 324
  recognition, 211, 214
Antigenic determinants, 317
Anti-GFRα4 CAR, 22
Anti-HlgC mAbs, 326
Anti-LukE antibody, 329
Anti-LukE3
  antibody, 329
  mAb, 324
AP-MS. See Affinity purification-mass spectrometry (AP-MS)
APC/C. See Anaphase promoting complex/cyclosome (APC/C)
AQUA. See Absolute quantification (AQUA)
Arg. See Arginine (Arg)
Arginine (Arg), 369
Artificial intelligence–based methods, 224

Asn. See Asparagine (Asn)
Asparagine (Asn), 370
ATP. See Adenosine 5′-triphosphate (ATP)
Autoimmune disease, 247, 250
Avidity effect, 55, 59
AviTag, 254

B
*Bacillus subtilis*, 455
Backbone, 13
Bacteria, expanding library in, 347
Bacterial physiology, limitations to phage display imposed by, 40–44
Barbas, Carlos, 2, 21
Base pairs (bp), 411
Bead-binding capacity of biotinylated target proteins for selection, determination of, 416–417
4-1BB protein, 22
B-cell epitopes, 323
  convalescent serum antibodies commonly recognize primary sequence–dependent, 327–329
  minimal B-cell epitopes, 328f, 329t
  human immunoproteome of S. aureus gene products, 321–322
  identification
    leucocidin-encoded B-cell epitopes by phage display, 324
    Luk neutralization epitope in β-hairpin loop in stem subdomain of leucocidin holoprotein, 326–327
  molecular characterization of cross-reactive leucocidin epitope, 324–325
  phagemid-based display system for toxin epitope recovery, 323–324
  SA infections associated with range of immunogenic antigens, 321
  staphylococcal infections pose substantial and increasing clinical burden, 318–321
  staphylococcal PFTS attractive antigens for phage display–based investigations, 322
  variations of cross-reactive epitope fold in conserved β-loop in leucocidin protein, 325–326
B-cell maturation process, 250
B-cell sorting, 248
β-hairpin loop in stem subdomain of leucocidin holoprotein, identification of Luk neutralization epitope in, 326–327
β-lactamase gene (*Bla*), 238
β-sheet, 18

β-strands, 329
Bicistronic Fab expression, 238
Binary code diversity, 206
Binary library, 361
Binder(s)
  binding assays, 388–393
  cellular assays, 397–399
  characterization of, 387–388
  diversity, 386
  functional assays, 393–394
  structural studies, 394–396
Binding assays, 386, 388
  BLI, 391–392
  ELISA, 388–390
  ITC, 390–391
  SPR, 392–393
Biohazardous materials, 453
Bioinformatics methods, 383
Biolayer interferometry (BLI), 388, 391–392
Biological safety
  fulfills, 455
  procedures, 455
Bio-panning, 192–194
Biosensors, 376–377
Biotechnology, 293
  companies, 250
Biotin, 51
Biotinylated antigen in solution, panning against, 135
Biotinylated target proteins for selection, determination of bead-binding capacity of, 416–417
Bivalent Fabs, 206, 238
BLI. See Biolayer interferometry (BLI)
Blood products, 455
BMMCs. See Bone marrow mononuclear cells (BMMCs)
Bone marrow mononuclear cells (BMMCs), 82, 91
  preparation from human blood or BMMCs from human bone marrow, 91–92
  total RNA preparation from, 92–93
Bone marrow, total RNA preparation from, 106–107
Bovine serum albumin (BSA), 11, 82, 87, 103, 125, 129, 139–140, 166, 227, 334, 388
Bragg's law, 395
1-bromo-3-chloropropane (BCP), 166
BSA. See Bovine serum albumin (BSA)
B2 medium, 448
Burton, Dennis, 2

C
Calorimeter, 390
CA-MRSA. See Community-associated MRSA (CA-MRSA)

458

Cancer, 250, 381
    diagnosis, 376
    diagnostics, 376–377
CAP. See Catabolic activator protein (CAP)
Capsular-alone vaccine approach, 320
Capture-related SUPs, 56
CAR T. See Chimeric antigen receptor T-cell (CAR T)
Carb. See Carbenicillin (Carb)
Carbenicillin (Carb), 87, 227, 260, 271, 281
Carboxyl termini, 30
Carriers, 11
CARs. See Chimeric antigen receptors (CARs)
Catabolic activator protein (CAP), 70
CCC. See Covalently closed circular (CCC)
CCC-dsDNA. See Covalently closed circular double-stranded DNA (CCC-dsDNA)
CD19, 21
cDNA. See Complementary DNA (cDNA)
CDRs. See Complementarity-determining regions (CDRs)
Cell(s), 398
    cell-based assays, 247
    cell-based display systems, 406
    cell-based panning, 255
    physiology, 40
Cellular assays, 397
    cellular deubiquitination assays, 398
    cellular ubiquitination assays, 397–398
    Co-IP, 397
    UbV characterization in organoids, 398–399
Cellular deubiquitination assays, 398
Cellular ubiquitination assays, 397–398
Centrifuge cells, 428
Chickens (*Gallus gallus domestica*), 161
    antibodies, 159–160
        advantages for biotechnological uses, 160–161
        antibody engineering, 162
        biomedical applications, 162
        diversity of chicken Ig repertoire, 161–162
        using phage display library technology generation, 162–163
        system, 10
    IgL diversity, 161
    Ig repertoire, diversity of, 161–162
    materials, 166
        equipment, 167
        reagents, 166–167
    method, 167

cDNA synthesis, 171
    immunization and titering immune serum by ELISA, 168–169
    organ harvesting, followed by RNA purification with optional LiCl precipitation, 169–171
    monoclonal antibodies, 162
    phage antibody libraries, 10
    scFv libraries, 10
        high-throughput sequencing analysis for in silico selection of antigen binders, 196–199
Chimeric Antigen Receptor Lymphocytes with Oncolytic Specificity project (C.A.R.L.O.S. project), 21–24
Chimeric antigen receptors (CARs), 21, 22b, 22f, 162
Chimeric antigen receptor T-cell (CAR T), 2, 21, 247
Chimeric rabbit
    materials, 102
        equipment, 105–106
        reagents, 102–105
    method, 106
        PCR amplification of rabbit $V_\kappa$, $V_\lambda$, and $V_H$ cDNA, and human $C_\kappa$ and $C_\lambda$ DNA, 108–110
        PCR assembly of rbV$_L$/huC$_L$/rbV$_H$ cassettes, 110–111
        phagemid and phage library generation, 112–114
        preparation of total RNA from rabbit spleen and bone marrow, 106–107
        reverse transcription of mRNA, 107–108
        SfiI digestion of rbV$_L$/huC$_L$/rbV$_H$ cassettes and phagemid pC3C, 111–112
        test ligation and transformation, 112
    troubleshooting, 114
Clonal diversity, 324
Clonal lineage
Clonal quantification, 308–309
Clone-by-clone sequencing, 18
Clone's reads, 17, 57
Cloning, preparation of phage display vector for, 342–344
CM. See Conserved motif (CM)
Coated biosensor, 391
Coat protein
    fusion, 1
    structure, 29
Coimmunoprecipitation (Co-IP), 388, 397
Co-IP. See Coimmunoprecipitation (Co-IP)

Cold Spring Harbor Laboratory Press (CSHLP), 456
Colorimetric immunoassay, 388
Community, 358
Community-associated MRSA (CA-MRSA), 318
Competency test, 119
Complementarity-determining regions (CDRs), 11, 18, 160, 202, 224, 250, 361
    canonical conformations of, 217–219
    CDR-H2, 217
    length and composition of CDRs L3 and H3, 216–217
    relative contributions of, 213f
Complementary DNA (cDNA), 10, 171
Computational intelligence–based methods, 224
Computational methods, 383
Conformational diversity, 217
Conserved motif (CM), 285
Conserved β-loop in leucocidin protein, variations of cross-reactive epitope fold in, 325–326
Contamination tests, 119–120
Convalescent serum antibodies commonly recognize primary sequence–dependent B-cell epitope, 327–329
Core positions, 214
Core structural paratope, 214–216
Coronavirus disease 2019 (COVID-19), 253
Counterselection, 56
Covalently closed circular (CCC), 230
    CCC-dsDNA
        conversion, 239
        conversion of CCC-dsDNA into phage-displayed Fab library, 232–234
        in vitro synthesis, 230–232
Covalently closed circular double-stranded DNA (CCC-dsDNA), 239, 384
COVID-19. See Coronavirus disease 2019 (COVID-19)
Cross-reactive epitope fold in conserved β-loop in leucocidin protein, variations of, 325–326
Cross-reactive leucocidin epitope, molecular characterization of, 324–325
Cross-reactivity, 160, 247, 329
Crude Fab ELISA
    with eukaryotic cells in solution, 145–146
    with immobilized antigen, 144–145
    library selection analysis by, 143–144
Cryoelectron microscopy (cryoEM), 5–6

# Index

cryoEM. *See* Cryoelectron microscopy (cryoEM)
Crystallization, 395
Crystal structures, 395
CSHLP. *See* Cold Spring Harbor Laboratory Press (CSHLP)
Cullin 1 (Cul1), 387
Cysteine residues, 406
Cytoplasmic proteins, 43, 64

## D

D11, 14
Data
    preprocessing, 301–302, 313–314
    sets, 215
        structures of synthetic and natural Ab–antigen complexes, 206–207
dATP. *See* Deoxyadenine triphosphate (dATP)
dCTP. *See* Deoxycytosine triphosphate (dCTP)
DD. *See* Dimerization domain (DD)
DeKosky, Brandon, 16
Deoxyadenine triphosphate (dATP), 384
Deoxycytosine triphosphate (dCTP), 384
Deoxyguanine triphosphate (dGTP), 384
Deoxynucleoside triphosphate (dNTP), 226
Deoxynucleotide (dNTP), 271, 281
Deoxythymine triphosphate (dTTP), 384
Deoxy-U (dU), 12
Deoxyuridine triphosphate (dUTP), 384
Designed FN3 molecules, 358
Deubiquitinases (DUBs), 383
Deubiquitination assay, 394, 398
dGTP. *See* Deoxyguanine triphosphate (dGTP)
Diabodies, 251
Diagnostic use, 317
Digested phagemid vector
    digested semisynthetic scFvs, ligation with, 289
    ligation of digested scFv libraries with, 267
Digested scFv libraries with digested phagemid vector, ligation of, 267
Digested semisynthetic scFvs, ligation with digested phagemid vector, 289
Dimerization domain (DD), 238
Dimethylsulfoxide (DMSO), 452
Discovery strategies, 254
Disease-relevant animal model, 248
Display
    bias, 41
    primers for display vector amplification, 411–412
    systems for monobody discovery, 362–364
Displayed peptide, 50
Dithiothreitol (DTT), 227
Diversity (D), 10, 13, 161
    chicken Ig repertoire, 161–162
DMF. *See* N,N-dimethylformamide (DMF)
DMSO. *See* Dimethylsulfoxide (DMSO)
DNA sequencing
    library preparation for, 353
    target-specific phage clones, 236–237
dNTP. *See* Deoxynucleoside triphosphate (dNTP)
Double-stranded DNA (dsDNA), 65, 79
Double-stranded DNA origin of replication (dsDNA ori), 238
DPBS. *See* Dulbecco's phosphate-buffered saline (DPBS)
dsDNA. *See* Double-stranded DNA (dsDNA)
dsDNA ori. *See* Double-stranded DNA origin of replication (dsDNA ori)
DTT. *See* Dithiothreitol (DTT)
dTTP. *See* Deoxythymine triphosphate (dTTP)
DUBs. *See* Deubiquitinases (DUBs)
Dulbecco's phosphate-buffered saline (DPBS), 150
dU-ssDNA. *See* Uracil-containing ssDNA (dU-ssDNA)
*dut* gene, 384
dUTP. *See* Deoxyuridine triphosphate (dUTP)

## E

EBV. *See* Epstein–Barr virus (EBV)
ECM. *See* Extracellular matrix (ECM)
EDTA. *See* Ethylenediamenetetraacetic acid (EDTA)
Efficacy, 247
EGFR. *See* Epidermal growth factor receptor (EGFR)
Egg yolk, 160
Electrocompetent *E. coli*
    materials, 116
        equipment, 117–118
        reagents, 116–117
    method, 118–119
        competency test, 119
        contamination tests, 119–120
Electrocompetent *S. carnosus* TM300 cells, preparation of, 444–446
Electron
    micrographs, 4
    microscopy, 3
Electroporation
    affibody libraries into *S. carnosus*, 446–447
    ligation mix, 267–268, 290
    system, 232
    test ligations, 345–346
ELISA. *See* Enzyme-linked immunosorbent assay (ELISA)
Eluted phage, amplification of, 275–276
Elution, 275
EMA. *See* European Medicines Agency (EMA)
Engineering binding proteins, 382
Engineering SH2 domains
    affinity, 371–373
    engineering SH2 domains for binding to noncanonical PTMs, 373–375
    using phage display, 371
    specificity, 373
Enrichment ratio (ER), 310
    calculation, 309
    mining clones from, 304
Enzyme-linked immunosorbent assay (ELISA), 249, 350, 386, 388–390
    immunization and titering immune serum by, 168–169
E1 isozyme, 382
Epidermal growth factor receptor (EGFR), 405
Epitope discovery, 17–18
Epitope identification
    materials, 334
        equipment, 336
        reagents, 334–336
    method, 337–340
        gene fragments preparation for cloning into phagemid library, 340–342
        generating and selecting from random gene fragment library, 339*f*
        libraries preparation for HTS, 351–353
        panning on 96-well high-binding-capacity microtiter plates, 348–349
        pComb-Opti8 vector, 337*f*
        PCR primers, 340*f*
        phage binding ELISA protocol, 350–351
        phage display vector preparation for cloning, 342–344
        protein G bead–based library selection by mAb, 349–350
        test and library ligations of gene fragment inserts into prepared vector, 344–348
        UDIs for oligonucleotide-based library labeling, 341*f*
    troubleshooting, 353

Epstein–Barr virus (EBV), 455
ER. *See* Enrichment ratio (ER)
*Escherichia coli*, 27, 63, 79, 224, 251, 358, 403, 455
    CJ236, 384
    Ff effect on *E. coli* physiology, 39–40
    host cell, 323
    *lac* promoter, 7
    materials, 436
        equipment, 437
        reagents, 436–437
    method, 437–438
        culture expansion and induction of protein expression, 438
        FACS of *E. coli*-displayed affibody libraries, 440–441
        MACS of *E. coli*-displayed affibody libraries, 438–440
Ethylenediamenetetraacetic acid (EDTA), 418, 430
E2 isozyme, 382
Eukaryotic cells, 382
    in solution
        crude Fab ELISA with, 145–146
        panning against, 135–136
        phage ELISA with, 143
European Medicines Agency (EMA), 245
ExbBD complex, 33
Excel-based analysis, 310
ExoI. *See* Exonuclease I (ExoI)
Exonuclease I (ExoI), 227
Exothermic reaction, 390
Extracellular matrix (ECM), 368

## F

Fab. *See* Antigen-binding fragment (Fab); Fragment antigen binding (Fab)
Fabs. *See* Synthetic antigen-binding fragments (Fabs)
FACS. *See* Fluorescence-activated cell sorting (FACS)
Far western blot analysis, 370
Fc. *See* Fragment crystallizable (Fc)
Fd fragment, 70
FDA. *See* U.S. Food and Drug Administration (FDA)
Fertility factor, 63–64
Ff
    effect on *E. coli* physiology, 39–40
    filamentous phages, 33
        physical properties of, 30–31
        limitations to phage display imposed by, 40–44
    phage, 27, 33, 63–64
        genes and proteins of, 32t
    production, 44
    virion proteins, 43
Fibronectin type III (FN3), 19, 358, 362

Filamentous bacteriophages
    Ff effect on *E. coli* physiology, 39–40
    filamentous phage
        assembly, 37–39
        infection, 33–34
        replication, 34–37
        structure, 27–31
    genome, genes, and transcripts, 31–33
    limitations to phage display imposed by Ff assembly and bacterial physiology, 40–44
Filamentous phage, 63, 68
    context of phage display, 63–64
    infection, 33–34
    particles, 79–80, 82
    phage assembly, 37–39
        pIV channel gate opening and interaction, 38f
    replication, 34–37
    structure, 27–29
        Ff virion structure, 28f
        major coat protein structure, 29
        minor virion proteins forming end caps of filament, 29–30
        physical properties of Ff filamentous phages, 30–31
    structure and function, 3–7
        Ff filamentous phage virion, 6f
        Ff minor coat protein pIII, 5t
        high-resolution negative-stained electron micrograph, 4f
        nascent virion, 7f
        rare antigen-binding antibody domains, 4f
        schematic diagram of phage antibody, 3f
Filamentous virions, 53
Filament, virion proteins forming end caps of, 29–30
Filtered libraries (FLs), 287
    competing interest statement, 277
    materials, 271
        equipment, 272
        reagents, 271–272
    method, 273
        amplification of eluted phage, 275–276
        elution, 275
        magnetic bead equilibration and blocking of nonspecific interactions, 274
        phage precipitation with PEG/NaCl, 273–274
        preparation of phage-displayed PLs for filtration, 273
        quality control of, 276–277
        selection for thermostable scFvs, 274–275

    PCR assembly of FLs with natural H3J fragments, 286–288
Filtrated libraries (FLs), 253
Filtration
    preparation of phage-displayed PLs for, 273
    process, 253
First-strand cDNA synthesis, 285
FL. *See* Fluorescein (FL)
FL-BSA. *See* Fluorescein conjugated to bovine serum albumin (FL-BSA)
FL-Ova. *See* Fluorescein-conjugated ovalbumin (FL-Ova)
FLs. *See* Filtered libraries (FLs)
Fluorescein (FL), 16
Fluorescein-conjugated ovalbumin (FL-Ova), 16
Fluorescein conjugated to bovine serum albumin (FL-BSA), 11
Fluorescence-activated cell sorting (FACS), 294, 406
    *E. coli*-displayed affibody libraries, 440–441
    *S. carnosus*-displayed affibody libraries, 447–448
FN3. *See* Fibronectin type III (FN3)
Fourier transformation, 395
Fragment antigen binding (Fab), 78, 224
    chimeric rabbit/human Fab libraries in phagemid pC3C, 81f
    cloning, 153–154
    expression, 154–155
    Fab–antigen complexes, 210
    Fab-phage library panning against immobilized antigen, 83f
    format, 78
    heterodimers, 238
    phage display antibody libraries, 79–80
        generation in, 81–82
        selection in, 82–84
    purification, 155–156
Fragment crystallizable (Fc), 78
Fragment variable (Fv), 78
Functional antibody, 257
Functional assays, 393
    deubiquitination assay, 394
    ubiquitination assay, 393
Functional clones, 360
Functional signal sequence, 43
Fv. *See* Fragment variable (Fv)
Fyn-SH2 domains, 376

## G

*Gallus gallus domestica*. *See* Chickens (*Gallus gallus domestica*)
Gel
    electrophoresis, 413
    filtration medium, 60

Index

Gene diversity analysis
  materials, 311
    equipment, 312
    reagents, 311–312
    sample data, 312t
  method, 312–313
    data preprocessing, 313–314
    interpret final results, 314–315
    paired $V_H:V_L$ repertoire lineage analysis, 313f
    perform automated $V_H:V_L$ data analysis, 314
Gene fragment
  libraries, 318, 329
  phage library, 323
  preparation for cloning into phagemid library, 340–342
  test and library ligations of gene fragment inserts into prepared vector, 344
    cleanup, 345
    electroporation of test ligations, 345–346
    expanding library in bacteria, 347
    generation of library based on optimized ratios, 346–347
    library recovery in phage format, 347–348
    ligations, 344–345
    titration of library, 346
Generic vocabulary, 15
Genetic diversity, 320
Gene(s), 31–33
  annotation, 308–309
  assignment and translation to amino acid sequences, 302
  conversion, 161
  I encodes, 37
Genome, 31–33
Genotype in phagemid-based phage display, physical linkage of, 69–70
GFRα4 protein, 22
Glycerol, 449
Glycine-HCl, 430
Glycosylation, 247
Grb2-domains, 376
gVIII, 41
gX gene, 31
gXI gene, 31

H

HA. See Hemagglutinin (HA)
hACE2. See Human angiotensin-converting enzyme 2 (hACE2)
Haemophilus, 455
Haptens, 11

Hazardous chemicals, general properties of, 455
HC. See Heavy chain (HC)
HCDR3. See Heavy chain complementarity determining region 3 (HCDR3)
Health, Education, and Welfare (HEW), 455
Heavy chain (HC), 10, 70, 161, 214, 250, 295
  CDRs, 206
  frameworks, 207
Heavy chain complementarity determining region 3 (HCDR3), 78
Heavy chain constant (CH), 160
Heavy-chain constant 1 (CH1), 362
Heavy-chain constant 3 (CH3), 362
Heavy-chain variable (VH), 362
Heavy metals, 455
HECT. See Homologous to E3-AP carboxyl terminus (HECT)
Helper phage, 7, 68–69, 256
  materials, 122
    equipment, 123
    reagents, 122–123
  method, 123–124
Helper plasmids, 39
Hemagglutinin (HA), 72
Hemolysin A (HlgA), 322–323
Hemolysins, 322
Hemophilus influenzae, 318
Hepatitis B virus, 455
HER2. See Human epidermal growth factor receptor 2 (HER2)
HEW. See Health, Education, and Welfare (HEW)
High-affinity, 143
  antibodies, 256
  selection of high-affinity synthetic Fabs, 234–235
  SH2 domains, 376
High-binding-capacity microtiter plate, panning on 96-well, 348–349
High productivity, 39
High stringency, 16
High-throughput
  fluorescence polarization assays, 370
  method, 392
High-throughput sequencing (HTS), 16, 21, 163, 293, 299, 306, 312, 323, 353
  analysis of antibody data, 294–295
  analysis of chicken scFv library for in silico selection of antigen binders, 196–199
  applications of antibody sequence analysis, 295–296
  library amplification, 352–353

library preparation for DNA sequencing, 353
PCR-based addition of oligonucleotides that document library source of selected clone, 352
preparation of libraries for, 351
as primary readout in affinity selection projects, 16–17
High yield, 16
HIV. See Human immunodeficiency virus (HIV)
HlgA toxins, 325
HlgC, 239–257
  peptides, 328
Homologous to E3-AP carboxyl terminus (HECT), 383
Horse radish peroxidase (HRP), 140
Host bacterial cells, 68
HRP. See Horse radish peroxidase (HRP)
HSA-AF647. See Alexa Fluor 647-conjugated human serum albumin (HSA-AF647)
HTS. See High-throughput sequencing (HTS)
huC$_L$ cassettes
  PCR assembly, 96–97, 110–111
  SfiI digestion, 97, 111–112
Human Ab repertoires, 220
Human angiotensin-converting enzyme 2 (hACE2), 253
Human blood, 455
  preparation of PBMCs from, 91–92
Human bone marrow, BMMCs from, 91–92
Human cutaneous, 318
Human C$_κ$ and C$_λ$ DNA, 93–96
  PCR amplification of, 108–110
Human epidermal growth factor receptor 2 (HER2), 405
Human error, 386
Human Fab format
  materials, 86, 102
    DNA sequencing, 88t–89t
    equipment, 90–91, 105–106
    PCR amplification, 88t–89t
    PCR assembly, 88t–89t
    reagents, 86–90, 102–105
  method, 91, 106
    PCR amplification of human V$_κ$, V$_λ$, and V$_H$ cDNA, and human C$_κ$ and C$_λ$ DNA, 93–96
    PCR amplification of rabbit V$_κ$, V$_λ$, and V$_H$ cDNA, and human C$_κ$ and C$_λ$ DNA, 108–110
    PCR assembly of huV$_L$/huC$_L$/huV$_H$ cassettes, 96–97
    PCR assembly of rbV$_L$/huC$_L$/rbV$_H$ cassettes, 110–111
    phagemid and phage library generation, 98–100, 112–114

preparation of PBMCs from human blood or BMMCs from human bone marrow, 91–92
preparation of total RNA from PBMCs or BMMCs, 92–93
preparation of total RNA from rabbit spleen and bone marrow, 106–107
reverse transcription of mRNA, 93, 107–108
SfiI digestion of huV$_L$/huC$_L$/huV$_H$ cassettes and phagemid pC3C, 97
SfiI digestion of rbV$_L$/huC$_L$/rbV$_H$ cassettes and phagemid pC3C, 111–112
test ligation and transformation, 97–98, 112
troubleshooting, 100, 114
Human Fyn-SH2 domain, 373
Human genome, 19
Human immune
responsiveness, 321
system, 321
Human immunodeficiency virus (HIV), 455
Human immunoproteome of *S. aureus* gene products, 321–322
Human tenascin, 359
Human V$_H$ (huV$_H$), 94
Human V$_κ$ (huV$_κ$), 94
PCR amplification of, 93–96
Human V$_λ$ (huV$_λ$), 94
PCR amplification of, 93–96
huV$_H$ cassettes
PCR assembly of, 96–97
SfiI digestion of, 97
huV$_L$ cassettes
PCR assembly of, 96–97
SfiI digestion of, 97
Hybridomas, 223
Hydrochloric acid (HCl), 227
Hyperconversion events, 161
Hyperphage, 360

### I

Ig. *See* Immunoglobulin (Ig)
IgG. *See* Immunoglobulin G (IgG)
IgG1. *See* Immunoglobulin G1 (IgG1)
*IGHV* genes, 251
*IGHV3-23* gene, 255
*IGKV* genes, 251
*IGKV1* gene, 252
*IGKV3* gene, 252
*IGKV3-20* gene, 252
*IGKV4* gene, 252
*IGKV4-01* gene, 252
IL-17A. *See* Interleukin-17A (IL-17A)

IMAC. *See* Immobilized metal affinity chromatography (IMAC)
IMGT. *See* International Immunogenetics Information System (IMGT)
Immobilization mode, 52
Immobilized antigen
crude Fab ELISA with, 144–145
panning against, 132–135
phage ELISA with, 142–143
Immobilized metal affinity chromatography (IMAC), 72, 368
Immobilized proteins, 391–392
Immune libraries, 250
Immune system, 357
Immunization, 159, 163
by ELISA, 168–169
Immunogenic antigens, SA infections associated with range of, 321
Immunogenic fitness, 18
Immunogenicity, 203, 318, 326
Immunoglobulin (Ig), 247, 358, 403
genes, 202
IgBLAST, 295
IgG-A7, 253
IgY, 160
Immunoglobulin G (IgG), 361–362
Immunoglobulin G1 (IgG1), 77, 78f
Immunoglobulin heavy chain (IgH chain), 160
Immunoglobulin light chain (IgL chain) gene, 161
Immunoglobulin M (IgM), 160
Immunology, 293
Immunoprecipitation MS (IP-MS), 369
IND. *See* Investigational New Drug (IND)
Influence solubility, 203
Inorganic acids, 455
Inorganic bases, 455
Inoviruses, 28
In silico selection of antigen binders, high-throughput sequencing analysis of chicken scFv library for, 196–199
Institutional safety office, 451
Integrated sensing and activating proteins (iSNAPs), 376
Interest, technical considerations to isolate antibodies of, 256–257
Intergenic region (IR), 6–7
phage f1 single-stranded viral DNA, 8f
Interleukin-17A (IL-17A), 405
International Immunogenetics Information System (IMGT), 315
Investigational New Drug (IND), 246
In vitro
antibody display methods, 223
methods, 201

selection methods, 223
synthesis of CCC-dsDNA, 230–232
IP-MS. *See* Immunoprecipitation MS (IP-MS)
IPTG. *See* Isopropyl β-D-1-thiogalactopyranoside (IPTG); Isopropyl-β-D-thiogalactoside (IPTG)
IR. *See* Intergenic region (IR)
iSNAPs. *See* Integrated sensing and activating proteins (iSNAPs)
Isolate antibodies of interest, technical considerations to, 256–257
Isopropyl β-D-1-thiogalactopyranoside (IPTG), 7, 70
Isopropyl-β-D-thiogalactoside (IPTG), 140, 150
Isothermal titration calorimetry (ITC), 388, 390–391
ITC. *See* Isothermal titration calorimetry (ITC)

### J

JH fragments by PCR, amplification of, 285–286
Joining segment (J segment), 10, 13, 161
June, Carl, 21

### K

Kanamycin (Kan), 227
κ-light-chain isotype, 219
Keyhole limpet hemocyanin (KLH), 326
Kinetic methods, 249
KLH. *See* Keyhole limpet hemocyanin (KLH)
Kunkel mutagenesis, 362

### L

Laboratory jargon, 79
Laboratory personnel, primary safety information resources for, 451
Laboratory waste, disposal of, 453–454
*Lac* operator (*lacO*), 69
LB liquid medium II, 418, 442
LCDR3. *See* Light chain complementarity determining region 3 (LCDR3)
LD. *See* Linear dichroism (LD)
Lead optimization, 248
*Legionella pneumophila*, 370
Leucocidin E (LukE), 324
Leucocidins (Luks), 322
identification
leucocidin-encoded B-cell epitopes by phage display, 324

Index

Leucocidins (Luks) (Continued)
    Luk neutralization epitope in β-hairpin loop in stem subdomain, 326–327
    LukS, 322–340
        peptides, 328
        toxins, 325
        monomers, 328
        neutralization epitope identification in β-hairpin loop in stem subdomain of leucocidin holoprotein, 326–327
        variations of cross-reactive epitope fold in conserved β-loop in, 325–326
Leukoplatelet concentrate, isolation of lymphocytes from, 283–284
Liang, Chen T., 20
Library
    amplification, 352–353
    construction, 238–239, 249, 384
        case study for therapeutic monoclonal antibody discovery, 251–253
    design, 405–406
        improvements, 387
        principles of, 202–203
    fragment length distribution, 416
    generation of library based on optimized ratios, 346–347
    library-amplification primers, 412
    ligation, 179–181
        gene fragment inserts into prepared vector, 344–348
    optimization, 220
    preparation, 255
        for DNA sequencing, 353
        for HTS, 351–353
    recovery in phage format, 347–348
    selection analysis
        by crude Fab ELISA, 143–146
        by phage ELISA, 142–143
    titration of, 346
Library reamplification, 181–182
    materials, 125
        equipment, 126–127
        reagents, 125–126
    method, 127–128
Ligate library, 414
Ligation
    efficiency, 416
    electroporation of ligation mix, 267–268, 290
Light chain ($C_L$), 77, 161, 295
Light chain complementarity determining region 3 (LCDR3), 78
Light-chain constant (CL), 362
Light-chain variable (VL), 362
Linear dichroism (LD), 30

Liquid crystalline, 30
Liquid–liquid diffusion method of crystallization, 395
LukE. See Leucocidin E (LukE)
Luks. See Leucocidins (Luks)
Lung cancer cell lines, 376
Lymphocytes
    isolation of lymphocytes from leukoplatelet concentrate, 283–284
    total RNA isolation from, 284
Lysogenization, 34

M

mAbs. See Monoclonal antibodies (mAbs)
MACS. See Magnetic-activated cell sorting (MACS)
Magnetic-activated cell sorting (MACS), 448
    E. coli-displayed affibody libraries, 438–440
Magnetic bead equilibration and blocking of nonspecific interactions, 274
Magnetic particle concentrator (MPC), 192
Mammalian cell
    lines, 397
    ubiquitination assays, 388
Manufacturability traits, 247
Mass spectrometry (MS), 375, 388, 396
    affinity purification tools for MS, 375–376
Mass-to-charge ratios, 396
Material Safety Data Sheets (MSDSs), 451
Mathematical modeling, 39
MCS. See Multiple cloning site (MCS)
Mechanism of action (MOA), 247
Medical Pathological Waste (MPW), 453
Medullary thyroid cancer (MTC), 21
Membrane proteins, 254
mESCs. See Mouse embryonic stem cells (mESCs)
Methicillin-resistant Staphylococcus aureus (MRSA), 18, 318
Microbatch crystallization, 395
Microcentrifuge tube, 192, 426
Microtome blades, 453
Minimalist libraries, 12
MOA. See Mechanism of action (MOA)
Molecular display library, 361
Molecular diversity, 357
Molecular dynamic simulations, 370
Molecular recognition, SH2 domain structure and role in, 369
Molecular weight (MW), 341
Molten agarose, 418, 431
Monobody
    crystal structures of, 363f
    display systems for monobody discovery, 362–364

library designs, 360–362
phage display of, 359–360
system, 358–359
    immunoglobulin heavy-chain variable and fibronectin type III domains, 359f
Monoclonal antibodies (mAbs), 27, 77, 201, 223, 250, 318, 348
    protein G bead–based library selection by, 349–350
Mouse embryonic stem cells (mESCs), 376
MPC. See Magnetic particle concentrator (MPC)
MPW. See Medical Pathological Waste (MPW)
mRNA
    isolation, 284–285
    RT of, 93, 107–108
MRSA. See Methicillin-resistant Staphylococcus aureus (MRSA)
MS. See Mass spectrometry (MS)
MSDSs. See Material Safety Data Sheets (MSDSs)
MTC. See Medullary thyroid cancer (MTC)
Mucosal microbiome, 318
Multipass transmembrane protein, 247
Multiple cloning site (MCS), 7, 70
Murine Abs, 204
Mutagenic oligonucleotides, 238–239

N

Naive libraries, 11–14, 250–251
Nanobody, 80
National Center for Biotechnology Information, 252
Natural antibody, 14, 203, 206, 213–214, 219–220, 224, 358
    complexes, 202
    data set, 207, 216
    structures of synthetic and natural Ab–antigen complexes, 206–207
    PDB structures selected for comparison of, 208t–209t
    repertoire, 79, 81
Natural CDR-L1 loops, 216
Natural data sets, 214
Natural FN3 molecules, 358
Natural H3J fragments, 253
    amplification of $V_L$–$V_H$_CM fragment, 287
    PCR assembly
        of filtered libraries with, 286
        of $V_L$–$V_H$_CM fragment with, 287–288
Natural HCDR3 by PCR, amplification of, 285–286
Natural immune system, 358

Natural immunity, 295
Natural naive libraries, 11
Natural peptide libraries (NPLs), 17
Natural sets, 217
NEDD4. *See* Neural precursor cell expressed developmentally down-regulated protein 4 (NEDD4)
NEDD4L, 399
Negative selection, 255, 427
*Neisseria meningitidis*, 318
Neural precursor cell expressed developmentally down-regulated protein 4 (NEDD4), 393
Neutravidin, 52
Next-generation sequencing (NGS), 57
　as primary readout, 57–59
NGS. *See* Next-generation sequencing (NGS)
NHPs. *See* Nonhuman primates (NHPs)
Nickel pull-down method, 398
*N,N*-dimethylformamide (DMF), 452
N1 domain, 4, 69
Nonantibody antigen-binding domains, display of, 18–19
Noncanonical PTMs, engineering SH2 domains for binding to, 373–375
Nonfat dried milk, 388
Nonhuman primates (NHPs), 247
Non-phosphorylated peptide, 373
Non-probe-based assay, 394
Nonspecific background yield, 15, 54
Nonspecific interactions, magnetic bead equilibration and blocking of, 274
NPLs. *See* Natural peptide libraries (NPLs)
N2 domain, 4, 69

## O

Occupational Safety and Health Administration (OSHA), 451
OD. *See* Optical density (OD)
Oligonucleotides
　oligonucleotide-directed mutagenesis method, 384
　PCR-based addition of oligonucleotides that document library source of selected clone, 352
Omicron, 253
OPALs. *See* Oriented peptide array libraries (OPALs)
Open reading frames (ORFs), 31, 321
Optical density (OD), 118, 169, 170, 181, 192
Optical density at 405 nm ($OD_{405}$), 169
Optical density at 600 nm ($OD_{600}$), 181, 186, 192–194, 415, 424

Optimized ratios, generation of library based on, 346–347
Optional LiCl precipitation, RNA purification with, 169–171
ORFs. *See* Open reading frames (ORFs)
Organ harvesting, followed by RNA purification with optional LiCl precipitation, 169–171
Organic compounds, 455
Organoids, UbV characterization in, 398–399
Oriented peptide array libraries (OPALs), 370
OSHA. *See* Occupational Safety and Health Administration (OSHA)
Output titration, 193

## P

pADL-23c vector
　digestion
　　cloning four synthetic scFvs, 266–267
　　cloning semisynthetic scFvs, 289
　　four synthetic scFv libraries for cloning into, 264–266
　　semisynthetic scFvs, for cloning into, 288–289
Panning, 49, 79
　approach, 256
　developing panning strategy, 254–255
　on 96-well high-binding-capacity microtiter plates, 348–349
　rounds, 256
Panning procedures
　materials, 129
　　equipment, 131–132
　　reagents, 129–131
　method, 132
　　panning against biotinylated antigen in solution, 135
　　panning against eukaryotic cells in solution, 135–136
　　panning against immobilized antigen, 132–135
　troubleshooting, 136–137
Panton–Valentine leucocidin (PVL), 321–322
Parameters, 395
Passive immunotherapy, 321
"Payloads", 69
pBluescript, 7
PBMCs. *See* Peripheral blood mononuclear cells (PBMCs)
PBS. *See* Phosphate-buffered saline (PBS)
PBSB. *See* Phosphate-Buffered Saline with BSA (PBSB)
PBSP buffer, 442, 449
PBST. *See* Phosphate-Buffered Saline with Tween (PBST)

PBSTB. *See* Phosphate-Buffered Saline with Tween and BSA (PBSTB)
pC3C, 82
　sort, 72–74
pComb-Opti8
　fragment libraries, 328
　phage display system, 318
　phagemid vector, 323
　vector, 17–18, 354
pComb3, 70
　descendants
　　pC3C, 72–74
　　pComb3H, 71–72
　family, 7–8, 19
　pComb3-bearing cell, 8
　pComb3X, 71–72
　pComb3XSS vector, 176–177
　"type 3 + 3" phage display vector, 70–71
pComb3 phagemid, 318
　expression cassettes of, 71f
　family, 65
　filamentous phage in context of phage display, 63–64
　for mining antibody repertoires, 66t
　pComb3 descendants pComb3H and pComb3X, 71–72
　phage display vectors
　　classification of, 64–65
　　phagemids as, 65–67
　phagemid-based phage display, 67–69
　　physical linkage of phenotype and genotype in, 69–70
　specialized pComb3 descendants pC3C and pC3Csort, 72–74
　"type 3 + 3" phage display vector pComb3, 70–71
PCRs. *See* Polymerase chain reactions (PCRs)
PDB. *See* Protein Data Bank (PDB)
PDGFRβ. *See* Platelet-derived growth factor receptor beta (PDGFRβ)
PEG. *See* Polyethylene glycol (PEG)
pelB-Fd-ΔpIII, 70
pelB-LC, 70
PepSpotDB. *See* Probabilistic interaction network database (PepSpotDB)
Peptide(s), 42, 65, 245
　display of peptides domains other than antibodies, 17–21
　libraries, 65
　peptide–selector interaction, 53
*Peptostreptococcus magnus*, 252
Peripheral blood mononuclear cells (PBMCs), 80, 91, 283–284
　preparation from human blood or BMMCs from human bone marrow, 91–92
　total RNA preparation from, 92–93

Index

Periplasm, 5
PET. *See* Positron-emission tomography (PET)
PFT. *See* Pore-forming toxin (PFT)
Phage antibody
  context, 15
  libraries, 8–9, 59
    affinity maturation, 14–15
    all-purpose phage antibody libraries, 11–14
    immunized humans, chickens, and other animals, 9–11
    phagemid vectors for, 7–8, 9f
    projects, 9
Phage-based affinity maturation campaigns, 257
Phage binding ELISA protocol, 350–351
Phage display, 1, 3, 79, 83, 245, 329, 357, 364, 382–383, 388, 406
  affinity selection, 15–17
  C.A.R.L.O.S. project, 21–24
  course, 2, 8
  display of peptides and protein domains other than antibodies, 17–21
  engineering SH2 domains using, 371–375
  expression cloning, 317
  filamentous phage in context of, 63–64
  filamentous phage structure and function, 3–7
  library, 17, 255, 360, 362, 382
    construction and selection for therapeutic monoclonal antibody identification, 249–251
    and defining characteristics, 249t
    design, 387
    generation of chicken antibody using, 162–163
    limitations to phage display imposed by Ff assembly and bacterial physiology, 40–44
  methodology, 2
  monobodies, 359–360
  panning campaigns, 255
  phage antibody libraries, 8–15
  phagemid vectors for phage antibody display, 7–8
  selection of UbV binders by, 386
  staphylococcal PFTS attractive antigens for phage display–based investigations, 322
  technology, 27, 318
  vectors
    classification of, 64–65
    phagemids as, 65–67
    preparation for cloning, 342–344
Phage display antibody libraries, 79–80, 82
Fab format
  generation in, 81–82
  selection in, 82–84
Phage display chicken single-chain variable fragment library
  materials, 172
    equipment, 173–174
    reagents, 172–173
  method, 174
    amplification of $V_H$ and $V_\lambda$ genes, 174–175
    assembling scFv genes, 175–176
    restriction digestion of scFv gene fragments and pComb3XSS vector, 176–177
    test ligation, 177–182
    troubleshooting, 182
Phage-displayed antibody libraries, 223
Phage-displayed Fab library, conversion of CCC-dsDNA into, 232–234
Phage-displayed PLs for filtration, preparation of, 273
Phage-displayed synthetic antibody libraries, 239
Phage display-selected Fab
  materials, 149
    equipment, 151–152
    reagents, 149–151
  method, 152–153
    Fab cloning, 153–154
    Fab expression, 154–155
    Fab purification, 155–156
Phage ELISA
  analysis of library selection by, 142
  binding analysis of selected Fabs by, 235–236
  with eukaryotic cells in solution, 143
  with immobilized antigen, 142–143
Phage format, library recovery in, 347–348
Phage library, 255, 424
  generation, 98–100, 112–114
Phage litmus, 30
Phagemid(s), 7, 63–64, 80, 98–100, 112–114
  design, 238
  DNA, 14
  pComb3, 70, 80
  pComb8, 65, 71
  as phage display vectors, 65–67
  phagemid-based display system for toxin epitope recovery, 323–324
  phagemid-based phage display, 67
    helper phage, 68–69
    host bacterial cells, 68
    monovalent Fab-phage display, 68f
    phagemid library, 67–68
    physical linkage of phenotype and genotype in, 69–70
  phagemid-transformed cells, 42
  preparation of gene fragments for cloning into phagemid library, 340–342
  SfiI digestion of phagemid pC3C, 97, 111–112
  ssDNA, 384
  system, 43
  vectors, 42
    for phage antibody display, 7–8, 9f
Phage nonantibody libraries, 18, 20
Phage panning campaign, operational considerations before starting, 255–256
Phage precipitation with PEG/NaCl, 273–274
Phage shock protein (Psp), 39
Pharmaceutical companies, 250
Phenotype in phagemid-based phage display, physical linkage of, 69–70
*PhoA*. *See* Alkaline phosphatase (*PhoA*)
Phosphate-buffered saline (PBS), 140, 189, 410, 418, 431
Phosphate-Buffered Saline with BSA (PBSB), 431
Phosphate-Buffered Saline with Tween (PBST), 418, 431
Phosphate-Buffered Saline with Tween and BSA (PBSTB), 432
Phosphorylated tyrosine (pTyr), 367
  peptides, 376
  pTyr-containing proteins, 376
  structure of pTyr-binding pocket, 369–370
Phosphotyrosine binding domains (PTBs domains), 368
pIII. *See* Protein III (pIII)
pIII fusions, 42
pIII molecules, 51
pIV, 38
pIX, 29
Plasmons, 392
Platelet-derived growth factor receptor beta (PDGFRβ), 405
Pleckstrin homology (PH), 368
PLs. *See* Primary libraries (PLs)
pmf. *See* Proton motive force (pmf)
Polyclonal chicken antibodies, 160
Polyclonal ELISA, 429
Polyethylene glycol (PEG), 335, 355
  PEG–NaCl, 432
  phage precipitation with, 273–274
Polymerase chain reactions (PCRs), 68, 412, 425
  amplification
    human $V_\kappa$, $V_\lambda$, and $V_H$ cDNA, and human $C_\kappa$ and $C_\lambda$ DNA, 93–96

natural HCDR3/JH fragments by, 285–286
rabbit $V_\kappa$, $V_\lambda$, and $V_H$ cDNA, and human $C_\kappa$ and $C_\lambda$ DNA, 108–110
assembly
filtered libraries with natural H3J fragments, 286–287
huV$_L$/huC$_L$/huV$_H$ cassettes, 96–97
rbV$_L$/huC$_L$/rbV$_H$ cassettes, 110–111
V$_L$–V$_H$_CM fragment with natural H3J fragment to generate SLs, 287–288
PCR-based addition of oligonucleotides that document library source of selected clone, 352
screening primers, 412
Polypeptides, 40
Pore-forming toxin (PFT), 321
Positron-emission tomography (PET), 405
Post data analysis, 302–303
Posttranslational modifications (PTMs), 367, 381
Potential cancer therapeutics, SH2 domains as, 377
PPIs. *See* Protein–protein interactions (PPIs)
Primary libraries (PLs), 252
construction, 264
digestion
four synthetic scFv libraries for cloning into pADL-23c vector, 264–266
pADL-23c vector for cloning the four synthetic scFvs, 266–267
electroporation of ligation mix, 267–268
ligation of digested scFv libraries with digested phagemid vector, 267
QC of, 268–269
Primary readout, 16
HTS as primary readout in affinity selection projects, 16–17
NGS as, 57–59
Primary safety information resources for laboratory personnel, 451
Primary sequence–dependent B-cell epitope, convalescent serum antibodies commonly recognize, 327–329
Primer design, 411
library-amplification primers, 412
PCR screening primers, 412
primers for display vector amplification, 411–412
sequencing primers, 412
Probabilistic interaction network database (PepSpotDB), 370

Production yields, 203
Progeny filamentous virions, 5
Proline, 406
Proof-of-concept studies, 324
"Proof-of-principle" project, 18
Propagation-related SUPs, 56
PROTAC. *See* Proteolysis-targeting chimera (PROTAC)
Protein Data Bank (PDB), 207
Protein III (pIII), 49
Protein–protein interactions (PPIs), 214, 323, 381
Protein(s), 42
complexes, 394
display, 294
library data, 293
protein domains other than antibodies, 17–21
drug targets, 202
G bead–based library selection by mAb, 349–350
immobilization, 391
library diversity, 65
microarrays, 370
Proteolysis-targeting chimera (PROTAC), 397
Proton motive force (pmf), 33
*Pseudomonas*, 29, 455
*P. aeruginosa*, 29
Psp. *See* Phage shock protein (Psp)
PTBs domains. *See* Phosphotyrosine binding domains (PTBs domains)
PTMs. *See* Posttranslational modifications (PTMs)
pTyr. *See* Phosphorylated tyrosine (pTyr)
Purpose-built antibody, 295
gene alignment software, 295
pVI, 30
pVII, 29
PVL. *See* Panton–Valentine leucocidin (PVL)
pV–ssDNA complex, 36

## Q

QC. *See* Quality control (QC)
Qualitative data, 33
Quality control (QC), 249
FLs, 276–277
PLs, 268–269
SLs, 290–291
Quantitative data, 33

## R

Rabbit spleen, total RNA preparation from, 106–107
Rabbit V$_H$ (rbV$_H$), 108
PCR amplification of, 108–110

Rabbit V$_\kappa$ (rbV$_\kappa$), 108
PCR amplification of, 108–110
Rabbit V$_\lambda$ (rbV$_\lambda$), 108
PCR amplification of, 108–110
Rader, Christoph, 8
Radioactive safety procedures, 454
Rakonjac, Jasna, 5
Random peptide libraries (RPLs), 17
Ranibizumab, 79
RBR. *See* Ring-between-ring (RBR)
RBS. *See* Ribosome binding site (RBS)
rbV$_H$ cassettes
PCR assembly of, 110–111
SfiI digestion of, 111–112
rbV$_L$ cassettes
PCR assembly of, 110–111
SfiI digestion of, 111–112
Rbx1. *See* Ring-box protein 1 (Rbx1)
RE. *See* Restriction enzyme (RE)
Really interesting new gene (RING), 383
Receptor tyrosine kinases (RTKs), 367–368
Reduced codes, 361
Regulate stringency, 59–62
Release-related SUPs, 56
Remaining diversity, 217
Response units (RUs), 392
Restriction enzyme (RE), 411
Reverse transcription (RT), 89, 93, 108
mRNA, 93, 107–108
Reverse transcription polymerase chain reaction (RT-PCR), 73
Ribosome binding site (RBS), 33, 264
RING. *See* Really interesting new gene (RING)
Ring-between-ring (RBR), 383
Ring-box protein 1 (Rbx1), 387
Rise per nucleotide ($h$), 29
RNA, 10
purification with optional LiCl precipitation, 169–171
Roscow, Olivia, 20
RPLs. *See* Random peptide libraries (RPLs)
RT. *See* Reverse transcription (RT)
RT-PCR. *See* Reverse transcription polymerase chain reaction (RT-PCR)
RTKs. *See* Receptor tyrosine kinases (RTKs)
RUs. *See* Response units (RUs)

## S

SA. *See* Staphylococcus aureus (SA)
SAbDab. *See* Structural Antibody Database (SAbDab)
*Salmonella*, 455
SAP. *See* Shrimp alkaline phosphatase (SAP)

# Index

SARS-CoV-2. *See* Severe acute respiratory syndrome coronavirus 2 (SARS-CoV-2)
SASA. *See* Solvent-accessible surface area (SASA)
SB. *See* Super Broth (SB)
Sbi. *See* Second immunoglobulinbinding protein (Sbi)
Scaffold protein, 382
SCF. *See* Skp1–Cul1–F-box (SCF)
scFv. *See* Single-chain variable fragment (scFv)
sdAbs. *See* Single-domain antibodies (sdAbs)
SDS-PAGE. *See* Sodium dodecyl-sulfate polyacrylamide gel electrophoresis (SDS-PAGE)
SEC. *See* Secretory (SEC)
Second immunoglobulinbinding protein (Sbi), 322
Secondary libraries (SLs), 14, 286
    construction of, 288
        digestion of pADL-23c vector for cloning semisynthetic scFvs, 289
        digestion of semisynthetic scFvs, for cloning into pADL-23c vector, 288–289
        electroporation of ligation mix, 290
        ligation of digested semisynthetic scFvs with digested phagemid vector, 289
    quality control of, 290–291
Secondary screenings, 248
Secretory (SEC), 70
Selected binders
    materials, 139
        equipment, 141–142
        reagents, 139–141
    method, 142
        analysis of library selection by crude Fab ELISA, 143–146
        analysis of library selection by phage ELISA, 142–143
        DNA fingerprinting and sequencing, 146–147
        troubleshooting, 147–148
Selected clone, PCR-based addition of oligonucleotides that document library source of, 352
Selected Fabs by phage ELISA, binding analysis of, 235–236
Selection, 249, 323
Selector, 49, 51
    concentrations, 60
    molecules, 59
    selector-coated substrate, 52
    selector-specific clones, 54

selector-specific yield, 54
selector–peptide binding, 53
Selector-unrelated phages (SUPs), 16, 21, 55
scourge, 55–57
Semisynthetic libraries, 250–251
Semisynthetic phage display library construction
    competing interest statement, 269
    materials, 260
        equipment, 261
        reagents, 260–261
    method, 261–264
        CDRs of $V_H$ and $V_L$ fragments, 262$t$
        construction of PLs, 264–269
        neutral H3J fragments, 263$t$
        synthesis of diversified scFv libraries, 264
    troubleshooting, 269
Semisynthetic scFvs
    digestion
        cloning into pADL-23c vector, 288–289
        pADL-23c vector for cloning semisynthetic scFvs, 289
SeqMan NGen Genome Assembly Software, 354
Sequencing primers, 412
Serum antibodies, titration of, 170$f$
Severe acute respiratory syndrome coronavirus 2 (SARS-CoV-2), 207, 253
SfiI digestion of huV$_L$/huC$_L$/huV$_H$ cassettes and phagemid pC3C, 97
SfiI sites, 72
SH2. *See* Src homology 2 (SH2)
SHM. *See* Somatic hypermutation (SHM)
Short peptides, 5
Shrimp alkaline phosphatase (SAP), 227, 343
Signal recognition particle (SRP), 70, 360
    phage display system, 360
Signal transduction, 363
    SH2 domains in, 368–369
Silverman, Gregg, 17
Single-chain antibody, 53
Single-chain variable fragment (scFv), 2–3, 162, 203, 249
    constructs, 9
    genes
        assembling, 175–176
        restriction digestion of, 176–177
    libraries
        ligation of digested scFv libraries with digested phagemid vector, 267
        synthesis of diversified, 264
    materials, 281

equipment, 282–283
reagents, 281–282
method, 283
    amplification of natural HCDR3/ JH fragments by PCR, 285–286
    construction of SLs, 288–290
    first-strand cDNA synthesis, 285
    isolation of lymphocytes from leukoplatelet concentrate, 283–284
    mRNA isolation, 284–285
    PCR assembly of FLs with natural H3J fragments, 286–288
    quality control of SLs, 290–291
    total RNA isolation from lymphocytes, 284
troubleshooting, 291
Single-domain antibodies (sdAbs), 9
Single-stranded DNA (ssDNA), 28, 64, 80, 228, 384
Skin and soft tissue infection (SSTI), 318
Skin tissue infection, 321
Skp1. *See* S-phase kinase-associated protein 1 (Skp1)
Skp1–Cul1–F-box (SCF), 387
SLs. *See* Secondary libraries (SLs)
Small molecule drugs, 377
Small-scale preparation of antibody fragment-displaying phage, 194–195
Small-scale transformation, 414–415
SMART database, 358
Smith, George, ix, 245
Sodium acetate, 419
Sodium azide (NaN$_3$), 105, 129
Sodium bicarbonate (NaHCO$_3$), 167
Sodium chloride (NaCl), 89, 105, 126, 194, 272
    phage precipitation with, 273–274
Sodium dodecyl-sulfate polyacrylamide gel electrophoresis (SDS-PAGE), 393, 410
Sodium hydroxide (NaOH), 128, 455
Sodium phosphate, 419, 433
"Soft randomization" strategy, 383
Soft tissue infection, 321
Software preparation, 301–302
Solid phase panning, 254
Soluble protein, 254
Solvent-accessible surface area (SASA), 207, 210
Somatic hypermutation (SHM), 316
SORT. *See* Sortase A (SORT)
Sortase A (SORT), 73
SpA. *See* Staphylococcal protein A (SpA)
Specialized protein domains, display of, 19–21
Specificity, 255

structure of specificity-determining
pockets, 370
Specific yield, 15
S-phase kinase-associated protein 1 (Skp1), 387
Spin columns, 60
Spin down culture, 429
SPR. See Surface plasmon resonance (SPR)
Src homology 2 (SH2), 19, 367
  biotechnological applications of SH2 superbinders, 375
    affinity purification tools for MS, 375–376
    biosensors and cancer diagnostics, 376–377
    domains as potential cancer therapeutics, 377
    synthetic biology, 376
  domains
    mutational studies of, 371t
    in signal transduction networks, 368–369
    structure and role in molecular recognition, 369
    structure of pTyr-binding pocket, 369–370
    structure of specificity-determining pockets, 370–371
    structure–function of, 368
Src homology 3 (SH3), 368
SRP. See Signal recognition particle (SRP)
ssDNA. See Single-stranded DNA (ssDNA)
ssDNA–pV complex, 37
SSTI. See Skin and soft tissue infection (SSTI)
Stahl, Stefan, 19
Staphylococcal display
  materials, 443
    equipment, 444
    reagents, 443–444
  method, 444
    electroporation of affibody libraries into S. carnosus, 446–447
    FACS of S. carnosus-displayed affibody libraries, 447–448
    preparation of electrocompetent S. carnosus TM300 cells, 444–446
Staphylococcal infections pose substantial and increasing clinical burden, 318–321
  gene fragment-encoded minimal epitopes, 320f
  pComb-Opti8 vector, 319f
Staphylococcal PFTS attractive antigens for phage display–based investigations, 322

Staphylococcal protein A (SpA), 322
Staphylococcus aureus (SA), 18, 252, 318, 353
  human immunoproteome of S. aureus gene products, 321–322
  infections associated with range of immunogenic antigens, 321
Staphylococcus carnosus, 19, 406
  electroporation of affibody libraries into S. carnosus, 446–447
  FACS of S. carnosus-displayed affibody libraries, 447–448
  preparation of electrocompetent S. carnosus TM300 cells, 444–446
State-of-the-art synthetic Ab libraries, 206
Stem subdomain of leucocidin holoprotein, identification of Luk neutralization epitope in β-hairpin loop in, 326–327
Strategic approach, 252
Streptavidin, 52
Streptococcus
  S. pneumoniae, 318
  S. pyogenes, 318
Streptomyces, 455
Stringency, 15–16, 54–55
Structural Antibody Database (SAbDab), 207
Structural diversity, 217
Structural homology, 322
Structural studies, 394
  MS, 396
  X-ray crystallography, 394–396
sTyr. See Tyrosine O-sulfation (sTyr)
Substrates, 20
  surface, 53
Subtraction, 255
Sucrose, 449
supE44, 72
Super Broth (SB), 89, 103, 117, 122, 126, 130, 140, 151
SUPs. See Selector-unrelated phages (SUPs)
Surface plasmon resonance (SPR), 388, 392–393
Synthetic antibodies, 204, 206, 213–214, 219–220, 224
  data sets, 207, 216
  structures of synthetic and natural Ab–antigen complexes, 206–207
  libraries, 202, 204, 220
    principles of library design, 202–203
    trastuzumab single-framework libraries, 203–206
  PDB structures selected for comparison of, 208t–209t
  structural analysis of, 206

analysis of Ab residues involved in antigen interaction, 207–214
  antibody residues to the structural paratope, 212f
  canonical conformations of CDR loops, 217–219
  core structural paratope, 214–216
  data set of structures of synthetic and natural Ab–antigen complexes, 206–207
  length and composition of CDRs L3 and H3, 216–217
  limitations, 219
Synthetic antigen-binding fragments (Fabs), 202
Synthetic binding proteins, 358
Synthetic biology, 376
Synthetic CDR-L1 loops, 216
Synthetic data sets, 214
Synthetic human antibody libraries via phage display
  CCC-dsDNA conversion, 239
  library construction, 238–239
  materials, 226
    equipment, 228
    reagents, 226–227
  method, 228
    binding analysis of selected Fabs by phage ELISA, 235–236
    conversion of CCC-dsDNA into phage-displayed Fab library, 232–234
    DNA sequencing of target-specific phage clones, 236–237
    preparation of dU-ssDNA template, 228–229
    selection of high-affinity synthetic Fabs, 234–235
    target-specific antibodies, 236f
    troubleshooting, 237–238
    in vitro synthesis of CCC-dsDNA, 230–232
  phagemid design, 238
  selection of high-affinity synthetic Fabs, 239
Synthetic IGHV gene, 255
Synthetic libraries, 11–14, 202–203, 220, 250–251, 362
  design, 213
Synthetic peptide studies, 326
Synthetic scFv
  digestion
    four synthetic scFv libraries for cloning into pADL-23c vector, 264–266
    pADL-23c vector for cloning for, 266–267
Synthetic sets, 217

Index

**T**

TAE. *See* Tris-acetate-EDTA (TAE)
Tailored diversity, 206
Target biology, 248
Target concentration, 256
Targeted host cell membrane, 329
Target protein, 391–392
Target specificity, 403
Target-specific phage clones, DNA sequencing of, 236–237
TBAB plates. *See* Tryptose blood agar base plates (TBAB plates)
TBS. *See* Tris-buffered saline (TBS)
T-cell receptor (TCR), 21, 293
TCR. *See* T-cell receptor (TCR)
TDPs. *See* Transducing particles (TDPs)
Termination codon, 362
Test ligation, 177–179
   electroporation of, 345–346
   gene fragment inserts into prepared vector, 344–348
   library ligation, 179–181
   library reamplification, 181–182
   and transformation, 97–98, 112
3,3′,5,5′-tetramethylbenzidine (TMB), 388, 422
Tetranomial diversity, 206
T4 DNA ligase, 384
Therapeutic antibody, 246
   antibody discovery process, 245–249
   developing panning strategy, 254–255
   discovery, 249
   operational considerations before starting phage panning campaign, 255–256
   programs, 257
   technical considerations to isolate antibodies of interest, 256–257
   therapeutic monoclonal antibody case study in library construction for, 251–253
   phage display library construction and selection for identification, 249–251
Therapeutic monoclonal antibody
   case study in library construction for, 251–253
   phage display library construction and selection for, 249–251
Therapeutic use, 317
Thermal stability, 253
Thermodynamics, 390
Thermostable scFvs, selection for, 274–275
Thioredoxin, 39
Tissues, 455
TKB. *See* Tyrosine kinase binding (TKB)
TKI. *See* Tyrosine kinase inhibitor (TKI)
TolQR component, 33

Total RNA
   isolation from lymphocytes, 284
   preparation
      from PBMCs or BMMCs, 92–93
      from rabbit spleen and bone marrow, 106–107
Toxic compounds, 456
Toxin epitope recovery, phagemid-based display system for, 323–324
Traditional hybridoma technology, 201
Transcripts, 31–33
Transducing particles (TDPs), 42
Translational studies, 293
Trastuzumab, 207
Trastuzumab single-framework libraries, 203–206
   Genentech's approved antibodies in 4D5 framework, 204
   minimalist libraries based on 4D5 framework, 205
TRIM. *See* Trinucleotide mutagenesis (TRIM)
Trinucleotide mutagenesis (TRIM), 250
Tris-acetate-EDTA (TAE), 102, 139, 149, 420, 434
Tris-buffered saline (TBS), 105, 126, 130, 140
Trp. *See* Tryptophan (Trp)
Trypsin release, 53
Tryptic soy broth (TSB), 420, 434
Tryptophan (Trp), 370
Tryptose blood agar base plates (TBAB plates), 420, 434, 442, 449
TSB. *See* Tryptic soy broth (TSB)
TSB + Y medium, 420, 434, 449
T7 DNA polymerase, 384
Titering immune serum by ELISA, 168–169
Tumors, 247
TUPs, 55
"Type 3 + 3" phage display vector pComb3, 70–71
Typhimurim, 39
Tyr/Ser binary code, 361
Tyrosine
   corner, 361
   phosphoproteome, 371
   phosphorylation, 367–368, 376
   protein kinase Fes, 373
   sulfation, 373
Tyrosine kinase binding (TKB), 370
Tyrosine kinase inhibitor (TKI), 376
Tyrosine O-sulfation (sTyr), 373

**U**

U.S. Environmental Protection Agency, 453
U.S. Food and Drug Administration (FDA), 21, 245

Ub. *See* Ubiquitin (Ub)
Ub-AMC. *See* Ubiquitin-7-amino-4-methylcoumarin (Ub-AMC)
Ub-AML. *See* Ubiquitin-aminoluciferin (Ub-AML)
UB-AQUA. *See* Ubiquitin absolute quantification (UB-AQUA)
UBDs. *See* Ubiquitin-binding domains (UBDs)
Ub-Rho110. *See* Ubiquitin-rhodamine110 (Ub-Rho110)
Ubiquitin (Ub), 20–21, 382
   signaling cascade, 382–383
      characterization of binders, 387–399
      design and generation of first UbV libraries, 383–385
      library design improvements, 387
      selection of UbV binders by phage display, 386
   Ub-activating enzyme, 382
   Ub-conjugating enzyme, 382
   Ub-protein ligase, 382
Ubiquitin absolute quantification (UB-AQUA), 396
Ubiquitin-7-amino-4-methylcoumarin (Ub-AMC), 394
Ubiquitin-aminoluciferin (Ub-AML), 394
Ubiquitination, 393
   assay, 393, 397
   interactions, 383
Ubiquitin-binding domains (UBDs), 20–21
Ubiquitin-interacting motif (UIM), 20–21
Ubiquitin-rhodamine110 (Ub-Rho110), 394
Ubiquitin-specific protease 7 (USP7), 383
   process, 382
Ubiquitin variant (UbV), 382, 393
   characterization in organoids, 398–399
   design and generation of first UbV libraries, 383–385
   selection of UbV binders by phage display, 386
UbV. *See* Ubiquitin variant (UbV)
UDI. *See* Unique dual index (UDI)
UIM. *See* Ubiquitin-interacting motif (UIM)
*ung* gene, 384
Unique dual index (UDI), 352
University of Pennsylvania (Penn), 21
Uracil-containing ssDNA (dU-ssDNA), 239
   preparation of dU-ssDNA template, 228–229
USP7. *See* Ubiquitin-specific protease 7 (USP7)
Utility, 83

## V

Variable (V) gene, 161
Variable heavy ($V_H$), 295
  amplification of, 174–175
Variable light ($V_L$), 295
Vascular endothelial growth factor receptor 2 (VEGFR2), 405
VCSM13 helper phage
  materials, 185
    equipment, 185–186
    reagents, 185
  method, 186–187
Vector DNA with, 414
VEGFR2. *See* Vascular endothelial growth factor receptor 2 (VEGFR2)
$V_H$ cDNA, 93–96
  PCR amplification of, 108–110
Virion, 1, 15, 49
  length, 28
  proteins forming end caps of filament, 29–30
  structure, 50f
  virion-borne peptide, 53
Virus-like particles (VLPs), 254
Visual spreadsheet-based analyses, 310
VLPs. *See* Virus-like particles (VLPs)
$V_L$–$V_H$_CM fragment, amplification of, 287
$V_\lambda$ genes, amplification of, 174–175

## W

Washing stringency, 256
Wild-type (WT), 253
  Ff infection, 39
World Health Organization, 318
WT. *See* Wild-type (WT)

## X

X-ray crystallographic structure, 4
X-ray crystallography, 388, 394–396

## Y

Yeast display, 362, 364
*Yersinia enterocolitica*, 39
Yield, 15–16, 54–55

## Z

Zhang, Wei, 20

www.ingramcontent.com/pod-product-compliance
Lightning Source LLC
Chambersburg PA
CBHW041918180526
45172CB00013B/1325